Late Cenozoic Structure and Evolution of the Great Basin–Sierra Nevada Transition

edited by

John S. Oldow
Department of Geosciences
University of Texas at Dallas
Richardson, Texas 75080-3021
USA

Patricia H. Cashman
Department of Geological Sciences and Engineering
University of Nevada
Reno, Nevada 89577
USA

THE
GEOLOGICAL
SOCIETY
OF AMERICA®

Special Paper 447

3300 Penrose Place, P.O. Box 9140 ▪ Boulder, Colorado 80301-9140, USA

2009

Published by The Geological Society of America, Inc.
3300 Penrose Place, P.O. Box 9140, Boulder, Colorado 80301-9140, USA
www.geosociety.org

Printed in U.S.A.

GSA Books Science Editors: Marion E. Bickford and Donald I. Siegel

Library of Congress Cataloging-in-Publication Data

Late Cenozoic structure and evolution of the Great Basin - Sierra Nevada transition / edited by John S. Oldow, Patricia H. Cashman.
    p. cm. — (Special paper ; 447)
  Includes bibliographical references.
  ISBN 978-0-8137-2447-8 (pbk.)
    1. Geology, Stratigraphic—Neogene. 2. Geology, Structural—Great Basin. 3. Geology—Great Basin. I. Oldow, John S. II. Cashman, Patricia Hughes.

QE693.5.L38 2009
551.7'860979—dc22
                                                                    2008051454

Cover, above: Mineral Ridge, Silver Peak Range, west-central, Nevada. Photo courtesy of John Oldow. **Below:** Soft-sediment debris flow deposit in 3 Ma sub-lacustrine fan sediments exposed at The Island, on the south edge of Honey Lake, California. Photo courtesy of Jim Trexler. **Back cover:** Deformation in 3 Ma old sub-lacustrine fan sediments exposed at The Island, on the south edge of Honey Lake. The rocks are deformed and uplifted in a compressive overlap, i.e., a left step between two strands of the Honey Lake fault. Photo courtesy of Pat Cashman.

10 9 8 7 6 5 4 3 2 1

Contents

The Geological Society of America
Special Paper 447
2009

Introduction

John S. Oldow

Department of Geosciences, University of Texas at Dallas, Richardson, Texas 75080-3021, USA

Patricia H. Cashman

Department of Geological Sciences and Engineering, University of Nevada, Reno, Nevada 89577, USA

The transition zone between the Sierra Nevada and the Great Basin accommodates nearly 25% of the motion between the Pacific and North American plates and is localized along a boundary that evolved during the late Cenozoic. Both contemporary and ancient displacement are concentrated in a relatively narrow belt of deformation that stretches from southern California northwesterly to southern Oregon (Fig. 1). From the Mojave Desert, the displacement extends north, east of the southern Sierra Nevada, in a narrow zone of transtensional deformation constituting the Eastern California shear zone (Wallace, 1984). From the latitude of the central Sierra Nevada, the zone broadens to include the Walker Lane (Locke et al., 1940) and the extensional deformation of the central Nevada seismic belt in the northwestern Great Basin (Fig. 1). Farther to the northwest, the locus of deformation becomes less distinct along the transition between the Great Basin and Blue Mountain block of central Oregon.

The tectonic boundary zone was initially recognized by Locke et al. (1940) from the abrupt change in the north-northeast trending physiography characteristic of the central Great Basin to the northwesterly trending mountains and valleys in the region lying east of the Sierra Nevada. They correctly surmised that the physiography was related to displacement on a diffuse system of right-lateral transcurrent faults whose regional complexity was summarized by Stewart (1988). Although the Walker Lane was accepted as the locus of active deformation in the western Great Basin (Wallace, 1984; Eddington et al., 1987; Rogers et al., 1991; Zoback, 1989; Oldow, 1992), the role of the tectonic belt in accommodating a significant component of the relative plate motion along the western margin of North America was most forcefully expressed by findings of space-geodesy studies (Savage et al., 1990; Dixon et al., 1995, 2000; Bennett et al., 1999, 2003; Thatcher et al., 1999; Oldow et al., 2001; Miller et al., 2001; Gan et al., 2000; Oldow, 2003; Ham-

mond and Thatcher, 2004). The recognition that the Eastern California shear zone and Walker Lane system are part of an evolving intra-continental plate boundary (Unruh et al., 2003; Faulds et al., 2005) has kindled interest in the Cenozoic evolution of the boundary zone, particularly in the context of how contemporary kinematics may or may not reflect longer term deformation patterns within the tectonic belt (Wernicke, 1992; Cashman and Fontaine, 2000; McQuarrie and Wernicke, 2005; Henry and Perkins, 2001; Oldow et al., 2008).

This volume contains new research that provides constraints and insights into the Neogene structure and evolution of the Great Basin–Sierra Nevada transition zone (Fig. 2). The studies presented here reflect the broad scope of topical studies underway in the region and range from biostratigraphic analysis of Tertiary basin sediments to calculation of velocity and strain fields determined from space geodesy.

We organized the chapters of this volume both geographically and topically. Chapters 1–3 are companion papers by Blewitt, Hammond, and Kreemer that address the Global Positioning System (GPS) velocity field, strain-rate field, and geodynamic implications of velocity transients induced by recent earthquakes for much of the western Great Basin. In chapter 4, Murphy et al. present a graphical method that documents the relation between contemporary velocity and strain fields in the spatially heterogeneous transtensional deformation characteristic of the central Walker Lane.

The remaining twelve chapters address geologic, geophysical, biostratigraphic, and seismological investigations of smaller regions within the boundary zone; these are arranged from north to south, from Oregon to southeastern California. In chapter 5, Liberty et al. use seismic reflection profiling in upper Klamath Lake to document the subsurface pattern of faulting within an extensional basin, which, if not recognized, could pre-

Oldow, J.S., and Cashman, P.H., 2009, Introduction, *in* Oldow, J.S., and Cashman, P.H., eds., Late Cenozoic Structure and Evolution of the Great Basin–Sierra Nevada Transition: Geological Society of America Special Paper 447, p. v–viii, doi: 10.1130/2009.2447(00). For permission to copy, contact editing@geosociety.org. ©2009 The Geological Society of America. All rights reserved.

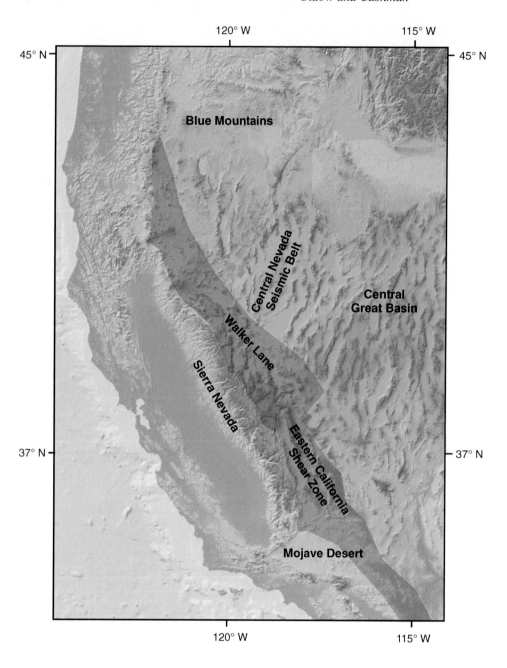

Figure 1. Physiographic map of the southwestern U.S. Cordillera showing major tectonic domains. Orange shaded region is the belt of active transtensional deformation marking the boundary zone between the Great Basin and Sierra Nevada. Yellow shaded regions are regions of greatest active extensional tectonism.

cipitate a mismatch between estimates of geologic, paleoseismological, and geodetic deformation rates. In chapter 6, Trexler et al. interpret the late Pliocene paleogeography of the southern Honey Lake basin from the lithofacies of the Neogene sedimentary rocks and document the subsequent (<3 Ma) deformation of the dextral Honey Lake fault system. In chapter 7, Hinz et al. use a combination of geological mapping, geochronology, and paleomagnetic analysis to establish piercing points across the Honey Lake fault system of the northern Walker Lane and conclude that deformation rates integrated over the last 3–6 million years are consistent with Quaternary fault studies and GPS geodetic rates. Kelly and Secord, in chapter 8, develop the

vertebrate biostratigraphy of the Hunter Creek Sandstone in the Verdi Basin west of Reno, Nevada; their work establishes that the area was an active depositional basin from late Miocene to late Pliocene time. Mass et al. (chapter 9) present a detailed investigation of the stratigraphy and structure of the Boca basin, within the eastern Sierra Nevada at the latitude of Reno, and conclude that the basin did not experience deformation until the Pliocene. To the south in the Gardnerville basin, between the Sierra Nevada frontal fault system and the through-going transcurrent faults of the northern Walker Lane, Cashman et al. (chapter 10) demonstrate that synextensional deposition (from >7 Ma to the present) occurred without accompanying strike-

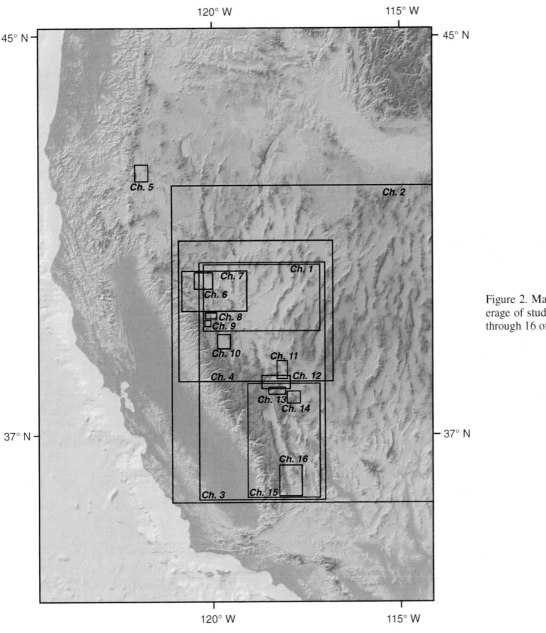

Figure 2. Map showing the spatial coverage of studies presented in chapters 1 through 16 of this volume.

slip motion in this region. In the central Walker Lane, Ferranti et al. (chapter 11) integrate the results of fault slip inversion, gravity modeling, and paleomagnetic analysis of rocks within and surrounding an extensional basin to demonstrate that displacement compatibility is retained around a curved array of extensional and strike-slip faults. In chapter 12, Petronis et al. use paleomagnetic analysis of Neogene volcanic rocks to show that pre-late Pliocene deformation in the central Walker Lane was accommodated by moderate vertical-axis rotation. Tincher and Stockli (chapter 13) provide new thermochronologic data and fault slip analysis to document reactivation of the White Mountain fault zone during right-transtension since 3 Ma. In

chapter 14, Oldow et al. provide a detailed analysis of the spatial and temporal pattern of late Neogene synextensional deposition in the upper plate of a large magnitude extensional complex formed in a stepover between right-lateral faults. Jayko (chapter 15) performs a regional analysis of Neogene erosional surfaces and concludes that they are regionally correlative and provide a regional vertical datum for reconstruction of post mid-Miocene deformation. In chapter 16, Unruh and Hauksson analyze seismicity in the southern Walker Lane belt or Eastern California shear zone and northern Mojave block to provide a detailed view of contemporary deformation in an evolving continental shear zone.

REFERENCES CITED

Bennett, R.A., Davis, J.L., and Wernicke, B.P., 1999, Present-day pattern of Cordilleran deformation in the western United States: Geology, v. 27, p. 371–374, doi: 10.1130/0091-7613(1999)027<0371:PDPOCD>2.3.CO;2.

Bennett, R.A., Wernicke, B.P., Niemi, N.A., Friedrich, A.M., and Davis, J.L., 2003, Contemporary strain rates in the northern Basin and Range province from GPS data: Tectonics, v. 22, doi: 10.1029/2001TC001355.

Cashman, P.H., and Fontaine, S., 2000, Strain partitioning in the northern Walker Lane, western Nevada and northeastern California: Tectonophysics, v. 326, p. 111–130, doi: 10.1016/S0040-1951(00)00149-9.

Dixon, T.H., Miller, M., Farina, F., Wang, H.Z., and Johnson, D., 2000, Present-day motion of the Sierra Nevada block and some tectonic implications for the Basin and Range province, North American Cordillera: Tectonics, v. 19, p. 1–24, doi: 10.1029/1998TC001088.

Dixon, T.H., Robaudo, S., Lee, J., and Reheis, M.C., 1995, Constraints on present-day basin and range deformation from space geodesy: Tectonics, v. 14, p. 755–772, doi: 10.1029/95TC00931.

Eddington, P.K., Smith, R.B., and Renggli, C., 1987, Kinematics of Basin and Range intraplate extension, *in* Coward, M.P., Dewey, J.F., and Hancock, P.L., eds., Continental extension tectonics: Geological Society, London, Special Publication 28, p. 371–392.

Faulds, J.E., Henry, C.D., and Hinz, N.H., 2005, Kinematics of the northern Walker Lane: An incipient transform fault along the Pacific–North American plate boundary: Geology, v. 33, p. 505–508, doi: 10.1130/G21274.1.

Gan, W., Svarc, J.L., Savage, J.C., and Prescott, W.H., 2000, Strain accumulation across the eastern California shear zone at latitude 36 degrees 30′N: Journal of Geophysical Research, v. 105, p. 16,229–16,236, doi: 10.1029/2000JB900105.

Hammond, W.C., and Thatcher, W., 2004, Contemporary tectonic deformation of the Basin and Range province, western United States: 10 years of observation with the Global Positioning System: Journal of Geophysical Research, v. 109, p. B08403, doi: 10.1029/2003JB002746.

Henry, C.D., and Perkins, M.E., 2001, Sierra Nevada–Basin and Range transition near Reno, Nevada: Two-stage development at 12 and 3 Ma: Geology, v. 29, p. 719–722, doi: 10.1130/0091-7613(2001)029<0719:SNBART>2.0.CO;2.

Locke, A., Billingsley, P., and Mayo, E.B., 1940, Sierra Nevada tectonic patterns: Geological Society of America Bulletin, v. 51, p. 513–540.

McQuarrie, N., and Wernicke, B.P., 2005, An animated tectonic reconstruction of southwestern North America since 36 Ma: Geosphere, v. 1, p. 147–172, doi: 10.1130/GES00016.1.

Miller, M.M., Johnson, D.J., Dixon, T.H., and Dokka, R.K., 2001, Refined kinematics of the Eastern California shear zone from GPS observations, 1993–1998: Journal of Geophysical Research, v. 106, p. 2245–2263, doi: 10.1029/2000JB900328.

Oldow, J.S., 1992, Late Cenozoic displacement partitioning in the northwestern Great Basin, *in* Craig, S.D., ed., Structure, tectonics, and mineralization of the Walker Lane, Walker Lane Symposium Proceedings Volume: Reno, Nevada, Geological Society of Nevada, p. 17–52.

Oldow, J.S., 2003, Active transtensional boundary zone between the western Great Basin and Sierra Nevada block, western US cordillera: Geology, v. 31, p. 1033–1036, doi: 10.1130/G19838.1.

Oldow, J.S., Aiken, C.L.V., Hare, J.L., Ferguson, J.F., and Hardyman, R.F., 2001, Active displacement transfer and differential block motion within the central Walker Lane, western Great Basin: Geology, v. 29, p. 19–22, doi: 10.1130/0091-7613(2001)029<0019:ADTADB>2.0.CO;2.

Oldow, J.S., Geissman, J.W., and Stockli, D.F., 2008, Evolution and strain reorganization within late Neogene structural stepovers linking the central Walker Lane and northern Eastern California shear zone, western Great Basin, *in* Dewey, J.F., ed., Lithospheric-Scale Transtension: International Geology Review, v. 50, p. 270–290, doi: 10.2747/0020-6814.50.3.270.

Rogers, A.M., Harmsen, S.C., Corbett, E.J., Priestley, K., and dePolo, D., 1991, The seismicity of Nevada and some adjacent parts of the Great Basin, *in* Slemmons, D.B., ed., Neotectonics of North America, Decade Map Volume 1: Boulder, Colorado, Geological Society of America, Geology of North America, p. 153–184.

Savage, J.C., Lisowski, M., and Prescott, W.H., 1990, An apparent shear zone trending north-northwest across the Mojave Desert into Owens Valley: Geophysical Research Letters, v. 17, p. 2113–2116, doi: 10.1029/GL017i012p02113.

Stewart, J.H., 1988, Tectonics of the Walker Lane belt, western Great Basin: Mesozoic and Cenozoic deformation in a shear zone, *in* Ernst, W.G., ed., Metamorphism and crustal evolution of the western United States: Englewood Cliffs, New Jersey, Prentice-Hall, p. 681–713.

Thatcher, W., Foulger, G.R., Julian, B.R., Svarc, J., Quilty, E., and Bawden, G.W., 1999, Present-day deformation across the Basin and Range Province, western United States: The Sciences, v. 283, p. 1714–1718.

Unruh, J.R., Humphrey, J., and Barron, A., 2003, Transtensional model for the Sierra Nevada frontal fault system, eastern California: Geology, v. 31, p. 327–330, doi: 10.1130/0091-7613(2003)031<0327:TMFTSN>2.0.CO;2.

Wallace, R.E., 1984, Patterns and timing of late Quaternary faulting in the Great Basin Province and relation to some regional tectonic features: Journal of Geophysical Research, v. 89, p. 5763–5769, doi: 10.1029/JB089iB07p05763.

Wernicke, B., 1992, Cenozoic extensional tectonics of the U.S. Cordillera, *in* Burchfiel, B.C., Lipman, P.W., and Zoback, M.L., eds., The Cordilleran Orogen: Conterminous U.S.: Boulder, Colorado, Geological Society of America, Geology of North America, v. G-3, p. 553–581.

Zoback, M.L., 1989, State of stress and modern deformation of the northern Basin and Range province: Journal of Geophysical Research, v. 94, p. 7105–7128, doi: 10.1029/JB094iB06p07105.

MANUSCRIPT ACCEPTED BY THE SOCIETY 21 JULY 2008

The Geological Society of America
Special Paper 447
2009

Geodetic observation of contemporary deformation in the northern Walker Lane: 1. Semipermanent GPS strategy

Geoffrey Blewitt*
William C. Hammond
Corné Kreemer
Nevada Bureau of Mines and Geology and Nevada Seismological Laboratory, University of Nevada, Reno, Nevada 89557-0178, USA

ABSTRACT

As of October 2005, the semipermanent Global Positioning System (GPS) network called MAGNET (Mobile Array of GPS for Nevada Transtension) included 60 stations and spanned 160 km (N-S) × 260 km (E-W) across the northern Walker Lane and central Nevada seismic belt. MAGNET was designed as a cheaper, higher-density alternative to permanent networks in order to deliver high-accuracy velocities more rapidly than campaigns. The mean nearest-neighbor spacing is 19 km (13–31 km range). At each site, the design facilitates equipment installation and pickup within minutes, with the antenna mounted precisely at the same location to mitigate eccentricity error and intersession multipath variation. Each site has been occupied ~50% of the time to sample seasonal signals. Using a custom regional filtering technique to process 1.5 yr of intermittent time series, the longest-running sites are assessed to have velocity accuracies of ~1 mm/yr. The mean weekly repeatability is 0.5 mm in longitude, 0.6 mm in latitude, and 2.1 mm in height. Within a few years, MAGNET will characterize strain partitioning in the northern Walker Lane to improve models of (1) geothermal activity, which is largely amagmatic in the Great Basin, (2) seismic hazard, (3) the ways in which northern Walker Lane accommodates strain between the Sierra Nevada block and the extending Basin and Range Province, and (4) Neogene development of the northern Walker Lane and its broader role in the ongoing evolution of the Pacific–North America plate-boundary system. MAGNET's design is generally applicable to regions with an abundance of vehicle-accessible rock outcrops and could be replicated elsewhere.

Keywords: GPS, Walker Lane, strain, tectonics.

INTRODUCTION

Scientific Objectives

By October 2005, the semipermanent global positioning system (GPS) network (Fig. 1) known as MAGNET (Mobile Array of GPS for Nevada Transtension) consisted of 60 stations built for the purpose of quantifying and characterizing crustal strain rates in the northern Walker Lane and central Nevada seismic belt. Funded by the Department of Energy, the primary objective of MAGNET is to improve our understanding of geothermal systems in the Great Basin that are known to be largely amagmatic. A working hypothesis is that transtensional tectonics is favorable to geothermal systems through enhanced heat flow (from crustal

*Corresponding author e-mail: gblewitt@unr.edu.

Blewitt, G., Hammond, W.C., and Kreemer, C., 2009, Geodetic observation of contemporary deformation in the northern Walker Lane: 1. Semipermanent GPS strategy, *in* Oldow, J.S., and Cashman, P.H., eds., Late Cenozoic Structure and Evolution of the Great Basin–Sierra Nevada Transition: Geological Society of America Special Paper 447, p. 1–15, doi: 10.1130/2009.2447(01). For permission to copy, contact editing@geosociety.org. ©2009 The Geological Society of America. All rights reserved.

thinning and lower-crustal magmatic intrusion) and permeability (from strike-slip faulting that penetrates the entire crust). Preliminary studies have indicated a correlation between the magnitude and style of crustal strain rates in the Great Basin and the location of economic and subeconomic amagmatic geothermal systems (Blewitt et al., 2003). Since MAGNET will constrain the rate and style of activity on faults in the northern Walker Lane and central Nevada seismic belt, it can also be used to improve seismic hazard assessment in this region.

Both these applications fundamentally depend on the ability of MAGNET to accurately resolve surface strain rates, and on using such kinematic data to improve tectonic models of the region (Kreemer et al., this volume). From this perspective, MAGNET is poised to assess how the northern Walker Lane accommodates strain between the blocklike motion of the Sierra Nevada and the extending Basin and Range Province. From an even broader perspective, MAGNET should provide important data to constrain models of Neogene development of the northern Walker Lane,

and the broader role of the Walker Lane in the ongoing evolution of the Pacific–North America plate-boundary system.

Here, we discuss the design and implementation of MAGNET, which is a new type of GPS network that offers advantages in accuracy and resolution time over traditional GPS campaigns, and yet it is far less expensive than permanent GPS networks. Moreover, the semipermanent design of MAGNET allows for a higher density of stations than permanent GPS for a fixed amount of available funding. Thus, MAGNET is primarily designed for the rapid resolution of highly accurate velocities with relatively high spatial sampling density, and thus it can be regarded as near-optimal for purposes of regional mapping of the strain-rate tensor. The emphasis of this paper is on the technological and logistical methodology employed by MAGNET and providing an initial assessment of MAGNET's performance. The goal of this paper is therefore to disseminate information that might be useful to other research groups who could benefit from applying the techniques described here.

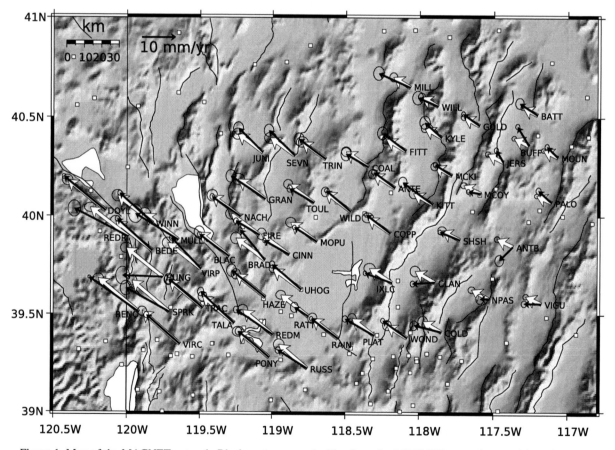

Figure 1. Map of the MAGNET network. Black vectors are velocities from the MAGNET network rotated into the model North America reference frame using only up to 1.5 yr of intermittent data. White vectors are predicted interpolated velocities at the MAGNET sites from the strain rate model of Kreemer et al. (this volume) obtained without the MAGNET results. Error ellipses represent 95% confidence but do not account for known systematic errors that can dominate for newly observed stations. White squares are locations of global positioning system (GPS) sites used in the strain-rate model. The reference frame is attached to stable North America, as defined by the stable North America reference frame (SNARF) Working Group (Blewitt et al., 2005).

Comparison of GPS Geodetic Survey Methods

Geodetic GPS surveys have mainly been conducted in two quite different modes (in the Walker Lane and more generally elsewhere): permanent GPS and campaign GPS. We now briefly describe and compare these methods in terms of advantages and limitations that might relate to specific requirements of a project.

Campaign GPS

Campaign GPS follows the recipe of traditional surveying with certain benefits over previous surveying technology, such as not having any need for line-of-sight between stations, no fundamental limitation on station separation, and all-weather capabilities. Campaigns are typically conducted every year or two over a period of ~10 yr, more or less. GPS antennas are typically mounted on tripods, centered over a permanent monument in the ground using a spirit-leveled optical tribrach. The height of the antenna above the monument thus varies with each setup of the tripod and thus is measured using a steel tape or measuring rod. The ability to center the antenna and measure its height accurately depends on the skill of the surveyor, on the calibration of the tribrach, and on the stability of the tripod, which can be subject to wind disturbance, ground moisture, and settling during the session. The fact that the antenna is physically located at a different point every campaign introduces another type of error. The phenomenon known as "multipath" occurs from the interference of phase arrivals from different paths that the satellite signal can take before reaching the antenna, for example, through ground reflections. This interference pattern can be very sensitive to the height of the antenna above the ground. In general, the outcome of multipath is a systematic bias in the estimated coordinates of the station. Having the antenna set up at different heights thus introduces a different bias each time the tripod is set up. A further possible problem at some GPS campaign sites is monument stability. Campaign GPS ultimately relies on a long time period (~10 yr) to produce accurate station velocities to reduce the effect of setup (or "eccentricity") error, variable multipath error, and other systematic and random errors in the epoch coordinate estimates. Another aspect of campaigns is budgetary; concern for security often demands that an operator be on site with the equipment during the session, thus adding to the cost of a campaign.

Permanent GPS

Permanent GPS is the ultimate method in terms of accuracy, and it mitigates many of the aforementioned problems with campaign GPS. The key advantages of permanent geodetic GPS are (1) the stability of the antenna, often mounted directly onto a very stable monument that is anchored deeply in bedrock (Langbein et al., 1995), and (2) the product is a continuous time series of station coordinates, which is important to characterize and possibly mitigate transient signals that may or may not be of tectonic origin. For example, seasonal signals can be filtered and removed (Blewitt and Lavallée, 2002), and nonlinear postseismic relax-

ation signals can be monitored over a broad range of time scales. The two main disadvantages of permanent GPS are (1) for a fixed amount of funding, fewer stations can be installed and have their velocities determined, and (2) siting and installation of a permanent station can be difficult, for example, requiring a permit to develop a permanent structure on the site. Since permanent GPS sites are typically easily accessible by automobiles (as a mobile drill rig is often required to build the monument), security might in some cases be a significant problem. Therefore, the location of a permanent station can in some cases be more of a compromise from a purely scientific point of view. For example, a station on a mountain top may only be permitted to be situated adjacent to other infrastructure, such as microwave towers and chain link fences, which might produce interference or increase the level of multipath. Often such environments (typically in enclosed areas on the tops of hills or mountains) are not static, but undergo sporadic development, further changing the electrical environment. These siting difficulties also add to the total cost, which is now typically in the range of $15,000–30,000 per station. For a relatively small, set amount of funding, permanent GPS may not be an option to meet the goals of a geodetic project, particularly if relatively high spatial sampling is required to map the variations in the strain-rate tensor across a region.

Semipermanent GPS

Semipermanent GPS is a recent concept. As the name suggests, the method involves moving a set of GPS receivers around a permanently installed network of monuments, such that each station is observed some fraction of the time. In practice, a set of GPS receivers can literally remain in the field for their entire life span, thus maximizing their usage. The monuments are designed with special mounts so that the GPS antenna is forced to the same physical location at each site. This has the advantage of mitigating errors (including possible blunders) in measuring the antenna height and in centering the antenna horizontally. This also has the advantage of reducing variation in multipath bias from one occupation session to another. The period of each "session" depends on the design of the operations. At one extreme, some stations might act essentially as permanent stations (though the equipment is still highly mobile), thus providing a level of reference frame stability, and some stations may only be occupied every year or two, in order to extend or increase the density of a network's spatial coverage.

We suggest that several sessions be planned per year if possible so as to sample potential seasonal signals. For example, if there are twice as many stations as receivers, then each station is occupied on average 50% of the time, and therefore sessions of anywhere in the region of 1–3 mo would suffice to sample the seasonal signal. The advantages of semipermanent GPS include enhanced spatial coverage as compared to permanent GPS, velocity accuracies closer to permanent GPS than campaign GPS, and time to achieve specified velocity accuracy also closer to permanent GPS, as often the seasonal signal is a fundamentally limiting factor (Blewitt and Lavallée, 2002). Another advantage is the lack of complexity in the station structure (only the monument is permanent), and therefore time is saved that would otherwise be needed for permitting (which could translate into more rapid scientific conclusions). A semipermanent station that uses a steel rod epoxied into a drill hole in bedrock can be completely installed and running within 20 min of arrival at the site. This simple setup produces a relatively low-multipath environment as compared to the chain-link enclosures commonly found at permanent sites. Also, the cost savings over permanent stations might be used to purchase more receivers. A disadvantage is that the time series is not continuous, but rather it is intermittent, so a transient geophysical signal might not be resolved very well, or perhaps missed entirely if it occurred between occupations. Another disadvantage over permanent GPS is that the mobile equipment may be more susceptible to damage and power failure due to transportation, wind, or rodents, which translates into a percentage of lost data and the need for higher maintenance costs (especially for solar panels). In contrast, permanent installations can easily be reinforced, as weight is not a design factor. Another advantage of permanent GPS as compared to semipermanent GPS is that a more robust monument can be constructed (Langbein et al., 1995), such as the Wyatt-type braced deep-anchored monuments (at a depth of ~10 m) that are now commonplace in western North America. To do this for semipermanent GPS would defeat the benefits: low cost, rapid installation, and minimal permitting.

Geodetic Requirements

For a general scientific goal that requires the mapping of crustal strain rates, a primary requirement is to sample sufficient stations spanning the region of interest with an appropriate spacing. Furthermore, the station velocities must be determined to within errors that are far smaller than the true (or anticipated) spatial variability in velocity signals of interest. It might also in some cases be important to monitor possible time variation in station velocity that might be geophysical in origin, or possibly due to some systematic error (such as seasonal variation in environmental conditions) that must be monitored and mitigated. There may also be a requirement on how quickly the velocity accuracy can be achieved to meet the goals of a specific project deadline. These considerations summarize the driving factors behind geodetic requirements. In turn, they can lead to very specific requirements for particular projects, which translate into details of experiment design. We shall now consider each of these main general areas that drive the geodetic requirements and relate them to the GPS survey methods discussed previously.

Spatial Resolution

In general, spatial resolution should be compatible with the characteristic distance scale over which geophysical signals vary. For volcano monitoring, this distance scale can be very small. For tectonic strain, the appropriate distance scale is typically much larger. Blewitt (2000) presented a rigorous solution for optimal network design (station placement) and applied it to idealized sample cases. A key feature of crustal strain associated with the earthquake cycle is that, typically, the characteristic distance scale corresponds to the thickness of the seismogenic zone within which faults are typically locked between earthquakes. Therefore, in the interseismic period, the surface strain rate tends to be smoothed and cannot change very much over distances <15 km. Maximum variation in strain rate is typically found adjacent to active faults that have a large slip rate at depth (corresponding to their geological slip rate), with the peak strain occurring in the region above the locked zone. However, there is no way to distinguish from the pattern of surface strain whether the strain field is due to a single very deeply locked fault, or due to a series of subparallel faults that are locked to a shallow depth, such as in a shear zone. An increased density of GPS stations does not help resolve the model in this case. We therefore suggest that there is a point of diminishing scientific return (for a fixed investment in equipment) to measure strain with a spatial resolution smaller than the thickness of the brittle crust. A reasonable nominal spacing for most geodetic networks measuring tectonic strain rates is therefore in the range of ~10–30 km. A spacing smaller than this is unlikely to enhance scientific return, and a spacing larger than this would only serve to constrain the level and style of tectonic activity within a region, but would be less capable of identifying the currently active Quaternary faults. However, it should be noted that a broadly spaced network, such as the BARGEN permanent GPS network with typical spacing ~90 km (Bennett et al., 2003) can serve an important role to provide a regional reference frame and a regional-scale tectonic context to more focused investigations, and it can help to identify regions that deserve more focus.

Velocity Accuracy

A common and simple objective of a GPS geodetic network is to resolve with adequate accuracy the velocities of all stations (within some reference frame). This raises two questions: (1) How long will it take to achieve the required velocity accuracy? (2) How often must the time series of positions be sampled in order to reduce random and systematic errors? This all assumes that the unknown geophysical motion of a site is linear (i.e., a constant velocity), and that any nonlinear motions can be either modeled (such as solid Earth tides) or adequately characterized (such as seasonal coordinate variation). In the case of campaigns where sites are sampled every one or two years, typically ~10 yr are required to achieve an accuracy of <<1 mm/yr in station velocity. However, permanent stations can typically achieve this accuracy within 1.5–2.5 yr, a critical time period when it first becomes possible to characterize and mitigate a seasonal signal (Blewitt and Lavallée, 2002). After 5 yr, permanent station velocity accuracy in a regional reference frame is typically ~0.1 mm (Davis et al., 2003). Stochastic models of permanent GPS time series show that monument stability can be a real issue with regard to how often it makes sense to sample the site position in order to resolve velocity (Langbein and Johnson, 1997; Williams, 2003). In the extreme case that monument instability is dominated by a random walk process it can be shown that, in effect, only the first and last data points provide information on site velocity, and so more frequent sampling does not help. Most GPS time series however, like most natural processes, can be characterized by a power-law noise, implying a finite time correlation (Agnew, 1992). This implies that frequent sampling will improve velocity estimation, but only to a limit, beyond which higher frequency sampling does not help. This suggests that a semipermanent observational strategy might be close to optimal in terms of resolving station velocity, assuming adequate sampling of two key cycles: the seasonal cycle and the diurnal cycle. For this reason, the fundamental epoch estimate ought to be based on a full 24 h session, and the station should be visited several times per year.

Temporal Resolution

The previous discussion addressed how often measurements should be made and for how long in order to resolve station velocities. However, if temporal variation in station velocity is expected, or if transient phenomena are important scientific objectives, then permanent GPS networks may be required. For example, the deep-crustal magma-injection event near Lake Tahoe in 2003 (Smith et al., 2004) produced an ~10 mm transient displacement over a 6 mo period at a permanent GPS station nearby. In principle, GPS campaign data might be used to detect this ~10 mm offset, though a 6 mo transient signal could not have been inferred. It is also possible that the 10 mm offset might be buried in the GPS campaign data noise, or it might have been attributed to a change in equipment or antenna setup error. Semipermanent GPS would not be temporally optimal in this scenario, though the chances of discovery would have actually been enhanced due to the higher spatial density that semipermanent GPS networks can provide. Thus, we can think of semipermanent GPS as trading off temporal resolution for spatial resolution. Indeed, it was fortunate that there was a permanent station that just happened to be within range of a detectable signal, given that the permanent GPS station spacing in this region is ~100 km. Perhaps the most obvious transient phenomenon that might be observed by GPS is a large earthquake. In this respect, permanent GPS is at a disadvantage compared to semipermanent GPS. In the "MAGNET-style" of deployment, all semipermanent GPS instruments are always operating somewhere, so they have no inherent disadvantage and can detect any transient signals as well as permanent stations. Additionally, semipermanent GPS have the advantage that more equipment can be purchased for the same cost, and they can be quickly redeployed to a more optimal configuration to map out the coseismic displacement (because all the monuments would have been presurveyed) and to monitor postseismic deformation. Once an earthquake has occurred, a decision can be made to keep the semipermanent stations in operation continuously (like permanent stations) for as long as the investigation requires. Moreover, instruments that might be borrowed from elsewhere could, in principle, be rapidly deployed to unoccupied monuments in the semipermanent network, or new semipermanent monuments could be quickly installed and inte-

grated seamlessly into the existing network. A disadvantage of semipermanent GPS is the lack of telemetering. The knowledge of a transient is delayed by data acquisition latency, and so the equipment might unfortunately be moved while the transient is in progress (if not associated with an obvious large event). Although this could be rectified with cellular or satellite data transfer, it impacts the cost efficiency (which will be outlined in a later section), and even so, a "silent" transient could still easily go undetected prior to moving the instrument(s).

METHODOLOGY

Semipermanent Network Design

Setting

The MAGNET network is designed to measure tectonic strain rates spanning the region bounded by the Sierra Nevada block to the west and the central Basin and Range to the east at the latitude of the northern Walker Lane and central Nevada seismic belt. This is a region of the Great Basin that has an enhanced number of known amagmatic geothermal systems, and it is a region of relatively high strain, with ~10 mm/yr of relative velocity across the network, mainly focused in the western Great Basin adjacent to the Sierra Nevada block. The central Nevada seismic belt is also expected to be undergoing considerable postseismic deformation as a result of a sequence of great earthquakes during the last century, where the postseismic deformation may exceed the magnitude of interseismic strain accumulation. Scientific goals require a velocity accuracy <1 mm/yr and a spatial resolution of 10–30 km.

Spatial Characteristics

As of October 2005, the MAGNET network consists of 60 stations, for which installation began in January 2004. The network spans 160 km N-S and 260 km E-W (roughly bounded by the quadrilateral defined by Carson City, Susanville, Battle Mountain, and Austin). The network is approximately uniform in distribution. Nearest-neighbor stations spacing ranges from 13.2 to 30.9 km, in accordance with the expected smoothness of the strain field imposed by crustal thickness (as previously noted). The mean nearest-neighbor station distance is 19.2 km. (Note added in proof: As of August 2008 (the time of this proof), the network has grown to 307 stations occupied by 56 receivers, spanning as far north as the Oregon border and as far south as the eastern California shear zone and northwestern Arizona.)

Tectonic Sensitivity

Locating GPS monuments in basins such as the Carson Sink would not be particularly useful, as hydrological effects in unconsolidated sediments can be expected to be significant compared to the underlying tectonic signals (e.g., Bell et al., 2002). Therefore, we have chosen to always site monuments in rock outcrops. Station spacing is in practice limited by availability of suitable and accessible rock outcrops. The largest gaps in the network span the largest basins, such as the Carson Sink. Sites located in bedrock are expected to be relatively immune to hydrological effects, except for possible elastic loading effects, which are computed to be less than one millimeter (relative) in the Great Basin. For example, Elósegui et al. (2003) identified perhaps the largest loading signals in the Great Basin of ~1 mm associated with seasonal loading of the Great Salt Lake. Atmospheric pressure loading and continental-scale hydrological loading may move the Great Basin as much as several millimeters; however, the relative motion within the Great Basin would be an order of magnitude smaller (van Dam et al., 2001). One remaining possible hydrological effect on bedrock sites relates to the sometimes substantial pumping of water from mines in the Great Basin. What is not yet understood is whether the underlying rock behaves purely elastically under such stress (in which case it would not be a significant problem), or whether failure might occur that could produce detectable surface deformation. Such a phenomenon, if it exists, might only be revealed by a sufficiently high-resolution technique such as interferometric synthetic aperture radar (InSAR), which has served as a useful tool to discriminate tectonic from hydrologic signals (Bawden et al., 2001; Bell et al., 2002).

Siting

Other than the criteria discussed previously on general network design to meet scientific goals, there are several practical factors considered in siting stations. The basic principle is to maximize the quantity, quality, and usefulness of the resulting data set, which includes a broad range of considerations, from how fast a site can be visited to securing a site from theft. In general, it is important to note that stringent requirements cannot generally be imposed if adequate spatial resolution (in our case, ~20 km) is to be maintained. This usually requires a site to be selected within a radius of ~5 km of a target candidate site. For comparison, the one standard deviation of our nearest-neighbor separation is 4.1 km. The exception to this rule is where the resulting data from any location in a given area would not be useful, or would divert resources that could be better spent elsewhere. We now consider each important aspect of siting, in a very approximate order of priority.

Accessibility

Of course, if we cannot access a site, it is of no use. Good accessibility means that more sites can be visited, which means more efficient use of funds, and more scientific return. The network has been designed so that operations can be sustained by day trips, where up to ~10 stations can be visited in one day. Typically one or two days of fieldwork per week by a technician is required, driving ~600 km. For this reason, all sites must be reasonably accessible by four-wheel drive all year-round, and the station should be no more than a 10 min hike from the truck. Due to time constraints, a hike should be less than 400 m in distance and less than 50 m in elevation. In most cases, a truck can be driven directly to the site. Whereas most of the time is spent driv-

ing on dirt roads, it is crucial to select looped routes that make the most use of the better roads. Seasonal conditions can vary greatly in the Great Basin, and contingency plans must be considered in the event of flash flooding and road washouts. Navigation to the sites (and to new candidate sites) is facilitated by a GPS-enabled field computer with digital maps (U.S. Geological Survey 1:24,000 series), or handheld GPS (which are also used for meter-level recording of the final selected site location).

Security

Nevada is the sixth least densely populated state in the United States, and most of the population lives in a few metropolitan areas. Approximately 83% of the land in Nevada is federally administered (the highest percentage in the United States), with some counties as high as 98%. All of our GPS sites are on open-access land managed by the Department of Interior's Bureau of Land Management. While this is good for access, it can be poor for security. Fences in most areas are rare or nonexistent, and, where they do exist, they are not intended to keep people out (but rather cattle in). Stations are selected so that they cannot be seen from the road, and they are situated away from populated areas and evidence of recent human activity. Needless to say, the impact of a stolen receiver is immense: the loss of data is minor; rather, the loss of the receiver to make future measurements is the main problem, plus the likelihood that the site will have to be abandoned, and so all previous data collected there may be wasted. Nevertheless, security cannot be overly cumbersome, otherwise no data would be collected at all. The issue therefore is risk management. For example, we have assessed that it would be an acceptable risk to lose an instrument once every year or two. A record of zero thefts might be achievable if we only selected areas that were totally unpopulated (and even this is doubtful), but such an approach would be inconsistent with our goals regarding seismic hazard in populated areas. As of 1.5 yr of network operations, we have lost only a $2000 insurance deductible due to the theft of one instrument out of a total of 34, representing ~0.4% loss per year on equipment costs.

Rock Stability

In the case of rock stability, the issue is whether the monument drilled into the rock accurately represents the motion of Earth's crust. If not, then at worst, an anomalous motion will be incorrectly interpreted as tectonic signal; at best, the problem will be detected eventually, but years of fieldwork getting to the site will have been wasted. Rock stability can therefore have the highest priority in cases where "some data" is actually worse than "no data." One exception might be in very sparsely sited regions, where it may be useful in the future to measure near-field coseismic displacement (but not for purposes of measuring interseismic strain). Within the concept of semipermanent networks, monument installation must be inexpensive and quick. Our choice (described in more detail in Station Design/Monumentation) is to epoxy a stainless 7/8-inch-diameter steel pin into ~15-cm-deep hole drilled in a rock outcrop. The selected rock outcrop must not be too weathered or fractured, and it must resound at a high pitch when struck with a sledge hammer. All else being equal, basement rocks such as granite are preferred. Volcanic rocks are sometimes chosen where they have few fractures. For sedimentary rocks, subvertical beds are often avoided because, where the bed planes intersect with the surface, they are susceptible to weathering and fracturing. Detached rocks and boulders are to be avoided, though clearly there are situations where this might be difficult to ascertain even to a trained geologist with only eyes and a hammer. The key therefore is to reject candidate sites that might well be good, but nevertheless are questionable.

Sky Visibility

Sites with the best sky visibility above 15° elevation are preferred. The tops of hills are preferred. Poor sky visibility to the north is acceptable, due to the fact that GPS satellites do not track within an ~35° cone around the north celestial pole. At the latitude of our network, this implies a large blank circle in the sky north of the station. Therefore south-facing slopes are often selected. South-facing slopes are also optimal for powering solar panels set on the ground. Areas with trees or other sky-blocking features, such as power-line towers or buildings, are to be avoided.

Multipath Environment

Good sky visibility also tends to mitigate multipath, because there will generally always be a direct GPS signal that is stronger than any reflected signal. Any significant (large, smooth) reflective surfaces that might create significant multipath should be avoided. Metallic structures such as fences and radio should also be avoided if possible. However, even relatively poor multipath environments can ultimately be acceptable; modern receivers do much to correct the problem through signal-processing techniques (exploiting the fact that reflected signals always arrive late and thus skew the phase-correlation pattern). For example, spatial sampling, accessibility, security, and rock condition took priority in one rare case (Rattlesnake Hill, Fallon).

Station Design

We have developed 12 key principles for the design of a semipermanent station. A station should be:

(1) mobile: to leave as little a permanent footprint as possible to minimize cost and permitting issues, and as a corollary, have as much of the station as possible be mobile;

(2) stable: to attach the monument effectively to the sited rock outcrop to maximize stability, using the same method at each site;

(3) easy: to monument the site as quickly as possible (accounting for the possible need to carry necessary equipment up to 400 m to the site) so that GPS data acquisition can begin immediately, and so that more sites can be installed in one day, and rapidly following a major event such as a large earthquake;

(4) repeatable: to ensure the antenna is force-mounted to precisely the same position at a station every session to minimize eccentricity and multipath error in the determination of velocity;

(5) fast: once monuments have been created, to be able routinely to install and move equipment while on site as quickly as possible, to increase the number of sites that can be visited in one day;

(6) modular: to have interchangeable components to facilitate ease of on-site system testing and repair, and swapping of parts;

(7) convenient: to minimize the time required to download data in the case that the equipment is not being moved (i.e., for stations acting as "permanent" for the time being, such as might be the case following a large earthquake);

(8) uniform: to have functionally identical stations, so that equipment is not specific to certain stations, and thus logistical efficiency is improved (as there are less planning constraints and fewer decisions to be made);

(9) invisible: to be difficult to discover accidentally unless in the immediate proximity to maximize security;

(10) secure: to deter opportunistic theft or vandalism if accidentally discovered;

(11) independent: to have adequate power and reliability while unattended for up to several months to maximize data collection; and

(12) robust: to minimize damage due to rodents, weather, and transportation.

We now describe the currently implemented design that meets these criteria.

Footprint

When the site is not in use, only two items remain on site: (1) the monument described next, and (2) an eye bolt epoxied into the rock as part of the security system. A 1 in. × 6 in. PV pipe with covering cap is placed over the monument for protection. The PV pipe also acts as a warning to avoid possible puncture of a random off-road vehicle (extremely unlikely, though theoretically possible).

Monumentation

The monument (Fig. 2A) is a 7/8-in.-diameter, 10-in.-long stainless steel pin that is epoxied into a hole drilled into a rock outcrop. We use a Hilti rock drill that is AC-powered using a hand-portable Honda generator. The hole is cleaned by blowing using a very light, powerful vacuum cleaner, also AC-powered. The monument design is based on the "Conquest Pin" by UNAVCO Inc. (http://www.unavco.org), with some modifications. Approximately 6 in. of the pin is epoxied below the ground in a hole drilled into the rock. The bottom half of the pin is milled to accommodate epoxy from the hole, and to better anchor the monument in place, withstanding vertical and rotational stresses. The hole itself is essentially friction tight, thus constraining lateral motion. The top of the monument has a 5/8 in. male screw thread for antenna mounting, as described in more detail later. The reference height of the monument is taken to be the collar immediately below the thread. As a backup, a groove is also milled into the side of the monument at a specific distance in case the top part of the monument for some reason gets damaged.

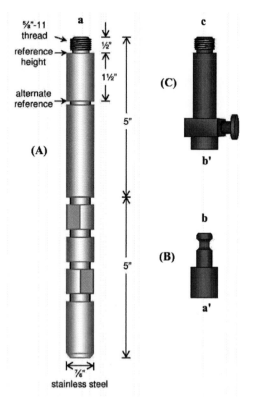

Figure 2. Diagram of monument and antenna mount design: (A) stainless steel monument, (B) monument adapter (Global Positioning System [GPS] quick-release adapter, Seco PN 5187-00), and (C) antenna adapter (GPS quick-disconnect adapter, Seco PN 5111-00). Parts that fit together are a–a' and b–b', and c-antenna (not shown).

Care must be taken to install the rod as vertically as possible, for which we use a spirit level several times during the drilling process. A tolerance of <3° is easily achieved, corresponding to <13 mm horizontal displacement between top and bottom of the rod. For example, this will ensure <1 mm horizontal displacement between different antenna types for which there is a 2 cm difference in phase center height. We use the same antenna types everywhere to further mitigate this potential source of error.

Antenna Mount

The international standard geodetic GPS antenna mount is a 5/8 in. male thread that inserts directly into the body of the GPS antenna. We have chosen a mount system that facilitates both rapid and accurate placement of the antenna. The system has two components manufactured by Seco originally intended for survey poles. The first part, the quick-release "monument adapter" (Fig. 2B) directly screws onto the 5/8 in. male thread of the monument down to the shoulder of the monument that defines the reference height. At the top of the monument adapter, there is a concave nipple. The second part, the quick-disconnect "antenna adapter" (Fig. 2C) screws into the 5/8 in. female thread of the antenna, where the bottom is designed to mount on the

first part, and it is secured using a quick-release spring-loaded button. This design has important features: (1) it facilitates quick (dis-)connection of the antenna itself, and quick (dis-)connection of the antenna cable; (2) it reduces the wear and tear that would arise from directly screwing the antenna onto the monument; and (3) it allows for arbitrary orientation of the antenna, and specifically, it allows for the antenna to be pointed toward geographic north (manually aided by a compass) so as to minimize the effect of azimuthal variation in antenna phase center. Furthermore, to minimize cost, the more-expensive antenna adapter moves with the receiver equipment rather than staying on site. No data logging by the field crew is required regarding the antenna location, because it is force-mounted to the same location every time. The Seco adapters are machined so that their heights are guaranteed to be consistent. The height between the reference mark (the shoulder on the monument at the bottom of the monument's male thread) and the reference point on the antenna (the base of the antenna's female thread) is precisely 100 mm.

Routine Installation

Speed of installation is partly facilitated by the antenna mount. A second important feature to improve the speed of routine installation is the mobile GPS system in a plastic carrying case, which also acts as a field enclosure. Finally, all equipment items, including solar panels, are light and easily portable. For example, the 32 W solar panels are foldable and fit easily into a backpack. The portable GPS equipment case is designed for portability. A 12 V, 80 Ah battery remains at each site in a battery box to facilitate routine fieldwork and can easily be augmented if more power is required (such as during weeks of snow cover), or replaced if depleted after a weeks of poor light conditions. When a receiver is picked up at a site, a compact flash card is swapped in the receiver rather than wasting time in the field downloading data to a computer.

Data Download

Data are recorded onto compact flash cards. The cards we use are manufactured by SanDisk and are of a industrial range, designed to withstand extreme cold and hot temperatures. It so happens that 64 MB cards can hold 3 mo of data. However, at the time of writing, 256 MB cards are the minimum capacity available, and so the card capacity is always going to be overengineered. We have never had a card fail. In the laboratory, data are automatically downloaded using a custom C-shell script running under Linux and converted to RINEX format. Files that are broken due to card swaps are automatically pieced back together using a database. The script connects to a custom database so as to automatically identify the station the card came from, without requiring any data entry from field notes. Data entry is required prior to the next download, so that the database can keep track of which receiver is at which site. The entire system is designed to minimize time in the field and minimize note-taking in the field. The only thing that needs to be recorded is the receiver ID number when it is installed at a site (which is required to identify filenames with sites).

Power and Reliability

We use a 12 V, 80 amp-hour lead-acid battery charged by a 32 W foldable solar panel connected to a 5 A SunWise controller. The battery sits in its own box to prevent possible corrosion of sensitive equipment from acid fumes. The battery also typically remains at the site unless it needs to be swapped out and taken back to the office for desulfation and recharge. We find that in winter months, sometimes the lack of sunlight or snow cover on the solar panels causes the battery to slowly discharge. In some cases, therefore, we double up the batteries in parallel. Another more sustainable option would be to double up the solar panels in winter (or use much more powerful solar panels all the time), though this requires a considerable investment. Even without any solar panels, we find that two 80 A-h batteries in parallel can power a Trimble 5700 receiver for ~3 wk, which is more than sufficient for a good epoch measurement. Given that batteries are subject to deep discharge on occasion, we use marine batteries, which are designed more for deep discharge than car batteries (which are designed rather to provide high current in short bursts). Car batteries would quickly break down if subject to a few deep discharges. By far the most likely cause for losing potential data is failure of the power system. However, by using a large capacity battery at a site, we are almost guaranteed to acquire at least a week of good data.

Station Similarity and Modularity

All stations are designed the same way to facilitate logistics, maintenance, reliability, and GPS accuracy. All systems are modular, which facilitates testing, repair, and swapping out of equipment. The heart of the system is the Trimble 5700 receiver. Three cables feed out from the carrying case: the antenna cable, a power cable to the external battery, and a power cable to connect a solar panel. All power cables are fused, and each segment can easily disconnect and be swapped out. All parts are numbered so that problems can be tracked, which is especially important when swapping out parts. All of the padlocks (240 in total) are keyed exactly the same. Simple repairs are very easy to do by a field technician at a moment's notice. If necessary, the GPS receiver can be reprogrammed or even have its firmware replaced in the field using a laptop computer. On very rare occasions this has been necessary, which we speculate may be associated with electrical storms that are common in the Basin and Range.

Damage Mitigation

We note three causes of damage that have led to failure. The most important is rodent damage, especially in early spring when rodents come out of hibernation. The most expensive loss is the antenna cable, worth approximately $100. Worse than this, by eating through either a battery or antenna cable, data is lost from that time forward. A very simple and inexpensive solution to this problem is to encase all cables in flexible plastic conduit. We use the type commonly used in automobiles that is ribbed, and is split lengthwise to facilitate rapid installation.

This actually also makes the cables easier to handle, further improving speed of installation.

The second most damaging factor is the wind. Solar panels are secured using nylon rope attached to available rocks. On rare occasions the rocks might be so sharp that the rope is cut, and the solar panels can then easily get destroyed by flapping in the wind. Detachment of the ropes has also happened if the site is accidentally discovered, then left alone once the warning label has been read (usually concealed by the solar panel which is laid on top of the box). Padlocks are made out of brass and are routinely lubricated. We have not yet suffered any significant damage due to transportation of equipment to the site, from water, extreme temperatures, or ultraviolet (UV) radiation. This is because our equipment is all commercially designed for the rigors of fieldwork. We would not, for example, recommend attempting to use GPS receivers more designed for permanent stations into such a regimen.

Security

Our approach to security is one of deterrence. The most important component to secure is the GPS receiver. This is achieved by binding the receiver inside a very tightly fitting bicycle U-lock (the Mini-Evolution by Kryptonite) that cannot be dislodged once locked. Typically, once locked, the U-lock never needs to be removed. Also, attached to the U-lock, there is a 5/8-in.-diameter, 4-ft-long braided steel cable looped on either end. The free end of the cable is passed through a hole in the carrying case and is then padlocked to an eye bolt epoxied into nearby rock. The cable is also passed through the padlocks holding the case shut, to make it more difficult to use a tool to break open the locks. The solar panel is also padlocked to the eye bolt. Someone sufficiently determined could break the padlocks and walk away with the box. However, removing the bicycle lock from the receiver itself would be extremely difficult without destroying the receiver. Finally, a large orange warning sticker is posted on each box to deliver a message to anyone discovering it, warning that the equipment is under GPS surveillance. While less important, a sticker on the battery box notes that the battery will only work with custom equipment.

Measurement Strategy

Receivers are set to log data every 15 s. Data files start and end on GPS midnight, which in Nevada corresponds to 4 p.m. local time in winter and 5 p.m. local time in summer. Typically, when a GPS receiver is installed at a station during the day, the few hours of data prior to GPS midnight are discarded in the subsequent analysis. In the office, data are subsequently processed in daily epoch batches. As a rule, daily files are only processed if they have at least 18 h of data, to minimize systematic biases associated with quasi-diurnal or semidiurnal signals such as multipath and tidal loading (Sanli and Blewitt, 2001). While in our time zone, the first day of a session never meets this criterion, but often the last day does.

Sites are visited several times per year so as to sample through the seasonal cycle. Typical sessions range from 2 to 12 wk. No attempt has been made to have a regular schedule (spatially or temporally), as we suspected that regularity can more likely lead to systematic artifacts that could be mitigated by a level of randomness in the measurement design. Random sampling leads to many more different ways through which the network becomes interconnected. From a practical point of view, a regular schedule would not be possible to keep anyway, especially due to extreme snow conditions in winter. Moreover, our equipment pool has been ramping up, and the network has been growing.

For velocity accuracy, the most important factor is the span of time since the very first and very last sessions, and, to a lesser extent, the percentage of time occupied. The reason for this is that the error in velocity estimation decreases linearly with the span of time (for a fixed amount of data), but at the most (in the white noise limit), it only decreases with the square root of the number of sessions (for a fixed span of time). Therefore, new station installations take clear priority over repeat occupations of existing sites.

Data Analysis

Data Processing

Starting with daily RINEX files, all data are processed using the GIPSY/OASIS II software from the Jet Propulsion Laboratory (JPL). The strategy used was precise point positioning (Zumberge et al., 1997), using dual-frequency carrier phase and pseudorange data, and precise orbit, clock, and reference frame transformation products publicly available from JPL. After automatic data editing for cycle slips and outliers (Blewitt, 1990), data were decimated to 300 s epochs. Estimated parameters included the three Cartesian coordinates of position as a constant over the day, a receiver clock parameter estimated stochastically as white noise, a zenith tropospheric parameter, and two tropospheric gradient parameters estimated stochastically as a random walk process (Bar Sever et al., 1998), and carrier phase biases to each satellite estimated as a constant. This was then followed by carrier-phase ambiguity resolution (Blewitt, 1989). The output station coordinate time series where then processed using a custom spatial filter, which we will now describe.

Spatial Filtering

Strain is a local quantity, and as such, it is relative station velocity that is important. Geocentric station velocity (referenced to the center of Earth) is not important in this context. However, the GPS data-processing system produces geocentric station coordinate time series. It is possible to filter these time series to optimize the estimation of relative velocities within a region. The key principle is to eliminate a common-mode bias in station coordinates at each epoch, which more reflects a variation in geocentric positioning rather than a variation at any particular station. This common-mode variation may be due to a variety of effects, such as satellite orbit and clock errors, global reference frame realization on that day, and real large-scale geophysical signals

such as atmospheric loading. Whether error or real signal, the common-mode variation is not the tectonic signal we seek. Elimination of the common-mode variation in the time series generally improves relative velocity determination, especially if different stations have different spans of data, or are occupied at different times. This is because the stations are sampling the common-mode signal at different times, and any difference caused by such sampling will map into the relative velocity estimates.

Methods to eliminate common-mode variations have been called regional filters or spatial filters, starting with Wdowinski et al. (1997). The basic concept is to estimate the average coordinate deviation over a regional network, and then subtract that deviation from the individual station coordinate time series. This procedure can then be iterated. The original method of Wdowinski et al. (1997) was to compute coordinate deviations with respect to each station's mean estimated position (for the purpose of analyzing data around the time of a large earthquake). For our purposes (to map strain rates), it is more appropriate to compute coordinate deviations with respect to each station's estimated constant velocity model.

A custom regional spatial filter was implemented as a series of C-shell scripts that utilize existing tools to transform data in the GIPSY/OASIS software, and it is available upon request. For this analysis, we implemented a filter that estimates the mean coordinate deviation from an initial model that uses only those sites with the best determined velocities (the longest-running sites). Specifically, stations were selected for which the formal error was <0.4 mm/yr in horizontal velocity components, which selects 34 data sets that span 1.0–1.5 yr. This innovative approach (versus averaging over the entire network) works better in principle because the shorter time series do not have a sufficiently adequate station motion model (epoch coordinate plus velocity) from which to infer the common-mode variation. As a result, the shorter time series tend to underestimate the magnitude of the common-mode variation, and thus bias the estimates.

A series of statistical tests, such as how well daily solutions within one week agree with each other, also serves to eliminate data outliers. The results discussed next demonstrate a factor of five improvement in the statistical quality of the station coordinate time series after performing spatial filtering, and considerable visual improvement in the smoothness of the resulting velocity field, especially for shorter time series, where one would predict the largest improvements to occur.

RESULTS

Network Performance

Table 1 shows a list of all stations, including their names, coordinates, time span of processed data, and "activity" percentage, defined as the fraction of possible weeks that have a valid (quality-assessed) solution (an upper bound on the percentage of weeks a site was occupied). These statistics therefore do not entirely reflect the intended measurement strategy, but

TABLE 1. STATION DATA ACQUISITION AND PERFORMANCE STATISTICS,
RANKED BY DATA SPAN

Station ID	Location			Span (yr)	Epochs (wk)	Active (%)	Weekly repeatability		
	Lat. (°N)	Long. (°W)	Height (m)				Lat. (mm)	Long. (mm)	H (mm)
RENO	39°31'	119°55'	1490	1.53	63	77	0.62	0.46	1.9
COPP	39°55'	118°12'	1280	1.53	35	42	0.42	0.43	1.7
WILD	40°01'	118°28'	1230	1.52	40	49	0.74	0.91	2.5
TOUL	40°04'	118°43'	1460	1.52	26	32	0.66	0.5	2.4
FIRE	39°54'	119°05'	1430	1.52	25	30	0.6	0.55	1.9
RAIN	39°23'	118°33'	1210	1.45	48	62	0.66	0.47	1.7
CINN	39°48'	118°53'	1290	1.45	41	53	0.69	0.55	2.0
MOPU	39°52'	118°42'	1250	1.45	37	47	0.5	0.36	2.5
ANTE	40°09'	118°10'	1450	1.42	29	38	0.43	0.55	1.5
BRAD	39°47'	119°03'	1330	1.42	37	49	0.43	0.46	1.7
BLAC	39°48'	119°18'	1280	1.42	35	46	0.36	0.46	1.8
NACH	40°00'	119°13'	1690	1.38	32	43	0.49	0.46	1.9
IXLC	39°40'	118°11'	1240	1.36	38	52	0.74	0.59	2.8
WOND	39°23'	118°05'	1590	1.36	35	48	0.49	0.45	2.4
WINN	39°58'	119°49'	1690	1.32	33	46	0.99	0.87	2.1
BEDE	39°49'	119°50'	1570	1.32	30	42	0.65	0.53	2.4
TALA	39°27'	119°16'	1640	1.32	16	22	0.39	0.46	1.1
GOLM	40°28'	117°36'	1470	1.32	39	55	0.46	0.49	1.8
MCKI	40°13'	117°47'	1590	1.32	36	51	0.57	0.3	1.7
MCOY	40°07'	117°35'	1180	1.32	34	48	0.39	0.35	2.4
DOYL	40°02'	120°09'	1690	1.28	39	57	0.62	0.49	2.2
VIRP	39°44'	119°30'	2080	1.27	38	56	0.49	0.44	1.7
VIRC	39°21'	119°38'	2010	1.27	29	42	0.59	0.55	3.7
SHSH	39°53'	117°44'	1210	1.26	43	64	0.43	0.38	1.7
JERS	40°17'	117°26'	1500	1.23	35	53	0.63	0.48	2.9
MOUN	40°18'	117°04'	1500	1.23	35	53	0.41	0.44	1.5
BUFF	40°22'	117°16'	1480	1.23	34	51	0.45	0.38	2.0
PLAT	39°24'	118°18'	1470	1.19	27	42	0.56	0.51	1.7
PALO	40°04'	117°07'	1650	1.13	26	42	0.37	0.31	1.4
VIGU	39°34'	117°11'	1950	1.13	20	32	0.37	0.28	1.8
RUSS	39°13'	118°46'	1380	1.10	27	45	0.62	0.38	1.5
REDM	39°24'	119°00'	1460	1.10	21	35	1.54	0.31	2.6
UHOG	39°38'	118°48'	1230	1.03	28	50	0.28	0.41	2.8
TRAC	39°32'	119°29'	1410	1.03	14	24	0.53	0.82	1.5
CLAN	39°40'	117°54'	1440	1.02	46	85	0.44	0.32	1.7
HAZE	39°35'	119°03'	1300	1.00	16	29	0.5	0.39	1.5
PONY	39°17'	119°01'	1310	0.96	24	46	0.39	0.42	2.6
COAL	40°15'	118°21'	1410	0.85	17	36	0.62	0.73	2.4
FITT	40°20'	118°06'	1450	0.85	16	34	0.71	0.45	2.8
KYLE	40°24'	117°51'	1530	0.85	16	34	1.02	0.36	2.4
MILL	40°40'	118°04'	1320	0.85	13	27	0.47	0.73	1.9
WILC	40°34'	117°53'	2140	0.85	10	20	0.8	0.53	2.6
SEVN	40°18'	118°51'	1220	0.84	18	39	1.13	0.37	2.0
GRAN	40°05'	119°03'	1520	0.84	18	39	0.88	0.32	2.1
JUNI	40°19'	119°04'	1410	0.84	17	36	0.45	0.73	1.5
TRIN	40°17'	118°38'	1660	0.84	9	18	0.72	0.56	1.7
RATT	39°29'	118°45'	1250	0.80	20	45	0.95	0.52	2.1
HUNG	39°41'	119°45'	1550	0.73	14	34	0.75	0.55	2.0
MULL	39°53'	119°39'	1430	0.73	12	29	0.41	0.49	0.9
SPRK	39°31'	119°43'	1360	0.71	19	48	0.65	0.53	2.1
COLD	39°25'	117°51'	1690	0.71	22	57	0.3	0.37	1.8
NPAS	39°35'	117°32'	1720	0.71	19	49	0.43	0.34	1.7
KITT	40°04'	117°55'	1660	0.71	17	43	0.43	0.22	2.3
BATT	40°31'	117°12'	1510	0.67	26	71	0.36	0.32	1.7
ANTB	39°51'	117°22'	1650	0.67	17	46	0.26	0.22	1.1
REDR	39°54'	119°59'	1490	0.63	14	39	0.39	0.68	1.6
MORG	40°27'	120°04'	1520	0.33	8	41	0.3	0.38	1.0
FLAN	40°09'	119°50'	1260	0.17	9	89	0.35	0.19	1.8
HONY	40°05'	119°56'	1310	0.17	9	89	0.27	0.32	2.2
SKED	40°17'	120°02'	1300	0.17	9	89	0.41	0.31	1.6

they also factor in the real-world problems of equipment failure and detected data problems. The activity of course will always start off high for a new station. For the 35 stations running longer than one year, the average activity was 48%, ranging from 22% to 85%. Therefore, the mean site occupation was around 50%, as planned.

GPS Time Series

Table 2 summarizes the data-quality statistics for the station coordinate time series for the 34 longest-running sites (>1 yr). These statistics are based on coordinate repeatability (defined by Dixon (1991). The table presents mean repeatability (averaged over the 34 stations), the standard deviation in the repeatability distribution, and the range (min–max) of the repeatability values. The statistics are given for each component (longitude, latitude, and height), and they are presented for the daily epoch time series (globally referenced and spatially filtered), and weekly epoch time series (spatially filtered). Table 1 shows weekly repeatability statistics for individual stations.

The longest-running sites were chosen to summarize the statistics because repeatability can be overly optimistic for short time series. The weekly averaged solutions are of interest because in most situations, the day to day variation within a week can be reasonably expected to be entirely due to errors. Slowly varying signals (or errors) are also easier to detect visually in the weekly time series, and therefore weekly solutions give a better visual indication of the quality of the resulting velocity estimates.

Regional spatial filtering on the daily epoch solutions produces about a factor of three improvement in the repeatability of all three coordinate components. Averaging the daily solutions down to weekly epoch solutions further improves the repeatability by about a factor of two in all components, indicating that daily solutions contain random (or high frequency) error that can be averaged down significantly. The resulting weekly repeatability statistics for spatially filtered coordinates are quite typical in magnitude as for the permanent BARGEN network (at 0.4 mm

for horizontal components, and 1.7 mm for height). We therefore conclude that GPS time series repeatability is not significantly degraded by the inherent difference in design between semipermanent and permanent networks (for example, that instruments are moved around a semipermanent network). In addition, we note that sampling the time series intermittently through the year produces similar repeatability statistics as sampling the time series continuously throughout the year.

Figure 3 presents typical examples of detrended coordinate time series (shown here with no regional spatial filtering) from three stations: CLAN, BLAC, and COPP (for which the height time series is also shown). In the case of CLAN, the station has been operating continuously since installation. CLAN serves as a "best case" scenario, since the antenna and receiver have not changed at all, where as for BLAC and COPP, the antenna and receivers changed randomly and were generally different for each several-week session. There is no obvious qualitative difference in the time series; weekly repeatability (spatially filtered) for all three stations is very similar at 0.4 mm (to within 0.05 mm) in both horizontal components (and 1.8 mm to within 0.2 mm for height).

From this, we conclude that the systematic effect of changing antennas and receivers is not a limiting factor, and it is likely to be ~0.1 mm (order of magnitude) for most, if not all, antennas. It is possible that the summary statistics hide a few antennas that do not meet this specification. If such antennas exist, they may be detected by a systematic deviation (in one direction) from the time series. This type of test would require more data than we have in hand, though it does indicate a possible method of calibration that could be used to fine tune the time series.

The longest-running stations have 1.53 yr of data that have been analyzed here, as of 20 August 2005. It so happens that for close to 1.5 yr of a continuous time series, any annual sinusoidal signal (no matter what the phase) will not significantly bias the velocity estimate (Blewitt and Lavallée, 2002). Figure 1 shows horizontal velocities for sites that have been operating more than 1 yr, with black arrows for >1.3 yr. The black arrows therefore represent velocities relatively less affected by seasonal signals. From now on, the longest-running stations will be contributing velocity solutions that might reasonably be interpreted in terms of their regional spatial pattern (Kreemer et al., this volume; Hammond et al., this volume). Remarkably, following the method of Davis et al. (2003), the smoothness of the velocity field shown by the black arrows in Figure 1 already indicates a degree of accuracy of ~1 mm/yr from only 1.3–1.5 yr of data. Further assessment of this velocity field is given by Kreemer et al. (this volume).

COST-BENEFIT ANALYSIS

Here, we present a first-order cost-benefit analysis to assess whether in fact the semipermanent network we have designed proves more cost-effective than other approaches. Specifically, we compare our semipermanent network with permanent GPS

TABLE 2. STATISTICS ON STATION COORDINATE REPEATABILITY

Coordinate component	Mean repeatability (mm)	Standard deviation (mm)	Range min–max (mm)
Daily: globally referenced			
Longitude	2.6	±0.3	2.0–3.2
Latitude	3.2	±0.5	2.3–3.9
Height	7.8	±1.1	5.0–11
Daily: spatially filtered			
Longitude	0.9	±0.2	0.6–1.5
Latitude	1.0	±0.2	0.7–1.6
Height	3.7	±0.7	2.6–5.8
Weekly: spatially filtered			
Longitude	0.47	±0.13	0.28–0.91
Latitude	0.57	±0.22	0.28–1.5
Height	2.1	±0.50	1.1–3.7

and campaign GPS with regard to their performance in delivering a set of accurate station velocities. The three key cost-driving assumptions that we make are that (1) permanent GPS requires significant site development, (2) campaign GPS requires an average of 3 d on site with an operator present for reasons of security and logistics, and (3) semipermanent GPS requires site visitations ~6 times per year, within day-trip driving distance of a central facility. So the conclusions of our analysis depend critically on the validity of these assumptions. Table 3 provides details on the calculation, which shows that for purposes of mapping strain rates, semipermanent networks outperform permanent networks by a factor of ~3 in spatial resolution, and they also outperform campaigns (with significantly higher accuracy within any specified time frame).

Perhaps surprisingly, annual operational costs for semipermanent GPS networks are less than either permanent or campaign GPS. This is because for semipermanent GPS, fieldwork days are scheduled such that up to nine sites are visited by one technician, thus the incremental cost per station visit is small in terms of transportation and labor. Permanent stations, on the other hand, are visited when there are problems, so no similar efficiency can be made, and therefore visits are relatively costly. Campaign GPS fieldwork is relatively expensive due to security and logistical issues. This extra cost for campaigns is not compensated by the savings that are gained by visiting many sites with one instrument during the year (eight different sites per campaign season are assumed here). Semipermanent GPS requires more up front cost in terms of equipment, but this is more than compensated by efficient (and therefore inexpensive) fieldwork. Adjusting the assumptions in Table 3 will clearly change the numbers, but it is difficult to change the relative ranking of the three techniques.

More competitive performance could in principle be obtained by permanent GPS if site-installation costs (specifically, monument installation) could be reduced. The trade-off would therefore be in terms of fewer more stable sites versus a denser network with potentially more local anomalies. The advantage of expending more on stable monuments is that the permanent network can serve as a stable "backbone" array that provides a good ground truth. Therefore, we view permanent networks as being essential and complementary to other types of networks.

More competitive performance could in principle be obtained by campaign GPS if fewer than 3 d were spent on each site. The best performance is obtained with one day on each site. Of course, the trade-off here is with regard to reliability, and the distinct possibility that a campaign might completely fail for specific sites due to the lack of redundancy. Another problem is the difficulty in the assessment of errors with only one daily epoch per campaign. Nevertheless, the number of extra sites that can be measured for the same cost might be preferable, and so identification of problem sites and the assessment of velocity errors might be done with a more spatial rather than temporal analysis.

Our cost-benefit analysis shows that for commonly assumed situations, semipermanent networks present an effective means

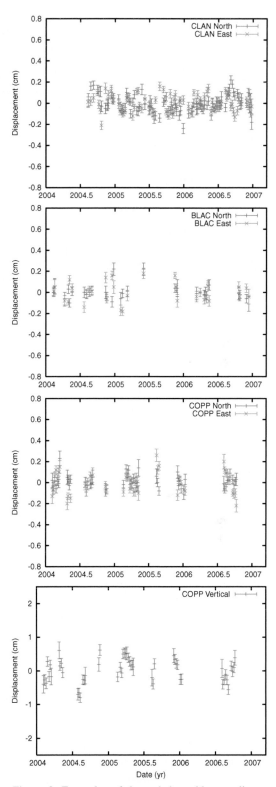

Figure 3. Examples of detrended weekly coordinate time series (unfiltered) from stations CLAN, BLAC, and COPP. Coordinate components are indicated as latitude, longitude, or height. Note that the half-scale is 8 mm for horizontal components and 25 mm for height, reflecting the approximate factor of three in relative scatter for height time series.

TABLE 3. COST-BENEFIT ANALYSIS OF GLOBAL POSITIONING SYSTEM (GPS) SURVEY METHOD

Cost-benefit factors	Permanent	Campaign	Semipermanent
Assumptions			
A. Velocity resolution time (yr)	2.50	10.00	3.50
B. GPS survey equipment set ($K)	8.00	8.00	8.00
C. GPS stations monitored per set	1.00	8.00	2.00
D. One-time installation cost ($K)	10.00	0.00	0.10
E. Site visits per year	1.00	1.00	6.00
F. Days on site per visit	0.50	3.00	0.15
G. Per diem meals and lodging ($K)	0.08	0.08	0.00
H. Transportation per visit ($K)	0.10	0.06	0.03
J. Field technician labor per day ($K)	0.20	0.20	0.20
K. Communications per year ($K)	0.30	0.00	0.00
Subtotals			
L. Capital (D + B/C) ($K)	18.00	1.00	4.10
M. Annual {E[H + F(G + J)] + K} ($K/yr)	0.54	0.90	0.36
Derived costs			
N. Per resolved station velocity	19.35	10.00	5.36
P. Per station over 10 yr	23.4	10.00	7.70
Benefit/cost ratios (relative to permanent)			
Number of resolved station velocities with fixed 10 yr funding (19.35/N)	1.00	1.94	3.61
Number of stations that can be routinely monitored with fixed 10 yr funding (23.4/P)	1.00	2.34	3.04

of mapping variations in crustal strain on the regional scale. Permanent networks retain a complementary utility in providing a backbone reference frame with velocities that can act as a ground truth. Permanent networks might also be essential if the goal is to monitor possible transient behavior. On the other hand, GPS campaigns appear (in comparison to semipermanent GPS) to have little redeeming qualities to justify their use on newly mapped terrain, though their use may be justified in specific cases to add value to existing data from previous campaigns.

CONCLUSIONS

We have embarked on an experiment to characterize strain partitioning in the Northern Walker Lane to improve models of (1) geothermal activity, which is largely amagmatic in the Great Basin, (2) seismic hazard, (3) the ways in which the northern Walker Lane accommodates strain between the Sierra Nevada block and the extending Basin and Range Province, and (4) Neogene development of the northern Walker Lane and its broader role in the ongoing evolution of the Pacific–North America plate-boundary system. To this end, we have designed, constructed, operated, and tested the 60-station semipermanent GPS network "MAGNET" spanning the northern Walker Lane and central Nevada seismic belt, with a mean nearest-neighbor spacing of 19 km (13–31 km range). Now at the watershed period of 1.5 yr (for the longest-running stations), when estimated GPS station velocities begin to converge and make sense (Blewitt and Lavallée, 2002), MAGNET is delivering data products that meet and even exceed the expectation and design specifications for the eventual objective of producing a uniform, high-resolution strain-rate map of the northern Walker Lane.

MAGNET is occupied by a total 34 receivers that are intermittently moved around the network, with a mean station occupancy of ~50% (after accounting for equipment failures). Assessment of the performance of MAGNET has proved that a semipermanent network can provide a cheaper, higher-density alternative to permanent networks, and a more accurate alternative to traditional GPS campaigns. The mean weekly repeatability (spatially filtered) is 0.5 mm in longitude, 0.6 mm in latitude, and 2.1 mm in height; these statistics are very similar to those of permanent GPS networks. This implies that antenna setup error has been mitigated, despite the fact that random antennas (of the same type) are assigned to station sessions (implying that antennas likely fall within the manufacturer's design specifications). MAGNET's design is generally applicable to regions with an abundance of vehicle-accessible rock outcrops, and it could be replicated elsewhere in western North America.

Other investigators may wish to install similar networks in a compatible way, creating a collaborative potential to form a dense (~20 km spacing), uniform, semipermanent GPS network spanning the western States. Such a network would complement permanent GPS networks, such as the Plate Boundary Observatory component of the National Science Foundation's EarthScope Program, with potentially much higher and more uniform resolution. Such a development would represent a major advance in earthquake preparedness, and it would provide a vital, spatially uniform data set to construct strain-rate maps that could be used for a multitude of purposes, including seismic hazard assessment, exploration of geothermal energy, understanding the dynamics of the Pacific–North America plate boundary, and understanding the structure and evolution of the North American continent.

ACKNOWLEDGMENTS

This research was funded in part by Department of Energy grant DE-FG36-02ID14311 to the Great Basin Center for Geothermal Energy and by the Department of Energy Yucca Mountain Project/Nevada System of Higher Education Cooperative Agreement DE-FC28-04RW12232. Jim Faulds provided expert assistance in the selection of sites on the basis of geological stability. We are grateful to Bret Pecoraro for being a model of efficiency on network installation, fieldwork operations, and equipment maintenance. This manuscript was improved by careful reviews from Jeff Freymueller and an anonymous reviewer.

REFERENCES CITED

Agnew, D.C., 1992, The time-domain behavior of power-law noises: Geophysical Research Letters, v. 19, p. 333–336, doi: 10.1029/91GL02832.

Bar Sever, Y., Kroger, P.M., and Borjesson, J.A., 1998, Estimating horizontal gradients of tropospheric path delay with a single GPS receiver: Journal of Geophysical Research, v. 103, p. 5019–5035, doi: 10.1029/97JB03534.

Bawden, G.W., Thatcher, W., Stein, R.S., Hudnut, K.W., and Peltzer, G., 2001, Tectonic contraction across Los Angeles after removal of groundwater pumping effects: Nature, v. 412, p. 812–815, doi: 10.1038/35090558.

Bell, J.W., Amelung, F., Ramelli, A.R., and Blewitt, G., 2002, Land subsidence in Las Vegas, Nevada, 1935–2000: New geodetic data show evolution, revised spatial patterns, and reduced rates: Environmental & Engineering Geoscience, v. 8, p. 155–174, doi: 10.2113/8.3.155.

Bennett, R.A., Wernicke, B.P., Niemi, N.A., Friedrich, A.M., and Davis, J.L., 2003, Contemporary strain rates in the northern Basin and Range Province from GPS data: Tectonics, v. 22, 1008, doi: 10.1029/2001TC001355.

Blewitt, G., 1989, Carrier phase ambiguity resolution for the global positioning system applied to geodetic baselines up to 2000 km: Journal of Geophysical Research, v. 94, p. 10,187–10,283, doi: 10.1029/JB094iB08p10187.

Blewitt, G., 1990, An automatic editing algorithm for GPS data: Geophysical Research Letters, v. 17, p. 199–202, doi: 10.1029/GL017i003p00199.

Blewitt, G., 2000, Geodetic network optimization for geophysical parameters: Geophysical Research Letters, v. 27, p. 3615–3618, doi: 10.1029/1999GL011296.

Blewitt, G., and Lavallée, D., 2002, Effects of annual signals on geodetic velocity: Journal of Geophysical Research, v. 107, no. B7, 2145, doi: 10.1029/2001JB000570.

Blewitt, G., Coolbaugh, M., Holt, W., Kreemer, C., Davis, J., and Bennett, R., 2003, Targeting of potential geothermal resources in the Great Basin from regional- to basin-scale relationships between geodetic strain and geological structures: Transactions of the Geothermal Resources Council, v. 27, p. 3–7.

Blewitt, G., Argus, D., Bennett, R., Bock, Y., Calais, E., Craymer, M., Davis, J., Dixon, T., Freymueller, J., Herring, T., Johnson, D., Larson, K., Miller, M., Sella, G., Snay, R., and Tamissiea, M., 2005, A stable North America reference frame (SNARF): First release, joint workshop: Stevenson, Washington, UNAVCO-IRIS.

Davis, J.L., Bennett, R.A., and Wernicke, B.P., 2003, Assessment of GPS velocity accuracy for the Basin and Range Geodetic Network (BARGEN): Geophysical Research Letters, v. 30, p. 1411, doi: 10.1029/2003GL016961.

Dixon, T.H., 1991, An introduction to the global positioning system and some geological applications: Reviews of Geophysics, v. 29, p. 249–276, doi: 10.1029/91RG00152.

Elósegui, P., Davis, J.L., Mitrovica, J.X., Bennett, R.A., and Wernicke, B.P., 2003, Measurement of crustal loading using GPS near Great Salt Lake, Utah: Geophysical Research Letters, v. 30, p. 1111–1114, doi: 10.1029/2002GL016579.

Hammond, W.C., Kreemer, C., and Blewitt, G., 2009, this volume, Geodetic constraints on contemporary deformation in the northern Walker Lane: 3. Central Nevada seismic belt postseismic relaxation, in Oldow, J.S., and Cashman, P.H., eds., Late Cenozoic Structure and Evolution of the Great Basin–Sierra Nevada Transition: Geological Society of America Special Paper 447, doi: 10.1130/2009.2447(03).

Kreemer, C., Blewitt, G., and Hammond, W.C., 2009, this volume, Geodetic constraints on contemporary deformation in the northern Walker Lane: 2. Velocity and strain rate tensor analysis, in Oldow, J.S., and Cashman, P.H., eds., Late Cenozoic Structure and Evolution of the Great Basin–Sierra Nevada Transition: Geological Society of America Special Paper 447, doi: 10.1130/2009.2447(02).

Langbein, J., and Johnson, H., 1997, Correlated errors in geodetic time series: Implications for time-dependent deformation: Journal of Geophysical Research, v. 102, p. 591–603, doi: 10.1029/96JB02945.

Langbein, J., Wyatt, F., Johnson, H., Hamann, D., and Zimmer, P., 1995, Improved stability of a deeply anchored geodetic monument for deformation monitoring: Geophysical Research Letters, v. 22, p. 3533–3536, doi: 10.1029/95GL03325.

Sanli, U., and Blewitt, G., 2001, Geocentric sea level trend using GPS and >100-year tide gauge record on a postglacial rebound nodal line: Journal of Geophysical Research, v. 106, p. 713, doi: 10.1029/2000JB900348.

Smith, K.D., von Seggern, D., Blewitt, G., Preston, L., Anderson, J.G., Wernicke, B.P., and Davis, J.L., 2004, Evidence for deep magma injection beneath Lake Tahoe, Nevada-California: Science, v. 305, no. 5688, p. 1277–1280, doi: 10.1126/science.1101304.

van Dam, T., Wahr, J., Milly, P.C.D., Shmakin, A.B., Blewitt, G., Lavallee, D., and Larson, K.M., 2001, Crustal displacements due to continental water loading: Geophysical Research Letters, v. 28, p. 651–654, doi: 10.1029/2000GL012120.

Wdowinski, S., Bock, Y., Zhang, J., Fang, P., and Gengrich, J., 1997, Southern California permanent GPS geodetic array: Spatial filtering of daily positions for estimating coseismic and postseismic displacements induced by the 1992 Landers earthquake: Journal of Geophysical Research, v. 102, p. 18,057–18,070, doi: 10.1029/97JB01378.

Williams, S.D.P., 2003, The effect of colored noise on the uncertainties of rates estimated from geodetic time series: Journal of Geodesy, v. 76, p. 483–494, doi: 10.1007/s00190-002-0283-4.

Zumberge, J.F., Heflin, M.B., Jefferson, D.C., Watkins, M.M., and Webb, F.H., 1997, Precise point positioning for the efficient and robust analysis of GPS data from large networks: Journal of Geophysical Research, v. 102, p. 5005–5018, doi: 10.1029/96JB03860.

MANUSCRIPT ACCEPTED BY THE SOCIETY 21 JULY 2008

The Geological Society of America
Special Paper 447
2009

Geodetic constraints on contemporary deformation in the northern Walker Lane: 2. Velocity and strain rate tensor analysis

Corné Kreemer*
Geoffrey Blewitt
William C. Hammond
Nevada Bureau of Mines and Geology, and Seismological Laboratory, University of Nevada, Reno, Nevada 89557-0178, USA

ABSTRACT

We present a velocity and strain rate model for the northern Walker Lane derived from a compilation of geodetic velocities and corrected for transient effects owing to historic earthquakes on the Central Nevada seismic belt. We find that from 37°N to 40°N, the Walker Lane is characterized by an ~100-km-wide zone with near-constant strain rates associated with ~10 mm yr⁻¹ total motion across the zone. The strain rates depict predominantly shear deformation, but south of 39°N, the extensional component of the strain rate tensor increases and thus reflects more of a transtensional domain there. We conclude that this transtension is a kinematic consequence of the motion of the Sierra Nevada–Great Valley block, which is not parallel to its eastern margin, i.e., the eastern Sierra front, south of 39°N. While the orientations of several normal and strike-slip faults in the Walker Lane region are consistent with the strain rate model results at several places, the mode and rate at which geologic structures accommodate the deformation are less clear. Left-lateral faulting and clockwise rotations there may contribute to the accommodation of the velocity gradient tensor field, and most normal faults are properly oriented to accommodate some component of the regional shear strain, but significant additional right-lateral strike-slip faulting is required to accommodate the majority of the 10 mm yr⁻¹ relative motion. Overall, the along-strike variation in the active tectonics of Walker Lane suggests that (1) various mechanisms are at play to accommodate the shear, (2) parts of the surface tectonics may (still) be in an early stage of development, and (3) inherited structural grain can have a dominant control on the strain accommodation mechanism.

Keywords: GPS, Walker Lane, strain, kinematics, tectonics.

INTRODUCTION

The Walker Lane belt is a structurally and kinematically complex zone of faulting in the Pacific–North America plate boundary zone between the Sierra Nevada Mountains and the Basin and Range Province (Stewart, 1988; Wesnousky, 2005a) (Fig. 1). It accommodates about a quarter of the total Pacific–North America plate motion (Bennett et al., 2003; Oldow et al., 2001; Thatcher et al., 1999). Numerous studies of its active and recent tectonics have revealed that the system is characterized by many relatively short northwest-trending right-lateral faults, north-trending normal faults, and a smaller number of ENE-trending left-lateral

*Corresponding author e-mail: Kreemer@unr.edu.

Kreemer, C., Blewitt, G., and Hammond, W.C., 2009, Geodetic constraints on contemporary deformation in the northern Walker Lane: 2. Velocity and strain rate tensor analysis, *in* Oldow, J.S., and Cashman, P.H., eds., Late Cenozoic Structure and Evolution of the Great Basin–Sierra Nevada Transition: Geological Society of America Special Paper 447, p. 17–31, doi: 10.1130/2009.2447(02). For permission to copy, contact editing@geosociety.org. ©2009 The Geological Society of America. All rights reserved.

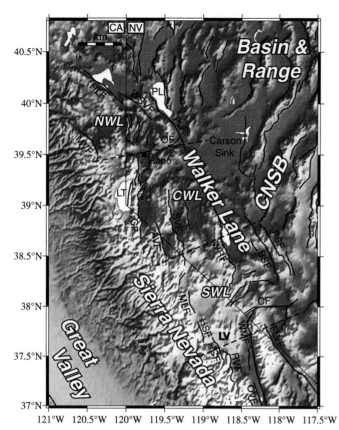

Figure 1. Overview map of the area of the main focus of this study. We divide the Walker Lane up into three segments: north (NWL), central (CWL), and south (SWL), separated by the dashed lines. AVF—Antelope Valley fault; BSF—Benton Springs fault; CA—California; CF—Coaldale fault; CNSB—Central Nevada seismic belt; FLVF—Fish Lake Valley fault; GF—Genoa fault; HCF—Hilton Creek fault; HLF—Honey Lake fault; HSF—Hartley Springs fault; LT—Lake Tahoe; LV—Long Valley; MLF—Mono Lake fault; MVF—Mohawk Valley fault; NV—Nevada; OF—Olinghouse fault; OVF—Owens Valley Fault; PL—Pyramid Lake; PLF—Pyramid Lake fault; PSF—Petrified Springs fault; RVF—Round Valley fault; SVF—Smith Valley fault; WMF—White Mountains fault; WRF—Wassuk Range fault; WSVF—Warm Springs Valley fault.

faults (Fig. 1) (e.g., Faulds et al., 2005b; Oldow, 2003; Slemmons et al., 1979; Stewart, 1988; Wesnousky, 2005a). It has been argued that the Walker Lane owes its tectonic complexity to (1) its location, i.e., it is situated between the approximately E-W–extending Basin and Range and the NW-moving Sierra Nevada–Great Valley microblock, (2) the fact that the current tectonic regime is geologically still young, and (3) the fact that many geologic features are inherited from pre–Walker Lane times. In general, it is now widely agreed upon that the Walker Lane is a young and developing transform or transtensional fault system (Faulds et al., 2005b; Wesnousky, 2005a, 2005b) characterized by the lack of long and well-developed faults, such as those present along, for example, the San Andreas fault system. At the latitude of ~39°N, significant clockwise rotations around vertical axes have been observed by paleomagnetic measurements (Cashman and Fontaine, 2000). It

has been argued that the clockwise rotations and the presence of left-lateral (rotated) conjugate faults are consistent with having a partially detached elastic-brittle crust that is being transported on a continuously deforming substratum (Wesnousky, 2005a). The recent recognition of the Walker Lane as a developing transform zone has expanded our general understanding of the plate boundary–wide partitioning of Pacific–North America plate motion and the development of the plate-boundary zone in general.

Adding to the complexity is the Central Nevada seismic belt, a zone of focused historic earthquakes extending from the Walker Lane northeastward into the Basin and Range Province (Wallace, 1984) (Fig. 1). These earthquakes and the geodetic observation of relatively large strain rates localized at the Central Nevada seismic belt (Hammond and Thatcher, 2004; Thatcher et al., 1999) have led (in part) to the speculation that this belt plays an important role in the strain accommodation in the western Basin and Range Province (Wesnousky, 2005a; Wesnousky et al., 2005). However, the high geodetic strain rate is probably not a permanent feature, since it can be largely explained through viscoelastic postseismic strain relaxation (Hammond et al., this volume; Hetland and Hager, 2003). Moreover, the long-term significance of the Central Nevada seismic belt as a zone of localized deformation is not supported by some geologic studies of the slip history of the faults and the fault slip rates within the Central Nevada seismic belt relative to other Great Basin faults (Bell et al., 2004; e.g., Wallace, 1987), although the region appears to have a relatively high recurrence rate of surface-rupture earthquakes (Wesnousky et al., 2005). In the analysis that follows, we use the results of a companion paper (Hammond et al., this volume) to correct the geodetic strain rate field for transient postseismic effects that can disrupt the comparison between geodetically estimated strain and fault slip.

Geodetic velocity measurements, particularly using the global positioning system (GPS), allow for the precise quantification of crustal strain rates and provide constraints on regional present-day kinematics. Such measures provide important constraints for understanding the role of observed faulting and seismicity in the region. Knowledge of the present-day kinematic framework is also important in understanding finite-strain markers and the recent evolution of the deformation field. Several GPS studies have recently been undertaken in the Walker Lane region (Hammond and Thatcher, 2004; Oldow et al., 2001; Svarc et al., 2002b; Thatcher et al., 1999). These studies all found that deformation in the Walker Lane is characterized by relatively large strain rates, although they are not all consistent on how the strain rate is accommodated geologically. Oldow et al. (2001) inferred from their results that the central Walker Lane acts as a distributed zone of displacements linking the Eastern California shear zone with the northern Walker Lane and the Central Nevada seismic belts. They concluded as well that the relative displacements are not accommodated by a spatially smooth transition, but rather as differential motions of tectonic blocks. Svarc at al. (2002a) used their GPS velocity results to calculate a strain rate tensor for the Walker Lane at the latitude of Reno. They concluded that their result is consistent with extension across and shear along a zone striking N35°W, which is the same orientation as the Pacific–North America small circle. Hammond and Thatcher (2004) obtained the same result, and, with Thatcher et al. (1999), found the strain rates in the Walker Lane to be significantly larger than elsewhere in the Great Basin, indicating to them that it is likely a zone of lithospheric weakness. Most of the geodetic results have illustrated the discrepancy that exists between geodetic deformation rates and those inferred from the slip activity of Quaternary faults (Hammond et al., this volume; Pancha et al., 2006). There are various explanations for that discrepancy, but the fact that the Walker Lane does not seem to behave as a mature fault system may provide a significant explanation.

In this study, we quantify the distributed strain rate tensor field in the Walker Lane region, taken here to be between ~37.5°N in the south and ~40.5°N in the north. For the purpose of this paper, we divide this part of the Walker Lane into three segments: a southern part south of ~38.75°N, a central part between 38.75°N and 39.5°N, and a northern part north of 39.5°N (Fig. 1). We address how faulting and other geologic and seismologic observations can be understood in terms of distributed strain. This is one of the first attempts to perform such analysis systematically along the entire Walker Lane. Because of the apparently diffuse nature of the deformation field, it is appropriate to analyze the deformation field in Walker Lane through modeling of the velocity gradient tensor field. From this, we can directly infer a continuous strain rate tensor field and interpolated velocities, as well as vertical-axis rotations. To do so, we combined published and updated GPS velocities into a synthesized observed velocity field, which has a much higher spatial resolution than similar previous attempts (Bennett et al., 2003; Oldow, 2003). To properly discuss and understand the context of contemporary deformation in the Walker Lane region, and to avoid modeling boundary effects, we briefly show and discuss our model results for the entire Great Basin region as well.

VELOCITY VECTOR SYNTHESIS

To perform the strain rate analysis, we required a spatially dense set of geodetic velocities estimates. Therefore, we synthesized and combined the velocities of several independent studies into one consistent reference frame. Within the Great Basin proper, we used campaign-style GPS velocities from published studies (Hammond and Thatcher, 2004, 2005, 2007; McClusky et al., 2001; Oldow et al., 2001; Svarc et al., 2002b), the Southern California Earthquake Center (SCEC) v. 3 velocity solution (which includes some velocities derived from the very long baseline interferometry (VLBI) technique; Shen et al., 2003), continuous GPS (CGPS) velocity estimates from the University of Utah's Eastern Basin and Range and Yellowstone Network (EBRY) (R. Smith, 2005, personal commun.), and U.S. Geological Society (USGS) campaign measurement results for the "Yucca Profile" (originally published by Gan et al. [2000], but we use a more recent solution from the USGS Web site). Crucial to providing a

robust regional frame, we include a CGPS velocity solution for the Basin and Range Geodetic Network (BARGEN) (e.g., Bennett et al., 1998, 2003) analyzed using the method of Blewitt et al. (this volume). Importantly, the permanent BARGEN network employs braced, deep, anchored monuments (down to ~10 m) into bedrock to ensure local stability (Langbein et al., 1995). We analyzed BARGEN data from 2000 to 2005.5, during which the GPS antenna/radome configuration was identical at each station (Smith et al., 2004). In addition, we include some vectors on the periphery of the Great Basin (including most of California's Central Valley) (d'Alessio et al., 2005; Freymueller et al., 1999; Mazzotti et al., 2003; Svarc et al., 2002a; Williams et al., 2006). The Bay Area Velocity Unification (BAVU) solution of d'Alessio et al. (2005) includes a large number of USGS campaign results as well as velocities from the Bay Area Regional Deformation (BARD) continuous GPS network. Our data set includes all available velocities prior to September 2005.

To include each set of velocities into this compilation, we estimated and applied a six-parameter Helmert transformation using the horizontal velocities at collocated sites between studies. In theory, the transformation involves a three-parameter translation rate and a three-parameter rotation rate. However, when the collocated sites are geographically close to one another, as is the case for most studies used here, there is a trade-off between the translation and rotation. We therefore only applied the translation if an F-test indicated that a translation in addition to a rotation would provide a statistically significant improvement to the velocity fit at the collocated sites compared to a case when only a rotation was applied. We used a global GPS velocity solution in the International Terrestrial Reference Frame (ITRF2000), known as GPSVEL (Holt et al., 2005; Kreemer et al., 2006)—a solution derived from a rigorous combination of International GNSS Service (IGS) solutions using the method of Davies and Blewitt (2000)—as the benchmark study into which we transformed the regional studies. Most velocity fields can only be transformed after others have been transformed so that the number of collocated sites is increased. All studies used and their transformation parameters are listed in Table 1. Next, to obtain velocities in a North American (NA) reference frame, we subtracted from the ITRF2000 velocities the values estimated from the NA-ITRF2000 angular velocity as defined by the Stable North America Reference Frame (SNARF) Working Group (Blewitt et al., 2005): the Euler pole of NA-ITRF2000 motion is 2.4°S, 83.6°W, 0.2° m.y.$^{-1}$. All 474 velocities (for 444 sites) in our study areas are shown in Figure 1 relative to the SNARF reference frame, and they are tabulated in a companion paper (Hammond et al., this volume).

To avoid spurious local strain rate anomalies, we discarded a small portion of GPS velocities that were significantly different from other nearby velocity estimates and that stood out from the regional pattern of velocity gradients. Often these anomalous velocities were from campaign-style measurements that only had measurements in two campaigns. Also, for some studies, we increased standard errors to be greater than those originally published (Table 1). Formal errors that are very small, particularly

TABLE 1. HELMERT TRANSFORMATION PARAMETERS

Study	k	Original ref.	ω_x (° m.y.$^{-1}$)	ω_y (° m.y.$^{-1}$)	ω_z (° m.y.$^{-1}$)	δ_x (mm yr^{-1})	δ_y (mm yr^{-1})	δ_z (mm yr^{-1})
BARGEN (this study)	2	ITRF00	−0.0141 ± 0.0381	−0.0490 ± 0.0909	0.0555 ± 0.0821	–	–	–
SCEC v. 3 (Shen et al., 2003)	1	N. Amer	0.0537 ± 0.0386	−0.1706 ± 0.0295	−0.0015 ± 0.0238	−2.60 ± 4.12	1.65 ± 3.44	3.56 ± 2.72
EBRY (R. Smith, 2005, personal commun.)	10	N. Amer	0.0198 ± 0.0019	−0.1988 ± 0.0043	−0.0125 ± 0.0042	–	–	–
Yucca profile	2	ITRF00	−0.0180 ± 0.1244	0.0031 ± 0.2399	0.0112 ± 0.1994	–	–	–
d'Alessio et al. (2005) (BAVU)	1	ITRF00	0.0061 ± 0.0009	0.0240 ± 0.0009	−0.0003 ± 0.0010	−1.41 ± 0.11	0.21 ± 0.11	1.82 ± 0.09
Freymueller et al. (1999)	1	Pacific	−0.3134 ± 0.1883	0.0022 ± 0.2988	−0.2924 ± 0.2918	–	–	–
Hammond and Thatcher (2004)	1	N. Amer	−0.0023 ± 0.0272	−0.2186 ± 0.0517	0.0211 ± 0.0482	–	–	–
Hammond and Thatcher (2005)	1	N. Amer	0.0173 ± 0.0285	−0.1847 ± 0.0525	−0.0138 ± 0.0519	–	–	–
Hammond and Thatcher (2007)	1	N. Amer	0.0303 ± 0.0143	−0.1894 ± 0.0278	−0.0113 ± 0.0258	–	–	–
Mazzotti et al. (2003)	1	ITRF00	0.0083 ± 0.0139	0.0247 ± 0.0233	−0.0157 ± 0.0309	–	–	–
McClusky et al. (2001)	2	N. Amer	0.0651 ± 0.0666	−0.0702 ± 0.1277	−0.1279 ± 0.1046	–	–	–
Oldow et al. (2001)	20	N. Amer	−0.0200 ± 0.2685	−0.2768 ± 0.4955	0.0488 ± 0.4559	–	–	–
Svarc et al. (2002a)	2	N. Amer	0.0588 ± 0.0890	−0.1289 ± 0.1572	−0.0669 ± 0.1506	–	–	–
Svarc et al. (2002b)	1	N. Amer	0.0093 ± 0.0616	−0.1667 ± 0.0970	−0.0468 ± 0.1103	–	–	–
Williams et al. (2006)	1	ITRF97	−0.0152 ± 0.0027	0.0271 ± 0.0053	−0.0349 ± 0.0055	–	–	–

Note: Cartesian components of the angular velocity (ω_x, ω_y, ω_z) and translation rate (δ_x, δ_y, δ_z) were solved for and applied to transform the velocities of each geodetic study from the original reference frame (Original ref.) into our ITRF2000 frame. The translation rate is only shown, and applied, when the transformation with translation led to a significantly better transformation compared to the case when only a rotation was applied (see text). k—factor with which the originally published formal velocity uncertainties are multiplied for use in our study. BARGEN—Basin and Range Geodetic Network; SCEC—Southern California Earthquake Center; EBRY—University of Utah's Eastern Basin and Range and Yellowstone Network.

with respect to those of nearby velocities, could lead locally to an overfit between the model and observed velocity, which in turn could potentially lead to locally spurious strain rate estimates.

STRAIN RATE ANALYSIS APPROACH

In this study, we characterize the regional deformation field on the assumption that most of the crust in the Great Basin deforms in a spatially continuous fashion. Given the geodetic data, which indicate a smooth velocity gradient (Fig. 2), likely as the result of elastic strain accumulation on nearby locked faults, a continuous modeling approach is appropriate. Moreover, although in the long-term, the velocity gradient is accommodated as discrete steps across faults, the large number of faults justifies a continuous approach, and the long-term and observed large-scale strain field are not expected to be significantly different. To derive a continuous velocity gradient tensor field, we applied a spline interpolation technique (e.g., Haines and Holt, 1993; Holt et al., 2000). In this method, model velocities are fitted to the observed geodetic velocities in a least-squares sense, using the full data covariance matrix. Model velocities are then interpolated using bicubic Bessel spline functions to derive a continuous velocity gradient tensor field, which provides estimates of strain rate, interpolated velocity, and vertical-axis rotation for any point in our model grid. Other studies using a similar model technique have been applied for the Pacific–North America plate-boundary zone (including most of the Basin and Range Province) using Quaternary faulting data, earthquake moment tensors, and early geodetic data (Flesch et al., 2000; Shen-Tu et al., 1998, 1999). However, the model resolution of those studies was limited by the use of relatively large grid cells in the inversion. Here, we use grid cells of $0.2° \times 0.2°$, which allow us to take advantage of the spatially dense velocity data that are now available in order to quantify the velocity gradients in higher detail than was previously possible.

We set up our model grid such that its northeastern edge is east of the Wasatch fault. No significant tectonic deformation appears to be present east of the Wasatch fault, as evidenced by the absence of Quaternary faults, seismicity, and insignificant

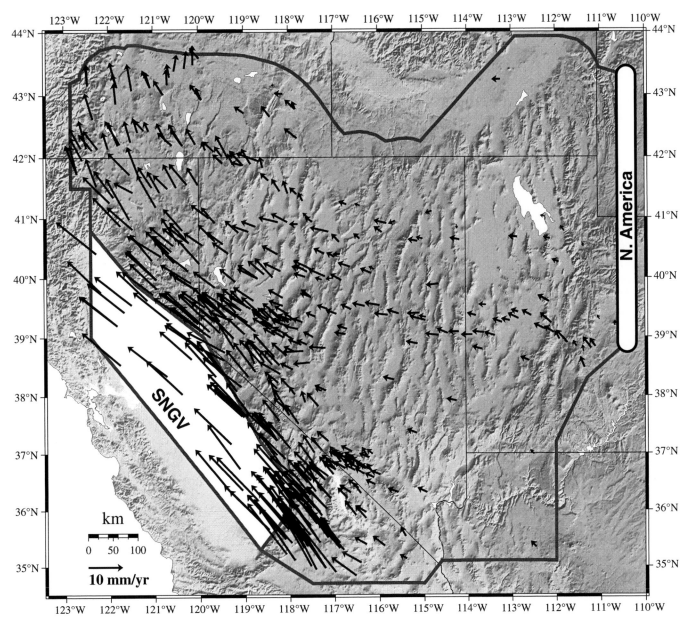

Figure 2. Geodetic velocities within the greater Great Basin, outlined by thick gray line. A 0.2° × 0.2° model grid is defined for our study area, with part of the grid defined a priori as the rigid Sierra Nevada–Great Valley block (SNGV, shown in white), and the eastern edge of the grid defined as North America. The geodetic velocities were compiled from multiple studies and are all shown in the same North America reference frame. Velocity uncertainties are omitted for clarity. In this figure, the correction for postseismic effects has not been applied.

GPS velocities of the easternmost sites in SNARF. Thus, we constrain the northeastern grid boundary to be equal to stable North America (Fig. 1). That is, we treat the velocities in our synthesized data set, which is in the NA reference frame, as being relative to this grid boundary. The southeastern edge of our model grid is near the western boundary of the Colorado Plateau (e.g., Bennett et al., 2003) and is free to deform, as is the northern grid boundary. The treatment of the western grid boundary is important to our analysis. We will present two models, one in which the Sierra Nevada–Great Valley is allowed to deform (the white

area in Fig. 2), and another where the Sierra Nevada–Great Valley moves as a rigid entity. Sierra Nevada–Great Valley motion is imposed as a rigid body rotation, which is estimated from the geodetic velocities, as discussed in the next section.

KINEMATIC BOUNDARY CONDITIONS

Rigidity and motion of the Sierra Nevada–Great Valley has been demonstrated by others (Argus and Gordon, 1991, 2001; Bennett et al., 2003; Dixon et al., 2000), and its motion provides

TABLE 2. ANGULAR VELOCITIES

	ω_x (° m.y.$^{-1}$)	±	ω_y (° m.y.$^{-1}$)	±	ω_z (° m.y.$^{-1}$)	±	Pole lat. (°)	Pole long. (°)	ω (° m.y.$^{-1}$)
SNGV-SNARF	−0.1469	0.0184	−0.0874	0.0311	−0.0025	0.0286	−0.9	−149.3	0.171
CGB-SNARF	−0.0001	0.0152	0.0148	0.0322	−0.0430	0.0292	−71.1	90.4	0.045
SNGV-CGB	−0.1468	0.0184	−0.1022	0.0311	0.0405	0.0286	12.6	−145.2	0.183

Note: SNGV—Sierra Nevada–Great Valley block; CGB—Central Great Basin; SNARF—Stable North America Reference Frame. Sites used to estimate SNGV motion: 0306, 0605, 0607, 0609, 0611, 0614, 1008, 3188, A300, CHO1, CMBB, CNDR, H112, ISLK, JAST, KMED, LUMP, MINS, MUSB, LIND, ORLA, ORVB, SUTB, UCD1, UU83. Sites used to estimate CGB motion: ALAM, ECHO, EGAN, ELKO, FOOT, GOSH, MINE, MONI, RAIL, RUBY.

the most important kinematic boundary condition to constrain and understand Walker Lane deformation. For the strain rate model that imposes Sierra Nevada–Great Valley motion, we estimated the angular velocity based on 25 velocities that are within the region that we define as being Sierra Nevada–Great Valley (Fig. 2; Table 2). We will show later that having the eastern margin of the Sierra Nevada–Great Valley closely follow the Sierra Nevada crest is in general agreement with the available geodetic velocities in the high Sierras. The estimated Sierra Nevada–Great Valley angular velocity (relative to SNARF) is somewhat sensitive to the choice of the velocities to use in the estimation, but resulting angular velocity differences are insignificant for the purposes of this paper.

The central Great Basin, which lies to the east of the Walker Lane, has been shown to behave as a geodetically rigid microblock (Bennett et al., 2003; Hammond and Thatcher, 2005). Our strain rate modeling results indicate ~4 nanostrain yr^{-1} for the central Great Basin (Fig. 3), the same order of magnitude as found for tectonic plates (e.g., Ward, 1998). Because later we would like to present our Walker Lane modeling results appropriately in an oblique Mercator projection around a Sierra Nevada–Great Valley–central Great Basin pole of rotation, we also estimate a central Great Basin–SNARF angular velocity based on 10 BARGEN sites (Table 2). We then combine that result with the Sierra Nevada–Great Valley (relative to SNARF) estimate to obtain the Sierra Nevada–Great Valley–central Great Basin angular velocity.

RESULTS

Great Basin

Before presenting our results for the Walker Lane, we briefly present the model strain rate field for the Great Basin as it is inferred directly from the geodetic velocity observations. Figure 3 shows the model velocity field and the contours of the second invariant of the model strain rates for a model that imposes rigid Sierra Nevada–Great Valley rotation. Formal uncertainties of the second invariant are on the order of 4–8 nanostrain yr^{-1}. As mentioned already, the central Great Basin shows very low strain rates and essentially moves as a geodetic microplate. Geologic strain rate estimates of ~1 nanostrain yr^{-1} corroborate the geodetic results, but Wesnousky et al. (2005) pointed out that the

uncertainty in the geologic estimate is possibly large, and, along with Bennett et al. (2003), they asserted that prevalent Quaternary faulting in the central Great Basin precludes the notion of a long-lived rigid "microplate." Strain rates are elevated east and west of the central Great Basin. The broad zone of elevated model strain rates in north-central Utah is at odds with studies that argued for localized strain along the Wasatch fault (Hammond and Thatcher, 2004; Martinez et al., 1998; Thatcher et al., 1999). However, a wide zone of elevated strain rate is consistent with an equally wide late Quaternary deformation field, derived from paleoseismic and seismic reflection data (Niemi et al., 2004), or the argument that the Wasatch fault is very late in the earthquake cycle (Malservisi et al., 2003). We show in a companion paper (Hammond et al., this volume) that the relatively high strain rates along the Central Nevada seismic belt are largely due to viscoelastic postseismic relaxation.

In the Eastern California shear zone and Walker Lane, strain rates are large as a consequence of the rapid east-to-west increase in velocity (as well as an ~20° clockwise change in direction). Strain rates along the Eastern California shear zone are ~30–130 nanostrain yr^{-1}, and are ~30–70 nanostrain yr^{-1} in the Walker Lane, slightly larger than the average strain rates inferred by Bennett et al. (2003). Our strain rates estimates for the area near Reno, Nevada, are also somewhat larger than those found by Svarc et al. (2002a), but they are consistent with the results of Hammond and Thatcher (2004). Contrary to Bennett et al. (2003), our solution does not indicate lower strain rates for the central Walker Lane compared to the northern and southern parts. This discrepancy cannot entirely be explained by having, for this model, Sierra Nevada–Great Valley block motion as a boundary condition.

Walker Lane

We show our results for the Walker Lane in Figure 4. Figure 4A contains the GPS velocities relative to Sierra Nevada–Great Valley. Figures 4B and 4C show the model velocities and principal strain rate axes (averaged for each grid cell) for a model that assumes a rigid Sierra Nevada–Great Valley and has Sierra Nevada–Great Valley–Stable North America Reference Frame motion applied as a boundary condition. Figure 4D shows the strain rate model for a case where the Sierra Nevada–Great Valley is part of the deforming grid, and no velocity boundary condition has been imposed. The strain results are corrected for the

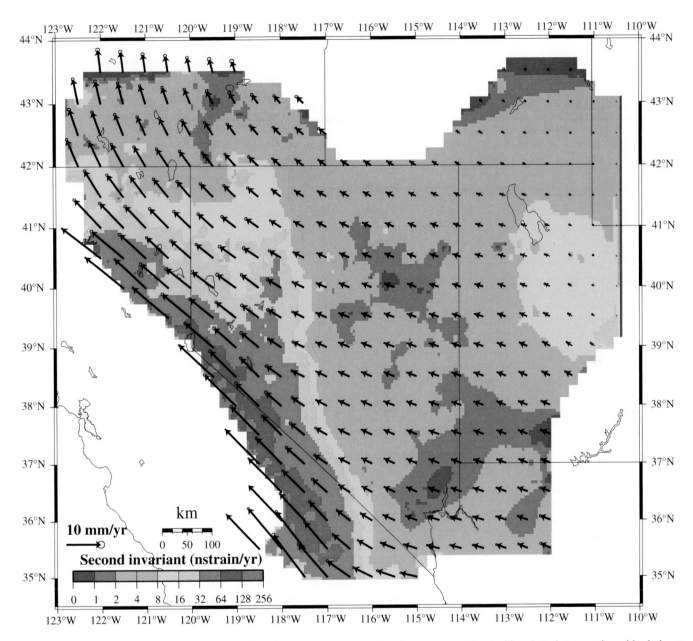

Figure 3. Color contour plot of second invariant of model strain rate tensor. Also shown are model velocities plotted at a regular grid relative to North America. Error ellipses represent one standard deviation.

effect of postseismic relaxation along the Central Nevada seismic belt (Hammond et al., this volume). To enhance interpretation of our results, we show the data results in an oblique Mercator projection with our Sierra Nevada–Great Valley–central Great Basin Euler pole (Table 2) as the pole of projection, and velocities are shown relative to our Sierra Nevada–Great Valley block. In this projection, any vector or feature aligned with the map's up-down direction is along the Sierra Nevada–Great Valley–central Great Basin small circle.

The azimuths of the GPS velocities (Fig. 4A) show some variation from the small circle orientation, but, on average,

observed velocities are aligned along the small circles associated with Sierra Nevada–Great Valley–central Great Basin motion. However, a slight clockwise rotation can be seen, particularly in the interpolated velocity field (Fig. 4B), and its effect is most profound north of 40°N. This rotation probably results from the fact that the figure is in an oblique projection around the Sierra Nevada–Great Valley–central Great Basin pole, while for the most northern Walker Lane region, the central Great Basin does not properly provide the appropriate bounding block (Hammond and Thatcher, 2005). The model velocity field is very consistent with the BARGEN velocities (white vectors in Fig. 4A), which

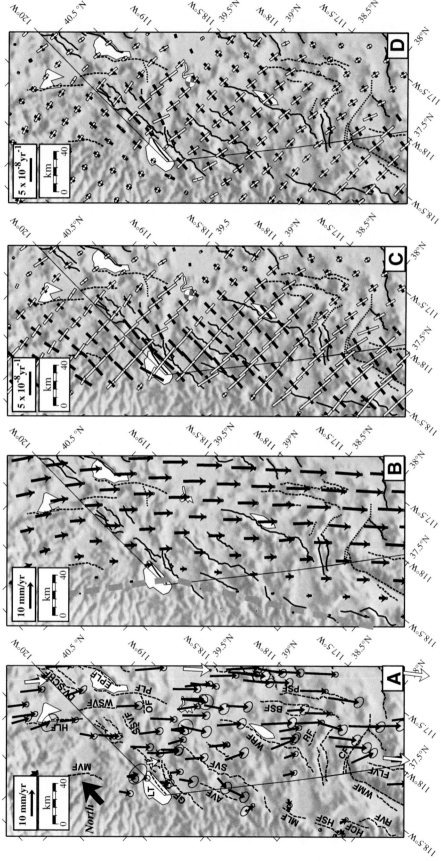

Figure 4. Oblique Mercator projection of the Walker Lane with the pole of projection equal to our obtained Sierra Nevada–Great Valley–central Great Basin Euler pole. (A) Global positioning system (GPS) velocities relative to the Sierra Nevada–Great Valley block that have been corrected for postseismic transient effects (white vectors for continuous BAR-GEN network, and black for all others). Error ellipses represent one standard deviation. Dashed lines are active faults: AVF—Antelope Valley fault; BSF—Benton Springs fault; CF—Coaldale fault; DV-SCRF—Dry Valley–Smoke Creek Ranch fault; EPLF—Eastern Pyramid Lake fault; FLVF—Fish Lake Valley fault; GF—Genoa fault; HCF—Hilton Creek fault; HLF—Honey Lake fault; HSF—Hartley Springs fault; LT—Lake Tahoe; MLF—Mono Lake fault; MVF—Mohawk Valley fault; OF—Olinghouse fault; PLF—Pyramid Lake fault; PSF—Petrified Springs fault; RF—Rattlesnake fault; RVF—Round Valley fault; SSVF—Spanish Springs Valley fault; SVF—Smith Valley fault; WMF—White Mountains fault; WRF—Wassuk Range fault; WSVF—Warm Springs Valley fault. (B) Model velocities relative to the Sierra Nevada–Great Valley block. Thick dashed line is eastern edge of our defined rigid Sierra Nevada–Great Valley block. Continuous black lines are normal faults, dashed lines are (dominantly) right-lateral faults, and dotted lines are (dominantly) left-lateral faults. Faults are indicated similarly as in B. Focal mechanism in western Carson Sink is for a 1992 $M_w = 4.1$ event (Ichinose et al., 2003). (D) Same as C, except for this model rigid motion of the Sierra Nevada–Great Valley block is not imposed as a velocity boundary condition.

provide strong constraints on our model solution because of their relatively small uncertainty. Despite the CGPS strong constraints, there are only a few CGPS velocities, and the campaign-style velocity measurements provide essential data in refining and confirming the deformation pattern implied by the CGPS velocities. Our results show that for the model with a rigid Sierra Nevada–Great Valley constraint, ~10 mm yr^{-1} is accommodated over a distance of ~135 km across the Walker Lane, and the width of this shear zone is roughly constant from north to south. Motion of 10 mm yr^{-1} of the eastern Walker Lane relative to the Sierra Nevada–Great Valley block is smaller than the earlier estimate by Thatcher et al. (1999), but it is in agreement with more recent estimates (Bennett et al., 2003; Hammond and Thatcher, 2004, 2007). Our results are different from Bennett et al. (2003) in that they found the width and total motion across the Walker Lane to vary along the Walker Lane, and we do not find that. The reasons for this difference can likely be found in the fact that we corrected for the effect of postseismic relaxation and in our treatment of the Sierra Nevada.

There is a significant difference between the strain magnitudes between the models with and without an imposed rigid Sierra Nevada–Great Valley (Fig. 4C vs. 4D). The difference in style of the tensor field is minimal. Without the assumption of a rigid Sierra Nevada–Great Valley, the model predicts significant strain rates in the Sierra Nevada and, in relation, lower strain rates in Walker Lane closest to the Sierra Nevada. Strain rate values toward the Basin and Range Province are roughly identical between the two models. For the most part, the difference in strain fields is the result of a lack of GPS velocity observations in the Sierra Nevada (particularly north of ~39°N), which causes a wide velocity gradient between the Walker Lane and the Great Valley. A secondary, and minor, reason for the difference is that the model results involves some level of smoothing, causing strain rates to diffuse into the Sierra Nevada in the case when the Sierras are not modeled as behaving rigidly.

The general alignment of the velocities with the small-circle orientation indicates that the deformation field is dominated by shear as a result of Sierra Nevada–Great Valley–central Great Basin motion. This is emphasized by the strain rate tensors, which generally have principal contractional and extensional strain rate axes of equal length, indicating shear (Figs. 4C–4D). The data and model results point out several general features of the kinematics in the Walker Lane. In the north (i.e., north of Lake Tahoe), shearing along the Sierra Nevada–Great Valley–central Great Basin small-circle direction dominates in the west, while smaller, dominantly extension, strain rates are present in the vicinity of Pyramid Lake. The Honey Lake (HL), Warm Springs Valley (WSV), Pyramid Lake (PL), and Mohawk Valley (MV) faults are optimally oriented to accommodate the inferred ~10 mm yr^{-1} total right lateral motion across these faults. Similarly, the Dry Valley–Smoke Creek Ranch, Eastern Pyramid Lake, and Spanish Springs Valley faults are well oriented to accommodate the extensional component of the strain tensor, and they are also consistent with the focal mechanisms of several earthquakes in

this area (Ichinose et al., 2003). Our model result shows that the region around Pyramid Lake is predominantly extending (Fig. 3C). Although this would agree with extension across the East Pyramid Lake and Spanish Springs Valley normal faults, it contradicts the geological findings that displacements in the region are predominantly right-lateral (Briggs and Wesnousky, 2004; Faulds et al., 2005b). We argue that this discrepancy may be an artifact of an insufficient number of reliable GPS velocities east of the Warm Springs Valley and Pyramid Lake faults.

South of the latitude of Lake Tahoe, the model strain rates indicate predominant shear, but the extensional strain component increases from the north to the south. The orientations of right-lateral faults, such as Fish Lake Valley, Petrified Springs (PS), and Benton Springs (BS) faults, are roughly along the small-circle orientation, indicating that they are well oriented to accommodate part of the 10 mm yr^{-1} of shear. Most other faults are oblique to the shear direction. In the central Walker Lane, the extensional component of the strain tensor is oriented consistently with the observed normal slip activity along normal faults such as the Genoa, Smith Valley, and Wassuk Range faults (e.g., Ramelli et al., 1999; Surpless et al., 2002). Our model results are not conclusive concerning the strain tensor style and orientation for Sierra range–bounding faults, such the Mono Lake and Hartley Springs faults, where the effect of choice of boundary condition is most profound. Further south, the strain tensor is more transtensional in style, and the orientation of the extensional axis deviates more from being perpendicular to faults such as, for example, the White Mountains (WM) fault. The latter is consistent with the observation that the White Mountains fault has accommodated oblique slip since ca. 3 Ma (Stockli et al., 2003; Wesnousky, 2005a). The strain rate tensors near the Rattlesnake and Coaldale faults are consistent with their behavior as left-lateral conjugate faults (with an extensional component) within a regional-scale right-lateral shear zone (Wesnousky, 2005a).

DISCUSSION

Transtension

The large-scale kinematics of the northern Walker Lane, as quantified by our strain rate model, the orientation of Sierra Nevada–Great Valley fixed velocities, and fault orientations and styles, are largely controlled by Sierra Nevada–Great Valley motion with respect to the Basin and Range Province. An earlier analysis of faulting along the entire eastern margin of the Sierra Nevada–Great Valley block came to the same conclusion (Unruh et al., 2003). Like Unruh et al. (2003), we find that the deformation field is not entirely characterized by simple shear strain, but rather by transtension. We note, however, that we observe transtension primarily south of 39°N. The transition from shear-dominated strain rates north of 39°N and transtension-dominated strain rate south of 39°N coincides with the change in the orientation of our defined eastern edge of the Sierra Nevada–Great Valley block. It could be argued that our transtensional strain rate results are simply a reflection of our imposed boundary condition. However, we see roughly the same pattern when we do not impose Sierra Nevada–Great Valley rigid body rotation as a boundary condition (Fig. 4D). Other studies also remarked that the transtension south of 39°N is due to the fact that the velocity field does not parallel the orientation of the eastern margin of the Sierra Nevada–Great Valley block (Oldow, 2003; Unruh et al., 2003).

Our results do not indicate fault-normal extension along the Genoa fault and within the Lake Tahoe basin. Yet, the orientations and locations of these faults, as well as the Quaternary faulting history along the Genoa fault (Ramelli et al., 1999), geophysical imaging (Kent et al., 2005), and Holocene seismic activity (Schweickert et al., 2004) near Lake Tahoe all indicate that these are, to first order, Sierra range–bounding normal faults. Moreover, the hypothesized lower crustal magma intrusion underneath Lake Tahoe in 2003 is consistent with crustal extension in a roughly E-W direction (Smith et al., 2004). Those observations would be consistent with the strain rate results if we would have imposed the eastern Sierra Nevada–Great Valley boundary to cut Lake Tahoe from south to north and then continue onward along the Mohawk Valley fault. Such geometry would create a large releasing bend across Lake Tahoe, with significant E-W extension across it. Unfortunately, the geodetic velocities northwest of Lake Tahoe that are available are rather unreliable kinematic indicators (Fig. 4A) and do not allow us to make conclusive inferences about the strain rate there and whether it could be considered part of the Sierra Nevada–Great Valley block. Tectonics near Lake Tahoe could be explained with having a large releasing bend in the eastern margin of the rigid Sierra Nevada–Great Valley block, but that does not reconcile the shear strain across normal faults such as the Smith Valley and Wassuk Range faults. If these faults play an important system role in the accommodation of the overall shear strain, then their simplest role would be of releasing or transfer steps between right-lateral faults. However such faults are not well documented. We discuss this in more detail in a section below.

"The Rate Debate"

At many places along the Walker Lane, a discrepancy appears to exist between the required total strike-slip motion imposed by GPS and that inferred geologically. Compare, for instance, the plots of the geodetic strain rate (Fig. 3) and geologic strain rates (Fig. 2 of Hammond et al., this volume). A similar discrepancy is found for the Basin and Range Province as a whole (Pancha et al., 2006). Next, we present a short overview of the geodetic and geologic values, largely in a regional context, as our study does not attempt to infer slip rates on individual faults.

In the most northern Walker Lane, the total range of geologic strike-slip rates (for the combined Honey Lake, Warm Springs Valley, and Pyramid Lake faults) is ~2–10 mm yr^{-1} (Faulds et al., 2005b). When adding a considerable, but still yet unknown, slip rate for the Mohawk Valley fault, the geologic rate is roughly consistent with the total geodetic rate of ~10 mm yr^{-1}. Ham-

mond and Thatcher (2007) used GPS campaign-style velocities (included in our velocities) to invert for fault slip rates and concluded that a preferred and maximum total of ~7 mm yr^{-1} and ~9 mm yr^{-1}, respectively, can be accounted for as right-lateral slip across the Mohawk Valley, Honey Lake, Warm Springs Valley, and Pyramid Lake faults.

Further south, in the central Walker Lane, no such consistency can be found. There, the dearth of significant observed Quaternary dextral strike-slip offsets (dePolo and Anderson, 2000) is puzzling. No right-lateral strike-slip fault is known to exist over a region ~100 km south of the Pyramid Lake fault. The only recognized significant strike-slip faults in the central Walker Lane, the Benton and Petrified Springs faults, have minimum slip rates of ~1 mm yr^{-1} (Wesnousky, 2005a).

More south, in the southern Walker Lane, the relative geodetic motion is roughly consistent with the relatively high geologic rates inferred for the Fish Lake Valley (FLV) fault (Reheis and Sawyer, 1997), as concluded elsewhere as well (Dixon et al., 2003). Predicted relative motion of 2 mm yr^{-1} over the White Mountains fault is higher than some observed geologic rates (dePolo, 1989; Reheis and Dixon, 1996) but more consistent with others (e.g., Schroeder et al., 2003).

The discrepancy between geologic and geodetic rates is most profound for the central Walker Lane. An outstanding question is: Does the discrepancy point to shortcomings or incompleteness in the paleoseismic slip estimation? The total cumulative slip on normal faults may approach the level of extension predicted as part of the inferred strain rate tensors. However, that would still require either a large amount of right-lateral slip between the normal faults on unknown strike-slip faults or a significant component of strike-slip motion on known normal faults. Does the discrepancy instead point to inappropriateness in the rate comparison? This would be the case, for instance, when a fault has a significant offset due to unrecognized aseismic slip, then the paleoseismic investigations are likely to underestimate the total offset. However, no documentation of "creeping" faults is available. Another case of an inappropriate comparison would be if only GPS velocities near a locked fault are considered, such that the full slip rate at depth may not be captured. However, most studies that have investigated the rate discrepancy consider slip rates over the entire Walker Lane belt or multiple faults, rather than one single fault, such that the geodetic velocity measurements should capture all of the interseismic strain. This leaves only one other explanation for the rate discrepancy: Can it be explained by strain accommodating processes other than slip along main faults? We investigate this third explanation below.

How Is Deformation Accommodated?

Right-Lateral Faulting

The simplest explanation of how the dextral shear is accommodated in the central Walker Lane is that it is simply done through right-lateral strike-slip faulting on undiscovered structures. Although no through-going dextral fault cuts the surface of the central Walker Lane between the Pyramid Lake and Benton Springs faults, some evidence of such a fault (zone) has recently emerged from field observations of possible dextral offsets and a topographic lineament in the western Carson Sink (J. Faulds, 2005, personal commun.). There is some additional evidence that NW-directed slip is currently being accommodated through strike-slip faulting. First, focal mechanism azimuths for the 1954 Fairview Peak and Rainbow Mountain earthquakes are ~N30°W (Doser, 1986), even though the surface breaks are oriented ~N17°E (Caskey et al., 2004). The slip direction of these events is close to the ~N35°W strike of the Benton Springs and Petrified Springs faults, which are located only ~25 km southeast of the Rainbow Mountain rupture. The 1932 Cedar Mountain earthquake displayed a similar discrepancy between seismic slip and surface faulting, as was seen for the 1954 events (e.g., Bell et al., 1999), and this discrepancy has been explained as the expression of Riedel shears above a fault at depth (Bell et al., 1999; Caskey et al., 2004). Secondly, while there is a dearth of mapped faults in the Carson Sink, a hint of its tectonic activity was revealed there by an $M_w = 4.1$ event in 1992 that had a strike-slip mechanism with one nodal plane trending N33°W (Ichinose et al., 2003). Thirdly, based on oblique striae rakes on the Wassuk Range fault (Stewart, 1988), evidence of small pull-apart basins and oblique dextral slip along this fault (Surpless et al., 2002), and orientations and arguable offsets of minor faults northwest of the mapped trace of the Wassuk Range front, called the White Mountain fault (Bell, 1981; Bingler, 1978; Wesnousky, 2005a), this region shows some evidence of right-lateral offsets.

Block Rotations

Crustal block rotation around a vertical axis, often involving "bookshelf faulting," is a well-described process in a zone of distributed homogeneous shear (e.g., McKenzie and Jackson, 1983; Nur et al., 1988). Through rotation, a significant part, or all, of the velocity gradient tensor can be accommodated. This process is well described for, for instance, the southern San Andreas fault system (Nicholson et al., 1986) and the northeastern Mojave Desert (e.g., Garfunkel, 1974; Nur et al., 1988), and the latter was recently corroborated with GPS velocity measurements (Savage et al., 2004).

It has been argued that clockwise rotations within the central Walker Lane accommodate, and have accommodated, a significant portion of the overall right-lateral shear, particularly south of the Pyramid Lake fault (known as the Carson domain) and near the Excelsior mountains, between the Benton Springs fault and the White Mountains and Fish Lake Valley faults (Cashman and Fontaine, 2000; Wesnousky, 2005a). Observed paleomagnetic rotations (Cashman and Fontaine, 2000) and observed (e.g., Olinghouse, Rattlesnake, and Coaldale faults), and inferred (i.e., the Carson and Wabuska lineaments; Stewart, 1988), left-lateral faults (or lineaments) hint at such an explanation. In this case, the crustal blocks between the left-lateral faults and lineaments act as the "books" in a "bookshelf faulting" scenario with the Pyramid Lake, Benton Springs, and Petrified Springs faults as the east-

ern "shelf." For this scenario, it is unclear what structures comprise the western "shelf." For the Carson domain, paleomagnetic clockwise rotation rates since 9–13 Ma have been estimated at 6 ± 2° m.y.[-1], with some indication that rotation has been slowing down over time (Cashman and Fontaine, 2000).

If currently present, can we observe crustal block rotations geodetically? Over timescales that are large compared to the seismic cycle, blocks translate and rotate with episodic slip along the fault zones between them. However, when measured over short timescales, and when unaffected by earthquakes, geodetic velocities cannot directly detect individual motions of blocks with small dimensions, i.e., 10–50 km. This is not only because of elastic loading along the block's bounding faults, but also because the lateral dimensions of the block under study are of the same order of magnitude as the elastic thickness of the block. As a consequence, the crust acts as a low-pass filter and geodetic velocities reflect only the strain accommodations over length scales many times the thickness of the crust. It is for these reasons, and the very low number of velocity measurements to be expected on any small crustal block, that we are cautious of studies that interpret geodetic velocities directly in terms of the kinematics of a small block, such as Oldow et al. (2001).

Even though geodetic velocities cannot directly constrain the rotation of small crustal blocks, we can infer from the velocity gradient model the contemporary rotation rate within the diffuse deformation zone, i.e., the vorticity of the velocity gradient tensor field, not a rigid block rotation. The relationship between these geodetic rotations and finite rotations is complicated because it is dependent on the size and orientations of crustal blocks (e.g., Lamb, 1994; McKenzie and Jackson, 1983). Nevertheless, we find relatively constant clockwise rotation rates of 1–2° m.y.[-1] for the Carson domain, and, regardless of the difficulties in comparing geodetic and finite rotations, we note that the geodetic rotation rate is significantly lower than the paleomagnetic estimates. Perhaps this is an expression of the deceleration of rotation since 9 Ma. If rotation is the only mechanism to accommodate the regional velocity gradient, a slowing down of rotation rate is surprising, because there is no indication that motion along the Walker Lane has slowed down since its inception at ca. 9 Ma. In fact, it has been argued that the Walker Lane is propagating northwestward as the Mendocino triple junction migrates northward along California's coast (Faulds et al., 2005b), with less cumulative slip in the north compared to the south (e.g., Faulds et al., 2005a; Wesnousky, 2005a). Thus, for the central Walker Lane region, the mechanism of rotation may have possibly been more significant between 9 and 3 Ma, compared to present-day, when the area was located ahead or near the "tip" of the propagating shear zone. This would be consistent with the latest timing estimate of onset of faulting (ca. 10 Ma) along the strike-slip faults of the central Walker Lane (Hardyman and Oldow, 1991).

Such a process can be seen at present, for instance, in central Greece, where significant vertical-axis rotation contributes significantly to the strain accommodation required in front of the North Anatolian fault as it propagates toward the Gulf of

Corinth (e.g., Armijo et al., 1996; Clarke et al., 1998; Kissel and Laj, 1988). Once a through-going shear zone has developed, block rotation will become less significant over time, in favor of strike-slip faulting.

Central Nevada Seismic Belt

It has been suggested that the Central Nevada seismic belt plays an important role in Walker Lane kinematics, transferring dextral shear (Faulds et al., 2005a; Oldow et al., 2001; Wesnousky et al., 2005) onto N-NE–striking normal faults. This idea has been based on several observations, but, most importantly, on the fact that earlier GPS velocity measurements suggested that the Central Nevada seismic belt separates ~6 ± 2 mm yr^{-1} of total motion across the northern Walker Lane and ~10 mm yr^{-1} across the southern portion (Gan et al., 2000; Thatcher et al., 1999). This would be consistent with the decrease of total offset from the southern to northern portion of Walker Lane (Faulds et al., 2005a; Wesnousky, 2005a). We show here that the motion over the northern Walker Lane is very similar to the motion over central and southern Walker Lane, in particular, when the effect of postseismic deformation from the Central Nevada seismic belt earthquakes is removed from the regional deformation field. Hammond et al. (this volume) assumed that when the time-dependent deformation is removed, extension across the Central Nevada seismic belt would be ~1 mm/yr, which is slightly elevated compared to the region further to the east. This is consistent with earthquake recurrence times being higher at the Central Nevada seismic belt than elsewhere in the central Great Basin (Wesnousky et al., 2005). Thus, although the twentieth-century seismicity along the Central Nevada seismic belt may have enhanced the apparent surface deformation, which could lead to an overemphasis of its role, the Central Nevada seismic belt likely plays some role in the regional strain accommodation. This idea is also supported by the orientation of the Central Nevada seismic belt relative to the Walker Lane trend and the fact that it connects to the Walker Lane where strike-slip faulting is not well established.

What Drives the Deformation?

We find that the strain rate field for the Walker Lane is, to first order, rather simple: the width of the zone is roughly constant, the strain rate magnitude in the zone is roughly constant, and the style of strain rate is dominated by shear/transtension throughout most of the region. The relative simplicity of the geodetic deformation field contrasts with the tectonic complexity inferred from the available (and, at places, lack of) geologic data discussed previously. Wesnousky (2005a) postulated that the complexity of faulting and the apparent rotation of crustal blocks are consistent with the concept of a partially detached elastic-brittle crust that is being transported on a continuously deforming anelastic substratum. Faulds et al. (2005a) noted that strike-slip faults in the northern Walker Lane appear to act as large-scale Riedel shears above a shear zone at depth. On physical arguments, it has been argued that the smoothness of the geodetic velocity field in a transform zone reflects the long-term continuous deformation in the lower

lithosphere (Bourne et al., 1998). Although Bourne et al. (1998) argued that the subcrustal shear zone drives the long-term movement of crustal blocks, similar to the floating block model of Lamb (1994), it should be noted that for a transform zone, the driving force of the instantaneous crustal deformation field is not uniquely resolvable from the geodetic velocity field and may come from the side or from beneath (Savage, 2000). Indeed, we observe that based on the orientation of the deformation zone, on the orientation of principal strain rate axes, and on the apparent influence of the geometry of the Sierra Nevada–Great Valley block on the Walker Lane strain rate tensor field, deformation in the Walker Lane is strongly controlled (and perhaps driven) by the Sierra Nevada–Great Valley block motion. Those strain features can be seen in Figures 4C and 4D, and our inference is thus not biased by whether or not we model the Sierra Nevada as a part of a rigid Sierra Nevada–Great Valley block.

Summary of Discussion

Irrespective of the actual driving mechanism, the surface complexity of the Walker Lane tectonic system is an expression of either its relative youth compared to, for example, the San Andreas fault system (Wesnousky, 2005b), and/or the influence of inherited structure. There is some suggestion that the southern Walker Lane region is more mature than the northern Walker Lane. Total dextral displacements are 48–75 km in the south (Ekren and Byers, 1984; Oldow, 1992) and 20–30 km in the north (Faulds et al., 2005b), and the inception of strike-slip faulting has been documented at ca. 6 Ma along the Fish Lake Valley fault (Reheis and Sawyer, 1997) and could be as recent as ca. 3 Ma in the northern Walker Lane (Henry et al., 2007). We show here that the difference in amount of finite strain between the northern and southern portions of Walker Lane is not matched by differences in the contemporary strain rate field (as was suggested by earlier GPS results), and we argue that Sierra Nevada–Great Valley block motion is the dominant kinematic boundary condition along the length of the entire Walker Lane. The discrepancy between cumulative offset and present-day strain rate distribution raises a few questions. Why does the northern Walker Lane appear as a more mature shear zone compared to the central Walker Lane? The Honey Lake and Warm Springs Valley faults (or at least the basement fault underneath them) are reactivated normal faults (Henry et al., 2007), and this reactivation may have occurred because those faults are optimally oriented to accommodate the shear. The normal faults in the central Walker Lane are more oblique to the shear orientation, which would inhibit them to reactivate as strike-slip faults (perhaps with the exception of the most northern Wassuk Range fault). Moreover, the process of crustal block rotations may have accommodated most of the strain without the need for those normal faults to take up a strike-slip component. If the rate of crustal rotations has slowed down, as suggested by the findings of Cashman and Fontaine (2000), one would expect oblique or dextral slip to become more prevalent on the normal faults, as has been documented further south along the White Mountains fault (Stockli et al., 2003), but concrete evidence for that is still missing.

Another point of consideration in understanding the discrepancy between finite strain and geodetic strain rate in the northern Walker Lane is that, kinematically, it is possible to have a shear zone with equal total displacement rates along its entire length that terminates rather abruptly. This can be seen, for example, in the northern Aegean, where the North Anatolian fault slips at a constant rate of ~24 mm yr^{-1} up to its abrupt termination near the Gulf of Evvia (Kreemer et al., 2004). In order to accommodate the deformation in the region in front of the propagating fault tip, relatively large contemporary vertical-axis rotation rates are expected, and can be seen, together with graben opening, in central Greece (Armijo et al., 1996; Clarke et al., 1998; Goldsworthy et al., 2002). The observed crustal rotations in the central Walker Lane ahead of the more established shear zone south of 38°N may be analogous. However, we would then also expect to see relatively large contemporary rotation northwest of the Honey Lake fault, where no active strike-slip fault has been documented. Moreover, we would expect to see relatively high activity along normal faults north of the Honey Lake–Warm Springs Valley fault system. Such activity is not documented, but given that migration of zones of active faulting has been documented before (Wallace, 1987), a future Central Nevada seismic belt–like zone of activity may be to the northwest of the Central Nevada seismic belt, connecting to the Walker Lane near 40.5°N.

CONCLUSIONS

We have presented a velocity and strain rate model for the northern part of the Walker Lane derived from a compilation of geodetic velocities. We find that the Walker Lane region from 37.5°N to 40.5°N is characterized by an ~135-km-wide zone with relatively constant strain rates associated with ~10 mm yr^{-1} total motion across the zone. These findings are consistent with most recent geodetic studies, from which came much of the data used in this study. The strain rates depict predominantly strike-slip deformation, but south of 39°N, the extensional component of the strain rate tensor increases, reflecting a more transtensional domain there. This transtension is the consequence of the motion of the Sierra Nevada–Great Valley block not being parallel to its eastern margin, i.e., the eastern Sierra front, south of 39°N. While several main faults in the northern and southern Walker Lane are consistent with the strain rate model results, the geologic mode and rate of deformation in the central Walker Lane are less clear. Left-lateral faulting and clockwise rotations there may contribute to the accommodation of the velocity gradient tensor field, and most normal faults are optimally oriented to accommodate some component of the regional shear strain. However, significant additional dextral strike-slip faulting is required to accommodate the majority of the 10 mm yr^{-1} relative motion. Several observations point at the accommodation of dextral shear in the Carson Sink, but no consistent pattern has emerged yet. The geologic complexity and variation along the length of the Walker Lane are

in contrast with the relatively simple strain rate field. This suggests that (1) various mechanisms are at play to accommodate the shear, (2) parts of the surface tectonics may (still) be in an early stage of development, and (3) inherited structural grain can have a dominant control on the strain accommodation mechanism.

ACKNOWLEDGMENTS

This research was funded in part by Department of Energy grant DE-FG36-02ID14311 to the Great Basin Center for Geothermal Energy and by the Department of Energy Yucca Mountain Project/Nevada System of Higher Education Cooperative Agreement DE-FC28-04RW12232. We thank D. Lavallée for providing his global velocity solution, T. Williams for providing us with his velocity solution before publication, R. Smith for making the unpublished results of the EBRY network available, and the U.S. Geological Survey for making all their campaign results publicly available. We thank J. Bell, S. Wesnousky, and J. Faulds for helpful discussions, and we are grateful to S. Wesnousky, P. LaFemina, and an anonymous reviewer for comments that improved this paper significantly.

REFERENCES CITED

Argus, D.F., and Gordon, R.G., 1991, Current Sierra Nevada–North America motion from very long baseline interferometry: Implications for the kinematics of the western United States: Geology, v. 19, p. 1085–1088, doi: 10.1130/0091-7613(1991)019<1085:CSNNAM>2.3.CO;2.

Argus, D.F., and Gordon, R.G., 2001, Present tectonic motion across the Coast Ranges and San Andreas fault system in central California: Geological Society of America Bulletin, v. 113, p. 1580–1592, doi: 10.1130/0016-7606 (2001)113<1580:PTMATC>2.0.CO;2.

Armijo, R., Meyer, B., King, G.C.P., Rigo, A., and Papanastassiou, D., 1996, Quaternary evolution of the Corinth Rift and its implications for the late Cenozoic evolution of the Aegean: Geophysical Journal International, v. 126, p. 11–53, doi: 10.1111/j.1365-246X.1996.tb05264.x.

Bell, J.W., 1981, Quaternary Fault Map of the Reno 1° by 2° Quadrangle: U.S. Geological Survey Open-File Report 81-982, 62 p.

Bell, J.W., dePolo, C.M., Ramelli, A.R., Sarna-Wojcicki, A.M., and Meyer, C.E., 1999, Surface faulting and paleoseismic history of the 1932 Cedar Mountain earthquake area, west-central Nevada, and implications for modern tectonics of the Walker Lane: Geological Society of America Bulletin, v. 111, p. 791–807, doi: 10.1130/0016-7606(1999)111<0791:SFAPHO>2.3.CO;2.

Bell, J.W., Caskey, S.J., Ramelli, A.R., and Guerrieri, L., 2004, Pattern and rates of faulting in the central Nevada seismic belt, and paleoseismic evidence for prior beltlike behavior: Bulletin of the Seismological Society of America, v. 94, p. 1229–1254, doi: 10.1785/012003226.

Bennett, R.A., Wernicke, B.P., and Davis, J.L., 1998, Continuous GPS measurements of contemporary deformation across the northern Basin and Range Province: Geophysical Research Letters, v. 25, p. 563–566, doi: 10.1029/98GL00128.

Bennett, R.A., Wernicke, B.P., Niemi, N.A., Friedrich, A.M., and Davis, J.L., 2003, Contemporary strain rates in the northern Basin and Range Province from GPS data: Tectonics, v. 22, 1008, doi: 10.1029/2001TC001355.

Bingler, E.C., 1978, Geologic Map of the Schurz Quadrangle: Nevada Bureau of Mines and Geology Map 60, scale 1:42,000.

Blewitt, G., Argus, D., Bennett, R., Bock, Y., Calais, E., Craymer, M., Davis, J., Dixon, T., Freymueller, J., Herring, T., Johnson, D., Larson, K., Miller, M., Sella, G., Snay, R., and Tamisiea, M., 2005, A stable North America reference frame (SNARF): First release, UNAVCO-IRIS Joint Workshop: Stevenson, Washington.

Blewitt, G., Hammond, W.C., and Kreemer, C., 2009, this volume, Geodetic observation of contemporary deformation in the northern Walker Lane: 1. Semipermanent GPS strategy, *in* Oldow, J.S., and Cashman, P.H., eds., Late Cenozoic Structure and Evolution of the Great Basin–Sierra Nevada Transition: Geological Society of America Special Paper 447, doi: 10.1130/2009.2447(01).

Bourne, S.J., England, P.C., and Parsons, B., 1998, The motion of crustal blocks driven by flow of the lower lithosphere and implications for slip rates of continental strike-slip faults: Nature, v. 391, p. 655–659, doi: 10.1038/35556.

Briggs, R.W., and Wesnousky, S.G., 2004, Late Pleistocene fault slip rate, earthquake recurrence, and recency of slip along the Pyramid Lake fault zone, northern Walker Lane, United States: Journal of Geophysical Research, v. 109, p. B08402, doi: 10.1029/2003JB002717.

Cashman, P.H., and Fontaine, S.A., 2000, Strain partitioning in the northern Walker Lane, western Nevada and northeastern California: Tectonophysics, v. 326, p. 111–130, doi: 10.1016/S0040-1951(00)00149-9.

Caskey, S.J., Bell, J.W., Ramelli, A.R., and Wesnousky, S.G., 2004, Historic surface faulting and paleoseismicity in the area of the 1954 Rainbow Mountain–Stillwater earthquake sequence, central Nevada: Bulletin of the Seismological Society of America, v. 94, p. 1255–1275, doi: 10.1785/012003012.

Clarke, P.J., Davies, R.R., England, P.C., Parsons, B., Billiris, H., Paradissis, D., Veis, G., Cross, P.A., Denys, P.H., Ashkenazi, V., Bingley, R., Kahle, H.-G., Muller, M.-V., and Briole, P., 1998, Crustal strain in central Greece from repeated GPS measurements in the interval 1989–1997: Geophysical Journal International, v. 135, p. 195–214, doi: 10.1046/j.1365-246X.1998.00633.x.

d'Alessio, M.A., Johanson, I.A., Bürgmann, R., Schmidt, D.A., and Murray, M.H., 2005, Slicing up the San Francisco Bay Area: Block kinematics and fault slip rates from GPS-derived surface velocities: Journal of Geophysical Research, v. 110, p. B06403, doi: 10.1029/2004JB003496.

Davies, P., and Blewitt, G., 2000, Methodology for global geodetic time series estimation: A new tool for geodynamics: Journal of Geophysical Research, v. 105, p. 11,083–11,100, doi: 10.1029/2000JB900004.

dePolo, C.M., 1989, Seismotectonics of the White Mountains fault system, east-central California and west-central Nevada [M.Sc. thesis]: Reno, University of Nevada, 354 p.

dePolo, C.M., and Anderson, J.G., 2000, Estimating the slip rates of normal faults in the Great Basin, USA: Basin Research, v. 12, p. 227–240, doi: 10.1046/j.1365-2117.2000.00131.x.

Dixon, T., Miller, M., Farina, F., Wang, H.Z., and Johnson, D., 2000, Present-day motion of the Sierra Nevada block and some tectonic implications for the Basin and Range Province, North American Cordillera: Tectonics, v. 19, p. 1–24, doi: 10.1029/1998TC001088.

Dixon, T., Norabuena, E., and Hotaling, L., 2003, Paleoseismology and global positioning system: Earthquake-cycle effects and geodetic versus geologic fault slip rates in the Eastern California shear zone: Geology, v. 31, p. 55–58, doi: 10.1130/0091-7613(2003)031<0055:PAGPSE>2.0.CO;2.

Doser, D., 1986, Earthquake processes in the Rainbow Mountain–Fairview Peak–Dixie Valley, Nevada region 1954–1959: Journal of Geophysical Research, v. 91, p. 12,572–12,586, doi: 10.1029/JB091iB12p12572.

Ekren, E.B., and Byers, F.M., Jr., 1984, The Gabbs Valley Range—A well-exposed segment of the Walker Lane in west-central Nevada, in Lintz, J., Jr., ed., Western Geologic Excursions, Volume 4: Boulder, Geological Society of America, Fieldtrip Guidebook, p. 204–215.

Faulds, J.E., Henry, C.D., Coolbaugh, M.F., and Garside, L.J., 2005a, Influence of the late Cenozoic strain field and tectonic setting on geothermal activity and mineralization in the northwestern Great Basin: Geothermal Resources Council Transactions, v. 29, p. 353–358.

Faulds, J.E., Henry, C.D., and Hinz, N.H., 2005b, Kinematics of the northern Walker Lane: An incipient transform fault along the Pacific–North American plate boundary: Geology, v. 33, p. 505–508, doi: 10.1130/G21274.1.

Flesch, L.M., Holt, W.E., Haines, A.J., and Shen-Tu, B., 2000, Dynamics of the Pacific–North American plate boundary in the western United States: Science, v. 287, p. 834–836, doi: 10.1126/science.287.5454.834.

Freymueller, J.T., Murray, M.H., Segall, P., and Castillo, D., 1999, Kinematics of the Pacific–North America plate boundary zone, northern California: Journal of Geophysical Research, v. 104, p. 7419–7441, doi: 10.1029/1998JB900118.

Gan, W.J., Svarc, J.L., Savage, J.C., and Prescott, W.H., 2000, Strain accumulation across the Eastern California shear zone at latitude 36°30′N: Journal of Geophysical Research, v. 105, p. 16,229–16,236, doi: 10.1029/2000JB900105.

Garfunkel, Z., 1974, Model for the late Cenozoic tectonic history of the Mojave Desert and its relation to adjacent areas: Geological Society of America Bulletin, v. 85, p. 1931–1944, doi: 10.1130/0016-7606(1974)85<1931:MFTLCT>2.0.CO;2.

Goldsworthy, M., Jackson, J., and Haines, J., 2002, The continuity of active fault systems in Greece: Geophysical Journal International, v. 148, p. 596–618, doi: 10.1046/j.1365-246X.2002.01609.x.

Haines, A.J., and Holt, W.E., 1993, A procedure for obtaining the complete horizontal motions within zones of distributed deformation from the inversion of strain rate data: Journal of Geophysical Research, v. 98, p. 12,057–12,082, doi: 10.1029/93JB00892.

Hammond, W.C., and Thatcher, W., 2004, Contemporary tectonic deformation of the Basin and Range Province, western United States: 10 years of observation with the global positioning system: Journal of Geophysical Research, v. 109, p. B08403, doi: 10.1029/2003JB002746.

Hammond, W.C., and Thatcher, W., 2005, Northwest Basin and Range tectonic deformation observed with the global positioning system, 1999–2003: Journal of Geophysical Research, v. 110, p. B10405, doi: 10.1029/2005JB003678.

Hammond, W.C., and Thatcher, W., 2007, Crustal deformation across the Sierra Nevada, Northern Walker Lane, Basin and Range transition, Western United States, measured with GPS, 2000–2004: Journal of Geophysical Research, v. 112, p. B05411, doi: 10.1029/2006JB004625.

Hammond, W.C., Kreemer, C., and Blewitt, G., 2009, this volume, Geodetic constraints on contemporary deformation in the northern Walker Lane: 3. Central Nevada seismic belt postseismic relaxation, in Oldow, J.S., and Cashman, P.H., eds., Late Cenozoic Structure and Evolution of the Great Basin–Sierra Nevada Transition: Geological Society of America Special Paper 447, doi: 10.1130/2009.2447(03).

Hardyman, R.F., and Oldow, J.S., 1991, Tertiary tectonic framework and Cenozoic history of the central Walker Lane, Nevada, in Raines, G.L., Lisle, R.E., Schafer, R.W., and Wilkinson, W.H., eds., Geology and Ore Deposits of the Great Basin, Symposium Proceedings: Reno, Nevada, Geological Society of Nevada, p. 279–301.

Henry, C.D., Faulds, J.E., and dePolo, C.M., 2007, Geometry and timing of strike-slip and normal faults in the northern Walker Lane, northwestern Nevada and northeastern California: Strain partitioning or sequential extensional and strike-slip deformation?, in Till, A.B., Roeske, S., Sample, J., and Foster, D.A., eds., Exhumation Associated with Continental Strike-Slip Fault Systems: Geological Society of America Special Paper 434, p. 59–79, doi: 10.1130/2007.2434(04).

Hetland, E.A., and Hager, B.H., 2003, Postseismic relaxation across the Central Nevada seismic belt: Journal of Geophysical Research, v. 108, 2394, doi: 10.1029/2002JB002257.

Holt, W.E., Shen-Tu, B., Haines, J., and Jackson, J., 2000, On the determination of self-consistent strain rate fields within zones of distributed deformation, in Richards, M.A., Gordon, R.G., and van der Hilst, R.D., eds., The History and Dynamics of Global Plate Motions: American Geophysical Union Geophysical Monograph 121, p. 113–141.

Holt, W.E., Kreemer, C., Haines, A.J., Estey, L., Meertens, C., Blewitt, G., and Lavallée, D., 2005, Project helps constrain continental dynamics and seismic hazards: Eos (Transactions, American Geophysical Union), v. 86, p. 383–387.

Ichinose, G.A., Anderson, J.G., Smith, K.D., and Zeng, Y.H., 2003, Source parameters of eastern California and western Nevada earthquakes from regional moment tensor inversion: Bulletin of the Seismological Society of America, v. 93, p. 61–84, doi: 10.1785/0120020063.

Kent, G.M., Babcock, J.M., Driscoll, N.W., Harding, A.J., Dingler, J.A., Seitz, G.G., Gardner, J.V., Mayer, L.A., Goldman, C.R., Heyvaert, A.C., Richards, R.C., Karlin, R., Morgan, C.W., Gayes, P.T., and Owen, L.A., 2005, 60 k.y. record of extension across the western boundary of the Basin and Range Province: Estimate of slip rates from offset shoreline terraces and a catastrophic slide beneath Lake Tahoe: Geology, v. 33, p. 365–368, doi: 10.1130/G21230.1.

Kissel, C., and Laj, C., 1988, The Tertiary geodynamical evolution of the Aegean arc: A paleomagnetic reconstruction: Tectonophysics, v. 146, p. 183–201, doi: 10.1016/0040-1951(88)90090-X.

Kreemer, C., Chamot-Rooke, N., and Le Pichon, X., 2004, Constraints on the evolution and vertical coherency of deformation in the northern Aegean from a comparison of geodetic, geologic and seismologic data: Earth and Planetary Science Letters, v. 225, p. 329–346, doi: 10.1016/j.epsl.2004.06.018.

Kreemer, C., Lavallée, D., Blewitt, G., and Holt, W.E., 2006, On the stability of a geodetic no-net-rotation frame and its implication for the International Terrestrial Reference Frame: Geophysical Research Letters, v. 33, doi: 10.1029/2006GL027058.

Lamb, S.H., 1994, Behavior of the brittle crust in wide plate boundary zones: Journal of Geophysical Research, v. 99, p. 4457–4483, doi: 10.1029/93JB02574.

Langbein, J., Wyatt, F., Johnson, H., Hamann, D., and Zimmer, P., 1995, Improved stability of a deeply anchored geodetic monument for deformation monitoring: Geophysical Research Letters, v. 22, p. 3533–3536, doi: 10.1029/95GL03325.

Malservisi, R., Dixon, T.H., La Femina, P.C., and Furlong, K.P., 2003, Holocene slip rate of the Wasatch fault zone, Utah, from geodetic data: Earthquake cycle effects: Geophysical Research Letters, v. 30, no. 13, p. 1673, doi: 10.1029/2003GL017408.

Martinez, L.J., Meertens, C.M., and Smith, R.B., 1998, Rapid deformation rates along the Wasatch fault zone, Utah, from first GPS measurements with implications for earthquake hazard: Geophysical Research Letters, v. 25, p. 567–570, doi: 10.1029/98GL00090.

Mazzotti, S., Dragert, H., Henton, J., Schmidt, M., Hyndman, R., James, T., Lu, Y., and Craymer, M., 2003, Current tectonics of northern Cascadia from a decade of GPS measurements: Journal of Geophysical Research, v. 108, p. 2554, doi: 10.1029/2003JB002653.

McClusky, S.C., Bjornstad, S.C., Hager, B.H., King, R.W., Meade, B.J., Miller, M.M., Monastero, F.C., and Souter, B.J., 2001, Present-day kinematics of the Eastern California shear zone from a geodetically constrained block model: Geophysical Research Letters, v. 28, p. 3369–3372, doi: 10.1029/2001GL013091.

McKenzie, D., and Jackson, J., 1983, The relationship between strain rates, crustal thickening, paleomagnetism, finite strain and fault movements within a deforming zone: Earth and Planetary Science Letters, v. 65, p. 182–202, doi: 10.1016/0012-821X(83)90198-X.

Nicholson, C., Seeber, L., Williams, P., and Sykes, L.R., 1986, Seismic evidence for conjugate slip and block rotation within the San Andreas fault system, southern California: Tectonics, v. 5, p. 629–648, doi: 10.1029/TC005i004p00629.

Niemi, N.A., Wernicke, B.P., Friedrich, A.M., Simons, M., Bennett, R.A., and Davis, J.L., 2004, BARGEN continuous GPS data across the eastern Basin and Range Province, and implications for fault system dynamics: Geophysical Journal International, v. 159, p. 842–862, doi: 10.1111/j.1365-246X.2004.02454.x.

Nur, A., Ron, H., and Scotti, O., 1988, Mechanics of distributed fault and block rotation, *in* Kissel, C., and Laj, C., eds., Paleomagnetic Rotations and Continental Deformation, Volume 254: Series C, Mathematical and Physical Sciences: Dordrecht, Kluwer, p. 209–228.

Oldow, J.S., 1992, Late Cenozoic displacement partitioning in the northwestern Great Basin, *in* Craig, S.D., ed., Structure, Tectonics and Mineralization of the Walker Lane: Reno, Nevada, Geological Society of Nevada, Geological Society of Nevada Symposium Proceedings Volume, p. 17–52.

Oldow, J.S., 2003, Active transtensional boundary zone between the western Great Basin and Sierra Nevada block, western U.S. Cordillera: Geology, v. 31, p. 1033–1036, doi: 10.1130/G19838.1.

Oldow, J.S., Aiken, C.L.V., Hare, J.L., Ferguson, J.F., and Hardyman, R.F., 2001, Active displacement transfer and differential block motion within the central Walker Lane, western Great Basin: Geology, v. 29, p. 19–22, doi: 10.1130/0091-7613(2001)029<0019:ADTADB>2.0.CO;2.

Pancha, A., Anderson, J.G., and Kreemer, C., 2006, Comparison of seismic and geodetic scalar moment rates across the Basin and Range Province: Bulletin of the Seismological Society of America, v. 96, p. 11–32, doi: 10.1785/0120040166.

Ramelli, A.R., Bell, J.W., dePolo, C.M., and Yount, J.C., 1999, Large-magnitude, late Holocene earthquakes on the Genoa fault, west-central Nevada and eastern California: Bulletin of the Seismological Society of America, v. 89, p. 1458–1472.

Reheis, M.C., and Dixon, T.H., 1996, Kinematics of the Eastern California shear zone: Evidence for slip transfer from Owens and Saline Valley fault zones to Fish Lake Valley fault zone: Geology, v. 24, p. 339–342, doi: 10.1130/0091-7613(1996)024<0339:KOTECS>2.3.CO;2.

Reheis, C., and Sawyer, T.L., 1997, Late Cenozoic history and slip rates of the Fish Lake Valley, Emigrant Peak, and Deep Springs fault zones, Nevada and California: Geological Society of America Bulletin, v. 109, p. 280–299, doi: 10.1130/0016-7606(1997)109<0280:LCHASR>2.3.CO;2.

Savage, J.C., 2000, Viscoelastic-coupling model for the earthquake cycle driven from below: Journal of Geophysical Research, v. 105, p. 25,525–25,532, doi: 10.1029/2000JB900276.

Savage, J.C., Svarc, J.L., and Prescott, W.H., 2004, Interseismic strain and rotation rates in the northeast Mojave domain, eastern California: Journal of Geophysical Research, v. 109, p. B02406, doi: 10.1029/2003JB002705.

Schroeder, J.M., Lee, J., Owen, L.A., and Finkel, R.C., 2003, Pleistocene dextral fault slip along the White Mountains fault zone, California: Geological Society of America Abstracts with Programs, v. 35, no. 6, p. 346.

Schweickert, R.A., Lahren, M.M., Smith, K.D., Howle, J.F., and Ichinose, G., 2004, Transtensional deformation in the Lake Tahoe region, California and Nevada, USA: Tectonophysics, v. 392, p. 303–323, doi: 10.1016/j.tecto.2004.04.019.

Shen, Z.K., Agnew, D.C., and King, R.W., 2003, The SCEC Crustal Motion Map, Version 3.0: Los Angeles, California, Southern California Earthquake Center.

Shen-Tu, B., Holt, W.E., and Haines, A.J., 1998, Contemporary kinematics of the western United States determined from earthquake moment tensors, very long baseline interferometry, and GPS observations: Journal of Geophysical Research, v. 103, p. 18,087–18,117, doi: 10.1029/98JB01669.

Shen-Tu, B., Holt, W.E., and Haines, A.J., 1999, Deformation kinematics in the western United States determined from Quaternary fault slip rates and recent geodetic data: Journal of Geophysical Research, v. 104, p. 28,927–28,955, doi: 10.1029/1999JB900293.

Slemmons, D.B., van Wormer, D., Bell, E.J., and Silberman, M., 1979, Recent crustal movements in the Sierra Nevada–Walker Lane region of California-Nevada: Part I. Rate and style of deformation: Tectonophysics, v. 52, p. 561–570, doi: 10.1016/0040-1951(79)90271-3.

Smith, K.D., von Seggern, D., Blewitt, G., Preston, L., Anderson, J.G., Wernicke, B.P., and Davis, J.L., 2004, Evidence for deep magma injection beneath Lake Tahoe, Nevada-California: Science, v. 305, p. 1277–1280, doi: 10.1126/science.1101304.

Stewart, J.H., 1988, Tectonics of the Walker Lane belt, western Great Basin: Mesozoic and Cenozoic deformation in a zone of shear, *in* Ernst, W.G., ed., Metamorphism and Crustal Evolution of the Western United States, Volume 7: Old Tappen, New Jersey, Prentice-Hall, p. 683–713.

Stockli, D.F., Dumitru, T.A., McWilliams, M.O., and Farley, K.A., 2003, Cenozoic tectonic evolution of the White Mountains, California and Nevada: Geological Society of America Bulletin, v. 115, p. 788–816, doi: 10.1130/0016-7606(2003)115<0788:CTEOTW>2.0.CO;2.

Surpless, B.E., Stockli, D.F., Dumitru, T.A., and Miller, E.L., 2002, Two-phase westward encroachment of Basin and Range extension into the northern Sierra Nevada: Tectonics, v. 21, p. 1002, doi: 10.1029/2000TC001257.

Svarc, J.L., Savage, J.C., Prescott, W.H., and Murray, M.H., 2002a, Strain accumulation and rotation in western Oregon and southwestern Washington: Journal of Geophysical Research, v. 107, p. 2087, doi: 10.1029/2001JB000625.

Svarc, J.L., Savage, J.C., Prescott, W.H., and Ramelli, A.R., 2002b, Strain accumulation and rotation in western Nevada, 1993–2000: Journal of Geophysical Research, v. 107, p. 2090, doi: 10.1029/2001JB000579.

Thatcher, W., Foulger, G.R., Julian, B.R., Svarc, J., Quilty, E., and Bawden, G.W., 1999, Present-day deformation across the Basin and Range Province, western United States: Science, v. 283, p. 1714–1718, doi: 10.1126/science.283.5408.1714.

Unruh, J., Humphrey, J., and Barron, A., 2003, Transtensional model for the Sierra Nevada fault system, eastern California: Geology, v. 31, p. 327–330, doi: 10.1130/0091-7613(2003)031<0327:TMFTSN>2.0.CO;2.

Wallace, R.E., 1984, Patterns and timing of late Quaternary faulting in the Great Basin Province and relation to some regional tectonic features: Journal of Geophysical Research, v. 89, p. 5763–5769, doi: 10.1029/JB089iB07p05763.

Wallace, R.E., 1987, Grouping and migration of surface faulting and variations in slip rates on faults in the Great Basin province: Bulletin of the Seismological Society of America, v. 77, p. 868–876.

Ward, S.N., 1998, On the consistency of earthquake moment release and space geodetic strain rates: Europe: Geophysical Journal International, v. 135, p. 1011–1018, doi: 10.1046/j.1365-246X.1998.t01-2-00658.x.

Wesnousky, S.G., 2005a, Active faulting in the Walker Lane: Tectonics, v. 24, p. TC3009, doi: 10.1029/2004TC001645.

Wesnousky, S.G., 2005b, The San Andreas and Walker Lane fault systems, western North America: Transpression, transtension, cumulative slip and the structural evolution of a major transform plate boundary: Journal of Structural Geology, v. 27, p. 1505–1512, doi: 10.1016/j.jsg.2005.01.015.

Wesnousky, S.G., Barron, A.D., Briggs, R.W., Caskey, S.J., Kumar, S., and Owen, L., 2005, Paleoseismic transect across the northern Great Basin: Journal of Geophysical Research, v. 110, p. B05408, doi: 10.1029/2004JB003283.

Williams, T.B., Kelsey, H.M., and Freymueller, J.T., 2006, GPS-derived strain in northwestern California: Termination of the San Andreas fault system and convergence of the Sierra Nevada–Great Valley block contribute to southern Cascadia forearc contraction: Tectonophysics, v. 413, p. 171–184, doi: 10.1016/j.tecto.2005.10.047.

MANUSCRIPT ACCEPTED BY THE SOCIETY 21 JULY 2008

The Geological Society of America
Special Paper 447
2009

Geodetic constraints on contemporary deformation in the northern Walker Lane: 3. Central Nevada seismic belt postseismic relaxation

William C. Hammond*
Corné Kreemer
Geoffrey Blewitt
Nevada Bureau of Mines and Geology, and Nevada Seismological Laboratory, University of Nevada, Reno, Nevada 89557-0178, USA

ABSTRACT

We combine horizontal Global Positioning System (GPS) velocities from a new compilation of published and new GPS velocities, results from an interferometric synthetic aperture radar (InSAR) study, and paleoseismic data to evaluate the postseismic response of historic earthquakes in the Central Nevada seismic belt. We assume that GPS velocity has contributions from time-invariant (i.e., steady permanent crustal deformation) and transient (i.e., time varying and associated with the seismic cycle) processes that are attributable to postseismic viscoelastic relaxation of the crust and upper mantle. In order to infer the viscosity structure of Basin and Range lower crust, η_{LC}, and upper mantle, η_{UM}, we apply three objective criteria to identify rheological models that fit both geodetic and geologic data. The model must (1) improve the apparent mismatch between geodetically and geologically inferred slip rates, (2) explain the InSAR-inferred vertical uplift rate, and (3) not imply time-invariant contractions anywhere in the extending province. It is not required for the postseismic deformation field to resemble the time-invariant velocity field in pattern, rate, or style. We find that the InSAR and horizontal GPS velocities form complementary constraints on the viscoelastic structure, excluding different parts of the model space. The best-fitting model has a lower crust that is stronger than the uppermost mantle, with $\eta_{LC} = 10^{20.5}$ Pa·s and $\eta_{UM} = 10^{19}$ Pa·s, a finding consistent with the majority of similar studies in the Basin and Range. The best-fitting viscosity model implies that the majority of Central Nevada seismic belt deformation is attributable to postseismic relaxation, and hence that western Basin and Range time-invariant deformation north of 39°N latitude is more tightly focused into the northern Walker Lane than would be inferred from uncorrected GPS velocities. However, significant deformation remains after correction for postseismic effects, consistent with Central Nevada seismic belt faults slipping at rates intermediate between the Walker Lane belt and the central Basin and Range.

Keywords: Basin and Range, GPS, earthquake, slip rates, seismic cycle.

*Corresponding author e-mail: whammond@unr.edu.

Hammond, W.C., Kreemer, C., and Blewitt, G., 2009, Geodetic constraints on contemporary deformation in the northern Walker Lane: 3. Central Nevada seismic belt postseismic relaxation, *in* Oldow, J.S., and Cashman, P.H., eds., Late Cenozoic Structure and Evolution of the Great Basin–Sierra Nevada Transition: Geological Society of America Special Paper 447, p. 33–54, doi: 10.1130/2009.2447(03). For permission to copy, contact editing@geosociety.org. ©2009 The Geological

INTRODUCTION

Rates of slip on active continental faults inferred from geodetic techniques sometimes disagree with the rates inferred from geologic studies. This discrepancy lies near the heart of the relationship between geodetic and geologic investigations of continental deformation (e.g., Friedrich et al., 2003). Similarity of space geodetic and geologic measurements of lithospheric plate motions (DeMets and Dixon, 1999; Sella et al., 2002) has led to anticipation that agreement might occur on the smaller scale of individual faults as well. Where these rates agree, geodetic measurement of relative motions across the fault match the rate inferred from coseismic rupture offset and time between major slip events. However, the comparison requires application of a correction of the geodetic data according to the buried dislocation model, composed of a faulted half space locked at the surface but slipping at depth (Savage and Burford, 1973; Freund and Barnett, 1976). In this case, the cumulative offset of many episodic surface ruptures from the largest earthquakes adds up to the slip predicted by far-field motion of crustal blocks. Some faults, such as the relatively linear and simple San Andreas in central California, exhibit a close agreement (e.g., Murray et al., 2001). This suggests that paleoseismic studies and geodesy measure the same physical processes at play in active faulting, albeit over greatly different time scales and different parts of the seismic cycle. Hence, disagreement between paleoseismic and geodetic slip rates suggests that an explanation is required for the deviation from this paradigm, and perhaps a modification of our physical model.

The Central Nevada seismic belt, which resides near the middle of the Basin and Range Province (Wallace, 1984b), is a well-documented example of disagreement between slip rates estimated with geodetic and geologic techniques. The Central Nevada seismic belt is a quasi-linear sequence of large-magnitude historical earthquakes that form an approximately north-south–trending belt (Fig. 1) (Caskey et al., 2000). The belt remains seismically active to this day, and it is responsible for a large proportion of the total historic seismic moment released in the Basin and Range Province (Pancha et al., 2006). Paleoseismically inferred slip rates for the set of faults that make up the belt near latitude 39°N total probably less than 1 mm/yr (Bell et al., 2004). Geodetic rates, however, inferred mostly from surveys with global positioning system (GPS) are closer to 3–4 mm/yr (Thatcher et al., 1999; Bennett et al., 2003; Hammond and Thatcher, 2004), a difference that is well outside the uncertainties in the measurements.

Recognition of the Central Nevada seismic belt as a zone of recent earthquakes is essential to explaining the discrepancy between geodetic and geologic rates because it allows for the possibility of geodetic strain rates being influenced by transient postseismic relaxation. A number of studies have identified a postseismic response to large earthquakes that has been explained by the viscoelastic properties of Earth's lower crust and upper mantle (starting with Nur and Mavko [1974] and Savage and Prescott [1978]). The transient postseismic part of the earthquake cycle may not be detectable in some cases because the response decreases to zero over time, and the time since the last earthquake may have been long. Recent progress in modeling the time-dependent surface deformation response (Pollitz, 1997) has made practicable the quantitative constraint of the viscous component of the rheology based on geodetic data.

In what follows, we present a new approach for constraining the viscoelastic properties of the lower crust and upper mantle. We apply three types of complementary data: horizontal GPS velocities, interferometric synthetic aperture radar (InSAR), which is predominantly sensitive to vertical motion, and paleoseismic estimates of fault slip rates and style. When comparing deformation measured with geodesy (e.g., GPS and InSAR) to deformation inferred from geology, we must account for the very different time scales measured with these techniques. In this study, we define time-invariant motion as the hypothetical steady tectonic deformation that occurs at a constant rate over many seismic cycles (e.g., associated with steady motion of the Sierra Nevada with respect to the Great Basin). We define transient motions as those that vary over time scales on the order of the seismic cycle (e.g., exponentially decaying postseismic relaxation of viscous material). Because motions that are measured with GPS at any given time are influenced by time-invariant and transient contributions, we cannot separate these processes if we assume that the predicted relaxation signal resembles the GPS velocity field. Instead, we here assume that the paleoseismic data constrain the time-invariant component of deformation, while physical models of viscoelastic relaxation constrain the complementary transient motion.

In our modeling, we calculate the postseismic velocities expected from the historic earthquakes in the Central Nevada seismic belt and many assumed viscoelastic Earth models, where we vary the viscosity of the lower crust and upper mantle. From each model, we infer the associated time-invariant deformation patterns by subtracting the transient component from the GPS velocities. To evaluate each model, we use misfit criteria based on three types of data: GPS velocities, interferometric synthetic aperture radar results (InSAR), and geologic estimates of fault slip rates. The best models must (1) explain the discrepancy between geologic and geodetic slip rate at the Central Nevada seismic belt, (2) explain the vertical motion observed with InSAR, and (3) not predict time-invariant contractions anywhere in the extensional Basin and Range Province. From the best-fitting viscoelastic Earth model, we obtain the most likely postseismic relaxation two-dimensional surface velocity pattern present inside the current snapshot of the geodetic velocity field. We then present the deformation patterns implied by the estimated time-invariant velocity field and the ways in which they differ from the apparent motion gleaned directly from GPS and other geodetic studies.

DATA

Historic Earthquakes of the Central Nevada Seismic Belt

We consider seismic events in central Nevada occurring in the past 150 yr that had magnitude of ~6.5 or above (Table 1).

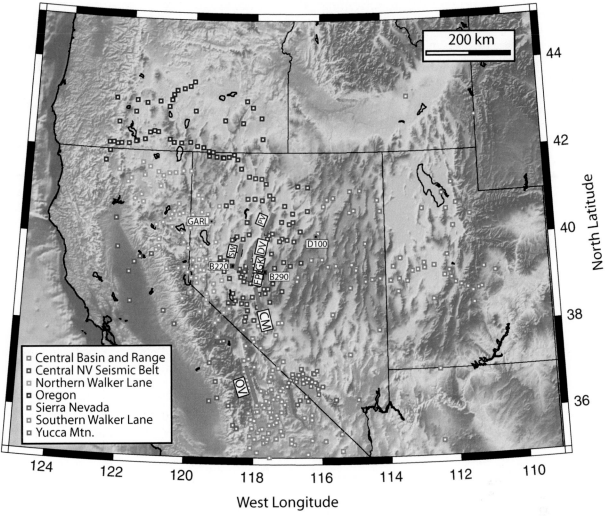

Figure 1. Horizontal velocities for global positioning system (GPS) sites used to constrain the viscosity of the Basin and Range lower crust and upper mantle. Site color designates the strain rate subnetwork discussed in text. Magenta line segments are the fault segments used to model the postseismic relaxation from the Central Nevada seismic belt historic earthquakes. See text for details of fault segments and coseismic slip parameters. Sites labeled GARL, B220, B290, and D100 and having black interiors are discussed in the text. Fault segments are: CM—Cedar Mountain; DV—Dixie Valley; GK—Gold King–Louderback sequence; FP—Fairview Peak; OV—Owens Valley; PV—Pleasant Valley; SW—Stillwater sequence.

TABLE 1. PARAMETERS USED FOR THE CENTRAL NEVADA SEISMIC BELT EARTHQUAKES

		Fault plane				Slip offset			
Earthquake	Date	Length (km)	Width (km)	Dip (°)	Strike (°)	Displ. (m)	Rake (°)	Mo (N·m)	M_W
Owens Valley	26 March 1872	100	15.0	90	339	6.0	180	2.7×10^{20}	7.6
Pleasant Valley	3 October 1915	59	21.2	45 W	210	4.0	−90	1.5×10^{20}	7.4
Cedar Mountain	21 December 1932	60	15.2	80 E	344	2.0	180	5.5×10^{19}	7.1
Combination of		70	19.6	50 E	15	1.0	−159	4.1×10^{19}	7.0
Rainbow Mtn.	6 July 1954								
Fourmile Flat	6 July 1954								
Stillwater	23 August 1954								
Fairview Peak	16 December 1954	32	17.3	60 E	15	2.4	−126	4.0×10^{19}	7.0
Combination of		22	17.3	60 W	170	0.6	−146	6.9×10^{18}	6.5
Gold King	16 December 1954								
Louderback	16 December 1954								
West Gate	16 December 1954								
Dixie Valley	16 December 1954	42	23.3	40 E	17	0.9	−90	2.6×10^{19}	6.9

The amount of postseismic motion that a GPS site might detect is a function of the size of the earthquake, its style and orientation, and the time and distance between the event and observation (Pollitz, 1997). For the earthquakes used here (simplified surface traces shown in Fig. 1), these parameters are constrained by geologic and paleoseismic investigation of the earthquake surface ruptures, seismic data, and eyewitness accounts of the earthquakes. Details of the earthquakes are discussed later, and the parameters used in the modeling are in listed Table 1. Our estimates of the moment magnitude, M_w, are, in all cases, within 0.2 of the estimates made by Pancha et al. (2006) for the largest earthquakes in the Basin and Range Province.

Earthquake Characteristics

Owens Valley (1872) M_w 7.6

This event was likely the largest to occur in the Basin and Range Province in historic time (Pancha et al., 2006), and it is the earliest and southernmost event considered. Its mechanism was predominantly right-lateral strike slip with a minor amount of normal, down-to-the-east slip (Beanland and Clark, 1994). The estimated times of previous events on this fault imply that its Holocene slip rate (2–4 mm/yr) (Lee et al., 2001) is among the fastest in the province. Geodetic rates (5–7 mm/yr) are even faster, however, and postseismic relaxation following the 1872 event has been invoked as a possible explanation for this apparent discrepancy (Miller et al., 2001a; Dixon et al., 2003).

Pleasant Valley (1915) M_w 7.4

Occurring at the northern end of the Central Nevada seismic belt, this event was recorded with early-generation seismic instrumentation that was sufficient to characterize its size and tensor moment (Doser, 1988), and it produced 59 km of surface rupture (Wallace, 1984a) that is clearly visible today. Its mechanism was predominantly normal, with extension oriented N65°W.

Cedar Mountain (1932) M_w 7.1

The rupture from this earthquake crossed through three valley fault systems, activating north-south–striking right-lateral strike-slip faults in Monte Cristo Valley, right-lateral and normal faults in Stewart Valley, and secondary north-northeast–striking normal faulting in Gabbs Valley (Bell et al., 1999). The seismic data favor a predominantly strike-slip earthquake on a fault dipping steeply to the east, containing two subevents separated by 10 s (Doser, 1988). In our modeling, the combined static stress changes of both events are what initiate and drive the postseismic response, so we simplify the calculations by assuming a single event that has the surface trace and combined moment of both subevents.

Rainbow Mountain, Fourmile Flat, Stillwater (1954) M_w 7.0

These events occurred in July and August in the same year, prior to the Fairview Peak–Dixie Valley earthquake sequence (Caskey et al., 2004). Their surface traces lie mostly in the eastern part of the Carson Sink, one valley west of the Fairview Peak–

Dixie Valley sequence (Caskey et al., 2004). The Stillwater (M_S 7.0) event was the largest (Doser, 1986), but it was similar to the others in style, consisting of a combination of right-lateral and down-to-the-west normal slip. Because of their similar style and close proximity in time, in our modeling, we combine these events into a single event with the combined moment of all three.

Fairview Peak (1954) M_w 7.0

The coseismic offset parameters for this event are constrained by a combination of paleoseismic (Slemmons, 1957; Caskey et al., 1996) and geodetic data (Savage and Hastie, 1969; Snay et al., 1985; Hodgkinson et al., 1996). Slip was right lateral and normal on an east-dipping fault. This rupture, and the ones discussed later, occupy the middle latitudes of the Central Nevada seismic belt and are hence transitional between the predominantly strike-slip faults to the south (e.g., Owens Valley and Cedar Mountain) and normal ruptures to the north (e.g., Pleasant Valley). The prior event on this fault was likely over 35,000 yr ago (Caskey et al., 2000).

Dixie Valley (1954) M_w 6.9

This event followed the Fairview Peak event by 4 min and ruptured the western range front fault in the valley immediately to the north (Slemmons, 1957). Paleoseismic, geodetic, and seismic data all indicate that it occurred on an east-dipping rupture plane accommodating mostly normal offset (Doser, 1986; Caskey et al., 1996; Hodgkinson et al., 1996). Thus, this event is similar to the Pleasant Valley earthquake, which was the next and northernmost event to the north.

Gold King, Louderback, West Gate (1954) M_w 6.5

These faults likely ruptured simultaneously with the Fairview Peak event. They are west-dipping and lie just north of the Fairview peak rupture. Their distribution is more complex than the other faults, but they are thought to have acted as a geometric link between the ruptures of the Dixie Valley and Fairview Peak earthquakes (Caskey et al., 2000). Individually and combined, they are smaller than the Fairview Peak and Dixie Valley events, so we represent these three ruptures in our modeling as a single event and fault plane consisting of normal and strike-slip offset. Collectively, these three events comprise the smallest event that we consider.

Examination of our final model shows that the smallest earthquakes we consider, the Gold King–Louderback–West Gate sequence (Table 1: M_w 6.5) contributes at most 0.15 mm/yr to the observed GPS velocity field (compared to 2.1 mm/yr for the Owens Valley event). Although some large earthquakes may have occurred in the past (prehistory) that could contribute to transient signals in our GPS data, we have used all the known earthquakes that can significantly affect our results. Given viscosities in the range of 10^{17}–10^{21} Pa·s (see Table 2 and references therein), a shear modulus of 3×10^{10} Pa, viscous Maxwell relaxation times are on the order of 0.1–1000 yr, and yet Central Nevada seismic belt recurrence intervals are on the order of 5000–35,000 yr (Caskey et al., 2000; Bell et al., 2004; Wesnousky et al., 2005).

TABLE 2. PUBLISHED RESULTS OF LOWER CRUST–UPPERMOST MANTLE VISCOSITIES
FOR THE BASIN AND RANGE PROVINCE

Study	Locality	η_{LC} (Pa·s)	η_{UM} (Pa·s)
Geodetic studies of modern* earthquakes			
Pollitz (2003)	ECSZ (Mojave)	3.2×10^{19}	4.6×10^{18}
Pollitz et al. (2001)	ECSZ (Hector Mine)	$\eta_{LC} \gg \eta_{UM}$	$3–8 \times 10^{17\dagger}$
Pollitz et al. (2000)	ECSZ (Landers)	$\eta_{LC} \gg \eta_{UM}$	$1–6 \times 10^{18}$
Geodetic studies of past earthquakes			
This study	CNSB/WL/EBR	$1–3 \times 10^{20}$	$1–3 \times 10^{19}$
Gourmelen and Amelung (2005)	CNSB	$>10^{20}$	$1–7 \times 10^{18}$
Chang and Smith (2005)	NBR (Hebgen Lk.)	$0.03–1.3 \times 10^{22}$	$1.3–3.2 \times 10^{19}$
Nishimura and Thatcher (2003)	NBR (Hebgen Lk.)	$>10^{20}$	$10^{18}–10^{20}$
Hetland and Hager (2003)	CNSB	$5–50 \times 10^{18}$	$>10^{19}$
Dixon et al. (2003)	SWL	1×10^{19}	
Geodetic studies of large lake loading/unloading			
Bills et al. (2007)	WL/CNSB	$2–10 \times 10^{20}$	$0.6–10 \times 10^{18}$
Kaufmann and Amelung (2000)	EBR (Lake Mead)	$>10^{20}$	$6–16 \times 10^{18}$
Adams et al. (1999)	WL/CNSB	n.p.	10^{18}
Bills et al. (1994)	EBR (Lk. Bonneville)	$1–10 \times 10^{20}$	$\sim 10^{19}$
Nakiboglu and Lambeck (1983)	EBR (Lk. Bonneville)	assumed elastic	$2.1–34 \times 10^{18}$
Studies of plate-boundary dynamics			
Flesch et al. (2000)	WL/CNSB	$1–5 \times 10^{21}$ (whole lithosphere)	

Note: n.p.—no estimate provided. CNSB—Central Nevada seismic belt, WL—Walker Lane, SWL—southern Walker Lane, ECSZ—Eastern California shear zone, EBR—eastern Basin and Range, NBR—northern Basin and Range.
*Modern earthquakes had coseismic deformation observed with space geodesy.
†Tends to increase toward $1–3 \times 10^{19}$ after 1–3 yr of relaxation, indicating transient rheology.

Thus, the GPS measurements were made very early in the seismic cycle for Central Nevada seismic belt faults. Because the expected relaxation times are so much shorter than the recurrence intervals, the effects from the penultimate earthquakes are almost certainly negligible and are not modeled.

Global Positioning System Data

We use the horizontal GPS velocities from the compilation of Kreemer et al. (this volume), which includes our own solutions for the continuously recording BARGEN network (Wernicke et al., 2000; Bennett et al., 2003), and also the published and updated velocities from about a dozen campaign networks (see Table 1 of Kreemer et al., this volume, and references therein). The individual velocity solution sets are transformed so that they refer to the Stable North America Reference Frame (SNARF) (Blewitt et al., 2005). The resulting velocity field spans the majority of the Basin and Range Province in California, Nevada, Utah, Arizona, Oregon, and Idaho. The sites are shown in Figure 1, and the velocities are shown in Kreemer et al. (this volume). While the velocity map has spatial density of sites that is highly variable (from less than tens to hundreds of kilometers), the coverage in the vicinity of the Central Nevada seismic belt has at least a dozen sites within 30 km of every fault segment.

Interferometric Synthetic Aperture Radar (InSAR)

We use the InSAR results of Gourmelen and Amelung (2005), which measure the change in range between the surface and a satellite in low Earth orbit over time. They stacked numerous interferograms constructed from repeat passes of the radar satellite over the Central Nevada seismic belt in order to cancel noise and reduce the uncertainty in the rate of movement. They argued that most of the surface motion observed at the Central Nevada seismic belt using this technique is attributable to vertical motion of the surface because the horizontal GPS signal in the same area (Hammond and Thatcher, 2004) cannot explain the InSAR data. Thus, the observations are most consistent with a domelike uplift with a rate of 2–3 mm/yr centered over the Fairview Peak–Dixie Valley–Pleasant Valley ruptures. This uplift is not well observed in the GPS measurements because the uncertainties in the vertical component of GPS are a factor of three to four larger than in the horizontal and exceed the signal observed by Gourmelen and Amelung (2005). Other GPS sites that are continuously recording, such as the BARGEN network (Bennett et al., 1998; Wernicke et al., 2000), or that have had significantly longer observation history, such as the Basin and Range Highway 50 network (Thatcher et al., 1999; Hammond and Thatcher, 2004), are not ideally positioned to observe the uplift seen with InSAR, since they lie to the south, east, or west of the maximum of the bulge. Thus, the InSAR measurements currently represent the best available source of relative vertical motion across the Central Nevada seismic belt.

The U.S. Geological Survey Quaternary Fault and Fold Database

We use data from the U.S. Geological Survey (USGS) Quaternary Fault and Fold database in the Basin and Range and

California (hereafter referred to as UQFD; Haller et al., 2002; Cao et al., 2003) to constrain the amount of permanent tectonic deformation that has occurred in the Basin and Range in the recent geologic past. A live internet accessible version of this database is available at http://earthquake.usgs.gov/qfaults/, but we used the specific tables associated with the Haller et al. (2002) and Cao et al. (2003) publications. It contains a compilation of fault geometries in the United States and associated estimates of slip direction and rate when available. In addition, we include results from the more recent study of paleoseismic slip rates for the Central Nevada seismic belt (Bell et al., 2004) and the compilation of dePolo and Anderson (2000), who used a reconnaissance technique to estimate rates of slip on 45 normal faults in the Basin and Range. Since their technique tends to identify an upper bound to slip rate, and we used these rates whenever available, our inferred rates of geologic moment release may be similarly high. The information contained in these databases is complementary to geodetic data that measure contemporary surface deformation. The geodetic velocity field may include transients that are not representative of time-invariant (i.e., many times greater than a seismic cycle) behavior or permanent deformation. In contrast, paleoseismic data constrain the age, location, and style of permanent surface deformation accommodated by slip on faults. If tectonic deformation progresses at a constant rate over time, then paleoseismology should constrain the same horizontal deformation field on Earth's surface that is measured by geodesy, when adjusted for transient effects.

The fault slip rate information is used to infer horizontal strain rates by using the moment tensor summation method of Shen-Tu et al. (1999), which is based on the approach of Kostrov (1974). We create a map-view grid of the Basin and Range Province with $0.4° \times 0.4°$ horizontal two-dimensional cells. Inside each cell with area A, we calculate the average horizontal strain rate tensor by summing over n fault segments within the cell:

$$\dot{\varepsilon}_{ij} = \frac{1}{2} \sum_{k=1}^{n} \frac{L_k \dot{u}_k}{A \sin \delta_k} \dot{m}_{ij}^k, \qquad (1)$$

where L_k is the length, δ_k is the fault dip angle, and $\dot{u}_k$ is the scalar slip rate of the kth fault. The unit moment tensor m_{ij}^k is defined as $m_{ij} = u_i n_j + u_j n_i$, where $\boldsymbol{n}$ is the horizontal unit vector normal to the fault trace, and $\boldsymbol{u}$ is the horizontal direction of slip across the fault (Shen-Tu et al., 1999; Kreemer et al., 2000). The horizontal tensor strain rate for each cell is obtained through the summation in Equation 1. The associated vector velocity field can be obtained through integration of the strains plus definition of a reference frame. The resulting map of permanent deformation associated with the fault database is shown in Figure 2.

The resulting strain field obtained from the UQFD (Fig. 2) is similar in spatial pattern and deformation style to that obtained from the geodetic velocity field (method discussed in Kreemer et al., this volume). Both have zones of more rapid deformation in the westernmost 100–200 km of the Basin and Range. Both have zones of right-lateral and extensional slip and have velocities oriented west/northwest that increase and rotate clockwise to the west. The similarity in these two deformation patterns suggests that the geodetic velocity field is dominated by the time-invariant component of deformation. There are, however, a number of differences in the details of the geologic and geodetic models. For example, in general, the geologic strain rates are considerably lower than the geodetic rates. The differences between them may be partly attributable to incompleteness of the catalog of surface faulting (which is spatially variable), and by the presence of transients in the geodetic velocity field caused by postseismic relaxation. Zones of higher rate deformation, associated with an east-west velocity gradient of 2–3 mm/yr, are also located in the vicinity of the Wasatch fault zone at the Great Basin eastern boundary. Note that a small appendage of 4–8 nanostrains/yr strain rate seen in the geologic strain rate map (Fig. 2) near 117°W longitude and 39°N latitude is not located at the Central Nevada seismic belt, but southeast of it. The slightly higher strain rate in this band is controlled by the proximity of several faults with <0.2 mm/yr slip rate estimates (Ione Valley fault, Southwest Reese River Valley, Western Toiyabe Range) and one with 0.22 mm/yr (Toiyabe Range fault zone).

Differences between the geodetic and geologic deformation fields can, at least partially, be explained by the incompleteness of the geologic catalog of prehistoric earthquakes. In a provincewide comparison of seismic, geologic, and geodetic moment rates in the Basin and Range, Pancha et al. (2006) have shown that the rate of geologically inferred moment release is less than that inferred from seismic and geodetic moment rates. They estimated a geodetic moment of between 4 and 7 × 10^18 N·m/yr and a geologic moment of 2.5 × 10^18 N·m/yr, implying a provincewide value for the ratio of geodetic moment over geologic moment R between 1.6 and 2.8. We calculate our own estimates of the ratio of geodetic over geologic moment rates R for each subdomain shown in Figure 1. To be consistent between our estimates of geodetic and geologic moment, we calculate the average moment for each inside the $0.4° \times 0.4°$ cells that cover each of the regions defined in Figure 1 from the continuum strain rate models. We assume that moment is proportional to the total strain rate defined as the second invariant of the strain rates tensor (Kreemer et al., this volume), and then we sum the moments in each region. The Central Basin and Range (which includes the Wasatch), northern Walker Lane, Yucca Mountain, Oregon, and southern Walker Lane domains have R values of 2.9, 4.9, 2.0, 3.9, and 3.9, respectively. For sites closest to the Central Nevada seismic belt (within ~70 km), $R = 9.5$, consistent with the geodetic deformation exhibiting elevated strain rates via transient effects. For a geographically broader selection of sites (within ~180 km) centered at the Central Nevada seismic belt, $R = 5.1$, consistent with the presence of enhanced geodetic moment that is focused near the Central Nevada seismic belt.

The relatively high values of R for the northern and southern Walker Lane are driven by geodetic strain rates that are highest at the westernmost boundary of the Basin and Range, adjacent to the Sierra Nevada microplate (Kreemer et al., this volume). In

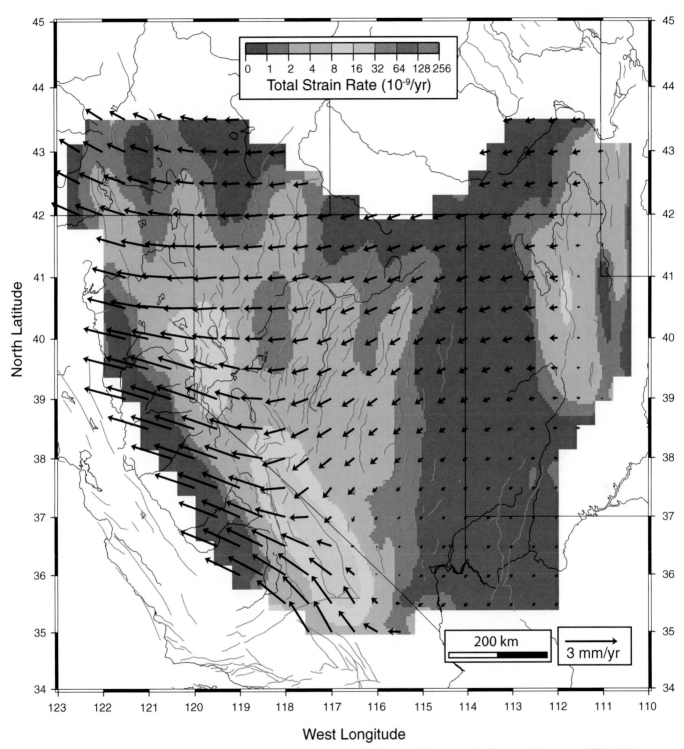

Figure 2. Strain rate field inferred from the paleoseismic databases of the U.S. Geological Survey and dePolo and Anderson (2000). Black arrows are velocities with respect to North America on a regular grid that result from integration of the strain rates. Colors are the second invariant of the strain rate (includes shear and dilatation). Magenta line traces are faults in the database. Note that the finger of higher strain rates near longitude 117°W and latitude 39°N lies east of the Central Nevada seismic belt.

this version of the UQFD, the Mohawk Valley fault is not present and thus effectively has a rate of 0.0 mm/yr. It has been shown recently to have 0.3–0.5 mm/yr slip on a single strand (Sawyer et al., 2005). Also in our version of the UQFD, the Owens Valley fault has a rate of 1.5 mm/yr, which is now estimated in the current version of the database to be 1.0–5.0 mm/yr (http://earthquake.usgs.gov/qfaults, accessed March 2007). Thus, it is likely that values for *R* would be smaller for the southern and northern Walker Lane if they were estimated from a more current UQFD database. The data are also consistent with a province-wide moment rate deficit in the paleoseismic earthquake catalog, of an approximate factor of 3 ± 1, since areas without significant historic seismic moment release also show a discrepancy.

Another difference between the geologic model (Fig. 2) and the geodetic model (Fig. 3) of Kreemer et al. (this volume) is seen in the azimuth of velocities in southern Nevada, western Utah, and northwest Arizona. In this area, the geologic model velocities have a southwest azimuth, whereas the geodetic model velocities have an azimuth north of west. However, in both models, the tensor strain rates are the primary solution, and velocities are obtained by integrating the strains while defining the velocity reference frame to be zero at the eastern edge of the model, east of the Wasatch fault zone in eastern Utah and southwest Wyoming. Uncertainties in this zero-velocity condition on this boundary allow for a small solid body rotational difference in the velocity fields, with a pole of rotation near the Wasatch. Adjusting for such a rotation could add a northwest component to velocities in southern Nevada, giving them a more western azimuth, without changing the fit to the geologic data. Furthermore, 1–3 mm/yr GPS velocities in southern Nevada and on the Colorado Plateau have a west azimuth and drive the azimuth in the geodetic model. No such constraint exists in the geologic model, since the rates on the Hurricane fault in southwest Utah and northwest Arizona are very low, and strain rates east of the Colorado Plateau are not taken into account. These differences, however, do not affect the strain rate patterns in either figure.

RELAXATION MODELING

There are several lines of evidence that suggest that postseismic processes are observable decades after the Central Nevada seismic belt earthquakes, and that these are dominated by the process of viscoelastic relaxation of the lower crust and/or upper mantle. These arguments fall into three classes. First, observations of similar- magnitude earthquakes that have occurred during the era of modern geodetic observation have been used to infer upper mantle and lower-crustal viscosities of 10^{17}–10^{21} Pa·s and rule out other processes, such as poroelastic rebound and afterslip (Pollitz et al., 2000, 2001; Hearn, 2003; Pollitz, 2003). Studies of other mid–twentieth-century Basin and Range earthquakes (e.g., Borah Peak, and Hebgen Lake) (Nishimura and Thatcher, 2003; Chang and Smith, 2008) and Lake Lahontan and Bonneville rebound (Bills and May, 1987; Bills et al., 1994; Adams et al., 1999) have found similar values

for lower crust–upper-mantle viscosities (Table 2). These suggest that the conditions for viscoelastic relaxation are present and will have the decades-long time scale needed to provide a signal in the year 2005. Other processes such as afterslip and poroelastic rebound are not expected to provide observable signals so long after the earthquakes. Second, the Central Nevada seismic belt exhibits a geodetic signature that is consistent with the presence of ongoing viscoelastic relaxation. A broad vertical uplift in the vicinity of the Dixie Valley–Fairview Peak and Pleasant Valley ruptures has been observed using stacks of InSAR scenes (Gourmelen and Amelung, 2005). This horizontal dimension and vertical rate, and decades-long relaxation time scale of this broad bulge, are consistent with viscosities obtained in other studies. Horizontal GPS measurements reveal anomalous dilatation at the Central Nevada seismic belt, which is consistent with this bulge (Hammond and Thatcher, 2004; Kreemer et al., this volume), and contractions east of the Central Nevada seismic belt that are otherwise difficult to reconcile with a region that is undergoing tectonic extension have been tentatively identified (Wernicke et al., 2000; Bennett et al., 2003; Friedrich et al., 2004; Hammond and Thatcher, 2004). Thirdly, the disagreement between geodetic and geologic slip and moment rates is large in the vicinity of the Central Nevada seismic belt ($R = 5.1$–9.5) and southern Walker Lane domains ($R = 3.9$), which contain the large historic earthquakes, compared to the adjacent domains. These large *R* values can be explained by an enhancement of geodetic strain resulting from viscoelastic relaxation. This argument alone is only consistent with relaxation rather than proof of it, however, since regions that have undergone less paleoseismic investigation can have relatively incomplete moment derived from the UQFD.

Modeling: Philosophy and Construction

Considering the previous arguments, we proceed under the assumption that transient deformation in the geodetic signal is caused by viscoelastic postseismic relaxation of the lower crust and upper mantle. However, measurements of the geodetic velocity field at the surface of the planet capture effects related to both time-invariant and transient processes that are a direct manifestation of the earthquake cycle. The transient effects can be related to the observed geodetic velocities via

$$v_{\text{geodesy}} = v_{\text{time-invariant}} + v_{\text{transient}}, \qquad (2)$$

where we constrain v_{geodesy} with GPS and InSAR observations, and we constrain $v_{\text{transient}}$ with the physics governing viscoelastic deformation of a layered medium. Equation 2 can also be expressed in terms of the spatial derivatives of velocity, i.e., strain rates, which we use in the criteria for model fit. We consider the time-invariant component of the geodetic velocity field as an unknown and model the transient portion of the velocity field by assuming a layered viscoelastic structure that is stressed by the Central Nevada seismic belt earthquake dislocations given in Table 1. Thus, for each viscosity structure, there is an associated postseismic transient velocity field and an associated time-invariant velocity field derived from Equation 2.

To model the evolution of the relaxation following the seismic events, we use the spherically layered viscoelastic modeling software VISCO1D (Pollitz, 1997). The software assumes a Newtonian Maxwell viscoelastic rheology and includes the effects of gravitation, which can have minor effects on the relaxation history at long intervals after the earthquake. Because we assume Newtonian rheology, we may sum the independently modeled earthquakes responses in the year 2005 to get the combined response. Although some studies have suggested that power-law, biviscous or transient rheologies are needed to explain the postseismic response, these studies are usually based on data obtained within days to a few years after the earthquake (Pollitz et al., 2001; Pollitz, 2003; Freed and Bürgmann, 2004), when multiple processes may have contributed to the postseismic response (Fialko, 2004). The Maxwell Newtonian rheology is simpler and is usually an adequate explanation of the data on time scales of 10,000–15,000 yr (e.g., Bills et al., 1994; Adams et al., 1999; Hetland and Hager, 2003; Nishimura and Thatcher, 2003; Pollitz et al., 2004). Furthermore, we expect that the mechanisms that have shorter relaxation times (e.g., poroelastic rebound; Fialko, 2004) will have dissipated early in the decades between the time of the GPS measurements and the events we consider. Based on previous structural and dynamic studies, we assume a laterally homogeneous 15-km-thick elastic uppermost layer that represents the upper crust, overlying a 15-km-thick viscoelastic lower crust (Fig. 3). This structure is a simplification based on studies that suggest a flat seismic Moho at 30–35 km depth (e.g., Allmendinger et al., 1987; Benz et al., 1990; Holbrook, 1990; Gilbert and Sheehan, 2004), and a rheological distinction between the lower and upper crust (see references in Table 2). The viscoelastic upper mantle extends from the bottom of the lower crust to a depth of 370 km. We tested models having a deeper bottom to the viscoelastic

upper mantle (down to a depth of 1088 km), and these gave nearly identical relaxation velocities (difference less than 0.1 mm/yr). The values for the shear and bulk elastic moduli are from the global one-dimensional seismic model PREM (Dziewonski and Anderson, 1981) mapped into discrete layers of 2–35 km thickness, thickening with depth. The values for the viscosity of the lower-crustal and upper-mantle layers are iteratively selected in a grid search, varying each from 10^{17} to 10^{21} Pa·s in logarithmic steps of one-half order of magnitude. For each model, the velocity in the year 2005 at each of the GPS sites shown in Figure 1 is computed, and these include the combined effects of each of the earthquakes listed in Table 1.

We use the characteristics of both the transient and time-invariant velocity fields to constrain the viscosities of the lower crust and upper mantle. To achieve this, we apply three objective criteria for the elimination of models that violate one or more of our three sources of data: GPS, InSAR, and paleoseismology. We cannot simply compare the results of our calculations to the geodetic observations because only the transient component of the velocity field is obtained in the viscoelastic modeling. Instead, we use the transient model and the observed GPS velocities to estimate the long-term velocity field using Equation 2, and then select a best model based on a synthesis of all of the following three objective, and independent, criteria:

1. The inferred time-invariant velocities in the vicinity of the Central Nevada seismic belt must imply a moment accumulation rate that makes a better match to the geologic moment release rate than do the raw GPS velocities. This criterion is based on the assumption that the mismatch between geodetic and geologic moment rates can be explained, at least in part, by the presence of postseismic viscoelastic relaxation.

2. The transient model must predict the uplift that is observed with InSAR. The vertical contribution from the time-invariant motion is presumed to be negligible at the Central Nevada seismic belt. This is consistent with slip rates, constrained by paleoseismology, that have an extensional component of slip that is less than 0.7 mm/yr total for four faults crossed by U.S. Highway 50 and less than 0.3 mm/yr for each of them (Bell et al., 2004). However, the observed motion is 2–3 mm/yr of surface uplift (Gourmelen and Amelung, 2005), which is at least several times greater than is expected based on strain accumulation. The wavelength of deformation expected from the transient response (>100 km) is much broader and is theoretically separable from the elastic strain response of individual faults (a few tens of kilometers). Thus, the spatial characteristics of the two processes could be used to separate the vertical component of the relaxation signal. However, that work is beyond the scope of this study, and we here assume that the entire vertical signal is attributable to relaxation.

3. The time-invariant model must not predict long-term contraction in regions characterized geologically by tectonic transtension. This is identical to, but more generally applied than, the assumption used by Hetland and Hager (2003) in an analysis of the horizontal GPS data of Thatcher et al. (1999) and Wernicke et al. (2000), where they assumed that contraction east of the

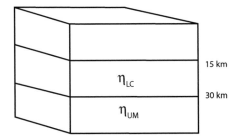

Figure 3. Rheological properties of our model as a function of depth. The uppermost layer (0–15 km) is assumed to be elastic; the lower crust (15–30 km) and upper mantle are assumed to be Maxwell viscoelastic. The elastic modulus μ is obtained from the one-dimensional seismic model PREM (Dziewonski and Anderson, 1981), and viscosities of the lower crust η_{LC} and upper mantle η_{UM} are varied in a grid search.

Central Nevada seismic belt needed to be explained by the post-seismic model.

These criteria can be more generally applied to tectonically deforming areas that have experienced recent earthquakes that may be affecting the modern geodetic velocity field. Also notice that none of these criteria requires that the postseismic deformation field resemble the time-invariant velocity field in pattern, style, or rate. For the Central Nevada seismic belt, the above criteria select a clearly defined best model, as described next.

Resolving Power of GPS, InSAR, and Geologic Data

The InSAR and horizontal GPS velocities form highly complementary constraints on the upper-mantle and lower-crustal viscoelastic structures. Each data type excludes very different portions of the model space. To evaluate the misfit of the viscosity models to the different types of data, we design the misfit as the combination of three different terms

$$\chi^2 = \alpha(R-1)^2 + \left(\frac{v_u - v_{InSAR}}{\sigma_{vu}}\right)^2 + \sum_{i=1}^{6}\left(\frac{\varepsilon_\Delta}{\sigma_{\varepsilon\Delta}}\right)^2, \qquad (3)$$

where R is the ratio of geodetic moment to geologic moment estimated in the Central Nevada seismic belt domain, and v_u is the maximum vertical velocity in the Central Nevada seismic belt domain obtained for each model in the grid search. The quantities v_{InSAR} and σ_{vu} are the vertical rate and its uncertainty observed with InSAR at the Central Nevada seismic belt taken to be 2.5 ± 0.5 mm/yr (Gourmelen and Amelung, 2005). The dilatational strain rate, ε_Δ, and its uncertainty, $\sigma_{\varepsilon\Delta}$, are obtained from the estimated time-invariant velocity field and the formal uncertainty obtained in the strain rate calculation. For the purposes of calculating the misfit, we assign $\varepsilon_\Delta = 0$ whenever $\varepsilon_\Delta \geq 0$, so that no extensional deformation, regardless of its rate, will be penalized. The summation in the third term sums the contributions from each of the domains. The Oregon domain is not included because contractions owing to Cascadia interseismic strain accumulation (McCaffrey et al., 2000; Miller et al., 2001b; Svarc et al., 2002a) and from the collision of the Sierra Nevada microplate with the Oregon domain (Hammond and Thatcher, 2005) are likely present and should not be penalized. The resulting misfit surfaces, for each of the terms in Equation 3, and for the combined constraint, are shown in Figure 4. The weight α is applied to control the relative importance of the first term, since it is not normalized by an uncertainty. We begin with a trial value of $\alpha = 1$, and then reduce α until models that are excluded by any single criterion are still excluded by the combined criteria, arriving at $\alpha = 0.7$. In other words, dark regions in Figures 4A–4C are also dark in Figure 4D. This approach gives approximately equal model exclusion power to the first term compared to the second and third terms in Equation 3. The minimum misfit taking all three criteria into account (Fig. 4D) identifies $\eta_{LC} = 10^{20.5}$ and $\eta_{UM} = 10^{19}$ as the best-fitting model. This result appears to

agree well with each of the individual criteria, since this model lies inside the region of minimal geologic/geodetic, InSAR, and contraction constraint misfit (Figs. 4A–4C). Models that fit similarly well can be found by decreasing η_{LC} and/or η_{UM} by one-half order of magnitude, but outside the zone defined by the contour where $\chi^2 = 2$, misfits increase rapidly. The misfit in the InSAR term increases most rapidly when η_{UM} is above 10^{19} Pa·s or reduced below $10^{18.5}$ Pa·s. When η_{LC} is decreased, the InSAR constraint yields a greater misfit; however, a more powerful limit on the lower bound of η_{LC} comes from the introduction of time-invariant contractions in the Sierra Nevada and Central Nevada seismic belt domains when η_{LC} is below 10^{20} Pa·s. The lower-crustal viscosity η_{LC} could be greater than $10^{20.5}$ and not strongly violate the InSAR result, but this would come at some cost to the fit between geologic and geodetic data. The precise amount of uncertainty in the estimates of η_{LC} and η_{UM} is difficult to quantify because Equation 3 is not a truly normalized measure of model misfit, and α is subjectively assigned. However, the models outside the $\chi^2 = 4$ combined constraint contour in Figure 4D are being excluded by at least one of the criteria. Thus, uncertainties in η_{LC} and η_{UM} are near one-half order of magnitude.

Results

To guide the evaluation of our results, the sites are divided into separate geographic domains so that different regions can be evaluated individually. These domains are: (1) the southern Walker Lane, (2) the northern Walker Lane, (3) the Sierra Nevada microplate, (4) the Central Nevada seismic belt, (5) southern Oregon, (6) the eastern Basin and Range, and (7) the Yucca Mountain area (Fig. 1). These domains are selected, somewhat subjectively, according to their first-order deformation characteristics based on previous geodetic and geological studies. For each viscoelastic model, we calculate the horizontal tensor strain rate inside each of the subdomains from the estimated time-invariant velocity field. We estimate the three horizontal strain rate parameters ($\varepsilon_{\Phi\Phi}$, $\varepsilon_{\lambda\lambda}$, $\varepsilon_{\Phi\lambda}$) simultaneously with three solid body rotation parameters (latitude [λ], longitude [Φ], and rotation rate [ω]) on the surface of a sphere according to the method of Savage et al. (2001). Figure 5A shows the dilatational component of the strain rate ε_Δ calculated for sites inside the Central Nevada seismic belt domain and corrected for the effects of postseismic relaxation based on each of the models in our grid search. The strain rate has been normalized by its uncertainty to illustrate the transition to models that imply no significant deformation in the Central Nevada seismic belt domain.

In order to evaluate the models according to our first objective criterion, we examine the ratio of geodetic moment rate corrected for postseismic effects to geologic moment rate in the Central Nevada seismic belt domain. The geologic moment is obtained from the UQFD and Equation 1, while the geodetic moment is obtained from the GPS velocities corrected for postseismic effects via the relaxation models. The modified ratio R is shown as a function of model viscosities in Figure 5B. Because

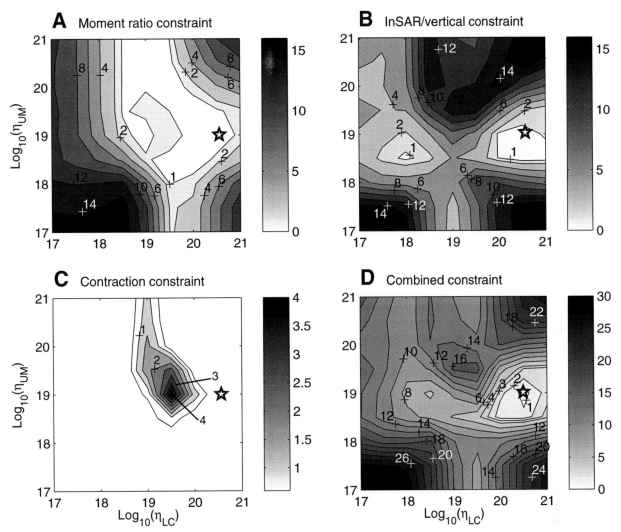

Figure 4. Model misfits as a function of lower-crustal (η_{LC}) and upper-mantle (η_{UM}) viscosity. (A) Contour of the first term of Equation 3, which depends on the ratio between geodetic and geologic scalar moment rate. Lower values indicate a better match between geologic and geodetic data. (B) Contour of the second term in Equation 3, which describes the misfit between the predicted vertical rate at the Central Nevada seismic belt and that measured by InSAR. (C) Contour of the third term in Equation 3, which is sensitive to the presence of undesirable time-invariant contractions in the areas surround the Central Nevada seismic belt. (D) The total constraint from all three data types combined using the method discussed in the text. Each panel has a star showing the preferred model ($\eta_{LC} = 10^{20.5}$, $\eta_{UM} = 10^{19}$ Pa·s).

the UQFD is not everywhere complete, we have speculated that there is a background disagreement rate between geodetic and geologic moment rates of a factor of 3 ± 1, which is near the lowest value for modified R (Fig. 5B). This assertion assumes, however, that the database is equally incomplete at all locations in the Basin and Range, and hence that the anomalously high ratio in the Central Nevada seismic belt is not the result of systematically less complete geologic record. In fact, it is more likely that the opposite is true, since the Central Nevada seismic belt has been the focus of numerous studies owing to its vigorous historic seismicity and proximity to the Reno/Carson metropolitan area. In Figure 5B, lower values for modified R occur in a part of

the model space that has intermediate values for lower-crust viscosity η_{LC} ($10^{18.5}$–$10^{20.5}$ Pa·s) and intermediate to high values for upper-mantle viscosity η_{UM} (>$10^{18.5}$ Pa·s), with the highest values for η_{UM} if η_{LC} is in a narrower band ($10^{18.5}$–$10^{19.5}$ Pa·s). The lowest values for modified R are near the speculated background level, and hence models inside the light central region of Figure 5B are more likely to be correct models according to the first objective criteria. The similarity between Figures 5A and 5B suggests that the misfit between geodetic and geologic moment rates is controlled by the dilatational component of the geodetic strain rate field in the vicinity of the Central Nevada seismic belt. This is consistent with the observation of anomalously high geodetic

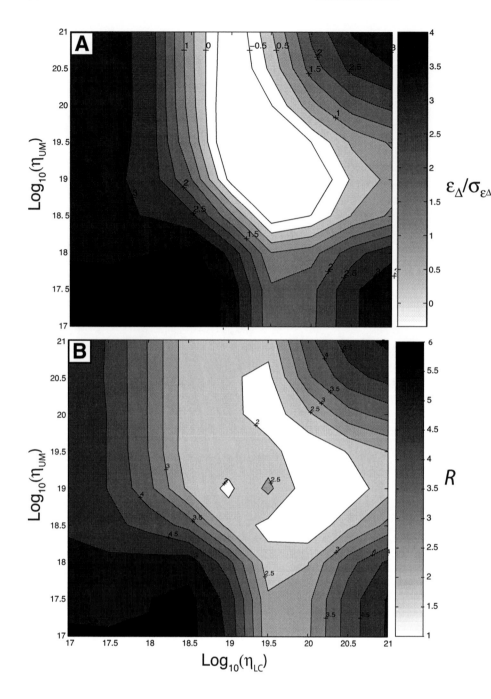

Figure 5. (A) Contour of the dilatational strain rate after correction for postseismic relaxation at the Central Nevada seismic belt normalized by its uncertainty as a function of lower-crustal (η_{LC}) and upper-mantle (η_{UM}) viscosity. Values greater than 2 indicate dilatation that is significantly larger than zero to 95% confidence. (B) Contour plot of R, the ratio of the scalar geodetic moment rate after correction for postseismic relaxation to the scalar geologic moment rate in the Central Nevada seismic belt.

dilatation at the Central Nevada seismic belt (Savage et al., 1995; Svarc et al., 2002b; Hammond and Thatcher, 2004) and the likely presence of dilatation in the postseismic deformation that follows earthquakes with a normal component of slip (i.e., Pleasant Valley and Dixie Valley).

The second objective criterion eliminates viscosity models based on misfit between the vertical motion observed with InSAR and that predicted by the relaxation calculations. The maximum vertical velocity inside the Central Nevada seismic belt domain as a function of the viscosity model is shown in Figure 6. It is immediately clear that the only models that can

produce a >2 mm/yr uplift are those with a relatively high lower-crustal viscosity ($\eta_{LC} > 10^{20}$ Pa·s) and an intermediate upper-mantle viscosity (η_{UM} between $10^{18.5}$ and 10^{19} Pa·s). The range of viscosities eliminated by this criterion is very large, and is complementary to that of models eliminated by the geologic/geodetic misfit, i.e., the overlap between regions of acceptable fit in Figure 5 and Figure 6 is small. What constitutes an acceptable fit will be discussed later.

The Basin and Range is characterized by (oblique) normal faulting, so the third objective criterion is designed to eliminate models that predict time-invariant contractions, in violation of

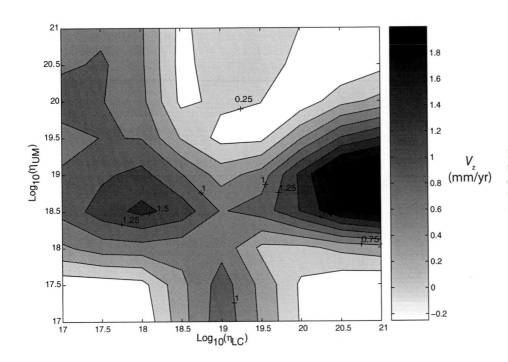

Figure 6. Contour of uplift rate (V_z) at the Central Nevada seismic belt predicted from the relaxation model as a function of lower-crustal (η_{LC}) and upper-mantle (η_{UM}) viscosity.

geological observation. The minimum and maximum time-invariant dilatational strain rates, ε_Δ, as a function of the viscosity model for each domain are shown in Table 3. None of the ε_Δ values has a normalized value below −2, which would indicate a contraction rate that has exceeded the 2σ uncertainty level. Only the Oregon domain exhibits a normalized contraction below −1, which indicates there may be some contraction of the entire domain. This is consistent with the presence of interseismic strain accumulation of the kind expected in the Cascadia backarc (McCaffrey et al., 2000; Miller et al., 2001b; Svarc et al., 2002a) and also with a north-south contraction of 2–3 mm/yr observed near the Oregon-California border (Hammond and Thatcher, 2004). The minimum and maximum ε_Δ values in Table 3 also provide an illustration of the sensitivity of strain rate field to the viscoelastic properties of the lower crust and upper mantle. The range between the maximum and minimum ε_Δ is greatest for the Central Nevada seismic belt domain, consistent with its location near the historic earthquakes. The northern Walker Lane domain sees a transition from marginally insignificant to marginally significant dilatation. This implies that geodetic results that show no dilatation (e.g., Hammond and Thatcher, 2007) may have been influenced by the presence of relaxation, which can mask a small amount of secular extension. This is discussed more completely later.

Table 4 shows that correcting the geodetic strain rate field with this model makes the deformation more similar to the geologic strain rate field with respect to geographic distribution of strain rate style, if not with respect to the total magnitude of strain (which is a factor of 2–5 smaller for the geologic model). The table shows a comparison between the strain rates inside the separate regions calculated from the geodetic strain rate model velocities, geologic data via the UQFD and Equation 1, and from the postseismic model. This table shows the similarity between the geologic and geodetic data sets with respect to the dominance of shear strain in western Basin and Range Province, particularly the southern Walker Lane, northern Walker Lane, and Yucca Mountain regions. The only region that shows significant geodetic dilatation is the Central Nevada seismic belt, in contrast to the geologic model, which shows generally less dilatation than shear everywhere. A property of the postseismic relaxation model is that all the dilatation and most of the shear is focused at the Central Nevada seismic belt, consistent with it being the locus of the historic earthquakes used in its construction. Thus, the geodetic strains minus the postseismic relaxation model will have much less dilatation, similar to the geologic model. Note that the geodetic minus the correction will not match the geologic model exactly, because of the provincewide lower magnitude of deformation rate seen in the geologic data. However, the corrected geodetic strains provide a model that is similar to the geologic model in the sense that most dilatation has been removed, and shear strain rates at the Central Nevada seismic belt are lower than in the Walker Lane belt. The very strong correlation between R and the dilatational component of the relaxation model suggests that the model has been appropriately built to compensate for the anomalous dilatation at the Central Nevada seismic belt.

DISCUSSION

Implications for Time-Invariant Deformation of the Northern Walker Lane

The postseismic strain rates predicted by our preferred viscosity model are from the postseismic velocity field obtained

TABLE 3. DILATATIONAL STRAIN RATE (ε_Δ) FOR GEODETIC VELOCITIES CORRECTED FOR
POSTSEISMIC RELAXATION

Domain	Strain rates				Normalized			
	Observed	Corrected			Observed	Corrected		
		Min.	Max.	Best		Min.	Max.	Best
Eastern Basin and Range	1.3 ± 1.3	0.8	1.8	1.3	1.0	0.6	1.4	1.0
CNSB	10.5 ± 2.8	−5.7	11.0	3.1	3.8	−2.0	3.9	1.1
Northern Walker Lane	7.2 ± 3.9	6.5	10.3	8.8	1.8	1.7	2.6	2.3
South Walker Lane	3.9 ± 2.9	3.5	7.1	6.3	1.3	1.2	2.4	2.2
Oregon	−3.1 ± 3.4	−3.7	−1.7	−2.5	−0.9	−1.1	−0.5	−0.7
Yucca Mountain	8.0 ± 11.3	5.3	13.3	13.3	0.6	0.5	1.2	1.2

Note: The uncorrected, minimum, maximum, and preferred dilatational strain rates ε_Δ and their 1σ formal uncertainties are given in units of nanostrains (10^{-9}) per year. Normalized strain rates are the strain rates divided by their formal uncertainties. Values that are greater than 2 or less than −2 indicate 95% significant area growth rate or contractions, respectively. The difference between the maximum and minimum is a measure of the sensitivity of ε_Δ to the viscosity of the lower crust and upper mantle. The "Best" column shows the estimated secular dilatation (corrected for postseismic relaxation) inside each domain given the model with $\eta_{LC} = 10^{20.5}$ Pa·s and $\eta_{UM} = 10^{19}$ Pa·s. CNSB—Central Nevada seismic belt.

TABLE 4. GEOLOGIC, GEODETIC, AND RELAXATION MODEL STRAIN RATES INSIDE REGIONS

Region	Geodetic				Postseismic		Geologic		
	ε_Δ	$\pm\sigma_\varepsilon$	ε_{xy}	$\pm\sigma_\varepsilon$	ε_Δ	ε_{xy}	ε_Δ	ε_{xy}	R
East Basin and Range	1.4	1.3	8.9	1.3	0.1	0.2	0.9	1.2	2.9
Walker Lane North	7.3	3.9	33.5	3.9	−0.7	1.1	2.3	7.3	4.9
Walker Lane South	3.9	2.9	79.0	2.9	−1.5	9.2	0.1	15.9	3.9
Oregon	−3.0	3.4	1.0	3.4	−0.1	0.8	1.3	0.9	3.9
Yucca Mountain	8.1	11.3	37.2	11.3	−4.1	1.8	−1.4	10.8	2.0
CNSB (within 180 km)	10.5	2.8	23.5	2.8	10.3	14.7	2.3	2.7	5.1
CNSB (within 70 km)	31.0	10.3	48.4	10.3	23.9	34.0	1.7	2.3	9.5

Note: Velocities used to calculate geodetic strain rates and uncertainties have not been corrected for postseismic relaxation and are from the strain rate model of Kreemer et al. (this volume). Geologic strain rates are from the UQFD (U.S. Geological Survey Quaternary Fault and Fold database), obtained in Equation 1 and shown in Figure 2. Postseismic strain rates come from the viscoelastic relaxation model. Strain rates are in nanostrains/yr. CNSB—Central Nevada seismic belt.

from the relaxation calculations. The velocities predicted from our preferred model (Fig. 7; Table DR1[1]) are consistent with northwest-southeast uniaxial extension across Pleasant Valley, and significant vertical uplift of ~2 mm/yr predicted north of the Fairview Peak rupture. To the south, progressively greater amounts of shear deformation appear where the style of the earthquakes becomes more strike-slip. The dilatational component of postseismic transient strain rate (Fig. 8A) shows enhanced extension in the vicinity of the Central Nevada seismic belt. This patch of dilatation is spatially coincident with the anomalously high Central Nevada seismic belt dilatation rates shown in the strain rate map in a companion article to this paper (Kreemer et al., this volume). This dilatation comes mostly in the form of uniaxial northwest-southeast–directed extension, and it is visible in the detail that includes the strain rate tensor axes (Fig. 8B). The relaxation model also contains lobes of low-intensity (1–4 nanostrain/yr) contraction to the northwest and east of the Cen-

tral Nevada seismic belt historic ruptures. The lobe to the east is consistent with GPS observations of geographically broad contraction east of the Central Nevada seismic belt (Hammond and Thatcher, 2004), but it does not explain a narrow zone of contraction inferred from the single continuous GPS site LEWI on Mt. Lewis, Nevada, near Crescent Valley (Wernicke et al., 2000; Bennett et al., 2003; Friedrich et al., 2004). The significant east-west asymmetry in these contractional lobes comes from summing earthquakes with different strikes, event times, and different ruptures styles. Shorter-wavelength variations are mainly the result of geographic irregularity in the GPS sites, which causes irregular sampling of the model relaxation velocity field that is used in the strain modeling. For example, near the Owens Valley rupture at the southern end of our model, small blotches of dilatation of alternating sign exist, but these contribute negligibly to the integrated dilatation over larger regions. The transition of tensor deformation style from south to north (Fig. 8B) can also be seen in the model velocities (Fig. 7). Where the earthquake styles to the south have a larger component of right-lateral slip, we see postseismic deformation that has close to equal parts of contraction and extension, indicating shear deformation. To the

[1]GSA Data Repository Item 2009001, Table DR1. Velocities from compilation of GPS, relaxation model, and secular motion, is available at www.geosociety.org/pubs/ft2009.htm, or on request from editing@geosociety.org, Documents Secretary, GSA, P.O. Box 9140, Boulder, CO 80301-9140.

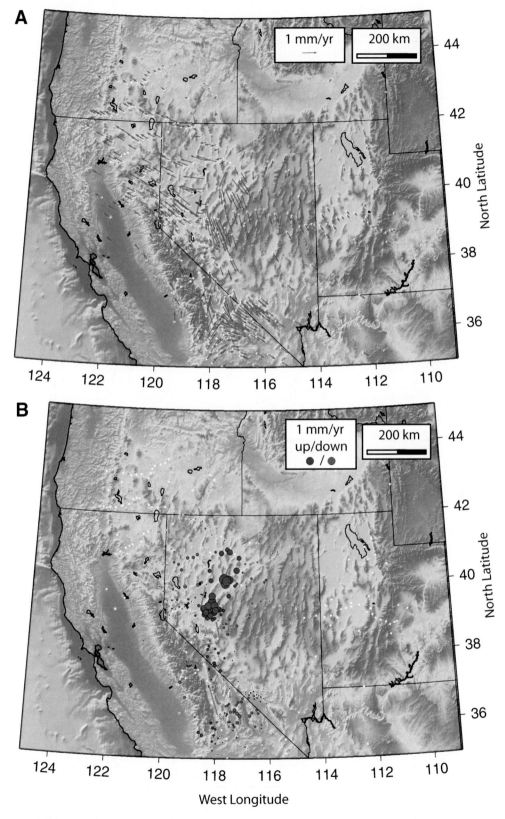

Figure 7. Postseismic relaxation velocities predicted from the preferred relaxation model ($\eta_{LC} = 10^{20.5}$, $\eta_{UM} = 10^{19}$ Pa·s), including the contributions from all the earthquakes considered in this study. (A) Red vectors show predicted horizontal velocity to the model faults (magenta line segments) at each global positioning system (GPS) site in the compilation of Kreemer et al. (this volume). (B) Vertical motion is shown blue for uplift, red for subsidence. Size of the circle indicates the vertical rate.

48

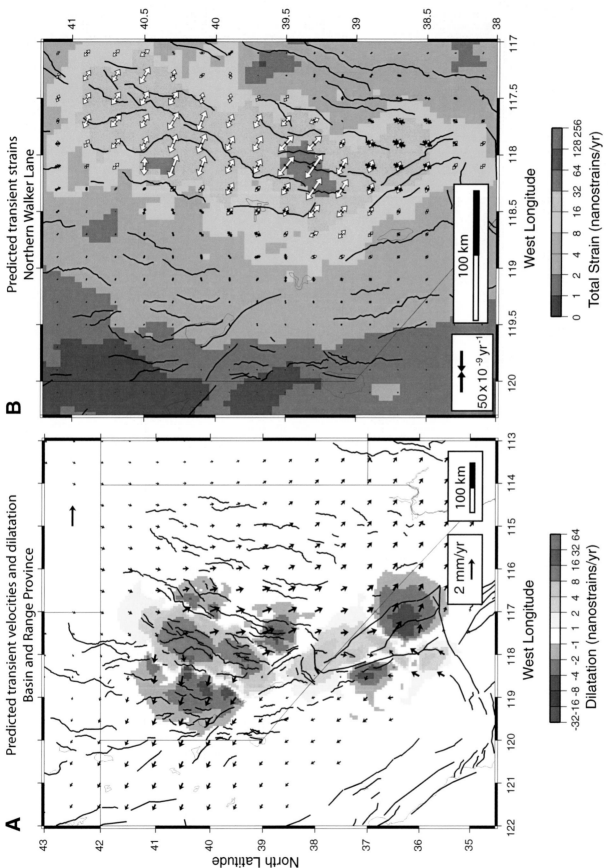

Figure 8. (A) Dilatational strain rates predicted from the preferred relaxation model velocities ($\eta_{LC} = 10^{20.5}$, $\eta_{UM} = 10^{19}$ Pa·s). This map was created using the same global position-ing system (GPS) site locations as the model of Kreemer et al. (this volume), so that the effects of station distribution on the strain modeling are included. Black arrows indicate the relaxation velocities interpolated onto a regular grid. Quaternary faults are shown with black line traces. State borders and major lakes are shown with faint lines. (B) A detailed view of the northern Walker Lane. The colors now represent the amount of total strain rate, which contains both horizontal shear and dilatation. Black arrow pairs indicate tensor contrac-tion; white arrows indicate tensor extension.

north, near the Pleasant Valley rupture, the style of extension is closer to northwest-southeast uniaxial extension, similar to the coseismic rupture there. In the vicinity of Pyramid Lake, roughly 2–4 nanostrain/yr of northwest-directed uniaxial contraction is predicted from the relaxation model.

The time-invariant deformation field is estimated by subtracting the transient velocities from the GPS velocities and then repeating the strain rate modeling of Kreemer et al. (this volume). Thus, the strain rate map of Figure 9B is expected to be in closer agreement with the deformations inferred from paleoseismological studies. A comparison of Figure 2 and Figure 9B shows that this is the case, since deformation in the time-invariant model and in the geologic model is more strongly focused into the northern Walker Lane, and there is relatively little deformation at the Central Nevada seismic belt. A single small zone of strain near longitude 118.5°W and latitude 40°N is caused by a single campaign GPS velocity (site WILD) that has a velocity that deviates by ~1 mm/yr from the smooth regional pattern. The finger of higher geologic strain rate near longitude 117°W and latitude 39°N and the Toiyabe Range (Fig. 2) is not observed in the uncorrected or corrected geodetic strain rate models (Fig. 9). It likely shows higher moment rate because of the close proximity of the Toiyabe Range fault (0.22 mm/yr), the Ione Valley fault (0.1 mm/yr), the Southwest Reese River Valley fault (0.1 mm/yr), and the Mahogany Mountain section of the Western Toiyabe Range fault zone (0.2 mm/yr).

After correction for postseismic effects, some deformation at the Central Nevada seismic belt remains, with rates that are still higher with respect to the eastern Basin and Range (Fig. 9B). In the vicinity of the Stillwater, Dixie Valley, and Fairview Peak faults, the total strain rates are 8–32 nanostrain/yr, and a bit lower near Pleasant Valley. Near the Cedar Mountain rupture, strain rates are higher, 16–32 nanostrain/yr, reduced from the 32–64 nanostrain/yr estimated before the correction. The velocity gradient across the Central Nevada seismic belt (between sites B220 and B290, Fig. 1) is 3.1 mm/yr in the GPS velocity compilation, but it has been reduced to 0.9 mm/yr in the time-invariant model. Thus, a significant fraction of the deformation at the Central Nevada seismic belt has been explained with postseismic relaxation, yet some remains and is consistent with the paleoseismological result of ~1 mm/yr extension across the Central Nevada seismic belt faults (Bell et al., 2004). The amount of velocity gradient that still exists across the Central Nevada seismic belt after the correction indicates that the belt may still have a higher rate of deformation than in the Central Basin and Range, where strain rates are close to zero (Bennett et al., 2003; Kreemer et al., this volume). The corrected rates are, however, less than what is observed in the Walker Lane to the west. Thus, these results are consistent with a greater frequency of paleoearthquakes at the Central Nevada seismic belt compared to the central Basin and Range, and the view that the Central Nevada seismic belt is a zone of deformation that extends to the northeast, possibly transferring Walker Lane dextral slip onto NNE-striking normal faults (Savage et al., 1995; Wesnousky et al., 2005; Faulds et al., 2005). For GPS sites further north and farther apart that span Pleasant

Valley, the difference owing to the correction is not as large. The relative velocity of GARL and D100 changes from 2.3 mm/yr to 1.3 mm/yr upon adjustment for postseismic deformation. In general, the magnitude of the correction decreases with distance from the Central Nevada seismic belt.

The time-invariant dilatation rate in the northern Walker Lane domain is 8.8 ± 3.9 nanostrains/yr (Table 3), just barely significant to 95% confidence, while its shear strain rate is 33.8 ± 3.9 nanostrains/yr (shear is defined as $\varepsilon_1 - \varepsilon_2$, where ε_1 and ε_2 are the maximum and minimum principal strain rates, respectively). This suggests that, as a whole, the northern Walker Lane acts as a right-lateral shear zone with a small component of extension. However, this extension becomes statistically significant only after the correction for postseismic relaxation from the Central Nevada seismic belt historic earthquakes has been made. Thus, in several respects, the northern Walker Lane behaves similarly to the central Walker Lane, where dextral shear rates are greater than extension rates, and the highest deformation rates are seen near the western boundary of the Walker Lane belt adjacent to the Sierra Nevada (Oldow et al., 2001; Oldow, 2003). Furthermore, these observations are consistent with campaign GPS data spanning Walker Lane between 37°N and 40°N, which show significant extension approximately normal to the trend of the shear zone but a different orientation from east to west (Oldow, 2003; Hammond and Thatcher, 2004). It is also consistent with the results of a profile of campaign GPS velocities across the northern Walker Lane in the vicinity of Pyramid Lake (Hammond and Thatcher, 2007) that did not correct for postseismic relaxation, and that observed no significant dilatation.

Implications for Basin and Range Lithospheric Rheology

The conclusions reached here are broadly consistent with Basin and Range lower-crustal and upper-mantle viscosity estimates from a number of other studies. We have summarized the values obtained by a sample of other studies in Table 2. The most common conclusion reached for the Basin and Range is that the lower crust has greater strength (i.e., higher viscosity) than the uppermost mantle on which it lies. This conclusion is consistent across studies that sample a great variety of temporal scales, from times immediately following (<3 yr) an earthquake (Pollitz et al., 2000, 2001; Pollitz, 2003) to over 10,000 yr of relaxation following the unloading owing to the draining of late Pleistocene postglacial lakes (e.g., Nakiboglu and Lambeck, 1983; Bills and May, 1987; Bills et al., 1994; Kaufmann and Amelung, 2000).

In their study of Central Nevada seismic belt rheological layering, Hetland and Hager (2003) relied on the criterion that time-invariant contraction should be explained by postseismic relaxation. Our modeling differs from theirs because we choose to exclude models that predict time-invariant contractions in any of the domains near the Central Nevada seismic belt. We do not explain the contraction east of the Central Nevada seismic belt directly, because we do not observe the contraction in our Central Basin and Range domain. This is likely owing to the fact that our domain is so large and also includes the Wasatch fault zone to the

50

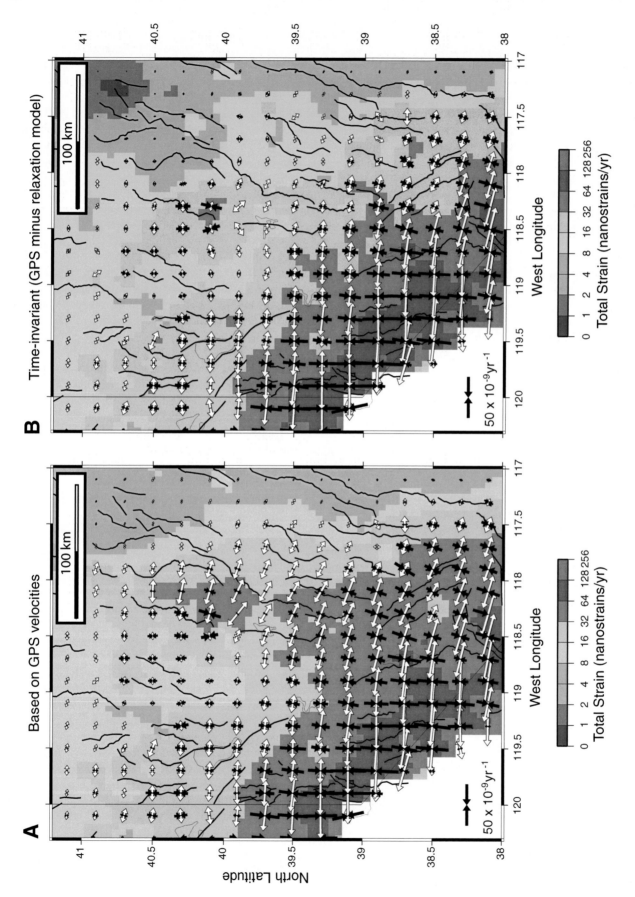

Figure 9. (A) Continuum strain rate model based on the global positioning system (GPS) velocity data. This is a detail of the map presented in Kreemer et al. (this volume). Other features are the same as B. (B) Same model except based on the time-invariant GPS velocity field obtained by subtracting the preferred relaxation model velocities ($\eta_{LC} = 10^{20.5}$, η_{UM} = 10^{19} Pa·s) from the GPS velocities. The zone of deformation has become significantly more focused to the western edge of the Walker Lane belt.

east. However, when we use just the criterion that Central Basin and Range domain extension should be positive, amounting to approximately the same model discrimination criterion that they use, we would have estimated values of $\eta_{LC} \approx 10^{19}$ and $\eta_{UM} \geq \eta_{LC}$, which are very similar to their results. Their reliance on this criterion may explain why their result is the only one of the survey of previous results (Table 2) that indicates a lower crust that is weaker than the upper mantle.

Our results, and most of the results listed in Table 2, appear to be in disagreement with the results of others that suggest that the lower crust behaves as a low-viscosity channel through which material flows. Sometimes called the "jelly sandwich" model (Jackson, 2002), this idea has been used to explain the scarcity of earthquakes in the lower crust, the existence of metamorphic core complexes (e.g., Block and Royden, 1990), and the flatness of Moho topography despite geologic extension that is, at the surface, localized to dipping normal faults and heterogeneous over scales of hundreds of kilometers (Gans, 1987). The geodetic networks considered in this study are geographically focused in the western Great Basin, where strain rates are one to two orders of magnitude higher and metamorphic core complexes are absent. Thus, variations in crust and mantle properties may partially account for the discrepancy. However, results from several studies in the eastern Great Basin (i.e., in the vicinity of Lakes Bonneville, Mead, and Hebgen) show no strong difference in η_{LC} compared to the western Great Basin (e.g., Central Nevada seismic belt, Walker Lane, or Mohave Desert) (Table 2). According to the two-layer theory of McKenzie et al. (2000), short-wavelength Moho relief is flattened on a time scale that is proportional to the sum of crust and upper-mantle viscosities. Thus, it is possible to flatten Moho topography if the crust is stronger than the mantle, as long as the crust is weak enough to allow flow on the time scale of Basin and Range extension. For example, in the two-layer viscous model with an upper layer $10^{20.5}$ Pa·s over a half-space of 10^{19} Pa·s (values of η_{LC} and η_{LC} in our preferred model), 10 km wavelength Moho topography will relax with a characteristic decay time of 2.6 m.y. This is short enough to flatten Moho topography over the history of Basin and Range extension, and thus a weak lower crust (compared to the mantle) is not an absolute requirement to flatten Moho topography. So our model is not in conflict with the observation of a flat Moho.

An additional possibility is that lateral contrasts of lithospheric viscosity exist that can affect our results and comparisons between the Central Nevada seismic belt and other parts of the Basin and Range. Modeling of the stresses driving deformation of the western United States in the context of a thin viscous sheet suggests that the focused deformation in the westernmost part of the province is attributable to lithosphere that has a lower viscosity compared to the adjacent areas (Flesch et al., 2000). East-west asymmetry in lithospheric effective viscosity is consistent with the presence of higher heat flow (Lachenbruch and Sass, 1978; Blackwell and Steele, 1992) and lower flexural rigidity to the west (Lowry and Smith, 1995). In this study, we assumed a laterally constant viscosity, and thus our modeling does not include the

effects of east-west asymmetry in the relaxation velocity field that would be expected from an increase in viscosity east of the Central Nevada seismic belt. This can affect the strain rates that are expected from the relaxation model far from the Central Nevada seismic belt. For example, our preferred model predicts some deformation in the southern Sierra Nevada microplate (Fig. 8A), which to date has not been definitively observed. Future modeling studies will have to be undertaken to understand the effect of postseismic relaxation in the presence of lateral variations in the strength of the Basin and Range lithosphere.

Transient or power-law rheology in the upper mantle may explain relaxations that occur more quickly immediately following a large earthquake than would be expected from the longer-term (>1 yr) response assuming a Newtonian Maxwell viscoelastic model (Pollitz, 2003; Freed and Bürgmann, 2004). Our study assumes a linear response, and hence the velocity fields from the individual earthquakes can be modeled separately and then summed to represent the velocity field in the year 2005. A rheology other than Newtonian violates this assumption. If, however, the mantle were characterized with a power-law rheology that increased the inferred viscosity over time, as has been observed in the Mojave (Pollitz et al., 2001; Pollitz, 2003), then our results would tend to overestimate mantle viscosity, and our conclusions about the relative strength between the crust and upper mantle would still hold. Similarly, if the viscosity is temperature dependent, and hence smoothly decreases with depth, our modeling would be most sensitive to the upper portion of each depth interval. This would result in a systematic overestimation of the viscosity in each layer, but it would not alter the relative strength of the lower crust and upper mantle.

SUMMARY AND CONCLUSIONS

We combine horizontal GPS velocities, InSAR, and paleoseismic data with modeling of the viscoelastic postseismic response of the Central Nevada seismic belt historic earthquakes to estimate the most likely viscoelastic structure of the western Basin and Range lower crust and upper mantle. From these, we develop a correction for postseismic effects that can be used to estimate secular crustal deformation from geodetic velocities near the Central Nevada seismic belt. Our preferred model has values of $\eta_{LC} = 10^{20.5}$ Pa·s and $\eta_{UM} = 10^{19}$ Pa·s. To obtain this result, we assume that the GPS velocity field is the sum of time-invariant and transient processes, and seek models that (1) explain the mismatch between geodetic and geologically inferred slip rates, (2) explain the vertical uplift that has been observed with InSAR and (3) do not imply time-invariant contractions anywhere in the Basin and Range. This model is consistent with other geodetic studies of earthquakes and lake loading of the lithosphere, and with the observation of flat Moho topography throughout much of the Basin and Range.

We evaluate the ability that different types of data have to constrain the viscoelastic structure. We find that InSAR, GPS, and geologic data exclude different parts of the model space and thus form complementary constraints on the viscoelastic structure. This suggests that studies that aim to constrain rheology of the lithosphere are best served by using techniques that measure both vertical and horizontal movements and that include geologic constraints on steady permanent tectonic deformation averaged over many seismic cycles.

The preferred viscosity model implies that the majority of the deformation observed at the Central Nevada seismic belt is attributable to postseismic relaxation, and that the time-invariant deformation is more tightly focused into northern Walker Lane than is inferred from the uncorrected GPS velocities. Whereas no significant dilatation was detectable before correcting for postseismic effects, following the correction, the dilatation is significant (8.4 ± 3.9 nanostrains/yr). Thus, the Central Nevada seismic belt postseismic relaxation masks time-invariant extension in the adjacent province to the west.

The preferred model decreases the apparent discrepancy between geologically and geodetically estimates slip rates on faults at the Central Nevada seismic belt. After using this model to correct the geodetic velocity field for transient postseismic effects from historic earthquakes, the amount of geodetically inferred steady permanent extension that has occurred at the Central Nevada seismic belt is reduced, making it less anomalous with respect to the Central Basin and Range, in agreement with paleoseismic data. However, after the correction, rates of deformation at the Central Nevada seismic belt are still greater than rates in the central Basin and Range to the east, and not as great as rates in the Walker Lane belt to the west.

ACKNOWLEDGMENTS

This work was supported by the Department of Energy Geothermal Program through grant DE-FG36-02ID14311 to the Great Basin Center for Geothermal Energy and Department of Energy Yucca Mountain Project/Nevada System of Higher Education Cooperative Agreement DE-FC28-04RW12232. Special thanks are due to Fred Pollitz for sharing his VISCO1D software and assisting us with installation. We are grateful to all who contribute to and maintain the U.S. Geological Survey Quaternary Fault and Fold paleoseismic database. This manuscript was improved by careful reviews from Dennis Harry and Craig Jones.

REFERENCES CITED

Adams, K.D., Wesnousky, S.G., and Bills, B.G., 1999, Isostatic rebound, active faulting, and potential geomorphic effects in the Lake Lahontan Basin, Nevada and California: Geological Society of America Bulletin, v. 111, p. 1739–1756, doi: 10.1130/0016-7606(1999)111<1739:IRAFAP>2.3.CO;2.

Allmendinger, R.W., Hauge, T.A., Hauser, E.C., Potter, C.J., Klemperer, S.L., Nelson, K.D., Knuepfer, P., and Oliver, J., 1987, Overview of the COCORP 40°N transect, western United States: The fabric of an orogenic belt: Geological Society of America Bulletin, v. 98, p. 308–319, doi: 10.1130/0016-7606(1987)98<308:OOTCNT>2.0.CO;2.

Beanland, S., and Clark, M.M., 1994, The Owens Valley fault zone, eastern California, and surface rupture associated with the 1872 earthquake: U.S. Geological Survey Bulletin 1982, p. 1–29.

Bell, J.W., dePolo, C.M., Ramelli, A.R., Sarna-Wojcicki, A.M., and Meyer, C.E., 1999, Surface faulting and paleoseismic history of the 1932 Cedar Mountain earthquake area, west-central Nevada, and implications for modern tectonics of the Walker Lane: Geological Society of America Bulletin, v. 111, p. 791–807, doi: 10.1130/0016-7606(1999)111<0791:SFAPHO>2.3.CO;2.

Bell, J.W., Caskey, S.J., Ramelli, A.R., and Guerrieri, L., 2004, Pattern and rates of faulting in the central Nevada seismic belt, and paleoseismic evidence for prior belt-like behavior: Bulletin of the Seismological Society of America, v. 94, p. 1229–1254, doi: 10.1785/012003226.

Bennett, R.A., Wernicke, B.P., and Davis, J.L., 1998, Continuous GPS measurements of contemporary deformation across the northern Basin and Range Province: Geophysical Research Letters, v. 25, p. 563–566, doi: 10.1029/98GL00128.

Bennett, R.A., Wernicke, B.P., Niemi, N.A., Friedrich, A.M., and Davis, J.L., 2003, Contemporary strain rates in the northern Basin and Range Province from GPS data: Tectonics, v. 22, no. 2, 1008, doi: 10.1029/2001TC001355.

Benz, H.M., Smith, R.B., and Mooney, W.D., 1990, Crustal structure of the northwestern Basin and Range Province from the 1986 Program for Array Seismic Studies of the Continental Lithosphere seismic experiment: Journal of Geophysical Research, v. 95, p. 21,823–21,842, doi: 10.1029/JB095iB13p21823.

Bills, B.G., and May, G.M., 1987, Lake Bonneville: Constraints on lithospheric thickness and upper mantle viscosity from isostatic warping of Bonneville, Provo and Gilbert stage shorelines: Journal of Geophysical Research, v. 92, p. 11,493–11,508, doi: 10.1029/JB092iB11p11493.

Bills, B.G., Currey, D.R., and Marshall, G.A., 1994, Viscosity estimates for the crust and upper-mantle from patterns of shoreline deformation in the Eastern Great Basin: Journal of Geophysical Research, v. 99, p. 22,059–22,086, doi: 10.1029/94JB01192.

Bills, B.G., Adams, K.D., and Wesnousky, S.G., 2007, Viscosity structure of the crust and upper mantle in western Nevada from isostatic rebound patterns of Lake Lahontan shorelines: Journal of Geophysical Research, v. 112, B06405, doi: 10.1029/2005JB003941.

Blackwell, D.D., and Steele, J.L., 1992, Geothermal Map of North America: Boulder, Colorado, Geological Society of America, scale 1:5000000.

Blewitt, G., Argus, D.F., Bennett, R.A., Bock, Y., Calais, E., Craymer, M., Davis, J.L., Dixon, T.H., Freymueller, J.T., Herring, T.A., Johnson, D., Larson, K.M., Miller, M.M., Sella, G.F., Snay, R.A., and Tamissiea, M., 2005, A stable North America reference frame (SNARF): First release: UNAVCO-IRIS Joint Workshop: Stevenson, Washington.

Block, L., and Royden, L.H., 1990, Core complex geometries and regional scale flow in the lower crust: Tectonics, v. 9, p. 557–567, doi: 10.1029/TC009i004p00557.

Cao, T., Bryant, W.A., Rowshandel, B., Branum, D., and Wills, C.J., 2003, The Revised 2002 California Probabilistic Seismic Hazard Maps: California Geological Survey Open-File Report 96-08, 12 p.

Caskey, J.S., Wesnousky, S.G., Zhang, P., and Slemmons, B.D., 1996, Surface faulting of the 1954 Fairview Peak (Ms=7.2) and Dixie Valley (Ms=6.9) earthquakes, central Nevada: Bulletin of the Seismological Society of America, v. 86, p. 761–787.

Caskey, J.S., Bell, J.W., Slemmons, B.D., and Ramelli, A.R., 2000, Historical surface faulting and paleoseismology of the central Nevada seismic belt, *in* Lageson, D.R., Peters, S.G., and Lahren, M.M., eds., Great Basin and Sierra Nevada: Boulder, Colorado, Geological Society of America Field Guide 2, p. 23–44.

Caskey, J.S., Bell, J.W., Ramelli, A.R., and Wesnousky, S.G., 2004, Historic surface faulting and paleoseismicity in the area of the 1954 Rainbow Mountain–Stillwater earthquake sequence, central Nevada: Bulletin of the Seismological Society of America, v. 94, p. 1255–1275, doi: 10.1785/012003012.

Chang, W.-L., and Smith, R.B., 2008, Lithospheric rheology from post-seismic deformation of a M7.5 normal faulting earthquake with implications for continental kinematics: Journal of Geophysical Research (in press).

DeMets, C., and Dixon, T.H., 1999, New kinematic models for Pacific–North America motion from 3 Ma to present: I. Evidence for steady motion and biases in the NUVEL-1A model: Geophysical Research Letters, v. 26, p. 1921–1924, doi: 10.1029/1999GL900405.

dePolo, C.M., and Anderson, J.G., 2000, Estimating the slip rates of normal faults in the Great Basin, USA: Basin Research, v. 12, p. 227–240, doi: 10.1046/j.1365-2117.2000.00131.x.

Dixon, T.H., Norabuena, E., and Hotaling, L., 2003, Paleoseismology and Global Positioning System: Earthquake-cycle effects and geodetic versus geologic fault slip rates in the eastern California shear zone: Geology, v. 31, p. 55–58, doi: 10.1130/0091-7613(2003)031<0055:PAGPSE>2.0.CO;2.

Doser, D.I., 1986, Earthquake processes in the Rainbow Mountain–Fairview Peak–Dixie Valley, Nevada, Region 1954–1959: Journal of Geophysical Research, v. 91, p. 12,572–12,586, doi: 10.1029/JB091iB12p12572.

Doser, D.I., 1988, Source parameters of earthquakes in the Nevada seismic zone, 1915–1943: Journal of Geophysical Research, v. 93, p. 15,001–15,015, doi: 10.1029/JB093iB12p15001.

Dziewonski, A.M., and Anderson, D.L., 1981, Preliminary reference Earth model: Physics of the Earth and Planetary Interiors, v. 25, p. 297–356, doi: 10.1016/0031-9201(81)90046-7.

Faulds, J.E., Henry, C.D., and Hinz, N.H., 2005, Kinematics of the northern Walker Lane: An incipient transform fault along the Pacific–North American plate boundary: Geology, v. 33, p. 505–508, doi: 10.1130/G21274.1.

Fialko, Y., 2004, Evidence of fluid-filled upper crust from observations of postseismic deformation due to the 1992 M_w7.3 Landers earthquake: Journal of Geophysical Research, v. 109, B08401, doi: 10.1029/2004JB002985.

Flesch, L.M., Holt, W.E., Haines, A.J., and Shen-Tu, B., 2000, Dynamics of the Pacific–North America plate boundary in the western United States: Science, v. 287, p. 834–836, doi: 10.1126/science.287.5454.834.

Freed, A.M., and Bürgmann, R., 2004, Evidence of power-law flow in the Mojave Desert mantle: Nature, v. 430, p. 548–551, doi: 10.1038/nature02784.

Freund, L.B., and Barnett, D.M., 1976, A two-dimensional analysis of surface deformation due to dip-slip faulting: Bulletin of the Seismological Society of America, v. 66, p. 667–675.

Friedrich, A.M., Wernicke, B.P., Niemi, N.A., Bennett, R.A., and Davis, G.A., 2003, Comparison of geodetic and geologic data from the Wasatch region, Utah, and implications for the spectral character of Earth deformation at periods of 10 to 10 million years: Journal of Geophysical Research, v. 108, no. B4, 2199, doi: 10.1029/2001JB000682.

Friedrich, A.M., Lee, J., Wernicke, B.P., and Sieh, K., 2004, Geologic context of geodetic data across a Basin and Range normal fault, Crescent Valley, Nevada: Tectonics, v. 23, p. TC2015, doi: 10.1029/2003TC001528.

Gans, P.B., 1987, An open-system, two-layer crustal stretching model for the eastern Great Basin: Tectonics, v. 6, p. 1–12, doi: 10.1029/TC006i001p00001.

Gilbert, H., and Sheehan, A.F., 2004, Images of crustal variations in the intermountain west: Journal of Geophysical Research, v. 109, B03306, doi: 10.1029/2003JB002730.

Gourmelen, N., and Amelung, F., 2005, Post-seismic deformation in the Central Nevada seismic belt detected by InSAR: Implications for Basin and Range dynamics: Science, v. 310, p. 1473–1476, doi: 10.1126/science.1119798.

Haller, K.M., Wheeler, R.L., and Rukstales, K.S., 2002, Documentation of Changes in Fault Parameters for the 2002 National Seismic Hazard Maps: Conterminous United States except California: U.S. Geological Survey Open-File Report 02-467, 34 p.

Hammond, W.C., and Thatcher, W., 2004, Contemporary tectonic deformation of the Basin and Range Province, western United States: 10 years of observation with the global positioning system: Journal of Geophysical Research, v. 109, p. B08403, doi: 10.1029/2003JB002746.

Hammond, W.C., and Thatcher, W., 2005, Northwest Basin and Range tectonic deformation observed with the global positioning system, 1999–2003: Journal of Geophysical Research, v. 110, p. B10405, doi: 10.1029/2005JB003678.

Hammond, W.C., and Thatcher, W., 2007, Crustal deformation across the Sierra Nevada, northern Walker Lane, Basin and Range transition, western United States measured with GPS, 2000–2004: Journal of Geophysical Research, v. 112, p. B05411, doi: 10.1029/2006JB004625.

Hearn, E.H., 2003, What can GPS data tell us about the dynamics of post-seismic deformation?: Geophysical Journal International, v. 155, p. 753–777, doi: 10.1111/j.1365-246X.2003.02030.x.

Hetland, E.A., and Hager, B.H., 2003, Postseismic relaxation across the Central Nevada seismic belt: Journal of Geophysical Research, v. 108, 2394, doi: 10.1029/2002JB002257.

Hodgkinson, K.M., Stein, R.S., and Marshall, G., 1996, Geometry of the 1954 Fairview Peak–Dixie Valley earthquake sequence from a joint inversion of leveling and triangulation data: Journal of Geophysical Research, v. 101, p. 25,437–25,457, doi: 10.1029/96JB01643.

Holbrook, S.W., 1990, The crustal structure of the northwestern Basin and Range Province, Nevada, from wide-angle seismic data: Journal of Geophysical Research, v. 95, p. 21,843–21,869, doi: 10.1029/JB095iB13p21843.

Jackson, J., 2002, Strength of the continental lithosphere; time to abandon the jelly sandwich?: GSA Today, v. 12, no. 9, p. 4–10, doi: 10.1130/1052-5173(2002)012<0004:SOTCLT>2.0.CO;2.

Kaufmann, G., and Amelung, F., 2000, Reservoir-induced deformation and continental rheology in vicinity of Lake Mead, Nevada: Journal of Geophysical Research, v. 105, p. 16,341–16,358, doi: 10.1029/2000JB900079.

Kostrov, V.V., 1974, Seismic moment and energy of earthquakes, and seismic flow of rocks: Izvestiya, Physics of the Solid Earth, v. 1, p. 23–44 (English translation).

Kreemer, C., Haines, J.A., Holt, W.E., Blewitt, G., and Lavallee, D., 2000, On the determination of the Global Strain Rate model: Earth Planets Space, v. 52, p. 765–770.

Kreemer, C., Blewitt, G., and Hammond, W.C., 2009, this volume, Geodetic constraints on contemporary deformation in the northern Walker Lane: 2. Velocity and strain rate tensor analysis, *in* Oldow, J.S., and Cashman, P.H., eds., Late Cenozoic Structure and Evolution of the Great Basin–Sierra Nevada Transition: Geological Society of America Special Paper 447, doi: 10.1130/2009.2447(02).

Lachenbruch, A.H., and Sass, J.H., 1978, Models of extending lithosphere and heat flow in the Basin and Range Province, *in* Smith, R.B., and Eaton, G.P., eds., Cenozoic Tectonics and Regional Geophysics of the Western Cordillera: Geological Society of America Memoir 152, p. 209–250.

Lee, J., Spencer, J., and Owen, L., 2001, Holocene slip rates along the Owens Valley fault, California: Implications for the recent evolution of the Eastern California shear zone: Geology, v. 29, p. 819–822, doi: 10.1130/0091-7613(2001)029<0819:HSRATO>2.0.CO;2.

Lowry, A.R., and Smith, R.B., 1995, Strength and rheology of the western U.S. Cordillera: Journal of Geophysical Research, v. 100, p. 17,947–17,963, doi: 10.1029/95JB00747.

McCaffrey, R., Long, M.D., Goldfinger, C., Zwick, P.C., Nabelek, J.L., Johnson, C.K., and Smith, C., 2000, Rotation and plate locking at the southern Cascadia subduction zone: Geophysical Research Letters, v. 27, p. 3117–3120, doi: 10.1029/2000GL011768.

McKenzie, D., Nimmo, F., Jackson, J.A., Gans, P.B., and Miller, E.L., 2000, Characteristics and consequences of flow in the lower crust: Journal of Geophysical Research, v. 105, no. B5, p. 11,029–11,046.

Miller, M.M., Johnson, D.J., and Dixon, T.H., 2001a, Refined kinematics of the eastern California shear zone from GPS observations, 1993–1998: Journal of Geophysical Research, v. 106, p. 2245–2263, doi: 10.1029/2000JB900328.

Miller, M.M., Johnson, D.J., Rubin, C.M., Dragert, H., Wang, K., Qamar, A., and Goldfinger, C., 2001b, GPS-determination of along-strike variation in Cascadia margin kinematics: Implications for relative plate motion, subduction zone coupling, and permanent deformation: Tectonics, v. 20, p. 161–176, doi: 10.1029/2000TC001224.

Murray, J.R., Segall, P., Cervelli, P., Prescott, W.H., and Svarc, J.L., 2001, Inversion of GPS data for spatially variable slip-rate on the San Andreas fault near Parkfield, CA: Geophysical Research Letters, v. 28, p. 359–362, doi: 10.1029/2000GL011933.

Nakiboglu, S.M., and Lambeck, K., 1983, A reevaluation of the isostatic rebound of Lake Bonneville: Journal of Geophysical Research, v. 88, p. 10,439–10,447, doi: 10.1029/JB088iB12p10439.

Nishimura, T., and Thatcher, W., 2003, Rheology of the lithosphere inferred from postseismic uplift following the 1959 Hebgen Lake earthquake: Journal of Geophysical Research, v. 108, no. B8, 2389, doi: 10.1029/2002JB002191.

Nur, A., and Mavko, G., 1974, Postseismic viscoelastic rebound: Science, v. 183, p. 204–206, doi: 10.1126/science.183.4121.204.

Oldow, J.S., 2003, Active transtensional boundary zone between the western Great Basin and Sierra Nevada block, western U.S. Cordillera: Geology, v. 31, p. 1033–1036, doi: 10.1130/G19838.1.

Oldow, J.S., Aiken, C.L.V., Hare, J.L., Ferguson, J.F., and Hardyman, R.F., 2001, Active displacement transfer and differential block motion within the central Walker Lane, western Great Basin: Geology, v. 29, p. 19–22, doi: 10.1130/0091-7613(2001)029<0019:ADTADB>2.0.CO;2.

Pancha, A., Anderson, J.G., and Kreemer, C., 2006, Comparison of seismic and geodetic scalar moment rates across the Basin and Range Province: Bulletin of the Seismological Society of America, v. 96, p. 11–32, doi: 10.1785/0120040166.

Pollitz, F.F., 1997, Gravitational-viscoelastic postseismic relaxation on a layered spherical Earth: Journal of Geophysical Research, v. 102, p. 17,921–17,941, doi: 10.1029/97JB01277.

Pollitz, F.F., 2003, Transient rheology of the uppermost mantle beneath the Mojave Desert, California: Earth and Planetary Science Letters, v. 215, p. 89–104, doi: 10.1016/S0012-821X(03)00432-1.

Pollitz, F.F., Peltzer, G., and Bürgmann, R., 2000, Mobility of continental mantle: Evidence from postseismic geodetic observation following the 1992

Landers earthquake: Journal of Geophysical Research, v. 105, p. 8035–8054, doi: 10.1029/1999JB900380.

Pollitz, F.F., Wicks, C.W., and Thatcher, W., 2001, Mantle flow beneath a continental strike-slip fault: Postseismic deformation after the 1999 Hector Mine earthquake: Science, v. 293, p. 1814–1818, doi: 10.1126/science.1061361.

Pollitz, F.F., Bukun, W.H., and Nyst, M., 2004, A physical model for strain accumulation in the San Francisco Bay region: Stress evolution since 1838: Journal of Geophysical Research, v. 109, B11408, doi: 10.1029/2004JB003003.

Savage, J.C., and Burford, R.O., 1973, Geodetic determination of relative plate motion in central California: Journal of Geophysical Research, v. 78, p. 832–845, doi: 10.1029/JB078i005p00832.

Savage, J.C., and Hastie, L.M., 1969, A dislocation model for the Fairview Peak, Nevada, earthquake: Bulletin of the Seismological Society of America, v. 59, p. 1937–1948.

Savage, J.C., and Prescott, W.H., 1978, Asthenosphere readjustment and the earthquake cycle: Journal of Geophysical Research, v. 83, p. 3369–3376, doi: 10.1029/JB083iB07p03369.

Savage, J.C., Lisowski, M., Svarc, J.L., and Gross, W.K., 1995, Strain accumulation across the central Nevada seismic zone, 1973–1994: Journal of Geophysical Research, v. 100, p. 20,257–20,269, doi: 10.1029/95JB01872.

Savage, J.C., Gan, W., and Svarc, J.L., 2001, Strain accumulation and rotation in the Eastern California shear zone: Journal of Geophysical Research, v. 106, p. 21,995–22,007, doi: 10.1029/2000JB000127.

Sawyer, T.L., Briggs, R.W., and Ramelli, A.R., 2005, Late Quaternary activity of the Southern Mohawk Valley fault zone, northeastern California: Seismological Research Letters, v. 76, p. 248.

Sella, G.F., Dixon, T.H., and Mao, A., 2002, REVEL: A model for recent plate velocities from space geodesy: Journal of Geophysical Research, v. 107, no. B4, 2081, doi: 10.1029/2000JB000033.

Shen-Tu, B., Holt, W.E., and Haines, J.A., 1999, Deformation kinematics in the western United States determined from Quaternary fault slip rates and recent geodetic data: Journal of Geophysical Research, v. 104, p. 28,927–28,955, doi: 10.1029/1999JB900293.

Slemmons, B.D., 1957, Geological effects of the Dixie Valley–Fairview Peak, Nevada earthquakes of December 16, 1954: Bulletin of the Seismological Society of America, v. 47, p. 353–375.

Snay, R.A., Cline, M.W., and Timmermann, E.L., 1985, Dislocation models for the 1954 earthquake sequence in Nevada: U.S. Geological Survey Open-File Report 89-290, p. 531–555.

Svarc, J.L., Savage, J.C., Prescott, W.H., and Murray, M.H., 2002a, Strain accumulation and rotation in western Oregon and southwestern Washington: Journal of Geophysical Research, v. 107, no. B5, 2087, doi: 10.1029/2001JB000625.

Svarc, J.L., Savage, J.C., Prescott, W.H., and Ramelli, A.R., 2002b, Strain accumulation and rotation in western Nevada, 1993–2000: Journal of Geophysical Research, v. 107, no. B5, 2090, doi: 10.1029/2001JB000579.

Thatcher, W., Foulger, G.R., Julian, B.R., Svarc, J.L., Quilty, E., and Bawden, G.W., 1999, Present-day deformation across the Basin and Range Province, western United States: Science, v. 283, p. 1714–1718, doi: 10.1126/science.283.5408.1714.

Wallace, R.E., 1984a, Faulting related to the 1915 earthquakes in Pleasant Valley, Nevada: U.S. Geological Survey Professional Paper 1274-A, p. A1–A33.

Wallace, R.E., 1984b, Patterns and timing of late Quaternary faulting in the Great Basin Province and relation to some regional tectonic features: Journal of Geophysical Research, v. 89, p. 5763–5769, doi: 10.1029/JB089iB07p05763.

Wernicke, B.P., Friedrich, A.M., Niemi, N.A., Bennett, R.A., and Davis, J.L., 2000, Dynamics of plate boundary fault systems from Basin and Range Geodetic Network (BARGEN) and Geologic Data: GSA Today, v. 10, no. 11, p. 1–7.

Wesnousky, S.G., Baron, A.D., Briggs, R.W., Caskey, J.S., Kumar, S.J., and Owen, L., 2005, Paleoseismic transect across the northern Great Basin: Journal of Geophysical Research, v. 110, p. B05408, doi: 10.1029/2004JB003283.

MANUSCRIPT ACCEPTED BY THE SOCIETY 21 JULY 2008

The Geological Society of America
Special Paper 447
2009

Spatially partitioned transtension within the central Walker Lane, western Great Basin, USA: Application of the polar Mohr construction for finite deformation

Justin J. Murphy*
A. John Watkinson
School of Earth and Environmental Science, Washington State University, Pullman, Washington 99163, USA

John S. Oldow
Department of Geosciences, University of Texas at Dallas, Richardson, Texas 75080-3021, USA

ABSTRACT

Deviation between velocity trajectories from global positioning system (GPS) networks and strain trajectories from earthquake focal mechanisms and fault-slip inversion within the central Walker Lane are reconciled as the consequence of non–plane strain (constriction) within a transtensional zone separating the Sierra Nevada and central Great Basin. Dextral transtension within the central Walker Lane is produced by differential displacement of the Sierra Nevada with respect to the central Great Basin, and it is partitioned into domains exhibiting simple shear–dominated and pure shear–dominated strain. From east to west across the central Walker Lane, GPS velocities change orientation from west-northwest to northwest and increase from 2–3 to 12–14 mm/yr as the incremental-strain elongation axis changes from west-northwest to west-southwest. The deviation between strain and velocity trajectories increases to 50° as the Sierra Nevada Range is approached from the east. This deviation in strain and velocity trajectories is consistent with analytical models linking kinematic vorticity, particle velocity paths, and incremental non–plane strain during transtensional deformation. We link field observations to the analytical models using a polar Mohr construction in a system that conserves kinematic boundary conditions to graphically demonstrate that the relationship between velocity and strain fields is a consequence of constrictional deformation.

Keywords: rotation, graphical link, vorticity, GPS velocities, strain orientation.

*Now at ExxonMobil Exploration Company, 233 Benmar, Houston, Texas 77060, USA; justin.j.murphy@exxonmobil.com.

Murphy, J.J., Watkinson, A.J., and Oldow, J.S., 2009, Spatially partitioned transtension within the central Walker Lane, western Great Basin, USA: Application of the polar Mohr construction for finite deformation, *in* Oldow, J.S., and Cashman, P.H., eds., Late Cenozoic Structure and Evolution of the Great Basin–Sierra Nevada Transition: Geological Society of America Special Paper 447, p. 55–70, doi: 10.1130/2009.2447(04). For permission to copy, contact editing@geosociety.org. ©2009 The Geological Society of America. All rights reserved.

INTRODUCTION

The first-order correspondence between geodetic and tectonic displacement fields across plate boundaries (DeMets et al., 1994) is in good agreement with modeled velocity and strain-rate fields based on geologic, geodetic, and seismological observations in diffuse deformation zones such as the western U.S. Cordillera (Shen-Tu et al., 1998, 1999; Holt et al., 2000; Thatcher, 2003). The general correspondence between strain and velocity field trajectories observed over large expanses of the U.S. Cordillera reflects both the scale of observation and, at least to some degree, the regularity of the deformation field.

As the spatial resolution of global positioning system (GPS) networks has improved over the years, however, deviations between velocity field and active strain field trajectories have become apparent, at least locally (Oldow, 2003). The seismic character of the boundary zone between the Sierra Nevada and central Great Basin (Fig. 1A) was interpreted as a product of non–plane strain deformation (Unruh et al., 2003). This interpretation was shared by Oldow (2003) as an explanation for the deviation between the trajectories of the regional velocity field and strain

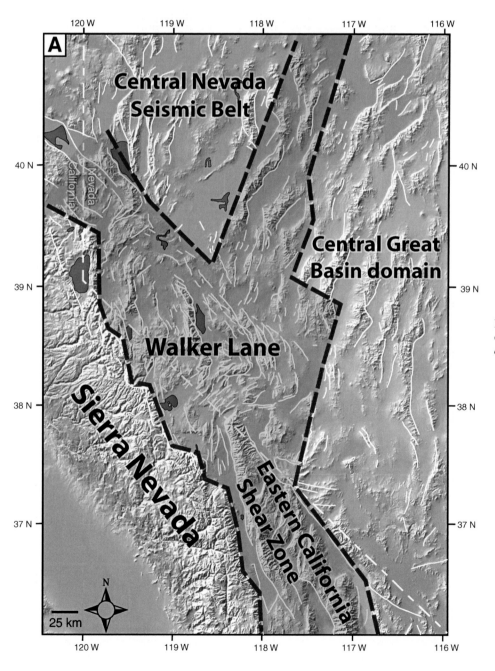

Figure 1. (A) Regional tectonic setting of the western Great Basin. (*Continued on following two pages.*)

field by as much as 50° (Fig. 1B). The interaction of regional displacement fields and crustal anisotropy was viewed as localizing the domains of plane and non–plane strain deformation (Fig. 1C).

Although the deviation between strain and velocity fields in the central Walker Lane is consistent with theoretical considerations, an improved understanding of the complexity within actively deforming zones can be gained by directly relating displacement fields to strain. Here, we expand on earlier works (McCoss, 1986; Withjack and Jamison, 1986; Wojtal, 1989; Jamison, 1991; Tron and Brun, 1991; Fossen and Tikoff, 1993; Horsman and Tikoff, 2005) by presenting a graphical tool, based on the polar Mohr construction (De Paor, 1981; Simpson and De Paor, 1993), to directly compare strain and velocity. This graphical approach provides (1) a visual link between the GPS velocity and the strain, (2) solutions that are spatially referenced, and (3) reflections of both incremental and finite strains, and (4) it captures rotation from a non–coaxial strain component.

GEOLOGIC SETTING

The northern part of the Eastern California Shear Zone (Dokka and Travis, 1990) and the central Walker Lane (Locke

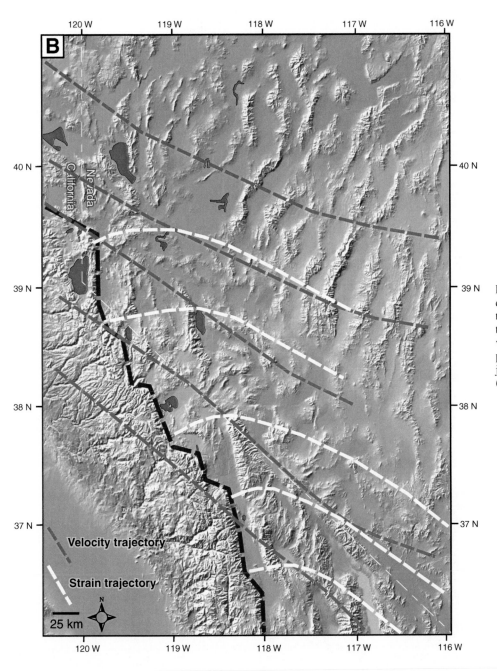

Figure 1 (*continued*). (B) Map of the central Walker Lane, Nevada, showing the systematic change in orientation of the strain trajectories (white) and deviation from parallelism with the global positioning system (GPS) velocity trajectories (red) toward the Sierra Nevada. (*Continued on following page.*)

et al., 1940; Stewart, 1988; Oldow, 1992) form a boundary zone that accommodates differential motion between the northwesterly migrating Sierra Nevada block and the central Great Basin (Fig. 1A). Differential displacement across the boundary zone is recorded by GPS velocities and seismicity and is accommodated on a complex array of normal and strike-slip faults (Burchfiel et al., 1987, 1992; Stewart, 1988; Oldow, 1992, 2003; Wesnousky et al., 2005). The seismically active faults have a recurrence interval on the order of 10,000 yr and exhibit normal and strike-slip focal mechanisms (Ryall and Priestly, 1975; Rogers et al., 1991) consistent with long-term displacements estimated from geologic

offsets and fault-slip inversion (Oldow, 1992, 2003; Wesnousky et al., 2005; Ferranti et al., this volume). GPS velocities increase from east to west across the boundary zone from 2–3 mm/yr in the central Great Basin to 12–14 mm/yr in the Sierra Nevada and are accompanied by a systematic clockwise change in orientation from west-northwest to northwest (Thatcher et al., 1999; Bennett et al., 1999, 2003; Dixon et al., 1995, 2000; Miller et al., 2001; Oldow et al., 2001; Oldow, 2003; Hammond and Thatcher, 2004). The orientations of incremental strain axes determined from fault-slip inversion and earthquake focal mechanisms also vary systematically across the central Walker Lane (Oldow, 2003) and

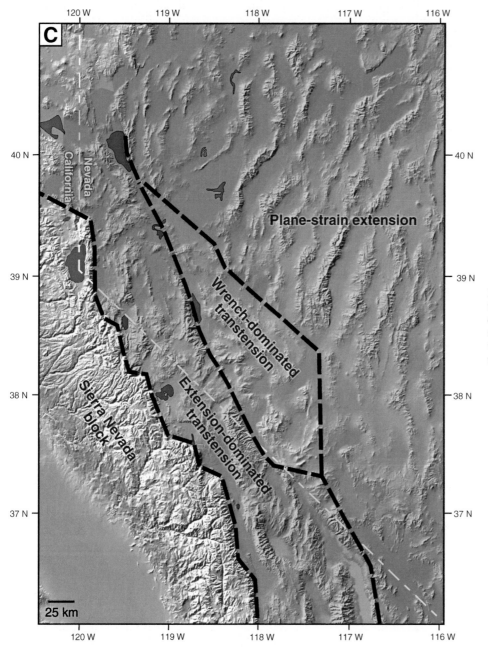

Figure 1 (*continued*). (C) Regional transtension is spatially partitioned into strain domains dominated by non–coaxial strain in the east and coaxial strain in the west of the central Walker Lane. All figures are after Oldow (2003).

are marked by a change in orientation from west-northwest to west-southwest as the Sierra Nevada Range is approached from the east (Oldow, 2003). From east to west, the velocity and strain trajectories deviate by as much as 50° (Fig. 1B).

Structures within the boundary zone are spatially segregated into domains dominated by dip-slip and strike-slip displacement. The domains are reflected in the regional physiography. The structures form a belt that narrows from 125 km in the south to ~75 km at the latitude of the central Sierra Nevada, where the belt broadens to ~175 km in the central Walker Lane (Fig. 1B). East of the southern Sierra Nevada, the northern part of the Eastern California Shear Zone is expressed by north-northwest–trending mountain ranges and basins separated by right-oblique normal faults (Burchfiel et al., 1987, 1992). This tectonic domain is separated from the central Great Basin (Fig. 1A) by the northwest-trending, right-lateral Furnace Creek fault system (Stewart, 1967, 1983; Reheis and McKee, 1991; Snow and Wernicke, 2000). Farther to the north, the central Walker Lane is divided into eastern and western physiographic domains. In the west, north-northwest–trending mountains and basins are separated by right-oblique normal faults (Oldow, 1992). In the eastern domain, a series of right-lateral transcurrent faults underlies a region of subdued topography bordered on the east by the central Great Basin. The transcurrent faults of the central Walker Lane are linked to the structures of the northern part of the Eastern California Shear Zone by a system of east-northeast left-lateral faults constituting a structural stepover known as the Mina deflection (Ryall and Priestly, 1975; Oldow, 1992).

DISPLACEMENT AND STRAIN IN THE CENTRAL WALKER LANE

In order to show how the strain and displacements are related within the central Walker Lane region, the general review here introduces the relationship between displacements and strain. Next, the particular assumptions and boundary conditions needed to employ the Mohr circle solution are outlined, followed by the graphical solutions for the domains within the central Walker Lane region.

When a rock mass is deformed, each particle within the rock is moved through a reference frame and creates a trace that records its path from an initial to a final position. The direction and length of these traces with respect to the reference frame give vectors describing the displacement of each particle. If the displacement is along a straight line, then each particle path vector will be parallel and the same length. Because each particle has moved the same amount, and in the same direction, the geometric properties of the object remain unchanged. This type of motion is considered a body translation (Ghosh and Ramberg, 1976; Ramsay and Huber, 1987). If one side of the object moves faster than the other, and volume is conserved, a velocity gradient will result in a strain, a change in length or angle from an original state (Ramsay, 1967).

If the previously described transformation is imposed on a square grid, strain can be recorded as changes in line lengths and angles. The grid is initially composed of squares geometrically

identical to each other. Following a homogeneous deformation, every square will be deformed into identical parallelograms. If the displacement vectors have differed in orientation and magnitude, heterogeneous strain (Ramsay, 1967), the squares would no longer be identical everywhere. An exception to this case is rigid body rotation, where rotation about a single point results in vectors of different orientation and magnitude but the geometric properties of the object remain unchanged. If all particle motion is contained within a single plane, then it is referred to as plane strain, while particle motion that cannot be fully described by a planar section is considered non–plane strain.

In plane strain conditions, pure shear and simple shear are end member forms of general shear deformation (Ramsay, 1967; Simpson and De Paor, 1993). Directions of elongation and shortening that remain in a constant orientation during deformation characterize pure shear. Simple shear is distinctly different from pure shear because the orientations of elongation and shortening rotate with increasing strain. In pure shear, an orthogonal set of lines parallel to the instantaneous principal strain axes will remain in this orientation throughout the deformation. Pure shear is considered coaxial because the principal axes of finite strain will remain parallel to the instantaneous principal strain axes. During simple shear, an orthogonal set of lines parallel to the instantaneous principal strain axes will rotate out of parallelism and become nonorthogonal. Simple shear is non–coaxial because the finite strain axes will immediately rotate out of parallelism with the instantaneous strain axes during progressive deformation (Ramsay, 1967).

The stability of lines differs in simple shear and plane general shear during progressive deformation. In simple shear, the orientations of lines oriented parallel to the shear couple are stable and do not change, whereas all other passively rotating lines will asymptotically approach the orientation of lines parallel to the shear-zone boundary. In contrast, in plane general shear, there are two directions of lines that remain stable. One is parallel to the shear-zone boundary and the other is inclined to the shear-zone boundary with an orientation dependent upon the relative components of pure and simple shear, or more specifically, the vorticity (Fossen and Tikoff, 1993).

Particle displacement paths within a deforming body can create curved traces (Ghosh and Ramberg, 1976; Weijermars and Poliakov, 1993; Tikoff and Fossen, 1995) except for those in stable orientations, which record motion that is either directly toward, away, or fixed with respect to the origin (Tikoff and Fossen, 1995). These linear traces are called eigenvectors, which represent a stable orientation throughout the deformational history. The eigenvectors within a velocity gradient field are known as the flow apophyses (Means et al., 1980; Bobyarchick, 1986; Fossen and Tikoff, 1993). In simple shear, there is only one flow apophysis, and in the case of two- (2D) and three-dimensional (3D) general shear, there are two and three flow apophyses, respectively (Tikoff and Fossen, 1995).

An understanding of the relationship between particle displacement paths in different strain regimes provides insight into the link between displacement fields and strain fields. Descriptions of this relationship typically require the use of quantitative tensor techniques; however, it can also be explored with the graphical method of De Paor (see Appendix 1) (De Paor, 1981; De Paor, 1987; Simpson and De Paor, 1993), used here for the analysis of the deforming tectonic zones in the central Walker Lane region.

Certain assumptions must be made in order to apply the polar Mohr construction to the Walker Lane region: (1) distributed deformation across the zone occurs at a scale that approximates a homogeneous state or can be generalized as a homogeneous domain within a heterogeneous or partitioned system (Thatcher, 1995; Horsman and Tikoff, 2005); (2) non–coaxial strain is restricted to a horizontal section, where the vorticity vector is perpendicular to this plane and parallel to one axis of the finite-strain ellipsoid; and (3) earthquake focal mechanisms and GPS velocities are not transient but represent steady-state conditions over modeled time scales. These three conditions are met, at least to a first approximation, within the Walker Lane region. The transcurrent and oblique-slip structures are distributed across the eastern and western domains, respectively, and the principal strain trajectories (Fig. 1B) are subparallel within each domain. Vertical gradients cannot be determined from GPS velocity fields, and any resultant components of non–coaxial strain can only be determined for the horizontal plane. Although unsatisfying, this assumption is axiomatic. The strain in the Walker Lane is not transient, as indicated by correspondence between the orientation

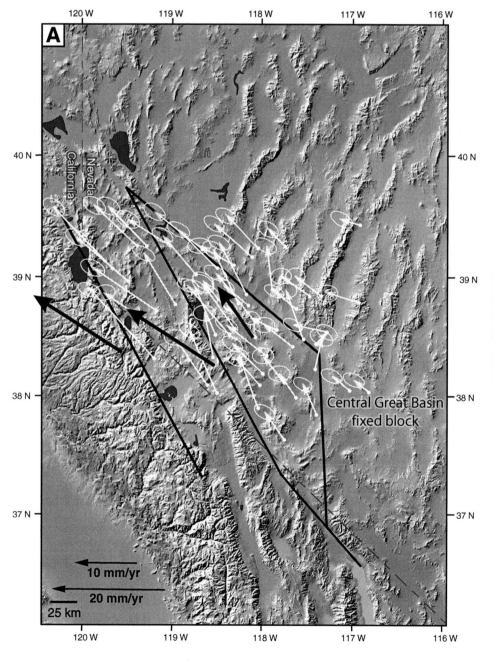

Figure 2. (A) Map of the central Walker Lane showing regional global positioning system (GPS) velocities (white arrows) with error ellipses (white circles) in a North American reference frame. The black arrows correspond to a visually determined best-fit orientation of the velocity field within each domain. (*Continued on following page.*)

of strain rate obtained from fault-slip inversion and from seismicity. This suggests that the state of strain has remained at a steady state over time scales of 10^6 yr (Oldow, 2003). The seismicity captures the strain rate that occurs over a similar time scale as the GPS velocities, and given the consistency of the strain over long time scales, the deviation between velocity and strain rate observed in the central Walker Lane region presumably is long-lived and cannot be attributed solely to transient strain.

The domainal character of deformation within the central Walker Lane region precludes a single construction. Rather, the relationship between strain-rate orientation and GPS velocity is calculated for the western and eastern domains separately.

For each of the domains, the GPS velocity recorded in a fixed North American frame does not show the differential displacement across domain boundaries and requires decomposition into a local reference frame.

The extension-dominated domain adjacent to and east of the Sierra Nevada (Fig. 1C) is a good candidate for the graphical construction because it has approximately parallel boundaries striking ~332° with GPS velocities that record a marked deviation between displacement and strain trajectories (Fig. 1B). Residual velocities for each domain were constructed by visually determining a best-fit velocity within each domain (Fig. 2A) and subtracting the relative velocity of the neighboring domains. The residual

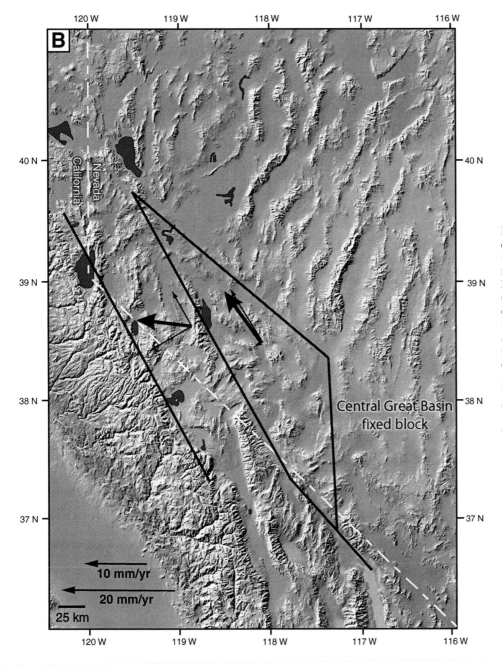

Figure 2 (*continued*). (B) Map of the central Walker Lane showing residual velocity of each domain with components (black arrows). The residual velocity of the wrench-dominated domain (right) was determined by taking the difference between the best-fit velocity within the wrench-dominated domain and the stable central Great Basin. The residual velocity within the extension-dominated domain (left) was determined by taking the difference between the best-fit velocities of the wrench-dominated domain and the motion of the Sierra Nevada. Figure was adapted from Oldow (2003).

GPS velocity in the extension-dominated domain trends ~276° and was determined by subtracting the average GPS velocity of the wrench-dominated domain to the east from the average GPS velocity of the domain to the west (Fig. 2B). Decomposition of the residual velocity into its boundary-parallel and boundary-perpendicular components yields rates of 5.7 mm/yr and 8 mm/yr, respectively (Fig. 2B). The width of this domain, measured orthogonal to its boundary, is ~60 km. The seismic recurrence interval of 10,000 yr determined by Rogers et al. (1991) is the period of time that the graphical construction is integrated for both the extension and wrench domains. This time period was chosen so the strain solution time scale matches the time scale of geologic strain data. The elongation direction is predicted to trend ~73° (θ) counterclockwise from the kinematic boundary, which is 16° more counterclockwise than the relative GPS velocity (Fig. 3A).

The wrench-dominated domain to the east can be analyzed in a similar way. The width of the deforming zone is taken as the distance between the boundary with the extension-dominated domain to the easternmost strike-slip fault within the wrench-dominated domain. The boundaries of this zone strike ~332°, and the relative GPS velocity trends ~324° (Fig. 1C). Because the residual GPS velocity is nearly parallel to the boundary, this domain is predominantly dextral simple shear with a small component of pure shear elongation (i.e., wrench-dominated transtension). The boundary-parallel and boundary-perpendicular components are 11.4 mm/yr and 1.5 mm/yr, respectively (Fig. 2B). The width is ~40 km. The maximum elongation direction is predicted to trend ~48° counterclockwise from the kinematic boundary, which is 40° more counterclockwise from the residual GPS velocity (Fig. 3B).

If these solutions are plotted on the map of the central Walker Lane (Fig. 4), they are consistent with the independently determined strain trajectories from first motion studies and fault-slip inversion described by Oldow (2003). The θ values predicted by the graphical solution also agree with the analytical solutions for strain obtained using the method of Tikoff and Fossen (1995), which predicted the maximum elongation direction in the extensional domain and the wrench dominated domain to be 73° and 48° from the kinematic boundary of each zone, respectively.

DISCUSSION

When deforming zones undergo contemporaneous coaxial and non–coaxial strains, structures in nonstable orientations will rotate over time at rates that vary as a function of both their orientation and the relative components of coaxial and non–coaxial strain (Ramsay, 1967; Tikoff and Fossen, 1995). The rotation of passive markers is dictated by the rotation and ellipticity of the finite-strain ellipse. Active markers may rotate at some different amount that can be a function of either the relative competence contrast between the marker and the country rock or the degree of anisotropy created by the structure.

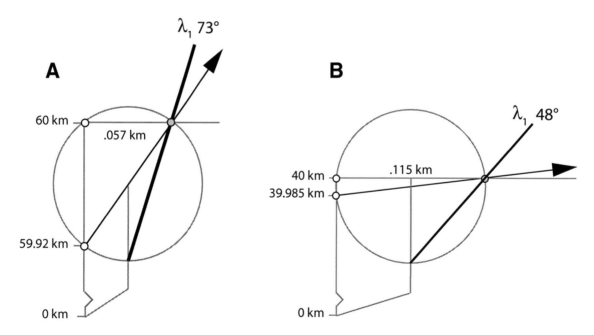

Figure 3. (A) Enlarged view of the completed graphical solution for the western, extension-dominated domain. (B) Enlarged view of the completed graphical solution for the eastern, wrench-dominated domain. Solid gray lines correspond to the vertical reference axis, horizontal baseline, the polar Mohr circle for strain (Simpson and De Paor, 1993), and the line projected from the center of the circle to the origin. The solid arrow corresponds to the orientation of the residual velocity, which is compared to the orientation of elongation shown by the solid black line. The vertical axis represents increasing distance perpendicular to the shear plane, while the horizontal axis represents increasing distance parallel.

To investigate the rotation of the principal strain axes over time, we graphically solved the orientation in 1 m.y. increments (Fig. 5). The rotation of the finite-strain ellipse within the extension-dominated domain (Fig. 5A) is slow (Simpson and De Paor, 1997). The coaxial component of transtension appears to counteract the rotation resulting from the non–coaxial strain component. It follows that the rotation of the principal axis of the finite-strain ellipse is much faster in the wrench-dominated domain (Fig. 5B) because the coaxial component is much smaller than the non–coaxial component. The non–coaxial component of deformation causes the principal strain axis to rotate at a rate close to but not equal to that in a simple shear condition.

The orientations of strain obtained using the graphical method agree well with the strain trajectories derived from fault-slip inversion and earthquake focal mechanism solutions (Oldow, 2003) within the central Walker Lane region. Furthermore, the predicted strain orientation is consistent with results obtained using existing analytical solutions (e.g., Fossen and Tikoff, 1993; Tikoff and Fossen, 1995). Because the path to a finite 3D strain is nonunique (Tikoff and Fossen, 1995), the orientation in 2D strain calculated in the Mohr construction cannot differentiate between

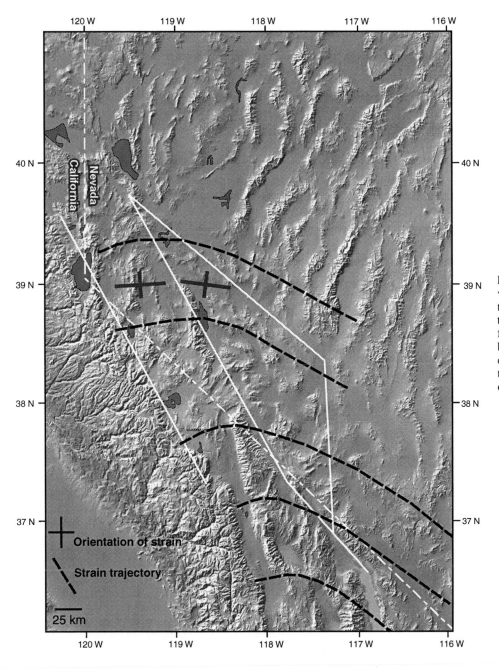

Figure 4. Map of the Walker Lane, Nevada, showing global positioning system (GPS) and strain trajectories with the predicted orientation of finite strain from the graphical method plotted in black. The dimensions of the strain axes do not correspond to the ratio of stretch; they indicate the long and short axes only. (Adapted from Oldow, 2003.)

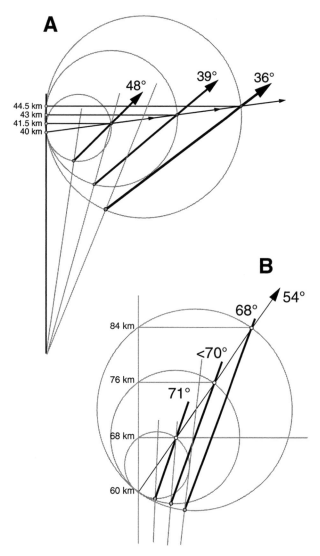

Figure 5. Rotation of the maximum elongation direction (solid line) integrated over 1, 2, and 3 m.y. for both the (A) wrench-dominated and (B) extension-dominated domains.

plane or non–plane strain conditions. Differentiation between plane and non–plane strain in the central Walker Lane is based on strain compatibility arguments.

Possible end members are: (1) the non–plane transtensional model (Sanderson and Marchini, 1984; Fossen and Tikoff, 1993), and (2) the horizontal–plane strain model, consisting of a broadening sub–simple shear zone (Simpson and De Paor, 1993). The non–plane strain model has boundary conditions where the length of the shear-zone boundary remains fixed and volume conservation is accommodated by vertical strain. In contrast, the plane strain boundary condition does not allow vertical displacement but requires the shear-zone boundary to shorten or elongate (Fig. 6).

The effects of the different boundary conditions and their relevance to the geologic evolution of the western Great Basin–

Sierra Nevada boundary zone can be assessed by simple geometric models. Given the present-day geometry of the boundary zone and observed kinematics, we can use an inverse strain model to restore the domain geometry over different time intervals. The extension-dominated zone is generalized as a deforming zone, 500 km in length and 60 km in width. The thickness of the crust is taken as the depth to the Mohorovicic discontinuity, which is estimated at 30 km (Gilbert et al., 2007; Gilbert and Sheehan, 2004; Wernicke et al., 1996; Fliedner et al., 1996). GPS velocity components are 8.0 mm/yr and 5.7 mm/yr perpendicular and parallel to the zone boundaries, respectively. The eastern, wrench-dominated domain has dimensions of ~500 km in length, 40 km in width, and 30 km in thickness, with displacement rate components perpendicular and parallel to the deforming zone of 1.5 mm/yr and 11.4 mm/yr, respectively. Velocities within each domain are integrated over 10^6 yr to model strain compatibility over a time scale comparable to the inferred life of the strain field discussed previously.

The inverse model for plane strain is characterized by a significant change in length of the shear-zone boundary. This results in a reduction of the width of the zone and elongation along the boundary but no change in the vertical thickness (Fig. 7). Using the contemporary conditions outlined here, the inverse model reduces the western domain width by a linear rate of 8 km/m.y. In this state of strain, there can be no vertical changes, and, assuming conservation of volume, the restoration of boundary length increases by a nonlinear rate. Over a period of 1 m.y., the boundary length increases from 500 to 577 km, over 2 m.y., the length increases to 682 km, and in 3 m.y., it increases to 833 km. For the eastern domain, the width decreases by 1.5 km/m.y. The length increases from 500 to 520 km in 1 m.y., to 542 km in 2 m.y. and to 563 km in 3 m.y. This model is untenable because it presents a striking mismatch in boundary length compatibility, both within the deformation zone and between the deformation zone and the Sierran block to the west and the central Great Basin to the east (Fig. 7).

The inverse model for non–plane transtension (Sanderson and Marchini, 1984; Tikoff and Teyssier, 1994; Dewey et al., 1998), however, is viable and is characterized by a fixed shear-zone boundary length and the movement of material in and out of the horizontal section. Strain components in this model consist of a non–coaxial strain in the horizontal plane and a vertically oriented coaxial shortening with an accompanying orthogonal coaxial elongation across the deforming zone (Fig. 7). Within the extension-dominated domain, non–plane transtension results in a decrease in domain width by 8 km/m.y. The model increases in thickness from 30 to ~35 km in 1 m.y., to ~41 km in 2 m.y., and to 50 km in 3 m.y. (Fig. 7) and exceeds 64 and 90 km in 4 and 5 m.y., respectively. The geometry of the wrench-dominated domain decreases in width by 1.5 km/m.y. as the thickness changes from 30 to ~31 km in 1 m.y., to ~32 km in 2 m.y., and ~34 km in 3 m.y. The deforming zone progressively thickens in order to accommodate width changes, but no changes occur in the length of the shear-zone boundary.

Non-plane strain transtension

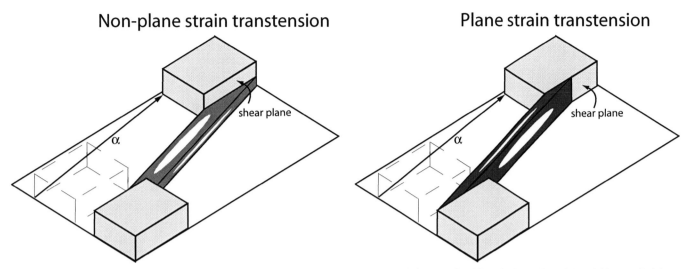

Figure 6. Cartoon of two end members of deformation: Non–plane strain transtension (left) has a fixed length shear plane. Material is supplied from the third dimension to accommodate stretching. The length of the shear-zone boundary in plane strain (right) changes, the deforming zone narrows to accommodate stretching, and no change in thickness occurs. Figure was adapted from Tikoff and Teyssier (1994).

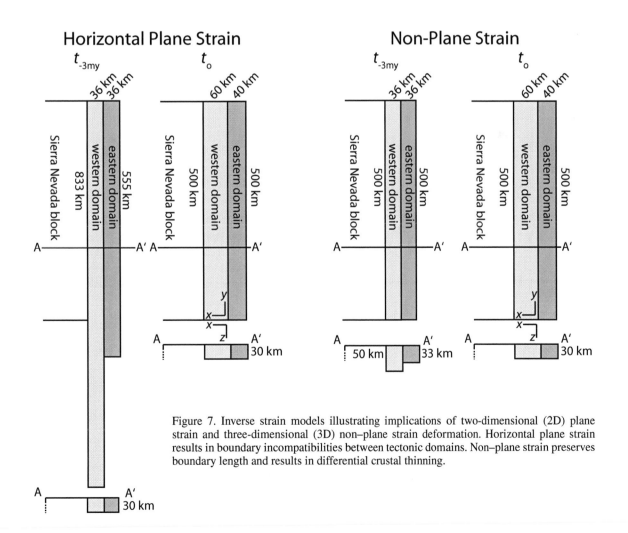

Figure 7. Inverse strain models illustrating implications of two-dimensional (2D) plane strain and three-dimensional (3D) non–plane strain deformation. Horizontal plane strain results in boundary incompatibilities between tectonic domains. Non–plane strain preserves boundary length and results in differential crustal thinning.

The rapid change of crustal thickness required in the non–plane transtensional strain state sets limits on the duration over which the model can be applied. The inverse model predicts a crustal thickness of 50 km for the extension-dominated domain of the transition zone at 3 Ma. This is consistent with the initial thickness of ~50 km for the Sierra Nevada and presumably the transition along the western Great Basin (Hauge et al., 1987; Louie et al., 2004; Gilbert and Sheehan, 2004; Heimgartner et al., 2005). The crustal thickness of the southern Sierra Nevada was dramatically changed during delamination of the lithospheric root at ca. 3 Ma (Gilbert et al., 2007), which coincides with a fundamental reorganization of kinematics in the northern Eastern California Shear Zone and central Walker Lane (Oldow et al., this volume). The inverse model clearly shows that the present-day kinematics recorded in earthquake seismology and GPS velocities cannot be extrapolated into the past beyond the mid-Pliocene.

Using the values of strain components from the inverse models presented here, the relative contributions from the non–coaxial and coaxial components to the overall strain can be expressed in terms of the vorticity number. This can be calculated using the value and orientation of the strains computed from the GPS rate components and the equation for the internal kinematic vorticity (Tikoff and Fossen, 1995).

The equation for elongation (Ramsay, 1967) gives the values of e_1, e_2, and e_3, the maximum, intermediate and minimum elongation respectively. Where l_1 is the current dimension of the deforming zone, and l_0 is the original.

$$e = \frac{l_1 - l_0}{l_0}. \qquad (1)$$

The principal values of stretch, K_1, K_2, and K_3 can be found by substituting the values of elongation determined from Equation 1 into the equation for stretch:

$$K_n = 1 + e_n. \qquad (2)$$

The angular shear, Ψ, can be found by constructing a right triangle where the short side is equal to the time-integrated parallel displacement, and the intermediate side is equal to the width of the zone:

$$\gamma = \tan(\Psi). \qquad (3)$$

The values of K_n and θ can be substituted into Equation 4 to determine the internal kinematic vorticity (Tikoff and Fossen, 1995).

$$Wk_i = \frac{\sqrt{\left(\gamma_{xy}\right)^2 + \left(\gamma_{yz}\right)^2 + \left(\gamma_{xz}\right)^2}}{\sqrt{2\left[\ln\left(k_1\right)^2 + \ln\left(k_2\right)^2 + \ln\left(k_3\right)^2\right] + \left(\gamma_{xz}\right)^2}} \qquad (4)$$

The calculated vorticity of the extension-dominated domain is ~0.36 in the non–plane transtension model and ~0.97 in the wrench-dominated domain. Now, we can compare both our Mohr graphical results and these vorticities computed from the inverse model with the analytical solutions of Tikoff and Fossen (1995). All methods yield a value for θ of ~73° for the extension-dominated domain and ~48° for the wrench-dominated domain of the non–plane strain model.

CONCLUSION

Velocity trajectories and strain trajectories deviate from parallelism within the central Walker Lane by as much as 50° toward the eastern side of the Sierra Nevada. This deviation has been attributed to spatial strain partitioning into domains of non–coaxially and coaxially dominated transtension (Oldow, 2003). A better understanding of the kinematics within these zones is gained by relating the GPS velocity and strain fields using a Mohr graphical solution (Simpson and De Paor, 1993) that estimates the orientations of the principal strain axes from the velocity field.

In the case of the central Walker Lane region, the solutions of the graphical method agree with both the strain trajectories determined independently (Oldow, 2003) and solutions obtained from existing analytical tools (Fossen and Tikoff, 1993; Tikoff and Fossen, 1995). Together, these tools illustrate that GPS velocities and strain trajectories are not parallel within deforming zones that exhibit a non–coaxial strain component. Furthermore, structures that behave as passive material lines rotate toward the oblique flow apophysis with increasing transtensional displacement.

The results of our graphical solutions combined with compatibility constraints document that the state of strain within the central Walker Lane region is characterized by 3D non–plane strain transtension. This strain state predicts significant crustal thinning along the transition zone between the Sierra Nevada and Walker Lane and is consistent with geologic observations in the region. When the strain model is inverted beyond 3–4 Ma, the results indicate that the contemporary deformation field is incompatible with accepted crustal thicknesses for the Sierra Nevada and western Great Basin. The implication is that this part of the North American plate margin experienced a profound kinematic reorganization in the mid-Pliocene.

APPENDIX 1

The De Paor construction (De Paor, 1981; Simpson and De Paor, 1993) was designed for the analysis of rotated porphyroclasts and passive material lines in ductile shear zones. Set in a polar coordinate reference system, the radius and polar angle of a line corresponds to the stretch and rotation, respectively, of a passive material line (Fig. A1). The circumference of a circle is created by the locus of points that describe the stretch and rotation of any line in any orientation (Simpson and De Paor, 1993). This plot can produce the orientation and value of the eigenvector(s), final and initial orientation of passive markers, maximum and minimum shear strain, two-dimensional kinematic vorticity, and the finite rotation of passive material lines.

Two integral pieces of data are needed to apply the graphical construction to tectonic problems: (1) the displacement rate components and orientation of the GPS velocity and/or NUVEL-1A plate motion (DeMets et al., 1994) to compute the velocity component parallel to the boundary, which captures the rotational strain, and (2) the width of the deforming zone. In interplate regimes, the width usually refers to the continental plate because exposure of the oceanic plate can be limited to island arcs. If partitioning has occurred within an arc, then the width of the transpressional terrane (Avé Lallemant and Oldow, 2000) defined by arc-parallel strike-slip faults could be used to define the boundary. Additional data that aid in the construction are first motion and fault-plane solutions, which give incremental strain directions that provide a powerful check on the graphical solution. Additionally, earthquake foci can delineate active fault planes, structural analysis gives the progressive superposition history, and fault-slip inversion techniques can give both the incremental strain axes and the vorticity, provided the data are discriminating enough (e.g., Unruh et al., 2003).

Certain assumptions must be made in order to apply this modified construction: (1) deformation across the zone is distributed, or can be generalized as a domain within a heterogeneous or partitioned system (Thatcher, 1995; Horsman and Tikoff, 2005), (2) non–coaxial strain is restricted to a horizontal section where the vorticity is perpendicular to this plane and parallel to one axis of the finite-strain ellipsoid, and (3) elastic strain accumulation recorded within the velocity field cannot be resolved from permanent deformation.

It is useful to analyze a kinematic cartoon for a brief walkthrough of the method before analyzing real-world tectonic settings. In a three-plate problem (Fig. A2), where plate A remains stationary relative to plate B, which is moving obliquely away from plate A at 23 mm/yr, the angle between the boundary and the plate motion vector is 30°. Decomposition of the plate motion vector yields a boundary-parallel velocity of 20 mm/yr and a boundary-normal velocity of 11.5 mm/yr, with a 16-km-wide deforming zone. The horizontal and vertical axes are scaled to the width of the zone.

The velocity field must be integrated over a period of time consistent with the available geologic data. For example, if the strain solutions from the graphical construction were to be compared with earthquake focal mechanism solutions, then the velocity would need to be integrated over the earthquake recurrence interval within the deforming zone. The integrated velocity gives the distance that the anchor point is translated (Fig. A3). One million years was chosen in this example, so the anchor point was moved 11.5 km perpendicular to the boundary and 20 km parallel to the boundary (Fig. A3A) from the height on the reference axis corresponding to the width of the deforming zone (Fig. A3A). Lines were drawn through the anchor point, both in the orientation of the velocity and parallel to the zone boundary (Fig. A3B). This method uses the motion vector of plate B as the orientation of the oblique stable orientation of flow (Weijermars and Poliakov, 1993).

Three points are created: (1) the anchor point, (2) the intercept on the reference axis, created by a line drawn from the anchor point to the reference axis in the orientation of the velocity field, and (3) the point where a line drawn from the anchor point, parallel to the kinematic boundary, intersects the reference axis (Fig. A3B). A circle can be constructed passing through these three points (Fig. A3C). A line drawn from the origin through the center of the circle will produce two intercepts on the circumference of the circle; these are points S_1 (greatest stretch) and S_2 (least stretch) (Fig. A3C).

Lines drawn from these two points back through the anchor point will produce the orientation of the principal strain axes with respect to the kinematic boundaries of the zone. Because the anchor point inverts the solution, the line drawn from S_2 through the anchor point yields the orientation of the long axis of the finite-strain ellipse (Fig. A3D). The orientations of the principal strain axes can be plotted on the kinematic model. In this example, the long axis of strain is predicted to be 55° from the kinematic boundary and 25° from the velocity direction (Fig. A3D).

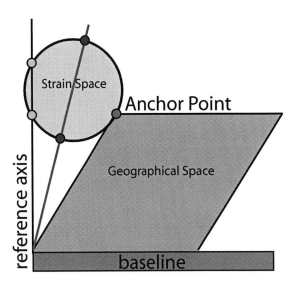

Figure A1. The graphical method of De Paor is based on a polar coordinate strain Mohr circle, where the Mohr circle is created by the locus of points that describe the stretch and rotation of all orientations of passive material lines. Solutions obtained within strain space are translated through the anchor point into geographical space. The reference axis and baseline axis are scaled to the width of the deforming zone, and the reference axis is perpendicular to the zone boundaries. Figure was adapted from Simpson and De Paor (1993).

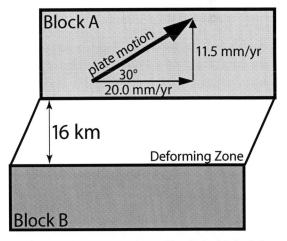

Figure A2. Three-plate problem: Plate B is fixed relative to plate A, which is moving obliquely away from plate B. Deformation is accommodated within the deforming zone (white), which is 16 km wide. The graphical construction uses the width of the deforming zone (white), the angle between the plate motion vector of plate A and the boundary, 30°, and the rate components of the plate motion vector.

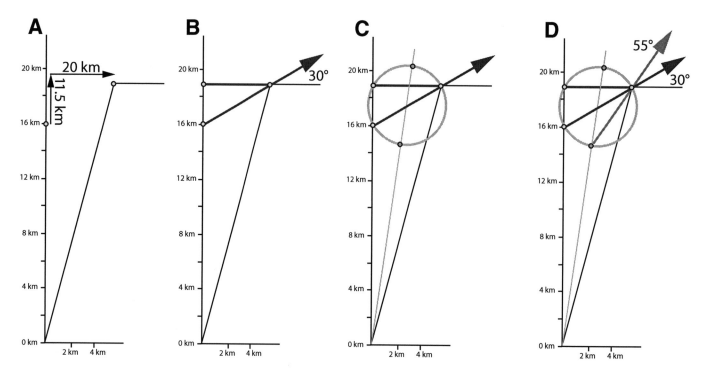

Figure A3. Stepwise construction of the graphical method. (A) The reference axis (vertical) and baseline (horizontal) are scaled to the width of the deforming zone being analyzed. The anchor point is translated in the direction of the velocity field by an amount equal to the velocity integrated over an amount of time consistent with the geologic data. In this example, the anchor point is translated 11.5 km perpendicular and 20 km parallel to the zone, with rates integrated over 1 m.y. (B) A line is drawn in the orientation of the velocity field (dark gray arrow) and parallel to the zone, which creates three points. (C) The Mohr circle is constructed to pass through these three points (dots). A line drawn from the origin through the center of the circle creates two intercepts, S_1 (upper intercept) and S_2 (lower intercept), which correspond to the greatest stretch (S_1) and least stretch (S_2). (D) A line drawn from S_2 through the anchor point gives the orientation of the long axis of the strain ellipse. Note the solutions are inverted through the anchor point.

ACKNOWLEDGMENTS

The authors thank Declan De Paor and Basil Tikoff for reviewing this manuscript and inspirational research into the intricacies of vorticity and strain. This research was partially supported by National Science Foundation grants (EAR-0126205 and EAR-0208219) to J.S. Oldow.

REFERENCES CITED

Avé Lallemant, H.G., and Oldow, J.S., 2000, Active displacement partitioning and arc-parallel extension of the Aleutian volcanic arc based on Global Positioning System geodesy and kinematic analysis: Geology, v. 28, p. 739–742, doi: 10.1130/0091-7613(2000)28<739:ADPAAE>2.0.CO;2.

Bennett, R.A., Davis, J.L., and Wernicke, B.P., 1999, Present-day pattern of Cordillera deformation in the western United States: Geology, v. 27, p. 371–374, doi: 10.1130/0091-7613(1999)027<0371:PDPOCD>2.3.CO;2.

Bennett, R.A., Wernicke, B.P., Niemi, N.A., Friedrich, A.M., and Davis, J.L., 2003, Contemporary strain rates in the northern Basin and Range province from GPS data: Tectonics, v. 22, no. 2, p. 1008, doi: 10.1029/2001TC001355.

Bobyarchick, A.R., 1986, The eigenvalues of steady flow in Mohr space: Tectonophysics, v. 122, p. 35–51, doi: 10.1016/0040-1951(86)90157-5.

Burchfiel, B.C., Hodges, K.V., and Royden, L.H., 1987, Geology of the Panamint Valley pull-apart system: Palinspastic evidence for low-angle geometry of a Neogene range-bounding fault: Journal of Geophysical Research, v. 92, p. 10,422–10,426, doi: 10.1029/JB092iB10p10422.

Burchfiel, B.C., Cowan, D.S., and Davis, G.A., 1992, Tectonic overview of the Cordilleran orogen in the western United States, *in* Burchfiel, B.C., et al., eds., The Cordilleran Orogen: Conterminous U.S.: Boulder, Colorado, Geological Society of America, Geology of North America, v. G-3, p. 407–479.

DeMets, C., Gordon, R.G., Argus, D.F., and Stein, S., 1994, Effect of recent revisions to the geomagnetic reversal time scale on estimates of current plate motions: Geophysical Research Letters, v. 21, p. 2191–2194, doi: 10.1029/94GL02118.

De Paor, D.G., 1981, Strain analysis using deformed line distributions: Tectonophysics, v. 73, p. T9–T14, doi: 10.1016/0040-1951(81)90217-1.

De Paor, D.G., 1987, Stretch in shear zones: Implications for section balancing: Journal of Structural Geology, v. 9, p. 893–895, doi: 10.1016/0191-8141(87)90089-7.

Dixon, T.H., Robaudo, S., Lee, J., and Reheis, M.C., 1995, Constraints on present-day Basin and Range deformation from space geodesy: Tectonics, v. 14, p. 755–772, doi: 10.1029/95TC00931.

Dixon, T.H., Miller, M., Farina, F., Wang, H., and Johnson, D., 2000, Present day motion of the Sierra Nevada block and some tectonic implications for the Basin and Range Province, North American Cordillera: Tectonics, v. 19, p. 1–24, doi: 10.1029/1998TC001088.

Dokka, R.K., and Travis, C.J., 1990, Role of the Eastern California shear zone in accommodating Pacific-North American plate motion: Geophysical Research Letters, v. 17, no. 9, p. 1323–1326, doi: 10.1029/GL017i009p01323.

Ferranti, L., Oldow, J.S., Geissman, J.W., and Nell, M.M., 2009, this volume, Flattening strain during coordinated slip on a curved fault array, Rhodes Salt Marsh extensional basin, central Walker Lane, west-central Nevada, *in* Oldow, J.S., and Cashman, P.H., eds., Late Cenozoic Structure and Evolution of the Great Basin–Sierra Nevada Transition: Geological Society of America Special Paper 447, doi: 10.1130/2009.2447(11).

Fliedner, M.M., Ruppert, S.D., Malin, P.E., Park, S.K., Jiracek, G.R., Phinney, R.A., Saleeby, J.B., Wernicke, B.P., Clayton, R.W., Keller, G.R., Miller, K.C., Jones, C.H., Luetgert, J.H., Mooney, W.D., Oliver, H.L., Klemperer, S.L., and Thompson, G.A., 1996, Three-dimensional crustal structure of the southern Sierra Nevada from seismic fan profiles and gravity modeling: Geology, v. 24, p. 367–370, doi: 10.1130/0091-7613(1996)024<0367: TDCSOT>2.3.CO;2.

Fossen, H., and Tikoff, B., 1993, The deformation matrix for simultaneous simple shearing, pure shearing and volume change, and its application to transpression-transtension tectonics: Journal of Structural Geology, v. 15, p. 413–422, doi: 10.1016/0191-8141(93)90137-Y.

Ghosh, S.K., and Ramberg, H., 1976, Reorientation of inclusions by combination of pure and simple-shear: Tectonophysics, v. 34, p. 1–70, doi: 10.1016/0040-1951(76)90176-1.

Gilbert, H.J., and Sheehan, A.F., 2004, Images of crustal variations in the intermountain west: Journal of Geophysical Research, v. 109, p. B03306, doi: 10.1029/2003JB002730.

Gilbert, H., Jones, C., Owens, T.J., and Zandt, G., 2007, Imaging Sierra Nevada lithospheric sinking: Eos (Transactions, American Geophysical Union), v. 88, no. 21, p. 225, 229, doi: 10.1029/2007EO210001.

Hammond, W.C., and Thatcher, W., 2004, Contemporary tectonic deformation of the Basin and Range province, western United States: 10 years of observation with the Global Positioning System: Journal of Geophysical Research, v. 109, p. B08403, doi: 10.1029/2003JB002746.

Hauge, T.A., Allmendinger, R.W., Caruso, C., Hauser, E.C., Klemperer, S.L., Opdyke, S., Potter, C.J., Sanford, W., Brown, L.D., Kaufman, S., and Oliver, J., 1987, Crustal structure of western Nevada from COCORP deep seismic-reflection data: Geological Society of America Bulletin, v. 98, p. 320–329, doi: 10.1130/0016-7606(1987)98<320:CSOWNF>2.0.CO;2.

Heimgartner, M., Scott, J.B., Thelen, W., Lozez, C.T., and Louie, J.N., 2005, Variable crustal thickness in the western Great Basin: A compilation of old and new refraction data: Geothermal Resources Council Transactions, v. 29, p. 239–242.

Holt, W.E., Shen-Tu, B., Haines, A., and Jackson, J., 2000, On the determination of self-consistent strain rate fields within zones of distributed continental deformation, *in* Richards, M.A., Gordon, R.G., and van der Hilst, R.D., eds., The History and Dynamics of Global Plate Motions: American Geophysical Union Geophysical Monograph 121, p. 113–142.

Horsman, E., and Tikoff, B., 2005, Quantifying simultaneous discrete and distributed deformation: Journal of Structural Geology, v. 27, p. 1168–1169, doi: 10.1016/j.jsg.2004.08.001.

Jamison, W.R., 1991, Kinematics of compressional fold development in convergent wrench terranes: Tectonophysics, v. 190, p. 209–232, doi: 10.1016/0040-1951(91)90431-Q.

Locke, A., Billingsley, P.R., and Mayo, E.B., 1940, Sierra Nevada tectonic patterns: Geological Society of America Bulletin, v. 51, no. 4, p. 513–539.

Louie, J.N., Thelen, W., Smith, S.B., Scott, J.B., Clark, M., and Pullammanappallil, S., 2004, The northern Walker Lane refraction experiment: Pn arrival and the northern Sierra Nevada root: Tectonophysics, v. 388, p. 253–269, doi: 10.1016/j.tecto.2004.07.042.

McCoss, A.M., 1986, Simple constructions for deformation in transpression/ transtension zones: Journal of Structural Geology, v. 8, p. 715–718, doi: 10.1016/0191-8141(86)90077-5.

Means, W.D., Hobbs, B.E., Lister, B.E., and Williams, P.F., 1980, Vorticity and non-coaxility in progressive deformations: Journal of Structural Geology, v. 2, p. 371–378, doi: 10.1016/0191-8141(80)90024-3.

Miller, M.M., Johnson, D.J., Dixon, T., and Dokka, R.K., 2001, Refined kinematics of the Eastern California shear zone from GPS observations, 1993–1998: Journal of Geophysical Research, v. 106, p. 2245–2264, doi: 10.1029/2000JB900328.

Oldow, J.S., 1992, Late Cenozoic displacement partitioning in the northwestern Great Basin, *in* Craig, S.D., ed., Structure, Tectonics, and Mineralization of the Walker Lane: Walker Lane Symposium Proceedings: Reno, Geological Society of Nevada, p. 17–52.

Oldow, J.S., 2003, Active transtensional boundary zone between the western Great Basin and Sierra Nevada block, western U.S. Cordillera: Geology, v. 31, p. 1033–1036, doi: 10.1130/G19838.1.

Oldow, J.S., Aiken, C.L.V., Ferguson, J.F., Hare, J.L., and Hardyman, R.F., 2001, Active displacement transfer and differential motion between tectonic blocks within the central Walker Lane, western Great Basin: Geology, v. 29, p. 19–22, doi: 10.1130/0091-7613(2001)029<0019:ADTADB> 2.0.CO;2.

Oldow, J.S., Elias, E.A., Ferranti, L., McClelland, W.C., and McIntosh, W.C., 2009, this volume, Late Miocene to Pliocene synextensional deposition in fault-bounded basins within the upper plate of the western Silver Peak–Lone Mountain extensional complex, west-central Nevada, *in* Oldow, J.S., and Cashman, P.H., eds., Late Cenozoic Structure and Evolution of the Great Basin–Sierra Nevada Transition: Geological Society of America Special Paper 447, doi: 10.1130/2009.2447(14).

Ramsay, J.G., 1967, Folding and Fracturing of Rocks: New York, McGraw Hill, 568 p.

Ramsay, J.G., and Huber, M.I., 1987, The Techniques of Modern Structural Geology: Volume 1. Strain Analysis: London Academic Press, 307 p.

Reheis, M.C., and McKee, E.H., 1991, Late Cenozoic history of slip on the Fish Lake Valley fault zone, Nevada and California: U.S. Geological Survey Open File Report 91-290, p. 26–45.

Rogers, A.M., Harmsen, S.C., Corbett, E.J., Priestly, K., and de Polo, D., 1991. The seismicity of Nevada and some adjacent parts of the Great Basin, *in* Slemmons, D.B., et al., eds., Neotectonics of North America: Boulder, Colorado, Geological Society of America, Decade Map Volume 1, p. 153–184.

Ryall, A.S., and Priestly, K., 1975, Seismicity, secular strain, and maximum magnitude in the Excelsior Mountains area, western Nevada and eastern California: Geological Society of America Bulletin, v. 86, p. 1585–1592, doi: 10.1130/0016-7606(1975)86<1585:SSSAMM>2.0.CO;2.

Sanderson, D.J., and Marchini, W.R.D., 1984, Transpression: Journal of Structural Geology, v. 6, p. 449–458, doi: 10.1016/0191-8141(84)90058-0.

Shen-Tu, B., Holt, W.E., and Haines, A., 1998, Contemporary kinematics of the western United States determined from earthquake moment tensors, very long baseline interferometry, and GPS observations: Journal of Geophysical Research, v. 103, no. B8, p. 18,087–18,118, doi: 10.1029/98JB01669.

Shen-Tu, B., Holt, W.E., and Haines, A., 1999, Deformation kinematics in the western United States determined from Quaternary fault slip rates and recent geodetic data: Journal of Geophysical Research, v. 104, no. B12, p. 28,927–28,956, doi: 10.1029/1999JB900293.

Simpson, C., and De Paor, D.G., 1993, Strain and kinematic analysis in general shear zones: Journal of Structural Geology, v. 15, p. 1–20, doi: 10.1016/0191-8141(93)90075-L.

Simpson, C., and De Paor, D.G., 1997, Practical analysis of general shear zones using the porphyroclast hyperbolic distribution method: An example from the Scandinavian Caledonides, *in* Sengupta, S., ed., Evolution of Geological Structures in Micro to Macro Scales: London, Chapman & Hall, p. 169–184.

Snow, J., and Wernicke, B., 2000, Cenozoic tectonism in the central Basin and Range: Magnitude, rate, and distribution of upper crustal strain: American Journal of Science, v. 300, p. 659–719, doi: 10.2475/ajs.300.9.659.

Stewart, J.H., 1967, Possible large right-lateral displacement along fault and shear zones in the Death Valley–Las Vegas areas, California and Nevada: Geological Society of America Bulletin, v. 78, p. 131–142, doi: 10.1130/ 0016-7606(1967)78[131:PLRDAF]2.0.CO;2.

Stewart, J.H., 1983, Cenozoic structure and tectonics of the northern Basin and Range Province, California, Nevada, and Utah, *in* Eaton, G., ed., The Role of Heat in the Development of Energy and Mineral Resources in the Northern Basin and Range Province: Geothermal Resources Council Special Report 13, p. 25–40.

Stewart, J.H., 1988, Tectonics of the Walker Lane belt, western Great Basin: Mesozoic and Cenozoic deformation in a shear zone, *in* Ernst, W.G., ed., Metamorphism and Crustal Evolution of the Western United States: Englewood Cliffs, New Jersey, Prentice-Hall, p. 681–713.

Thatcher, W., 1995, Microplate versus continuum descriptions of active tectonic deformation: Journal of Geophysical Research, v. 100, p. 3885–3894, doi: 10.1029/94JB03064.

Thatcher, W., 2003, GPS constraints on the kinematics of continental deformation: International Geology Review, v. 45, no. 3, p. 191–212, doi: 10.2747/0020-6814.45.3.191.

Thatcher, W., Foulger, G.R., Julian, B.R., Svarc, J., Quilty, E., and Bawden, G.W., 1999, Present-day deformation across the Basin and Range Province, western United States: Science, v. 283, p. 1714–1718, doi: 10.1126/ science.283.5408.1714.

Tikoff, B., and Fossen, H., 1995, The limitations of three-dimensional kinematic vorticity analysis: Journal of Structural Geology, v. 17, p. 1771–1784, doi: 10.1016/0191-8141(95)00069-P.

Tikoff, B., and Teyssier, C., 1994, Strain modeling of displacement-field partitioning in transpressional orogens: Journal of Structural Geology, v. 16, p. 1575–1588, doi: 10.1016/0191-8141(94)90034-5.

Murphy et al.

Tron, V., and Brun, J.P., 1991, Experiments on oblique rifting in brittle-ductile systems: Tectonophysics, v. 188, p. 71–84, doi: 10.1016/0040-1951(91)90315-J.

Unruh, J., Humphrey, J., and Barron, A., 2003, Transtensional model for the Sierra Nevada frontal fault system, eastern: California Geology, v. 31, p. 327–330.

Weijermars, R., and Poliakov, A., 1993, Stream functions and complex potentials: Implications for development of rock fabric and the continuum assumption: Tectonophysics, v. 220, p. 33–50, doi: 10.1016/0040-1951 (93)90222-6.

Wernicke, B., Clayton, R., Ducea, M., Jones, C.H., Park, S., Ruppert, S., Saleeby, J., Snow, J.K., Squires, L., Fliedner, M., Jiracek, G., Keller, G.R., Klemperer, S., Luetgert, J., Malin, P., Miller, K., Mooney, W., Oliver, H., and Phinney, R., 1996, Origin of high mountains in the continents: The southern Sierra Nevada: Science, v. 271, p. 190–193, doi: 10.1126/science.271.5246.190.

Wesnousky, S.G., Barron, A.D., Briggs, R.W., Caskey, S.J., Kumer, S., and Owen, L., 2005, Paleoseismic transect across the northern Great Basin: Journal of Geophysical Research, v. 110, B05408, doi: 10.1029/2004JB003283.

Withjack, M.O., and Jamison, W.R., 1986, Deformation produced by oblique rifting: Tectonophysics, v. 126, p. 99–124, doi: 10.1016/0040-1951(86)90222-2.

Wojtal, S., 1989, Measuring displacement gradients and strain in faulted rock: Journal of Structural Geology, v. 11, p. 669–678, doi: 10.1016/0191-8141 (89)90003-5.

MANUSCRIPT ACCEPTED BY THE SOCIETY 21 JULY 2008

The Geological Society of America
Special Paper 447
2009

Neotectonic analysis of Upper Klamath Lake, Oregon: New insights from seismic reflection data

Lee M. Liberty*
Center for Geophysical Investigation of the Shallow Subsurface, Boise State University, Boise, Idaho 83725-1536, USA

Thomas L. Pratt
U.S. Geological Survey and School of Oceanography, University of Washington, Seattle, Washington 98195, USA

Mitchell Lyle
Department of Oceanography, Texas A&M University, College Station, Texas 77840-3146, USA

Ian P. Madin
Oregon Department of Geology and Mineral Industries, Portland, Oregon 97232, USA

ABSTRACT

We present marine high-resolution seismic reflection data from Upper Klamath Lake, Oregon, to discern the underlying structure and estimate Quaternary slip rates in this actively extending Basin and Range system. The sediment patterns and structures imaged on our seismic profiles reveal a complex geologic system that reflects a changing climate record, shallow water conditions, growth faulting, contrasting sediment sources, and high slip rates. We observe that Upper Klamath Lake is a sediment-saturated environment, and sediment accumulation rates are therefore controlled by basin subsidence rather than sediment supply. Published slip rates for Holocene extension are greater than our determined late Quaternary slip rates, assuming reasonable rates of deposition. The apparent increased Holocene fault-slip rates may be in part an artifact of long recurrence intervals between major earthquakes, with recent seismicity accommodating long-term strain. The quantity of observed faults below the lake is at least an order of magnitude greater than those mapped outside the lake, suggesting that many hidden faults throughout the region may be unaccounted for when estimating Basin and Range extension rates.

Keywords: seismic reflection, neotectonics, Basin and Range Province.

INTRODUCTION

Upper Klamath Lake is located in the seismically active region of southwestern Oregon between the Oregon Cascade Range and the northwestern margin of the Basin and Range

*Corresponding author e-mail: lml@cgiss.boisestate.edu.

Province (Fig. 1). Normal faults with Holocene motion have been mapped throughout the region from northern California to central Oregon (McKee et al., 1983; Sherrod and Pickthorn, 1992; Muffler et al., 1994; Bacon et al., 1999; Personius, 2002). Oregon's largest historical earthquakes were located along the Lake of the Woods fault zone, ~25 km west of Upper Klamath Lake (Wiley et al., 1993; Braunmiller et al., 1995; Dreger et al., 1995). In 1993,

Liberty, L.M., Pratt, T.L., Lyle, M., and Madin, I.P., 2009, Neotectonic analysis of Upper Klamath Lake, Oregon: New insights from seismic reflection data, *in* Oldow, J.S., and Cashman, P.H., eds., Late Cenozoic Structure and Evolution of the Great Basin–Sierra Nevada Transition: Geological Society of America Special Paper 447, p. 71–82, doi: 10.1130/2009.2447(05).

Liberty et al.

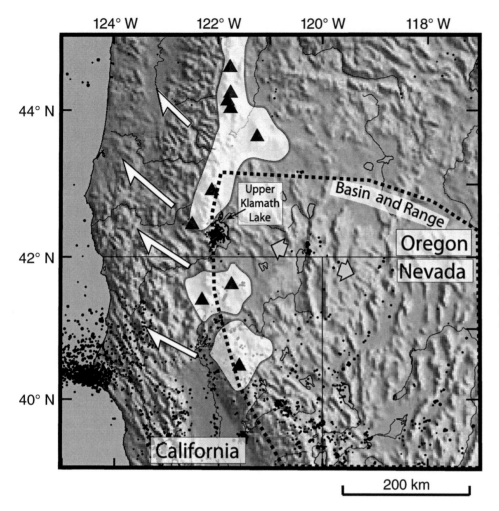

Figure 1. Regional topographic map showing the approximate boundary of the Basin and Range Province (dotted line), Quaternary Cascade arc (light region) with volcanoes (triangles), direction of rotating Oregon block and Basin and Range extension (arrows), and seismicity since 1973 (circles). Upper Klamath Lake, southern Oregon, is located along the northwest margin of the Basin and Range.

M5.9 and M6.0 main shocks were followed by hundreds of aftershocks along a north-northwest–striking, east-dipping normal fault system at 5–12 km depths. Although no surface displacement resulted from these events, surface cracking, landslides, and considerable damage to the nearby city of Klamath Falls did occur (Fig. 2; Sherrod, 1993; Wiley et al., 1993).

In addition to active faulting, the Klamath Lake area has experienced a complex late Quaternary climate history as indicated by marked changes in sedimentation rates, sediment type, and sediment sources (Bradbury et al., 2004; Colman et al., 2004; Rosenbaum and Reynolds, 2004). These changing climate conditions are critical to understanding the neotectonic history of the basin (e.g., estimating past slip rates), and, conversely, the neotectonic history is important to understanding basin evolution and past lake conditions.

We collected more than 200 km of seismic reflection profiles from the very shallow Upper Klamath Lake to map Neogene and younger strata, identify and characterize hidden faults, and measure vertical slip rates. Our findings suggest that fluctuating lake levels within Upper Klamath Lake, combined with changing climate, alternating sediment sources, and active tec-

tonics, have led to a complex basin evolution. These conditions lead to a basin where sedimentation rates have been largely controlled by subsidence. We imaged a large number of faults, and here we describe the depositional environment and estimate Quaternary slip rates. Our analysis suggests that the Quaternary slip rate for the Upper Klamath Lake region is less than that calculated from Holocene deposits, yet the quantity of faults is significantly greater than geologic mapping may suggest. However, complex depositional conditions, large uncertainties in depositional rates, and limitations of seismic imaging capabilities do not allow us to constrain Quaternary slip rates beyond first-order estimates.

GEOLOGIC SETTING

A series of normal faults in southwestern Oregon along the western margin of the Modoc Plateau comprises the Klamath graben fault system (Figs. 1 and 2). Faults are mapped every few kilometers throughout the fault system, which extends more than 50 km (MacDonald, 1966; McKee et al., 1983; Bacon et al., 1999; Personius, 2002). The area lies in the transition

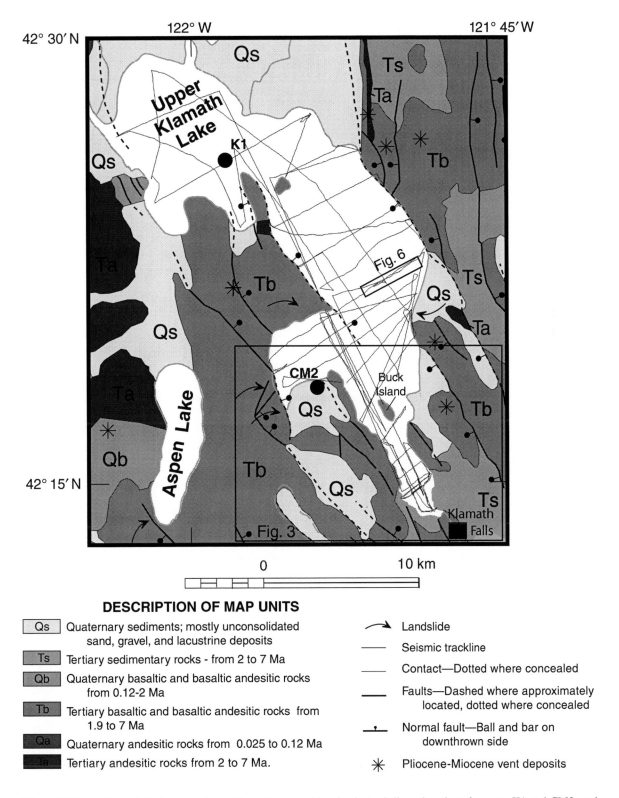

DESCRIPTION OF MAP UNITS

Qs	Quaternary sediments; mostly unconsolidated sand, gravel, and lacustrine deposits
Ts	Tertiary sedimentary rocks - from 2 to 7 Ma
Qb	Quaternary basaltic and basaltic andesitic rocks from 0.12-2 Ma
Tb	Tertiary basaltic and basaltic andesitic rocks from 1.9 to 7 Ma
Qa	Quaternary andesitic rocks from 0.025 to 0.12 Ma
Ta	Tertiary andesitic rocks from 2 to 7 Ma.

Landslide

Seismic trackline

Contact—Dotted where concealed

Faults—Dashed where approximately located, dotted where concealed

Normal fault—Ball and bar on downthrown side

Pliocene-Miocene vent deposits

Figure 2. Upper Klamath Lake geologic and tectonic map with seismic track lines, locations for cores K1 and CM2, and locations of seismic data presented in Figures 3–6. The box located in southeast corner appears on Figure 3.

between zones of dominantly north-striking normal faults of the Cascade volcanic arc and north-northwest–striking (N10W to N50W) normal faults of the Basin and Range Province of southwest Oregon. Using regional gravity data, Blakely et al. (1997) suggested that magmatism caused by thermal expansion of the Cascade volcanic arc resulted in a thermally weakened crust and east-west extension, accompanied by volcanic vents and north-south–oriented faults. Where amagmatic conditions exist, the region has responded to regional stresses associated with right-lateral shear strain of the Walker Lane belt (e.g., Oldow, 2003). The Klamath graben represents a 1500 km² area that is seismically active, and it extends along a series of sub-parallel faults typical of the Modoc Plateau portion of the Basin and Range Province (MacDonald, 1966).

Sherrod and Pickthorn (1992) identified probable Holocene faulting in and around Upper Klamath Lake based on steep fault scarps that cut late Pliocene volcanic rocks. Based on late Quaternary measured vertical slip rates of 0.3 mm/yr and estimated fault lengths from offset of ca. 35–300 ka lava flows, Bacon et al. (1999) estimated regional recurrence intervals for M7 earthquakes at 3–7 k.y. Hawkins et al. (1989) measured slightly smaller slip rates of 0.17 mm/yr from vertical displacements of nearby late Quaternary and early Holocene glacial and fluvial deposits.

Faults beneath Upper Klamath Lake have been previously imaged using a high-resolution 3.5 kHz seismic system (Colman et al., 2000). Offsets of the 7550 calendar yr B.P. Mazama tephra and younger layers indicate growth faulting and vertical slip rates of up to 0.43 mm/yr. This high slip rate and variable throw on layers of different ages suggest that multiple events within the past 7 k.y. have activated the faults below the lake floor. Measured vertical slip rates on many nearby faults on land are similar to Basin and Range estimates of ~0.1–0.4 mm/yr (e.g., Bell et al., 2004). However, these slip rates and estimates of extension are anomalously low when compared to geodetic measurements (e.g., Savage et al., 1995; Martinez et al., 1998; Thatcher et al., 1999; Bennett et al., 2003; Hammond and Thatcher, 2004), and many studies suggest a tremendous spatial and temporal variation of slip rates throughout the Basin and Range Province (e.g., Wallace, 1984; Oldow, 1992; Hammond and Thatcher, 2004; Hetzel and Hampel, 2005; Personius and Mahan, 2005).

Open lake, diatomaceous soft sediments dominate cores recovered from the center of Upper Klamath Lake, and numerous dated ash layers, including the Mazama tephra, constrain deposition rates (Colman et al., 2004). The lake-bottom sediments are mostly Quaternary fluvial and lacustrine sediments (Sherrod and Pickthorn, 1992). Holocene sediment accumulation rates have been measured to be 1.0 mm/yr at core site K1 in the north end of the lake and 0.5 mm/yr at core site CM2 on the west margin of the lake (Fig. 2; Colman et al., 2004). Interbedded and massive basalt ranging in age from 1.9 to 6.9 Ma lies adjacent to, and beneath, the lake sediments (Sherrod and Pickthorn, 1992) and acts as acoustic basement for our high-resolution seismic studies.

SEISMIC REFLECTION SURVEYS

We acquired digital seismic reflection data using both uniboom and airgun high-resolution sources (Figs. 3–6). We triggered the uniboom seismic source every 0.5 s and recorded the returns with a 9 m oil-filled streamer containing 12 equally spaced elements combined into a single channel. Spatial sampling for the uniboom data ranged from 0.5 to 1.0 m, depending on boat speed. The uniboom seismic source produces ~300 J of energy with frequencies up to 1500 Hz and signal penetration to ~100 m depth. We deployed the airgun source with a 12 channel, single-element, solid-state hydrophone string. Hydrophones were spaced at 3 m, with a 3 m minimum source-receiver offset. The airgun source was equipped with interchangeable 41, 82, and 164 cm³ air chambers, equivalent to 4000, 6800, and 13,500 J of energy (McQuillin et al., 1979). During our survey, we primarily used 82 and 164 cm³ air guns pressured from 1900 to 2000 psi. We fired the air gun every 4 s, for a shot interval of ~6 m at our boat speed. We recorded frequencies of 800 Hz or more and signal penetration to more than 200 m depth.

The uniboom seismic data consist of single-channel records that required minimal signal processing. Processing consisted of a true amplitude recovery and predictive deconvolution to attenuate water-bottom multiples. We stacked the airgun data using 3 m common midpoint (CMP) spacing to produce 6-fold data. Processing of the airgun seismic data included: geometry, bandpass filter (50–1000 Hz), spiking deconvolution, normal moveout (NMO) velocity corrections, CMP stack, automatic gain correction, and time-depth conversion based on stacking velocities and core information (Colman et al., 2000, 2004). Both the uniboom and airgun data were migrated using a phase-shift migration method (Gazdag, 1978).

The sublake stratigraphy and structural style of the region are clear on many of our seismic profiles, especially in the southern portions of the lake. However, the geology and data quality vary considerably over a surprisingly small area. Where seismic energy penetrated the lake floor, we observe clear images that suggest complex structures and stratigraphy. Although the soft Quaternary sediments beneath the lake often contain ideal seismic reflectors, a blue-green algal mat dominates much of the lake floor and traps gas near the water bottom (Sanville et al., 1974). This trapped gas creates a strong lake-floor impedance contrast that severely inhibits seismic energy from reaching deeper targets. Colman et al. (2000) mapped the gassy regions of the lake and documented where 3.5 kHz seismic energy penetrated the lake bottom. We surveyed throughout the lake using both the airgun and uniboom seismic sources (Fig. 2) and found a pattern of seismic penetration similar to that of Colman et al. (2000). We therefore concentrate our discussion on the central and southern portions of Upper Klamath Lake (Fig. 2) where Colman et al. (2000) clearly identified the Mazama tephra reflector from 3.5 kHz records and where deeper uniboom and airgun images were obtained to estimate slip rates and characterize the style of faulting and depositional patterns.

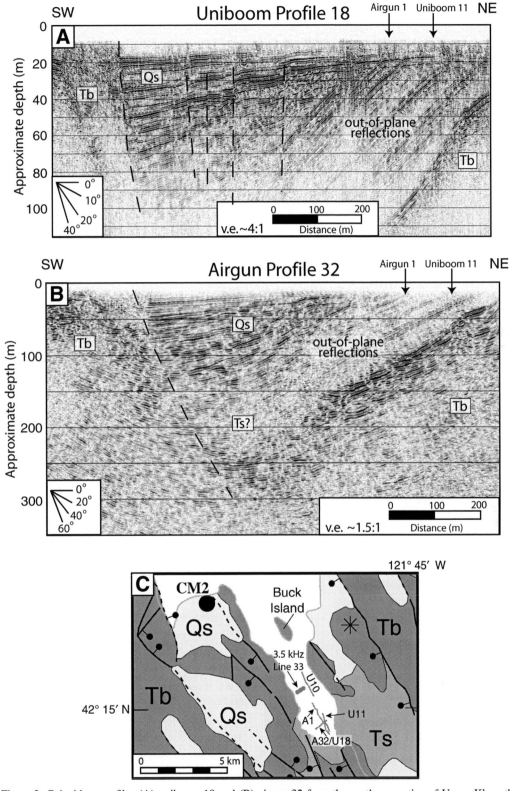

Figure 3. Coincident profiles (A) uniboom 18 and (B) airgun 32 from the southern portion of Upper Klamath Lake (Fig. 2). These data show a graben controlled by northwest-striking normal faults (dashed lines). The west side of the graben is presumably formed by a 60° northeast-dipping normal fault that may connect, at depth, with southwest-dipping normal faults within the basin. Note the steep southwest-dipping out-of-plane arrivals from south-dipping basalt bedrock (see Fig. 4). (C) Geologic map of seismic profiles in southern Upper Klamath Lake. v.e.—vertical exaggeration.

Uniboom Profile 11

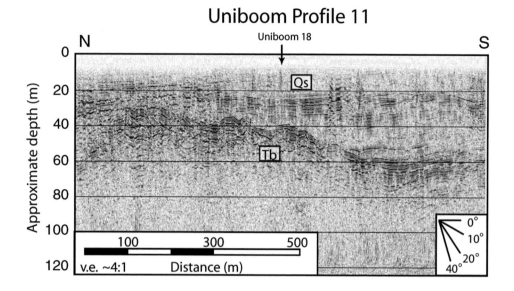

Airgun Profile 1

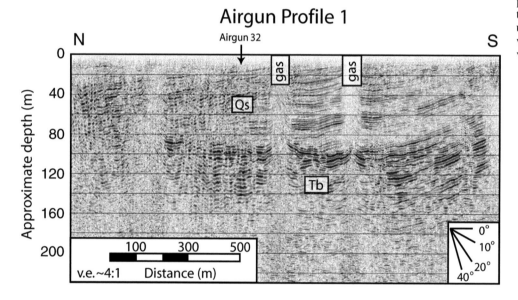

Figure 4. Uniboom profile 11 and airgun profile 1 from the southern portions of Upper Klamath Lake (Fig. 2). These parallel profiles cross uniboom profile 18 and airgun profile 32 (Fig. 3) and show gently north-dipping strata and angular unconformities within a small basin with complex bedrock geometry. v.e.—vertical exaggeration.

SEDIMENTATION RATES

Upper Klamath Lake is shallow (3–6 m depth), fills much of the Klamath graben, and is strongly influenced by changing climatic and tectonic conditions (Sherrod and Pickthorn, 1992; Colman et al., 2004). The lake has drained to the south into the Klamath River since at least late Pliocene or early Pleistocene (Donnelly-Nolan and Nolan, 1986). Historically, the Klamath graben was filled with marshland areas that seasonally changed in area as the lake levels rose and fell. The lake level is now controlled by a dam and kept near its minimum predam level, and sediment has accumulated at a rate of 1.5–2.0 mm/yr over the past ~100 yr (Colman et al., 2004). This sedimentation rate is much faster than longer-term averages, which may be due in part to increased sediment influx when some of the marshes that fringed the lake were drained (Bradbury et al., 2004). The lake floor appears undisturbed by faulting, suggesting a relatively rapid modern deposition rate or erosion rate with respect to vertical slip rates (Colman et al., 2000).

Estimates of deposition rates are based on two cores in the north and west portions of the lake (K1 and CM2 in Fig. 2), from which a chronology has been deduced to 45,000 yr B.P. (Colman et al., 2004). Late Holocene sedimentation rates have been measured by radiometric, tephrochronology, and $^{210}Pb/^{137}Cs$ measurements at 1.0 (K1) and 0.5 (CM2) mm/yr, while early to middle Holocene rates have been measured at roughly half of the late Holocene values (Colman et al., 2004; Rosenbaum and Reynolds, 2004). Changes in the late Pleistocene deposition

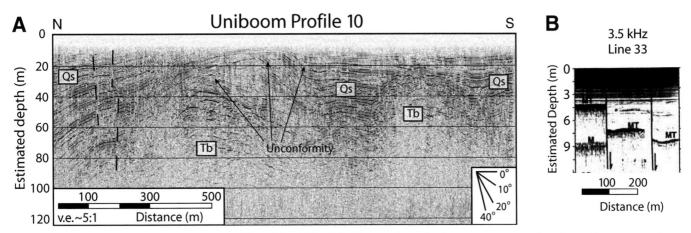

Figure 5. (A) Uniboom profile 10, located south of Buck Island (Fig. 3C). The complex reflection patterns suggest changing sediment source directions, faulting, and subaerial exposures. (B) 3.5 kHz line 33 showing deposition rates and offset above Mazama tephra (MT). Mazama tephra offsets on line 33 document the largest vertical slip that Colman et al. (2000) identified beneath the lake. M—multiple; v.e.—vertical exaggeration.

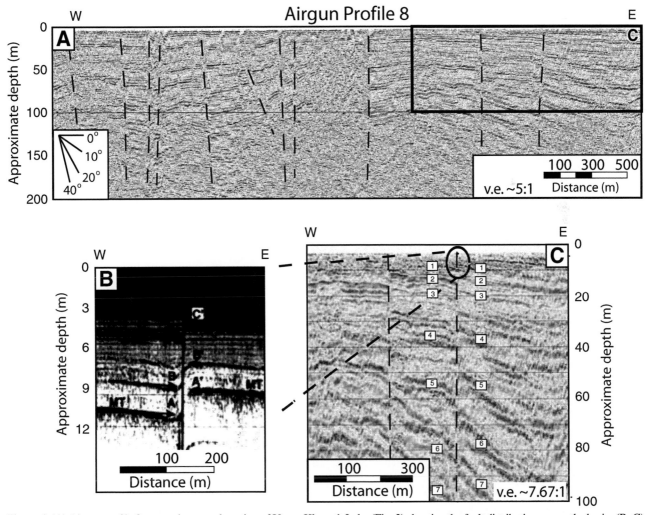

Figure 6. (A) Airgun profile 8 across the central portion of Upper Klamath Lake (Fig. 2) showing the fault distribution across the basin. (B–C) Expanded airgun profile 8 with the circled location and adjacent image of 3.5 kHz line 23 (Colman et al., 2000). Note that the offsets documented on line 23 match the offsets seen at similar depths on the airgun profile (see Fig. 7). Increased offsets on deeper reflections (numbers) document a growth fault pattern. MT—Mazama tephra; A,B,C—3.5 kHz reflectors; numbers 1–7—airgun reflectors. v.e.—vertical exaggeration.

rate correlate with glacial cycles, in which wetter interglacial periods caused increased deposition rates due to increased sediment load (Rosenbaum and Reynolds, 2004).

The cores provide sedimentation rates at only two locations, but a sedimentation rate can also be inferred by the thickness of sediment that has accumulated since the 7540 B.P. Mazama tephra was deposited throughout the lake. The tephra appears as a strong reflector in 3.5 kHz data sets (Colman et al., 2000). In the central lake basin, there is ~3.5 m of sediment overlying the tephra (Colman et al., 2000), which implies a sedimentation rate of ~0.46 mm/yr, similar to core CM1 along the west margin of the lake. In the southern and northern lake basins, the seismic sections shown in Colman et al. (2000) show 6–8.5 m of strata above the tephra, which requires sedimentation rates of 0.8–1.1 mm/yr. This latter rate is consistent with the 1 mm/yr rate derived from the core in the north part of the lake (Colman et al., 2004).

The seismic sections shown in Colman et al. (2000) suggest that sediment accumulation rates are not determined only by the sediment supply, however. The seismic sections show faults with up to 3.1 m of displacement on the tephra, yet the overlying lake bottom is flat (Fig. 4). The lack of water-bottom relief above the faults suggests that there is ample sediment supply to quickly fill the depressions and hide the scarps, and that another factor likely determines the highest elevation of the sediment accumulation. The obvious culprit is the water level, in which the top of the sediments is determined by the lowest lake levels, with wave erosion or wind-driven currents and waves removing sediments above this level (Laenen and LeTourneau, 1996). This erosion would create the flat lake bottom evident on the seismic profiles, it would explain the numerous unconformities observed on our profiles, and it would explain the different thicknesses of sediment above the Mazama tephra.

If the sedimentation rate is controlled by the lake level, the total sediment thickness must be a measure of the subsidence rate rather than the sedimentation rate. Indeed, Bradbury et al. (2004) suggested that long-term sedimentation in the lake has been nearly balanced by subsidence of the basin. This implies that the post-Mazama 0.5–1.1 mm/yr sediment accumulation reflects a Holocene subsidence rate rather than a sedimentation rate.

BASIN STRATIGRAPHY AND STRUCTURE

Southern Upper Klamath Lake

Seismic reflection data and geologic mapping from Upper Klamath Lake suggest that Tertiary and younger sediments fill fault-bounded subbasins within the active Klamath graben (Figs. 3–6; Sherrod and Pickthorn, 1992; Personius, 2002). South of Buck Island (Fig. 2), we identified and characterized reflections to depths greater than 200 m on four seismic profiles. Exposures of Tertiary basalt appear along the adjacent lake shore and are dated as young as 2.3 Ma (Sherrod and Pickthorn, 1992). Mapped normal faults project into the lake and trend approximately N20W to N50W (Fig. 2).

Coincident N60E-oriented uniboom profile 18 and airgun profile 32 show coherent southwest-dipping, laterally discontinuous reflections to depths of ~250 m (Fig. 3). Reflector dips increase with increasing depths on both profiles. The southwest-dipping reflection package terminates downward at a strong-amplitude basal reflection package that ranges from ~30 m to 250 m depth. Near the southwest end of these profiles, the subparallel reflections laterally terminate. The northeast portions of the profiles contain anomalous, steeply dipping reflectors that cut shallower dipping reflectors and do not migrate predictably.

We interpret reflectors on uniboom profile 18 and airgun profile 32 to represent a package of southwest-dipping lake sediments contained within an asymmetric fault-bounded half graben (Fig. 3). Reflector dips range from approximately 1° at 10 m depth to 22° on the basement horizon. Reflector discontinuities that offset basin strata likely define steeply dipping normal faults that postdate the Pliocene basalt emplacement. The southwestern boundary of this small graben appears to be formed by a steep northeast-dipping normal fault, bounded to the southwest by a bedrock ridge. We identified faults within the basin about every 100 m across the width of the graben, and we measured a trend of N56W on the basin-bounding fault from parallel seismic profiles (Fig. 2). The steeply dipping reflectors that appear to cut basin strata along the eastern portions of the profiles are best explained by out-of-plane basement reflectors in the northward shallowing part of the graben (see Fig. 4).

Reflectors in the upper 20–30 m drape over the mapped faults on uniboom profile 18, whereas deeper basin strata show clear fault offsets. This deformation pattern and the increasing dip on basin strata are characteristics of growth faulting, in which the faults have been active throughout basin formation. The lack of obvious offsets on strata in the upper 20 m indicates relatively long recurrence times for earthquakes on these faults relative to deposition rates. The steeply dipping faults dip toward the central portion of the graben and may connect with the border fault at depths greater than imaged on our profiles. Assuming a relatively small component of strike-slip motion typical of Basin and Range extension, a single master fault likely controls this graben's evolution. Assuming that the master fault has vertically offset estimated 2.5 Ma basalt by ~250 m, the long-term vertical slip rate is 0.1 mm/yr. The estimated 250 m of strata above Pliocene bedrock, combined with the measured Holocene sedimentation rates from lake cores (Colman et al., 2004), suggest that the deepest sediments in the subbasin are as old as 250,000–500,000 yr. Assuming slower Pleistocene sedimentation rates (Colman et al., 2004; Rosenbaum and Reynolds, 2004), the deepest sediments may be older. However, a more transparent reflection zone at the deeper portions of the graben above bedrock may represent a change in lithology. Tertiary sedimentary rocks overlie Pliocene basalt immediately south of Upper Klamath Lake (Fig. 2; Sherrod and Pickthorn, 1992) and may overlie bedrock beneath the lake. If Tertiary strata overlie older basalt at the base of the graben, and faulting has been active since basalt emplacement, it implies that sedimentation rates have fluctuated greatly, large portions of the basin strata have been removed through erosion, or that sediments were not deposited during portions of the Quaternary history.

The N20W-oriented uniboom profile 11 in southern Upper Klamath Lake shows a sequence of predominantly flat-lying to gently northwest-dipping reflectors unconformably deposited upon a rugged, high-amplitude reflector at 30–60 m depth (Fig. 4). Uniboom profile 11 crosses the profiles shown in Figure 3 near the eastern boundary of the basin, and the rugged basement reflector topography supports our earlier interpretation of out-of-plane arrivals as the cause of the steeply dipping events on uniboom 18 (Fig. 3). The 82 cm³ airgun profile 1 is located parallel to, and ~100 m west of, uniboom profile 11 (Fig. 3C). Airgun profile 1 shows continuous reflections to 100 m depth and the characteristic high-amplitude reflection package below (Fig. 4).

We interpret the top of Pliocene basalt as acoustic basement on both uniboom profile 11 and airgun profile 1 at depths of 30–100 m (Fig. 4). These depths are consistent with the Figure 3 profiles, which cross near the centers of uniboom profile 11 and airgun profile 1. We observe angular unconformities in the central portion of the basin on both airgun and uniboom profiles that suggest either a hiatus in past sedimentation or erosion prior to subsequent deposition. The increase in reflector dip with shallowing depths may be depositional or tectonic in origin. If the dip on Quaternary strata noted on all four seismic profiles from southern Upper Klamath Lake is, in part, depositional in origin, a sediment supply from the east or southeast is required, contrasting with present-day conditions, where sediments are derived mostly from the north (Donnelly-Nolan and Nolan, 1986). Improved age control on these sediments may better constrain the basin's evolution.

Farther north, but still within the southern portions of Upper Klamath Lake, we show uniboom profile 10 to highlight the complex nature of the structures and stratigraphy that dominate the Klamath graben system (Fig. 5). This N20W-oriented profile shows a large variation in the attitudes of strata in the upper 100 m depth. Along the northernmost portions of the profile, we observe a package of more than 80 m of north-dipping, laterally discontinuous reflectors, while reflectors farther south either dip to the south or are flat-lying.

Reflector discontinuities along the northern portion of the profile suggest that near-vertical normal faults disrupt the basin strata. We cannot trace individual faults between our coarsely spaced parallel profiles, but 3.5 kHz mapping by Colman et al. (2000) suggested a N60W fault orientation for these faults, and we do not see evidence to the contrary. However, this fault orientation is not prominent on geologic or hazard maps (Sherrod and Pickthorn, 1992; Personius, 2002). An increase in reflector offset with depth across these faults indicates growth faulting or changes in sedimentation rate through time.

Near the center of uniboom profile 10, reflectors above ~10 m depth unconformably overlie deeper strata (Fig. 5). We map other angular unconformities noted on Figure 5. These unconfor-mities suggest that sediment ages based on constant deposition rates underestimate the age of deeper sediments because of missing strata. Reflectors immediately below the 10-m-deep unconformity are draped over deeper strata in a foreset pattern that may be explained by a southward-prograding delta sequence. The deposition rate in the north part of the lake, if applicable here, suggests that sediments in the upper 10 m are Holocene strata. The unconformity beneath these shallow sediments may correspond to a change in Holocene sedimentation rates, as documented by Colman et al. (2004), perhaps related to increased wetness of the climate in the late Holocene. Water depths that average 3 m south of Buck Island may support subaerial conditions and perhaps erosion during drier periods, while wetter time periods resulted in deposition of lacustrine sediments. Thus, sedimentation rates and the nature of the unconformities could hold significant information regarding past climate for the region surrounding Upper Klamath Lake if deeper cores are collected to determine the sediment chronology.

A high-amplitude, arcuate reflector at depths of 35–60 m near the center of uniboom profile 10 likely represents a buried bedrock surface. Draping and tilting of sediment over this surface produce opposing reflector dips that suggest differing Quaternary fault motion (Fig. 5). Along the southern portion of uniboom profile 10, flat-lying reflectors surround a seismically transparent arch, the appearance of which is typical of a buried bedrock surface. Reflectors within the basin strata unconformably lap onto on this bedrock surface, suggesting that this bedrock surface is not bounded by active faults.

Central Upper Klamath Lake

Airgun profile 8 crosses the central portion of Upper Klamath Lake (Fig. 2) and is nearly coincident with a 3.5 kHz seismic profile that shows Mazama tephra and younger sediments offset as much as 1.9 m across a growth fault (Fig. 5; Colman et al., 2000). The east-west–oriented 164 cm³ airgun profile shows strong, discontinuous reflectors in the upper 200 m across the width of the basin and no evidence for a strong reflection from acoustic basement (Fig. 6). As observed by Colman et al. (2000) in the upper 10 m, we see increasing reflector offsets to depths of more than 100 m.

We interpret the central part of Upper Klamath Lake as containing a thick sequence of sediments that are deformed into a broad arch (Fig. 6). The arched shape to the sedimentary sequence suggests two opposing subbasins with border faults near each side of the lake (Sherrod and Pickthorn, 1992). Colman et al. (2000) showed a northwest trend to these faults, which is consistent with our profiles. We identified many normal faults cutting the strata, and increasing offset with depth suggests growth faulting. Although reflectors are discontinuous across the faults, we can nonetheless trace specific marker horizons across much of the profile (Fig. 6C).

If we assume a constant sediment accumulation rate through time for central Upper Klamath Lake, offsets of deeper strata indicate smaller slip rates than offsets across the Mazama tephra and younger sediments (Fig. 7). We measured long-term, late Quaternary slip rates on this fault at 0.14 mm/yr when incorporating deeper strata and assuming a sedimentation rate of 1.0 mm/yr, as opposed to 0.25 mm/yr when looking only at the Mazama tephra and overlying units. Slower sedimentation rates for late Pleistocene and early Holocene drier periods (Colman et al., 2004) would further reduce long-term slip-rate estimates. Unconformities evident in some seismic profiles also suggest that substantial periods of time are not represented by strata, further decreasing the calculated slip rates. An anomalously high 2.0 mm/yr long-term deposition rate would be required to match post-Mazama slip rates. Nearby cores (Colman et al., 2004), post-Mazama deposition rates (Colman et al., 2000), and discontinuous deposition do not support such a high sedimentation rate in the basin center. These calculations (Fig. 7) suggest that late Holocene slip rates documented by Colman et al. (2000) are greater than the longer-term late Quaternary slip rates, given any reasonable rate of deposition. A reduced longer-term slip rate on these faults suggests Upper Klamath Lake may be more consistent with other documented vertical slip rates for the region (Bacon et al., 1999; Hawkins et al., 1989) and for Basin and Range structures in general (e.g., Bell et al., 2004).

Although different measured offsets across faults suggest differing slip rates for Upper Klamath Lake faults, these

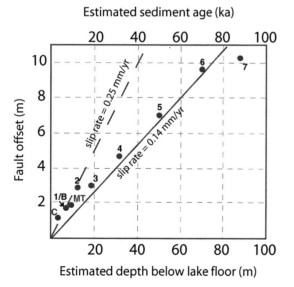

Figure 7. Vertical slip rates measured from 3.5 kHz line 23 (letters) and airgun profile 8 (numbers) based on an estimated deposition rate of 1.0 mm/yr (Fig. 6). Note the increased slip rate for late Holocene sediments (letters) compared to late Pliostocene strata (numbers), perhaps due to the young age of the Mazama tephra (MT) not reflecting a long-term slip rate because of the 3–7 k.y. recurrence intervals of large earthquakes in the region.

seismic profiles show that the quantity of faults present below Upper Klamath Lake is significantly greater than those shown on geologic and tectonic maps (Sherrod and Pickthorn, 1992; Personius, 2002). Slip rates on individual fault strands vary significantly across the lake, yet these faults likely connect at depth to a few controlling master faults. Additionally, slip rates likely vary through time, and measurements here suggest that late Holocene rates may be anomalously high. This may suggest an increase in late Holocene seismicity, but it could also represent a short-term anomaly measured on a few faults. Increased rates of Holocene fault activity within the Basin and Range have been linked to glacial unloading and lithospheric rebound (Hetzel and Hampel, 2005). In contrast, anomalously low late Holocene slip rates have been observed across the nearby Santa Rosa Range fault zone (e.g., Personius and Mahan, 2005).

DISCUSSION

Although we derive slip rates in Upper Klamath Lake close to normal Basin and Range estimates, all of these rate estimates have potentially large errors and are based on only a few measurements. We mentioned the potential error due to changes in the sedimentation rate through time, but there are two other potential sources of error. First, the analyses are based on the assumption of dip-slip motion on the faults, but even a small component of strike-slip motion could significantly increase slip rates.

A second, less obvious source of error in Holocene slip-rate estimates is the use of the relatively young Mazama tephra as a marker for measuring slip rates. The 7550 yr age of the Mazama tephra is near the recurrence time for earthquakes on nearby fault systems (Bacon et al., 1999), so the offsets we see on the Mazama tephra may not reflect a long-term average. Many of the active faults may not yet have ruptured the tephra, while at least one fault has evidence for three earthquakes since the tephra was deposited (Colman et al., 2000). The timing of earthquakes with respect to the deposition of the tephra creates large uncertainties in calculated slip rates. For example, a fault with a 6500 yr recurrence interval could have had two earthquakes in the past 7500 yr, which would result in an artificially high slip rate consistent with a 3750 yr recurrence interval (two events in 7500 yr). The distribution of earthquakes in time and across faults could therefore have a dramatic affect on slip rates estimated from displacements on the Mazama tephra alone.

The latter point may explain the observation that there are many more faults in the pre-Holocene section than there are cutting the Mazama tephra. If these faults have a recurrence interval greater than 7500 yr, many of them could be potentially active faults despite the lack of displacement on the tephra. Likewise, the apparent draping of sediments across many of the faults we imaged (e.g., Fig. 3) could be indicating long recurrence intervals, where slip is widely distributed across a large number of faults. Slip appears artificially concentrated on a fewer number of faults in the Holocene (post-Mazama) because of the small number of Holocene earthquakes.

Our interpretations have several implications for earthquake hazard assessments in the region. First, the half graben imaged on the profiles (Fig. 3) confirms that the lake is formed by extension and therefore appears to be part of the Basin and Range Province. Second, the profiles show us that displacements, and therefore measured vertical slip rates, can vary widely on sets of faults in the region. Seismic hazard assessments must therefore consider the fault density or cumulative slip across subsurface faults as well as surface ruptures. The identification of an order-of-magnitude greater subsurface fault density compared to that inferred from surface mapping may provide an explanation for the disparities between estimates of extension based on geodetic measurements (e.g., Thatcher et al., 1999; Bennett et al., 2003) and measurements based on fault slip. Slip-rate estimates based on a single fault seem to be of limited value; it is only by integrating displacements on all the faults in the region that a reliable extension rate can be computed. This could most easily be done using a more extensive set of seismic reflection profiles coupled with additional borehole information to allow for the reconstruction of horizons of different ages. Until such an exercise is carried out to measure the average slip over long time periods, the slip rates computed from these studies will be influenced by the incomplete sampling only of faults reaching the surface or cutting the Mazama tephra.

CONCLUSIONS

Our seismic results show a history of sedimentation and faulting that suggests that Upper Klamath Lake is situated in a complex, extending basin with numerous, steeply dipping normal faults. Upper Klamath Lake appears to be a sediment-saturated lake that has a history of varying sediment sources and subaerial exposures, which, coupled with the active tectonics, have created complex sedimentologic and tectonic structures. Our data suggest that the 0.43 mm/yr maximum vertical slip rate computed on post-Mazama strata using previous 3.5 kHz seismic data overestimates the long-term slip rate. The style of faulting and the slip rates vary across the basin, but measured slip suggests that the basin is extending at a comparable rate to other Basin and Range structures. Growth faulting appears to have been active throughout the late Quaternary.

ACKNOWLEDGMENTS

We would like to thank the U.S. Bureau of Recreation for the use of their boat, and Greg Olsen from the U.S. Geological Survey for expertly piloting it. Also, we would like to thank Frank Hladky, Tom Wiley, Margi Jenks, and Jerry Black for assisting in the fieldwork. Dick Sylwester of Golder and Associates provided some of the seismic equipment, and other equipment came from the University of Washington's School of Oceanography. Partial funding was provided by the Oregon Department of Geology and Mineral Industries. The manuscript has benefited from reviews by Joseph Rosenbaum, David Dinter, Derek Lerch, and Simon Klemperer.

REFERENCES CITED

Bacon, C.R., Lanphere, M.A., and Champion, D.E., 1999, Late Quaternary slip rate and seismic hazards of the West Klamath Lake fault zone near Crater Lake, Oregon Cascades: Geology, v. 27, no. 1, p. 43–46, doi: 10.1130/0091-7613(1999)027<0043:LQSRAS>2.3.CO;2.

Bell, J.W., Caskey, S.J., Ramelli, A.R., and Guerrieri, L., 2004, Pattern and rates of faulting in the central Nevada seismic belt, and paleoseismic evidence for prior beltlike behavior: Bulletin of the Seismological Society of America, v. 94, no. 4, p. 1229–1254, doi: 10.1785/012003226.

Bennett, R.A., Wernicke, B.P., Niemi, N.A., Friedrich, A.M., and Davis, J.L., 2003, Contemporary strain rates in the northern Basin and Range Province from GPS data: Tectonics, v. 22, no. 2, p. 1008, doi: 10.1029/2001TC001355.

Blakely, R.J., Christiansen, R.L., Guffanti, M., Wells, R.E., Donnelly-Nolan, J.M., Muffler, L.J.P., Clynne, M.A., and Smith, J.G., 1997, Gravity anomalies, Quaternary vents, and Quaternary faults in the southern Cascade Range, Oregon and California; implications for arc and backarc evolution: Journal of Geophysical Research, v. 102, no. B10, p. 22,513–22,527, doi: 10.1029/97JB01516.

Bradbury, J.P., Colman, S.M., and Dean, W.E., 2004, Limnologic and climatic environments at Upper Klamath Lake, Oregon, during the past 45,000 years: Journal of Paleolimnology, v. 31, no. 2, p. 151–165, doi: 10.1023/B:JOPL.0000019233.12287.18.

Braunmiller, J., Nabelek, J., Leitner, B., and Qamar, A., 1995, The 1993 Klamath Falls, Oregon, earthquake sequence; source mechanisms from regional data: Geophysical Research Letters, v. 22, no. 2, p. 105–108, doi: 10.1029/94GL02844.

Colman, S.M., Rosenbaum, J.G., Reynolds, R.L., and Sarna-Wojcicki, A.M., 2000, Post-Mazama (7 ka) faulting beneath Upper Klamath Lake, Oregon: Bulletin of the Seismological Society of America, v. 90, no. 1, p. 243–247, doi: 10.1785/0119990033.

Colman, S.M., Bradbury, J.P., McGeehin, J.P., Holmes, C.W., Edginton, D., and Sarna-Wojcicki, A.M., 2004, Chronology of sediment deposition in Upper Klamath Lake, Oregon: Journal of Paleolimnology, v. 31, no. 2, p. 139–149, doi: 10.1023/B:JOPL.0000019234.05899.ea.

Donnelly-Nolan, J.M., and Nolan, K.M., 1986, Catastrophic flooding and eruption of ash-flow tuff at Medicine Lake Volcano, California: Geology, v. 14, no. 10, p. 875–878, doi: 10.1130/0091-7613(1986)14<875:CFAEOA>2.0.CO;2.

Dreger, D., Ritsema, J., and Pasyanos, M., 1995, Broadband analysis of the 21 September, 1993, Klamath Falls earthquake sequence: Geophysical Research Letters, v. 22, no. 8, p. 997–1000, doi: 10.1029/95GL00566.

Gazdag, J., 1978, Wave equation migration with the phase-shift method: Geophysics, v. 43, no. 7, p. 1342–1351, doi: 10.1190/1.1440899.

Hammond, W.C., and Thatcher, W., 2004, Contemporary tectonic deformation of the Basin and Range Province, western United States: 10 years of observation with the global positioning system: Journal of Geophysical Research, v. 109, p. B08403, doi: 10.1029/2003JB002746.

Hawkins, F.F., Foley, L.L., and LaForge, R.C., 1989, Seismotectonic study for Fish Lake and Fourmile Lake dams, Rogue River Basin Project, Oregon: Denver, Colorado, U.S. Bureau of Reclamation Seismotectonic Report 89-3, 26 p.

Hetzel, R., and Hampel, A., 2005, Slip rate variations on normal faults during glacial-interglacial changes in surface loads: Nature, v. 435, p. 81–84, doi: 10.1038/nature03562.

Laenen, A., and LeTourneau, A.P., 1996, Upper Klamath Lake Nutrient Loading Study—Estimate of Wind-Induced Resuspension of Bed Sediment during Periods of Low Lake Elevation: U.S. Geological Survey Open-File Report 95-414, 11 p.

MacDonald, G.A., 1966, Geology of the Cascade Range and Modoc Plateau, *in* Bailey, E.H., ed., Geology of Northern California: California Division of Mines and Geology Bulletin 190, p. 65–96.

Martinez, L.J., Meertens, C.M., and Smith, R.B., 1998, Rapid deformation rates along the Wasatch fault zone, Utah, from first GPS measurements with implications for earthquake hazard: Geophysical Research Letters, v. 25, no. 4, p. 567–570, doi: 10.1029/98GL00090.

McKee, E.H., Duffield, W.A., and Stern, R.J., 1983, Late Miocene and early Pliocene basaltic rocks and their implications for crustal structure, northeastern California and south-central Oregon: Geological Society of America Bulletin, v. 94, p. 292–304, doi: 10.1130/0016-7606(1983)94<292:LMAEPB>2.0.CO;2.

McQuillin, R., Bacon, M., and Barclay, W., 1979, An Introduction to Seismic Interpretation: Houston, Texas, Gulf Publishing Co., 191 p.

Muffler, L.J.P., Clynne, M.A., and Champion, D.E., 1994, Late Quaternary normal faulting of the Hat Creek Basalt, northern California: Geological Society of America Bulletin, v. 106, p. 195–200.

Oldow, J.S., 1992, Late Cenozoic displacement partitioning in the northwestern Great Basin, *in* Craig, S.D., ed., Structure, Tectonics, and Mineralization of the Walker Lane: Walker Lane Symposium: Reno, Geological Society of Nevada, p. 17–52.

Oldow, J.S., 2003, Active transtensional boundary zone between the western Great Basin and Sierra Nevada block, western U.S. Cordillera: Geology, v. 31, p. 1033–1036, doi: 10.1130/G19838.1.

Personius, S.F., 2002, Fault number 843c, Klamath graben fault system, South Klamath Lake section, *in* Quaternary fault and fold database of the United States, ver. 1.0: U.S. Geological Survey Open-File Report 03-417.

Personius, S.F., and Mahan, S.A., 2005, Unusually low slip rates on the Santa Rosa fault zone, northern Nevada: Bulletin of the Seismological Society of America, v. 95, no. 1, p. 319–333, doi: 10.1785/0120040001.

Rosenbaum, J.G., and Reynolds, R.L., 2004, Record of late Pleistocene glaciation and deglaciation in the southern Cascade Range: II. Flux of glacial flour in a sediment core from Upper Klamath Lake, Oregon: Upper Klamath Lake, Oregon, USA: Journal of Paleolimnology, v. 31, no. 2, p. 235–252, doi: 10.1023/B:JOPL.0000019229.75336.7a.

Sanville, W.D., Powers, C.F., and Gahler, A.R., 1974, Sediments and sediment-water nutrient interchange in Upper Klamath Lake, Oregon: U.S. Environmental Protection Agency Report EPA-660/3-74-015, 45 p.

Savage, J.C., Lisowski, M., Svarc, J.L., and Gross, W.K., 1995, Strain accumulation across the central Nevada seismic zone, 1973–1994: Journal of Geophysical Research, v. 100, no. B10, p. 20,257–20,269, doi: 10.1029/95JB01872.

Sherrod, D.R., 1993, Historic and prehistoric earthquakes near Klamath Falls, Oregon: Earthquakes and Volcanoes, v. 24, no. 3, p. 106–120.

Sherrod, D.R., and Pickthorn, L.G., 1992, Geologic Map of the West Half of the Klamath Falls 1° × 2° Quadrangle, South-Central Oregon: U.S. Geological Survey Miscellaneous Investigations Map I-2182, scale 1:250,000.

Thatcher, W., Foulger, G.R., Julian, B.R., Svarc, J., Quilty, E., and Bawden, G.W., 1999, Present-day deformation across the Basin and Range Province, western United States: Science, v. 283, p. 1714–1718, doi: 10.1126/science.283.5408.1714.

Wallace, R.E., 1984, Patterns and timing of late Quaternary faulting in the Great Basin Province and relation to some regional tectonic features: Special section; fault behavior and the earthquake generation process: Journal of Geophysical Research, v. 89, no. B7, p. 5763–5769, doi: 10.1029/JB089iB07p05763.

Wiley, T.J., Sherrod, D.R., Keefer, D.K., Qamar, A., Schuster, R.L., Dewey, J.W., Mabey, M.A., Black, G.L., and Wells, R.E., 1993, Klamath Falls earthquakes, September 20, 1993; including the strongest quake ever measured in Oregon: Oregon Geology, v. 55, no. 6, p. 127–134.

MANUSCRIPT ACCEPTED BY THE SOCIETY 21 JULY 2008

The Geological Society of America
Special Paper 447
2009

Late Neogene basin history at Honey Lake, northeastern California: Implications for regional tectonics at 3 to 4 Ma

J.H. Trexler Jr.*
H.M. Park[†]
P.H. Cashman
K.B. Mass[§]
Department of Geological Sciences and Engineering, University of Nevada, Reno, Nevada 89577, USA

ABSTRACT

Neogene sediments in a structural and geomorphic high in the southwestern Honey Lake basin represent lacustrine deposition from 3.7 to 2.9 Ma, interrupted once by a significant lowstand. Tephras in the upper section are 3.26 Ma and 3.06 Ma. A thick debris-flow bed, truncated by an erosional surface and overlain concordantly by a thin interval of subaerial sediments, is evidence for lake-level fall at ca. 3.4 Ma. The dominant structure is a broad east-southeast–plunging anticline cut by several sets of faults. These include northwest-striking dextral and northeast-striking sinistral strike-slip faults and a conjugate set of west-northwest–striking thrust faults; all are consistent with north-south shortening. Mutually crosscutting relationships between faults, and tilt fanning of the dextral faults, indicate that tightening of the anticline was synchronous with faulting. A Quaternary strand of the dextral Honey Lake fault crops out near the northern end of the exposure, suggesting that the cause of the local shortening and uplift was a contractional stepover between two strands of the Honey Lake fault. The Neogene section limits this faulting to some time after 2.9 Ma. The Honey Lake basin lies at the intersection of the Walker Lane with the Sierran frontal fault system. Although the timing of tectonic disruption was roughly consistent with passage of the triple junction to the west and with uplift and exhumation of several nearby basins, the described deformation seems to be directly related to dextral faulting, dating the propagation of a strand of the Honey Lake fault through the southwestern Honey Lake basin.

Keywords: Neogene, Walker Lane, strike-slip tectonics, eastern Sierra Nevada.

INTRODUCTION

The Sierra Nevada–Basin and Range transition zone exhibits two different styles of intraplate deformation, east-west extension, and northwest-directed dextral slip. East-west extension, primarily middle Miocene in age, characterizes the central Basin and Range Province and has propagated westward with time into the adjacent Sierra Nevada (e.g., Dilles and Gans, 1995; Surpless et al., 2002). In contrast, extension did not start until 10–12 Ma in the northern Basin and Range (e.g., Colgan and Dumitru, 2003;

*Corresponding author e-mail: Trexler@mines.unr.edu.
[†]Now at Department of Geology, University of South Carolina, Columbia, South Carolina 29208, USA.
[§]Now at Black and Veatch Special Projects Corp., Irvine, California 92618, USA.

Trexler, J.H., Jr., Park, H.M., Cashman, P.H., and Mass, K.B., 2009, Late Neogene basin history at Honey Lake, northeastern California: Implications for regional tectonics at 3 to 4 Ma, *in* Oldow, J.S., and Cashman, P.H., eds., Late Cenozoic Structure and Evolution of the Great Basin–Sierra Nevada Transition: Geological Society of America Special Paper 447, p. 83–100, doi: 10.1130/2009.2447(06). For permission to copy, contact editing@geosociety.org. ©2009 The Geological Society of America. All rights reserved.

Lerch et al., 2004; Whitehill et al., 2004), suggesting that extension also may be propagating, or stepping, northward with time. The Sierran frontal fault system forms the prominent western topographic boundary of the Basin and Range Province, but it is not the structural boundary: young extensional faulting is occurring within the eastern part of the Sierra "block" (e.g., Hawkins et al., 1986; Mass, 2005; Mass et al., this volume). Strike-slip faulting in the Walker Lane, along the western edge of the Basin and Range Province, has occurred at several times since the initiation of the transform plate boundary (e.g., Hardyman and Oldow, 1991; Oldow, 1992; Cashman and Fontaine, 2000). The most recent dextral slip episode has been attributed to the opening of the Gulf of California and the development of the Eastern California shear zone, starting ca. 6 Ma (e.g., Oskin and Stock, 2003; Faulds et al., 2003). The northwest-trending strike-slip fault system of the Walker Lane intersects the north-trending Sierran frontal fault system just south of Honey Lake, California (Fig. 1).

Neogene sedimentary basins are present in all of these domains: the relatively coherent Sierra Nevada block, the Walker Lane, and the extended regions east and west of the Walker Lane. These basins record the time interval during which the structural domains were developing (Muntean, 2001; Park, 2004; Mass, 2005). Sedimentary facies, provenance, and paleocurrent indicators record the evolving paleogeography during deposition. Interbedded volcanic ash and fossils provide age control. Many of the sedimentary basins in the Sierra Nevada–Basin and Range transition zone are now exhumed; tilting, faulting, and folding of these sedimentary rocks record the postdepositional deformation. Comparison of individual basin histories allows us to constrain the evolution of the Sierra Nevada–Basin and Range transition zone in space and time.

The Honey Lake basin lies near the intersection of Walker Lane dextral faults and the Sierran frontal fault system, and also marks the change from primarily exhumed Neogene sedimentary basins, to the south, to still-active depositional basins on the east side of the Cascade arc, to the north. The Honey Lake basin forms a prominent gravity low (e.g., Cerón, 1992), and most of it is still an active depositional basin. However, part of the Neogene sedimentary section at Honey Lake is exposed at the southwest edge of a topographic and structural high near the south end of the basin, locally known as The Island (Figs. 2, 3). Most of the exposures occur on a small peninsula at the southern end of The Island called the South Island. We have examined these rocks for information regarding the depositional history, age of exhumation, and style of deformation at this unique area in the Sierra Nevada–Basin and Range transition zone.

In this paper we present new sedimentological and structural data from the Neogene rocks at Honey Lake and use these to interpret part of the tectonic evolution of this part of the Sierra Nevada–Basin and Range transition zone. We start with the stratigraphy, sedimentology, and age of the exposed section, the subject of an M.S. thesis by H. Park (Park, 2004; Park et al., 2004). Then we present the deformational record in the local structural high that exposes the sediments; this structural study was part of

an M.S. thesis by K. Mass (Mass et al., 2003; Mass, 2005). We conclude with the implications of our results for the style and timing of intraplate deformation in the region.

BACKGROUND

Tectonic Setting

Several important tectonic events in Miocene–Pliocene time influenced the geologic evolution of the Sierra Nevada–Basin and Range transition zone at the latitude of the northern Sierra. At the plate margin the triple junction migrated northward, accompanied by the cessation of subduction and Cascadian volcanism, and changes in the sublithospheric mantle; it passed the latitude of Reno, Nevada, ca. 4 Ma (Atwater and Stock, 1998). Basin and Range extension propagated westward into the Sierra block, possibly related to side heating from the Basin and Range (e.g., Dilles and Gans, 1995; Surpless et al., 2002). Sierran root delamination sometime between 10 and 4 Ma caused uplift and a change in composition of basaltic volcanism in the central Sierra (e.g., Ducea and Saleeby, 1996). Tilting and incision of Neogene sedimentary rocks as young as 2 or 2.5 Ma are evidence for a significantly more recent uplift event in northwestern Nevada and northeastern California (e.g., Trexler et al., 2000; Muntean, 2001; Park, 2004; Cashman and Trexler, 2004; Mass, 2005; Cashman et al., this volume; Mass et al., this volume). Although dextral shear across the continental margin started with the inception of the transform boundary ca. 30 Ma (Atwater, 1970), the accommodation of the shear changed with the opening of the Gulf of California ca. 6 Ma (Oskin and Stock, 2003) and development of the eastern California shear zone. This latest dextral slip may have propagated northward with time, providing an opportunity to examine the development of a major intracontinental fault zone or transform plate boundary (e.g., Faulds et al., 2003, 2004). There is little direct evidence for the age of onset of dextral faulting in the northern Walker Lane; estimates include between 9 and 4 Ma (Cashman and Fontaine, 2000), 6 Ma (Faulds et al., 2003), and 3 Ma (Henry et al., 2002). At the latitude of the White Mountains in central-eastern California, dextral faulting, like normal faulting, has propagated westward with time (e.g., Stockli et al., 2003). It is not yet possible to determine whether this has been the case at the latitude of the northern Sierra.

Because these tectonic events (and the structures that result from them) overlap in space and in time, it can be difficult to determine the precise age, extent, or regional significance of any one event. Our approach has been to study Neogene sedimentary basins in different structural domains, thus isolating specific structural styles and tectonic driving forces. For example, the Honey Lake basin, specifically the Neogene sedimentary section exposed at the South Island, lies within the Walker Lane zone of dextral faults. The postdepositional deformation of these rocks therefore may represent the initiation of dextral faulting in this area, and the age of the youngest deformed sediments constrain its timing. In addition, the Neogene sedimentary record prior to

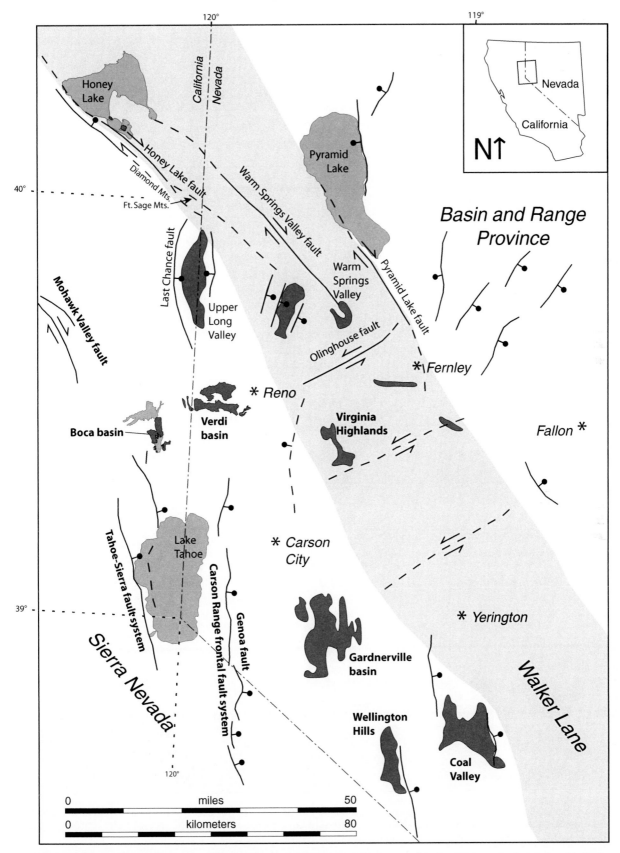

Figure 1. Map of the northern Sierra Nevada–Basin and Range transition zone, showing Neogene basins and major faults. Light-gray shading—Walker Lane. Modified from Mass (2005).

Trexler et al.

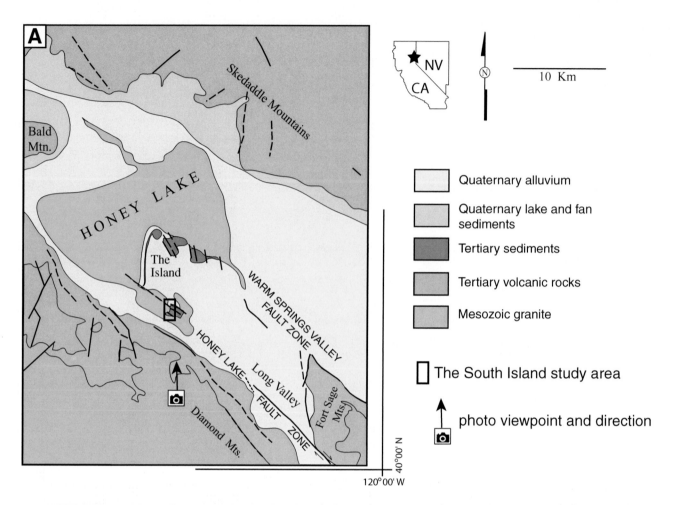

Figure 2. (A) Regional geologic map, showing the geologic and geographic setting of The Island, and the location of the study area on the South Island (geology after Lydon et al., 1960). (B) Photograph taken from the crest of the Diamond Mountains, looking north across Honey Lake. The South Island field area is in the middle distance. Photo credit: Nick Hinz.

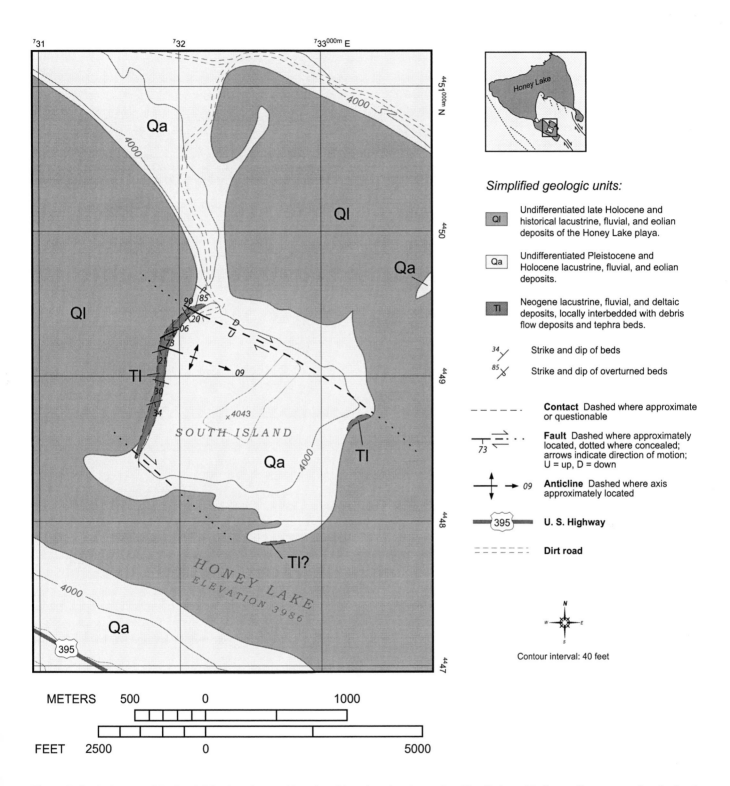

Figure 3. Geologic map of the South Island study area. Note that although regional mapping (Fig. 2) shows Tertiary sediments covering the South Island, in fact most of the sediment at the surface there is Pleistocene or Holocene. Tertiary sediments are well exposed only along wave-cut cliff exposures along the west side of the South Island. Poorly exposed and patchy outcrops are present at the southeast corner of the South Island and the northeast edge of The Island that were not included in this study.

deformation may reveal changes in regional paleogeography that reflect other nearby tectonic events.

Regional Geology

The Honey Lake basin lies within the "Pyramid Lake domain" of the Walker Lane; this northernmost structural domain of the Walker Lane is characterized by northwest-trending, left-stepping dextral faults (Stewart, 1988). In the original definition of the domain the major dextral faults were, from southeast to northwest, the Pyramid Lake, Warm Springs Valley, and Honey Lake faults (Fig. 1). The Mohawk Valley fault, farther west and within the Sierra Nevada, is now commonly included in the domain (e.g., Faulds et al., 2003; Hinz, 2005; Hinz et al., this volume). The Honey Lake fault is mapped along the western edge of the Honey Lake basin. A Quaternary scarp of the Honey Lake fault cuts the eastern edge of the exposed Neogene section we studied at the South Island (Grose et al., 1984; Wills and Borchard, 1993) but cannot be traced in the topography northwestward into the modern lake or playa. A second dextral fault of the Pyramid Lake domain, the Warm Springs Valley fault, projects into the eastern part of the Honey Lake basin from the southeast, along the northeastern boundaries of the Fort Sage Mountains and the South Island (e.g., Grose et al., 1989; Wills and Borchard, 1993). Warm Springs Valley, along fault strike to the southeast, contains Neogene sediments that range from ca. 10 Ma (TenBrink et al., 2000) to 3 Ma (Henry et al., 2003), but little is known about the depositional system(s) or the deformation history of these rocks. The Warm Springs Valley fault has deformed the youngest known part of the Neogene section and latest Pleistocene strata at the northwest end of Warm Springs Valley (e.g., Henry et al., 2004) and has a Holocene scarp in the central part of the valley (Bell, 1984). It is not clear whether the fault continues northwestward under the Quaternary alluvium and lake deposits in the eastern Honey Lake Valley, but the shape of The Island, Honey Lake, and the steep-sided gravity low at the Honey Lake basin (e.g., Cerón, 1992) support this interpretation.

Recent research using paleochannels in the Oligocene tuff section as piercing points has led to better estimates of dextral slip in the Pyramid Lake domain, although the time period over which these dextral faults have acted is not well constrained. Correlation of paleovalley segments between the Diamond and Fort Sage Mountains implies 9.5–17 km of dextral displacement across the Honey Lake fault zone; other constraints favor ~10 km of displacement (Hinz, 2005; Hinz et al., this volume). Interestingly, although the topography suggests significant east-side-down offset across the mountain front along the southwest edge of Honey Lake, the paleovalleys are at virtually the same elevation in the Diamond and Fort Sage Mountains (Hinz, 2005; Hinz et al., this volume), implying little or no dip-slip motion across the Honey Lake fault or a range-front fault during this time. For the Warm Springs Valley fault, the preliminary estimate of dextral displacement based on offset of paleovalleys is 8–10 km (Delwiche et al., 2002; Henry et al., 2004).

The intersection of the normal Sierran frontal fault system with the dextral faults of the Walker Lane is at the northern end of Upper Long Valley, directly southwest of the Honey Lake basin (Fig. 1). Upper Long Valley contains an exhumed, west-tilted Neogene section, with a normal fault—the Upper Long Valley fault—forming its western boundary (Koehler, 1989; Grose and Mergner, 1992); this basin is the northernmost of the exhumed Neogene basins. The Neogene deposits, known as the Hallelujah Formation (e.g., Koehler, 1989; Grose and Mergner, 1992), include ca. 8 Ma rocks low in the exposed section (C.D. Henry, 2000, personal commun.). However, preliminary gravity studies suggest a thick Neogene section along the *east* side of the basin that underlies this dated locality, indicating that basin initiation was well before 8 Ma (TenBrink et al., 2002). Sedimentation included both lacustrine deposition and periods of axial drainage that probably flowed north into the Honey Lake basin (Koehler, 1989; Park, 2004). Coarse, monolithologic debris-flow conglomerates occur throughout the Hallelujah Formation (Koehler, 1989; Grose and Mergner, 1992); these conglomerates have a local source and appear to record syndepositional faulting at several times during the history of the Long Valley basin. The upper age limit for Neogene deposition in Long Valley—and therefore the timing of exhumation—is poorly constrained. Age control comes from Late Blancan (<5 Ma) fossils high in the section (Koehler, 1989), but we do not know how much younger than 5 Ma the youngest of these rocks are.

The dextral faults may be working in combination with normal faults to accommodate dextral slip in the northern Walker Lane (Faulds et al., 2003). Slip on the Pyramid Lake fault appears to transfer northward onto several west-dipping range-front normal faults, and slip on the Warm Springs Valley, Honey Lake, and Mohawk Valley faults may transfer southward onto a series of east-dipping normal faults north of Reno, Nevada (Faulds et al., 2003). One of these normal faults is the Upper Long Valley fault.

STRATIGRAPHY AT HONEY LAKE

Neogene sediments of the Honey Lake basin are exposed in scattered outcrops around the north end of Honey Lake, although mapping there has not been published (Roberts, 1985), and in an excellent and continuous exposure of a broad anticline at a low topographic rise on the South Island at the southwest end of the lake (Fig. 3). More than 600 m of section at the South Island was studied and described in detail by Park (2004), whose thesis is the basis for the following sedimentary descriptions. Other than these limited outcrops, little is exposed of the Neogene Honey Lake basin stratigraphy, although strata are presumed to be extensive underneath the modern playa.

Investigations at the north end of Honey Lake (Roberts, 1985) document two sedimentary sections, at Bald Mountain and Rice's Canyon, that are Miocene and Pliocene in age; these establish an older limit for the Neogene section. The lower section is intercalated with 10 Ma andesitic volcanic rocks, and the upper part preserves Miocene–Pliocene flora. The upper section includes diatomite and tephra, and is capped by 4.9 Ma basalt. These sections are older than the exposed sections at the South Island. There is no reported sedimentary section in the northern part of the basin younger than 4.9 Ma and older than Pleistocene. We did not include these older Tertiary sediments in our investigation. Small and disconnected patches of presumed Tertiary sedimentary rocks are also exposed on the northern shore of The Island, and reconnaissance investigation suggests that these rocks are part of the younger Honey Lake section. These limited exposures were not included in this study.

Two long and detailed stratigraphic sections were measured at the South Island, one on each limb of the anticline. The north limb yields 185 m of section, and the south limb, 435 m. Although faults cut the center of the anticline, the two limbs can be correlated (Fig. 4). The result is two overlapping sections from slightly different positions in the lake's paleogeography.

Lithologic Description

Neogene sediments are silt- to sand-sized grains of quartz, feldspar, mica, and lithics, with minor amounts of clay and diatoms. Virtually all clasts larger than coarse sand are unlithified rip-up clasts derived from previously deposited Neogene lacustrine sediments. Some of these are quite large, up to several meters in size. No large clasts of granite or volcanic rock were observed in the section.

Two sediment sources are implied by the clastic material in the South Island section. The provenance of coarse clasts (pebble size and larger), consisting of unlithified rip-ups of subjacent Neogene sediments, is entirely within the basin. In contrast, coarse sand and finer material is consistent with a granitic or felsic volcanic source. The north-sloping fan (see discussion of paleoslope, below) was geometrically aligned with the axis of the Neogene Honey Lake basin. The Neogene Upper Long Valley basin, preserved in an uplifted valley southwest of Honey Lake, also shares this axial alignment and trend (Fig. 1). Mesozoic granitic rocks and Oligocene tuffs are widespread south of the Honey Lake basin, so sediment composition does not identify a unique provenance for these rocks.

Diatomaceous sediments are common in the South Island section, especially in finer-grained intervals. However, clean diatomite was not deposited in this section. The finer parts of the section are dominated by hemipelagic silt. Diatoms identified by Park (2004) are consistent with deep-water lacustrine conditions. The most common assemblage is *Aulacoseira* (or *Meloseira*), *Stephanodiscus carconensis*, *Stephanodiscus niagarae*, *Stephanodiscus elgeri*, *Cyclotella stylorum,* and *Cyclostephanos*, which are tychoplanktonic and euplanktonic diatoms (Fig. 5). In the diatom study of Nakayamadiara palaeolake, Japan (Yamaoka and Shimada, 1962), a depth of 150 m is suggested from the diatom assemblage for *Aulacoseira, Stephanodiscus, and Cyclotella*, which is the same diatom assemblage found in the Honey Lake Neogene sediments.

Thin tephras occur throughout the section; 65 were sampled, of which 10 were analyzed. Of these, two were identified by Mike Perkins (2004, personal commun.) and are used for age control in this study.

Description and Interpretation of Sedimentology

Sedimentary structures commonly include graded beds with scoured bed bases and diffuse tops. Lamination is uncommon, and cross-lamination is rare. The coarse clasts of the graded beds are rip-up clasts, and most were unlithified at the time of deposition. All sedimentary structures in these sections are consistent with a depositional-environment interpretation of a turbidite-dominated, sublacustrine fan.

The sedimentary section is interpreted to have been dominated by density-modified flow deposits, primarily turbidites and debris-flow sediments. Density-modified flow deposits with a grain-size range and a mud/sand ratio of this type are described by Gani (2004). Most of these sandy turbidites are unchannelized, but in a few localities eroded channels are filled with turbidites. In some intervals, bed-thickness trends indicate coarsening and thickening, lobe-building relationships (Fig. 6). The fine-grained intervals in between turbidites are hemipelagic silt and mud. These hemipelagic parts of the section commonly have a significant diatomite component. Detailed sedimentology and facies analysis are documented by Park (2004).

The north limb section contains a single, thick, and very coarse debris-flow deposit low in the section. Blocks in the debris flow are up to 10 m in long dimension, and all are rip-up clasts of the underlying lacustrine sediment. Most clasts and blocks are deformed, and some are folded (Fig. 7). At the base of the debris bed are partially detached turbidite beds that are rolled into the flow. The sense of shear in the debris flow is tops-north, as it is in all paleoslope indicators here (discussed below).

Soft-sediment deformation is common throughout the Honey Lake section. Structures that record soft-sediment failure include load casts, flame structures, detached and distorted bedding, and ball-and-pillow structures. A continuum of deformed beds can be defined, from simple loading to debris slides and flows. Most deformed or broken bedding shows soft-fold vergence and rip-up imbrication, which is evidence of downslope movement toward the north (Fig. 8).

In both north and south limb sections an important unconformity breaks the stratigraphy. In the north limb section this unconformity caps the thick debris flow. In the south limb section it caps a thick, channelized sand unit that is correlated with the debris flow and presumed to be a mass-flow channel-filling sand. A distinctive, pyroclastic (lapilli)–dominated bed just above the unconformity in both sections confirms the correlation. In both sections the unconformity is scoured and is overlain by cross-laminated, coarse, micaceous sandstone that leveled out micro-paleotopography. We interpret this unconformity as a lacustrine lowstand that formed an erosional surface capped by a thin, fluvial conglomerate and sandstone interval

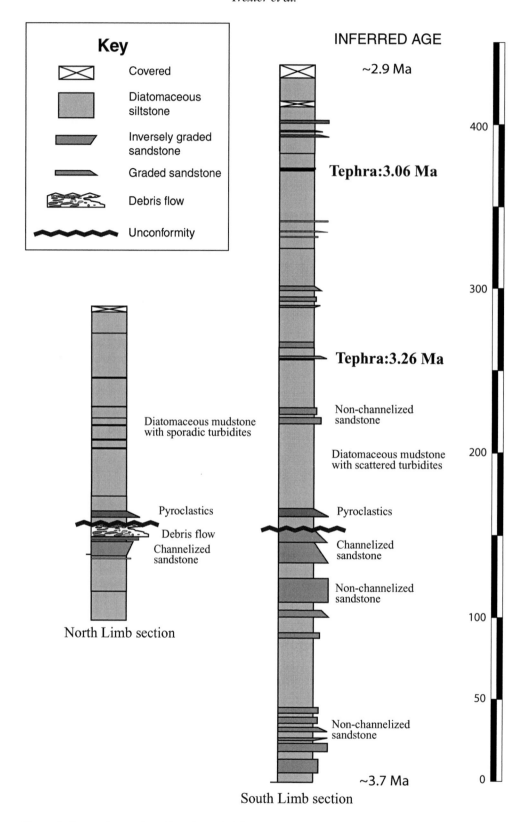

Figure 4. Composite stratigraphic column for Neogene sediments at The Island, from Park (2004). The sections from the north and south limbs of the anticline are correlated on the basis of the erosional surface, underlying channelized sandstone, and distinctive overlying pyroclastic (lapilli-dominated) zones. The time scale is based on the deposition rate calculated from the two dated tephras in the south limb section. Vertical scale at right in meters.

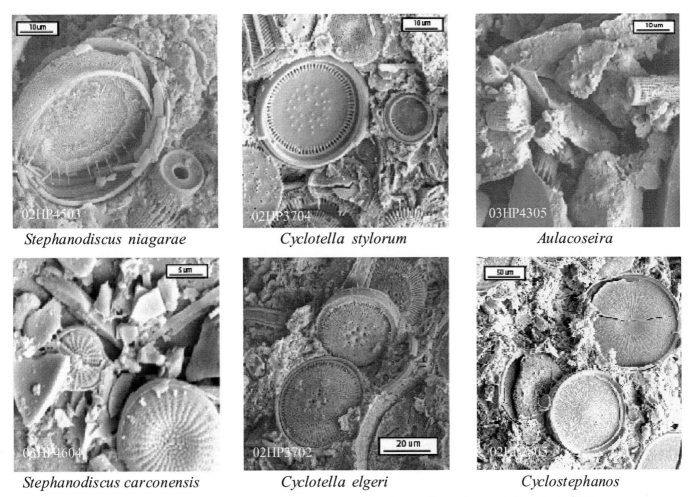

Figure 5. Diatoms common in the Honey Lake diatomaceous siltstone. Photomicrographs from Park (2004).

Figure 6. A coarsening-thickening-upward succession of turbidites, interpreted as a sublacustrine fan-lobe sequence.

Figure 7. View, looking east, of the thick debris-flow deposit directly below the 3.45 Ma unconformity in the north limb section. Blocks in the debris flow are rip-up clasts of the underlying lacustrine sediment; folding of clasts indicates downslope-to-the-north displacement. The debris flow may be related to instability during a lacustrine lowstand, or to tectonic activity possibly as a coseismic event. Note students for scale.

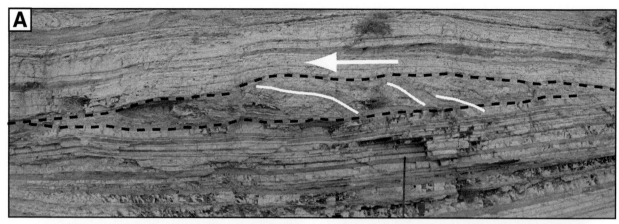

Figure 8. Several examples of soft-sediment deformation, showing sense of motion downslope to the
north (left in all photos). (A) Soft-sediment folds;
axial surfaces, shown by white lines, indicate
vergence toward the north. (B) Imbricated rip-up
clasts, showing flow toward the north. (C) Roll-up
structure, showing flow toward the north. (D) Ten-
sional fracture filled with the overlying sand, suggesting extension downslope toward the north.

that is unique in the sections. The overlying turbidites record a
return to lacustrine conditions.

Age Control

Two tephras ~100 m apart in the Honey Lake section have
been identified and correlated with regional ash beds: one at
3.26 Ma, and another at 3.06 Ma (Fig. 4; Perkins, 2004, written
commun.). By interpolation, these tephras allow an estimation
of lake sedimentation rate at 550 m per million years. This rate,
in turn, allows estimation of the age range of the exposed south-
limb section of ca. 3.7 Ma to 2.9 Ma.

**Summary: Depositional Environment of the South Island
Section**

The Neogene section at the South Island represents ~800,000
yr of lacustrine deposition, interrupted once by a significant lowstand. These sediments were deposited on a north-sloping axial
ramp at the south end of the Honey Lake basin, with a significant sediment source to the south. There are no paleoslope indicators or any coarse clastic debris that would be evidence of a
source-area uplift to the west (modern Diamond Mountains) or
east (modern Skedaddle Mountains). Sedimentation was dominated by sandy turbidites and hemipelagic mud, with some silty

diatomaceous intervals. The dominance of fine clastic sediments and the lack of clean diatomite intervals are consistent with deposition as a prodelta slope rather than a part of the basin isolated from active sedimentation. Diatoms identified by Park (2004) are consistent with deepwater (>150 m) lacustrine conditions.

The Neogene sediments were very unstable, as recorded by the abundant evidence of sediment failure and collapse. Everywhere in the South Island section the paleoslope was steep enough so that once mobilized, sediments sheared downslope to the north. The sediment instability and failure here can be attributed to three possible explanations (e.g., see discussion by Leeder, 1999):

1. A high sedimentation rate contributed to autocyclic failure. This phenomenon occurs as rapid sedimentation overpressures a saturated sediment column, resulting in fluid escape. Common evidence in known examples includes sand volcanoes and dewatering structures extending upward through the section to the coeval sediment-water interface.

2. A dramatic lake-level drop destabilized sediments, causing dewatering and allocyclic failure. As water level and hydrostatic pressure fell, interstitial water pressure became greater, and subsequent fluid escape caused sediment deformation and failure.

3. Co-seismic shaking caused intergranular instability and loss of material strength, resulting in allocyclic failure. Such failure can also result from shock caused by storm waves in open-marine environments, but this cause can be discounted for lakes the size of the Honey Lake basin.

Any of these mechanisms is possible in the South Island section.

The dramatic combination of the thick debris-flow bed, the erosional surface that truncates it, and the thin overlying interval of subaerial sediments is evidence for a rapid lake-level fall at ca. 3.4 Ma. Rothwell et al. (1998) invoke low sea level as a possible trigger for large submarine landslides (for the Pleistocene case they document, buried organic matter and methane hydrate also played a role). The debris flow may have been triggered by subaerial exposure of the upper (southern) part of the delta as lake level dropped, resulting in erosion and resedimentation of inner-delta sediments down the delta slope. Alternatively, seismic shaking may have triggered the debris flow, although there is no direct evidence for this. In any case, although subaerial exposure followed the debris-flow event, the lake soon returned, and lacustrine sedimentation continued much as before. There is no evidence of angular discordance across the lowstand boundary, so local tectonism is not required as a cause for the lake-level drop. Lake-level response to climate change also is a viable possibility.

POST-DEPOSITIONAL DEFORMATION RECORDED AT HONEY LAKE

The Neogene section exposed at the South Island is in a structural high formed by a broad east-southeast–plunging anti-cline (Figs. 3 and 9). This anticline is cut by two strike-slip faults that have tens of meters of offset, and by numerous mesoscopic faults. A northwest-striking dextral fault near the north end of the exposure displaces the ground surface and alluvial gravels as well as the Neogene section (Fig. 10), and has been mapped as an active strand of the Honey Lake fault zone (Grose et al., 1984; Wills and Borchard, 1993). Bedding is locally vertical in the vicinity of this fault; this amount of deformation suggests significant displacement on the fault. A northeast-striking sinistral fault near the core of the anticline is also associated with strong localized drag folding, and bedding cannot be matched across the fault; both suggest relatively large offset (tens of meters or greater?). Mesoscopic faults are common throughout the anticline, but fault offsets are small enough (most are of centimeter to meter scale) so that the stratigraphy can be reconstructed across the faults. Kinematic indicators are present on ~40% of the small faults; sense of slip on the remainder is inferred from the combination of stratigraphic separation and analogy with similarly oriented faults of known slip sense. The faults can be divided into three main sets, in order of abundance: northwest- to north-northwest–striking dextral faults, east-northeast- to east-southeast–striking sinistral faults, and thrust faults that are most commonly west-northwest striking. The following structural descriptions are summarized from Mass (2005).

The most abundant faults in the exposed Neogene section are subvertical northwest- to north-northwest–striking dextral faults. Slickenlines and grooves on these fault surfaces plunge <10° to the northwest or southeast, recording pure strike-slip motion (Fig. 11). Riedel shears and stratigraphic separation indicate dextral slip. The dextral faults appear to have been active throughout the deformation history of the Neogene section, and this deformation continues with the Quaternary fault that offsets the ground surface at the north end of the outcrop on the South Island. There appears to be some tilt-fanning of the dextral faults across the core of the anticline, suggesting that some of these faults were present prior to formation of the anticline and that tightening of the anticline has been synchronous with dextral faulting. This dextral fault set, not including the Quaternary fault that offsets the ground surface, accommodates at least 10 m of cumulative displacement.

The northeast- to east-striking sinistral faults are more variable in both strike and dip than the dextral faults (Fig. 12). The few kinematic indicators (other than stratigraphic separation) that are preserved are subhorizontal on northeast-striking surfaces. Slip indicators are slightly steeper on east-striking surfaces and indicate oblique sinistral-reverse slip. The sinistral faults have mutually crosscutting relationships with the thrust faults, and both seem to have formed to accommodate space problems during tightening of the anticline (see below) (Figs. 9 and 10). This sinistral fault set, not including the major sinistral fault near the core of the anticline, accommodates at least 5 m of cumulative displacement.

Mesoscopic west-northwest–striking faults constitute a conjugate set that dips gently northeast or southwest, and stratigraphic separations consistently show that the hanging wall moved up

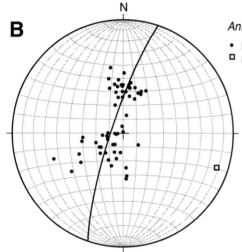

Anticline geometry

- Pole to bedding (N=59)
- ▫ Pole to calculated best fit line
  (trend and plunge): 108.5°, 08.7°

Figure 9. Geometry of the anticline exposed along the west shore of The Island at Honey Lake. T—toward; A—away. (A) Mosaic of photos, showing the core of the anticline with sinistral and thrust faults (students for scale). The area inside the box shows an example of a thrust fault truncated by a sinistral strike-slip fault. (B) Great circle best fit, calculated from 59 poles to bedding across the ~1 km exposure, indicates that the fold plunges gently to the east-southeast.

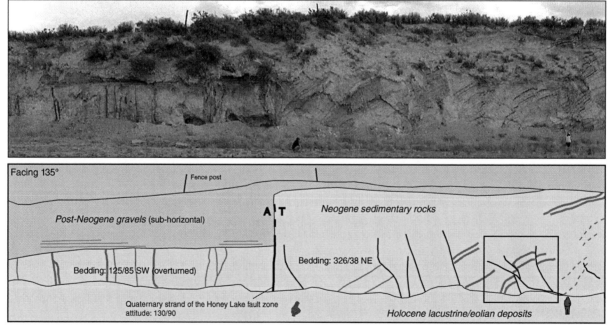

Figure 10. A Quaternary strand of the Honey Lake fault zone is exposed at the northern termination of the anticline. Sandstone beds (gray) in silty diatomite (light gray) change dip dramatically across the fault. Minor faults include dextral and sinistral strike-slip faults and thrust faults, and show crosscutting relationships locally. The area inside the box shows an example of a sinistral strike-slip fault cut by a thrust fault. Note students for scale. T—toward; A—away.

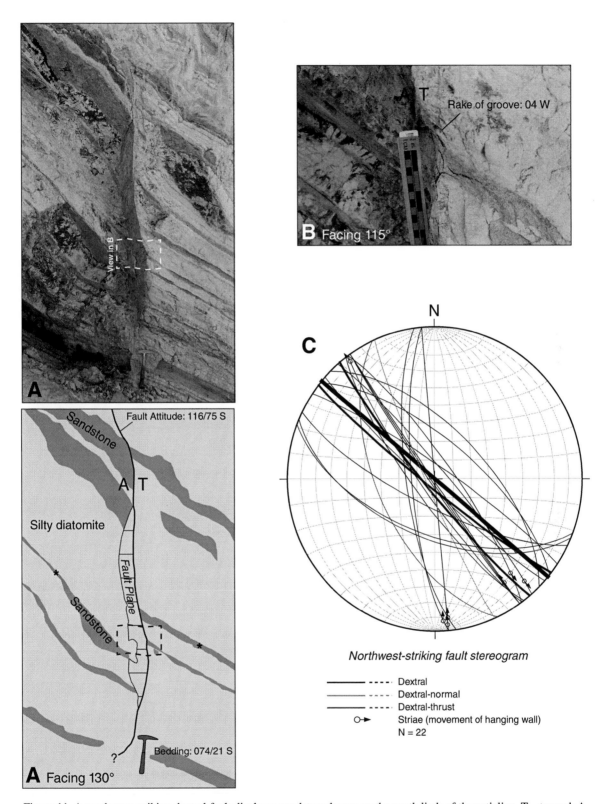

Figure 11. A northwest-striking dextral fault displaces sandstone layers on the south limb of the anticline. T—toward; A—away. (A) The thin gray sandstone at center (*) has a total vertical stratigraphic separation of ~0.2 m. (B) Subhorizontal grooves on the fault plane indicate strike-slip motion. This particular fault has ~2.3 m of total displacement. (C) Equal-area, lower hemisphere stereographic projection of northwest-striking faults interpreted to have dextral motion (22 planes). Thicker lines indicate faults with greater displacement. Although striae are not preserved on most fault surfaces, where they are present they are consistently subhorizontal.

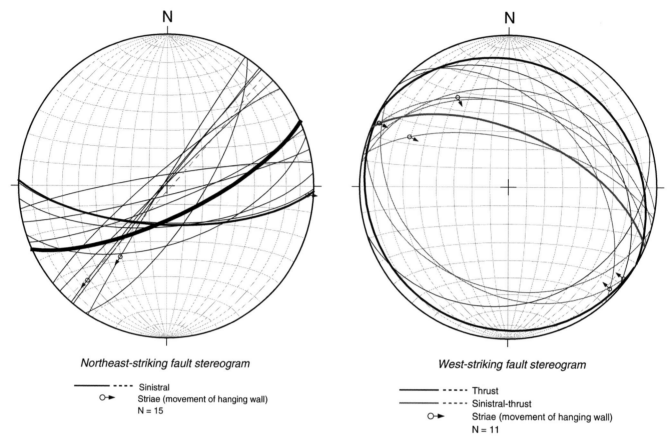

Figure 12. Equal-area, lower hemisphere stereographic projection of northeast-striking faults interpreted to have sinistral motion (15 planes). Thicker lines indicate faults with greater displacement. Striae are not preserved on most of these faults, but drag folding and offset bedding indicate sinistral motion.

Figure 13. Equal-area, lower hemisphere stereographic projection of west-striking conjugate faults interpreted to have thrust motion (11 planes). Thicker lines indicate faults with greater displacement. Note that most striae are not consistent with thrust motion. These striae are interpreted to indicate reactivation of these fault surfaces during strike-slip events on nearby faults.

relative to the footwall (Figs. 9, 10, and 13). These faults are interpreted to be thrust faults, resulting from north-northeast–south-southwest shortening, which is consistent with the strain recorded by the broad anticline. The thrust faults show mutually crosscutting relationships with sinistral faults, both in the core and on the north limb of the anticline, indicating that the two were active simultaneously (Figs. 9 and 10). Multiple generations of thrusting—oblique thrust and sinistral faults, particularly in the core of the anticline—attest to continuing activity of the faults during folding. The few striae preserved on the thrust faults record northwest-southeast slip at a low angle to the fault strike, which is an unlikely initial slip direction for gently dipping faults. The northwest-trending striae probably represent reactivation of some of the thrust-fault surfaces during northwest-directed dextral slip along the dominant northwest-striking dextral fault set, possibly as coseismic deformation during motion on the nearby Quaternary fault.

Three-dimensional strain can be determined using a fault inversion method, based on individual faults that have kinematic indicators (Marrett and Allmendinger, 1990). The method

assumes that regional strain is uniform and invariant over time. The fault data for the analysis must form over a relatively short time interval, and all faults must be included in the data set. The Honey Lake fault data presented here conform to these criteria, with one caveat: Only faults with kinematic indicators can be used, and these make up a relatively small (but on the basis of the consistent slip directions they show, probably a representative) percentage of the faults (Figs. 8, 10–12). Although the maximum and minimum strain directions determined from the analysis may not necessarily correspond to the extensional and contractional axes for individual faults, the inversion method is reasonable for many faults distributed over a large area (Pollard et al., 1993), such as the ~1-km-length cliff exposures on the South Island.

The fault inversion shows that the faults have accommodated approximately east-west extension and north-south contraction, with total extension smaller than contraction; motion on the three main fault sets is consistent with this strain field (Fig. 14A, B). The dextral nodal plane from the fault plane solution strikes 313°, which corresponds well with the measured 310° strike of the

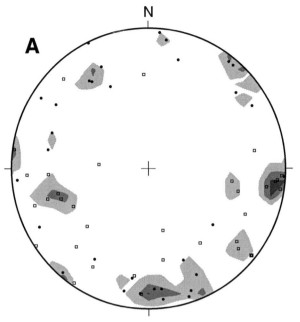

Contour interval 4.0% axes per 2% area

● P-axis
□ T-axis

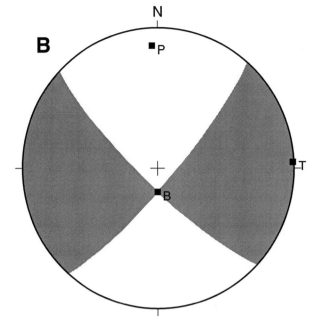

Nodal planes
133.2°, 80.0° and 041.6°, 80.9°

Eigenvalue; Eigenvector (trend, plunge)
T) 0.1132; 087.5°, 00.6°
B) 0.0579; 180.0°, 76.4°
P) -0.1711; 357.3°, 13.5°

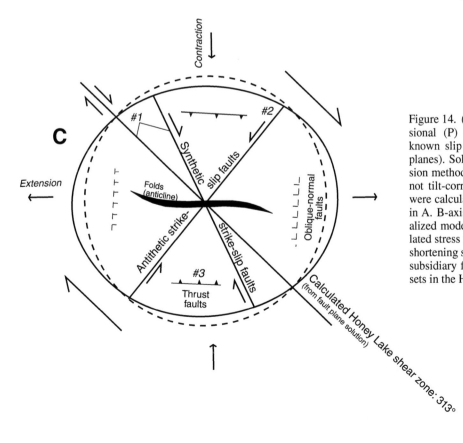

Figure 14. (A) Contoured tensile (T) and compressional (P) stress-axis orientations for faults with known slip direction and displacement >0.1 m (47 planes). Solutions were modeled using a fault-inversion method (program of R. Allmendinger). Data are not tilt-corrected. (B) Nodal planes and stress axes were calculated by combining all of the faults shown in A. B-axis is the intermediate stress axis. (C) Idealized model of structures associated with the calculated stress field (east-west extension and north-south shortening stress is predicted). Note that the predicted subsidiary fault sets are similar to the observed fault sets in the Honey Lake basin.

Quaternary fault strand of the Honey Lake fault zone (Fig. 10). Predicted subsidiary fault sets for this strike-slip shear zone are similar to the observed fault sets of the Neogene Honey Lake basin (cf. Sylvester, 1988) (Fig. 14C). The synthetic dextral faults have a northwest strike (fault set no. 1), while antithetic sinistral faults have a northeast strike (fault set no. 2) and are less abundant than the synthetic faults. The thrust and oblique-thrust faults (fault set no. 3) have a west strike, which is perpendicular to the maximum contraction direction. The trend and plunge of the anticline (109°/09°) is also consistent with the orientation of the maximum contraction direction (Fig. 9). The few sinistral-normal faults on the south limb of the anticline are compatible with the maximum extensional direction.

In summary, the deformation of the Neogene section at the South Island is consistent with north–south shortening and with the regional northwest-striking dextral faults of the Walker Lane. The broad anticline and conjugate set of thrust faults indicate north-south shortening; northwest-striking dextral and northeast-striking sinistral faults are consistent with this strain field. Mutually crosscutting relationships between the thrust and strike-slip fault sets, and possible tilt fanning of the dextral faults, suggest that tightening of the anticline was synchronous with faulting. Faulting at the core of the anticline may in part have accommodated space problems that developed during tightening of the fold. The prevalence of the northwest-striking dextral-fault set, as well as the presence of a Quaternary dextral fault, suggests that these faults are accommodating some of the regional dextral slip of the Walker Lane in addition to local north-south shortening at the South Island. A fault-inversion analysis, using only the faults with kinematic indicators, shows that these faults accommodated subhorizontal north-south shortening and subhorizontal east-west extension (Fig. 14). The nodal planes correspond to many of the measured (northwest-striking dextral and northeast-striking sinistral) strike-slip faults. The northwest-striking nodal plane closely parallels the active strand of the dextral Honey Lake fault at the northeast end of the exposures at the South Island.

DISCUSSION—IMPLICATIONS FOR REGIONAL TECTONICS

The lacustrine Neogene section exposed at the South Island represents only a small, and relatively young, portion of the depositional history of the Honey Lake basin. In exposures northwest of Honey Lake, fluvial lithic wacke and lacustrine diatomite are interbedded with Miocene (10.1 Ma) andesite and Pliocene ash deposits, suggesting that deposition has been active in this basin since at least middle Miocene time (Roberts, 1985). An estimate of depth to basement based on gravity modeling yielded an independent but consistent age of basin initiation ca. 10 Ma (Cerón, 1992). The initiation of sedimentation in the basin does not necessarily represent a tectonic event; the local volcanism could have created the enclosed basin.

The South Island section records the paleogeography at the south end of the Honey Lake basin from ca. 3.7 to 2.9 Ma. The section consists of lake-slope or delta deposits, both pelagic and hemipelagic; diatoms in the finer-grained layers suggest lake depths of ~150 m. The calculated sedimentation rate (550 m per million years) is high for lacustrine deposits but is consistent with this setting, because the hemipelagic sediments and diatoms were supplemented by, and could have been overwhelmed by, the clastic input from the rivers feeding the delta systems. The composition of the clastic material is consistent with derivation from an intermediate to felsic igneous source; lithic clasts are rare in these generally fine-grained sediments, so a more specific provenance description is not possible.

Interpretation of the Honey Lake sediments as deltaic is consistent both geometrically and geographically with possibly coeval fluvial sediments of the Hallelujah Formation of Upper Long Valley (e.g., Koehler, 1989; Grose and Mergner, 1992), an argument supported by both paleocurrent and slope-direction indicators at Honey Lake. The Hallelujah Formation fluvial system may have been the upstream part of a river-delta-fan drainage. Alternatively, the Hallelujah Formation may have been exhumed and eroded, and that debris redeposited in the Honey Lake basin. Because of insufficient age control for the Hallelujah Formation, it is not yet possible to distinguish between these two possibilities, thereby testing the model of kinematic linkage between dextral and normal faults (Faulds et al., 2003) on the basis of the section at the South Island.

A significant lake-lowering event occurred at ca. 3.4 Ma. Age control suggests that the lowstand had a short duration and that lacustrine deposition had resumed after a brief period of subaerial erosion and fluvial deposition. There is no evidence of angular discordance across the lowstand boundary, so local faulting is not required as a cause for the lake-level drop. Because lacustrine sedimentation was quickly reestablished, large-scale topographic rearrangement is not indicated. A climatically driven lowstand cannot be ruled out. However, the coincidence of the thick debris-flow deposits and the drop in lake level are suggestive of tectonic processes.

The anomalous structural high that exposes the Neogene section is interpreted to have formed in a contractional stepover between two strands of the Honey Lake fault. Folding and faulting of the Neogene section are consistent with north-south shortening, and our fault-inversion analysis gives similar results. Mutually crosscutting relationships between the thrust and strike-slip fault sets, and possible tilt fanning of the dextral faults, suggest that tightening of the anticline was synchronous with mesoscopic faulting. An active strand of the dextral Honey Lake fault crops out at the northeast end of the South Island exposures; it offsets the ground surface as well as the sedimentary section. Dextral slip on this surface is consistent with the strain field represented by the postdepositional structures. The coincidence of this fault with the end of Neogene exposures suggests that the cause of the local shortening and uplift is a contractional stepover—i.e., a left step between two strands of the dextral Honey Lake fault. The tephra dates from the South Island section, and the sedimentation rate calculated from them, indicate that the currently active strands of the Honey Lake fault did not propagate through this area until some time after 2.9 Ma.

By analogy with the local structural high that exposes the Neogene section, The Island as a whole appears to be a broad structural high formed by a contractional stepover between the Warm Springs Valley and Honey Lake fault systems. The parallel, northwest-trending boundaries of The Island are along strike with the Honey Lake and Warm Springs Valley faults (Fig. 2). This interpretation also explains both the presence of The Island as a topographic feature and its abrupt termination along trend to the northwest, as shown in gravity data (Cerón, 1992). If correct, this interpretation implies that the Warm Springs Valley fault extends northwest of Warm Springs Valley and the Fort Sage Mountains into the Honey Lake basin.

Like several other Neogene basins along or within the eastern edge of the Sierra, the Honey Lake basin was a lacustrine environment until <3 Ma, but the similarity of timing between basins may be coincidental. Sedimentation (primarily lacustrine) continued in the Verdi basin, directly west of Reno, and the Boca basin, west of Verdi within the Sierra, until <2.6 and ca. 2.7 Ma, respectively (e.g., Mass et al., this volume). The tectonism that terminated deposition in these two basins also exhumed them. In contrast, the tectonism that exposed the section at the South Island was local; most of the Honey Lake basin is still structurally low. The top of the exposed Honey Lake section is slightly <3 Ma; a younger deformed section may exist but is not exposed. Another contrast with the Verdi and Boca basins is that the Honey Lake basin may also record tectonism ca. 3.4 Ma (the lake-lowering event associated with the thick debris flow), and there is no similar evidence at Verdi or Boca.

In summary, the short geologic history recorded in the sedimentary section at the South Island contains evidence of only one of the important Miocene–Pliocene tectonic events in the region. Although the timing of tectonic disruption is roughly consistent with passage of the triple junction to the west and with uplift and exhumation of several nearby basins (the latter possibly related to Sierran root delamination), the deformation at the South Island appears to be directly related to dextral faulting. It therefore appears to date the propagation of a strand of the Honey Lake fault through southwestern Honey Lake. The fact that the rest of the Honey Lake basin remains structurally low is consistent with a contractional-stepover origin for the South Island and suggests that the forces responsible for exhumation of nearby Neogene basins are not active here. The Neogene section at the South Island contains no direct record of regional normal faulting, so it does not place additional constraints on northward or westward propagation of normal faulting, or on kinematic linkage between normal and dextral faulting in the northern Walker Lane.

ACKNOWLEDGMENTS

Dave Wagner at the California Geological Survey originally called our attention to the spectacular exposures at the South Island and has always been available and willing to discuss the regional geology with us. The Biology Department at the University of Nevada, Reno (UNR), provided access to the South Island, which is part of the University of Nevada Research Station at Honey Lake. Carl Lee, station caretaker, was generous with his time and resources. Gary Oppliger discussed gravity data with us and helped us interpret the data. John McCormack helped with photographs of diatoms on the UNR scanning electron microscope. Work with diatoms was partially funded through Alan Wallace and the U.S. Geological Survey. Mike Perkins at the University of Utah helped with collection of tephras and provided identification and age data. The authors also wish to thank Alan Wallace, Becky Dorsey, and John Oldow for very helpful reviews.

REFERENCES CITED

Atwater, T.M., 1970, Implications of plate tectonics for the Cenozoic tectonic evolution of western North America: Geological Society of America Bulletin, v. 81, p. 3513–3536, doi: 10.1130/0016-7606(1970)81[3513:IOPTFT]2.0.CO;2.

Atwater, T., and Stock, J., 1998, Pacific–North American plate tectonics of the Neogene southwestern United States: An update, *in* Ernst, W.G., and Nelson, C.A., eds., Integrated Earth and Environmental Evolution of the Southwestern United States; the Clarence A. Hall Jr. Volume: Columbia, Maryland, Bellwether Publishing, p. 393–420.

Bell, J.W., 1984, Reno Sheet, Quaternary Fault Map of Nevada: Nevada Bureau of Mines and Geology Map 79, scale 1:250,000.

Cashman, P.H., and Fontaine, S.A., 2000, Strain partitioning in the northern Walker Lane, western Nevada and northeastern California: Tectonophysics, v. 326, p. 111–130, doi: 10.1016/S0040-1951(00)00149-9.

Cashman, P., and Trexler, J.H.J., 2004, The Neogene Verdi basin records <2.0–2.5 Ma dextral faulting west of the Walker Lane near Reno, Nevada: Geological Society of America Abstracts with Programs, v. 36, no. 4, p. 55.

Cashman, P.H., Trexler, J.H., Jr., Muntean, T.W., Faulds, J.E., Louie, J.N., and Oppliger, G.L., 2009, this volume, Neogene tectonic evolution of the Sierra Nevada–Basin and Range transition zone at the latitude of Carson City, Nevada, *in* Oldow, J.S., and Cashman, P.H., eds., Late Cenozoic Structure and Evolution of the Great Basin–Sierra Nevada Transition: Geological Society of America Special Paper 447, doi: 10.1130/2009.2447(10).

Cerón, J.F., 1992, Gravity modeling of the Honey Lake basin [M.S. thesis]: Golden, Colorado School of Mines, 145 p.

Colgan, J.P., and Dumitru, T.A., 2003, Reconstructing Late Miocene Basin and Range extension along the oldest part of the Yellowstone hot spot track in northwestern Nevada: Geological Society of America Abstracts with Programs, v. 35, no. 6, p. 347.

Delwiche, B., Faulds, J.E., and Henry, C.D., 2002, Structural framework and late Oligocene paleotopography of the Pah Rah Range, western Nevada: Implications for estimating offset across the Warm Springs Valley fault zone in the northern Walker Lane: Geological Society of America Abstracts with Programs, v. 36, no. 4, p. 28.

Dilles, J.H., and Gans, P.B., 1995, The chronology of Cenozoic volcanism and deformation in the Yerington area, western Basin and Range and Walker Lane: Geological Society of America Bulletin, v. 107, p. 474–486, doi: 10.1130/0016-7606(1995)107<0474:TCOCVA>2.3.CO;2.

Ducea, M., and Saleeby, J.B., 1996, Buoyancy sources for a large, unrooted mountain range, the Sierra Nevada, California: Evidence from xenolith thermobarometry: Journal of Geophysical Research, v. 101, no. B4, p. 8229–8244.

Ducea, M.N., and Saleeby, J., 2003, Trace element enrichment signatures by slab derived carbonate fluids in the continental mantle wedge: An example from the Sierran-Nevada, California: Geological Society of America Abstracts with Programs, v. 35, no. 6, p. 138.

Faulds, J.E., Henry, C.D., and Hinz, N.H., 2003, Kinematics and cumulative displacement across the northern Walker Lane: An incipient transform fault, northwest Nevada and northeast California: Geological Society of America Abstracts with Programs, v. 35, no. 6, p. 305.

Faulds, J.E., Coolbaugh, M., Henry, C.D., and Blewitt, G., 2004, Why is Nevada in hot water? Relations between plate boundary motions, the Walker Lane, and geothermal activity in the northern Great Basin: Geological Society of America Abstracts with Programs, v. 36, no. 4, p. 27.

Gani, M.R., 2004, From turbid to lucid: A straightforward approach to sediment gravity flows and their deposits: Sedimentary Record, v. 2, p. 2–8.

Grose, T.L.T., and Mergner, M., 1992, Geologic map of the Chilcoot 15′ quadrangle, Lassen and Plumas counties, CA: Sacramento, CA, California Division of Mines and Geology, scale 1:62,000.

Grose, T.L.T., Saucedo, G.J., and Wagner, D.L., 1984, Geologic Map of the Susanville Quadrangle, California: Sacramento, CA, California Division of Mines and Geology, scale 1:100,000.

Grose, T.L.T., Wagner, D.L., Saucedo, G.J., and Medrano, M.D., 1989, Geologic Map of the Doyle 15′ Quadrangle, Lassen and Plumas Counties, California: California Geological Survey Open-File Report 89-31, scale 1:62,500.

Hardyman, R.F., and Oldow, J.S., 1991, Tertiary tectonic framework and Cenozoic history of the central Walker Lane, Nevada, in Raines, G.L., et al., eds., Geology and Ore Deposits of the Great Basin: Reno, Geological Society of Nevada, p. 279–301.

Hawkins, F.F., LaForge, R., and Hansen, R.A., 1986, Seismotectonic Study of the Truckee/Lake Tahoe Area, Northeastern Sierra Nevada, California, for Prosser Creek, Stampede, Boca and Lake Tahoe Dams: Denver, U.S. Bureau of Reclamation Seismotectonic Report 85-4, 210 p.

Henry, C.D., Faulds, J.E., and DePolo, C.M., 2002, Structure and evolution of the Warm Springs Valley fault, northern Walker Lane, NV: Post–3 Ma initiation(?): Geological Society of America Abstracts with Programs, v. 34, no. 5, p. 84.

Henry, C.D., Faulds, J.E., Garside, L.J., and Hinz, N.H., 2003, Tectonic implications of ash-flow tuffs and paleovalleys in the western US: Geological Society of America Abstracts with Programs, v. 35, no. 6, p. 138.

Henry, C.D., Faulds, J.E., Garside, L.J., Castor, S.B., DePolo, C.M., Davis, D.A., Hinz, N.H., and Delwiche, B., 2004, A geologic mapping transect across the northern Walker Lane, NW Nevada: Determining the location, displacement and timing of major, plate boundary strike-slip faults: Geological Society of America Abstracts with Programs, v. 36, no. 4, p. 94.

Hinz, N.H., 2005, Tertiary volcanic stratigraphy of the Diamond and Fort Sage Mountains, northeastern California and western Nevada—Implications for development of the northern Walker Lane [M.S. thesis]: Reno, University of Nevada, 110 p.

Hinz, N.H., Faulds, J.E., and Henry, C.D., 2009, this volume, Tertiary volcanic stratigraphy and paleotopography of the Diamond and Fort Sage Mountains: Constraining slip along the Honey Lake fault zone in the northern Walker Lane, northeastern California and western Nevada, in Oldow, J.S., and Cashman, P.H., eds., Late Cenozoic Structure and Evolution of the Great Basin–Sierra Nevada Transition: Geological Society of America Special Paper 447, doi: 10.1130/2009.2447(07).

Koehler, B.M., 1989, Stratigraphy and depositional environments of the Late Pliocene (Blancan) Hallelujah Formation, Long Valley, Lassen County, California, Washoe County, Nevada [M.S. thesis]: Reno, University of Nevada, 120 p.

Leeder, M.L., 1999, Sedimentology and Sedimentary Basins, from Turbulence to Tectonics: London, Blackwell Science, 592 p.

Lerch, D.W., McWilliams, M.O., Miller, E.L., and Colgan, J.P., 2004, Structure and magmatic evolution of the northern Blackrock Range, Nevada: Preparation for a wide-angle refraction/reflection survey: Geological Society of America, Rocky Mountain Section, Abstracts with Programs, v. 36, no. 4, p. 37.

Lydon, P.A., Gay, T.E., and Jennings, C.W., 1960, Geologic Map of California, Westwood Sheet: California Division of Natural Resources, scale 1:250,000.

Marrett, R., and Allmendinger, R.W., 1990, Kinematic analysis of fault-slip data: Journal of Structural Geology, v. 12, no. 8, p. 973–986, doi: 10.1016/0191-8141(90)90093-E.

Mass, K.B., 2005, The Neogene Boca basin, northeastern California: Stratigraphy, structure and implications for regional tectonics [M.S. thesis]: Reno, University of Nevada, 195 p.

Mass, K.B., Cashman, P.H., Trexler, J.H., Jr., Park, H., and Perkins, M.E., 2003, Deformation history in Neogene sediments of Honey Lake basin, northern Walker Lane, Lassen County, California: Geological Society of America Abstracts with Programs, v. 36, no. 6, p. 26.

Mass, K.B., Cashman, P.H., and Trexler, J.H., Jr., 2009, this volume, Stratigraphy and structure of the Neogene Boca basin, northeastern California: Implications for Late Cenozoic tectonic evolution of the northern Sierra Nevada, in Oldow, J.S., and Cashman, P.H., eds., Late Cenozoic Structure and Evolution of the Great Basin–Sierra Nevada Transition: Geological Society of America Special Paper 447, doi: 10.1130/2009.2447(09).

Muntean, T.W., 2001, Evolution and stratigraphy of the Neogene Sunrise Pass Formation of the Gardnerville sedimentary basin, Douglas County, Nevada [M.S. thesis]: Reno, University of Nevada, 250 p.

Oldow, J.S., 1992, Late Cenozoic displacement partitioning in the northern Great Basin, in Craig, S.D., ed., Walker Lane Symposium; Structure, Tectonics, and Mineralization of the Walker Lane: Reno, Geological Society of Nevada, p. 17–52.

Oskin, M., and Stock, J., 2003, Pacific–North America plate motion and opening of the Upper Delfin basin, northern Gulf of California, Mexico: Geological Society of America Bulletin, v. 115, p. 1173–1190, doi: 10.1130/B25154.1.

Park, H., 2004, Honey Lake basin: Depositional settings and tectonics [M.S. thesis]: Reno, University of Nevada, 259 p.

Park, H., Trexler, J.H.J., Cashman, P., and Mass, K.B., 2004, Local initiation of Walker Lane tectonism prior to 3.6 Ma recorded in Neogene sediments at Honey Lake basin, northeastern California: Geological Society of America Abstracts with Programs, v. 36, no. 4, p. 52.

Pollard, D.D., Saltzer, S.D., and Rubin, A.M., 1993, Stress inversion methods: Are they based on faulty assumptions? Journal of Structural Geology, v. 15, no. 3, p. 245–248, doi: 10.1016/0191-8141(93)90176-B.

Roberts, C.T., 1985, Cenozoic evolution of the NW Honey Lake Basin, Lassen County, California: Colorado School of Mines Quarterly, v. 80, p. 1–64.

Rothwell, R.G., Thomas, J., and Kaehler, G., 1998, Low-sea-level emplacement of very large Pleistocene "megaturbidite" in the western Mediterranean Sea: Nature, v. 392, p. 377–380, doi: 10.1038/32871.

Stewart, J.H., 1988, Tectonics of the Walker Lane belt, western Great Basin—Mesozoic and Cenozoic deformation in a zone of shear, in Ernst, W.G., ed., Metamorphism and Crustal Evolution of the Western United States: Englewood Cliffs, New Jersey, Prentice Hall, p. 683–713.

Stockli, D.F., Dumitru, T.A., McWilliams, M.O., and Farley, K.A., 2003, Cenozoic tectonic evolution of the White Mountains, California and Nevada: Geological Society of America Bulletin, v. 115, p. 788–816, doi: 10.1130/0016-7606(2003)115<0788:CTEOTW>2.0.CO;2.

Surpless, B.E., Stockli, D.F., Dumitru, T.A., and Miller, E.L., 2002, Two-phase westward encroachment of Basin and Range extension into the northern Sierra Nevada: Tectonics, v. 21, p. 2-1–2-13.

Sylvester, A.G., 1988, Strike-slip faults: Geological Society of America Bulletin, v. 100, no. 11, p. 1666–1703, doi: 10.1130/0016-7606(1988)100<1666:SSF>2.3.CO;2.

TenBrink, A., Cashman, P.H., and Trexler, J.H., Jr., 2000, Neogene depositional and deformational history of Warm Springs Valley, northern Walker Lane: Geological Society of America, Cordilleran Section, Abstracts with Programs, v. 32, no. 6, p. A-71.

TenBrink, A.L., Cashman, P.H., Trexler, J.H., Jr., Louie, J., and Smith, S., 2002, Active tectonism since 8 Ma in the Sierra Nevada–Basin and Range transition zone, Lassen County, CA: Geological Society of America Abstracts with Programs, v. 34, no. 5, p. A100.

Trexler, J.H., Jr., Cashman, P.H., Henry, C.D., Muntean, T.W., Schwartz, K., TenBrink, A., Faulds, J.E., Perlins, M., and Kelly, T.S., 2000, Neogene basins in western Nevada document tectonic history of the Sierra Nevada–Basin and Range transition zone for the last 12 Ma, in Lageson, D.R., et al., eds., Great Basin and Sierra Nevada: Geological Society of America Field Guide 2, p. 97–116.

Whitehill, C.S., Miller, E.L., Colgan, J.P., Dumitru, T.A., Lerch, D.W., and McWilliams, M.O., 2004, Extent, style, and age of Basin and Range faulting east of Pyramid Lake: Geological Society of America, Rocky Mountain Section, Abstracts with Programs, v. 36, no. 4, p. 37.

Wills, C.J., and Borchard, G., 1993, Holocene slip rate and earthquake recurrence on the Honey Lake fault zone, northeastern California: Geology, v. 21, p. 853–856, doi: 10.1130/0091-7613(1993)021<0853:HSRAER>2.3.CO;2.

Yamaoka, K., and Shimada, I., 1962, On fossil diatoms from the Onikobe, Narugo-Cho, Miyagi Prefecture: Mineral Resources for Industry in Tohoju District, v. B-6, p. 259–262.

MANUSCRIPT ACCEPTED BY THE SOCIETY 21 JULY 2008

The Geological Society of America
Special Paper 447
2009

Tertiary volcanic stratigraphy and paleotopography of the Diamond and Fort Sage Mountains: Constraining slip along the Honey Lake fault zone in the northern Walker Lane, northeastern California and western Nevada

Nicholas H. Hinz
James E. Faulds*
Christopher D. Henry
Nevada Bureau of Mines and Geology, MS 178, University of Nevada, Reno, Nevada 89557, USA

ABSTRACT

The Honey Lake fault zone is one of four major, northwest-striking dextral faults that constitute the northern Walker Lane in northwestern Nevada and northeastern California. Global positioning system (GPS) geodetic data indicate that the northern Walker Lane accommodates ~10%–20% of the dextral motion between the North American and Pacific plates. Regional relations suggest that dextral movement in the Honey Lake area began ca. 6–3 Ma. Five 31.3–25.3 Ma ash-flow tuffs, totaling ~250 m in thickness, were distinguished in a paleovalley in the Black Mountain area of the Diamond Mountains, southwest of the Honey Lake fault. Four of these tuffs, totaling ~200 m in thickness, also occupy a paleovalley in the Fort Sage Mountains northeast of the fault. On the basis of the similar tuff sequences, we infer that the Diamond and Fort Sage Mountains contain offset segments of a once-continuous, westerly trending late Oligocene paleovalley. Paleomagnetic data from the 25.3 Ma Nine Hill Tuff indicate negligible vertical-axis rotation in the Diamond and Fort Sage Mountains.

Correlation of the paleovalley segments in the Diamond and Fort Sage Mountains suggests 10–17 km of dextral displacement across the Honey Lake fault. About 10 km of offset is favored on the basis of constraints near the southeast end of the fault. The spread of possible offset values implies long-term slip rates of ~1.7–2.8 mm/yr for a 6 Ma initiation, and ~3.3–5.7 mm/yr for a 3 Ma initiation. These rates are comparable to slip rates inferred from Quaternary fault studies and GPS geodesy.

Keywords: Walker Lane, paleovalley, Honey Lake fault, ash-flow tuffs, strike-slip faults, Nevada.

---

*Corresponding author e-mail: jfaulds@unr.edu.

Hinz, N.H., Faulds, J.E., and Henry, C.D., 2009, Tertiary volcanic stratigraphy and paleotopography of the Diamond and Fort Sage Mountains: Constraining slip along the Honey Lake fault zone in the northern Walker Lane, northeastern California and western Nevada, *in* Oldow, J.S., and Cashman, P.H., eds., Late Cenozoic Structure and Evolution of the Great Basin–Sierra Nevada Transition: Geological Society of America Special Paper 447, p. 101–131, doi: 10.1130/2009.2447(07).

INTRODUCTION

It has long been known that the Pacific–North American plate motion is primarily taken up by right-lateral slip on the San Andreas fault system and that extension in the Basin and Range Province is intimately linked to the evolving transform boundary (Atwater, 1970; Wernicke, 1992). Global positioning system (GPS) geodetic studies across the western United States (Feigl et al., 1993; Williams et al., 1994; Bennett et al., 1999; Antonellis et al., 1999; Argus and Gordon, 2001) have shown that the San Andreas fault system accounts for only 75%–80% of the motion between the Pacific and North American plates. The remaining 20%–25% of that motion is taken up primarily by systems of strike-slip faults within the western Great Basin, largely on the east side of the Sierra Nevada block (Fig. 1; Dixon et al., 1995, 2000; Bennett et al., 1997, 1998, 2003; Thatcher et al., 1999; Miller et al., 2001; Oldow et al., 2001; Hammond and Thatcher, 2004; Kreemer et al., 2006). These strike-slip fault

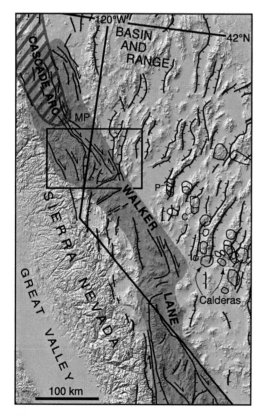

Figure 1. Faults of the Walker Lane (modified from Stewart, 1988) on a shaded elevation model (modified slightly from Faulds et al., 2005a). Circles in central Nevada are calderas, including sources of ash-flow tuffs studied in this project. C—Campbell Creek caldera, source of tuff of Campbell Creek; MP—Modoc Plateau; P—Poco Canyon caldera, source of tuff of Chimney Spring (John, 1995). Box shows area of Figure 2.

systems have been referred to as the Walker Lane in northern reaches (Locke et al., 1940; Stewart, 1988; Oldow, 1992) and the eastern California shear zone in the south (Dokka and Travis, 1990; Dixon et al., 1995).

The magnitude and timing of deformation is poorly defined across much of the Walker Lane and eastern California shear zone. Thus, the kinematic role of this strike-slip fault system in the evolving transform boundary between the Pacific and North American plates is not well understood. Determining the offset and timing of deformation is notoriously difficult, however, for most strike-slip faults. Strike-slip faults within the Walker Lane and eastern California shear zone are no exception. Lacking fortuitously offset volcanic centers or discrete volcanic fields (e.g., Anderson, 1973; Graymer et al., 2002), the complex Tertiary volcanic stratigraphy that characterizes the western Great Basin hinders determination of strike-slip displacement. However, tuff-filled Oligocene paleovalleys have recently been identified and used as piercing lines across parts of the Walker Lane (Henry et al., 2003; Faulds et al., 2003c, 2005a).

Based on analysis of Oligocene paleovalleys, this paper constrains cumulative dextral offset across the Honey Lake fault zone, one of the major strike-slip faults in the northern part of the Walker Lane. Detailed and reconnaissance geologic mapping, paleomagnetic and anisotropy of magnetic susceptibility investigations, and $^{40}Ar/^{39}Ar$ geochronology were used to determine the Tertiary stratigraphy and paleotopography in the Diamond and Fort Sage Mountains (Fig. 2), which abut the Honey Lake fault zone in northeastern California and northwestern Nevada, respectively (Hinz, 2004). Constraining the magnitude of dextral slip on the Honey Lake fault zone is crucial for estimating cumulative displacement across the northern Walker Lane and relating dextral shear in this region to the evolving plate boundary.

NORTHERN WALKER LANE

In northwestern Nevada and adjacent northeastern California the Walker Lane consists of four major left-stepping, northwest-striking right-lateral faults (Fig. 1). From west to east, these are the Mohawk Valley, Honey Lake, Warm Springs Valley, and Pyramid Lake fault zones (Fig. 2). We refer to this series of faults as the northern Walker Lane. GPS geodetic studies indicate that the northern Walker Lane currently accommodates ~4–10 mm/yr of dextral shear (Thatcher et al., 1999; Bennett et al., 1999, 2003; Oldow et al., 2001; Hammond and Thatcher, 2004; Kreemer et al., 2006). Strike-slip faults in the northern Walker Lane are marked by Quaternary scarps (Anderson and Hawkins, 1984; Wills and Borchardt, 1993; Briggs and Wesnousky, 2004, 2005) and appear to be intimately linked to normal-fault systems in the western part of the Basin and Range (Faulds et al., 2003c, 2005a, b). Quaternary fault scarps (e.g., Bell, 1984; Briggs and Wesnousky, 2004, 2005), seismological data (Ichinose et al., 1998), and kinematic relationships (Faulds et al., 2005b) indicate that northwest-striking dextral and northerly striking normal faults are both active in the contemporary strain field.

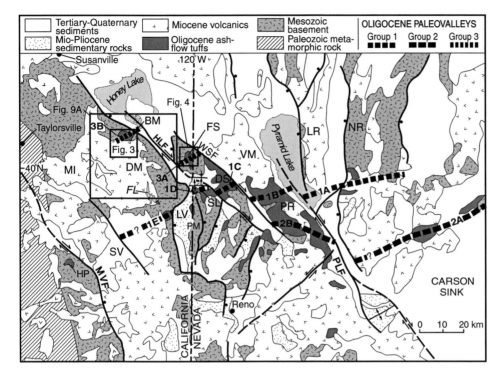

Figure 2. Generalized geologic map of the northern Walker Lane, showing the four primary left-stepping, NW-striking dextral faults and offset paleovalley axes. The paleovalleys are shown as three distinct groups, based on probable correlations (modified slightly from Faulds et al., 2005a, 2005b). BM—Black Mountain; DM—Diamond Mountains; DS—Dogskin Mountain; FL—Frenchman Lake; FS—Fort Sage Mountains; HLF—Honey Lake fault; HP—Haskell Peak; LR—Lake Range; LV—Long Valley; MI—Mount Ingalls; MVF—Mohawk Valley fault; NR—Nightingale Range; PLF—Pyramid Lake fault; PM—Peterson Mountain; PR—Pah Rah Range; SL—Seven Lakes Mountain; SV—Sierra Valley; VM—Virginia Mountains; WSF—Warm Springs Valley fault. Boxes outline study areas in Diamond and Fort Sage Mountains and correspond to areas shown in Figures 3, 4, and 9A, respectively.

Although intimately linked in the current strain field, normal and strike-slip faulting may have begun at different times in the northern Walker Lane region. Studies of Neogene basins near the northern Walker Lane have established the onset of Basin and Range normal faulting at ca. 12–10 Ma (Trexler et al., 2000; Henry and Perkins, 2001). The onset of strike-slip faulting is more poorly defined. On the basis of clockwise vertical-axis rotations gleaned from paleomagnetic data, strike-slip faulting began ca. 9–5 Ma at the latitude of Carson City (Cashman and Fontaine, 2000). Estimates range from ca. 3–6 Ma for the onset of strike-slip faulting farther north, where 3.5 Ma sedimentary rocks are as highly deformed as Oligocene strata along some of the strike-slip faults (Henry et al., 2002, 2007; Faulds et al., 2003c, 2005a, 2005b). The Mendocino triple junction began migrating north through the latitudes of the northern Walker Lane ca. 6 Ma (Atwater and Stock, 1998) and is currently at 40.3°N latitude, which corresponds to the northern part of the Honey Lake basin (Fig. 2).

The northern Walker Lane has undergone episodic volcanism from the early Oligocene through the Quaternary. Three major, overlapping episodes of volcanism include (1) ca. 31–23 Ma rhyolitic ash-flow tuffs associated with calc-alkaline magmatism of the "ignimbrite flareup," (2) ca. 22–4 Ma mafic to dacitic flows associated with the ancestral Cascade arc, and (3) middle Miocene to present bimodal volcanism associated with ~east-west Basin and Range extension (McKee and Silberman, 1970; McKee, 1971; Deino, 1985; McKee and Noble, 1986; Best et al., 1989; Christiansen and Yeats, 1992; Garside et al., 2000). In addition to volcanic rocks, thick sedimentary sections started

accumulating in north-trending basins at ca. 12 Ma (Trexler et al., 2000; Henry and Perkins, 2001). Northward migration of the Mendocino triple junction has coincided with a change from arc-to extension-related volcanic signatures (Wilson, 1989; Schwartz, 2001; Schwartz et al., 2002; Cousens et al., 2003) and is possibly associated with a 3 Ma episode of extension in the region (Trexler et al., 2000; Henry and Perkins, 2001; Henry et al., 2002).

Of particular importance to this study are the ash-flow tuffs that were erupted during the ignimbrite flareup, a period of voluminous felsic volcanism that swept southward across Nevada and western Utah from Oligocene to early Miocene time (Best et al., 1989; Christiansen and Yeats, 1992) before strike-slip faulting began in the northern Walker Lane. Multiple ash-flow tuffs erupted from calderas in central Nevada (Best et al., 1989) and were deposited in paleovalleys that extended across western Nevada, through northeast California, to the Pacific Ocean (Lindgren, 1911; Henry et al., 2003; Faulds et al., 2005a, 2005b), which at that time occupied the Central Valley in California. Many of the tuffs therefore serve as regional structural markers. Moreover, the tuff-filled paleovalleys can be used as piercing lines across strike-slip faults in the Walker Lane (e.g., Faulds et al., 2005a).

HONEY LAKE FAULT ZONE

The ~80-km-long Honey Lake fault zone extends from the southern part of Warm Springs Valley, Nevada, on the southeast to near Susanville, California, on the northwest (Fig. 2). Its surface trace is marked by fault-related landforms, including Quaternary scarps up to 6 m high, offset stream channels and alluvial-fan surfaces, and sag ponds at right steps. The best exposures of the Honey Lake fault zone are found in the Honey Lake basin, a large composite basin that started forming in middle to late Miocene time (Roberts, 1985; Ceron, 1992). Gravity modeling (Ceron, 1992) indicated that up to 3 km of Neogene sedimentary rocks fill the basin. The Neogene sedimentary rocks are locally well exposed and highly deformed between major strands of the Honey Lake fault zone (Mass et al., 2003, 2005; Park et al., 2004). Based on an offset stream bank along Long Valley Creek in the southeastern part of the Honey Lake basin, Wills and Borchardt (1993) established a Quaternary slip rate of 1.1–2.6 mm/yr for the Honey Lake fault zone, which is similar to estimated Quaternary slip rates for other major strike-slip faults in the northern Walker Lane (Briggs and Wesnousky, 2004). Minor strike-slip faults have also been noted in the Susanville, California, area (Fig. 2; Roberts, 1985; Grose and Porro, 1989; Grose et al., 1990, 1992; Colie et al., 2002). These faults may be a northwesterly extension of either the Honey Lake or Warm Springs Valley fault zones. The Modoc Plateau of northern California (Fig. 1) also contains many other minor, strike-slip faults related to the northern Walker Lane (e.g., Pease, 1969; Page et al., 1993; Grose, 2000b) and possibly to the Honey Lake fault zone.

DIAMOND AND FORT SAGE MOUNTAINS

Black Mountain (2368 m elevation) lies along the crest of the ~750–900-m-high Diamond Mountains escarpment along the southwest side of the Honey Lake basin (Fig. 2). The escarpment exposes a large area of Cretaceous Sierra Nevadan batholithic basement (Oldenburg, 1995). Nearly flat-lying sections of Tertiary volcanic rocks dominate the interior of the Diamond Mountains (Figs. 2 and 3). The northeastern escarpment of the central Diamond Mountains coincides with a normal fault that parallels the Honey Lake fault zone (Grose et al., 1990; Henry et al., 2007).

Similar in elevation and overall composition to the Diamond Mountains, the Fort Sage Mountains (1800–2150 m peak elevations) consist largely of Cretaceous batholithic basement (Oldenburg, 1995), with the upper, central part of the mountains covered by Tertiary volcanic rocks (Figs. 2 and 4). Bounded by a system of strike-slip, normal, and oblique-slip faults, the Fort Sage Mountains constitute a fault block between the Honey Lake and Warm Springs Valley fault zones. A normal fault with Quaternary scarps and a 1950 surface rupture bounds the west side of the Fort Sage Mountains (Gianella, 1957). Previous studies interpreted a strand of the Warm Springs Valley fault zone along the base of the steep northeast escarpment (Grose, 1984; Grose et al., 1990). Significant vertical relief along the southwest and southeast sides of the Fort Sage Mountains indicates a large normal component along at least some of the faults.

Several possible Oligocene paleovalley segments straddle the Honey Lake fault zone (Fig. 2). These segments are best exposed (1) in the Black Mountain area of the Diamond Mountains directly west of the Honey Lake fault zone, (2) at the crest of the Fort Sage Mountains to the east of the Honey Lake fault

zone, (3) at the north end of Dogskin Mountain also on the east side of the Honey Lake fault zone, and (4) at Seven Lakes Mountain in a restraining bend in the Honey Lake fault zone. Grose et al. (1990) documented ~200–250 m of undefined ash-flow tuffs in both the Diamond and Fort Sage Mountains. An ~625-m-thick section containing 13 ash-flow tuffs was distinguished in the paleovalley segment at Dogskin Mountain (Henry et al., 2004), and 14 ash-flow tuffs were observed at Seven Lakes Mountain (Henry et al., 2006).

This study focused on the previously undefined ash-flow-tuff sections in the Diamond and Fort Sage Mountains. The Mesozoic basement and Miocene volcanic rocks were not studied in detail.

Stratigraphy

To define the ash-flow-tuff stratigraphy, as well as the pre-tuff paleotopography, areas of ~50 km^2 and 25 km^2 were mapped in the Black Mountain area and Fort Sage Mountains, respectively (Figs. 3 and 4). The Black Mountain area was mapped in detail, whereas mapping of the Fort Sage Mountains involved field reconnaissance transects and analysis of aerial photos. In both areas the Cenozoic stratigraphic section includes Oligocene ash-flow tuffs, Miocene volcanic rocks and intrusions, and surficial Quaternary deposits.

Diamond Mountains

Five ash-flow tuffs were distinguished in the Black Mountain area (Tables 1, 2; Figs. 3 and 5). From oldest to youngest, these include the 31.3 Ma tuff of Axehandle Canyon, the 31.0 Ma tuff of Rattlesnake Canyon, the 30.4 Ma tuff of Sutcliffe, the 28.8 Ma tuff of Campbell Creek, and the 25.3 Ma Nine Hill Tuff. All five tuffs correlate with units found elsewhere in the northern Walker Lane and Sierra Nevada region (Deino, 1985; Garside et al., 2003; Faulds et al., 2003a, 2003b, 2005b; Henry et al., 2004, 2006). Correlations were based on stratigraphic position, phenocryst assemblages, and $^{40}Ar/^{39}Ar$ ages (Table 1; Hinz, 2004).

The primary exposure of tuff in the Black Mountain area is ~4.5 km wide by 7 km long, elongated N70°E, and up to 250 m thick (Fig. 3). Here, the tuffs of Rattlesnake Canyon and Campbell Creek are the most extensive, with thicknesses up to 125 m each and moderately to densely welded zones forming cliffs with 10–40 m of relief. The tuff of Sutcliffe is also as thick as 125 m but crops out principally in the western half of this primary exposure in low-relief, blocky masses. The tuff of Axehandle Canyon is up to 70 m thick along the western and northeastern margins of the Black Mountain area, locally forming ledges with up to 8 m of relief. The Nine Hill Tuff forms one small, 30-m-thick, low-relief outcrop along the southeastern margin of this primary exposure.

A secondary area of exposure includes several outcrops directly south of Black Mountain aligned along a N75°W trend. These outcrops are composed almost entirely of the tuff of

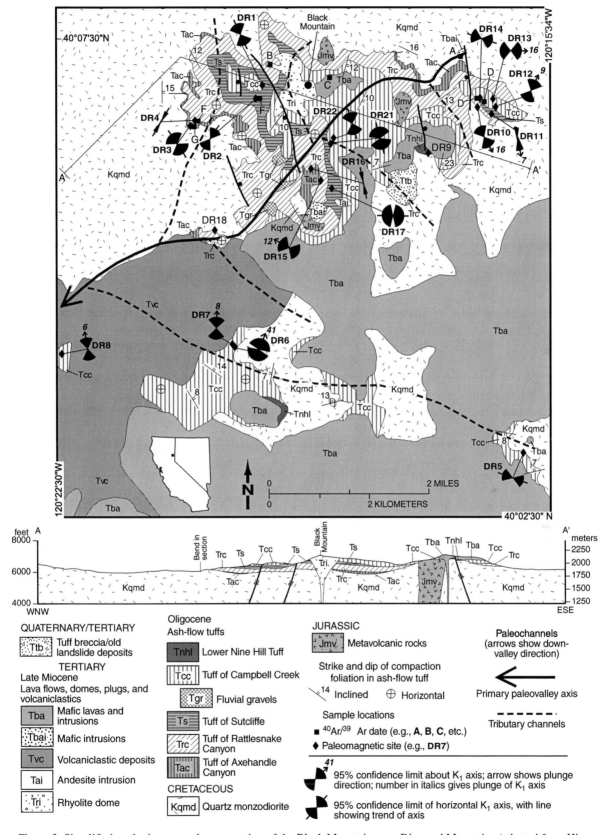

Figure 3. Simplified geologic map and cross section of the Black Mountain area, Diamond Mountains (adapted from Hinz, 2004), showing interpreted paleovalleys developed on Oligocene erosion surface, as well as anisotropy of magnetic susceptibility (AMS) data (bow-tie symbols).

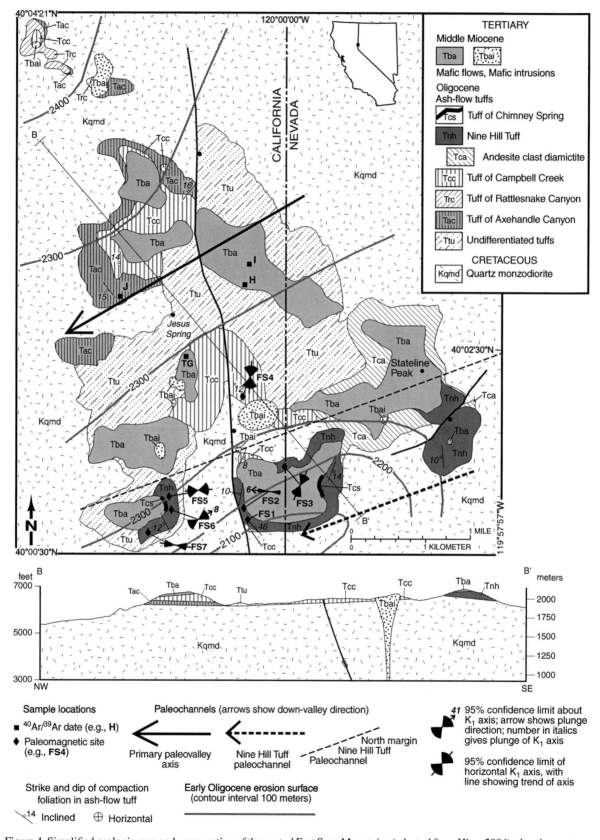

Figure 4. Simplified geologic map and cross section of the central Fort Sage Mountains (adapted from Hinz, 2004), showing contours of the early Oligocene erosion surface and anisotropy of magnetic susceptibility (AMS) data (bow-tie symbols). Contours are corrected for a 6° west tilt about a north-trending axis (N0°E) and fault offsets. TG—14.3 ± 0.4 Ma K/Ar date (Grose et al., 1990).

TABLE 1. $^{40}Ar/^{39}Ar$ AGES OF VOLCANIC ROCKS AND INTRUSIONS

Sample	Sample ID #	Rock type	N Latitude	W Longitude	Material					Ages (Ma)									

Black Mountain area, Diamond Mountains

Step heating						mg	Plateau	±2σ	%³⁹Ar	Steps	Isochron	±2σ	$^{40}Ar/^{36}Ari$	±2σ	MSWD	Steps	Total gas	±2σ
A	NH02-103	Andesite intrusion	40° 7.317'	120° 16.850'	Matrix	26.93	**8.80**	0.08	61	2/9	**8.79**	0.12	296.2	0.9	2.4	3/9	9.60	0.30
B	NH02-81	Rhyolite intrusion	40° 7.233'	120° 19.550'	Plagioclase	26.84	**11.33**	0.35	58	4/10	**11.36**	0.12	291.0	6.0	5.2	5/10	11.99	0.28
C	NH02-86	Basaltic andesite intrusion	40° 7.083'	120° 18.683'	Matrix	25.06	**11.48**	0.16	64	5/9	**11.49**	0.06	294.4	0.8	2.3	6/9	**11.54**	0.35

Single crystal		Rhyolite ash-flow tuffs				Weighted mean	±2σ	K/Ca	±2σ	n
D	H02-90	Tuff of Campbell Creek, upper cooling unit	40° 6.873'	120° 16.482'	Sanidine	**28.68**	0.09	53.4	8.3	15
D	H02-91	Tuff of Campbell Creek, lower cooling unit	40° 6.870'	120° 16.498'	Sanidine	**28.80**	0.08	52.5	8.0	10
E	H02-80	Tuff of Sutcliffe	40° 6.843'	120° 19.775'	Sanidine	**30.33**	0.08	32.2	7.4	15
F	H02-79	Tuff of Rattlesnake Canyon	40° 6.627'	120° 20.585'	Sanidine	**31.00**	0.09	39.0	5.1	14
G	H02-78	Tuff of Axehandle Canyon	40° 6.582'	120° 20.620'	Sanidine	**31.28**	0.08	46.4	13.3	15

Fort Sage Mountains

Step heating						mg	Plateau	±2σ	%³⁹Ar	Steps	Isochron	±2σ	$^{40}Ar/^{36}Ari$	±2σ	MSWD	Steps	Total gas	±2σ
H	NH02-22	Basalt lava	40° 2.450'	120° 00.400'	Plagioclase	26.50	**14.78**	0.29	56	6/10	**14.92**	0.10	223.0	4.0	1.6	6/10	13.97	0.21
I	NH02-26	Basalt lava	40° 2.617'	120° 00.367'	Matrix	25.43	**14.82**	0.15	100	8/8	**14.78**	0.04	293.6	0.9	1.3	6/8	**14.52**	0.27

Single crystal		Rhyolite ash-flow tuffs				Weighted mean	±2σ	K/Ca	±2σ	n
J	NH02-18	Tuff of Axehandle Canyon	40° 2.317'	120° 01.050'	Sanidine	**31.27**	0.10	46.2	5.3	14

Note: Ages in bold are best estimates of eruption or emplacement age; ages in regular type are alternative age calculations; mg—weight of each step heating sample; %³⁹Ar—percentage of ³⁹Ar used to define plateau age; Steps—number of incremental heating steps out of total used in age calculation; MSWD—mean square of weighted deviates; n—number of individual grains used to define weighted mean age. Decay constants and isotopic abundances after Steiger and Jäger (1977). Minerals were separated from crushed, sieved samples by standard magnetic and density techniques; sanidine and plagioclase were leached with dilute HF to remove matrix and handpicked. Analyses performed at the New Mexico Geochronological Research Laboratory (methods in McIntosh et al., 2003). Samples were irradiated in Al discs for 7 hours in D-3 position, Nuclear Science Center, College Station, Texas. Neutron flux monitor, Fish Canyon Tuff sanidine (FC-1); assigned age, 28.02 Ma (Renne et al., 1998). Single sanidine grains were fused with a CO_2 laser operating at 10 W. Other samples were heated in a low blank, molybdenum resistance furnace. Extracted gases were purified with SAES GP-50 getters. Argon was analyzed with a Mass Analyzer Products (MAP) model 215-50 mass spectrometer operated in static mode. Weighted mean $^{40}Ar/^{39}Ar$ ages calculated by the method of Samson and Alexander (1987).

TABLE 2. PHENOCRYST ASSEMBLAGES IN TERTIARY VOLCANIC ROCKS

Unit	Age (Ma)	Total phenocrysts	Sanidine	Plagioclase	Quartz	Biotite	Anorthoclase	Hornblende	Augite	Olivine	Hypersthene
Diamond Mountains											
Tba	11.5	<1–8	0	0–1	0–1	0	0	0	0–4	0–4	0–1
Tdi	~7–11	12–24	6–12	6–12	<1	<1	0	0	0	0	0
Tbap	~7–11	4	0	0	0	0	0	4	0	0	0
Tri	11.3	6–10	0	5–8	1	<1	0	0	0	0	0
Tnh	25.3	3–5	2–2.5	<1	<1–1.5	<1	<1	0	0	0	0
Tcc	28.8	7–12	3–4	2–4	2–4	<1	0	0	0	0	0
Ts	30.4	10–18	3–9	5–8	<1	1–2	0	0	0	0	0
Trc	31.0	8–10	5–7	2–5	<1	<1	0	0	0	0	0
Tac	31.3	10–30	1–2	7–22	1–2.5	<1–3	0	trace	0	0	0
Fort Sage Mountains											
Tba	14.8	5–25	0	5–25	0	0	0	0	trace	<1–3	trace
Tbai	~14–15	20–35	0	20–29	0	0	0	0	0–4	0–3	0–4
Tcs	25.1	25	11	3	11	<1	0	0	0	0	0
Tnh	25.3	5–15	4–11	<1–2	<1	<1	<1–2	0	0	0	0
Tcc	28.8	8–12	3–4	2–4	2–4	<1	0	0	0	0	0
Trc	31.0	8–10	5–7	2–5	<1	<1	0	0	0	0	0
Tac	31.3	10–15	3–4	6–8	1	1–2	0	0	0	0	0

Note: Phenocryst content of the Miocene lavas and intrusions and Oligocene ash-tuffs in the Diamond and Fort Sage Mountains. Phenocryst quantities are percentage of total volume of each unit. Unit abbreviations same as in Figures 3 and 4. Age estimates of Tdi and Tbap from the Diamond Mountains are based on this study and Grose et al. (1990). Age of Tbai from the Fort Sage Mountains is based on probable correlation with Tba and the Pyramid sequence.

Campbell Creek but include one small outcrop of Nine Hill Tuff. The tuffs approach a thickness of 60 m in this area.

The 25.3 Ma Nine Hill Tuff is a distinctive unit that is widely distributed in the western Great Basin and northern Sierra Nevada (Deino, 1985). It is characterized by abundant, large (to 40 cm) fiamme and three feldspars (sanidine, plagioclase, and anorthoclase). Two phases of Nine Hill Tuff, a lower sparsely porphyritic zone and an upper, moderately porphyritic zone, are found in the region (Deino, 1985). Only the lower phase crops out in the Diamond Mountains, at two localities.

Thin lenses of fluvial gravels crop out locally at the base of the tuff section and between tuffs, but these were generally too thin to map at a scale of 1:24,000. Only a 20-m-thick gravel that fills a narrow, inter-tuff paleochannel was large enough to map (Tgr, Fig. 3).

Miocene basaltic andesite and lesser andesite and low-silica rhyolite form plugs, domes, and flows that respectively intrude and cap Mesozoic basement and Oligocene ash-flow tuffs. Intercalated volcaniclastic rocks are composed of clasts derived from the Miocene intrusions and volcanic rocks (Tvc, Fig. 3). Three individual intrusions yielded ^{40}Ar/^{39}Ar ages ranging from 11.5 to 8.8 Ma (Table 1). Similar volcanic rocks in other parts of the Diamond Mountains and nearby areas yielded K-Ar or ^{40}Ar/^{39}Ar dates ranging from 16 to 7 Ma (Grose et al., 1990; Grose, 2000b).

Other ash-flow-tuff outcrops shown on published geologic maps (Grose et al., 1989; Saucedo and Wagner, 1990; Wagner and Saucedo, 1990; Grose and Mergner, 2000) in the Diamond Mountains were examined to determine tuff stratigraphy and interpret pre-ash-flow-tuff topography. These supposed Oligocene tuff outcrops include six specific localities that range from tens of meters to 2 km across. The tuff of Campbell Creek and the Nine Hill Tuff were found at the south end of Frenchman Lake with a maximum combined thickness of 60 m (Fig. 2). The tuff of Campbell Creek also appears south of Mount Ingalls, but

any underlying units are covered. The four other published tuff outcrops are either concentrations of ash-flow-tuff clasts within fluvial gravels intercalated with, or large blocks (up to 5 m long) contained within, middle Miocene andesitic volcaniclastic deposits (Hinz, 2004).

Fort Sage Mountains

Five ash-flow tuffs with a maximum combined thickness of ~200 m were observed in the Fort Sage Mountains. From oldest to youngest, these include the 31.3 Ma tuff of Axehandle Canyon, the 31.0 Ma tuff of Rattlesnake Canyon, the 28.8 Ma tuff of Campbell Creek, the 25.3 Ma Nine Hill Tuff, and the 25.1 Ma tuff of Chimney Spring (Tables 1, 2; Figures 4 and 5). On the basis of stratigraphic position, phenocryst assemblages, and one new ^{40}Ar/^{39}Ar date, the lower four units correlate with tuffs in the Diamond Mountains.

The tuffs of Axehandle Canyon and Campbell Creek are the most widespread tuffs in the Fort Sage Mountains, with thicknesses up to 100 and 125 m, respectively. Aerial photo analysis and reconnaissance mapping indicate that they likely account for much of the undifferentiated tuffs shown in Figure 4. The thickest Oligocene sections, which include the tuffs of Axehandle Canyon and Campbell Creek and undifferentiated tuff, crop out in the prominent ridges northeast, northwest, and southwest of Jesus Spring (Fig. 4). These tuffs thin southward, where they pinch out beneath a series of N70°E-trending outcrops of Nine Hill Tuff up to 200 m thick along the southern edge of the map area (Fig. 6A). The Nine Hill Tuff in the Fort Sage Mountains is mostly the lower phase, but as much as 5 m of the upper phase crops out locally. The contact between the lower, sparsely porphyritic phase and the upper, more abundantly porphyritic phase is sharp but does not mark a cooling break. The tuff of Rattlesnake Canyon was only observed in three small outcrops up to 40 m thick at the north end of the Fort Sage Mountains. The tuff

109

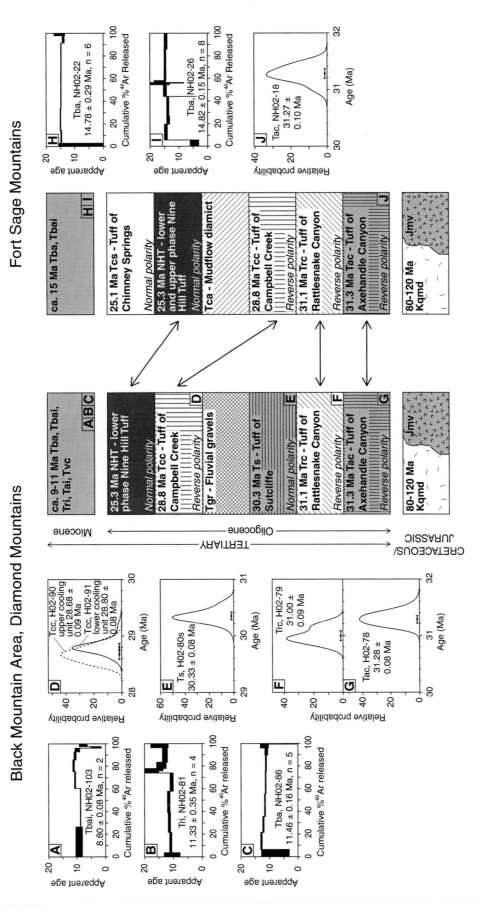

Figure 5. Generalized stratigraphic columns for the Diamond and Fort Sage Mountains, showing correlation of the ash-flow tuffs and ^{40}Ar/^{39}Ar age spectra and ideograms for dated units. Letters for dated samples correspond to identically lettered sample locations in Figures 3 or 4 and lettered samples in Table 1; *n* is number of incremental heating steps used to calculate weighted mean plateau ages. Patterns and labels are the same as in Figures 3 and 4.

of Chimney Spring forms a thin, weakly welded and nonresistant layer in two locations in the Fort Sage Mountains. Although this tuff does not crop out in the Diamond Mountains, it is found at many other localities in the northern Walker Lane (Deino, 1985; Garside et al., 2003; Henry et al., 2003, 2004, 2006; Faulds et al., 2003a, 2003b). As at Black Mountain, lenses of fluvial gravels (too thin to map) locally reside between individual tuffs and at the base of the Oligocene section.

As many as 15 basaltic andesite flows, collectively ranging up to ~150 m thick, cap many ridges in the Fort Sage Mountains. Several aphyric to abundantly porphyritic basaltic andesite plugs also intrude the area. These plugs are petrographically similar to, and may therefore have been, sources for the ridge-capping flows.

Lower and upper flows from a sequence of 10 flows yielded statistically identical ^{40}Ar/^{39}Ar ages of 14.78 ± 0.29 Ma and 14.82 ± 0.15 Ma (Table 1), respectively, which coincide with a 14 Ma K-Ar date from another basaltic andesite sequence within the Fort Sage Mountains (Grose et al., 1990). The petrography and age of these mafic volcanic rocks suggest a correlation with the ca. 15–13 Ma Pyramid sequence in the Virginia Mountains (Faulds and Henry, 2002; Faulds et al., 2003b). It is important to note that the Miocene volcanic rocks in the Fort Sage Mountains predate those exposed in the Black Mountain area of the Diamond Mountains. Thus, the Miocene sections in the two areas do not correlate and therefore cannot be used to gauge offset across the Honey Lake fault zone.

Figure 6. (A) Outcrops of the 25.3 Ma Nine Hill Tuff (Tnh) in an interpreted tributary paleochannel segment in the Fort Sage Mountains. View is looking west-southwest from the easternmost outcrop (foreground) of Nine Hill Tuff toward two other exposures outlined in black on ridges in the middle ground. The Honey Lake fault zone (HLFZ) is approximated by the dashed black line in northern Long Valley at left, and the Diamond Mountains rise in the background. (B) View of the southern flank of the Fort Sage Mountains, looking north-northwest from the Seven Lakes Mountain area. Westward slope of the early Oligocene erosion surface (outlined by solid black line) illustrates the 6° west tilt of the Fort Sage Mountains. Small ridges outlined in dashed black lines interrupt the view of the early Oligocene erosion surface.

Structural Framework

The Diamond Mountains form a relatively coherent structural block bounded by the Honey Lake basin on the northeast and Sierra Valley on the southwest. A prominent ~750–900-m-high escarpment marks the northeastern margin of the Diamond Mountains along the southwest side of the Honey Lake basin. The Mohawk Valley fault zone and minor normal faults near Taylorsville, California, bound this structural block on the west side (Fig. 2). Several north- to northwest-striking, both east- and west-dipping normal faults with up to ~75 m of throw cut the nearly flat-lying volcanic units in the Black Mountain area. Miocene and Oligocene rocks show the same amount of displacement along these faults.

The Diamond Mountains are not measurably tilted (<~3°). The clustering of the 139 poles to compaction foliations of ash-flow tuffs in the Black Mountain area (Fig. 7A) suggests a subhorizontal average foliation and negligible regional tilting. Compaction foliations that are greater than or equal to 15° (Fig. 7B) predominantly strike northwest-southeast, with subequal partitioning between northeast and southwest dips, and are evenly distributed throughout the Black Mountain area.

In the Fort Sage Mountains, north- to northwest-striking, east- and west-dipping normal faults cut the volcanic rocks. As with the Diamond Mountains, displacements of middle Miocene rocks are identical to those of the Oligocene ash-flow tuffs. Displacement of the base of the Miocene lava and the base of Oligocene tuff sections constrains the maximum stratigraphic throw (150 m) across a down-to-east normal fault in the north-central part of the Fort Sage Mountains (Fig. 4). Several west-dipping normal faults are shown on published geologic maps (Grose et al., 1989, 1990) of the western slope of the Fort Sage Mountains.

Several features indicate a gentle west tilt (~6°) of the Fort Sage Mountains. For example, the nonconformity at the base of the Oligocene section slopes gently westward throughout the mountain range (Fig. 6B). In addition, the base of the Nine Hill Tuff and the contact between the Miocene mafic lavas and the Oligocene tuff section dip gently westward. Collectively, these features suggest an ~6° westward tilt of the Fort Sage Mountains about a north- to northeast-trending axis. The gentle westward tilt of the range may have been accommodated by east- to northeast-dipping normal and dextral-normal faults along the east and northeast flanks of the Diamond Mountains and southwest margin of the Honey Lake basin (Fig. 2).

Oligocene Erosion Surface

Relief on the nonconformity at the base of the Oligocene ash-flow-tuff section largely reflects paleotopography developed on an Eocene–early Oligocene erosion surface. From the early to middle Tertiary (Stewart, 1980), river systems flowed west through paleovalleys and paleocanyons from central Nevada across California to the Pacific Ocean. Auriferous gravels in the Sierra Nevada of California accumulated in these paleovalleys during the Eocene (Lindgren, 1911; Jenkins, 1932; Garside et al., 2005). During the Oligocene to early Miocene, as magmatism swept southward in Nevada, thick sequences of ash-flow tuffs were deposited in these paleovalleys across much of the northern

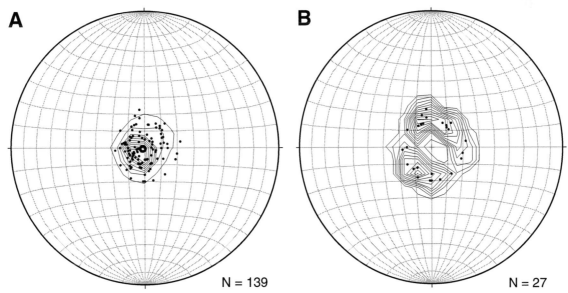

Figure 7. Lower-hemisphere equal-area stereographic projections of poles to compaction foliation from ash-flow tuffs in the Black Mountain area (Hinz, 2004). N is number of measurements (shown as dots). (A) Mean orientation of pole is 237°, 89° with 95% cone of confidence equal to 1.6° (shown as black circle). Density contour interval equals 5% of the data per 1% area. (B) Only attitudes greater than or equal to 15°. Density contour interval equals 2% of the data per 1% area.

Walker Lane, effectively filling some paleovalleys and preserving the pre-tuff erosion surface. Locally, these river channels migrated into less resistant basement rock following deposition of resistant ash-flow-tuff units. In other areas, these rivers cut into parts of the ash-flow-tuff stratigraphy. Subsequent tuffs were then confined to relatively narrow channels. Many of the paleovalleys were sufficiently broad and deep to localize depositional patterns throughout the ignimbrite flareup. Our mapping allows for detailed analysis of the configuration of the early Oligocene erosion surface in both the Diamond and Fort Sage Mountains (e.g., Fig. 8).

Diamond Mountains

The configuration of the early Oligocene erosion surface in the Black Mountain area of the Diamond Mountains reflects an ~5-km-wide, west-southwest–trending paleovalley, here referred to as the Black Mountain paleovalley. Contours of the erosion surface (Fig. 9B) were derived from elevation points along the tuff-basement contacts (Fig. 8). For parts of the erosion surface

either covered by or between exposed tuff-basement contacts, contours were extrapolated between the nearest exposures, with current basement outcrops providing minimum elevations of the early Oligocene erosion surface. Because faulting is minor and tilting is negligible, these contours closely approximate the early Oligocene paleotopography and reflect an ~5-km-wide by ~10-km-long paleovalley, trending 240° (S60°W). Several tributaries intersect the primary, west-southwest–trending paleovalley at high angles, producing a dendritic drainage pattern. At the base of the Oligocene tuff section the tuff of Axehandle Canyon crops out at the northeast and southwest exposures of the primary paleovalley near Black Mountain and along the northern and southern margins of the north-trending tributary paleochannel in the northwest part of the map area, essentially residing in the lowest parts of the Black Mountain paleovalley (Figs. 3, 9B). The thickest parts of the ash-flow-tuff section also lie along the primary paleovalley axis. Two relatively resistant metamorphic roof pendants within the Cretaceous granite generate the most

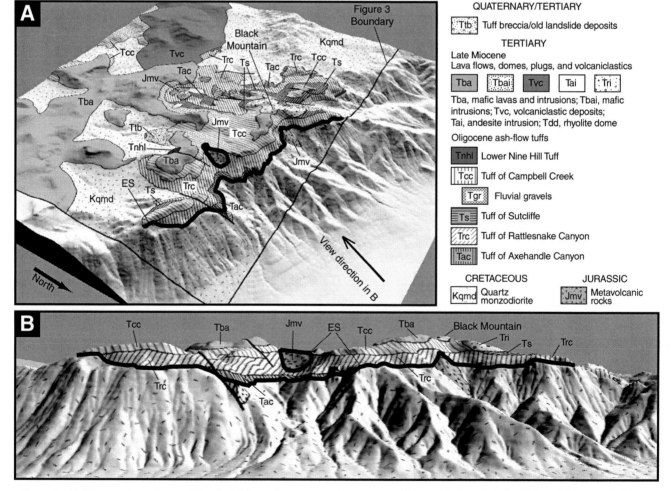

Figure 8. (A) Oblique view of part of the Black Mountain map area draped over a 10-m-resolution shaded-relief model (no vertical exaggeration). A segment of the early Oligocene erosion surface (ES) is highlighted by the thick black line. (B) View southwest, perpendicular to the Diamond Mountain range front (1.5× vertical exaggeration). Units in these figures are the same as in Figure 3.

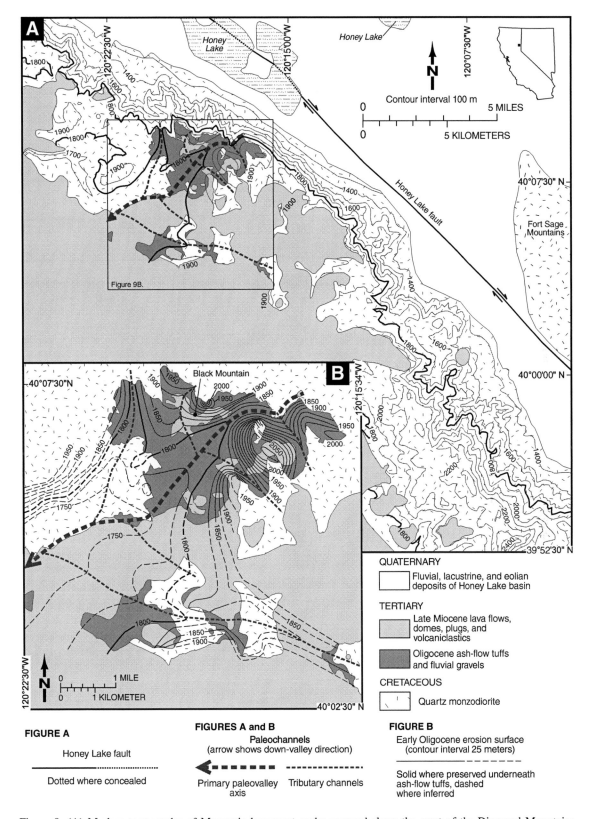

Figure 9. (A) Modern topography of Mesozoic basement rocks exposed along the crest of the Diamond Mountains. Contours of the modern basement rocks are merged with contours of the early Oligocene erosion surface for the Black Mountain map area (inset, Fig. 9B). Geologic units adapted from Figure 4 (this study), Grose et al. (1990), and Grose and Mergner (2000). (B) Paleotopography of the early Oligocene erosion surface for the Black Mountain map area. Geologic units simplified from Figure 3.

dramatic basement relief, with >250 m of vertical relief over a 1 km horizontal distance. The roof pendants also define a narrow axial part of the paleovalley, although they do not appear to delineate paleovalley margins. A 10-km-long southern tributary channel primarily contains outcrops of the tuff of Campbell Creek and one small outcrop of the Nine Hill Tuff. Potential down-valley gradients of 0.5° over 10 km for both the primary paleovalley and the southernmost tributary channel are in accord with the 0.05° to 0.5° gradients of the 50–10 Ma ancestral San Joaquin River in the central Sierra Nevada (Huber, 1981).

The distribution of Mesozoic basement and Tertiary volcanic rocks and current elevations of the Tertiary–Mesozoic basement contact indicate that the Black Mountain paleovalley is the lowest point of the early Oligocene erosion surface along a 45-km-long northwest-trending span of the Diamond Mountains. Elevations of the paleovalley bottom at Black Mountain range from ~1750 to 1800 m (Fig. 9B), whereas current elevations of exposed Mesozoic basement rocks along the crest of the Diamond Mountains are 30–600 m greater than the paleovalley axis (Fig. 9A). Thus, the Black Mountain paleovalley was likely the primary early Oligocene paleochannel in the Diamond Mountains area.

Fort Sage Mountains

The configuration of the early Oligocene erosion surface in the Fort Sage Mountains reflects an ~7-km-wide, west-southwest–plunging paleovalley, here referred to as the Fort Sage Mountains paleovalley. Contours of the erosion surface (Fig. 4) were derived from elevation points along the tuff-basement contacts in a process similar to that for the Black Mountain area. However, because our work in the Fort Sage Mountains was largely of a reconnaissance nature, contours for the Fort Sage Mountains were generated at 100 m intervals in contrast to 25 m intervals for the Black Mountain area.

Analysis of the early Oligocene erosion surface in this area required adjustments for the inferred 6° of westward tilt (Fig. 6B) and displacement along the central, north-striking normal fault (Fig. 4). The throw on this fault reaches a maximum of 150 m in the northern Fort Sage Mountains and decreases to <20 m in the southern part. The resulting contours approximate the early Oligocene paleotopography and reflect an ~6–7.5-km-wide by ~4-km-long southwest-plunging paleovalley, with a southern 5.5 km-long tributary channel trending 250° (S70°W; Fig. 4). The paleovalley axis transects the thickest parts of the tuff of Axehandle Canyon, the tuff of Campbell Creek, and undifferentiated tuffs, and the southern tributary channel primarily contains the Nine Hill Tuff and lesser amounts of the tuff of Campbell Creek and undifferentiated tuffs. Elevations of the early Oligocene erosion surface rise ~100 m directly to the north and south of the primary paleovalley axis. The southern tributary channel provides the most dramatic paleotopographic relief of >200 m. Paleotopographic features do not clearly constrain the orientation of the short paleovalley segment beyond "generally west-to-southwest trending." A relatively small area of the early Oligocene erosion surface is preserved north of the inferred paleovalley axis.

Thus, the northern extent of the paleovalley is not well defined. However, on the basis of the relatively high elevation of parts of the early Oligocene erosion surface exposed to the north of the inferred axis, and similarities in exposed widths and internal stratigraphy between the Fort Sage Mountains paleovalley and other paleovalleys in the region (e.g., Faulds et al., 2005a), it is unlikely that the axis of the Fort Sage Mountains paleovalley was much farther north than that inferred (Fig. 4).

It is noteworthy that both the Fort Sage Mountains and Black Mountain paleovalleys include southern tributary paleochannels. In the Diamond Mountains the 10-km-long southern tributary is filled primarily with the tuff of Campbell Creek with one small outcrop of Nine Hill Tuff. Along the southern edge of the Fort Sage Mountains the Nine Hill Tuff crops out in a 1-km-wide by 5.5-km-long strip that trends 250° (S70°W; Figs. 4 and 6A). This strip thins against a paleohigh preserved along its northern margin (Fig. 4), filling a paleochannel eroded primarily into basement and partly into andesite-clast diamictite (Tca, Fig. 4), tuff of Campbell Creek, and undifferentiated tuff. This paleochannel is largely filled by the Nine Hill Tuff, with minor undifferentiated tuff and tuff of Campbell Creek in its western third. Thus, the southern tributaries to both the Black Mountain and Fort Sage Mountains paleovalleys began forming prior to deposition of the 28.8 Ma tuff of Campbell Creek, but the Fort Sage Mountains tributary continued to develop headwardly before deposition of the Nine Hill Tuff. These tributary channels appear to subparallel the main paleovalleys and thus provide additional evidence of the overall westerly trend of the paleovalley system. They may also indicate slight rearrangement of the drainage systems, as late Oligocene rivers may have shifted toward the margins of the tuff-filled paleovalleys and incised into softer basement rocks. This is especially evident in the Fort Sage Mountains, where the Nine Hill Tuff and underlying andesite-clast diamictite fill the southern tributary but are largely absent in the main paleovalley to the north. Ultimately, the N70°E trend of this 5.5-km-long tributary provides the best paleotopographic constraint for the orientation of the Fort Sage Mountains paleovalley.

It is also important to note that the basal elevations of the Nine Hill Tuff in the tributary paleochannel in the Fort Sage Mountains decrease steadily west-southwest along an ~5° slope (Hinz, 2004). Presumably, these tributary channels were originally subhorizontal (<1°) in accord with other studied paleochannels flowing west into California from Nevada (Huber, 1981). The ~5° west slope of the southern tributary paleochannel therefore supports the inference that the Fort Sage Mountains are tilted gently westward.

ROCK MAGNETISM

The rock magnetic investigations included analysis of characteristic remanent magnetization (ChRM) and anisotropy of magnetic susceptibility (AMS) from 29 sites in the 5 ash-flow tuffs in the Diamond and Fort Sage Mountains (Tables 3 and 4). ChRMs facilitated both the correlation of ash-flow tuffs and

TABLE 3. CHARACTERISTIC REMANENT MAGNETIZATIONS

Site	Unit	Location	Latitude	Longitude	Declination	Inclination	k	α_{95}	N/N_0	Attitude	Corrected for foliation dec., inc.	Corrected for regional tilt dec., inc.	Comments
DR1	Tcc	Black Mountain area	040° 07' 01"	120° 19' 43"	205.7	−33.8	9.3	21.0°	7/9	000, 21 W	190.4, −40.4	--	Poor statistical mean
DR2	Ts	Black Mountain area	040° 06' 54"	120° 19' 49"	014.9	38.1	58.5	9.8°	8/8	subhorizontal	14.9, 38.1	--	
DR3	Tac	Black Mountain area	040° 06' 34"	120° 20' 42"	117.3	−61.4	39.0	14.9°	4/9	subhorizontal	117.3, −61.4	--	
DR4	Trc	Black Mountain area	040° 06' 37"	120° 20' 38"	150.9	−68.0	260.7	5.7°	4/8	275, 06 N	157.6, −62.8	--	
DR5	Tcc	Black Mountain area	040° 03' 10"	120° 15' 53"	204.8	−48.3	114.4	5.2°	8/9	305, 07 NE	206.0, −41.4	--	
DR6	Tcc	Black Mountain area	040° 04' 18"	120° 20' 01"	209.8	−44.5	102.9	9.1°	4/9	010, 05 W	205.0, −46.0	--	
DR7	Tcc	Black Mountain area	040° 04' 20"	120° 20' 04"	208.8	−47.7	328.7	3.7°	6/8	010, 05 W	203.4, −49.1	--	
DR8	Tcc	Black Mountain area	040° 04' 10"	120° 22' 28"	---	---	---	---	0/7	320, 13 W	--	--	Poor statistical mean
DR9	Tnhl	Black Mountain area	040° 06' 18"	120° 17' 15"	348.7	59.1	134.1	4.8°	8/13	333, 10 NE	3.0, 55.2	--	
DR10	Trc	Black Mountain area	040° 06' 47"	120° 16' 34"	161.5	−72.5	392.1	2.8°	8/8	293, 10 NE	175.8, −64.2	--	
DR11	Ts	Black Mountain area	040° 06' 44"	120° 16' 22"	014.5	40.0	84.6	6.6°	7/8	subhorizontal	14.5, 40.0	--	
DR12	Tcc	Black Mountain area	040° 06' 47"	120° 16' 19"	204.7	−42.0	392.1	2.8°	8/8	073, 10 NW	199.7, −34.2	--	
DR13	Tcc	Black Mountain area	040° 06' 52"	120° 16' 21"	196.6	−43.7	341.7	3.0°	8/8	350, 25 W	170.1, −49.7	--	
DR14	Tcc	Black Mountain area	040° 06' 52"	120° 16' 32"	204.2	−42.4	173.2	7.0°	4/8	340, 10 NE	209.7, −35.1	--	
DR15	Tac	Black Mountain area	040° 06' 01"	120° 18' 57"	061.7	−74.9	159.4	4.4°	8/8	305, 15 SW	48.9, −60.7	--	
DR16	Trc	Black Mountain area	040° 06' 07"	120° 19' 04"	161.5	−69.7	335.4	3.3°	7/10	080, 07 N	163.6, −62.8	--	
DR17	Ts	Black Mountain area	040° 05' 53"	120° 18' 48"	012.2	38.1	106.1	5.4°	8/8	subhorizontal	12.2, 38.1	--	
DR18	Trc	Black Mountain area	040° 05' 30"	120° 20' 22"	136.7	−70.8	194.5	3.7°	9/11	350, 10 NE	167.1, −74.0	--	
DR19	Tcc	Frenchman Lake	039° 51' 25"	120° 10' 10"	207.4	−43.2	491.5	2.5°	8/9	022, 08 W	199.8, −43.4	--	
DR20	Tnhl	Frenchman Lake	039° 51' 20"	120° 10' 29"	332.4	40.6	266.2	3.4°	8/8	285, 14 SW	321.8, 50.1	--	
DR21	Tcc	Black Mountain area	040° 06' 30"	120° 18' 53"	---	---	---	---	0/9	355, 13 E	--	--	Poor statistical mean
DR22	Tcc	Black Mountain area	040° 06' 13"	120° 18' 51"	208.0	−38.7	180.5	4.5°	7/8	345, 14 E	214.5, −28.5	359.9, 47.6	
FS1	Tnhl	Fort Sage Mountains	040° 00' 46"	120° 00' 30"	006.4	47.3	779.7	2.4°	6/8	296, 44 NE	12.8, 4.9	356.4, 54.3	
FS2	Tnh	Fort Sage Mountains	040° 00' 52"	120° 00' 31"	004.7	54.2	513.1	2.4°	8/8	280, 25 N	6.4, 29.3		
FS3	Tnh	Fort Sage Mountains	040° 01' 08"	120° 00' 07"	--	--	--	--	--	013, 15 SE	--	--	Lightning
FS4	Tcc	Fort Sage Mountains	040° 01' 38"	120° 00' 31"	211.5	−45.0	405.7	3.3°	6/6	000, 12 E	220.1, −37.9	206.0, −47.9	
FS5	Tnhl	Fort Sage Mountains	040° 01' 43"	120° 01' 11"	353.6	57.6	1047.5	1.7°	8/8	355, 13 SW	334.3, 55.1	344.5, 56.5	
FS6	Tnh	Fort Sage Mountains	040° 00' 47"	120° 01' 11"	321.4	42.1	297.0	3.2°	8/8	050, 13 NW	321.2, 29.1	317.5, 38.2	
FS7	Tnh	Fort Sage Mountains	040° 00' 39"	120° 01' 24"	344.4	44.9	188.0	4.1°	8/8	333, 12 SW	332.2, 46.0	338.9, 43.0	
DR Trc mean	Trc	Diamond Mountains	--	--	152.6	−70.5	352.4	4.9°	4/4	--	--	--	Uncorrected
DR Trc mean	Trc	Diamond Mountains	--	--	165.8	−66.1	165.8	7.2°	4/4	--	--	--	Compaction foliation tilt corrected
DR Ts mean	Ts	Diamond Mountains	--	--	014.4	38.1	1053.0	3.8°	3/3	--	--	--	Uncorrected
DR Tcc mean	Tcc	Diamond Mountains	--	--	205.5	−43.9	341.7	3.0°	8/11	--	--	--	Uncorrected, without DR1, 8, 21
DR Tcc mean	Tcc	Diamond Mountains	--	--	201.5	−42.8	43.4	8.5°	8/11	--	--	--	Compaction foliation tilt corrected, without DR1, 8, 21
DR–FS Tnh mean	Tnh	Diamond & Fort Sage Mountains	--	--	346.2	50.5	37.6	10.0°	7/8	--	--	--	Uncorrected
DR–FS Tnh mean	Tnh	Diamond & Fort Sage Mountains	--	--	346.1	40.8	40.7	10.4°	7/8	--	--	--	Compaction foliation tilt corrected
DR–FS Tnh mean	Tnh	Diamond & Fort Sage Mountains	--	--	341.3	49.3	39.4	9.7°	7/8	--	--	--	Regional 6° tilt correction for Fort Sage Mountains sites and uncorrected for Diamond Mountains
Tnh reference	Tnh		--	--	335.1	57.5	121	3.0°	20/20	--	--	--	Deino (1985)

Note: Paleomagnetic data, Diamond and Fort Sage Mountains. Unit abbreviations same as in Figures 3 and 4. Location of paleomagnetic sites are shown in Figures 3 and 4. Frenchman Lake sites (location not shown) are at south end of Frenchman Lake (Fig. 2). k—precision parameter (Fisher, 1953); α_{95}—cone of 95% confidence about the site mean; N/N_0—number of demagnetized samples used in statistical analysis versus number of total samples analyzed, or number of sites used in site mean calculations versus total number of sites analyzed. Regional tilt correction of 6°W about a N0°E axis was applied for sites from the Fort Sage Mountains. Corrections for compaction foliations generally increased dispersion and therefore were not used in assessing vertical-axis rotations or in determining reference direction for Tcc.

TABLE 4. ANISOTROPY OF MAGNETIC SUSCEPTIBILITY DATA

Site	Unit	Location	Latitude	Longitude	D_{K1}	I_{K1}	D_{K2}	I_{K2}	D_{K3}	I_{K3}	K_1/K_2	K_2/K_3	K-mean (10^{-3} SI)	N/N_0	Attitude	Regional tilt corrected D_{K1}	Regional tilt corrected I_{K1}	VDAB (m)
DR1	Tcc	Black Mountain area	040° 07' 01"	120° 19' 43"	155±45	01±13	246±48	21±04	063±30	69±12	1.011	1.017	0.914	5/8	000, 21 W	--	--	3-5
DR2	Ts	Black Mountain area	040° 06' 54"	120° 19' 49"	034±35	03±07	124±35	09±10	289±12	81±07	1.012	1.068	2.412	12/14	Subhorizontal	--	--	65
DR3	Tac	Black Mountain area	040° 06' 34"	120° 20' 42"	326±58	03±18	056±58	03±14	188±20	86±13	1.007	1.019	1.266	10/10	Subhorizontal	--	--	5
DR4	Trc	Black Mountain area	040° 06' 37"	120° 20' 38"	031±21	09±10	301±21	06±11	178±13	80±10	1.008	1.019	2.484	10/10	275, 06 N	--	--	3-5
DR5	Tcc	Black Mountain area	040° 03' 10"	120° 15' 53"	109±33	04±17	200±32	04±12	329±19	84±08	1.004	1.023	7.739	12/12	305, 07 NE	--	--	10-15
DR6	Tcc	Black Mountain area	040° 04' 18"	120° 20' 01"	023±79	40±12	281±79	15±16	174±19	47±10	1.001	1.020	0.727	8/9	010, 05 W	--	--	10-15
DR7	Tcc	Black Mountain area	040° 04' 20"	120° 20' 04"	002±45	11±08	095±45	16±14	238±16	70±09	1.005	1.018	3.945	11/11	010, 05 W	--	--	30-35
DR8	Tcc	Black Mountain area	040° 04' 10"	120° 22' 28"	350±35	01±06	260±36	06±11	087±14	84±07	1.006	1.019	0.565	7/9	320, 13 W	--	--	10
DR9	Tnhl	Black Mountain area	040° 06' 18"	120° 17' 15"	--	--	--	--	--	--	--	--	0.038	0/13	333, 10 NE	--	--	12
DR10	Trc	Black Mountain area	040° 06' 47"	120° 16' 34"	150±41	10±10	057±40	17±09	269±12	71±07	1.004	1.026	4.235	10/11	293, 10 NE	--	--	3
DR11	Ts	Black Mountain area	040° 06' 44"	120° 16' 22"	167±20	07±04	076±28	07±18	301±27	80±06	1.023	1.087	0.155	6/11	Subhorizontal	--	--	5
DR12	Tcc	Black Mountain area	040° 06' 47"	120° 16' 19"	026±44	16±04	118±45	09±09	238±10	72±04	1.005	1.027	3.706	8/9	073, 10 NW	--	--	15-40
DR13	Tcc	Black Mountain area	040° 06' 52"	120° 16' 21"	271±52	08±06	001±52	01±10	098±82	82±07	1.002	1.040	6.697	11/13	350, 25 W	--	--	35-45
DR14	Tcc	Black Mountain area	040° 06' 52"	120° 16' 32"	288±39	06±14	198±40	08±21	053±24	80±14	1.006	1.017	1.251	11/11	340, 10 NE	--	--	12-25
DR15	Tac	Black Mountain area	040° 06' 01"	120° 18' 57"	300±49	14±06	034±49	13±05	164±08	71±04	1.005	1.036	6.204	12/12	305, 15 W	--	--	25-30
DR16	Trc	Black Mountain area	040° 06' 07"	120° 19' 04"	348±11	10±06	079±24	02±10	180±25	80±06	1.011	1.008	1.426	8/8	080, 07 N	--	--	5
DR17	Ts	Black Mountain area	040° 05' 53"	120° 18' 48"	271±77	03±12	181±77	08±09	020±14	82±10	1.002	1.058	0.214	7/14	Subhorizontal	--	--	20
DR18	Trc	Black Mountain area	040° 05' 30"	120° 20' 22"	--	--	--	--	--	--	--	--	0.499	4/12	350, 10 NE	--	--	3-5
DR19	Tcc	Frenchman Lake	039° 51' 25"	120° 10' 10"	222±63	03±05	132±63	03±07	354±08	86±05	1.003	1.039	6.647	9/9	022, 08 W	--	--	5-10
DR20	Tnhl	Frenchman Lake	039° 51' 20"	120° 10' 29"	151±50	12±08	241±51	01±09	335±14	78±09	1.003	1.013	2.221	10/12	285, 14 SW	--	--	10-15
DR21	Tcc	Black Mountain area	040° 06' 30"	120° 18' 53"	345±69	04±30	253±70	23±47	084±55	67±20	1.004	1.012	0.471	7/8	355, 13 E	--	--	10-12
DR22	Tcc	Black Mountain area	040° 06' 13"	120° 18' 51"	343±45	03±07	073±45	00±14	167±15	87±06	1.008	1.055	0.120	7/12	345, 14 E	--	--	25-30
Group 1	All	Black Mountain area	--	--	353±74	01±16	083±74	02±19	234±20	87±16	1.002	1.030	3.005	(185)	--	--	--	--
FS1	Tnhl	Fort Sage Mountains	040° 00' 46"	120° 00' 30"	--	--	--	--	--	--	--	--	0.059	0/9	296, 44 E	--	--	15-30
FS2	Tnh	Fort Sage Mountains	040° 00' 52"	120° 00' 31"	275±10	04±03	005±12	01±07	106±11	86±04	1.007	1.014	1.429	7/11	280, 25 N	277±10	06±03	40-70
FS3	Tnh	Fort Sage Mountains	040° 01' 08"	120° 00' 07"	026±30	06±06	117±30	11±08	266±15	78±04	1.002	1.017	10.37	9/9	013, 15 E	027±30	03±06	35
FS4	Tcc	Fort Sage Mountains	040° 01' 38"	120° 00' 31"	005±41	01±13	095±42	01±23	245±28	89±10	1.010	1.014	1.153	7/7	000, 12 E	005±41	00±13	15-20
FS5	Tnh	Fort Sage Mountains	040° 00' 43"	120° 01' 11"	263±32	17±03	165±33	23±10	026±19	61±06	1.007	1.014	0.122	6/9	355, 13 W	263±32	04±04	60
FS6	Tnh	Fort Sage Mountains	040° 00' 47"	120° 01' 11"	060±28	06±04	327±28	16±07	171±08	73±06	1.003	1.018	8.169	9/9	050, 13 NW	058±28	08±04	70
FS7	Tnh	Fort Sage Mountains	040° 00' 39"	120° 01' 24"	278±18	10±06	184±17	07±06	059±10	77±05	1.004	1.015	9.129	10/10	333, 12 SW	277±18	00±06	50
Group 2	Tnh	Fort Sage Mountains	--	--	266±76	04±16	356±76	01±18	102±20	86±15	1.005	1.004	5.004	(54)	--	--	--	--

Note: Anisotropy of magnetic susceptibility data, Diamond and Fort Sage Mountains. Directional statistics same as in Figures 3 and 4. D and I are the declination and inclination of the maximum (K_1), intermediate (K_2), and minimum (K_3) axes of susceptibility, respectively, with error ranges derived from α_{95} cones of 95% confidence. K_1/K_2 and K_2/K_3 are the axial magnitude ratios. K-mean is the mean susceptibility of all samples measured per site (N_0). N is the number of samples that were used to calculate directional statistics per site. Regional tilt correction of 6°W about a N0°E axis was applied for sites from the Fort Sage Mountains. VDAB—vertical distances above base of tuff, with ranges indicating best estimates for buried or poorly exposed basal contacts. Group 1 includes all the samples from the Black Mountain area and corresponds to Figure 12A. Group 2 includes all the Tnh samples from the Fort Sage Mountains and corresponds to Figure 12C.

evaluation of vertical-axis rotation in the Diamond and Fort Sage Mountains. AMS was employed to assess flow directions in ash-flow tuffs, which in turn were used to constrain orientations of paleovalley segments. In the Diamond Mountains, samples from 20 sites were collected in the Black Mountain area and from 2 sites near Frenchman Lake. Samples from the remaining seven sites were gathered in the Fort Sage Mountains. Most of the sites contained the tuff of Campbell Creek and the Nine Hill Tuff.

Remanent Magnetizations

Conventional paleomagnetic methods (e.g., Knight et al., 1986; Hillhouse and Wells, 1991; Hudson et al., 2000) were employed in this study. A portable drill and a Pomeroy orienting fixture were used to collect oriented cores from each site. Samples were prepared for analyses using nonmagnetic saw blades. All analyses were carried out at the Keck Paleomagnetic Laboratory at the University of Nevada, Reno. Remanent magnetizations were generally measured on an Agico JR-5A magnetometer, but a few odd-shaped samples were analyzed on a Schonstedt magnetometer. To isolate components of natural remanent magnetization, all samples were subjected either to alternating field or thermal demagnetization. Demagnetization trajectories were evaluated on orthogonal demagnetization diagrams (Fig. 10). ChRMs were calculated using standard methods such as the multivariate technique of principal component analysis (e.g., Kirschvink, 1980). Conventional statistical analyses on a sphere (e.g., Fisher, 1953) were employed to determine site means and dispersion parameters (Table 3).

Twenty-two of the 29 sites yielded well-grouped site means, with an $\alpha_{95} < 10°$ and $k > 100$ (k is a precision parameter denoting the concentration of the distribution about the mean direction; Fisher, 1953). Marginal site means, with $\alpha_{95} \sim 10°–15°$ and $k > 35$, were obtained at three other sites. Four sites yielded highly dispersed site means, probably induced by lightning strikes.

Most samples behaved simply during demagnetization (Fig. 10). After removal of magnetizations with variable directions at low inductions or temperatures, a single ChRM was first recognized between 5 and 40 mT (milliteslas) and 300 to 500 °C. At higher temperatures and inductions, demagnetization trajectories typically continued straight to the origin.

Demagnetization behaviors suggest that fine-grained, pseudo–single domain titanomagnetite is the principal carrier of the ChRMs in the ash-flow tuffs. Alternating field demagnetization commonly removed nearly all magnetization by 100 mT. Thermal demagnetization effectively removed all magnetization by 590 °C for all samples from the tuff of Rattlesnake Canyon and about half of the samples from the tuffs of Axehandle Canyon and Campbell Creek. Titanomagnetite typically loses most magnetization by 100 mT and 590 °C (Butler, 1992). However, some samples from the tuffs of Sutcliffe and Campbell Creek and most samples of Nine Hill Tuff retained a minor amount of magnetization at 100 mT. In addition, about half of the samples from the tuffs of Axehandle Canyon and Campbell Creek, along with

all samples from the tuff of Sutcliffe and Nine Hill Tuff, retained a small level of magnetization at 590 °C that was completely removed by 610 to 625 °C. This demagnetization behavior indicates that small amounts of hematite or fine-grained maghemite reside in some of the tuffs.

Anisotropy of Magnetic Susceptibility

In recent decades, AMS in ash-flow tuffs has been shown to reflect hydrodynamic orientation of ferromagnetic mineral grains, thus allowing interpretation of flow direction during deposition. Flow directions of ash-flow tuffs obtained through field data and petrographic analyses generally correlate well with flow directions inferred from AMS (Ellwood, 1982; Knight et al., 1986; Wolff et al., 1989; MacDonald and Palmer, 1990; Seaman et al., 1991). AMS has been used to help identify source calderas for ash-flow tuffs (Hillhouse and Wells, 1991; Palmer et al., 1991; MacDonald et al., 1998), document the variation in flow turbulence relative to distance from the source calderas, and demonstrate the control of topography on flow directions (Fisher et al., 1993; Baer et al., 1997; Ort et al., 1999, 2003). Because flow directions are intimately related to paleotopography, it follows that AMS measurements may be used to infer orientations of paleovalley segments in this study.

AMS is expressed in terms of a triaxial ellipsoid with K1, K2, and K3 representing the maximum, intermediate, and minimum axes of susceptibility, respectively. Bulk susceptibility (K) of a rock is defined as (K1+ K2 + K3)/3. F-test statistics produced by SUSAM software for each sample were used to exclude samples that were not statistically anisotropic and samples that did not have statistically unique K-values for each of the three axes such that K1 > K2 > K3. Lineations defined by K1 axes define true flow directions, whereas K3 axes define the inclination of the K1/K2 plane, which is useful in evaluating imbrication of the ellipsoid (Hillhouse and Wells, 1991; Baer et al., 1997; Ort et al., 2003).

Flow directions most commonly parallel the long axes of magnetic grains or the K1 axes of the AMS ellipsoid (Fisher et al., 1993; Baer et al., 1997; Le Pennec et al., 1998; Ort et al., 2003). However, several processes can induce prolate grains to align perpendicular to flow. For example, sedimentological studies of detrital rocks have shown that prolate-shaped clasts can roll along substrate, with their long axes orthogonal to the flow direction (Mills, 1984; Best, 1992). This process has also been linked to transverse-to-flow, K1 orientations in sandstones (Ellwood and Ledbetter, 1977; Taira and Scholle, 1979). In addition, transverse magnetic fabrics from an ignimbrite in the central Andes have been interpreted to result from the rolling of prolate-shaped clasts along a progressively aggrading surface (Ort, 1993). Other circumstances by which nonflow-parallel AMS fabrics can be produced include fixed magnetic grains in pumice and lithic fragments (Le Pennec et al., 1998), domain state of magnetite grains (Cagnoli and Tarling, 1997), and secondary iron-titanium oxide mineral-emplacement in pore spaces (Thomas et al., 1992; Le Pennec et al., 1998).

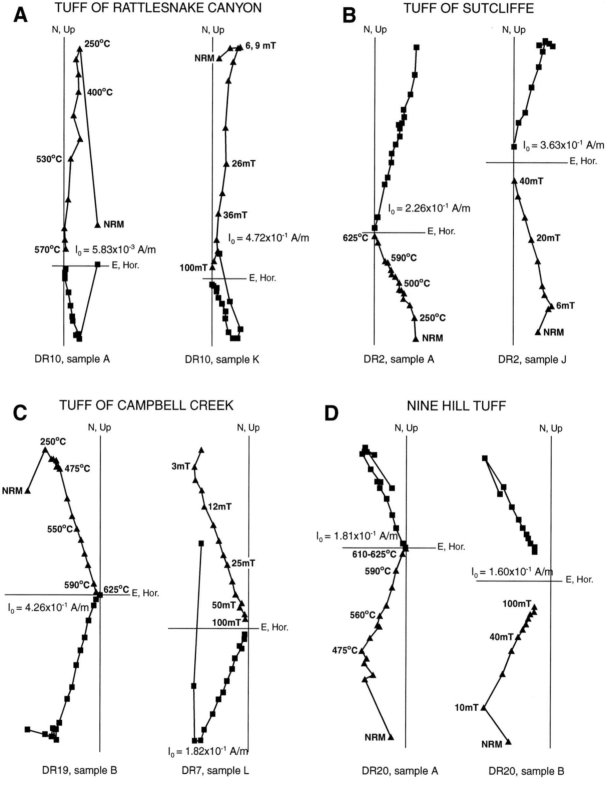

Figure 10. Representative orthogonal demagnetization plots of representative samples. Squares and triangles are projections on the horizontal (declination) and vertical (inclination) planes, respectively. C—Celsius; mT—millitesla; I_0—intensity of the natural remanent magnetization (NRM) prior to demagnetization. Coordinates are not corrected for tilt of strata.

Where K1 axes are flow parallel, imbricated K1 axes can be used to infer actual flow directions, assuming an upstream plunge. Plunging or imbricated K1 axes are defined by comparing K1 inclinations to apparent dips of the compaction foliation, which generally mimics paleotopography (Chapin and Lowell, 1979) and therefore facilitates identifying K1 axes that are imbricated relative to paleotopography.

AMS analyses for this study were conducted with a Kappabridge KLY-3S susceptibility bridge operated with SUSAM 3.0 software. All 29 sites from the Diamond and Fort Sage Mountains, including 7–14 samples per site, were analyzed for AMS. Directional statistics (Fig. 11; Table 4) were acquired with Anisoft 3.0 software using multivariate analysis (Jelinek, 1978). The mean tensor was determined through averaging the individual components of the normalized tensors of specimens, and 95% confidence regions were calculated around the principal directions of the mean tensor.

Twenty-six of the 29 total sites produced statistically viable AMS results (Table 4), with most samples yielding anisotropic results. However, all samples from two Nine Hill Tuff sites (DR9 and FS1) were isotropic. In addition, only four samples from site DR18 in the tuff of Rattlesnake Canyon were statistically anisotropic, whereas five samples are needed to compute directional statistics with Anisoft 3.0 software. Sites from the Fort Sage Mountains were corrected for regional tilt.

The observed magnetic fabrics are characteristic of ash-flow tuffs, with most of the K3 axes clustering near vertical, and moderately to widely dispersed K1 and K2 axes near horizontal (Fig. 11; Table 4). Similar to other studies (e.g., Ort et al., 2003), highly dispersed but geologically reasonable results were obtained. Twenty-six (90%) of the K3 site averages are within 20° of vertical (e.g., DR2 and DR19), two (7%) are within 20° to 30° of vertical (e.g., FS5), and one (3%) is near 45° (DR6). Declinations of K1 and K2 site averages have α_{95} cones of confidence that range from ±10° to ±79° and average near ±40°. Ranges in orientation for K1 axes exhibit greater variability for the Black Mountain area than for the Fort Sage Mountains (Fig. 12). In the Fort Sage Mountains, K1 axes display a bimodal distribution with both north-south and east-west trends, with the majority oriented east-west. In the Black Mountain area, K1 axes show prominent north-south and east-west trends but also exhibit NW-SE and NE-SW trends. For 11 sites from the Diamond and Fort Sage Mountains, the mean K1 inclinations differ from the apparent dip of the compaction foliation at the 95% confidence level, indicating statistically significant imbrication of the ellipsoid at the 95% confidence level.

DISCUSSION

Correlating Oligocene Paleovalleys

The similar sequences and thicknesses of Oligocene ash-flow tuffs suggest that the Black Mountain and Fort Sage Mountains paleovalleys were originally part of one continuous topo-graphic depression. The 31.3–25.3 Ma ash-flow-tuff sequence of the Black Mountain area in the Diamond Mountains correlates well with the 31.3–25.1 Ma tuffs of the Fort Sage Mountains. Four ash-flow tuffs, the tuff of Axehandle Canyon, the tuff of Rattlesnake Canyon, the tuff of Campbell Creek, and the Nine Hill Tuff, are found in both the ~250-m-thick section in the Black Mountain area and the ~200-m-thick section in the Fort Sage Mountains. Although the 25.1 Ma tuff of Chimney Spring crops out in the Fort Sage Mountains, its poor welding indicates a relatively distal exposure. The absence of the tuff of Chimney Spring at Black Mountain is therefore not surprising. Reconnaissance of outlying areas in and around the Diamond Mountains indicates that the Black Mountain paleovalley is the only Oligocene paleovalley preserved in the Diamond Mountains. Both mountain ranges are relatively coherent structural blocks, separated by the Honey Lake fault zone. Only minor faulting and tilting disrupt Oligocene strata in both areas. Both paleovalleys also have a definitive southern tributary filled with the tuff of Campbell Creek and the Nine Hill Tuff. These relations suggest that the Black Mountain and Fort Sage Mountains paleovalleys correlate and therefore can be used as piercing lines to determine the magnitude of dextral displacement across the Honey Lake fault zone. The configuration of the early Oligocene erosion surface indicates west-southwest trends for both the Black Mountain and Fort Sage Mountains paleovalleys. However, inferring offset on the Honey Lake fault zone is contingent on the original orientation of the paleovalleys and the amount of subsequent vertical-axis rotation of the Fort Sage and Diamond Mountain structural blocks, as constrained by paleomagnetic data.

It is important to note that the paleovalley in the Diamond and Fort Sage Mountains may not directly correlate with other paleovalley segments in the northern Walker Lane (Faulds et al., 2005a). The best exposed paleovalley in the northern Walker Lane extends from the Nightingale Mountains across the Pyramid Lake and Warm Springs Valley faults and into the Honey Lake fault zone at Seven Lakes Mountain (Fig. 2). The Black Mountain–Fort Sage Mountains paleovalley appears to be a separate, more northerly system that probably connects westward to other small exposures of ash-flow tuff in the Diamond Mountains (Durrell, 1959; Grose, 2000a), such as an isolated outcrop of the tuff of Campbell Creek south of Mount Ingalls (Fig. 2). East of the Fort Sage Mountains this paleovalley is likely buried under younger rocks.

Rock Magnetism Interpretations

Remanent Magnetizations

For both the Diamond and Fort Sage Mountains, uncorrected-site means for ChRMs from individual ash-flow tuffs cluster more tightly than those corrected for the attitudes of compaction foliation (assuming an original subhorizontal foliation). Before tilt corrections, respective group means tightly cluster for the tuffs of Rattlesnake Canyon, Sutcliffe, and Campbell Creek

Hinz et al.

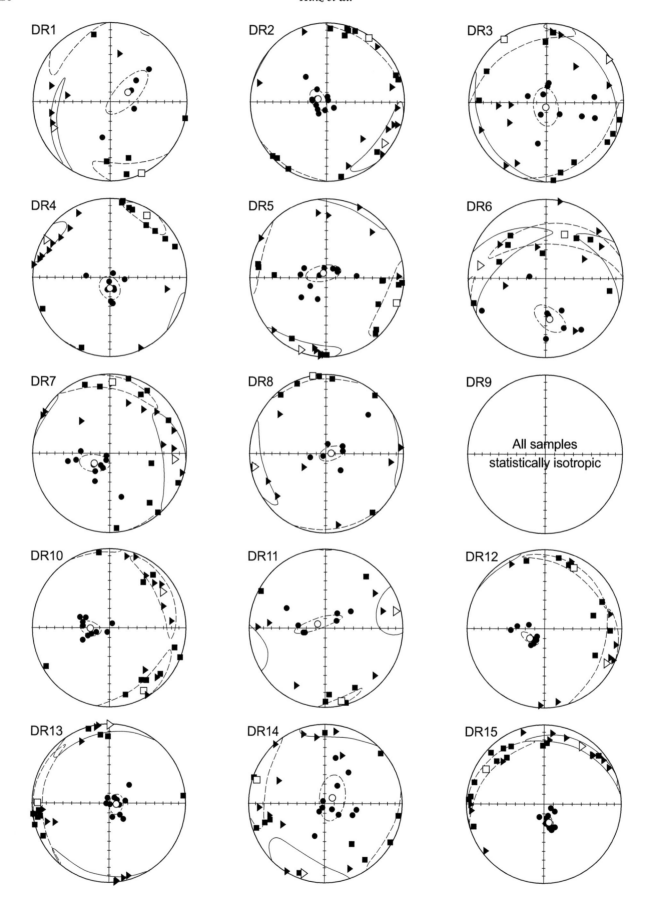

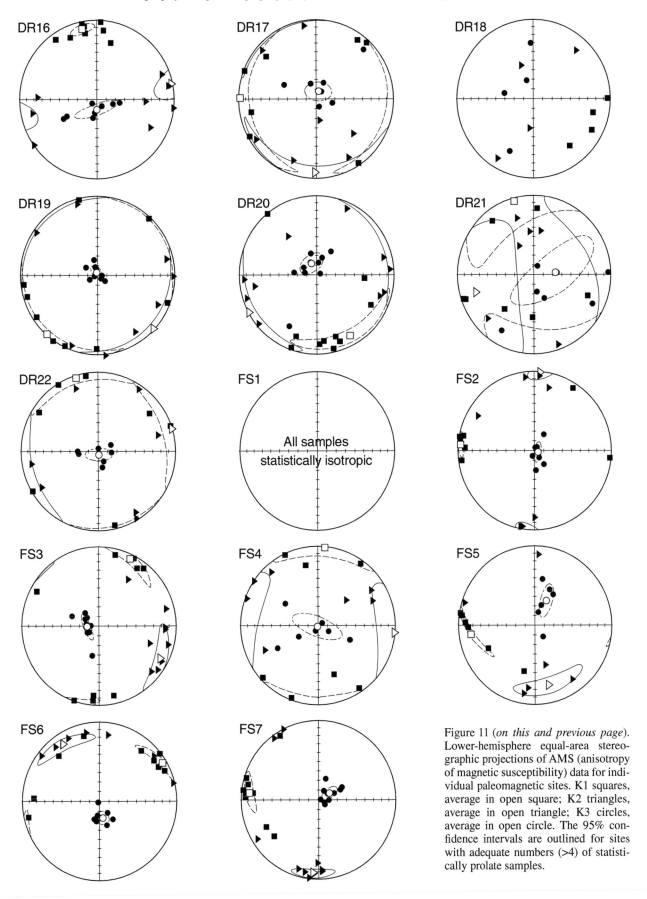

Figure 11 (*on this and previous page*). Lower-hemisphere equal-area stereographic projections of AMS (anisotropy of magnetic susceptibility) data for individual paleomagnetic sites. K1 squares, average in open square; K2 triangles, average in open triangle; K3 circles, average in open circle. The 95% confidence intervals are outlined for sites with adequate numbers (>4) of statistically prolate samples.

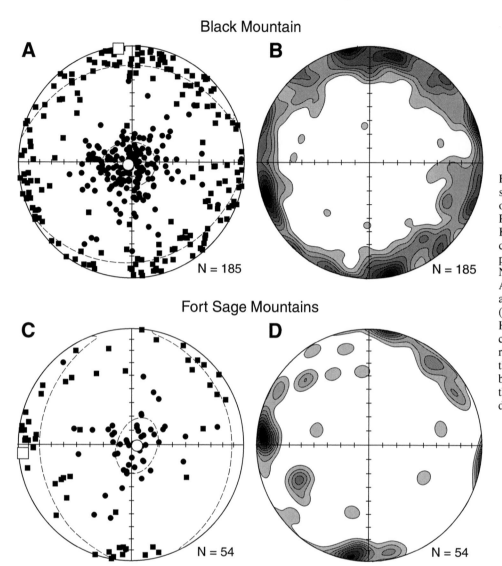

Figure 12. Lower-hemisphere equal area stereographic projections of anisotropy of magnetic susceptibility (AMS) data. Projections in left column (A and C): K1 squares (average in open square); K3 circles (average in open circle). Contour plots of K1 in right column (B and D). N equals number of samples. (A and B) All AMS samples from Black Mountain area, corresponding to group 1, Table 4. (C and D) All AMS data from the Nine Hill Tuff in the Fort Sage Mountains, corresponding to group 2, Table 4. With respect to the data displayed in D and the corresponding "group 2" row in Table 4, correction for the 6° west tilt of the Fort Sage Mountains was not done due to software limitations.

(Fig. 13A, B, C; Table 3). Group means for the Nine Hill Tuff cluster moderately well (Fig. 13D). Tilt corrections, based on the orientation of compaction foliations in the ash-flow tuffs, consistently increased dispersion of group means for all units in both the Diamond and Fort Sage Mountains (Fig. 13; Table 3). This suggests that the gently dipping compaction foliations observed in the Diamond and Fort Sage Mountains are generally primary and reflect irregularities in Oligocene erosion surfaces. In addition, these relations further support our inference of negligible tilting of the Diamond and Fort Sage Mountains. Nonetheless, uncorrected-site means for the Nine Hill Tuff from the Fort Sage Mountains (Fig. 13D) show more dispersion than would be expected from a single coherent fault block, suggesting minor variations in tilting that have not been accounted for in our reconnaissance study of this area.

Comparison of the group average for sites from the Nine Hill Tuff in the Diamond and Fort Sage Mountains with estab-

lished reference directions (Deino, 1985; Faulds et al., 2004) indicates negligible vertical-axis rotation of these areas. In making this assertion, the group mean of five sites in the Fort Sage Mountains was corrected for 6° of westward tilt about a north-trending axis (N0°E; Fig. 6B). The group average from the Diamond and Fort Sage Mountains overlaps the established reference directions for the Nine Hill Tuff at the 95% confidence level (Fig. 14).

As a result of the inferred negligible vertical-axis rotation, a reference direction can be established for the 28.8 Ma tuff of Campbell Creek, using the group site mean from the Diamond Mountains (Fig. 14; Table 3). The reference direction has a declination of 205.5° and an inclination of −43.9°. Comparison of this newly established reference direction with the lone tuff of Campbell Creek site (corrected for the 6° west tilt) in the Fort Sage Mountains (FS4) supports the premise of negligible vertical-axis rotation of the Fort Sage Mountains.

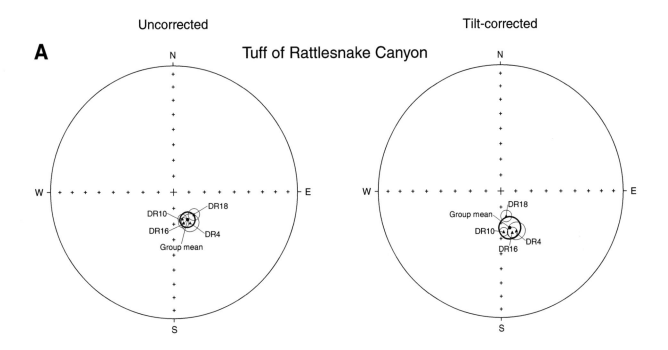

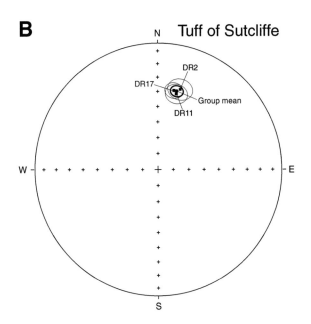

Figure 13 (*on this and following page*). Equal-area projections of site mean directions and their 95% confidence cones before (left figure) and after (right figure) tilt corrections for compaction foliations from representative ash-flow tuffs. Site numbers correspond to Table 3. Triangles and squares are projections on the upper and lower hemispheres, respectively. Bold 95% confidence cone denotes group site mean for each plot. (A) Tuff of Rattlesnake Canyon, Diamond Mountains; (B) tuff of Sutcliffe (sites have subhorizontal compaction foliation and thus are not tilt-corrected), Diamond Mountains; (C) tuff of Campbell Creek, Diamond Mountains; (D) Nine Hill Tuff, Diamond and Fort Sage Mountains.

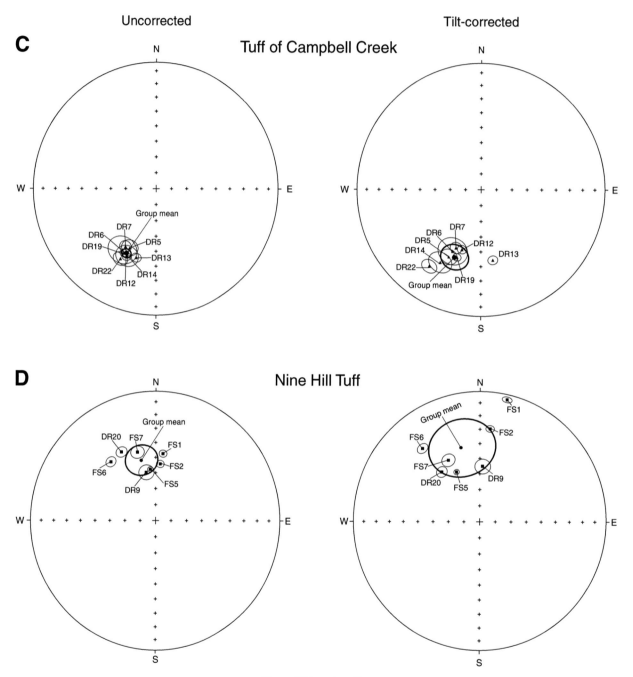

Figure 13 (*continued*).

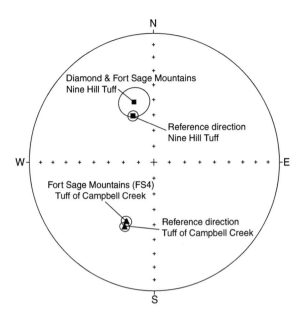

Figure 14. Equal-area stereographic projection, comparing the Nine Hill Tuff group average from the Diamond and Fort Sage Mountains with the Nine Hill Tuff reference direction (Deino, 1989; Faulds et al., 2004), as well as the tuff of Campbell Creek reference direction, established in the Diamond Mountains (this study), with the tuff of Campbell Creek site (FS4) from the Fort Sage Mountains. Triangles and squares are projections on the upper and lower hemispheres, respectively. Data from Fort Sage Mountains corrected for regional tilt.

AMS Interpretations

AMS data were used to interpret flow directions in the ash-flow tuffs and ultimately to enhance the understanding of paleovalley geometries. Principal-axis orientations of AMS from the Diamond Mountains were evaluated in situ because geologic relations and ChRMs indicate no significant rotations or tilting of the ash-flow tuffs. Principal axis orientations of AMS from the Fort Sage Mountains were corrected for the inferred 6° west tilt. Considering the aforementioned studies of prolate grains in both detrital rocks and ignimbrites, as well as the bimodal distribution of much of our AMS data (Fig. 11), horizontal (i.e., nonimbricated) K1 axes from the Diamond and Fort Sage Mountains are interpreted as bidirectional lineations oriented either parallel or transverse to the flow direction (Figs. 3 and 4). Where K1 axes are flow parallel, imbricated K1 axes are used to infer actual flow directions.

A compilation of all AMS samples in the Black Mountain area in the Diamond Mountains (group 1, Table 4; Figs. 3, 11, and 12A) reveals near bimodal northerly and westerly trends with high angular dispersions about the horizontal, which probably indicate a combination of flow within the main-stem paleovalley, flow up into tributary valleys, and rolling of prolate grains transverse to flow. The clusters of easterly and westerly oriented K1 axes (Figs. 3 and 12A) probably relate to down-valley main-

stem flow, whereas northerly and southerly oriented K1 axes may reflect a combination of flow up and down tributary valleys (e.g., sites DR1, 2, 3, and 4) and transverse-to-flow K1 orientations (e.g., sites DR15, 16, 20, and 21). The configuration of the Oligocene erosion surface supports the influence of complex paleotopography on flow directions, as several tributary paleovalleys appear to intersect the primary channel at high angles (Hinz, 2004; Fig. 9B). In the northeast, sites DR10, 11, 12, 13, and 14 (e.g., K1 orientations and imbrications) indicate net west-northwestward flow in accord with the northwest-trending ridge shoulder protruding into the south side of the primary paleovalley axis (Figs. 3 and 9B). In the northwest, sites DR1, 2, 3, and 4 generally indicate north and/or southward flow, which probably reflects flow up and/or back down a north-trending tributary. In the central part of the Black Mountain area, sites DR15, 16, 17, 20, and 21 vary widely in flow directions, which may reflect turbulence imparted by upstream channel sinuosity and steep, confining topography, as illustrated by the erosion-surface contours. Alternatively, some or all of the sites in the central Black Mountain area may contain transverse-to-flow K1 axes produced by the rolling of prolate clasts during deposition. Ultimately, these correlations of flow directions and paleotopography are in accord with other studies documenting topographic influences on flow directions (e.g., Baer et al., 1997; Ort et al., 2003).

Similar to the Black Mountain area, a compilation of AMS samples from the Fort Sage Mountains (Fig. 12B; Table 4) reveals bimodal northerly and westerly trends of K1 axes with high angular dispersions about the horizontal (note tilt correction information specific to Fig. 12B in caption). In the Fort Sage Mountains, paleomagnetic sites are concentrated in the Nine Hill Tuff in the aforementioned southern tributary channel (Fig. 4). Five of these sites yielded high quality AMS data, including only moderate variation in the trend of K1 axes (Figs. 4 and 11). The K1 axes for these sites trend predominantly west-northwest to west-southwest, with one site (FS3) yielding a southwest-trending axis. Sites FS3 in the Nine Hill Tuff and FS4 in the tuff of Campbell Creek may record transverse-to-flow K1 orientations. Two sites, FS2 and FS6, exhibit opposing imbrication orientations. The opposite imbrication direction at site FS2 may have resulted from locally complex patterns rather than from actual eastward flow. For the Fort Sage Mountains the overall westward trend of K1 axes probably reflects a down-valley flow direction, which is consistent with the ~240° (S60°W) paleovalley axis inferred from analysis of the early Oligocene erosion surface and with the inferred source of the tuffs in central Nevada (e.g., Deino, 1985; Henry et al., 2003). The northerly trends may reflect local turbulence, possibly induced by interaction with local topographic irregularities, and the rolling of prolate clasts with long axes oriented transverse to flow.

We conclude that high scatter in AMS data may characterize many of the paleovalleys in the northern Walker Lane region, where irregularities or roughness in the underlying topography likely disrupted flow patterns. This conclusion contrasts, however, with those of several other studies, whereby the more distal parts of ash-flow tuffs are presumably deposited under conditions of minor turbulence or laminar flow, thereby producing more highly clustered AMS orientations in comparison with areas proximal to calderas (Fisher et al., 1993; Baer et al., 1997; Ort et al., 2003). In the distal northern Walker Lane region (relative to source calderas in central Nevada), however, we suggest that topographic roughness and paleovalley sinuosity may have generated turbulent flow patterns throughout the stratigraphic column. It would appear that drainage systems in the Diamond and Fort Sage Mountains did not fill with outflow sheets, thereby preventing smoothing of irregular erosion surfaces and ultimately imparting relatively high scatter in the AMS data within the tuffs.

Constraining Offset on the Honey Lake Fault Zone

Delineating the orientation of Oligocene paleovalleys is crucial to constraining displacement on the Honey Lake fault zone. In the Diamond Mountains the bottom and central axis of the Black Mountain paleovalley transects the central, thickest part of the ash-flow-tuff sequence. Based on the configuration of the early Oligocene erosion surface, the axis of the Black Mountain paleovalley trends ~240° (S60°W). AMS data from the Diamond Mountains vary considerably but are consistent with local variations in topography along a main-stem west-southwest–trending paleovalley within a dendritic drainage system, as supported by the configuration of the early Oligocene erosion surface. However, the margins of the Black Mountain paleovalley are poorly preserved, as the present elevation of Mesozoic basement does not surpass the maximum modern-day elevation of ash-flow tuffs (2070 m or 6830 ft) within a 5 km radius of the map area. Thus, paleovalley margins expressed by basement-rock paleohighs have largely been eroded.

In the Fort Sage Mountains, paleotopographic features and AMS data are consistent with a general west-southwest–trending paleovalley. However, the small size of the fault block and resulting limited extent of ash-flow-tuff exposures limit accuracy in defining the orientation of the Fort Sage Mountains paleovalley. The margins of the Fort Sage Mountains paleovalley are also poorly preserved, similar to the Black Mountain paleovalley. Nonetheless, the bottom and probable central axis of an ~250° (S70°W) trending, 4-km-long paleovalley transects the north-central part of the range. The average K1 orientation from AMS data for the Nine Hill Tuff sites (group 2, Table 4; Fig. 12B) and the trend of the paleochannel containing this tuff are consistent with a west-southwest down-valley slope of the main-stem paleovalley.

The inferred axes of the Black Mountain and Fort Sage Mountains paleovalleys provide piercing lines with which to gauge offset across the Honey Lake fault zone. Paleotopographic constraints indicate a probable orientation of ~240° for the Black Mountain paleovalley and ~250° for the Fort Sage Mountains paleovalley. Other, well-defined paleovalley segments at Dogskin Mountain, in the Virginia Mountains, and in the Nightingale Range trend ~255° to 270° (Faulds et al., 2003c, 2005a; Henry et al., 2004). Because the paleotopographic constraints identified in this study may represent local variations in an overall approximately west-trending paleovalley system, the inferred range in paleovalley orientations in the Diamond and Fort Sage Mountains is expanded to include 270°. Correlation of the Black Mountain and Fort Sage Mountains paleovalleys, with orientations ranging from 240° to 270° for the former and 250° to 270° for the latter, implies 10–17 km of right-lateral displacement across the Honey Lake fault zone (Fig. 15).

Tectonic Implications

The 10–17 km of inferred dextral offset along the Honey Lake fault zone between the Fort Sage and Diamond Mountains is roughly compatible with estimates of displacement along the southeastern part of this fault zone. The probable correlation of paleovalley margins between Seven Lakes Mountain and Dogskin Mountain constrains cumulative dextral displacement to ~3–8 km for strands of the Honey Lake fault zone to the southeast of the Fort Sage Mountains (Henry et al., 2006; Fig. 2). In addition, as much as 3 km of dextral offset on the Honey Lake fault zone may have been transferred onto north-northeast–striking normal faults at the north end of Long Valley. West-tilted ca. 11.7–2.7 Ma sedimentary rocks in Long Valley record extension beginning by 11.7 Ma and continuing to 2.7 Ma (TenBrink et al., 2002; Henry et al., 2007). Strands of the Honey Lake fault zone between Seven Lakes Mountain and Dogskin Mountain and normal faults in Long Valley may accommodate 6–11 km of cumulative displacement, thus overlapping with the 10–17 km of dextral offset inferred in this study. Collectively, these constraints support ~10 km of offset on the Honey Lake fault zone but also imply possible differential offset between the southeastern and central parts of the fault zone.

The range of 10–17 km of inferred dextral offset along the Honey Lake fault zone is also compatible with the presumed onset of strike-slip faulting in the northern Walker Lane, which ranges from ca. 3 Ma (Henry et al., 2002, 2007) to 3–9 Ma (Faulds et al., 2003c, 2005a). If the Walker Lane is intimately linked to the San Andreas fault system, the late Miocene to Holocene northward track of the Mendocino triple junction (Atwater and Stock, 1998) favors post–6 Ma onset of strike-slip faulting in the northern Walker Lane (Faulds et al., 2003c). Long-term slip rates necessary for 10–17 km of cumulative displacement on the Honey Lake fault zone are ~1.7–2.8 mm/yr for a 6 Ma initiation and ~3.3–5.7 mm/yr for a 3 Ma initiation. These long-term slip rates are compatible with both GPS geodetic data (Thatcher et al., 1999; Bennett et al., 1999, 2003; Oldow et al., 2001; Hammond and Thatcher, 2004) and inferred Quaternary slip rates derived from fault studies (Anderson and Hawkins, 1984; Wills and Borchardt, 1993; Briggs and Wesnousky, 2004), which imply ~1–3 mm/yr on each of the four primary right-lateral strike-slip faults of the northern Walker Lane.

The 10 km value is also consistent with slip estimates of other major strike-slip faults in the northern Walker Lane. Faulds et al. (2005a) inferred ~10 km of offset for both the Warm Springs Valley and Pyramid Lake faults. Prior to this study, cumulative offset on the central part of the Honey Lake fault zone was unknown.

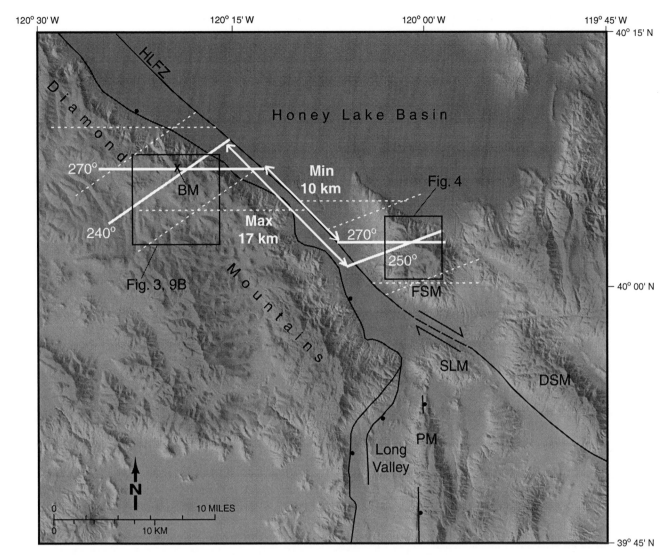

Figure 15. Inferred displacement across the Honey Lake fault zone (HLFZ) of the Black Mountain (BM) and Fort Sage Mountains (FSM) paleovalley segments. Solid white lines represent the probable ranges in orientation of the paleovalley axes as they project to the fault zone. Azimuths represent inferred down-paleovalley directions. Although paleovalley margins are commonly difficult to define, dotted white lines illustrate theoretical paleovalley margins based on a hypothetical 8 km width, which is the approximate width of a nearby paleovalley at Seven Lakes Mountain (e.g., Henry et al., 2006). Lines with arrows illustrate minimum and maximum displacement of the paleovalley. DSM—Dogskin Mountain; PM—Peterson Mountain; SLM—Seven Lakes Mountain.

Similarities in the apparent magnitude of slip on each of these faults support a Riedel shear model for the major left-stepping, dextral faults in the northern Walker Lane and attest to the youthfulness of this strike-slip fault system (see discussion in Faulds et al., 2005a, 2005b). Thus, the inferred offset along the Honey Lake fault zone represents an important step in unraveling the development of the northern Walker Lane within the context of the evolving plate boundary.

The inferred constraints on offset across the Honey Lake fault zone will facilitate estimates of cumulative dextral displacement and long-term slip rates across the northern Walker Lane, which in turn will permit comparisons with other parts of the

Walker Lane and the eastern California shear zone. Cumulative offset on any of the overlapping segments of the northern Walker Lane probably does not exceed 25 km. This suggests long-term slip rates of ~4–8 mm/yr, which are compatible with rates (~4–10 mm/yr) inferred from GPS geodetic data (Thatcher et al., 1999; Bennett et al., 1999, 2003; Oldow et al., 2001; Hammond and Thatcher, 2004; Kreemer et al., 2006).

CONCLUSIONS

The similar sequences and thicknesses of Oligocene ash-flow tuffs support correlation of the Black Mountain and Fort

Sage Mountains paleovalley segments. The 31.3–25.3 Ma ash-flow-tuff stratigraphy of the Black Mountain area in the Diamond Mountains closely matches the 31.3–25.1 Ma tuffs of the Fort Sage Mountains. Four ash-flow tuffs, the 31.3 Ma tuff of Axehandle Canyon, the 31.0 Ma tuff of Rattlesnake Canyon, the 28.8 Ma tuff of Campbell Creek, and the 25.3 Ma Nine Hill Tuff are found in both the ~250-m-thick section in the Black Mountain area and the ~200-m-thick section in the Fort Sage Mountains.

The configuration of an Oligocene erosion surface and rock magnetic data constrain the orientations of the paleovalley segments as they project toward the Honey Lake fault zone, thus allowing for interpretation of dextral offset. Both the Diamond and Fort Sage Mountains are relatively coherent structural blocks. ChRMs from the ash-flow tuffs indicate negligible tilting of the Diamond Mountains; geologic data suggest only 6° of west tilt in the Fort Sage Mountains. The configuration of an early Oligocene erosion surface indicates paleovalley orientations of ~240° (S60°W) for the Diamond Mountains and ~250° (S70°W) in the Fort Sage Mountains. AMS data from the Diamond and Fort Sage Mountains are consistent with westerly trending dendritic paleovalley segments, thus supporting the paleotopographic analyses. Comparison of the group site mean for CHRMs from the Nine Hill Tuff in the Diamond and Fort Sage Mountains with established reference directions (Deino, 1985; Faulds et al., 2004) indicates negligible vertical-axis rotation of these areas. The new reference direction established for the tuff of Campbell Creek (this study) also implies negligible vertical-axis rotation of the Fort Sage Mountains. Therefore, the estimated trends of the paleovalley segments reflect their original orientations.

In addition to local constraints on paleovalley orientation inferred from paleomagnetic, structural, and paleotopographic analyses, regional relations advocate consideration of a 270° down-valley direction. The inferred 240° to 270° orientation for the paleovalley in the Diamond Mountains and 250° to 270° for the paleovalley in the Fort Sage Mountains imply ~10–17 km of cumulative dextral displacement across the Honey Lake fault zone. Resulting long-term slip rates, assuming initiation of strike-slip faulting ca. 6–3 Ma, are compatible with strain rates derived from Quaternary faults and GPS geodetic data in other studies.

The 10–17 km range of displacement inferred in this study along the central part of the Honey Lake fault zone overlaps the interpreted cumulative strain (6–11 km) along the southeast part of the fault zone, which includes 3–8 km of dextral offset (Henry et al., 2006, 2007) and ~3 km of normal displacement on northerly striking faults bordering Long Valley. Collectively, these relations favor ~9–10 km of offset on the central part of the Honey Lake fault zone (i.e., between the Fort Sage Mountains and Black Mountain paleovalleys). The 10 km value is also consistent with slip estimates of other major strike-slip faults in the northern Walker Lane (Faulds et al., 2005a, b). The inferred offset along the Honey Lake fault zone represents an important step in constraining cumulative offset across the northern Walker Lane and evaluating kinematic models for this youthful strike-slip fault system.

ACKNOWLEDGMENTS

This work was supported by the U.S. National Science Foundation (grant EAR0124869). Constructive reviews by Mark Hudson and Elizabeth Schermer greatly improved this manuscript. We thank Rich Briggs, Woody Brooks, Patricia Cashman, Larry Garside, Trobe Gross, George Saucedo, and Dave Wagner for fruitful discussions. ^{40}Ar/^{39}Ar ages were determined at the New Mexico Geochronology Research Laboratory under the guidance of Bill McIntosh, Matt Heizler, Rich Esser, and Lisa Peters. We are also grateful for cartographic assistance from Robert Chaney and preparation of several figures by Kris Pizarro.

REFERENCES CITED

Anderson, L.W., and Hawkins, F.F., 1984, Recurrent Holocene strike-slip faulting, Pyramid Lake fault zone, western Nevada: Geology, v. 12, p. 681–684, doi: 10.1130/0091-7613(1984)12<681:RHSFPL>2.0.CO;2.

Anderson, R.E., 1973, Large-magnitude late Tertiary strike-slip faulting north of Lake Mead, Nevada: U.S. Geological Survey Professional Paper 794, 18 p.

Antonelis, K., Johnson, D.J., Miller, M.M., and Palmer, R., 1999, GPS determination of current Pacific–North American plate motion: Geology, v. 27, p. 299–302, doi: 10.1130/0091-7613(1999)027<0299:GDOCPN>2.3.CO;2.

Argus, D.F., and Gordon, R.G., 2001, Present tectonic motion across the Coast Ranges and San Andreas fault system in central California: Geological Society of America Bulletin, v. 113, p. 1580–1592, doi: 10.1130/0016-7606(2001)113<1580:PTMATC>2.0.CO;2.

Atwater, T., 1970, Implications of plate tectonics for the Cenozoic tectonic evolution of western North America: Geological Society of America Bulletin, v. 81, p. 3513–3536, doi: 10.1130/0016-7606(1970)81[3513:IOPTFT]2.0.CO;2.

Atwater, T., and Stock, J., 1998, Pacific–North America plate tectonics of the Neogene southwestern United States: An update: International Geology Review, v. 40, p. 375–402.

Baer, E.M., Fisher, R.V., Fuller, M., and Valentine, G., 1997, Turbulent transport and deposition of the Ito pyroclastic flow; determinations using anisotropy of magnetic susceptibility: Journal of Geophysical Research, v. 102, p. 22,565–22,586, doi: 10.1029/96JB01277.

Bell, J.W., 1984, Quaternary fault map of Nevada, Reno sheet: Nevada Bureau of Mines and Geology Map 79, scale 1:250,000.

Bennett, R.A., Wernicke, B.P., Davis, J.L., Elósegui, P., Snow, J.K., Abolins, M.J., House, M.A., Stirewalt, G.L., and Ferrill, D.A., 1997, Global Positioning System constraints on fault slip rates in the Death Valley region, California and Nevada: Geophysical Research Letters, v. 24, p. 3073–3076, doi: 10.1029/97GL02951.

Bennett, R.A., Wernicke, B.P., and Davis, J.L., 1998, Continuous GPS measurements of contemporary deformation across the northern Basin-Range province: Geophysical Research Letters, v. 25, p. 563–566, doi: 10.1029/98GL00128.

Bennett, R.A., Davis, J.L., and Wernicke, B.P., 1999, Present-day pattern of Cordilleran deformation in the western United States: Geology, v. 27, p. 371–374, doi: 10.1130/0091-7613(1999)027<0371:PDPOCD>2.3.CO;2.

Bennett, R.A., Wernicke, B.P., Niemi, N.A., Friedrich, A.M., and Davis, J.L., 2003, Contemporary strain rates in the northern Basin and Range province from GPS data: Tectonics, v. 22, 1008, doi: 10.1029/2001TC001355.

Best, J.L., 1992, Sedimentology and event timing of a catastrophic volcaniclastic mass flow, Volcano Hudson, southern Chile: Bulletin of Volcanology, v. 54, p. 299–318, doi: 10.1007/BF00301484.

Best, M.G., Christiansen, E.H., Deino, A.L., Gromme, C.S., McKee, E.H., and Noble, D.C., 1989, Eocene through Miocene Volcanism in the Great Basin of the Western United States: New Mexico Bureau of Mines and Mineral Resources Memoir 131, 22 p.

Briggs, R.W., and Wesnousky, S.G., 2004, Late Pleistocene fault slip rate, earthquake recurrence, and recency of slip along the Pyramid Lake fault zone, northern Walker Lane, United States: Journal of Geophysical Research, v. 109, p. B08402, doi: 10.1029/2003JB002717.

Briggs, R.W., and Wesnousky, S.G., 2005, Late Pleistocene paleoearthquake activity of the Olinghouse fault zone, Nevada: Bulletin of the Seismological Society of America, v. 95, p. 1301–1313, doi: 10.1785/0120040129.

Butler, R.F., 1992, Paleomagnetism: Magnetic Domains to Geologic Terranes: Boston, Blackwell Scientific Press, 319 p.

Cagnoli, B., and Tarling, D.H., 1997, The reliability of anisotropy of magnetic susceptibility (AMS) data as flow direction indicators in friable base surge and ignimbrite deposits: Italian examples: Journal of Volcanology and Geothermal Research, v. 75, p. 309–320, doi: 10.1016/S0377-0273(96)00038-8.

Cashman, P.H., and Fontaine, S.A., 2000, Strain partitioning in the northern Walker Lane, western Nevada and northeastern California: Tectonophysics, v. 326, p. 111–130, doi: 10.1016/S0040-1951(00)00149-9.

Ceron, J.F., 1992, Gravity modeling of the Honey Lake basin [M.S. thesis]: Golden, Colorado School of Mines, 145 p.

Chapin, C.E., and Lowell, G.R., 1979, Primary and secondary flow structures in ash-flow tuffs of the Gribbles Run paleovalley, central Colorado, *in* Chapin, C.E., and Elston, W.E., eds., Ash-Flow Tuffs: Geological Society of America Special Paper 180, p. 137–154.

Christiansen, R.L., and Yeats, R.S., 1992, Post Laramide geology of the U.S. Cordilleran region, *in* Burchfiel, B.C., et al., eds., The Cordilleran Orogen: Conterminous U.S.: Boulder, Colorado, Geological Society of America, Geology of North America, v. G-3, p. 261–406.

Colie, E.M., Roeske, S.M., and McClain, J., 2002, Strike-slip motion on the northern continuation of the late Cenozoic to Quaternary Honey Lake fault zone, Eagle Lake, northeast California: Geological Society of America Abstracts with Programs, v. 34, no. 5, p. 108.

Cousens, B.L., Prytulak, J., Henry, C.D., and Wise, W., 2003, The relationship between late Cenozoic tectonics and volcanism in the northern Sierra Nevada, California/Nevada: The roles of the upper mantle, subducting slab, lithosphere and Basin and Range extension: Geological Association of Canada, Mineralogical Association of Canada Abstracts, v. 28, p. 183.

Deino, A.L., 1985, Stratigraphy, chemistry, K-Ar dating, and paleomagnetism of the Nine Hill Tuff, California-Nevada; Miocene/Oligocene ash-flow tuffs of Seven Lakes Mountain, California-Nevada; improved calibration methods and error estimates for potassium-40–argon-40 dating of young rocks [Ph.D. thesis]: Berkeley, University of California, 457 p.

Deino, A.L., 1989, Single-Crystal ^{40}Ar/^{39}Ar Dating as an Aid in Correlation of Ash Flows: Examples from the Chimney Spring/New Pass Tuffs and the Nine Hill/Bates Mountain Tuffs of California and Nevada: New Mexico Bureau of Mines and Mineral Resources Bulletin 131, 70 p.

Dixon, T.H., Robaudo, S., Lee, J., and Reheis, M.C., 1995, Constraints on the present day Basin and Range deformation from space geodesy: Tectonics, v. 14, p. 755–772, doi: 10.1029/95TC00931.

Dixon, T.H., Miller, M., Farina, F., Wang, H., and Johnson, D., 2000, Present-day motion of the Sierra Nevada block and some tectonic implications for the Basin and Range province, North America Cordillera: Tectonics, v. 19, p. 1–19, doi: 10.1029/1998TC001088.

Dokka, R.K., and Travis, C.J., 1990, Late Cenozoic strike-slip faulting in the Mojave Desert, California: Tectonics, v. 9, p. 311–340, doi: 10.1029/TC009i002p00311.

Durrell, C., 1959, Tertiary stratigraphy of the Blairsden Quadrangle, Plumas County, California: University of California Special Publications in Geological Sciences, v. 34, p. 161–192.

Ellwood, B.B., 1982, Estimates of flow directions for calc-alkaline welded tuffs and paleomagnetic data reliability from anisotropy of magnetic susceptibility measurements: Central San Juan Mountains, southwest Colorado: Earth and Planetary Science Letters, v. 59, p. 303–314, doi: 10.1016/0012-821X(82)90133-9.

Ellwood, B.B., and Ledbetter, M.T., 1977, Antarctic bottom water fluctuations in the Vema channel: Effects of velocity changes on particle alignment and size: Earth and Planetary Science Letters, v. 35, p. 189–198, doi: 10.1016/0012-821X(77)90121-2.

Faulds, J.E., and Henry, C.D., 2002, Tertiary stratigraphy and structure of the Virginia Mountains, western Nevada: Implications for development of the northern Walker Lane: Geological Society of America Abstracts with Programs, v. 34, no. 5, p. 84.

Faulds, J.E., Henry, C.D., and dePolo, C.M., 2003a, Preliminary geologic map of the Sutcliffe quadrangle, Nevada: Nevada Bureau of Mines and Geology Open-File Report OF03-17, scale 1:24,000.

Faulds, J.E., Henry, C.D., and dePolo, C.M., 2003b, Preliminary geologic map of the Tule Peak quadrangle, Washoe County, Nevada: Nevada Bureau of Mines and Geology Open-File Report OF03-10, scale 1:24,000.

Faulds, J.E., Henry, C.D., and Hinz, N.H., 2003c, Kinematics and cumulative displacement across the northern Walker Lane, an incipient transform fault, northwest Nevada and northeast California: Geological Society of America Abstracts with Programs, v. 35, no. 6, p. 305.

Faulds, J.E., Henry, C.D., Hinz, N.H., Delwiche, B., and Cashman, P.H., 2004, Kinematic implications of new paleomagnetic data from the northern Walker Lane, western Nevada: Counterintuitive anticlockwise vertical-axis rotation in an incipient dextral shear zone: Eos (Transactions, American Geophysical Union), v. 85, Fall Meeting Supplement, Abstract GP42A-08.

Faulds, J.E., Henry, C.D., and Hinz, N.H., 2005a, Kinematics of the northern Walker Lane: An incipient transform fault along the Pacific–North American plate boundary: Geology, v. 33, p. 505–508, doi: 10.1130/G21274.1.

Faulds, J.E., Henry, C.D., Hinz, N.H., Drakos, P.S., and Delwiche, B., 2005b, Transect across the northern Walker Lane, northwest Nevada and northeast California: An incipient transform fault along the Pacific–North American plate boundary, *in* Pederson, J., and Dehler, C.M., eds., Interior Western United States: Geological Society of America Field Guide 6, p. 129–150, doi: 10.1130/2005.fld006(06).

Feigl, K.L., Agnew, D.C., Bock, Y., Dong, D., Donnellan, A., Hager, B.H., Herring, T.A., Jackson, D.D., Jordan, T.H., King, R.W., Larsen, S., Larson, K.M., Murray, M.H., Shen, Z., and Webb, F.H., 1993, Space geodetic measurements of crustal deformation in central and southern California: Journal of Geophysical Research, v. 98, p. 21,677–21,712, doi: 10.1029/93JB02405.

Fisher, R.A., 1953, Dispersion on a sphere: Proceedings of the Royal Society of London, v. A217, p. 295–305.

Fisher, R.V., Orsi, G., Ort, M., and Heiken, G., 1993, Mobility of large-volume pyroclastic flow—Emplacement of the Campanian Ignimbrite, Italy: Journal of Volcanology and Geothermal Research, v. 56, p. 205–222, doi: 10.1016/0377-0273(93)90017-L.

Garside, L.J., Castor, S.B., Henry, C.D., and Faulds, J.E., 2000, Structure, Volcanic Stratigraphy, and Ore Deposits of the Pah Rah Range, Washoe County, Nevada: Geological Society of Nevada Symposium 2000, Field Trip Guidebook no. 2, 132 p.

Garside, L.J., Castor, S.B., dePolo, C.M., and Davis, D., 2003, Geologic map of the Fraser Flat quadrangle and the west half of the Moses Rock quadrangle, Washoe County, Nevada: Nevada Bureau of Mines and Geology Map M146, scale 1:24,000.

Garside, L.J., Henry, C.D., Faulds, J.E., and Hinz, N.H., 2005, The upper reaches of the Sierra Nevada auriferous gold channels, California and Nevada, *in* Rhoden, H.N., et al., eds., Geological Society of Nevada Symposium 2005: Window to the World: Reno, p. 209–236.

Gianella, V.P., 1957, Earthquake and faulting, Fort Sage Mountains, California, December 1950: Bulletin of the Seismological Society of America, v. 47, p. 173–177.

Graymer, R.W., Sarna-Wojcicki, A.M., Walker, J.P., McLaughlin, R.J., and Fleck, R.J., 2002, Controls on timing and amount of right-lateral offset on the East Bay fault system, San Francisco Bay region, California: Geological Society of America Bulletin, v. 114, p. 1471–1479, doi: 10.1130/0016-7606(2002)114<1471:COTAAO>2.0.CO;2.

Grose, T.L.T., 1984, Geologic map of the State Line Peak quadrangle, Nevada-California: Nevada Bureau of Mines and Geology Map 82, scale 1:24,000.

Grose, T.L.T., 2000a, Geologic map of the Blairsden 15′ quadrangle, Lassen County, California: California Division of Mines and Geology Open-File Report 2000-21, scale 1:62,500.

Grose, T.L.T., 2000b, Volcanoes in the Susanville region, Lassen, Modoc, Plumas counties, northeastern California: California Geology, v. 53, September/October, p. 4–23.

Grose, T.L.T., and Mergner, M., 2000, Geologic map of the Chilcoot 15′ quadrangle, Lassen and Plumas Counties, California: California Division of Mines and Geology Open-File Report 2000-23, scale 1:62,500.

Grose, T.L.T., and Porro, C.T.R., 1989, Geologic map of the Litchfield 15-Minute quadrangle, Lassen County, California: California Division of Mines and Geology Open-File Report 89-32, scale 1:62,500.

Grose, T.L.T., Saucedo, G.J., Wagner, D.L., and Medrano, M.D., 1989, Geologic map of the Doyle 15-minute quadrangle, Lassen and Plumas Counties, California: California Division of Mines and Geology Open-File Report 89-31, scale 1:62,500.

Grose, T.L.T., Saucedo, G.J., and Wagner, D.L., 1990, Geologic map of the Susanville quadrangle, Lassen and Plumas Counties, California: California Division of Mines and Geology Open-File Report 91-1, scale 1:100,000.

Grose, T.L.T., Saucedo, G.J., and Wagner, D.L., 1992, Geologic map of the Eagle Lake quadrangle, Lassen County, California: California Division of Mines and Geology Open-File Report 92-14, scale 1:100,000.

Hammond, W.C., and Thatcher, W., 2004, Contemporary tectonic deformation of the Basin and Range province, western United States: 10 years of observation with the Global Positioning System: Journal of Geophysical Research, v. 109, p. B08403, doi: 10.1029/2003JB002746.

Henry, C.D., and Perkins, M.E., 2001, Sierra Nevada–Basin and Range transition near Reno, Nevada; two-stage development at 12 and 3 Ma: Geology, v. 29, p. 719–722, doi: 10.1130/0091-7613(2001)029<0719:SNBART>2.0.CO;2.

Henry, C.D., Faulds, J.E., and dePolo, C.M., 2002, Structure and evolution of the Warm Springs Valley fault (WSF), northern Walker Lane, Nevada: Post-3-Ma initiation?: Geological Society of America Abstracts with Programs, v. 34, no. 5, p. 84.

Henry, C.D., Faulds, J.E., Garside, L.J., and Hinz, N.H., 2003, Tectonic implications of ash-flow tuffs and paleovalleys in the western US: Geological Society of America Abstracts with Programs, v. 35, no. 6, p. 346.

Henry, C.D., Faulds, J.E., dePolo, C.M., and Davis, D.A., 2004, Geologic map of the Dogskin Mountain quadrangle, Northern Walker Lane, Nevada: Nevada Bureau of Mines and Geology Map 148, scale 1:24,000, with accompanying text.

Henry, C.D., Ramelli, A.R., and Faulds, J.E., 2006, Preliminary geologic map of the Seven Lakes Mountain quadrangle, Nevada and California: Nevada Bureau of Mines and Geology Open-File Report 06-14, scale 1:24,000.

Henry, C.D., Faulds, J.E., and dePolo, C.M., 2007, Geometry and timing of strike-slip and normal faults of the northern Walker Lane, northwestern Nevada and northeastern California: Strain partitioning or sequential extensional and strike-slip deformation? *in* Till, A.B., et al., eds., Exhumation Associated with Continental Strike-Slip Fault Systems: Geological Society of America Special Paper 434, p. 59–80.

Hillhouse, J.W., and Wells, R.E., 1991, Magnetic fabric, flow directions, and source areas of lower Miocene Peach Springs Tuff in Arizona, California, and Nevada: Journal of Geophysical Research, v. 96, p. 12,443–12,460, doi: 10.1029/90JB02257.

Hinz, N.H., 2004, Tertiary volcanic stratigraphy of the Diamond and Fort Sage Mountains, northeastern California and western Nevada—Implications for development of the northern Walker Lane [M.S. thesis]: Reno, University of Nevada, 110 p.

Huber, N.K., 1981, Amount and timing of late Cenozoic uplift and tilt of the central Sierra Nevada, California: Evidence from the upper San Joaquin River basin: U.S. Geological Survey Professional Paper 1197, 28 p.

Hudson, M.R., John, D.A., Conrad, J.E., and McKee, E.H., 2000, Style and age of late Oligocene–early Miocene deformation in the southern Stillwater Range, west central Nevada: Paleomagnetism, geochronology, and field relations: Journal of Geophysical Research, v. 105, p. 929–954, doi: 10.1029/1999JB900338.

Ichinose, G.A., Smith, K.D., and Anderson, J.G., 1998, Moment tensor solutions of the 1994 to 1996 Double Spring Flat, Nevada, earthquake sequence and implications for local tectonic models: Bulletin of the Seismological Society of America, v. 88, p. 1363–1378.

Jelinek, V., 1978, Statistical processing of anisotropy of magnetic susceptibility measured on groups of specimens: Studia Geophysica et Geodaetica, v. 22, p. 50–62, doi: 10.1007/BF01613632.

Jenkins, O.P., 1932, Geologic map of northern Sierra Nevada showing Tertiary river channels and Mother Lode belt, *in* Bradley, W.W., ed., 28th Report of the State Mineralogist Accompanying July–October Quarterly Chapter.

John, D.A., 1995, Tilted middle Tertiary ash-flow calderas and subjacent granitic plutons, southern Stillwater Range, Nevada: Cross sections of an Oligocene igneous center: Geological Society of America Bulletin, v. 107, p. 180–200, doi: 10.1130/0016-7606(1995)107<0180:TMTAFC>2.3.CO;2.

Kirschvink, J.L., 1980, The least-squares line and plane and the analysis of paleomagnetic data: Geophysical Journal of the Royal Astronomical Society, v. 62, p. 699–718.

Knight, M.D., Walker, G.P.L., Ellwood, B.B., and Diehl, J.F., 1986, Stratigraphy, paleomagnetism, and magnetic fabric of the Toba Tuffs: Constraints on the source and eruptive styles: Journal of Geophysical Research, v. 91, p. 10,355–10,382, doi: 10.1029/JB091iB10p10355.

Kreemer, C., Hammond, W.C., and Blewitt, G., 2006, On the motion and geometry of the Sierra Nevada–Great Valley micro-plate: Implications for

Walker Lane tectonics: Eos (Transactions, American Geophysical Union), v. 87, Fall Meeting Supplement, Abstract G42A-05.

Le Pennec, J.L., Chen, Y., Diot, H., Froger, J.L., and Gourgaud, A., 1998, Interpretation of anisotropy of magnetic susceptibility fabric of ignimbrites in terms of kinematic and sedimentological mechanisms: An Anatolian case-study: Earth and Planetary Science Letters, v. 157, p. 105–127, doi: 10.1016/S0012-821X(97)00215-X.

Lindgren, W., 1911, The Tertiary gravels of the Sierra Nevada of California: U.S. Geological Survey Professional Paper 73, 226 p.

Locke, A., Billingsley, P.R., and Mayo, E.B., 1940, Sierra Nevada tectonic patterns: Geological Society of America Bulletin, v. 51, p. 513–540.

MacDonald, W.D., and Palmer, H.C., 1990, Flow directions in ash-flow tuffs: A comparison of geological and magnetic susceptibility measurements, Tshirege member (upper Bandelier Tuff), Valles caldera, New Mexico, USA: Bulletin of Volcanology, v. 53, p. 45–59, doi: 10.1007/BF00680319.

MacDonald, W.D., Palmer, H.C., and Hayatsu, A., 1998, Structural rotation and volcanic source implications of magnetic data from Eocene volcanic rocks, SW Idaho: Earth and Planetary Science Letters, v. 156, p. 225–237, doi: 10.1016/S0012-821X(98)00017-X.

Mass, K.B., 2005, The Neogene Boca basin, northeastern California: Stratigraphy, structure and implications for regional tectonics [M.S. thesis]: Reno, University of Nevada, 195 p.

Mass, K.B., Cashman, P.H., Trexler, J.H., Park, H., and Perkins, M.E., 2003, Deformation history in Neogene sediments of Honey Lake basin, northern Walker Lane, Lassen County, California: Geological Society of America Abstracts with Programs, v. 35, no. 6, p. 26.

McIntosh, W.C., Heizler, M., Peters, L., and Esser, R., 2003, ^{40}Ar/^{39}Ar Geochronology at the New Mexico Bureau of Geology and Mineral Resources: New Mexico Bureau of Geology and Mineral Resources Open File Report OF-AR-1, 7 p.

McKee, E.H., 1971, Tertiary igneous chronology of the Great Basin of the western United States—Implications for tectonic models: Geological Society of America Bulletin, v. 82, p. 3497–3502, doi: 10.1130/0016-7606(1971)82[3497:TICOTG]2.0.CO;2.

McKee, E.H., and Noble, D.C., 1986, Tectonic and magmatic development of the Great Basin of western United States during late Cenozoic time: Modern Geology, v. 10, p. 39–49.

McKee, E.H., and Silberman, M.L., 1970, Geochronology of Tertiary igneous rocks in central Nevada: Geological Society of America Bulletin, v. 81, p. 2317–2318, doi: 10.1130/0016-7606(1970)81[2317:GOTIRI]2.0.CO;2.

Miller, M., Johnson, D., Dixon, T., and Dokka, R.K., 2001, Refined kinematics of the eastern California shear zone from GPS observations 1993–1998: Journal of Geophysical Research, v. 106, p. 2245–2263, doi: 10.1029/2000JB900328.

Mills, H.H., 1984, Clast orientation in Mount St. Helens debris-flow deposits, North Fork Toutle River, Washington: Journal of Sedimentary Petrology, v. 54, p. 626–634.

Oldenburg, E.A., 1995, Chemical and petrologic comparison of Cretaceous plutonic rocks from the Diamond Mountains, Fort Sage Mountains and Yuba Pass region of northeastern California [M.S. thesis]: Arcata, California, Humboldt State University, 99 p.

Oldow, J.S., 1992, Late Cenozoic displacement partitioning in the northwest Great Basin, *in* Stewart, J.H., ed., Structure, Tectonics, and Mineralization of the Walker Lane; Walker Lane Symposium: Reno, Geological Society of Nevada, p. 15–52.

Oldow, J.S., Aiken, C.L.V., Hare, J.L., Ferguson, J.F., and Hardyman, R.F., 2001, Active displacement transfer within the central Walker Lane, western Great Basin: Geology, v. 29, p. 19–24, doi: 10.1130/0091-7613(2001)029<0019:ADTADB>2.0.CO;2.

Ort, M.H., 1993, Eruptive processes and caldera formation in a nested downsag-collapse caldera: Cerro Panizos, central Andes Mountains: Journal of Volcanology and Geothermal Research, v. 56, p. 221–252, doi: 10.1016/0377-0273(93)90018-M.

Ort, M.H., Rose, M., and Anderson, C.D., 1999, Correlation of deposits and vent locations of proximal Campanian Ignimbrite deposits, Campi Flegrei, Italy, based on natural remanent magnetization and anisotropy of magnetic susceptibility characteristics: Journal of Volcanology and Geothermal Research, v. 91, p. 167–178, doi: 10.1016/S0377-0273(99)00034-7.

Ort, M.H., Orsi, G., Pappalardo, L., and Fisher, R.V., 2003, Anisotropy of magnetic susceptibility studies of depositional processes in the Campanian Ignimbrite, Italy: Bulletin of Volcanology, v. 65, p. 55–72.

Page, W.D., Sawyer, T.L., McLaren, M., Savage, W.U., and Wakabayashi, J., 1993, The Quaternary Tahoe–Medicine Lake trough: The western margin of the Basin and Range transition, NE California: Geological Society of America Abstracts with Programs, v. 25, no. 5, p. 131.

Palmer, H.C., MacDonald, W.D., and Hayatsu, A., 1991, Magnetic, structural and geochronologic evidence bearing on volcanic sources and Oligocene deformation of ash flow tuffs, northeast Nevada: Journal of Geophysical Research, v. 96, p. 2185–2202, doi: 10.1029/90JB02224.

Park, H., Trexler, J., Cashman, P., and Mass, K.B., 2004, Local initiation of Walker Lane tectonism prior to 3.6 Ma recorded in Neogene sediments at Honey Lake basin, northeastern California: Geological Society of America Abstracts with Programs, v. 36, no. 4, p. 16–17.

Pease, R.W., 1969, Normal faulting and lateral shear in northeastern California: Geological Society of America Bulletin, v. 80, p. 715–720, doi: 10.1130/0016-7606(1969)80[715:NFALSI]2.0.CO;2.

Renne, P.R., Swisher, C.C., Deino, A.L., Karner, D.B., Owens, T.L., and DePaolo, D.J., 1998, Intercalibration of standards, absolute ages and uncertainties in $^{40}Ar/^{39}Ar$ dating: Chemical Geology, v. 145, p. 117–152, doi: 10.1016/S0009-2541(97)00159-9.

Roberts, C.T., 1985, Cenozoic evolution of the northwestern Honey Lake Basin, Lassen County, California: Colorado School of Mines Quarterly, v. 80, no. 1, 64 p., geologic map scale 1:24,000.

Samson, S.D., and Alexander, E.C., Jr., 1987, Calibration of the interlaboratory $^{40}Ar/^{39}Ar$ dating standard, MMhb-1: Chemical Geology, v. 66, p. 27–34.

Saucedo, G.J., and Wagner, D.L., 1990, Reconnaissance geologic map of part of the Kettle Rock 15-minute quadrangle, Plumas and Lassen Counties, California: California Division of Mines and Geology Open-File Report 90-7, scale 1:62,500.

Schwartz, K.M., 2001, Evolution of the middle to late Miocene Chalk Hills basin in the Basin and Range–Sierra Nevada transition zone, western Nevada [M.S. thesis]: Reno, University of Nevada, 160 p.

Schwartz, K.M., Faulds, J.E., and Henry, C.D., 2002, Cenozoic magmatic evolution in the western Virginia Range, western Nevada: Transition from subduction- to extension-related magmatism in the western Great Basin: Geological Society of America Abstracts with Programs, v. 34, no. 5, p. 4.

Seaman, S.J., McIntosh, W.C., Geissman, J.W., Williams, M.L., and Elston, W.E., 1991, Magnetic fabrics of the Bloodgood Canyon and Shelley Peak Tuffs, southwestern New Mexico: Implications for emplacement and alteration processes: Bulletin of Volcanology, v. 53, p. 460–476, doi: 10.1007/BF00258185.

Steiger, R.H., and Jäger, E., 1977, Subcommission on geochronology: Convention on the use of decay constants in geo- and cosmochronology: Earth and Planetary Science Letters, v. 36, p. 359–362, doi: 10.1016/0012-821X(77)90060-7.

Stewart, J.H., 1980, Geology of Nevada: A Discussion to Accompany the Geologic Map of Nevada: Nevada Bureau of Mines and Geology Special Publication 4, 136 p.

Stewart, J.H., 1988, Tectonics of the Walker Lane belt, western Great Basin: Mesozoic and Cenozoic deformation in a zone of shear, *in* Ernst, W.G., ed., The Geotectonic Development of California: Englewood Cliffs, New Jersey, Prentice Hall, p. 683–713.

Taira, A., and Scholle, P.A., 1979, Deposition of resedimented sandstone beds in the Pico Formation, Ventura Basin, California, as interpreted from magnetic fabric measurements: Geological Society of America Bulletin, v. 90, p. 952–962, doi: 10.1130/0016-7606(1979)90<952:DORSBI>2.0.CO;2.

TenBrink, A.L., Cashman, P.H., Trexler, J.H., Louie, J., and Smith, S., 2002, Active tectonism since 8 Ma in the Sierra Nevada—Basin and range transition zone, Lassen County, CA: Geological Society of America Abstracts with Programs, v. 34, no. 5, p. 100.

Thatcher, W., Foulger, G.R., Julian, B.R., Svarc, J., Quilty, E., and Bawden, G.W., 1999, Present-day deformation across the Basin and Range province, western United States: Science, v. 283, p. 1714–1718, doi: 10.1126/science.283.5408.1714.

Thomas, I.M., Moyer, T.C., and Wikswo, J.P., Jr., 1992, High resolution magnetic susceptibility imaging of geological thin sections: Pilot study of a pyroclastic sample from the Bishop Tuff, California: Geophysical Research Letters, v. 19, p. 2139–2142, doi: 10.1029/92GL02322.

Trexler, J.H., Cashman, P.H., Henry, C.D., Muntean, T.W., Schwartz, K., Ten-Brink, A., Faulds, J.E., Perlins, M., and Kelly, T.S., 2000, Neogene basins in western Nevada document the tectonic history of the Sierra Nevada–Basin and Range transition zone for the last 12 Ma, *in* Lageson, D.R., et al., eds., Great Basin and Sierra Nevada: Geological Society of America Field Guide 2, p. 97–116.

Wagner, D.L., and Saucedo, G.J., 1990, Reconnaissance geologic map of the Milford 15-minute quadrangle, Lassen and Plumas Counties, California: California Division of Mines and Geology Open-File Report 90-08, scale 1:62,500.

Wernicke, B., 1992, Cenozoic extensional tectonics of the U.S. Cordillera, *in* Burchfiel, B.C., et al., eds., The Cordilleran Orogen: Conterminous U.S.: Boulder, Colorado, Geological Society of America, Geology of North America, v. G-3, p. 553–581.

Williams, S.D.P., Svarc, J.L., Lisowski, M., and Prescott, W.H., 1994, GPS measured rates of deformation in the northern San Francisco Bay region, California, 1990–1993: Geophysical Research Letters, v. 21, p. 1511–1514, doi: 10.1029/94GL01227.

Wills, C.J., and Borchardt, G., 1993, Holocene slip rate and earthquake recurrence on the Honey Lake fault zone, northeastern California: Geology, v. 21, p. 853–856, doi: 10.1130/0091-7613(1993)021<0853:HSRAER>2.3.CO;2.

Wilson, D.S., 1989, Deformation of the so-called Gorda plate: Journal of Geophysical Research, v. 94, p. 3065–3075, doi: 10.1029/JB094iB03p03065.

Wolff, J.A., Ellwood, B.B., and Sachs, S.D., 1989, Anisotropy of magnetic susceptibility in welded tuffs: Application to a welded-tuff dyke in the Tertiary Trans Pecos Texas volcanic province, USA: Bulletin of Volcanology, v. 51, p. 299–310, doi: 10.1007/BF01073518.

MANUSCRIPT ACCEPTED BY THE SOCIETY 21 JULY 2008

The Geological Society of America
Special Paper 447
2009

Biostratigraphy of the Hunter Creek Sandstone, Verdi Basin, Washoe County, Nevada

Thomas S. Kelly*

*Vertebrate Paleontology Section, Natural History Museum of Los Angeles County, 900 Exposition Blvd.,
Los Angeles, California 90007, USA*

Ross Secord

Department of Geosciences and Nebraska State Museum, University of Nebraska, 200 Bessey Hall, Lincoln, Nebraska 68588, USA

ABSTRACT

The Hunter Creek Sandstone of the Verdi Basin, Nevada, yielded a succession of superposed continental faunal assemblages ranging in age from the late Clarendonian (late Miocene) through the late Blancan (late Pliocene) in the North American land mammal age framework, or ca. 10.5–2.5 Ma. We describe two new local faunas from the Hunter Creek Sandstone: the East Verdi local fauna, of late-medial to late Clarendonian age, which includes *Dinohippus* cf. *D. leardi*, Camelidae, ?Antilocapridae, and Mammutidae or Gomphotheriidae; and the Mogul local fauna, of Hemphillian age, which includes *Dinohippus* sp., Rhinocerotidae, Camelidae (at least two species), *Mammut* sp., and possibly Gomphotheriidae. A third unnamed assemblage, of latest Hemphillian or earliest Blancan age, is represented by a small sample of fossils from W.M. Keck Museum locality P-105. The only taxa recovered from this locality are cf. *Megatylopus* and Gomphotheriidae or Mammutidae. A single late Blancan locality, the Byland locality, yielded *Equus idahoensis*. The recognition of this faunal succession provides a biostratigraphic framework for the Hunter Creek Sandstone that corroborates and is consistent with the previous chronostratigraphy based on radioisotopic and tephrochronologic dating methods.

Keywords: Clarendonian, Hemphillian, Blancan, biostratigraphy, Nevada.

INTRODUCTION

The Verdi Basin is a structural basin that extends along the valley of the Truckee River and surrounding foothills from the west side of the community of Reno, Nevada, to the vicinity of Boca Reservoir, California (Trexler et al., 2000). Neogene sedimentary rocks are well exposed within the Verdi Basin (Fig. 1). These sedimentary rocks were originally mapped by King (1878)

as Truckee Formation, the type section of which occurs ~90 km to the northeast in Churchill County, Nevada. Axelrod (1956, 1958) reevaluated this sedimentary unit and referred it instead to the Coal Valley Formation, the type section of which occurs ~120 km to the southeast in Coal Valley, Lyon County, Nevada. Bingler (1975) described the Verdi Basin sedimentary rocks and informally referred to them as the "Sandstone of the Hunter Creek." More recent investigators have recognized that these rocks were depos-

*Corresponding author e-mail: tom@tskelly.gardnerville.nv.us.

Kelly, T.S., and Secord, R., 2009, Biostratigraphy of the Hunter Creek Sandstone, Verdi Basin, Washoe County, Nevada, *in* Oldow, J.S., and Cashman, P.H., eds., Late Cenozoic Structure and Evolution of the Great Basin–Sierra Nevada Transition: Geological Society of America Special Paper 447, p. 133–146, doi: 10.1130/2009.2447(08). For permission to copy, contact editing@geosociety.org. ©2009 The Geological Society of America. All rights reserved.

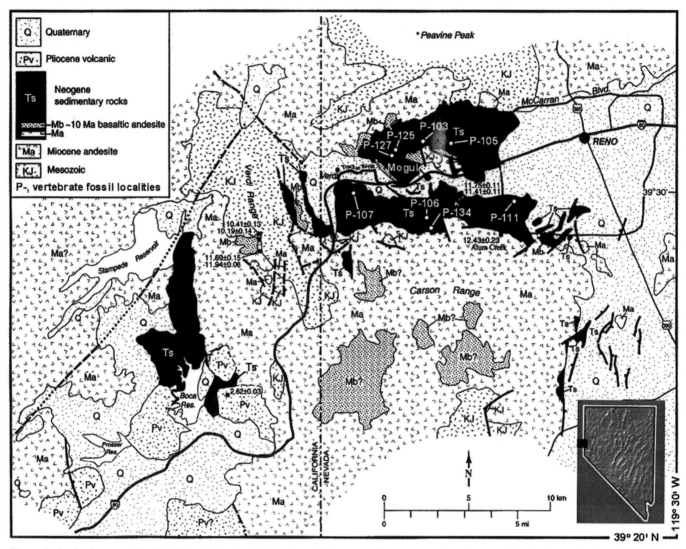

Figure 1. Geologic map of the Verdi Basin showing distribution of Neogene sedimentary rocks and geographic locations of selected W.M. Keck museum vertebrate fossil localities (P-) representing different stratigraphic levels within the Hunter Creek Sandstone (map modified from Trexler et al., 2000). Nevada map insert shows geographic location of Verdi Basin.

ited within a separate depositional basin from those of the type Truckee and Coal Valley Formations and should not be referred to either of these units (e.g., Kelly, 1998a; Trexler et al., 2000). Trexler et al. (2000) demonstrated that the Neogene sedimentary rocks of the Verdi Basin represent a continuously deposited rock unit that should be regarded as a distinct formation. For the purpose of this report, the informal name Hunter Creek Sandstone will be used for the Neogene sedimentary rocks of the Verdi Basin.

The Hunter Creek Sandstone is composed of fluvial, delta and fan-delta, and lacustrine deposits, with minor debris-flow lahars at several stratigraphic levels within the formation (Trexler et al., 2000). Trexler et al. (2000) recognized four facies within the Hunter Creek Sandstone: (1) fluvial facies consisting of conglomerate, sandstone, and intercalated mudstone; (2) deltaic facies consisting of minor conglomerate, coarse- to fine-grained

sandstone, and mudstone; (3) lacustrine facies consisting of diatomite and silty diatomite; and (4) lahar deposits consisting of distinctive coarse-grained breccia. Whole-rock and single-crystal (plagioclase and hornblende) ^{40}Ar/^{39}Ar radioisotopic dating, tephrochronologic dating of volcanic tuffs, and biostratigraphic data indicate that the Hunter Creek Sandstone was deposited from ca. 11 to 2.5 Ma (Garside et al., 2000; Trexler et al., 2000).

Ninety-five years ago, Buwalda (1914) described the first fossil mammal from the Hunter Creek Sandstone, a partial mastodon tooth from the Verdi area. Axelrod (1958) stated that this tooth was found ~1.2 km (~0.75 mile) southeast of the community of Verdi, but the exact location was unknown. Subsequently, two additional fragmentary mammalian fossils from the Hunter Creek Sandstone were reported in the literature. Axelrod (1958) noted that a second mastodon tooth was discovered from exposures near the community of Mogul, and a partial horse tooth, which he regarded as representing either *Hipparion* or *Neohipparion*, was discovered ~1.6 km (~1 mile) north of Mogul. Axelrod (1958) also described a diverse flora from University of California plant locality 102 (the Verdi Flora), which occurs in a railroad cut along the Southern Pacific Railroad near the Truckee River, just west of Mogul. The Verdi Flora is regarded as Hemphillian (late Miocene) in age based on paleobotanical correlation and a K-Ar radioisotopic age of 5.85 Ma (recalculated; see Methods) from the locality (Axelrod, 1958; Evernden and James, 1964; Schorn et al., 1994). During this study, we relocated and surface prospected the two known vertebrate fossil localities within the Hunter Creek Sandstone and discovered many new localities. The most productive localities occur on private lands to the north of the Truckee River and the community of Mogul. Additional localities were discovered south of the Truckee River and within the Toiyabe National Forest. The purposes of this report are to: (1) document the new fossil localities; (2) identify and describe the new fossil specimens; and (3) construct a biostratigraphic framework for the Hunter Creek Sandstone. The biostratigraphic framework provides independent age constraints for the Hunter Creek Sandstone, and it will be used to test the existing chronostratigraphy, which is based on radioisotopic and tephrochronologic methods, for congruence.

METHODS

The two previously known fossil localities were relocated using published records and locality data on file at the W.M. Keck Museum. Exposures of the Hunter Creek Sandstone were examined and surface prospected for additional vertebrate fossil localities. All specimens discovered during this study were deposited in the W.M. Keck Museum, Mackay School of Mines, University of Nevada, Reno. Specimens collected within the Toiyabe National Forest were collected under the U.S. Department of Agriculture Forest Service Special Use Permit CAR37. All locality data are on file at W.M. Keck Museum, University of Nevada, Reno.

Trexler et al. (2000) recently provided a summary of the geology and stratigraphy of the Hunter Creek Sandstone of the Verdi Basin. They provided whole-rock and single-crystal ^{40}Ar/^{39}Ar (hornblende and plagioclase) radioisotopic ages from three different stratigraphic levels in the basin. Plagioclase from the "Kate Peak"–type andesitic lavas that directly underlie the Hunter Creek Sandstone at Alum Creek yielded a ^{40}Ar/^{39}Ar age of 12.43 ± 0.23 Ma (plagioclase), and similar andesite capping in the Verdi Range yielded ages of 11.69 ± 0.15 Ma (plagioclase) and 11.94 ± 0.06 Ma (hornblende). Clasts from an andesite breccia, which were shown to be contemporaneous with additional "Kate Peak"–type volcanism at the base of the Hunter Creek Sandstone (Trexler et al., 2000), provided ^{40}Ar/^{39}Ar ages of 11.41 ± 0.11 Ma (plagioclase) and 11.75 ± 0.10 Ma (hornblende). A group of mafic lavas, which lie within the lower part of the Hunter Creek Sandstone in many areas of the Verdi Basin, provided five ^{40}Ar/^{39}Ar (whole-rock) ages within a narrow range

of 10.19 ± 0.14 Ma to 10.41 ± 0.13 Ma. In addition, they identified several ash beds within the thick lacustrine facies of the upper half of the formation that were correlated geochemically with known, regional tephras dated at 3.0–3.1 and 4.4–4.8 Ma (±2σ, not provided). Their work helped to provide a basis for correlating the stratigraphic positions of the fossil localities within the Verdi Basin.

Measurements of teeth and appendicular elements were made to the nearest 0.1 mm with a Vernier caliper. All horse specimens were measured following the standards set forth by Eisenmann et al. (1988), and horse dental terminology follows MacFadden (1984). Definitions of the wear stages for horse cheek teeth follow Kelly (1998b). All other specimens were measured at their greatest dimensions. Upper teeth are designated by uppercase letters, and lower teeth are designated by lowercase letters. Metric abbreviations and dental formulae follow standard usage. Definition of taxon-range chron (= range chron of Walsh, 1998) follows Woodburne (2004), and definition of chronostratigraphy follows Aubry et al. (1999). Subzones or subages (e.g., Cl2, Cl3, Hh1, Hh2, Hh3) of the Clarendonian and Hemphillian North American land mammal ages follow Tedford et al. (2004), and Blancan V arvicoline division of the Blancan land mammal age follows Repenning (1987) and Bell et al. (2004). All taxonomic identifications were determined by the authors using published accounts and comparative material in the vertebrate paleontology collections of the Natural History Museum of Los Angeles County, the Museum of Paleontology, University of California, and the W.M. Keck Museum, University of Nevada, Reno.

Older published K-Ar radioisotopic ages presented herein were recalibrated using International Union of Geological Sciences constants following the method of Dalrymple (1979). Older published ^{40}Ar/^{39}Ar ages were recalibrated relative to the Fish Canyon Tuff sanidine interlaboratory standard at 28.02 Ma. All radioisotopic ages include ±2σ (deviation), except those that were published without any citation for their deviations.

Abbreviations and institutional acronyms are as follows: A-P—anteroposterior; L—left; lf—local fauna; LACM—Natural History Museum of Los Angeles County; Ma—million yr B.P.; R—right; ROC—radius of curvature; TR—transverse; UCMP—Museum of Paleontology, University of California, Berkeley; WMK—W.M. Keck Museum, Mackay School of Mines, University of Nevada, Reno; WMK P—vertebrate fossil locality.

SYSTEMATIC PALEONTOLOGY

Class Mammalia Linnaeus, 1758
Order Perissodactyla Owen, 1848
Family Rhinocerotidae Owen, 1845
Rhinocerotidae, genus and species indeterminate

Referred specimens: From locality WMK P-103: partial lower premolar, WMK 6623. From locality WMK P-126: cheek tooth fragment, WMK 6718. From locality WMK P-104: cheek tooth fragment, WMK 6635.

Age and fauna: Hemphillian, Mogul lf.

Figure 2. *Dinohippus* cf. *D. leardi* from East Verdi lf. (A) RP4, WMK 6712, posterior view. (B) RP4, WMK 6712, occlusal view. Bar scale = 5 mm.

Discussion: WMK 6623 is a well-worn, partial premolar, possibly deciduous that appears to represent the Rhinocerotidae. A small labial cingulid is present. WMK 6635 and 6718 are cheek tooth enamel fragments. Due to their fragmentary states, a familial identification is all that is possible. However, they do indicate that representatives of the Rhinocerotidae were present in the Mogul lf.

Family Equidae Gray, 1821
Subfamily Equinae Gray, 1821
Tribe Equini Gray, 1821
Genus *Dinohippus* Quinn, 1955
Dinohippus cf. *D. leardi* (Drescher, 1941)

Referred specimen (Fig. 2): From locality WMK P-125: RP4, WMK 6712.

Age and fauna: Clarendonian, East Verdi lf.

Description: The RP4 (WMK 6712) (Fig. 2) is complete and in moderate wear (stage of wear after Kelly, 1998b). It is characterized by having the following: (1) the fossette enamel borders are very simple; (2) the protocone is well connected to the protoloph, oval in occlusal outline, well separated from the metaloph, and the anterior portion is extended only slightly anteriorly; (3) a distinct pli caballin is present; (4) the hypoconal groove is closed with only an incipient plication at its position along the posterior enamel border of the tooth, and a hypoconal lake is present; (5) the crown is moderately tapered, with the A-P dimension at the occlusal surface greater than the A-P dimension at the base of the crown; (6) the crown is slightly curved (ROC = 80 mm); and (7) the cement is thick (~2 mm). The measurements for WMK 6712 are: A-P = 29.5 mm, TR = 29.2 mm, protocone A-P = 8.7 mm, protocone TR = 6.7 mm, and mesostylar crown height = 46.0 mm.

Discussion: WMK 6712 exhibits characters that are shared with certain species of *Dinohippus* sensu lato (see Kelly [1998b]

for detailed discussion of *Dinohippus* and *Pliohippus*). It is most similar to *Dinohippus leardi* and shares the following dental characters: (1) simple fossette enamel borders; (2) a closed hypoconal groove that does not extend down to the base of the crown; (3) an oval protocone connected to the protoloph; (4) a single pli caballin; (5) thick cement; and (6) a tapered crown. It differs from *D. leardi* by having a greater ROC, a slightly more persistent pli caballin, and a hypoconal lake. Although hypoconal lakes are usually absent in *D. leardi*, they are occasionally present (Kelly, 1998b). In *Pliohippus*, hypoconal lakes usually form when the hypoconal groove closes (Kelly, 1998b). WMK 6712 differs from the upper cheek teeth of *Pliohippus* by having much less curvature (ROC of *Pliohippus* < 40 mm), a more persistent pli caballin, a protocone that is well separated from the hypocone, and larger size. These differences preclude assignment of WMK 6712 to *Pliohippus*.

WMK 6712 represents *D. leardi* or a species closely related to it that is at a similar evolutionary stage within the *Dinohippus* lineage (see Kelly, 1998b), and therefore we refer it to *Dinohippus* cf. *D. leardi*.

Dinohippus sp.

Referred specimens (Fig. 3; Table 1): From locality WMK P-102: upper cheek tooth partial fossette, WMK 6790; partial Lp3 or 4, WMK 6789; lower cheek tooth postflexid, WMK 6791. From locality WMK P-103: upper left cheek tooth fossette, WMK 6618; upper cheek tooth partial protoloph and protocone, WMK 6617; upper cheek tooth partial fossette, WMK 6619; upper cheek tooth partial ectoloph, WMK 6620; upper cheek tooth partial fossette, WMK 6769; upper cheek tooth partial fossette, WMK 6811; partial Lp2, WMK 6768; associated partial Rp3–4, WMK 6809; lower left cheek tooth postfossettid, WMK 6621; lower left cheek tooth prefossettid, WMK 6770; lower right cheek tooth prefossettid, WMK 6771; astragalus, WMK 6616. From locality WMK P-104: upper cheek tooth partial fossette, WMK 6741; upper left cheek tooth fossettes, WMK 6628; partial Rm1 or 2, WMK 6627; partial lower right cheek tooth, probably m1, WMK 6629; partial left lower cheek tooth, WMK 6725; partial lower cheek tooth, WMK 6750; astragalus, WMK 6734; astragalus, WMK 6735; astragalus, WMK 6736; first phalanx, WMK 6655; partial third phalanx (hoof), WMK 6726. From locality WMK P-107: partial right upper cheek tooth, WMK 6820; upper cheek tooth fossette, WMK 6637. From locality WMK P-127: upper left cheek tooth ectoloph and partial fossette, WMK 6721; partial Rm1 or 2, WMK 6720.

Age and fauna: Hemphillian, Mogul lf.

Description: The horse material in the Mogul lf consists of several almost complete lower cheek teeth and numerous upper and lower cheek teeth fragments (Fig. 3). Although only a few specimens are relatively complete, the morphologies exhibited by these and all of the other tooth fragments are rather consistent. This horse dental material is characterized by having the following: (1) the upper cheek teeth have only slight curvature (average ROC = ~80 mm); (2) the upper cheek teeth fossette enamel borders are very simple, with either weakly developed single plis

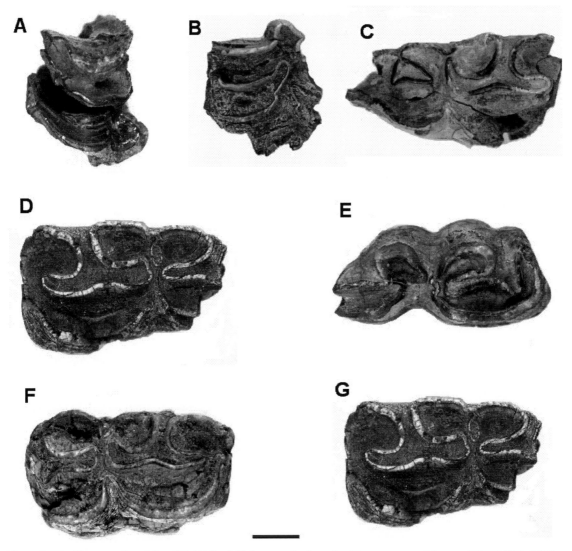

Figure 3. *Dinohippus* sp. from Mogul lf. (A) Partial left upper cheek tooth with protocone and fossette, WMK 6618. (B) Partial right upper cheek tooth, WMK 6820. (C) Partial Lp2, WMK 6768. (D) Partial Rm1 or 2, WMK 6720. (E) Partial Rm1 or 2, WMK 6627. (F) Partial Lp3 or 4, WMK 6789. (G) Partial Rp4, WMK 6809. All occlusal views; bar scale = 5 mm.

prefossette and postfossette or lacking any plications; (3) the protocones are oval and well connected to the protolophs; (4) the plis caballin are single and weakly developed; (5) the lower cheek teeth are almost straight, with little curvature; (6) the prefossettid and postfossettid borders are very simple and lack plications; (7) the ectoflexids in the lower premolars do not penetrate the isthmuses between the metaconids and metastylids; (8) the linguaflexids are generally U-shaped; (9) the plis caballinid are either absent or represented by a single, very slight indentation along the ectoflexid; (10) the ectoflexids in the lower molars generally do not penetrate the isthmuses between the metaconids and metastylids (only one partial lower molar shows penetration); (11) all the lower cheek teeth in which the anterolabial portion of the tooth is preserved lack protostylids; and (12) the cement covering is thick (>1.5 mm). The teeth appear to have been hypsodont because even teeth in which the lower portion of the crown is broken off or teeth that are in a moderate or well-worn stage have mesostylar or metastylar heights greater than 50 mm. In WMK 6721, most of the mesostyle is present, and it measures 64.7 mm in height. Measurements of selected teeth are presented in Table 1.

Appendicular elements are represented by several astragali, a first phalanx, and a partial hoof. The dimensions (Table 1) of the appendicular and dental specimens indicate that the horse of the Mogul lf was of moderate size as compared with contemporary Hemphillian Equini from other localities within the Great Basin (Macdonald, 1959; Shotwell, 1970; Azzaroli, 1988; Downs and Miller, 1994; Kelly, 2000).

TABLE 1. MEASUREMENTS OF SELECTED DENTAL AND
APPENDICULAR SPECIMENS OF *DINOHIPPUS* SP. FROM
MOGUL LOCAL FAUNA

Specimen	Dimension measured	Measurement (mm)
WMK 6618	Fossette height	59.6
	ROC	82.0
	Protocone A-P	5.3
	Protocone TR	4.9
WMK 6811	ROC	83.0
WMK 6628	ROC	80.5
WMK 6721	Mesostylar height	64.6
WMK 6768	Lp2 A-P	33.4 est.
	Lp2 TR	16.9
WMK 6789	Lp3 or 4 A-P	30.5
	Lp3 or 4 TR	16.9
	Lp3 or 4 metastylid height	47.0
WMK 6809	Rp3 or 4 TR	16.0 est.
WMK 6627	?Rm1 TR	12.1
	?Rm1 metastylid height	52.0
WMK 6720	Rm1 or 2 A-P	30.1
	Rm1 or 2 TR	15.5
	Rm1 or 2 metastylid height	51.0
WMK 6734	Astragalus height	53.5
	Astragalus width	53.6
WMK 6735	Astragalus height	57.0
	Astragalus width	54.4
WMK 6726	First phalanx length	80.5

Note: A-P–anteroposterior; est.–estimated; ROC–radius of curvature; TR–transverse.

Discussion: The horse material in the Mogul lf can be confidently assigned to the subfamily Equinae and tribe Equini (see Hulbert and MacFadden, 1991; MacFadden, 1992) based on the dental characters listed previously. The teeth differ from those of the tribe Hipparionini by having the following: (1) simple upper cheek teeth fossette enamel borders; (2) protocones that are well connected to the protolophs; (3) protostylids absent; (4) plis caballinid absent or only very weakly developed; and (5) simple prefossettid and postfossettid enamel borders. They differ from those of *Pliohippus* by having the following: (1) cheek teeth with much less curvature (ROC of *Pliohippus* < 40 mm); (2) protocones that are well separated from the hypocones during moderate-wear stage; (3) shallower lower molar ectoflexids that generally do not penetrate the isthmuses during moderate-wear stage; and (4) larger size.

All the dental characters exhibited by the horse material in the Mogul lf from the Hunter Creek Sandstone are seen also in species of the genus *Dinohippus* sensu lato (Kelly, 1998b). Specifically, the Hunter Creek equid teeth are indistinguishable in dental occlusal morphology and size to those of *Dinohippus* cf. *D. spectans* from the Coal Valley Formation of Smith Valley, Nevada (Macdonald, 1959; Kelly, 1998b). They differ from *D.* cf. *D. spectans* in having slightly less curvature of the upper cheek teeth, which is generally regarded as a derived character (Hulbert, 1988; Kelly, 1998b). However, until additional material is discovered, a specific diagnosis cannot be made based on the fragmentary nature of the sample from the Hunter Creek Sand-

stone. Therefore, we refer this material to *Dinohippus* sp., recognizing that it represents a species closely related to, or possibly conspecific with, *D.* cf. *D. spectans*.

An interesting note is that Axelrod (1958) referred a partial horse tooth from the Hunter Creek Sandstone, which was collected ~1.6 km (~1 mile) north of Mogul, to either *Hipparion* or *Neohipparion*. Axelrod stated that this partial tooth was in the collection at the W.M. Keck Museum. During a complete curation of all vertebrate fossil specimens in the collection by one of us (Kelly), we were unable to relocate the tooth and must assume that it has been lost or misplaced. All the horse partial teeth and numerous tooth fragments in the collection from the Hunter Creek Sandstone exhibit characters typical of *Dinohippus*. Thus, we cannot confirm Axelrod's report that hipparionines were present in the Hunter Creek Sandstone.

Genus *Equus* Linnaeus, 1758
Equus idahoensis Merriam, 1918

Referred specimen (Fig. 4; Table 2): From locality WMK P-111: associated partial RP3-M1 and partial LM1–3, WMK 6634.

Age and fauna: Blancan, Byland locality.

Description: The individual teeth making up WMK 6634 are assumed to represent a single individual and were cataloged as a single specimen because of the following facts: (1) they were recovered from a small isolated pocket (~20 × 30 cm) within the Hunter Creek Sandstone, and no other fossil material was evident in the area; (2) they are all at the same occlusal wear stage (moderate wear); and (3) the anterior and posterior surfaces (interstitial wear facets) of successive teeth fit precisely together. The premolars are represented by a partial RP3 and RP4, which exhibit the following characters: (1) the fossette enamel borders are relatively simple, with single plis protoloph, single plis hypostyle, rounded plis protoconule, and two small plis postfossette; (2) the protocones are elongated anteroposteriorly and indented along their lingual borders; (3) the protocones are well connected to the protolophs by narrow isthmuses, the anterior portions of the protocones extend anteriorly well past the isthmuses, and the posterior portions of the protocones extend to about the middle of the metalophs; (4) a single, well-developed pli caballin is present; (5) the hypoconal grooves are well developed and extend down to near the base of the crown; and (6) there is very little curvature, and the crown is almost straight (ROC = 126 mm). The mesostylar height of the RP4 is 53.9 mm.

The upper molars are represented by a RM1, a partial LM1, and LM2–3. The M1–2 are morphologically very similar to the upper premolars and exhibit all of the characters listed for the premolars. The M3 differs from the M1–2 by being slightly more elongated anteroposteriorly and having a less distinct hypoconal groove. The mesostyle height of the RM1 is 53.4 mm and that of the LM2 is 52.6 mm. Additional measurements for the premolars and molars are presented in Table 2.

Discussion: The cheek teeth of WMK 6634 can be confidently assigned to *Equus* based on the characters listed above. They differ from those of *Dinohippus* by having the following:

Figure 4. *Equus idahoensis* from Byland locality. (A) Partial RP3-M1, WMK 6634. (B) Partial LM1-M3, WMK 6634. All occlusal views; bar scale = 5 mm.

(1) the protocones are very elongated anteroposteriorly with a well-developed indentation along the lingual borders; (2) the protocones extend anteriorly well past the isthmuses that connect the protocones to the protolophs; and (3) the crowns are straighter, with only very slight curvature. In size and dental morphology, the associated teeth of WMK 6634 are indistinguishable from those of *Equus idahoensis* (see Merriam, 1918; Shotwell, 1970; Azzaroli and Voorhies, 1993; Downs and Miller, 1994; Repenning et al., 1995; Kelly, 1997), and we refer them to this species.

Order Artiodactyla Owen, 1848
Family Camelidae Gray, 1821
Genus *Hemiauchenia* Gervais and Ameghino, 1880
cf. *Hemiauchenia* sp.

Referred specimens: From locality P-107: partial metapodial, WMK 6830. From locality WMK P-103: partial astragalus, WMK 6772.

Age and fauna: Hemphillian, Mogul lf.

Description: The partial metapodial is missing the distal condyles. The measurements of the metapodial are as follows; the TR proximal articular surface = 40.7 mm, TR mid-shaft = 24.6 mm, A-P at mid-shaft = 24.4 mm; and the broken length (missing the distal condyles) = 290.3 mm. The partial right astragalus is missing the lateral trochlear crest and distal astragalar facet. The height from the medial trochlear crest to the navicular facet is 65.3 mm.

Discussion: The elongated and slender proportions of the metapodial are typical of those of *Hemiauchenia*. The size of the

partial astragalus is within the size range of those of *Hemiauchenia* from the Hemphillian Smith Valley Fauna of the Coal Valley Formation, Smith Valley, Nevada (Macdonald, 1959; unpublished specimens in the LACM and UCMP). The specimens probably represent *Hemiauchenia*, but without additional diagnostic material, their assignment to this taxon is tentative.

Genus *Megatylopus* Matthew and Cook, 1909
cf. *Megatylopus* spp.

TABLE 2. MEASUREMENTS OF *EQUUS IDAHOENSIS* FROM THE BYLAND LOCALITY

Position/dimension	A-P (mm)	TR (mm)
RP3	33 est.	29 est.
RP3 protocone	13.0	5.1
RP4	35.4	31.2
RP4 protocone	13.4	5.8
RM1	34.6	32.6
RM1 protocone	13.1	5.9
LM1	n.d.	n.d.
LM1 protocone	n.d.	n.d.
LM2	31.6	n.d.
LM2 protocone	n.d.	n.d.
LM3	36.3	27.8
LM3 protocone	14.1	5.5

Note: A-P—anteroposterior; n.d.—no data; TR—transverse; est.—estimated.

Referred specimens: From locality WMK P-104: fibular tarsal, WMK 6738. From locality WMK P-106: navicular tarsal, WMK 6865. From locality WMK P-105: partial distal metapodial, WMK 6632.

Age and fauna: Hemphillian, Mogul lf and latest Hemphillian or earliest Blancan, unnamed assemblage from WMK P-105.

Description: The tarsals and partial metapodial are characterized by their large size. The fibular tarsal is complete with the following dimensions; height = 55.5 mm and width = 31.7 mm. The navicular tarsal is missing small portions of bone from the anterior medial and posterior medial aspects. It has the following dimensions: height = 34.0 mm and A-P length = 53.3 mm. The partial distal metapodial is missing the medial condylar surface. The lateral condylar A-P dimension is 40.5 mm.

Discussion: The robust proportions of the tarsals and partial metapodial indicate the presence of a large camel within the Mogul lf and the unnamed assemblage from locality WMK P-105. These elements are significantly larger than those of *Hemiauchenia* and *Alforjas*, but smaller than those of *Gigantocamelus* (see Harrison, 1979; Kelly, 1994, 1997). It cannot be determined whether the specimens from WMK P-104 and P-106 represent the same species as the specimen from WMK P-105. In size and morphology, they all compare well with those of *Megatylopus* from the Hemphillian Smith Valley Fauna of the Coal Valley Formation, Smith Valley, Nevada (Macdonald, 1959; unpublished specimens in the collections of the LACM and UCMP). Thus, we tentatively assign this sample to cf. *Megatylopus* spp.

Camelidae, genus and species indeterminate

Referred specimen: From locality WMK P-132: partial upper molar ectoloph, WMK 6852.

Age and fauna: Clarendonian, East Verdi lf.

Figure 5. *Mammut* sp. from Mogul lf. LM2, WMK 6968, occlusal view. Bar scale = 10 mm.

Discussion: Although the specimen is too fragmentary for generic identification, it documents the presence of the Camelidae in the East Verdi lf.

Family Antilocapridae Gray, 1966
?Antilocapridae, genus and species indeterminate

Referred specimen: From locality WMK P-110: distal metapodial condyle, WMK 6654.

Age and fauna: Clarendonian, East Verdi lf.

Description: The specimen consists of the distal portion of a metapodial with one condyle preserved. The condyle measures 24.3 mm A-P and 12.8 mm TR.

Discussion: Morphologically, the metapodial condyle is typical of the Antilocapridae. In size, it is comparable to those of the Hemphillian to Rancholabrean Antilocaprinae, and it is larger than those of the Barstovian to Clarendonian Merycodontinae. However, even a familial assignment is difficult, so we refer it questionably to the Antilocapridae.

Order Proboscidea Illiger, 1811
Family Mammutidae Cabrera, 1929
Mammut Blumenbach, 1799
Mammut (= *Pliomastodon* Osborn, 1926) sp.

Referred specimen (Fig. 5): From locality WMK 107: LM2, WMK 6968.

Age and fauna: Hemphillian, Mogul lf.

Description: The upper molar is in early wear and has small portions of the anterior and posterior enamel missing. There are three primary transverse lophs that are well separated by complete valleys that are open labially and lingually (Fig. 5). The transverse lophs lack any trefoiling. Small, but distinct anterior and posterior cingula are present. Small, intermittent cingula are also present between the labial aspects of the transverse lophs. The measurements of WMK 6968 are 113 mm A-P and 86 mm TR.

Discussion: Buwalda (1914) described a partial mastodon tooth that was later determined by Axelrod (1958) to have been recovered from an interval estimated to be ~400–500 ft (~120–150 m) above the Verdi Flora locality. He questionably assigned this tooth to *Tetrabelodon*. Stirton (1940) noted that this tooth was similar to *Gomphotherium simpsoni* (Stirton, 1939) but stressed that the tooth was not diagnostic. Subsequently, a second, more complete, mastodon tooth (WMK 6968, Fig. 5) was recovered from an exposure that was reported to occur ~600 ft (~180 m) stratigraphically below the Verdi Flora and 400 ft (~120 m) above mafic lavas (Axelrod, 1958). Based on limited WMK locality data, this second tooth appears to have come from the middle part of the formation at locality WMK P-107. WMK 6968 was reported to be a left M2 of a true mammutid mastodon (personal commun. from D.E. Savage *in* Axelrod, 1958, p. 158). Savage thought it might represent *Miomastodon* or *Pliomastodon*. Lambert and Shoshani (1998) now regard *Miomastodon* as a junior synonym of *Zygolophodon* and *Pliomastodon* a junior synonym of *Mammut*. Based on size and the simple, trilophodont occlusal pattern with open valleys between the lophs, WMK 6968 can be confidently referred to *Mammut* (= *Pliomastodon*). However, because of morphological similarities between bunolophodont proboscideans (Lambert and Shoshani, 1998), a specific diagnosis is not possible.

Family Gomphotheriidae Hay, 1922
or Mammutidae Cabrera, 1929
Genus and species indeterminate

Referred specimens: From locality WMK P-102: cheek tooth enamel fragments, WMK 6607–6609, 6719, 6796–6801. From locality WMK P-103: cheek tooth enamel fragments, WMK 6614, 6624, 6774–6778, 6816–6817, 6819; partial tusks, WMK 6614–6615, 6786; magnum carpal, WMK 6773. From locality WMK P-104: cheek tooth enamel fragments, WMK 6626, 6728–6732, 6751–6767, 6787–6788. From locality WMK P-105: cheek tooth enamel fragments, WMK 6782–6783. From locality WMK P-106: cheek tooth enamel fragments, WMK 6633, 6866–6866; associated tusk fragments, WMK 6864; humerus, WMK 6967. From locality WMK P-107: cheek tooth enamel fragments, WMK 6640–6647, 6649, 6651–6653, 6833–6851, 6858. From locality WMK P-108: cheek tooth enamel fragment, WMK 6853. From locality WMK P-109: cheek tooth enamel fragment, WMK 6717. From locality WMK P-110: associated cheek tooth fragments, WMK 6781. From locality WMK P-125: cheek tooth enamel fragments, WMK 6713–6715. From locality WMK P-127: cheek tooth enamel fragments, WMK 6723–6724. From locality WMK P-128: cheek tooth enamel fragments, WMK 6779–6780. From locality WMK P-129: cheek tooth enamel fragments, WMK 6806–6807. From locality WMK P-131: cheek tooth enamel fragments, WMK 6854–6857. From locality WMK P-132: associated cheek tooth fragments, WMK 6863. From locality WMK P-133: cheek tooth enamel fragments, WMK 6859–6862. From locality WMK P-134: partial cervical vertebra, WMK 6868; partial scapula (glenoid fossa), WMK 6869. From locality WMK P-167: partial tusk, WMK 6966.

Age and faunas: Clarendonian, East Verdi lf; Hemphillian, Mogul lf; and latest Hemphillian or earliest Blancan, unnamed assemblage from WMK P-105.

Discussion: Proboscidean cheek tooth fragments are by far the most common vertebrate fossils throughout the section of the Hunter Creek Sandstone below the thick lacustrine facies in the upper part of the formation. Almost every locality yielded specimens. The enamel morphology of the cheek tooth fragments indicates that they could represent either the Gomphotheriidae or Mammutidae.

BIOSTRATIGRAPHY

At least three North American land mammal ages (Woodburne, 1987, 2004) are represented by fossils from the Hunter Creek Sandstone (Fig. 6). These include the Clarendonian (late Miocene), the Hemphillian (late Miocene to earliest Pliocene), and the Blancan (Pliocene).

The localities that occur from ~75 m below the mafic lavas to ~500 m above these lavas (−75–500 m levels in Fig. 6) yielded

the East Verdi lf, which includes *Dinohippus* cf. *D. leardi*, Camelidae, ?Antilocapridae, and Mammutidae or Gomphotheriidae. The most productive localities within this stratigraphic interval occur on private lands north of the Truckee River and east the community of Verdi (Fig. 1). These localities can be accurately placed within the stratigraphic section relative to the dated mafic lavas and the lacustrine facies. The stratigraphic positions of WMK P-133 and P-134, south of the Truckee River, are less confident because they cannot be directly correlated relative to the mafic lavas. However, WMK P-133 and P-134 do occur very low in the section in silty sandstones that directly overlie a conglomerate that appears to correlate with the conglomerate facies interbedded with mafic lavas dated at ca. 10 Ma in the Steamboat Ditch section (Trexler et al., 2000). The only fossils recovered from WMK P-133 and P-134 are fragments of cheek teeth representing Equidae and Gomphotheriidae or Mammutidae. The stratigraphic position of locality WMK P-167, which occurs along Steamboat Ditch, is more confident because it can be tied to the measured Steamboat Ditch section (Trexler, et al., 2000).

The most biostratigraphically informative species in the East Verdi lf is *Dinohippus* cf. *D. leardi*. *Dinohippus leardi* is known only from the following Clarendonian faunas: upper localities of the Iron Canyon Fauna (Cl2) and Ricardo Fauna (Cl3), Dove Spring Formation, California; North Tejon Hills Fauna (Cl3), Chanac Formation, California; and the Black Hawk Ranch Quarry (Cl3), Green Valley Formation, California (Savage, 1955; Drescher, 1941; Richey, 1948; Whistler and Burbank, 1992; Kelly, 1998b; Tedford et al., 2004). Based on biostratigraphic correlation, radioisotopic and paleomagnetic data constraints, the taxon-range chron for *D. leardi* in these formations is late-medial to late Clarendonian (Cl2–Cl3), or ca. 10.8–9.0 Ma (Savage, 1955; Drescher, 1941; Whistler and Burbank, 1992; Wilson and Prothero, 1997; Kelly, 1998b; Prothero and Tedford, 2000; Tedford et al., 2004). Kelly (1998b) provided cladistic analyses that documented the progressive acquisition of apomorphies within the *Dinohippus* clade. These analyses provided also an evolutionary and chronological framework for the succession of dental characters within the clade. *Dinohippus* cf. *D. leardi* is morphologically very similar to *D. leardi* and either represents *D. leardi* or a species closely related to *D. leardi* that is at a similar stage of evolution within the *Dinohippus* clade (see Kelly, 1998b), suggesting a similar age constraint for the East Verdi lf. The stratigraphic positions of the localities that yielded the East Verdi lf, relative to the dated mafic lavas (ca. 10 Ma), are also consistent with a late-medial to late Clarendonian age (ca. 10.8–9 Ma) for the fauna.

The Mogul lf was collected from localities that occur from ~850–1350 m in the measured section (Fig. 6), and it includes *Dinohippus* sp., Rhinocerotidae, Camelidae (at least two species), *Mammut* (= *Pliomastodon*) sp., and possibly Gomphotheriidae. The most productive localities within this interval were WMK P-103, P-104, P-127 (north of the community of Mogul), and WMK P-107 (southwest of Mogul). *Dinohippus* sp. is the most biostratigraphically informative taxon and is most similar to *Dinohippus* cf. *D. spectans* of the Smith Valley Fauna, Coal

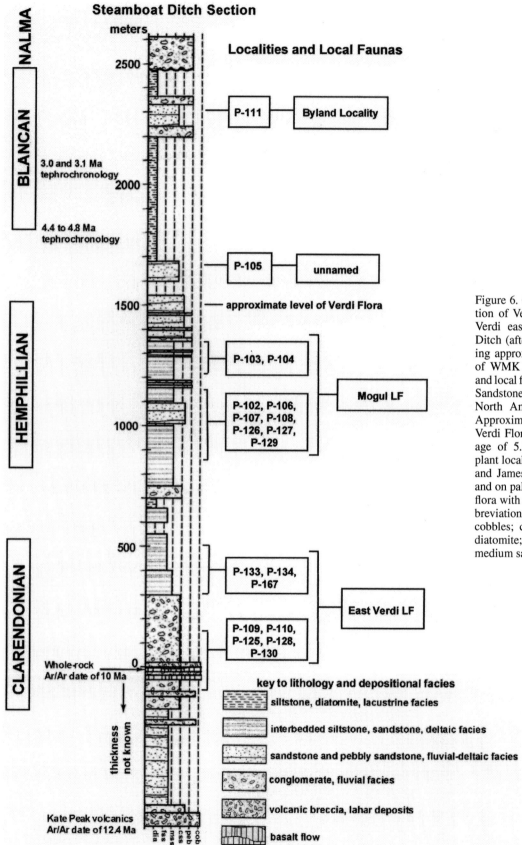

Figure 6. Generalized stratigraphic section of Verdi Basin in a transect from Verdi east to Reno along Steamboat Ditch (after Trexler et al., 2000) showing approximate stratigraphic positions of WMK P vertebrate fossil localities and local faunas within the Hunter Creek Sandstone (lf—local fauna; NALMA—North American land mammal age). Approximate stratigraphic position of Verdi Flora is based on a radioisotopic age of 5.85 Ma (corrected) from the plant locality (Axelrod, 1958; Evernden and James, 1964; Schorn et al., 1994) and on paleobotanical correlation of the flora with other Hemphillian floras. Abbreviations are: BT—basal tuff; cob—cobbles; css—coarse sandstone; dia—diatomite; fss—fine sandstone; mss—medium sandstone; peb—pebbles.

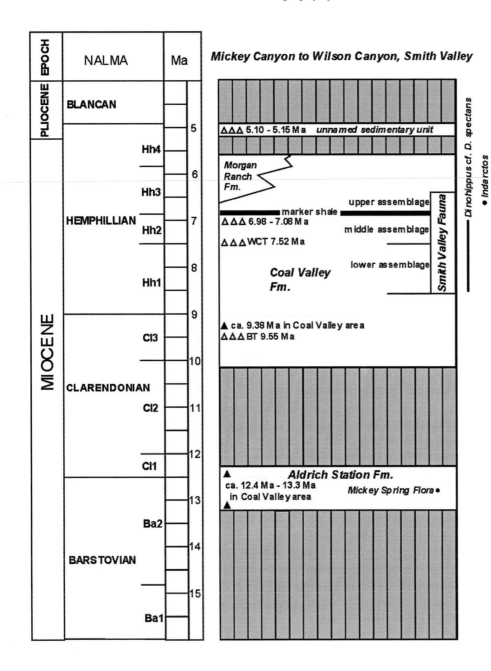

Figure 7. Schematic chart of generalized section from Mickey Canyon to Wilson Canyon, Smith Valley, Nevada, showing the biostratigraphy, chronostratigraphy, and stratigraphic relationships of formations exposed in this area. Open triangles represent tuffs with radioisotopic ages from the section (see text), and filled triangles (black) represent tuffs with radioisotopic ages (corrected) that are extrapolated onto the section from the type Coal Valley and Aldrich Station Formations, Coal Valley, Nevada (Gilbert and Reynolds, 1973). Shaded, vertically lined intervals represent unconformities. The stratigraphic levels that yielded *Indarctos* and the taxon-range chron for *Dinohippus* cf. *D. spectans* are shown on the right. Subzones of North American land mammal ages (NALMA) follow Tedford et al. (2004). WCT—Wilson Canyon Tuff.

Valley Formation, Smith Valley, Nevada (Macdonald, 1959; Kelly, 1998b).

The age of the Smith Valley Fauna is constrained biostratigraphically and radioisotopically. Three marker horizons have been previously recognized in the Smith Valley section (Fig. 7). Using the lowest marker horizon as a datum (basal tuff, 0 m) and field data with estimated stratigraphic levels, they are: (1) a basal tuff that occurs ~350 m below the lowest occurrence of mammals on the eastern edge of Smith Valley, which yielded a K-Ar (biotite) age of 9.55 Ma (Evernden et al., 1964; 2σ errors not provided); (2) the Wilson Canyon Tuff at ~650 m, which yielded a ^{40}Ar/^{39}Ar (plagioclase) age of 7.52 ± 0.08 Ma (Swisher, 1992);

and (3) a marker shale (Gilbert and Reynolds, 1973) at ~850 m. Chris Henry (2005, personal commun.) recently dated two additional tuffs (CV-427 and CV-517) that were collected by one of us (Kelly) from the Coal Valley Formation in upper Petrified Canyon, Smith Valley. These tuffs occur in the interval between the Wilson Canyon Tuff and the marker shale at 130 m (CV-427) and 160 m above (CV-517) the Wilson Canyon Tuff. The tuffs yielded ^{40}Ar/^{39}Ar ages of 7.08 ± 0.01 Ma (CV-427, sanidine) and 6.98 ± 0.01 Ma (CV-517, sanidine) and are consistent with the ^{40}Ar/^{39}Ar age of the underlying Wilson Canyon Tuff. Furthermore, the basal tuff in Smith Valley may be a correlative of a tuff with a K-Ar (hornblende) age of 9.38 ± 0.44 Ma that occurs in the type Coal Valley Formation, Coal Valley, Nevada, and overlies the type Coal Valley Fauna of Clarendonian age (Gilbert and Reynolds, 1973). Gilbert and Reynolds (1973) reported two K-Ar ages of 5.15 ± 0.26 Ma (hornblende) and 5.10 ± 0.35 Ma (biotite) for a tuff from an unnamed sedimentary unit that unconformably overlies the Coal Valley Formation in Smith Valley. Taxa included in the Smith Valley Fauna were collected from three stratigraphic intervals (Macdonald, 1959; locality data at the LACM and UCMP): (1) from ~350 m above the basal tuff to just below the Wilson Canyon Tuff at ~650 m; (2) from the interval between the Wilson Canyon Tuff and the marker shale at ~650–850 m; and (3) from just above the marker shale to ~950 m. A schematic chart of the stratigraphic relationships of the section in the vicinity of Mickey Canyon northward to Wilson Canyon, Smith Valley, is shown in Figure 7.

Tedford et al. (2004, 6.2) regarded the Smith Valley Fauna as early Hemphillian in age (Hh1, ca. 9.0–7.5 Ma) based on its faunal content and on the K-Ar age from the basal tuff (9.55 Ma). However, the bear *Indarctos*, one of several taxa whose first occurrences were used by Tedford et al. (2004) to define the beginning of the late-early Hemphillian (Hh2, ca. 7.5–6.8 Ma), was collected from the upper assemblage of the Smith Valley Fauna (Macdonald, 1959; Hunt, 1998), above the marker shale (>850 m, Fig. 7) and new dated tuffs. The taxon-range chron for *Indarctos* is from the late-early Hemphillian to early-late Hemphillian (Hh2–Hh3), or ca. 7.5–6.0 Ma (Hunt, 1998). The occurrence of *Indarctos* above the marker shale and the $^{40}Ar/^{39}Ar$ ages for the Wilson Canyon Tuff (7.52 Ma) and the two tuffs (7.08 and 6.98 Ma) between the Wilson Canyon Tuff and the marker shale strongly suggest, contrary to Tedford et al. (2004, their Fig. 6.2), a late-early Hemphillian or younger age for strata above the Wilson Canyon Tuff. A detailed systematic analysis of all taxa comprising the middle assemblage is required to determine an unequivocal biochron subage assignment for this assemblage, and, because this is not yet available, we provisionally refer it to the late-early Hemphillian (Hh2).

Based on the overall taxonomic composition of the Smith Valley Fauna (Macdonald and Pelletier, 1958; Macdonald, 1959; Hunt, 1998), the stratigraphic distribution of fossils relative to the radioisotopically aged tuffs within the Smith Valley section, and the presence of *Inarctos* above the marker shale, the Smith Valley Fauna probably spans the early to late-early Hemphillian or younger (Hh1 to possibly Hh3, ca. 8.5 to at 6.5 Ma). Kelly (1998b) determined that the dental morphology of *Dinohippus* cf. *D. spectans* of the Coal Valley Formation is more derived than that of the Clarendonian *D. leardi*, but less derived than those of the late Hemphillian *D. leidyanus* and latest Hemphillian *D. mexicanus*. In the Smith Valley section, *Dinohippus* cf. *D. spectans* occurs in the lower and middle faunal assemblages and is restricted to a stratigraphic interval (Fig. 7) from ~300 m below the Wilson Canyon Tuff to ~30 m below the marker shale (Macdonald, 1959; specimen and locality data from the LACM and UCMP). Following the previous discussion, this interval is mostly early Hemphillian (Hh1),

but it also appears to include part of the late-early Hemphillian (Hh2) above the Wilson Canyon Tuff.

East of Smith Valley, *Dinohippus* cf. *D. spectans* has been recorded in the Yerington lf of the Coal Valley Formation (Macdonald and Pelletier, 1958; specimen and locality data from the LACM and UCMP) from a stratigraphic interval ~100 m below the marker shale to ~43 m below the marker shale (the base of the section in this area is covered by alluvium), which is consistent with its stratigraphic range in Smith Valley.

The Smith Valley Fauna correlates well with the Thousand Creek Fauna from the Thousand Creek Formation, Washoe County, Nevada, indicating that these faunas are of similar age (Furlong, 1932; Shotwell, 1955; Korth, 1999; Tedford et al., 2004). Perkins et al. (1998) determined that an ash overlying the Thousand Creek Fauna correlated geochemically with the Rattlesnake tuff, which elsewhere has a $^{40}Ar/^{39}Ar$ (sanidine) age of 7.05 ± 0.01 Ma (Streck and Grunder, 1995). Perkins et al. (1998) identified also two ashes from the fossil-producing part of the section at Thousand Creek that were correlated geochemically to the Rush Valley ash and Alamo ash, which they extrapolated to be 7.90 ± 0.50 Ma and 8.00 ± 1.00 Ma in age, respectively. These ash dates are consistent with the geochronology of Smith Valley.

All of the above data indicate that the taxon-range chron for *Dinohippus* cf. *D. spectans* in the Coal Valley Formation in Smith Valley and east of Smith Valley is ca. 8.5–7.0 Ma, or early to late-early Hemphillian (Hh1 to Hh2). The close morphological similarity between *D.* cf. *D. spectans* from the Coal Valley Formation and *Dinohippus* sp. from the Hunter Creek Sandstone, including similar derived dental character states relative to the Clarendonian *D. leardi* (see Kelly, 1998b), suggests they are closely related species or are conspecific. This, in turn, suggests that the Mogul lf may be of a similar age. In addition, the relative stratigraphic positions of the localities that yielded the Mogul lf, which occur between $^{40}Ar/^{39}Ar$ ages of ca. 10 Ma and tephrochronologic ages of 4.4–4.8 Ma, are also consistent with a Hemphillian age.

At least two species of camel are represented in the Mogul lf, one very large-sized camel and a medium-sized camel. Based on size and morphology, the large camel probably represents *Megatylopus*, whereas the medium-sized camel probably represents *Hemiauchenia*. Both of these camels are found commonly in Hemphillian faunas of the Great Basin (Macdonald, 1959; Shotwell, 1970; Kelly, 1997, 1998a, 2000), and their presence is consistent with a Hemphillian age estimate for the Mogul lf.

North of the community of Mogul, a single locality (WMK P-105) was discovered high in the section in a dark brown sandstone bed that underlies the thick lacustrine facies of diatomaceous shale and siltstone (Fig. 6). WMK P-105 occurs ~75 m stratigraphically below ash beds within the lacustrine facies that were tephrochronologically dated at 4.4–4.8 Ma (Trexler et al., 2000). This locality yielded only a partial camel metapodial questionably referred to cf. *Megatylopus* and several bunolophodont proboscidean cheek tooth fragments. The occurrence of *Megatylopus* is consistent with a Hemphillian or Blancan age, and the stratigraphic position of WMK P-105 relative to the ash dates suggests either a latest Hemphillian or earliest Blancan age. Because the sample from WMK P-105 is meager, we do not assign a faunal name.

The Blancan is confidently represented by one locality, the Byland locality (WMK P-111), which occurs on the east side of the Verdi Basin and is named in honor of Mr. Al Byland of Reno, Nevada, for allowing us access to his property. The single specimen (WMK 6634) from the Byland locality consists of associated horse upper cheek teeth that we refer to *Equus idahoensis*. *Equus idahoensis* is a common taxon in late Blancan faunas of the Pacific Coast and Great Basin that range from ca. 2.8 to 2.2 Ma in age (Shotwell, 1970; Downs and Miller, 1994; Azzaroli and Voorhies, 1993; Kelly, 1994; Repenning et al., 1995), and it indicates a late Blancan age for the Byland locality. Tephrochronologic ages of 3.0 and 3.1 Ma have been reported from ashes that occur ~100 m stratigraphically below the Byland locality (Trexler et al., 2000) within the thick lacustrine facies composed of diatomite and siltstone (Fig. 6), and these corroborate a Blancan age.

CONCLUSIONS

Twenty-three WMK localities yielding mammalian fossils are now known from the Hunter Creek Sandstone in the Verdi Basin, including 21 localities found during the course of this study. Although, many of the fossils are fragmentary, several specimens preserve morphology adequate for generic identification and specific conferral. These specimens, which are primarily horses, allow for the recognition of three North American land mammal ages, including the Clarendonian (late Miocene), the Hemphillian (late Miocene to earliest Pliocene), and the Blancan (Pliocene). We recognize two mammalian assemblages from the Hunter Creek Sandstone, the East Verdi and Mogul lfs. Based on faunal content, the East Verdi and Mogul lfs are late-medial to late Clarendonian (Cl2–Cl3) and early to possibly late-early Hemphillian (Hh1–Hh2) in age, respectively. In addition, a specimen from the Byland locality, higher in the section, indicates a late Blancan (V) age for this level. The biostratigraphy of the local faunas is consistent with the chronostratigraphy determined by tephrochronologic and radioisotopic methods for the Hunter Creek Sandstone. A meager sample of material from WMK P-105 may represent an additional unnamed assemblage of latest Hemphillian or earliest Blancan age.

The most productive vertebrate fossil localities occur north of the Truckee River and the community of Mogul on private lands. South of the Truckee River in the Humbolt-Toiyabe National Forest, most of the vertebrate fossil localities yielded only fragmentary specimens, and because of local faulting and incomplete exposure, they are subject to greater stratigraphic uncertainty.

The consistent presence of both browsing (Camelidae and bunolophodont proboscideans) and grazing (Equidae) taxa through the late Clarendonian and Hemphillian portions of the Hunter Creek Sandstone suggests a relatively stable local paleoenvironment for at least 5 m.y. The combination of browsers and grazers suggests that the paleoenvironment was characterized by grasslands or open woodlands, probably with thicker woodlands or riparian vegetation along perennial streams and lakes.

ACKNOWLEDGMENTS

We are indebted to Jim Trexler and Pat Cashman of the University of Nevada, Reno, and Larry Garside of the Nevada Bureau of Mines and Geology, Reno, for their considerable help in providing information on the geology and stratigraphy of the Hunter Creek Sandstone. We are particularly grateful to Chris Henry of the Nevada Bureau of Mines and Geology, Reno, for his many constructive comments and for allowing us to cite his unpublished radioisotopic data for Smith Valley. We are also grateful to the late Tom Lugaski for his help with specimen curation at the W.M. Keck Museum and to Joseph Lintz for help in relocating a fossil locality. We give special thanks to Terry Birk of the Carson Ranger District, U.S. Forest Service, for his support of this study and for his considerable help in obtaining a collecting permit. Christopher Bell of the Department of Geological Sciences, University of Texas, Austin, and Donald Prothero of Department of Geology, Occidental College, Los Angeles, provided constructive comments and advice on the original draft of this report. Collections at the University of California, Berkeley, Museum of Paleontology, and the Natural History Museum of Los Angeles County were made available by Patricia Holroyd and Samuel McLeod, respectively.

REFERENCES CITED

Aubry, M.P., Berggren, W.A., Van Couvering, J.A., and Steininger, F., 1999, Problems in chronostratigraphy: Stages, series, unit and boundary stratotypes, global stratotype section and point and tarnished golden spikes: Earth-Science Reviews, v. 46, p. 99–148, doi: 10.1016/S0012-8252(99)00008-2.

Axelrod, D.I., 1956, Mio-Pliocene Floras from West-Central Nevada: University of California Publications in Geological Sciences, v. 33, 322 p.

Axelrod, D.I., 1958, The Pliocene Verdi Flora of Western Nevada: University of California Publications in Geological Sciences, v. 34, p. 91–160.

Azzaroli, A., 1988, On the equid genera *Dinohippus* Quinn 1955 and *Pliohippus* Marsh 1874: Gollettino della Societa Palaeontologica Italiana, v. 27, p. 61–72.

Azzaroli, A., and Voorhies, M.R., 1993, The genus *Equus* in North America. The Blancan species: Palaeontographia Italica, v. 80, p. 175–198.

Bell, C.J., Lundelius, E.L., Jr., Barnosky, A.D., Graham, R.W., Lindsay, E.H., Ruez, D.R., Jr., Semken, H.A., Jr., Webb, S.D., and Zakrzewski, R.J., 2004, The Blancan, Irvingtonian, and Rancholabrean mammal ages, *in* Woodburne, M.O., ed., Late Cretaceous and Cenozoic Mammals of North America, Biostratigraphy and Geochronology: New York, Columbia University Press, p. 232–314.

Bingler, E., 1975, Guidebook to the Quaternary geology along the western flank of the Truckee Meadows, Washoe County, Nevada: Nevada Bureau of Mines and Geology Report, v. 22, p. 1–14.

Buwalda, J.P., 1914, A proboscidean tooth from the Truckee beds of western Nevada: University of California Publications, Bulletin of the Department of Geological Sciences, v. 8, p. 305–308.

Dalrymple, G.B., 1979, Critical tables for conversion of K-Ar ages from old to new constants: Geology, v. 7, p. 558–560, doi: 10.1130/0091-7613(1979)7<558:CTFCOK>2.0.CO;2.

Downs, T., and Miller, G.J., 1994, Late Cenozoic equids from the Anza-Borrego Desert of California: Natural History Museum of Los Angeles County: Contributions in Science, no. 440, p. 1–90.

Drescher, A.B., 1941, Later Tertiary Equidae from the Tejon Hills, California: Carnegie Institution of Washington Publication, v. 530, p. 1–23.

Eisenmann, V., Alberdi, M.T., De Giuli, C., and Staesche, U., 1988, Methodology, *in* Woodburne, M.O., and Sondarr, P., eds., Studying Fossil Horses: Leiden, E.J. Brill, p. 1–9.

Evernden, J.F., and James, G.T., 1964, Potassium-argon dates and the Tertiary floras of North America: American Journal of Science, v. 262, p. 945–974.

Evernden, J.F., Savage, D.E., Curtis, G.H., and James, G.T., 1964, Potassium-argon dates and the Cenozoic mammal chronology of North America: American Journal of Science, v. 262, p. 145–198.

Furlong, E.L., 1932, Distribution and description of skull remains of the Pliocene antelope *Sphenophalos* from the northern Great Basin Province: Carnegie Institution of Washington Publication, v. 418, p. 27–38.

Garside, L.J., Castor, S.B., Henry, C.D., and Faulds, J.E., 2000, Structure, volcanic stratigraphy, and ore deposits of the Pah Pah Range, Washoe County, Nevada: Reno, Nevada, Geological Society of Nevada, Symposium 2000 Field Trip Guidebook no. 2, p. 1–132.

Gilbert, C.M., and Reynolds, M.W., 1973, Character and chronology of basin development, western margin of the Basin and Range Province: Geological Society of America Bulletin, v. 84, p. 2489–2510, doi: 10.1130/0016-7606(1973)84<2489:CACOBD>2.0.CO;2.

Harrison, J.A., 1979, Revision of the Camelinae (Artiodactyla, Tylopoda) and description of the new genus *Alforjas*: University of Kansas: Paleontological Contributions, v. 95, p. 1–20.

Hulbert, R.C., 1988, Phylogenetic interrelationships and evolution of North American late Neogene Equinae, *in* Prothero, D.R., and Schoch, R.M., eds., The Evolution of Perissodactyls: Oxford Monographs on Geology and Geophysics 15, p. 178–196.

Hulbert, R.C., and MacFadden, B.J., 1991, Morphological transformation and cladogenesis at the base of the adaptive radiation of Miocene hypsodont horses: American Museum Novitates, no. 3000, p. 1–61.

Hunt, R.M., Jr., 1998, Ursidae, *in* Janis, C.M., Jacobs, L.L., and Scott, K.M., eds., Evolution of Tertiary Mammals of North America: Cambridge, Cambridge University Press, p. 174–195.

Kelly, T.S., 1994, Two Pliocene (Blancan) vertebrate faunas from Douglas County, Nevada: PaleoBios, v. 16, p. 1–23.

Kelly, T.S., 1997, Additional late Cenozoic (latest Hemphillian to earliest Irvingtonian) mammals from Douglas County, Nevada: PaleoBios, v. 18, p. 1–31.

Kelly, T.S., 1998a, New Miocene mammalian faunas from west central Nevada: Journal of Paleontology, v. 72, p. 137–149.

Kelly, T.S., 1998b, New middle Miocene equid crania from California and their implications for the phylogeny of the Equini: Natural History Museum of Los Angeles County: Contributions in Science, no. 473, p. 1–44.

Kelly, T.S., 2000, A new Hemphillian (late Miocene) mammalian fauna from Hoye Canyon, west central Nevada: Natural History Museum of Los Angeles County: Contributions in Science, no. 481, p. 1–21.

King, C., 1878, The Comstock lode, *in* Hague, J.D., ed., Mining Industry, U.S. Geological Exploration of the 40th Parallel: Washington, D.C., U.S. Geological Survey, p. 10–91.

Korth, W.W., 1999, *Hesperogaulus*, a new genus of mylagaulid rodent (Mammalia) from the Miocene (Barstovian to Hemphillian) of the Great Basin: Journal of Paleontology, v. 73, p. 945–951.

Lambert, W.D., and Shoshani, J., 1998, Proboscidea, *in* Janis, C.M., et al., eds., Evolution of Tertiary Mammals of North America: Cambridge, Cambridge University Press, p. 606–621.

Macdonald, J.R., 1959, The middle Pliocene mammalian fauna from Smiths Valley, Nevada: Journal of Paleontology, v. 33, p. 872–887.

Macdonald, J.R., and Pelletier, W.J., 1958, The Pliocene mammalian faunas of Nevada, U.S.A.: Session 20: Paleontology, Taxonomy, and Evolution, Section VII: International Geologic Congress, p. 365–388.

MacFadden, B.J., 1984, Systematics and phylogeny of *Hipparion, Neohipparion, Nannippus,* and *Cormohipparion* (Mammalia, Equidae) from the Miocene and Pliocene of the New World: Bulletin of the American Museum of Natural History, v. 179, 196 p.

MacFadden, B.J., 1992, Fossil Horses: Systematics, Paleobiology, and Evolution of the Family Equidae: New York, Cambridge University Press, p. xii + 369.

Merriam, J.C., 1918, New Mammalia from the Idaho Formation: University of California Publications: Bulletin of the Department of Geology, v. 10, p. 523–530.

Perkins, M.E., Brown, F.H., Nash, F.H., McIntosh, W.P., and Williams, S.K., 1998, Sequence, age, and source of silicic fallout tuffs in middle to late Miocene basins of the northern Basin and Range Province: Geological Society of America Bulletin, v. 110, p. 344–360, doi: 10.1130/0016-7606(1998)110<0344:SAASOS>2.3.CO;2.

Prothero, D.R., and Tedford, R.H., 2000, Magnetic stratigraphy of the type Montediablan stage (late Clarendonian, late Miocene), Black Hawk Ranch, Contra Costa County, California: Implication for Clarendonian correlations: PaleoBios, v. 20, p. 1–37.

Repenning, C.A., 1987, Biochronology of the microtine rodents of the United States, *in* Woodburne, M.O., ed., Cenozoic Mammals of North America, Geochronology and Biostratigraphy: Berkeley, University of California Press, p. 236–368.

Repenning, C.A., Weasma, T.R., and Scott, G.R., 1995, The early Pleistocene (latest Blancan–earliest Irvingtonian) Froman Ferry Fauna and history of the Glenns Ferry Formation, southwestern Idaho: U.S. Geological Survey Bulletin 2105, p. 1–86.

Richey, K.A., 1948, Lower Pliocene horses from Black Hawk Ranch, Mount Diablo, California: University of California Publications in Geological Sciences, v. 28, p. 1–44.

Savage, D.E., 1955, Nonmarine lower Pliocene sediments in California: University of California Publications in Geological Sciences, v. 31, p. 1–26.

Schorn, H.E., Bell, C.J., Starratt, S.W., and Wheeler, D.T., 1994, A computer-assisted annotated bibliography and preliminary survey of Nevada paleobotany: U.S. Geological Survey Open-File Report 94-441A, 180 p.

Shotwell, J.A., 1955, Review of the Pliocene beaver *Dipoides*: Journal of Paleontology, v. 29, p. 129–144.

Shotwell, J.A., 1970, Pliocene mammals of south-east Oregon and adjacent Idaho: Bulletin of the Museum of Natural History, University of Oregon, no. 17, p. 1–103.

Stirton, R.A., 1939, Cenozoic mammal remains from the San Francisco Bay region: University of California Publications: Bulletin of the Department of Geological Sciences, v. 24, p. 339–410.

Stirton, R.A., 1940, The Nevada Miocene and Pliocene mammalian faunas as faunal units: Proceedings of the VI Pacific Science Congress, no. 2, p. 627–640.

Streck, M.J., and Grunder, A.L., 1995, Crystallization and welding variations in a widespread ignimbrite sheet; the Rattlesnake Tuff, eastern Oregon, USA: Bulletin of Volcanology, v. 57, p. 151–169.

Swisher, C.C., III, 1992, ^{40}Ar/^{39}Ar dating and its application to the calibration of the North American land mammal ages [Ph.D. thesis]: Berkeley, University of California, 239 p.

Tedford, T.H., Albright, L.B., III, Barnosky, A.D., Ferrusquia-Villafranca, I., Hunt, R.M., Jr., Storer, J.E., Swisher, C.C., III, Voorhies, M.R., Webb, S.D., and Whistler, D.P., 2004, Mammalian biochronology of the Arikareean through Hemphillian interval (late Oligocene through early Pliocene epochs), *in* Woodburne, M.O., ed., Late Cretaceous and Cenozoic Mammals of North America: New York, Columbia University Press, p. 169–231.

Trexler, J.H., Jr., Cashman, P.H., Henry, C.D., Muntean, T., Schwartz, K., Ten-Brink, A., Faulds, J.E., Perkins, M., and Kelly, T., 2000, Neogene basins in western Nevada document the tectonic history of the Sierra Nevada–Basin and Range transition zone for the last 12 Ma, *in* Lageson, D.R., et al., eds., Great Basin and Sierra Nevada: Boulder, Colorado: Geological Society of America Field Guide 2, p. 97–116.

Walsh, S.L., 1998, Fossil datum and paleobiological event terms, paleontostratigraphy, chronostratigraphy, and the definition of land mammal "age" boundaries: Journal of Vertebrate Paleontology, v. 18, p. 150–179.

Whistler, D.P., and Burbank, D.W., 1992, Miocene biostratigraphy and biochronology of the Dove Spring Formation, Mojave Desert, California, and characterization of the Clarendonian mammal age (late Miocene) in California: Geological Society of America Bulletin, v. 104, p. 644–658, doi: 10.1130/0016-7606(1992)104<0644:MBABOT>2.3.CO;2.

Wilson, E.L., and Prothero, D.R., 1997, Magnetic stratigraphy and tectonic rotation of the middle-upper Miocene "Santa Margarita" and Chanac Formations, north-central Transverse Ranges, California, *in* Girty, G.H., Hanson, R.E., and Cooper, J.D., eds., Geology of the Western Cordillera: Perspectives from Undergraduate Research: Pacific Section, Society of Economic Paleontologists and Mineralogists Special Publication 82, p. 35–48.

Woodburne, M.O., 1987, Cenozoic Mammals of North America, Geochronology and Biostratigraphy: Berkeley, University of California Press, p. xv + 336.

Woodburne, M.O., 2004, Late Cretaceous and Cenozoic Mammals of North America, Biostratigraphy and Geochronology: New York, Columbia University Press, p. xv + 391.

Manuscript Accepted by the Society 21 July 2008

The Geological Society of America
Special Paper 447
2009

Stratigraphy and structure of the Neogene Boca basin, northeastern California: Implications for late Cenozoic tectonic evolution of the northern Sierra Nevada

Kevin B. Mass*
Patricia H. Cashman
James H. Trexler Jr.
Department of Geological Sciences and Engineering, University of Nevada, Reno, Nevada 89557, USA

ABSTRACT

The Neogene Boca basin, located 15 km northeast of Truckee, California, records the depositional and deformational history for the late Miocene–Pliocene period in this part of the northern Sierra Nevada. This study consists of fine-scale analysis of the well-exposed Neogene sedimentary rocks in an otherwise poorly exposed area of the northern Sierra Nevada. The Neogene Boca basin sedimentary section is >500 m thick and dips generally west to southwest. Four distinct lithologic intervals are deposited unconformably over lahars and intermediate lavas of the Miocene Kate Peak Formation. An ~180-m-thick section of conglomerate and conglomeratic litharenite represents a generally southwest directed fluvial system that existed from at least 4.4 Ma (interval I). This is overlain by and locally interfingered with a ca. 4.38 Ma basalt flow of Boca Hill. Above this basalt, an ~107-m-thick section of quartz wacke and siltstone deposits represents a deltaic system controlled by local volcanic topography from ca. 4.4 to 4.1 Ma (interval II). Conformably above interval II, an ~122-m-thick section of silty diatomite deposits with interbedded tephra and litharenite represents a lacustrine environment from ca. 4.1 to 2.7 Ma (interval III). Overlying the diatomite along a disrupted surface, a >91-m-thick section of medium- to coarse-grained litharenite and cobble conglomerate represents an abrupt change in depositional environment, to a west directed fluvial system (interval IV). Pliocene westward tilting and change in base level began during deposition of interval IV (ca. 2.7 Ma) and prior to eruption of the Boca Ridge Formation (ca. 2.61 Ma).

Four orientations of large faults (>0.1 m displacement) are distributed evenly across the basin: (1) northeast to north-northeast striking sinistral faults; (2) northwest to north-northwest striking dextral faults; (3) west to west-northwest striking oblique-reverse faults; and (4) other fault orientations that have apparent motions not included in these categories. Strike-slip faulting is thought to have occurred during

*Present address: Black & Veatch, 15615 Alton Parkway, Suite 300, Irvine, California 92618, USA; masskb@gmail.com.

Mass, K.B., Cashman, P.H., and Trexler, J.H., Jr., 2009, Stratigraphy and structure of the Neogene Boca basin, northeastern California: Implications for late Cenozoic tectonic evolution of the northern Sierra Nevada, *in* Oldow, J.S., and Cashman, P.H., eds., Late Cenozoic Structure and Evolution of the Great Basin–Sierra Nevada Transition: Geological Society of America Special Paper 447, p. 147–170, doi: 10.1130/2009.2447(09). For permission to copy, contact editing@geosociety.org. ©2009 The Geological Society of America. All rights reserved.

tilting of the Neogene section. The distributed conjugate strike-slip faults in the rocks of Boca basin accommodated east-southeast directed extension and south-southwest directed contraction.

These new stratigraphic and structural data provide information on late Miocene–Pliocene deformation at the eastern edge of the Sierra Nevada. The Boca basin appears to have been an isolated basin controlled by volcanic topography. A late Miocene deformation event is not recorded in Boca basin; however, a Pliocene event is recorded in the termination of deposition and deformation of the section through tilting, incision, and distributed faulting. Pliocene deformational style is consistent with generally east-west extension associated with westward encroachment of the Basin and Range or northward migration of normal faults at Lake Tahoe. The structural data cannot disprove migration of Walker Lane deformation into the Sierra Nevada but merely show that this did not occur in the area occupied by the Neogene Boca basin. The Pliocene deformation event coincided with local eruption of high-potassium lavas and a regional base-level change, and it may represent rollback of the Juan de Fuca plate after ca. 3 Ma.

Keywords: Sierra Nevada, extension, sedimentary basin, Neogene, tephrochronology.

INTRODUCTION

Active deformation in western North America is not confined to the Pacific–North American plate boundary, but it is distributed across the western third of the continent (e.g., Atwater, 1970; Wright, 1976; Zoback and Zoback, 1980). East-west extension occurs in the Basin and Range Province. A complex zone of dextral shear along the west edge of the Basin and Range, known as the Walker Lane, accommodates some of the northwest directed dextral shear of the plate boundary (Fig. 1). West of the Walker Lane, the Sierra Nevada block, or microplate, is moving as a relatively coherent unit northwestward relative to the North American craton (e.g., Dixon et al., 2000).

In the Sierra Nevada, both extensional and dextral faulting are propagating westward from the Basin and Range into the Sierra Nevada, but the timing, mechanisms, and implications for this deformation are poorly understood. Stockli et al. (2003) documented post–middle Miocene dextral transtensional deformation in the White Mountains, west of the Walker Lane and east of the central Sierra Nevada. Farther north, in the Yerington area (Fig. 1), Dilles and Gans (1995) documented westward stepping of normal faulting with time and initiation of strike-slip faulting at ca. 25 Ma. Both may be facilitated by rheologic changes due to side-heating from the Basin and Range (e.g., Surpless et al., 2002). Henry and Perkins (2001) also documented normal faulting into the northern Sierra Nevada near the latitude of the study area and suggested it stepped west with time. The Walker Lane converges with the Sierra Nevada frontal fault system north of Reno, Nevada, and recent studies have suggested that Walker Lane dextral slip steps westward into the Sierra Nevada at Mohawk Valley, just north of the study area (e.g., Sawyer and Briggs, 2001). Active deformation is also occurring within the Sierra Nevada "block," as shown by seismicity (e.g., Tsai and Aki, 1970; Hawkins et al., 1986; Ichinose et al., 2003; Schweickert et al., 2004) and global positioning system (GPS) data (e.g.,

Argus and Gordon, 1991; Dixon et al., 1995, 2000; Thatcher et al., 1999; Bennett et al., 2003); however, these record only instantaneous strain and may not reflect the longer-term (million-year-scale) strain in the northern Sierra Nevada.

The Neogene sedimentary rocks of the Boca basin were deposited within the Sierra Nevada "block" during a time period when both dextral and extensional faulting were occurring in the Sierra Nevada–Basin and Range transition zone just to the east. This study examines these rocks to determine whether either of these faulting styles developed within the Sierra Nevada, and if so, when. The late Miocene–Pliocene Boca basin is located entirely within the Sierra Nevada "block," and it is one of the few well-exposed areas of Neogene sedimentary rocks in the Sierra Nevada (Fig. 1). The section is well exposed along the wave-cut shores of Boca Reservoir and incised streams nearby, and it records both syndepositional paleogeography and the post-depositional deformation. Interbedded tephra provide age control through tephrochronology or $^{40}Ar/^{39}Ar$ isotopic dating.

In this paper, we present a detailed stratigraphic and structural analysis of the Neogene sedimentary section at Boca Reservoir, here named the Neogene Boca basin. First, we summarize the internal stratigraphy of the Neogene Boca basin, with emphasis on paleogeographic information that relates to tectonism. Next, we present a structural analysis of the postdepositional faulting, culminating with three-dimensional local stress determined by 9 fault-slip inversion method. We conclude with a discussion of the results and their implications for the nature and timing of late Miocene–Pliocene deformation within the northern Sierra Nevada.

GEOLOGIC SETTING

The late Cenozoic volcanic and sedimentary rocks overlying the Mesozoic Sierra Nevada batholith record the late Miocene–Pliocene evolution of the northern Sierra Nevada (e.g., Bateman and Wahrhaftig, 1966). Subduction-related intermediate volcanic

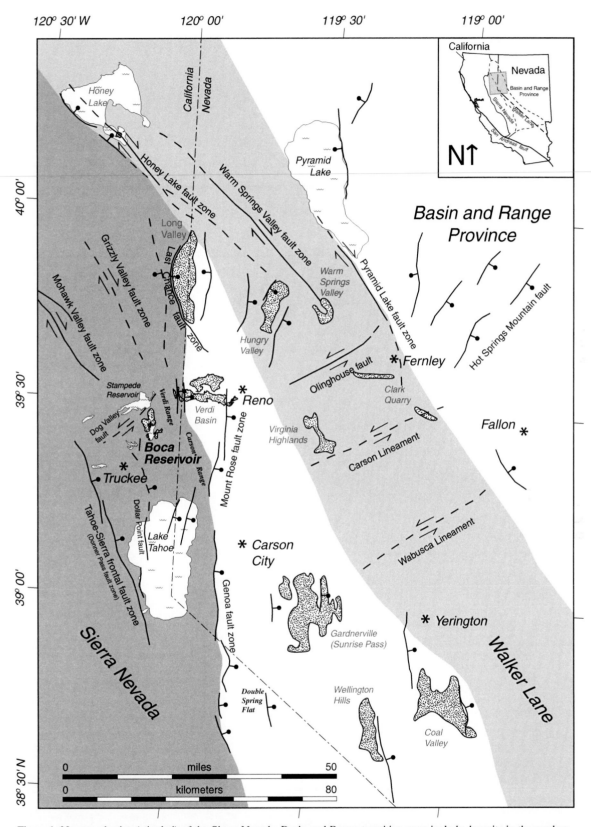

Figure 1. Neogene basins (stippled) of the Sierra Nevada–Basin and Range transition zone include deposits in the modern Sierra Nevada microplate (dark gray), Walker Lane belt (light gray), and between the two provinces. Major faults (black) distinguish the Sierra Nevada, Walker Lane, and Basin and Range. Neogene Boca basin sedimentary rocks are exposed near Boca Reservoir. Map was modified from Kelly (1998) and Trexler et al. (2000); regional tectonics are after Stewart (1988).

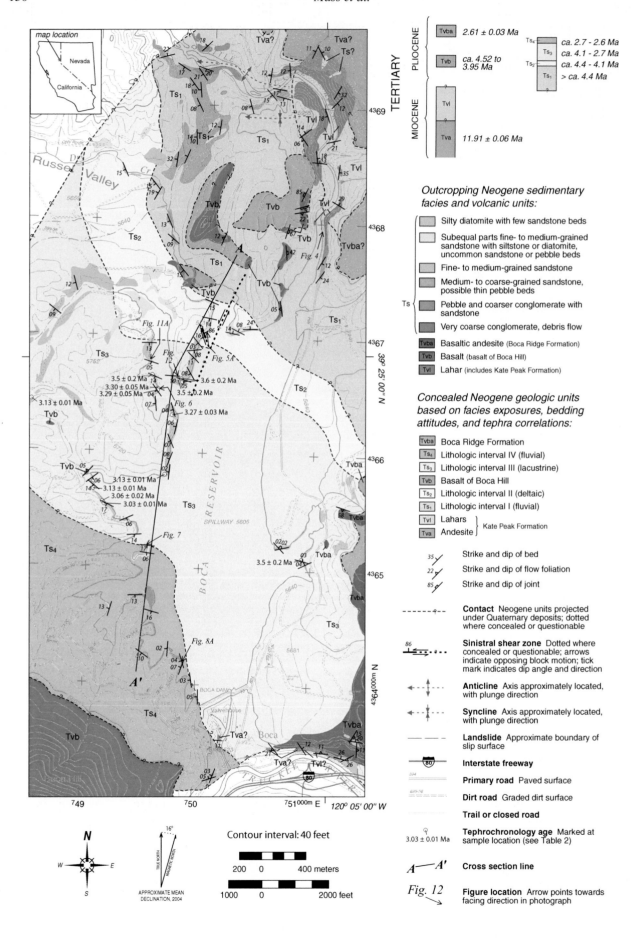

rocks dominate the Miocene section. In the study area, these rocks include the Kate Peak Formation (11.91 ± 0.06 Ma; Henry and Perkins, 2001), which underlies much of the modern topography, including the Verdi Range, Boca Ridge, and the Carson Range (Fig. 1). Subduction ceased with the northward migration of the Mendocino triple junction, starting with the inception of a transform boundary ca. 30 Ma (Atwater, 1970). The triple junction passed the latitude of the study area ca. 4 Ma, bringing an end to subduction (Atwater and Stock, 1998). In the Miocene and early Pliocene, intermediate volcanism was followed by basaltic volcanism at both the eastern Sierra Nevada (e.g., Henry et al., 2004) and south-central Great Basin (e.g., Best and Christiansen, 1991), and this progression may have been related to extension tapping a deep-seated source. In the Boca Reservoir area, intermediate Kate Peak Formation rocks were followed by the basalt of Ladybug Peak (10.17 ± 0.14 Ma; Henry and Perkins, 2001) and the basalt of Boca Hill (4.38 ± 0.03–3.95 ± 0.29 Ma; Henry et al., 2004). In addition, Henry et al. (2004) documented a compositional change in local volcanism between the basalt of Boca Hill and the relatively less mafic, potassium-rich basaltic andesite of the Boca Ridge Formation (2.61 ± 0.03 Ma; Latham, 1985; Henry and Perkins, 2001). A similar compositional change to high-potassium basalt at 4–3 Ma is documented farther south in the Sierra Nevada, and it has been attributed to delamination of the Sierra Nevada root and uplift of the southern Sierra Nevada (Ducea and Saleeby, 1996; Manley et al., 2000; Farmer et al., 2002), and possible related subsidence in the southwestern Sierra Nevada (Zandt, 2003; Saleeby and Foster, 2004).

The Sierra Nevada frontal fault system forms a dramatic topographic and structural break at the eastern boundary of the Sierra Nevada; however, the time at which the high topography formed is a subject of debate. One explanation is that the Sierra Nevada became a topographic high fairly recently (late Cenozoic), with most of the relief developing since 10 Ma (Lindgren, 1911; Hudson, 1955; Axelrod, 1962; Huber, 1981; Unruh, 1991; Wakabayashi and Sawyer, 2001; Jones et al., 2004). Another explanation is that the Sierra Nevada became topographically high in the Cretaceous (e.g., House et al., 1998) or early Cenozoic time as part of a large plateau that included the region now occupied by the Great Basin. In this model, the Sierra Nevada gradually sub-

sided throughout the late Cenozoic (e.g., Wernicke et al., 1996; Wolfe et al., 1997). In northwestern Nevada, deep erosion of Oligocene tuffs suggests an uplift event in late Oligocene or early Miocene time (e.g., Deino, 1985; Stockli et al., 2003; Hinz, 2004; Faulds et al., 2005). In addition, late Pliocene uplift, tilting, and incision in northwestern Nevada and northeastern California are recorded in rocks as young as 2.5 Ma (e.g., Trexler et al., 2000; Muntean, 2001; Park, 2004; Cashman and Trexler, 2004; Mass, 2005; Trexler et al., this volume; Cashman et al., this volume). Near Boca basin, uplift is accommodated along the Verdi, Mount Rose, and the Genoa fault zones (e.g., Bell, 1981; Hawkins et al., 1986; Schweickert et al., 2000b), which presently define the steep eastern edges of the Verdi Range and Carson Range (Fig. 1).

Neogene sedimentary basins document evolution of the Sierra Nevada–Basin and Range transition zone since late Miocene time; age control is available from mammalian fossils (e.g., Kelly, 1994, 1997, 1998), plant fossils (e.g., Axelrod, 1956, 1958, 1962), isotopic dating (e.g., Henry and Perkins, 2001), and tephrochronology (e.g., Perkins et al., 1998). Both extensional faulting and strike-slip faulting are documented in Neogene sedimentary rocks east of the Sierra Nevada since late Miocene time. Between the Sierra Nevada and the Walker Lane, continuous extensional faulting from ca. 7 Ma to present is documented in the Gardnerville basin (Muntean, 2001; Cashman et al., this volume). In contrast, distinct periods of deformation are documented between ca. 12.5 and 3 Ma at Coal Valley Basin west of the Walker Lane (Gilbert and Reynolds, 1973; Golia and Stewart, 1984). The Neogene Verdi basin records basin formation prior to 10.3 Ma (Trexler et al., 2000; Henry and Perkins, 2001), including some faulting early in the basin history, but then no faulting occurred until <2.5 Ma when the deformation was dominated by distributed conjugate strike-slip faulting (Cashman and Trexler, 2004). In the northern Walker Lane, strike-slip deformation is documented at Honey Lake Basin after 6 Ma (Mass et al., 2003; Park, 2004; Faulds et al., 2005; Trexler et al., this volume).

The Neogene Boca basin includes some of the few exposures of Neogene sedimentary rocks in the modern Sierra Nevada; however, it is one of the least-studied basins in the entire transition zone. Previous studies have documented the aerial extent of Neogene sedimentary rocks at Boca basin as a single mapped

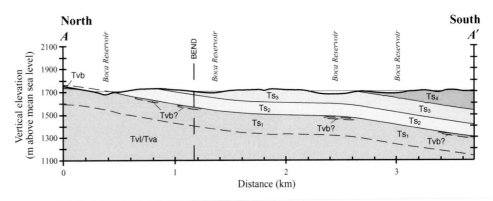

Figure 2 (*on this and previous page*). Neogene geologic map and cross section (A–A′) for the southwestern part of the Boca quadrangle, northeastern California. Outcropping Neogene sedimentary facies are the surface exposures of Neogene sedimentary rocks. Concealed geologic units are the projected Neogene intervals under Quaternary alluvium. Extent of projected geologic units is confirmed using facies exposures, bedding attitudes, and tephra correlations.

unit but did not distinguish internal stratigraphy of the Neogene section (Lindgren, 1911; Birkeland, 1962; Latham, 1985; Saucedo and Wagner, 1992). Only recently, Trexler et al. (2000) and Henry and Perkins (2001) have provided general descriptions of exposed Neogene Boca basin sedimentary rocks, mainly for a comparison to the better-studied Verdi basin (Fig. 1). The total extent of the exposed Boca basin is ~10 km north-south by at least 4 km east-west (Fig. 1). Neogene sedimentary rocks are exposed along the wave-cut shores of Boca Reservoir and along incised channels draining into it, such as the Little Truckee River Canyon (Fig. 2). Neogene rocks are locally covered by Quaternary alluvium in low-lying areas, such as Russel Valley (Fig. 2). Sedimentary rocks might continue in the subsurface as far west as the Tahoe-Sierra frontal fault (Donner Pass fault zone), extending 25 km southwest of the Verdi Range (Fig. 1).

At the eastern edge of the Sierra Nevada block, the Boca basin is near three fault zones with different geometries and kinematics (Fig. 1): (1) the northwest trending, dextral-normal Mohawk Valley fault zone (e.g., Hawkins et al., 1986; Sawyer and Briggs, 2001), (2) the northeast trending, sinistral-normal Dog Valley fault zone (Hawkins et al., 1986), and (3) north trending, east-side-down normal faulting around Lake Tahoe, including the Dollar Point and Donner Pass fault zones (Schweickert et al., 2000a, 2000b; Henry and Perkins, 2001). This faulting represents the propagation of both normal and dextral faulting into the Sierra Nevada; evidence from the Neogene sediments of the Boca basin will help to constrain the age(s) of the faulting.

RESULTS: STRATIGRAPHY

Boca basin Neogene stratigraphy consists of a >500-m-thick section that is divided here into four distinct stratigraphic intervals (I–IV, Fig. 3; Table 1), each of which is defined by a discrete texture, age, and depositional environment (cf. Miall, 1996, 2000). Stratigraphic intervals I and IV are further subdivided into lithofacies that have slightly different texture, sedimentary structure, or depositional environment. These intervals and lithofacies, each with a distinct mineralogy (according to the methods of Okada, 1971), refer only to the Neogene Boca basin sedimentary rocks. The base of the sedimentary section is defined as the base of a coarse cobble conglomerate that contains significantly less primary volcanic material than the underlying Miocene lahars and lavas (cf. Fisher, 1971; Fisher and Schmincke, 1984; Miller, 1989). The top of the Neogene

section is defined at the unconformity of the overlying Boca Ridge Formation and Quaternary glacial deposits and alluvium (Birkeland, 1962). Each of the stratigraphic intervals is described below, and descriptions are followed by an interpretation of the depositional environment.

Intervals I, II, and III consist of a fining-upward section that has been dated using basalt flows near the top of interval I and tephra beds in interval III; the unconformably overlying interval IV coarsens upward (Fig. 3). Interval I fines upward from coarse conglomerate debris flow to fluvial conglomerate and pebbly litharenite. Lava flows from the basalt of Boca Hill (ca. 4.38 Ma; Henry et al., 2004) are intercalated with the conglomeratic litharenite near the top of interval I. Interval II, immediately overlying the uppermost basalt flow, is a fining-upward sequence of quartz wacke and diatomite (locally organic-rich) with <2-m-thick litharenite beds. Its upper contact, with the diatomite and silty diatomite of interval III, is gradational. The diatomite of interval III is interbedded with <1-m-thick litharenite and tephra beds; the age of the section was determined from 14 tephra samples (from eight beds) dated from 3.03 ± 0.01 Ma to 3.6 ± 0.2 Ma (Table 2). Tephra ages were determined by Michael Perkins at the Department of Geology and Geophysics, University of Utah, using the methods described in Appendix 1. Interval III is unconformably overlain by a coarsening-upward sequence of cross-bedded, coarse-grained litharenite and pebble to small cobble conglomerate (interval IV). Interval IV is in turn unconformably overlain by basaltic andesite of the Boca Ridge Formation (ca. 2.61 Ma; Henry and Perkins, 2001) and Quaternary glacial deposits and alluvium (Birkeland, 1962).

Interval I

Interval I is a fining-upward sequence of debris flows, conglomerate, and conglomeratic litharenite up to 181 m thick (Fig. 3). The base is defined by a >4-m-thick deposit of boulder and large cobble conglomerate that unconformably overlies the highest lahar in the Kate Peak Formation. This conglomerate is ungraded, very coarse grained (clasts up to 2 m diameter), and matrix-rich; it is interpreted as a debris flow. Local conglomerate horizons also occur within the lahar section, but they are not laterally continuous like the debris flow (Mass, 2005). The ca. 4.38 Ma basalt of Boca Hill is interbedded near the top of interval I. Best exposures of this interval occur in the southern Little Truckee River Canyon and along incised channels near Russel Valley (Fig. 2).

Figure 3. Summary of the Neogene Boca basin sedimentary section compiled from individual stratigraphic columns and map relationships (Mass, 2005). Dashed sections are estimated thicknesses from map relationships. Section thickness starts at the base of interval I. Lithologic blocks represent dominant percentage of rock type rather than thickness of individual beds. Tephra ages and names in the column are obtained from tephra samples sent to M. Perkins at the Department of Geology and Geophysics, University of Utah. Average paleocurrent directions are grouped in each stereogram; arrows point toward flow direction. Intervals at right refer to the number scheme for individual interval descriptions. Lithology abbreviations: CRBN DIAT—carbonaceous diatomite; VF SS—very fine–grained sandstone; F SS—fine-grained sandstone; M SS—medium-grained sandstone; C SS—coarse-grained sandstone; VC SS—very coarse–grained sandstone; CGL SS—conglomeratic sandstone; PBL CGL—pebble conglomerate; CBL CGL—cobble conglomerate; BLDR/VOLC—boulder conglomerate, debris flow, or volcanic unit.

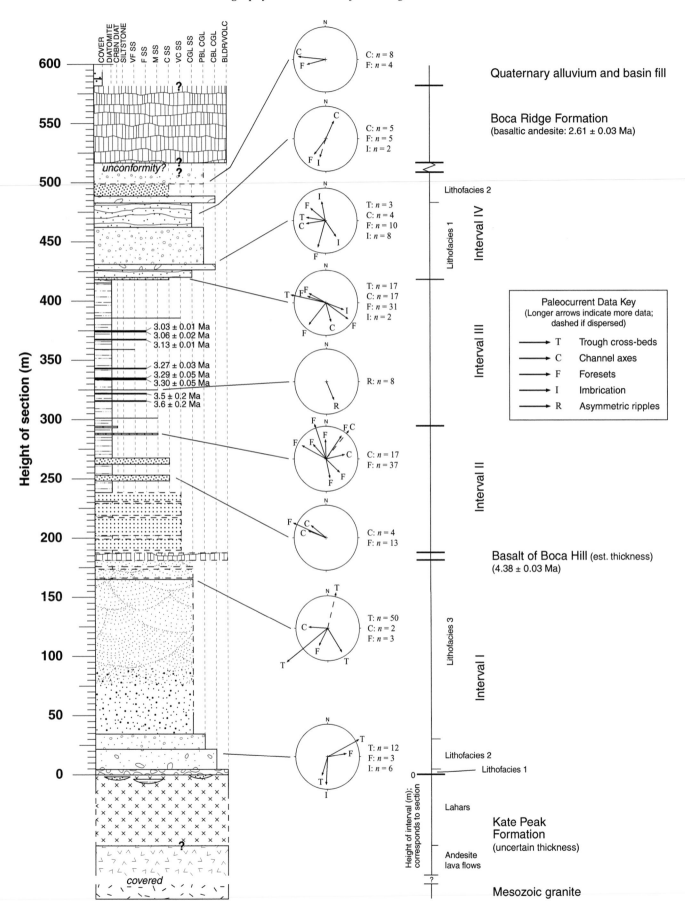

TABLE 1. SUMMARY OF BOCA BASIN NEOGENE LITHOLOGIC INTERVALS

	Interval	Thickness (m)	Lithofacies	Dominant rock type	Depositional environment
	Interval IV (ca. 2.7 to 2.6 Ma)	>91	2 (>26 m)	Fine- to medium-grained litharenite and conglomerate lenses with broad channels filled with conglomerate	High bed load, meandering fluvial system
			1 (65 m)	Very coarse-grained litharenite and matrix- and clast-supported conglomerate	Shallow, broad-channel fluvial system with increasing strength and incision by deeper channels
	Interval III (ca. 4.1 to 2.7 Ma)	~122		Silty diatomite with litharenite and tephra layers	Lacustrine
	Interval II (ca. 4.4 to 4.1 Ma)	~107		Fine-grained quartz wacke grading to silty diatomite with litharenite layers	Weakening low-velocity, meandering fluvial-deltaic system
Upsection→	Interval I (<11.9 to ca. 4.4 Ma)	~166–181	3 (~147 m)	Conglomeratic litharenite	Shallow, broad stream channels
			2 (15–30 m)	Pebble and cobble conglomerate	Strong to moderate velocity, high bed load, shallow fluvial system
			1 (~4 m)	Clast-supported conglomerate with isolated large boulders of lahar and layered sandstone	Debris flow

TABLE 2. TEPHROCHRONOLOGY AGES AND LOCATIONS: NEOGENE BOCA BASIN

Tephra sample no.	Sample location (UTM, Zone 10S, NAD27)	Age (Ma)*	Position in measured section (m)	Depositional rate for mean age (m/m.y.)	Range of depositional rate using age uncertainty (m/m.y.)
03KM00201	4365554 m E / 749145 m N	3.03 ± 0.01 (Ishi tephra)	375.1	16.7	8.3–16.7[†]
03KM00401	4365572 m E / 749128 m N	3.06 ± 0.02 (Upper Horse Hill tephra)	374.6	98.6	69.0–172.5
03KM00402	4365572 m E / 749128 m N				
03KM07701	4366445 m E / 748469 m N	3.13 ± 0.01 (Chalk Hills Fm. tephra)	367.7	174.3	135.6–244.0
03KM08102	4365670 m E / 749044 m N				
03KM08601	4365810 m E / 749678 m N				
03KM10502	4366322 m E / 749719 m N	3.27 ± 0.03 (Chalk Hills Fm. tephra)	343.3	415.0	83.0–415.0[†]
03KM11201	4366528 m E / 749586 m N	3.29 ± 0.05 (Nomlaki tephra)	335.0	80.0	7.3–80.0[†]
03KM11202	4366532 m E / 749591 m N	3.30 ± 0.05 (Putah tephra)	334.2	60.5	26.9–60.5[†]
03KM02601	4366635 m E / 749756 m N	3.5 ± 0.2 (Honey Lake tephra)	322.1	59.0	11.8–59.0[†]
03KM11501	4366661 m E / 749543 m N				
04KM06401	4365104 m E / 750926 m N				
03KM03401	4366639 m E / 749794 m N	3.6±0.2 (Honey Lake tephra)	316.2	n.d.	n.d.
03KM03601	4366607 m E / 749799 m N				

Note: n.d.—depositional rate not available; no adjacent age range for calculation.
*Ages and ash bed correlations were determined by M. Perkins at the Department of Geology and Geophysics, University of Utah, U.S.A.
[†]No upper rate calculated; youngest age uncertainty of older tephra overlaps with oldest age uncertainty of younger tephra.

The interval is divided into three lithofacies from bottom to top: (1) boulder and large cobble conglomerate (debris flow; lowest 4 m of interval I), (2) pebble and cobble conglomerate (~4–34 m), and (3) conglomeratic litharenite with thin to medium diatomite beds near the top (34–~181 m; partly estimated from map relationships). Lithofacies 1 contains lenses of clast-supported conglomerate with medium to large boulders from the underlying basement (layered Miocene litharenite and lahar) in a matrix of poorly sorted, conglomeratic volcanic and lithic clasts (Fig. 4). The debris flow is overlain by lithofacies 2, a clast- and matrix-supported pebbly conglomerate with a pebbly litharenite matrix, 15–30 m thick. Clasts range from 0.5 to 4.0 cm in diameter (most are 1 cm) with isolated boulders; >95% are intermediate volcanics. Smaller clasts have imbrication with an interpreted south directed flow (Mass, 2005; Fig. 3). At the Little Truckee River valley (Fig. 2), the interval contains 1–4-m-thick beds of conglomerate and <0.5-m-thick beds of fine- or coarse-grained litharenite. Subtle trough cross-beds in the litharenite indicate an average south-southwest directed flow (Mass, 2005; Fig. 3). Lithofacies 2 fines upward into lithofacies 3; the latter is ~147 m thick. Lithofacies 3 is fine-grained litharenite interbedded with medium- to coarse-grained litharenite and lithic wacke. Individual litharenite beds are >0.4 m thick and separated by up to 0.2 m beds of scoured silty, blocky diatomite. This lithofacies crops out on the northeast side of Russel Valley and along the Dry Creek stream incision (Fig. 2). At Dry Creek, the coarse-grained litharenite contains very coarse sand lenses with well-defined trough cross-beds. Individual troughs are 5–20 cm deep and have an open concave-upward shape. Troughs in the Dry Creek litharenite exposures indicate an average southwest directed flow, but individual troughs range from west to south-southwest directed (Mass, 2005; Fig. 3).

The top of interval I is well dated at ca. 4.4 Ma; the age of the base is poorly constrained, but it is thought to be younger than ca. 11.9 Ma. Interval I overlies a lahar that is probably a correlative of the upper Kate Peak Formation. This lahar is undated, but a Kate Peak andesite flow on the top of the Verdi Range just east of the study area has an age of ca. 11.91 Ma (Henry and Perkins, 2001). This relationship is the basis for the estimated oldest possible age for the base of interval I. At the northwestern shore of Boca Reservoir, the uppermost part of interval I is interbedded with the basalt of Boca Hill (Mass, 2005); this basalt is dated at ca. 4.38 Ma (Henry et al., 2004), providing the younger age bracket. For interval I, there are no changes of clast composition, no evidence of angular unconformities, and no departure from the overall fining-upward pattern, any of which would suggest change or reactivation in the source area during deposition. Therefore, it appears that interval I represents a single depositional cycle of response to creation of accommodation space. Since the dated basalt flow is interbedded with the top of interval I, it is most likely that all of interval I was deposited in the younger part of the potential age range.

Interval II

Interval II is at least 107 m thick and consists of interbedded litharenite and diatomite (Fig. 3). The lower contact is defined near the top of an exposed basalt flow of the Boca Hill series. The upper contact is defined at the uppermost carbonaceous diatomite and coincides with a significant decrease in litharenite and increase in diatomite. Interval II has two main lithic types: (1) beds of fine-grained quartz wacke that grade upward to silty diatomite, with or without carbonaceous material, and (2) distinct beds of medium- to coarse-grained lith-

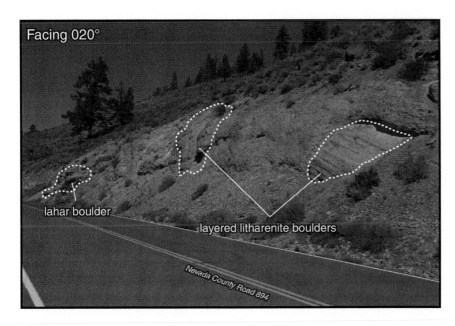

Facing 020°

lahar boulder

layered litharenite boulders

Nevada County Road 894

Figure 4. An oblique view north along Nevada County Road 894 of lithofacies 1 from interval I in the Boca basin section. Lithofacies 1 is composed of large boulders of layered litharenite and lahar in a conglomeratic matrix (0–4 m of the composite section).

arenite and lithic wacke. There is a general fining-upward trend throughout interval II. The best exposures of interval II occur at the cliff exposures along the north and north-northwest edge of Boca Reservoir and in southern Russel Valley (Fig. 2).

Fine-grained quartz wacke grading to silty diatomite is the most common rock type in interval II. Typical beds of quartz wacke are 1–2 m thick (rarely 4 m thick) and separated by 1–3-m-thick diatomite beds. The diatomite commonly has inverse grading, coarsening upward into the overlying quartz wacke. Carbonaceous material, mostly unidentifiable woody fragments <2 cm length, constitutes <10% by volume of some diatomite intervals in the uppermost 5 m of interval II. Medium- to coarse-grained litharenites and lithic wackes generally occur in <1-m-thick beds and have abrupt, often convoluted contacts with the surrounding diatomite. Rare cobble conglomerates are also present as medium beds within muddy diatomite. Channels, some with lateral accretion bars, are common in litharenite layers (Fig. 5A) and indicate an average northeast directed flow (Fig. 5B). The channels are generally broad, with widths <1 m and average thickness of 0.15 m. Channels are typically three times as wide as they are deep. Bar foresets, tabular and typically 5–15 cm thick, are also common in the top 15 m of interval II. These foresets have flow directions toward the northeast, south-southeast, and northwest (Mass, 2005; Fig. 3).

The bottom of interval II is ca. 4.4 Ma, based the age of the basalt of Boca Hill. An approximate depositional rate for the overlying interval III was determined using the thickness of section between two dated tephra beds (discussed later). This depositional rate was then projected down to the top of interval II, at the last carbonaceous diatomite layer, to obtain an estimated age of ca. 4.1 Ma for the upper contact.

Interval III

Interval III is 122 m thick and consists of diatomite and silty diatomite with interbedded tephra and litharenite beds (Fig. 3). The bottom of the interval is defined by the top of the highest carbonaceous layer in interval II. The top of the interval is well defined by the scoured erosional surface at the base of interval IV. The best exposures of interval III are along the wave-cut shores of Boca Reservoir.

The dominant rock type is blocky, massive silty diatomite. However, several very thin to medium litharenite beds are deposited with the diatomite. These beds are generally <0.5 m thick, medium- to coarse-grained pebbly lithic wacke with a few beds of poorly sorted lithic arenite. Some beds grade upward into the overlying diatomite. Many of the medium (>0.1 m) beds have uneven upper and lower contacts, load casts, and flame structures. One very thick clastic bed consists of a 3.5 m fining-upward sequence of matrix-supported cobble conglomerate grading to pebbly litharenite with ~10% volume of woody fragments.

The diatomite in the lowest 20 m of interval III contains evidence for syndepositional slumping. Slump features commonly occur in individual litharenite beds and produce minor scours in underlying bedding. In contrast, postdepositional slumps

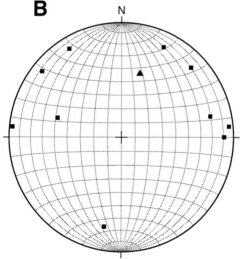

▲ Average of channel axes: 017°, 41° (trend and plunge)

■ Individual channel axis: *n* = 10 (trend and plunge)

Figure 5. Meandering streams in lithologic interval II, interpreted as a deltaic fan system, commonly incise into silty floodplain or lacustrine sediments. (A) An oblique view of an ~20-cm-deep incised channel is exposed along the wave-cut shore of Boca Reservoir. This particular paleochannel has an east-northeast flow direction (at 269 m of composite column). Hammer is shown for scale. (B) Stereogram (equal area projection, lower hemisphere) of channel axes in lithologic interval II, rotated to horizontal. The dispersed channel axes, common of meandering stream systems, have a mean north-northeast flow direction.

are characterized by >1-m-thick sections of irregular-shaped, pebble- to boulder-sized clasts of diatomite in a sandy diatomite matrix. Postdepositional slumps commonly sole into diatomite beds. The slip surfaces and slump blocks are more permeable than unbroken diatomite and are therefore commonly intruded by modern plant roots.

Interval III is the best dated interval in the Neogene Boca basin section. Numerous tephra beds, composed of loosely consolidated ash with distinguishable glass and crystalline matrix, provide age control for this interval. Very fine to coarse crystalline tephra are interbedded in the diatomite and are often deposited with coarse litharenite (Fig. 6). Within interval III, 25 distinct tephra samples were obtained for analysis. Of these, 14 tephra samples yielded eight distinct age estimates that conform to the measured section based on tephrochronology (Table 2; Appendix 1). The oldest tephra dated in this study is 28.8 m from the bottom of this interval and has an age of 3.6 ± 0.2 Ma. The youngest dated tephra is 87.1 m from the bottom of this interval and has an age of 3.03 ± 0.01 Ma. The abundance of tephra throughout interval III allows age calibration of the entire diatomite section. Depositional rates between two tephra beds were determined from the two tephra ages and the thickness of undisturbed diatomite section using the equation:

Depositional rate (m/m.y.) =

$$\frac{\text{Thickness of undisturbed sedimentary section (m)}}{\text{Age of tephra A (Ma)} - \text{Age of tephra B (Ma)}}, \quad (1)$$

where the thickness is the stratigraphic section between tephra A and tephra B, and tephra A is older than tephra B. This method was used to calculate depositional rates for interval III

(Table 2). Uncertainties in stratigraphic thickness were noted, particularly in areas of slump or fault displacement. Postdepositional slumps were not included in total thickness of the section. The highest accumulation rates in interval III occurred at ca. 3.3 Ma (Table 2).

Projected ages from the tephra, based on mean accumulation rates, indicate that the lower boundary of interval III (at the last carbonaceous diatomite layer of interval II) is ca. 4.1 Ma and the upper boundary (beginning of deposition of interval IV) is ca. 2.7 Ma. The lower boundary was calculated using the mean depositional rate between the 3.5 ± 0.2 Ma and 3.6 ± 0.2 Ma tephra beds (Table 2). A slightly older age for the lower boundary is possible with the tephra age uncertainties, but it is not likely considering the relative stratigraphic position of an interbedded ca. 4.38 Ma basalt flow near the top of interval I. The upper boundary was calculated using the mean depositional rate between the 3.06 ± 0.02 Ma and 3.13 ± 0.02 Ma tephra beds. The uppermost tephra (3.03 ± 0.01 Ma) was not used for calculation because only a small portion of section (~0.5 m) is represented below this bed. The range for the upper boundary, considering tephra age uncertainties, is ca. 2.6–2.9 Ma. Since the tilted sedimentary section is unconformably overlain by a basaltic andesite dated at ca. 2.61 Ma (Henry and Perkins, 2001; discussed later), the 2.7 Ma age from the mean accumulation rate is reasonable for the top of interval III and will be used throughout this paper. However, a slightly younger or older age for the top of interval III cannot be discounted.

Interval IV

Interval IV is >91 m thick and consists of pebbly litharenite and pebble to small cobble conglomerate (Fig. 3). The bottom

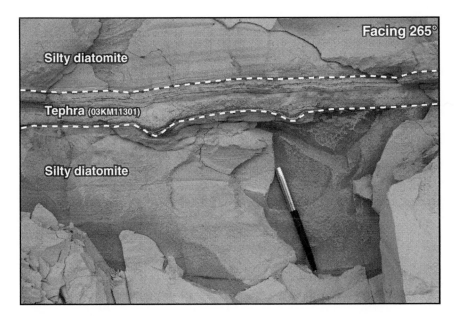

Figure 6. Dominant stratigraphy of lithologic interval III is silty diatomite with litharenite and tephra interbeds. This 3.5-cm-thick tephra bed exhibits load casts in the silty diatomite (333 m of composite column). Pen is shown for scale. Of the sampled tephra, 16 likely regional correlations and one possible correlation were determined by tephrochronology (cf. Perkins et al., 1995, 1998).

158   *Mass et al.*

of this interval is defined by a scoured surface overlain by 0.3 m of disturbed, reworked diatomite (Fig. 7). The upper contact is obscured by Quaternary alluvium. The best exposures of this interval occur along the southwestern shore of Boca Reservoir.

The interval is divided into two lithofacies: (1) very coarse-grained litharenite and matrix- and clast-supported conglomerate (lowest 65 m of interval IV), and (2) fine- to medium-grained litharenite and conglomerate lenses with broad channels filled with conglomerate (65 to ~91 m of interval IV). Lithofacies 1 has several distinct sedimentary textures. The base of lithofacies 1 is an ~0.3-m-thick section of disrupted diatomite (Fig. 7), overlain by pebbly, silty litharenite; these are deposited on a scoured surface. From 0.3 to 10 m above the base, lithofacies 1 is very coarse-grained, trough cross-bedded litharenite and lithic wacke grading with pebble conglomerate. Troughs are ~8 cm deep, are commonly separated by up to 4 cm of planar bedding, and indicate west-northwest directed flow (Mass, 2005; Fig. 3). Irregularly shaped silty lenses fill many of the troughs. From 10 to 50 m above the interval base, lithofacies 1 is conglomerate and

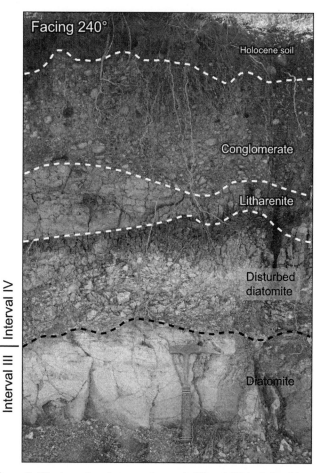

Figure 7. The boundary between intervals III and IV is defined by a scoured surface below ~0.3 m of disrupted diatomite. Dashed lines represent scoured surfaces or unconformities (409–410 m of composite column). Hammer is shown for scale.

litharenite in >1-m-thick beds. Many beds of conglomerate and litharenite in this part of the interval have scoured bases; diatomite and siltstone rip-up clasts are present in some conglomerate layers (Fig. 8A). Clast composition in the conglomerate layers is dominantly intermediate volcanics with 10%–25% metamorphic clasts. Imbrication in the conglomerates is subtle and nonconclusive about flow direction. Other structures, such as trough cross-beds, small channels, and tabular foresets in litharenite layers, indicate an average west or southwest directed flow for this part of the lithofacies (Fig. 8B). From 50 to 65 m above the interval base, lithofacies 1 is dominated by medium sand-supported conglomerate with conglomeratic litharenite lenses. This conglomerate has larger cobble clasts than other conglomerates of lithofacies 1. The remainder of interval IV exposures, 65 to ~91 m above the interval base, is lithofacies 2, which consists of medium- to coarse-grained litharenite that is incised by broad, conglomerate-filled channels, up to 2 m wide and 0.5 m in depth, that have an average west directed flow (Mass, 2005; Fig. 3). Incised channels are commonly filled with clast-supported conglomerate.

The age of this interval is projected from the uppermost tephra in the underlying diatomite and bracketed by an overlying basaltic andesite flow. According to tephra ages in interval III, deposition of interval IV began sometime after ca. 2.7 Ma. Clasts of mafic volcanics are rare in interval IV, and no parts of interval IV are found interbedded with the Boca Ridge Formation (ca. 2.61 Ma; Henry and Perkins, 2001). Therefore, deposition of interval IV of the Neogene Boca basin occurred between ca. 2.7 and 2.6 Ma.

Interpretation

Depositional environments of the four Neogene Boca basin lithologic intervals shed light on the evolution of this part of the Sierra Nevada. The Boca basin records nearly continuous deposition from before ca. 4.4 Ma to 2.6 Ma: an important time interval in the evolution of the Sierra Nevada–Basin and Range transition zone. Depositional environments changed several times during this period (illustrated in Fig. 9) and each records details of local tectonism.

The timing of basin inception is uncertain, but it was sometime after 11.9 Ma, and it may not have been much before 4.4 Ma. The initial volcaniclastic debris-flow deposition in interval I (Fig. 4) was followed by moderate- to high-velocity, shallow streams. These streams had low to moderate sinuosity, shallow gravel bar bed forms, and an average flow direction to the southwest. The fining-upward sequence in interval I indicates a gradual decrease in streamflow velocity with time. The highest part, lithofacies 3, contains abundant shallow trough cross-beds, thin-bedded, fine-grained overbank deposits, and diatomaceous floodplain deposits (Fig. 9A).

Flows from the basalt of Boca Hill (ca. 4.38 Ma) are interbedded near the boundary between interval I and interval II and may be the reason for the change in depositional environment at this boundary. Basalt flowed northward from the volcanic source

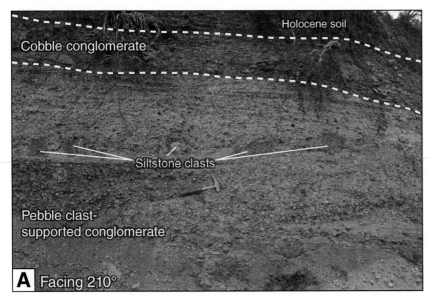

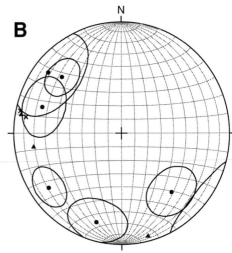

▲ Average of channel axes; *n* = 21, from 3 outcrops
✕ Average of trough cross-bed axes; *n* = 20, from 2 outcrops
● Average of foresets with 95% confidence circle;
   n = 38, from 6 outcrops

Figure 8. The lowest 15 m in lithofacies 1 of interval IV is composed of (A) pebbly conglomerate with siltstone rip-up clasts (421–422 m of composite column). Hammer is shown for scale. (B) Stereogram (equal-area projection, lower hemisphere) of channel axes, trough cross-bed axes, and foresets within the lowest 15 m of interval IV, rotated to horizontal. Averages calculated from measurements in same outcrop. The axes indicate a dominantly westward flow direction in troughs and most channels. Distributed foresets have flow directions between northwest and south, which may indicate local stream meanders during deposition of interval IV.

(Boca Hill; Fig. 2) to interfinger with the sedimentary section, indicating the formation of a topographic high associated with the volcanism. Basalt flows formed local topographic highs and redirected or blocked stream channels (Fig. 9B). The changes in paleocurrent direction, grain size, and depositional environment reflect this change in paleotopography.

In interval II (ca. 4.4–4.1 Ma), the abundance of channels and bar foresets indicates moderate to weak velocity and a broad, meandering fluvial-deltaic fan system. Rare conglomerate layers in muddy diatomite are interpreted as fan-delta deposits. Paleocurrent directions are widely dispersed for this interval, which is common for meandering channels (Miall, 1996). These paleocurrent directions, toward the northeast, south-southeast, and northwest, indicate that these fan systems flowed away from topographic highs in the southwest (Boca Hill) and southeast (Carson Range?) and toward a topographic low near the modern Boca Reservoir. Several channels are incised into silt deposits (e.g., Fig. 5A), indicating that the deltaic system probably terminated into a shallow-lake environment or migrated over a low-gradient floodplain (Fig. 9B). A developing lacustrine system is suggested near the top of interval II by the thick silty diatomite beds.

In interval III (ca. 4.1–2.7 Ma), the abundance of diatomite suggests a quiet lake basin (Fig. 9C). Intermittent volcanic eruptions deposited ash-fall tephra and lithics directly into the water. Several tephra and litharenite beds show load casts and differential settling in the underlying diatomite. This suggests that the diatomite was still soft and underwater at the time of ash or

litharenite deposition. Slump features in litharenite layers suggest that minor syndepositional slope failure occurred in the low-density material. Depositional rates gradually increased after ca. 4.1 Ma and reached maximum at ca. 3.3 Ma (Table 2).

Interval IV (ca. 2.7–2.6 Ma) documents a dramatic change from lacustrine to west directed fluvial depositional environments. The sedimentary structures in lithofacies 1 of interval IV indicate an increase in fluvial strength and stream size. Trough cross-beds, low in lithofacies 1, indicate shallow, broad streams with silty overbank floodplain deposits. Starting above 10 m in lithofacies 1, scoured litharenite surface and coarse conglomerate (e.g., Fig. 7) are interpreted as an upsection increase in the strength of streams. These streams incised into the fine-grained floodplain deposits of the earlier shallow streams and locally incised into the silty diatomite of interval III (e.g., Fig. 8A). Paleocurrent directions indicate a dominantly west directed flow (Fig. 8B); metamorphic clasts in conglomerate indicate exhumation of basement rocks from an eastern source (Verdi Range and Boca Ridge). Lithofacies 2 documents a meandering stream system with high-volume bed load that developed after the broad streams in lithofacies 1. Dispersed paleocurrent directions in channels and bar foresets of lithofacies 2 indicate shallow, sinuous multichannel fluvial systems (Fig. 9D). Compared to lithofacies 1, the larger channels and more consistent paleocurrents of lithofacies 2 indicate that rivers were broad, slightly deeper, and more sinuous higher in interval IV; however, they maintained the generally west directed paleocurrent direction.

160

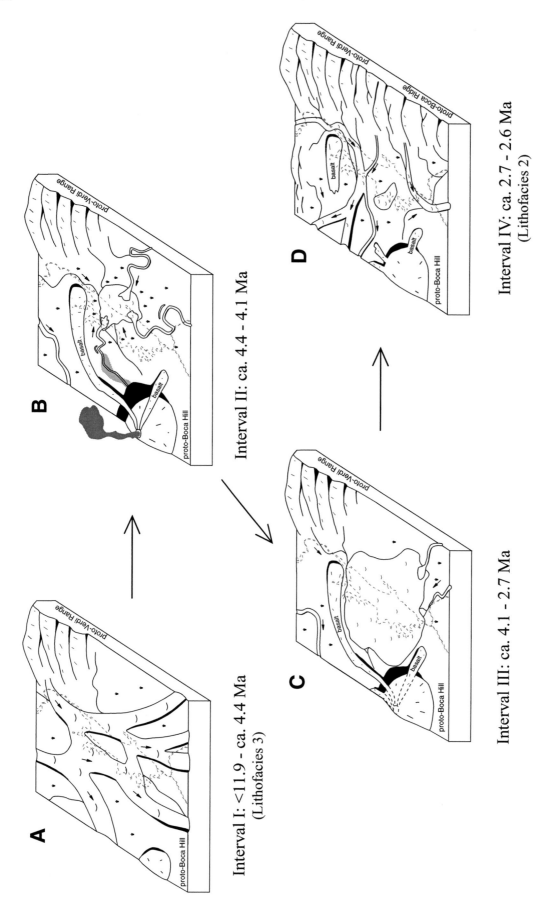

Interval I: <11.9 - ca. 4.4 Ma
(Lithofacies 3)

Interval II: ca. 4.4 - 4.1 Ma

Interval III: ca. 4.1 - 2.7 Ma

Interval IV: ca. 2.7 - 2.6 Ma
(Lithofacies 2)

Figure 9. Late Miocene–Pliocene evolution of the Neogene Boca basin is characterized by changes in depositional environments. The block diagrams represent conceptual models for discrete stages of basin evolution. Outlines of Boca Reservoir and surrounding rivers and ranges are shown for reference. Arrows indicate paleocurrent flow directions. Light source is from the southwest (bottom left). (A) Lithofacies 1 of interval I consists of west to southwest directed, shallow, broad stream channels. Uplift of the northern Verdi Range occurred during this time. (B) Interval II consists of a weakening, meandering fluvial-deltaic system. Erupted basalt flows from Boca Hill greatly influenced the local drainage pattern. (C) Interval III consists of a quiet lacustrine environment. (D) Lithofacies 2 of interval IV consists of a broad, locally meandering fluvial system. Deposition of this interval is thought to be coeval with westward tilting of the basin and uplift of the southern Verdi Range and Boca Ridge.

The basaltic andesite of the Boca Ridge Formation (ca. 2.61 Ma) overlies the Boca basin sedimentary section along an angular unconformity. Basaltic andesite flows are deposited directly on top of the diatomite of interval III along the eastern shore of Boca Reservoir, yet the flows are not interbedded with interval IV. Map-scale trends of Neogene intervals (Fig. 2) show that the section was tilted to the west or southwest and eroded prior to eruption of the Boca Ridge Formation. Intervals I and II generally have steeper dips than those of interval IV and the Boca Ridge Formation. Syndepositional tilting of the section was the catalyst for deep incision of the channel system in interval IV. However, minor tilt (~5°) in interval IV and the Boca Ridge Formation suggest that tilting continued after deposition/eruption of these units.

RESULTS: STRUCTURE

Postdepositional deformation of the Neogene Boca basin includes tilting, distributed small-scale faulting, and broad folding. Neogene sedimentary rocks exposed along the wave-cut cliffs of Boca Reservoir dip primarily west and southwest. Bedding is locally tilted south at the western shore of Boca Reservoir and northeast at the eastern and northern shores (Fig. 2). In addition, broad, west-plunging folds deform rocks in the northern parts of the study area. However, the dominant postdepositional structures in the Boca basin are pervasive, usually small-scale, faults. In the following section, we describe the four fault sets and the minor folds, the timing of the deformation, and the three-dimensional stress of postdepositional structures.

Several factors control the distribution and quality of fault data. Finer-grained sedimentary rocks, such as the silty diatomite of lithologic intervals II and III, preserve fault planes better than coarse-grained clastic sedimentary rocks. Hence, most of the faults analyzed in this study are in the vertical cliff exposures from the central western shore of Boca Reservoir (Fig. 2). Faults of all orientations are visible due to the irregular shape of the reservoir shoreline. Faults displace the entire Neogene section and are not truncated by interval boundaries. The majority of planes strike at a high angle to the bedding and have well-defined stratigraphic separation best seen in litharenite and tephra interbeds. Diatomite does not consistently preserve kinematic indicators on fault surfaces, and only ~50% of the faults in this study have measurable kinematic indicators. Stratigraphic separation can be measured for the remaining faults, but slip sense can only be interpreted by analogy with faults of similar orientation.

Four orientations of faulting are distributed across the basin: (1) northeast to north-northeast striking sinistral faults; (2) northwest to north-northwest striking dextral faults; (3) west to west-northwest striking sinistral-reverse or dextral-reverse faults; and (4) other fault orientations that have apparent motions not included in these categories (Fig. 10). Only faults with >0.1 m stratigraphic separation or displacement, referred to as "large faults," were analyzed individually (91 of 380 measured planes). The smaller faults generally agree with large fault trends. Sense of motion on large faults was determined from slickenlines,

grooves, Riedel shears, and other kinematic indicators where available. Displacement was calculated for each fault plane based on rake of slickenlines or grooves and total dip separation:

$$\text{Total displacement (m)} = \frac{\text{Dip separation (m)}}{\sin(\text{Rake}°)}. \quad (2)$$

Few crosscutting relationships were found to help determine the relative ages of the faults. Fault sets appear mutually crosscutting where observed in a few subhorizontal, wave-cut outcrop surfaces.

Northeast Striking Fault Set (No. 1)

The most abundant fault orientations in the Boca basin, 37% of the fault data, are north-northeast striking (000°–030°) and northeast striking (045°–065°), usually steeply dipping planes (most dip >70°). The planes strike at a high angle to the bedding and have well-defined stratigraphic separation. Viewed along strike, stratigraphic separation on these faults is such that bedding on the northwest side of a fault is displaced upward relative to the southeast side (e.g., Fig. 11A). Subhorizontal slickenlines and grooves (0°–15° northeast or southwest) indicate almost pure strike-slip motion. Riedel shears, mainly R1 fractures, are commonly present at outcrop scale (Fig. 11B). North striking, steeply dipping Riedel shears form acute angles with the main north-northeast or northeast striking fault surface.

Stratigraphic separation and kinematic indicators for the north-northeast and northeast striking planes indicate sinistral motion (Fig. 10A). The few sinistral faults with moderate dips (<70°) have gently plunging striae, suggesting possible strike-slip reactivation of preexisting surfaces. The sinistral fault set accommodates at least 160 m of cumulative displacement. Approximately 128 m of this total displacement is accommodated along one plane at the northeastern shore of Boca Reservoir near the transition of intervals II and III (Fig. 2). The fault set controls much of the drag folding along the west and northwest shores of Boca Reservoir.

Northwest Striking Fault Set (No. 2)

The second most abundant fault orientations in the Boca basin, 36% of the fault data, are northwest striking (300°–000°), steeply dipping planes (most dip >75°). The planes commonly strike near parallel to bedding strike, which results in minimal stratigraphic separation, even on the largest faults. Viewed along strike, stratigraphic separation on these faults is such that bedding on the northeast side of a fault is displaced upward relative to the southwest side. However, this separation is dependent on the strike of the fault and rake of the slip, and it is not uncommon to have a reverse stratigraphic separation where the strike of the fault nearly parallels the strike of the bedding. Rakes of slickenlines and grooves are subhorizontal (0°–20° northwest or southeast) and indicate strike-slip motion. Riedel shears, mainly R1

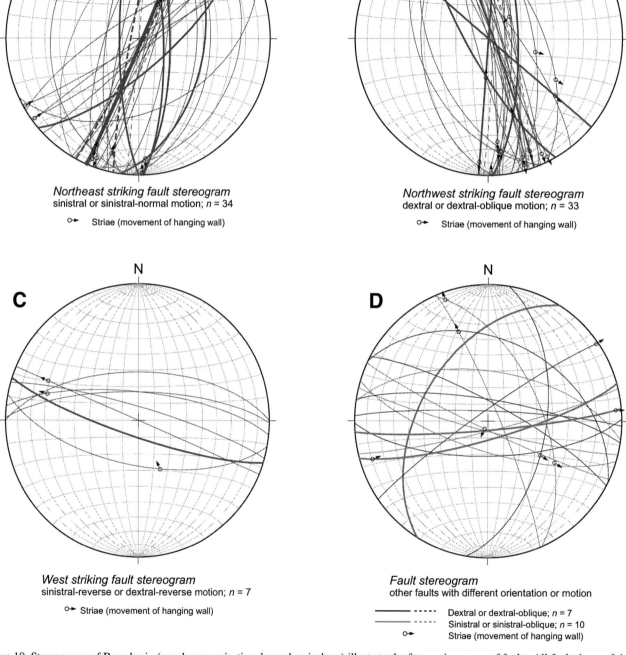

Figure 10. Stereograms of Boca basin (equal area projection, lower hemisphere) illustrate the four main groups of faults. All faults have >0.1 m of stratigraphic separation or displacement. Thicker lines indicate faults with greater stratigraphic separation or displacement. Dashed lines represent an average attitude of a fault surface. Fault motion is determined by striae, grooves, or other kinematic indicators. (A) Northeast striking faults of the Boca basin are interpreted to have sinistral motion. Greater than 160 m displacement is accommodated by this set. (B) Northwest striking faults of the Boca basin are interpreted to have dextral motion. Greater than 24 m displacement is accommodated by this set. (C) West striking faults of the Boca basin are interpreted to have oblique-reverse motion. Greater than 2 m displacement is accommodated by this set. (D) Other faults with orientation of motion different from the main sinistral, dextral, or oblique-normal fault sets. Greater than 22 m displacement is accommodated by these faults.

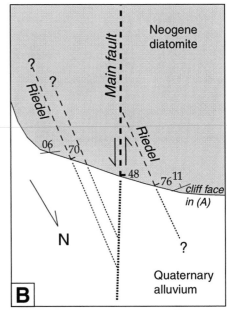

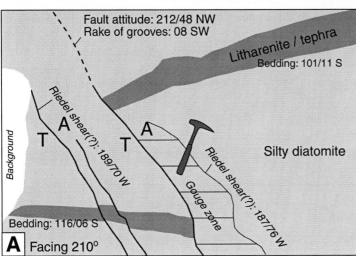

Figure 11. A northeast striking sinistral fault has >1 m of vertical stratigraphic separation (>7 m displacement), measured in a litharenite/tephra layer dipping into the cliff. Hammer is shown for scale. (A) Although the fault exhibits apparent reverse separation, the actual motion is predominantly sinistral, as verified by subhorizontal grooves on the fault surface (T—toward; A—away). Possible Riedel shears also indicate sinistral motion. (B) Map view schematic interpretation of faults in the vertical cliff face.

fractures, are locally present at outcrop scale, are north-northeast striking and steeply dipping, and form acute angles with the main northwest or north-northwest striking fault surface (Fig. 12).

Stratigraphic separation and kinematic indicators on the northwest striking set indicate dextral motion (Fig. 10B). A few dextral faults with moderate dip (<70°) and moderate rake (>20°) suggest a significant component of oblique-normal motion; these may be reactivated older surfaces. The dextral fault set accommodates at least 24 m of cumulative displacement, but it may accommodate much more where the bedding/fault intersection is close to the slip direction. The northwest striking faults commonly displace bedding along strike, and the total displacement along these faults could be drastically underestimated if no slip indicators are available.

West Striking Oblique-Reverse Fault Set (No. 3)

A small portion of the fault data set (8%) is west to west-northwest striking (270°–295°) and moderately to steeply dip-

ping (dips >60°). The planes strike near parallel to the bedding dip direction and have well-defined stratigraphic separation. Viewed across strike, planes have apparent reverse motion. Measured rakes of slickenlines and grooves are moderately to steeply plunging (25°–75°) west or east.

The few kinematic indicators for the west and west-northwest striking planes indicate oblique sinistral-reverse motion on these fault sets (Fig. 10C). Although a small data set, it has internal consistency and is distinct from other faults. The few oblique-reverse faults accommodate >2 m of displacement. These faults commonly occur at the northwestern shore of Boca Reservoir and may result from compressive overlap between numerous small-scale strike-slip faults in this part of the study area.

Other Faults (No. 4)

The remaining faults (19%) have strikes that differ from the three main fault groups (Fig. 10D). Fault motion is dex-

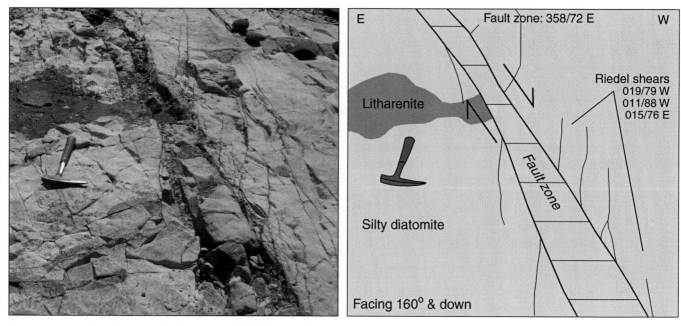

Figure 12. A northwest trending fault zone is exposed along a wave-cut diatomite shore. Although striae are not well preserved on slip surfaces in diatomite, the sense of slip on this surface is shown by well-developed Riedel shears. The thin litharenite bed on the east side of the fault zone is not found on the west side, which indicates >2 m stratigraphic separation. Hammer is shown for scale.

tral, dextral-oblique, sinistral, or sinistral-oblique; however, orientations are different from the previous fault groups. Those that have strikes within the described groups have stratigraphic separation or kinematic indicators that suggest motion different from the other main groups. Rakes on these faults range from subhorizontal to ~60°, when present. Although these planes are a significant percentage of the total faults, they do not have significant displacement. Cumulative displacement on these faults is >22 m. These faults are most common at the northwestern shore of Boca Reservoir and may have resulted from reactivation of older surfaces between numerous small-scale strike-slip faults.

Folds

Map-scale folds occur at the northern Little Truckee River Canyon and along the southeast shore of Boca Reservoir (Fig. 2). Folding is subtle at outcrop scale, but map-scale trends of bedding attitudes show the folds. Stereograms of bedding illustrate that these folds plunge gently (10°) to the west (Fig. 13). Like the west striking reverse faults, the folds accommodate minor north-south contraction.

Summary

In summary, the postdepositional deformation includes southwestward tilting, several sets of distributed (dominantly strike-slip) faults, and broad folding around gently west plunging axes. Weighting the faults by total displacement (according to the method of Angelier et al., 1985; Fig. 14) establishes that the conjugate (northeast striking sinistral and northwest striking dextral) strike-slip faults are the dominant style of deformation in the basin. The broad folding accommodates minor north-south shortening, as do the relatively insignificant west and west-northwest striking oblique reverse faults. Some additional fault planes appear to be incompatible with local strain and may represent reactivated or deformed fault surfaces (cf. Marrett and Allmendinger, 1990).

Timing of Deformation

Tilting and erosion of the Neogene Boca basin section occurred between the end of lacustrine deposition (interval III) at ca. 2.7 Ma and the eruption of Boca Ridge Formation basaltic andesite at ca. 2.61 Ma. Map relationships indicate that the Neogene sedimentary section was already tilted to the southwest prior to eruption of the Boca Ridge Formation (Fig. 2). The abrupt change from lacustrine to fluvial deposition (interval III to interval IV), and the downcutting associated with this contact, may signal the same tectonic event.

Distributed small-scale faulting postdates deposition, and may, at least in part, have been coeval with tilting and folding of the section. Few crosscutting relationships are exposed, but the available data suggest that strike-slip fault sets are mutually crosscutting and, therefore, were active at the same time. There is no significant difference between the tilted and tilt-corrected fault data (Fig. 14), which can be interpreted to show that fault-

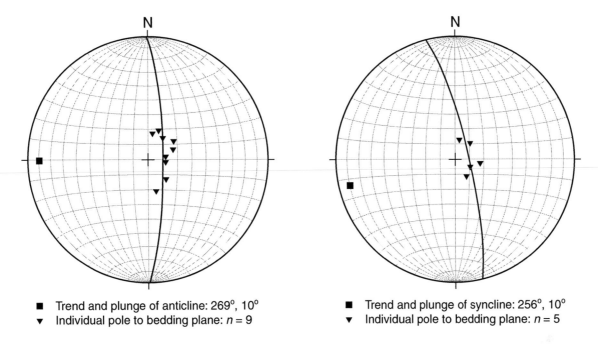

■ Trend and plunge of anticline: 269°, 10°	■ Trend and plunge of syncline: 256°, 10°
▼ Individual pole to bedding plane: *n* = 9	▼ Individual pole to bedding plane: *n* = 5

Figure 13. Stereograms (equal area projection, lower hemisphere) of two map-scale folds in the Little Truckee River valley, north of Boca Reservoir (see Fig. 2). Bedding attitudes indicate the west trending folds plunge shallowly to the west. The fold orientations support minor north-south contractional deformation at the Neogene Boca basin.

ing was coeval with tilting and folding of the section (cf. Marrett and Allmendinger, 1990). However, the low dip angles in the Boca basin section may make this comparison statistically invalid. Therefore, the relative ages of tilting and faulting cannot be established unequivocally.

Estimating Local Stress

Tensile and compressional stress directions can be estimated using a fault-slip inversion method for individual faults that have kinematic indicators (Marrett and Allmendinger, 1990). The methodology assumes regional stress is uniform and invariant over time. The faults included in the data set must have formed over a relatively short time interval and all fault orientations must be included in the data set. The Boca basin fault data presented here conform to these criteria. A stress-axis contour plot was modeled using faults with known slip direction indicators (Fig. 15A). Fault-slip inversion data from the many planes were then combined to model the overall local stress field for the Neogene Boca basin. The resulting fault-plane solution consists of two nodal planes and a maximum and minimum stress direction (Fig. 15B). The maximum and minimum stress directions may not necessarily correspond to the tensile and compressional axes for individual faults, but the inversion method is reasonable for many faults distributed over a large area (Pollard et al., 1993), such as the cliff exposures surrounding Boca Reservoir.

Fault-slip inversion analysis shows that distributed faults in the rocks of Boca basin accommodate east-southeast extension and south-southwest contraction. The modeled tensile stress axis plunges gently to the east-southeast, and the compressional stress axis plunges to the south-southwest (Fig 15B). The sinistral nodal plane from the fault-plane solution strikes 063° (Fig. 15B); this is the modeled shear zone for the Boca basin area. Predicted fault sets, simplified for the modeled strike-slip shear zone (cf. Sylvester, 1988), are similar to three main observed fault sets at the Neogene Boca basin (Fig. 15C). The synthetic sinistral faults have a northeast strike (fault set no. 1), while antithetic dextral faults have a northwest strike (fault set no. 2). Moderately dipping, north-northeast striking sinistral faults, oriented perpendicular to the maximum extension direction, partly accommodate the overall east-southeast directed extension through oblique sinistral-normal motion. Minor south-southwest directed contraction is accommodated by west-northwest striking oblique-reverse faults (fault set no. 3) and west trending folds. Faults of other orientations or slip directions probably result from small perturbations in the local stress field.

In the study area, extensional deformation appears to be greater than contractional deformation. The oblique-reverse faults and map-scale folds that record contraction might result from local compressive overlaps or drag folding from strike-slip faults. Compressive overlaps are common in areas with closely spaced strike-slip faults of similar orientation.

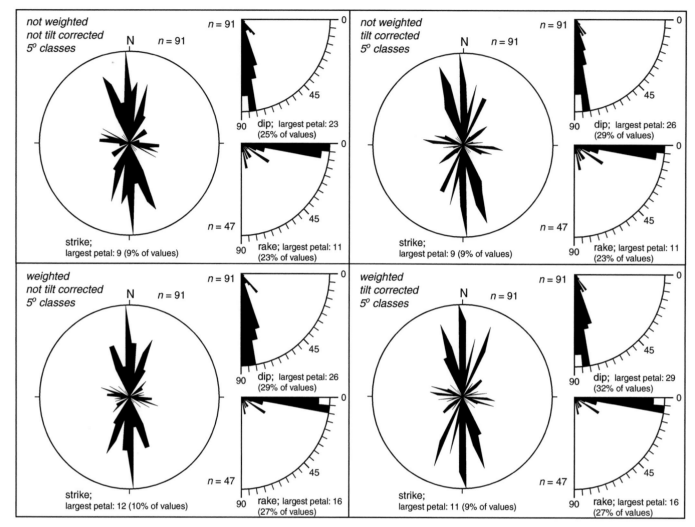

Figure 14. Rose diagrams for faults with >0.1 m stratigraphic separation are used to compare raw data with tilt-corrected and weighted results. Faults are weighted according to the methodology of Angelier et al. (1985). Tilt-corrected data appear comparable to non–tilt-corrected data; however, shallow dip angles in the Boca basin section may make this comparison invalid. Weighting the data emphasizes a few large, subvertical, northeast trending faults.

DISCUSSION: REGIONAL TECTONIC SIGNIFICANCE

This study provides new information about the style and timing of late Miocene–Pliocene deformation history at the eastern edge of the Sierra Nevada. Although late Miocene deformation is documented elsewhere in this part of the Sierra Nevada (e.g., Trexler et al., 2000; Henry and Perkins, 2001), this deformation is not recorded in the Boca basin section. There is probably a significant unconformity below, or possibly within, the lowest Boca basin rocks (interval I, lithofacies 1), which may be the local expression of this deformation. The age of basin initiation is not well constrained, but it predates a ca. 4.38 Ma basalt that is ~181 m above the base of the section. Late Pliocene deformation is recorded by the termination of deposition in the Boca basin. Tilting, incision, and probable initiation of distributed

faulting of the Boca basin rocks occurred between 2.7 Ma and 2.6 Ma. This period of tectonism is represented both by the initiation of interval IV in the Boca basin section and by the subsequent deformation of the section. Most of the tilting occurred prior to the eruption of the Boca Ridge Formation (2.61 Ma). This age control refines previous estimates for the timing of deformation (Latham, 1985; Henry and Perkins, 2001) and is consistent with the interpretation that the Sierra Nevada became a topographic high in the late Cenozoic, with most of the relief developing since 10 Ma (Lindgren, 1911; Hudson, 1955; Axelrod, 1962; Huber, 1981; Unruh, 1991; Wakabayashi and Sawyer, 2001; Jones et al., 2004). The distributed faulting in the Boca basin rocks may have accompanied, as well as followed, southwestward tilting; it records east-southeast extension and minor north-northeast contraction.

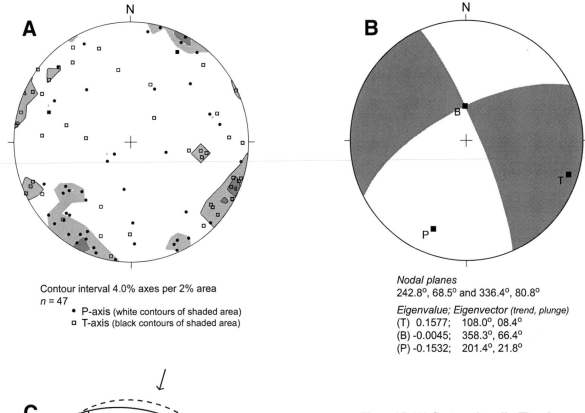

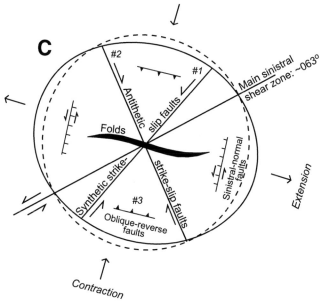

Figure 15. (A) Contoured tensile (T) and compressional (P) stress axis orientations for faults with >0.1 m displacement and known slip-direction indicators (47 planes). Solutions were modeled using a fault-slip inversion method (program of R. Allmendinger). Data are not tilt-corrected. (B) Nodal planes and stress axes were calculated by combining all the faults shown in A. B-axis is the intermediate stress axis. (C) Idealized model of structures associated with the calculated stress field (modified from Sylvester, 1988). West-northwest directed extension and north-northeast directed extension is predicted. The predicted fault sets are similar to those observed at the Boca basin. Fault sets have been labeled 1, 2, and 3 to correspond with the text.

To determine the regional significance of the depositional system recorded in Boca basin, we compare the Neogene Boca basin section to another approximately coeval basin, the Neogene Verdi basin (Fig. 1). Exposures of the Verdi basin are superficially similar to the Boca basin, and several tephra beds are correlated between the two (Trexler et al., 2000; Henry and Perkins, 2001). However, the Neogene Verdi basin consists of a longer-lived and thicker Neogene section (>2000 m), while the Boca basin appears to be an isolated, smaller basin (~500 m). The change in depositional environment and paleocurrent direction where ca. 4.38 Ma basalt flows are interbedded in the Boca basin section suggests that the Boca basin depositional system was controlled by volcanic topography rather than being part of a regional basin system. It is unlikely that the two basins were part of the same depositional system.

The motivation for this study of style and timing of deformation in the Boca basin was to determine whether any of the regional tectonic influences were expressed *within* the Sierra

Nevada, and if so, when each occurred. We have shown that at Boca basin, a Pliocene deformational event is expressed as southwestward tilting, erosional downcutting by west directed streams, and diffuse faulting throughout the basin. Well-defined through-going faults are lacking within the study area. We tested two potential regional tectonic mechanisms for the Pliocene deformation observed at Boca basin: (1) Walker Lane dextral slip encroaching westward into the Sierra Nevada at the latitude of Boca basin, and (2) extensional deformation stepping westward from the Basin and Range or propagating northward from normal faulting at Lake Tahoe. In addition, we examined evidence from the study area that might pertain to the timing and origin of the Sierra Nevada frontal fault system in the northern Sierra Nevada.

There is no evidence of large-scale northwest striking dextral slip in the Boca basin in the last 5 m.y. The deformation found in this study has occurred since ca. 2.7 Ma and consists of southwestward tilting and distributed small-scale faulting. The faulting includes northwest striking dextral faults, but these seem to be conjugate to equally (or better) developed northeast striking sinistral faults; together, these faults accommodate east-southeast extension and minor north-northeast contraction. However, the coexistence of extensional and contractional structures, which we see in the distributed faulting in the Boca basin, is typical of a continental strike-slip setting (e.g., van der Pluijm and Marshak, 1997). So the evidence from the study area does not disprove the westward stepping of large-scale dextral (Walker Lane) deformation to the Mohawk Valley or Grizzly faults within the Sierra Nevada (Hawkins et al., 1986; Sawyer and Briggs, 2001) but rather shows merely that this did not occur through the region occupied by the Neogene Boca basin section.

The deformation style (and local stress field it documents) within the Boca basin is consistent with generally east-west extension propagating from the Basin and Range Province into the Sierra Nevada. Focal mechanism studies along the Sierra Nevada frontal fault system show that the along-strike terminations of large normal faults are characterized by distributed conjugate strike-slip faulting (e.g., the Double Spring Flat earthquake sequence at the south end of the Genoa fault; Ichinose et al., 1998; Fig. 1 herein). The 1983 aftershock sequence of the 1966 M_L 6.0 Truckee earthquake on the sinistral Dog Valley fault (Hawkins et al., 1986; Fig. 1 herein) was very similar to the migration of aftershocks of the M 5.8 Double Spring Flat earthquake. Both 1966 and 1983 aftershock sequences of the Truckee earthquake indicated west-northwest–east-southeast tension, compatible with the stress calculated for distributed faulting in the Boca basin and with motion on normal faults to the south (e.g., Lake Tahoe; Schweickert et al., 2004) and east (e.g., Sierra Nevada range front; Dixon et al., 2000; Bennett et al., 2003). The southwestward tilt that characterizes the deformation of the Boca basin section suggests the presence of an east dipping normal fault to the west or southwest. In summary, the deformation recorded in the rocks of the Neogene Boca basin is consistent with extensional faulting within the Sierra Nevada and with the pattern of distributed conjugate strike-slip faulting at the terminations of large normal faults. In this case, the Boca basin is north of normal faults at Lake Tahoe and could lie near the along-strike termination of one normal fault system and in the (tilted) hanging wall of another, en echelon normal fault. Furthermore, the tight timing control at the Boca basin indicates that extensional deformation started ca. 2.7 Ma in this part of the Sierra Nevada.

Pliocene deformation at Boca basin, characterized by tilting of the section and an abrupt change in base level, was closely followed by eruption of high-potassium mafic lavas of the Boca Ridge Formation ca. 2.6 Ma (Latham, 1985; Henry et al., 2004). A similar compositional change to high-potassium volcanic rocks in the southern Sierra Nevada (4–3 Ma) is attributed to delamination of the Sierra Nevada root and uplift of the southern Sierra Nevada (Ducea and Saleeby, 1996; Manley et al., 2000; Farmer et al., 2002) or subsidence of the southwestern Sierra Nevada (Zandt, 2003; Saleeby and Foster, 2004). The mechanism for producing high-potassium volcanism in the northern Sierra Nevada is still under debate, but it may have been the result of rollback of the Juan de Fuca plate after 3 Ma (Benz and Zandt, 1993; Zandt and Carrigan, 1993). Discussion of the relationship between basalt compositional changes and crustal delamination or regional uplift is beyond the scope of this paper. However, the eruption of high-potassium lavas in the northern Sierra Nevada at ca. 2.61 Ma (Boca Ridge Formation) suggests a northward migration of this phenomenon with time, and it may have been genetically linked to regional uplift of the Sierra Nevada in the vicinity of the study area.

APPENDIX 1. TEPHROCHONOLOGY LABORATORY METHODS

Boca basin tephra were analyzed on a Cameca 50SX electron microprobe at the Department of Geology and Geophysics, University of Utah, USA. On average, 18 glass shards were analyzed for each of the 25 tephra samples. The analyses of these glass shards were compared with those in the University of Utah Western U.S. Tephra Database. This database currently contains the analyses of ~720 different tephra that range in age from recent to ca. 16 Ma. From the database, 16 likely and 1 possible compositional matches to the Boca basin tephra were identified using methods discussed in Perkins et al. (1995, 1998). Of the identified tephra, 14 of the 16 likely matches stratigraphically conformed to the Boca basin measured section (Table 2); all were Pliocene tephra in the age range 3.6–3.0 Ma. Correlative tephra to the Boca basin tephra are found in the Chalk Hills Formation (Idaho), in the subsurface of the Bonneville basin (Utah), in the Honey Lake basin (California), in the Verdi Formation (Nevada), and in the Tuscan Formation (California).

ACKNOWLEDGMENTS

This manuscript is based on a master of science thesis by Mass (2005). This project was supported by grants to K. Mass from the Nevada Petroleum Society, and the Graduate Student Association and the Larry T. Larson Graduate Field Geology scholarship of the University of Nevada, Reno. Special thanks are due to Michael Perkins for providing his time, effort, and knowledge to correlate the tephra samples presented in this study. This manuscript greatly benefited from constructive reviews from Andrew Hanson and Jeffrey Lee, and from discussions with Christopher Henry, Brian Cousens, and Michael Perkins. We appreciate U.S. Forest Service, Truckee Ranger District, for granting access near a bald eagle nesting site along the northern shore of Boca Reservoir.

REFERENCES CITED

Angelier, J., Colletta, B., and Anderson, R.E., 1985, Neogene paleostress changes in the Basin and Range: A case study at Hoover Dam, Nevada-Arizona: Geological Society of America Bulletin, v. 96, no. 3, p. 347–361, doi: 10.1130/0016-7606(1985)96<347:NPCITB>2.0.CO;2.

Argus, D.F., and Gordon, R.G., 1991, Current Sierra Nevada–North American plate motion from very long baseline interferometry: Implications for kinematics of the western United States: Geology, v. 19, no. 11, p. 1085–1088, doi: 10.1130/0091-7613(1991)019<1085:CSNNAM>2.3.CO;2.

Atwater, T., 1970, Implications of plate tectonics for the Cenozoic tectonic evolution of western North America: Geological Society of America Bulletin, v. 81, no. 12, p. 3513–3535, doi: 10.1130/0016-7606(1970)81[3513:IOPTFT]2.0.CO;2.

Atwater, T., and Stock, J., 1998, Pacific–North America plate tectonics of the Neogene southwestern United States: An update: International Geology Review, v. 40, p. 375–402.

Axelrod, D.I., 1956, Mio-Pliocene floras from west-central Nevada: University of California Publications in Geological Sciences, v. 33, 321 p.

Axelrod, D.I., 1958, The Pliocene Verdi flora of west central Nevada: University of California Publications in Geological Sciences, v. 34, no. 2, p. 91–160.

Axelrod, D.I., 1962, Post-Pliocene uplift of the Sierra Nevada, California: Geological Society of America Bulletin, v. 73, p. 183–198, doi: 10.1130/0016-7606(1962)73[183:PUOTSN]2.0.CO;2.

Bateman, P.C., and Wahrhaftig, C., 1966, Geology of the Sierra Nevada: California Division of Mines and Geology Bulletin, v. 190, p. 107–172.

Bell, J.W., 1981, Quaternary Fault Map of the Reno 1° by 2° Quadrangle: United States Geological Survey Open File Report 81-0982, 63 p.

Bennett, R.A., Wernicke, B.P., Niemi, N.A., Friedrich, A.M., and Davis, J.L., 2003, Contemporary strain rates in the northern Basin and Range Province from GPS data: Tectonics, v. 22, no. 2, 1008, doi: 10.1029/2001TC001355.

Benz, H.M., and Zandt, G., 1993, Teleseismic tomography: Lithospheric structure of the San Andreas fault system in northern and central California, *in* Iyer, H.M., and Hirahara, K., eds., Seismic Tomography: Theory and Practice: New York, Chapman and Hall, p. 440–465.

Best, M.G., and Christiansen, E.H., 1991, Limited extension during peak volcanism, Great Basin of Nevada and Utah: Journal of Geophysical Research, v. 96, no. B8, p. 13,509–13,528, doi: 10.1029/91JB00244.

Birkeland, P.W., 1962, Pleistocene volcanism and deformation in the Truckee area, north of Lake Tahoe, California [Ph.D. thesis]: Palo Alto, California, Stanford University, 126 p.

Cashman, P.H., and Trexler, J.H., Jr., 2004, The Neogene Verdi basin records <2.0–2.5 Ma dextral faulting west of the Walker Lane near Reno, NV: Geological Society of America Abstracts with Programs, v. 36, no. 4, p. 28.

Cashman, P.H., Trexler, J.H., Jr., Muntean, T.W., Faulds, J.E., Louie, J.N., and Oppliger, G.L., 2009, this volume, Neogene tectonic evolution of the Sierra Nevada–Basin and Range transition zone at the latitude of Carson City, Nevada, *in* Oldow, J.S., and Cashman, P.H., eds., Late Cenozoic Structure and Evolution of the Great Basin–Sierra Nevada Transition: Geological Society of America Special Paper 447, doi: 10.1130/2009.2447(10).

Deino, A.L., 1985, I. Stratigraphy, chemistry, K-Ar dating, and paleomagnetism of the Nine Hill tuff, California-Nevada; II. Miocene/Oligocene ash-flow tuffs of Seven Lake Mountain, California-Nevada; III. Improved calibration methods and error estimates for ⁴⁰K-⁴⁰Ar dating of young rocks [Ph.D. thesis]: Berkeley, University of California, 498 p.

Dilles, J.H., and Gans, P.B., 1995, The chronology of Cenozoic volcanism and deformation in the Yerington area, western Basin and Range and Walker Lane: Geological Society of America Bulletin, v. 107, no. 4, p. 474–486, doi: 10.1130/0016-7606(1995)107<0474:TCOCVA>2.3.CO;2.

Dixon, T.H., Robaudo, S., Lee, J., and Reheis, M.C., 1995, Constraints on present-day Basin and Range deformation from space geodesy: Tectonics, v. 14, p. 755–772, doi: 10.1029/95TC00931.

Dixon, T.H., Miller, M., Farina, F., Wang, H., and Johnson, D., 2000, Present-day motion of the Sierra Nevada block and some tectonic implications for the Basin and Range Province, North American Cordillera: Tectonics, v. 19, no. 1, p. 1–24, doi: 10.1029/1998TC001088.

Ducea, M.N., and Saleeby, J.B., 1996, Buoyancy sources for a large, unrooted mountains range, the Sierra Nevada, California: Evidence from xenolith thermobarometry: Journal of Geophysical Research, v. 101, p. 8229–8244, doi: 10.1029/95JB03452.

Farmer, G.L., Glazner, A.F., and Manley, C.R., 2002, Did lithospheric delamination trigger the late Cenozoic potassic volcanism in the southern Sierra Nevada?: Geological Society of America Bulletin, v. 114, no. 6, p. 754–768, doi: 10.1130/0016-7606(2002)114<0754:DLDTLC>2.0.CO;2.

Faulds, J.E., Henry, C.D., and Hinz, N.H., 2005, Kinematics of the northern Walker Lane: An incipient transform fault along the Pacific–North American plate boundary: Geology, v. 33, no. 6, p. 505–508, doi: 10.1130/G21274.1.

Fisher, R.V., 1971, Features of coarse-grained, high-concentration fluids and their deposits: Journal of Sedimentary Petrology, v. 41, no. 4, p. 916–927.

Fisher, R.V., and Schmincke, H., 1984, Pyroclastic Rocks: Berlin, Springer-Verlag Inc., 472 p.

Gilbert, C.M., and Reynolds, M.W., 1973, Character and chronology of basin development, western margin of the Basin and Range Province: Geological Society of America Bulletin, v. 84, no. 8, p. 2489–2510, doi: 10.1130/0016-7606(1973)84<2489:CACOBD>2.0.CO;2.

Golia, R.T., and Stewart, J.H., 1984, Depositional environments and paleogeography of the Upper Miocene Wassuk Group, west-central Nevada, *in* Nilsen, T.H., ed., Fluvial Sedimentation and Related Tectonic Framework, Western North America: Sedimentary Geology, v. 38, no. 1–4, p. 159–180.

Hawkins, F.F., LaForge, R., and Hansen, R.A., 1986, Seismotectonic Study of the Truckee/Lake Tahoe Area, Northeastern Sierra Nevada, California, for Prosser Creek, Stampede, Boca, and Lake Tahoe Dams: Denver, Colorado, United States Bureau of Reclamation Seismotectonic Report no. 85-4, 210 p.

Henry, C.D., and Perkins, M.E., 2001, Sierra Nevada–Basin and Range transition near Reno, Nevada: Two-stage development at ca. 12 and 3 Ma: Geology, v. 29, no. 8, p. 719–722, doi: 10.1130/0091-7613(2001)029<0719:SNBART>2.0.CO;2.

Henry, C.D., Cousens, B.L., Castor, S.B., Faulds, J.E., Garside, L.J., and Timmermans, A., 2004, The ancestral Cascades arc, northern California/western Nevada: Spatial and temporal variations in volcanism and geochemistry: American Geophysical Union, Fall Meeting 2004, abstract no. V13B-1478.

Hinz, N.H., 2004, Tertiary volcanic stratigraphy of the Diamond and Fort Sage Mountains, northeastern California and western Nevada—Implications for development of the northern Walker Lane [M.S. thesis]: Reno, University of Nevada, 110 p.

House, M.A., Wernicke, B.P., and Farley, K.A., 1998, Dating topography of the Sierra Nevada, California, using apatite (U-Th)/He ages: Nature, v. 396, p. 66–69, doi: 10.1038/23926.

Huber, N.K., 1981, Amount and Timing of Late Cenozoic Uplift and Tilt of the Central Sierra Nevada, California—Evidence from the Upper San Joaquin River Basin: U.S. Geological Survey Professional Paper 1197, 28 p.

Hudson, F.S., 1955, The deformation of the Sierra Nevada, California, since middle Eocene: Geological Society of America Bulletin, v. 66, no. 7, p. 835–869, doi: 10.1130/0016-7606(1955)66[835:MOTDOT]2.0.CO;2.

Ichinose, G.A., Smith, K.D., and Anderson, J.G., 1998, Moment tensor solutions of the 1994 to 1996 Double Spring Flat, Nevada, earthquake sequence and implications for local tectonic models: Bulletin of the Seismological Society of America, v. 88, no. 6, p. 1363–1378.

Ichinose, G.A., Anderson, J.G., Smith, K.D., and Zeng, Y., 2003, Source parameters of eastern California and western Nevada earthquakes from regional moment tensor inversion: Bulletin of the Seismological Society of America, v. 93, no. 1, p. 61–84, doi: 10.1785/0120020063.

Jones, C.H., Farmer, G.L., and Unruh, J., 2004, Tectonics of Pliocene removal of lithosphere of the Sierra Nevada, California: Geological Society of America Bulletin, v. 116, no. 11/12, p. 1408–1422, doi: 10.1130/B25397.1.

Kelly, T.S., 1994, Two Pliocene (Blancan) vertebrate faunas from Douglas County, Nevada: PaleoBios, v. 16, no. 1, p. 1–23.

Kelly, T.S., 1997, Additional late Cenozoic (latest Hemphillian to earliest Irvingtonian) mammals from Douglas County, Nevada: PaleoBios, v. 18, no. 1, p. 1–31.

Kelly, T.S., 1998, New Miocene mammalian faunas from west central Nevada: Journal of Paleontology, v. 72, p. 137–149.

Latham, T.S., Jr., 1985, Stratigraphy, structure, and geochemistry of Plio-Pleistocene volcanic rocks of the western Basin and Range Province near Truckee, California [Ph.D. thesis]: Davis, University of California, 341 p.

Lindgren, W., 1911, Tertiary Gravels of the Sierra Nevada, California: U.S. Geological Survey Professional Paper P-0073, 226 p.

Manley, C.R., Glazner, A.F., and Farmer, G.L., 2000, Timing of volcanism in the Sierra Nevada of California: Evidence for Pliocene delamination of the batholithic root?: Geology, v. 28, no. 9, p. 811–814, doi: 10.1130/0091-7613(2000)28<811:TOVITS>2.0.CO;2.

Marrett, R., and Allmendinger, R.W., 1990, Kinematic analysis of fault-slip data: Journal of Structural Geology, v. 12, no. 8, p. 973–986, doi: 10.1016/0191-8141(90)90093-E.

Mass, K.B., 2005, The Neogene Boca basin, northeastern California: Stratigraphy, structure, and implications for regional tectonics [M.S. thesis]: Reno, University of Nevada, 195 p.

Mass, K.B., Cashman, P.H., Trexler, J.H., Jr., Park, H., and Perkins, M.E., 2003, Deformation history in Neogene sediments of Honey Lake basin, northern Walker Lane, Lassen County, California: Geological Society of America Abstracts with Programs, v. 35, no. 6, p. 26.

Miall, A.D., 1996, The Geology of Fluvial Deposits: Sedimentary Facies, Basin Analysis and Petroleum Geology: Heidelberg, Springer Verlag Inc., 582 p.

Miall, A.D., 2000, Principles of Sedimentary Basin Deposits: Berlin-Heidelberg, Springer Verlag Inc., 616 p.

Miller, C.D., 1989, Potential Hazards from Future Eruptions in California: U.S. Geological Survey Bulletin 1847B, 17 p.

Muntean, T.W., 2001, Evolution and stratigraphy of the Neogene Sunrise Pass Formation of the Gardnerville sedimentary basin, Douglas County, Nevada [M.S. thesis]: Reno, University of Nevada, 250 p.

Okada, H., 1971, Classification of sandstone: Analysis and proposal: The Journal of Geology, v. 79, p. 509–525.

Park, H., 2004, Honey Lake basin: Depositional settings and tectonics [M.S thesis]: Reno, University of Nevada, 197 p.

Perkins, M.E., Nash, W.P., Brown, F.H., and Fleck, R.J., 1995, Fallout tuffs of Trapper Creek, Idaho—A record of Miocene explosive volcanism in the Snake River Plain volcanic province: Geological Society of America Bulletin, v. 107, no. 12, p. 1484–1506, doi: 10.1130/0016-7606(1995)107<1484:FTOTCI>2.3.CO;2.

Perkins, M.E., Brown, F.H., Nash, W.P., McIntosh, W., and Williams, S.K., 1998, Sequence, age, and source of silicic fallout tuffs in middle to late Miocene basins of the northern Basin and Range Province: Geological Society of America Bulletin, v. 110, no. 3, p. 344–360, doi: 10.1130/0016-7606(1998)110<0344:SAASOS>2.3.CO;2.

Pollard, D.D., Saltzer, S.D., and Rubin, A.M., 1993, Stress inversion methods: Are they based on faulty assumptions?: Journal of Structural Geology, v. 15, no. 8, p. 1045–1054, doi: 10.1016/0191-8141(93)90176-B.

Saleeby, J., and Foster, Z., 2004, Topographic response to mantle lithosphere removal in the southern Sierra Nevada region, California: Geology, v. 32, no. 3, p. 245–248, doi: 10.1130/G19958.1.

Saucedo, G.J., and Wagner, D.L., 1992, Geologic Map of the Chico Quadrangle, California: California Division of Mines and Geology Regional Geological Map Series 7A, scale 1:250,000.

Sawyer, T.L., and Briggs, R., 2001, Kinematics and late Quaternary activity of the Mohawk Valley fault zone: Pacific Cell, Friends of the Pleistocene Field Trip Guidebook, p. 49–62.

Schweickert, R.A., Lahren, M.M., Karlin, R.E., Howie, J.F., and Smith, K.D., 2000a, Lake Tahoe active faults, landslides, and tsumanis, in Lageson, D.R., Peters, S.G., and Lahren, M.M., eds., Great Basin and Sierra Nevada: Boulder, Colorado, Geological Society of America Field Guide 2, p. 1–22.

Schweickert, R.A., Lahren, M.M., Karlin, R., Howle, J., and Smith, K., 2000b, Preliminary Map of the Pleistocene to Holocene Faults in the Lake Tahoe Basin: Nevada Bureau of Mines and Geology Open-File Report 2000-4, scale 1:1,000,000.

Schweickert, R.A., Lahren, M.M., Smith, K.D., Howle, J.F., and Ichinose, G., 2004, Transtensional deformation in the Lake Tahoe region, California

and Nevada, USA: Tectonophysics, v. 392, no. 1–4, p. 303–323, doi: 10.1016/j.tecto.2004.04.019.

Stewart, J.H., 1988, Tectonics of the Walker Lane belt, western Great Basin: Mesozoic and Cenozoic deformation in a zone of shear, in Ernst, W.G., ed., The Geotectonic Development of California: Englewood Cliffs, New Jersey, Prentice Hall, p. 71–86.

Stockli, D.F., Dumitru, T.A., McWilliams, M.O., and Farley, K.A., 2003, Cenozoic tectonic evolution of the White Mountains, California and Nevada: Geological Society of America Bulletin, v. 115, no. 7, p. 788–816, doi: 10.1130/0016-7606(2003)115<0788:CTEOTW>2.0.CO;2.

Surpless, B.E., Stockli, D.F., Dumitru, T.A., and Miller, E.L., 2002, Two-phase westward encroachment of Basin and Range extension into the northern Sierra Nevada: Tectonics, v. 21, no. 1, 13 p.

Sylvester, A.G., 1988, Strike-slip faults: Geological Society of America Bulletin, v. 100, no. 11, p. 1666–1703, doi: 10.1130/0016-7606(1988)100<1666:SSF>2.3.CO;2.

Thatcher, W., Foulger, G.R., Julian, B.R., Svarc, J., Quilty, E., and Bawden, G.W., 1999, Present-day deformation across the Basin and Range Province, western United States: Science, v. 283, p. 1714–1718, doi: 10.1126/science.283.5408.1714.

Trexler, J.H., Jr., Cashman, P.H., Henry, C.D., Muntean, T., Schwartz, K., Ten-Brink, A., Faulds, J.E., Perkins, M., and Kelly, T., 2000, Neogene basins in western Nevada document the tectonic history of the Sierra Nevada–Basin and Range transition zone for the past 12 Ma, in Lageson, D.R., Peters, S.G., and Lahren, M.M., eds., Great Basin and Sierra Nevada: Boulder, Colorado, Geological Society of America Field Guide 2, p. 97–116.

Trexler, J.H., Jr., Park, H.M., Cashman, P.H., and Mass, K.B., 2009, this volume, Late Neogene basin history at Honey Lake, northeastern California: Implications for regional tectonics at 3 to 4 Ma, in Oldow, J.S., and Cashman, P.H., eds., Late Cenozoic Structure and Evolution of the Great Basin–Sierra Nevada Transition: Geological Society of America Special Paper 447, doi: 10.1130/2009.2447(06).

Tsai, Y., and Aki, K., 1970, Source mechanism of the Truckee, California, earthquake of September 12, 1966: Bulletin of the Seismological Society of America, v. 60, p. 1199–1208.

Unruh, J.R., 1991, The uplift of the Sierra Nevada and implications for late Cenozoic epeirogeny in the western Cordillera: Geological Society of America Bulletin, v. 103, p. 1395–1404, doi: 10.1130/0016-7606(1991)103<1395:TUOTSN>2.3.CO;2.

van der Pluijm, B.A., and Marshak, S., 1997, Earth Structure and Introduction to Structural Geology and Tectonics: United States of America: New York, McGraw-Hill, 495 p.

Wakabayashi, J., and Sawyer, T.L., 2001, Stream incision, tectonics, uplift, and evolution of topography of the Sierra Nevada, California: The Journal of Geology, v. 109, p. 539–562, doi: 10.1086/321962.

Wernicke, B., Clayton, R., Ducea, M., Jones, C.H., Park, S., Ruppert, S., Saleeby, J., Snow, J.K., Squires, L., Fliedner, M., Jiracek, G., Keller, R., Klemperer, S., Luetgert, J., Malin, P., Miller, K., Mooney, W., Oliver, H., and Phinney, R., 1996, Origin of high mountains in the continents: The southern Sierra Nevada: Science, v. 271, p. 190–193, doi: 10.1126/science.271.5246.190.

Wolfe, J.A., Schorn, H.E., Forest, C.E., and Molnar, P., 1997, Paleobotanical evidence for high altitudes in Nevada during the Miocene: Science, v. 276, p. 1672–1675, doi: 10.1126/science.276.5319.1672.

Wright, L.A., 1976, Late Cenozoic fault pattern in the Great Basin and westward displacement of the Sierra Nevada block: Geology, v. 4, no. 8, p. 489–494, doi: 10.1130/0091-7613(1976)4<489:LCFPAS>2.0.CO;2.

Zandt, G., 2003, The southern Sierra Nevada drip and mantle wind direction beneath the southwestern United States: International Geology Review, v. 45, no. 3, p. 213–223, doi: 10.2747/0020-6814.45.3.213.

Zandt, G., and Carrigan, C.R., 1993, Small-scale convective instability and upper mantle viscosity under California: Science, v. 261, p. 460–463, doi: 10.1126/science.261.5120.460.

Zoback, M.L., and Zoback, M., 1980, State of stress in the conterminous United States: Journal of Geophysical Research, v. 85, no. B11, p. 6113–6156, doi: 10.1029/JB085iB11p06113.

Manuscript Accepted by the Society 21 July 2008

The Geological Society of America
Special Paper 447
2009

Neogene tectonic evolution of the Sierra Nevada–Basin and Range transition zone at the latitude of Carson City, Nevada

Patricia H. Cashman*
James H. Trexler Jr.
Thomas W. Muntean[†]
Department of Geological Sciences and Engineering, University of Nevada, Reno, Nevada 89577, USA

James E. Faulds
Nevada Bureau of Mines and Geology, University of Nevada, Reno, Nevada 89557, USA

John N. Louie
Department of Geological Sciences and Engineering, University of Nevada, Reno, Nevada 89577, USA, and Nevada Seismological Laboratory, University of Nevada, Reno, Nevada 89557, USA

Gary L. Oppliger
Department of Geological Sciences and Engineering, University of Nevada, Reno, Nevada 89577, USA, and Arthur Brant Laboratory for Exploration Geophysics, University of Nevada, Reno, Nevada 89557, USA

ABSTRACT

Sedimentary rocks of the Neogene Gardnerville Basin record a complex normal faulting history from mid-Miocene to the present; this record bridges an important gap between contemporary tectonics and the older geologic record. The upper Neogene sediments are preserved in a west-dipping half graben, and fanning of dips within the section shows that the basin-bounding Carson Range frontal fault system to the west has been active since at least 7 Ma. In addition, the sedimentary history clearly shows that several north-striking normal faults within the basin have been active at different times during deposition. Gravity data enable us to extend the faulting history back beyond what is exposed at the surface and reveal mid-Miocene(?) normal faults that are no longer active below the western part of the basin. Gravity modeling suggests that the underlying fault-bounded basin is structurally symmetric. These faults have accommodated extension within Sierran crystalline rocks west of the Walker Lane, in the eastern part of the Sierra Nevada microplate.

The Neogene Gardnerville Basin documents the tectonic evolution of a distinctive part of the Sierra Nevada–Basin and Range transition zone. It lies west of the Walker Lane at this latitude, and, during its history from >7 Ma to the present, it shows no evidence of the distributed dextral slip that characterizes that zone. The field relation-

*Corresponding author e-mail: pcashman@mines.unr.edu.
[†]Present address: Department of Geosciences, University of Nevada, Las Vegas, Nevada 89154, USA.

Cashman, P.H., Trexler, J.H., Jr., Muntean, T.W., Faulds, J.E., Louie, J.N., and Oppliger, G.L., 2009, Neogene tectonic evolution of the Sierra Nevada–Basin and Range transition zone at the latitude of Carson City, Nevada, *in* Oldow, J.S., and Cashman, P.H., eds., Late Cenozoic Structure and Evolution of the Great Basin–Sierra Nevada Transition: Geological Society of America Special Paper 447, p. 171–188, doi: 10.1130/2009.2447(10). For permission to copy, contact editing@geosociety.org. ©2009 The Geological Society of America. All rights reserved.

ships, combined with sedimentology of the Neogene strata, document a multistage intrabasin faulting history during deposition; several intrabasin normal faults have acted in concert with the Carson Range frontal fault system to accommodate extension. This could be an analog for other normal fault systems in the vicinity, e.g., the Lake Tahoe Basin, immediately west of our study area.

Keywords: Walker Lane, syntectonic, basins, Neogene, structure.

INTRODUCTION

The northward migration of the Mendocino triple junction marks the progressive change from Cascadian subduction to dextral transform tectonics along the western margin of North America. Although dextral shear across the continental margin started with the inception of the transform boundary ca. 30 Ma (Atwater, 1970), the accommodation of the shear changed with the opening of the Gulf of California ca. 6 Ma (Oskin and Stock, 2003). Currently, in addition to the dextral offset along the plate boundary itself, 20%–25% of Pacific–North America relative plate motion is accommodated by dextral strain *within* the continent (e.g., Dixon et al., 1995; Thatcher et al., 1999; Miller et al., 2000; Oldow et al., 2001). This occurs in the Walker Lane, a zone of complex dextral faulting along the boundary between the Sierra Nevada and the Basin and Range Province.

In eastern California and western Nevada, several additional tectonic influences are superimposed on those resulting from the northward-migrating transition from a convergent to a transform margin:

1. East-west extension, primarily mid-Miocene in age, characterizes the Basin and Range Province; this extension is thought to have propagated westward with time, into the previously coherent Sierra Nevada block (e.g., Dilles and Gans, 1995; Surpless et al., 2002). The extension may also have stepped northward, into previously unextended northwestern Nevada, in the last 10–12 m.y. (e.g., Colgan and Dumitru, 2003; Lerch et al., 2004; Whitehill et al., 2004).

2. Changes in the direction or amount of Pacific–North America relative plate motion appear to have influenced deformation within the continent (e.g., Atwater and Stock, 1998; Wernicke and Snow, 1998). These include an increase in rate of motion at 12 Ma and a change in direction of motion (from WNW to NNW) at 8 Ma (Atwater and Stock, 1998).

3. Formation of the high topography in the Sierra Nevada was a significant tectonic event, but the timing of and explanation for this event are subject to debate: Most of the uplift may have occurred in Cretaceous time (e.g., House et al., 1998). Discontinuous preservation of Oligocene tuffs in northwestern Nevada records deep erosion (suggesting uplift) in late Oligocene or early Miocene time (e.g., Deino, 1985; Faulds et al., 2003; Henry et al., 2003; Hinz et al., 2003; Stockli et al., 2003; Hinz,

2004). Tilting and incision of Neogene sedimentary rocks as young as 2 or 2.5 Ma provide evidence for a late Pliocene uplift event in northwestern Nevada and northeastern California (e.g., Trexler et al., 2000; Muntean, 2001; Mass et al., 2003; Park, 2004; Park et al., 2004; Cashman and Trexler, 2004; Mass, 2005; Mass et al., this volume; Trexler et al., this volume).

4. Delamination of the Sierran root in late Miocene or Pliocene time has been proposed to explain both uplift and petrologic changes in the southern Sierra (e.g., Ducea and Saleeby, 2003; Feldstein and Lange, 1999; Manley et al., 2000; Farmer et al., 2002). The northward extent of this event is not yet known.

5. Motion of the Sierra Nevada microplate relative to North America (rather than Basin and Range extension, or Sierran uplift due to some other cause) has been proposed to explain the strain recorded by earthquakes along the Sierran frontal fault system (Unruh et al., 2003). Focal mechanism analysis reveals a significant component of dextral motion, and seismogenic deformation primarily characterized by horizontal shearing and oblique crustal thinning. Modern seismicity may not be representative, however; the long-term slip history on most of these faults is not well known.

It is difficult to recognize and separate the effects of these tectonic influences. Our approach to this problem is to examine localities that allow us to isolate individual factors, and therefore determine the contribution of each factor to the complex evolution of the continental margin.

The Neogene sedimentary basins of northwest Nevada and northeast California provide important constraints on the development of intraplate deformation along the eastern side of the Sierra Nevada. Geographically, some of these basins occur within the Basin and Range Province and others are in the Walker Lane. Still others are confined to a faulted region west of the Walker Lane; these include examples both west and east of the present Sierran topographic front (Fig. 1). Faulting west of the Walker Lane is thought to have started more recently than the extension in the Basin and Range to the east (e.g., Dilles and Gans, 1995; Surpless et al., 2002), and it has been suggested that this western faulting is related to rheological changes in the Sierran crystalline rocks, as a result of "side-heating" from the Basin and Range Province (e.g., Surpless et al., 2002). Temporally, the Neogene sedimentary rocks were deposited during the time period when both the Walker Lane and the Sierran frontal fault system were

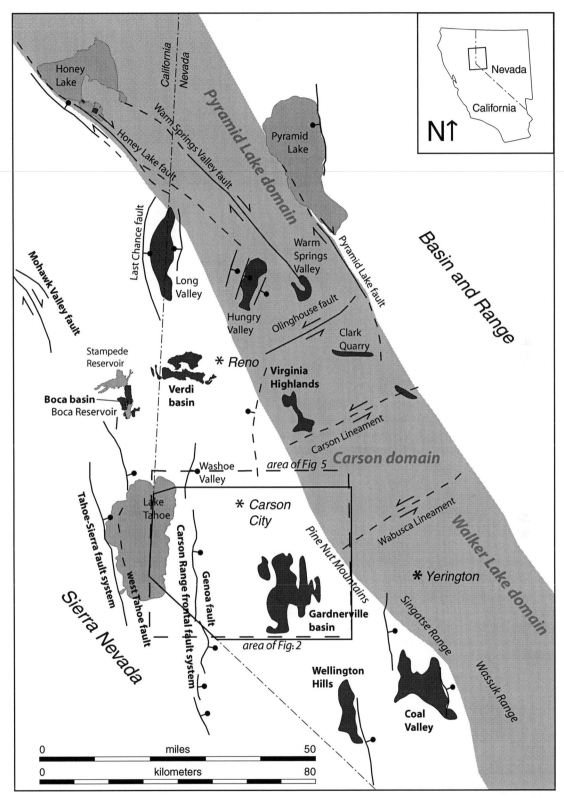

Figure 1. Regional location map, showing the Neogene basins along the Sierra Nevada–Basin and Range transition zone. Map was modified from Cashman and Fontaine (2000). Light gray is the Walker Lane; dark gray is outcrop areas of Neogene sedimentary rocks. Buried Neogene sedimentary rocks are presumed to be more extensive than these surface exposures. "Northern Walker Lane," as used herein, contains the Pyramid Lake domain (characterized by northwest-striking dextral faults) and Carson domain (characterized by northeast-striking sinistral faults) of Stewart (1988). "Central Walker Lane," as used herein, corresponds to the Walker Lake domain (characterized by northwest-striking dextral faults) of Stewart (1988).

becoming active at this latitude. The sedimentary rocks preserve evidence of the evolution of these major tectonic features. They complement other kinds of geologic data and fill an important gap between the older geologic record and contemporary kinematics revealed by neotectonics, geodesy, and seismicity.

The Neogene Gardnerville Basin, preserved in the present-day Carson Valley of northwest Nevada, is ideally situated to record the age, style, and magnitude of tectonism in the crystalline rocks west of the Walker Lane and to test models for the evolution of this structural domain in the context of the evolving North America–Pacific plate boundary. The sedimentary section depositionally overlies Mesozoic crystalline rocks of the Pine Nut Mountains on the east and terminates against Sierran granite of the Carson Range along the Carson Range frontal fault system (here, the Genoa fault) on the west. To the east of the Pine Nut Mountains, there lie the Yerington extensional domain and the central Walker Lane; the style and timing of faulting in these areas are well documented (see Background section). In addition, the Gardnerville Basin occupies an intermediate position between older Tertiary basins in the Walker Lane to the east (e.g., Stewart, 1992; Perkins et al., 1998; Kelly, 1998) and younger sedimentary deposits *within* the Carson Range in the Lake Tahoe and Boca Reservoir area to the west and northwest (e.g., Mass et al., 2003; Mass, 2005; Mass et al., this volume).

In this paper, we present new sedimentological, structural, geochronological, and geophysical data that record the development of the Neogene Gardnerville Basin. We summarize the depositional history of the Gardnerville Basin from older than 7 Ma to younger than 2 Ma, focusing on paleogeographic information that relates to tectonism. This is based on new mapping and sedimentological analysis of the Neogene sedimentary section—the Sunrise Pass Formation, defined in Muntean (2001). We then discuss the evidence for the style and timing of syndepositional deformation, including map relationships, age control, and a short seismic-reflection sounding. We present a new gravity map of the entire Gardnerville Basin, correlate the gravity anomalies with the surface geology, and show how they modify our interpretation of basin development to include older faulting. We conclude with a summary of the evolution of the Gardnerville Basin and the implications of this evolution for the development of the broad zone of intracontinental deformation along the eastern edge of the Sierra Nevada.

BACKGROUND

Neogene Sedimentary Rocks along the Sierra Nevada–Basin and Range Transition Zone

In the Sierra Nevada–Basin and Range transition zone, a few names have been widely (and loosely) applied to Neogene sedimentary rocks of similar appearance; this stratigraphic terminology has obscured important tectonic relationships. "Truckee Formation," a term still in use today, was the name originally applied to Tertiary sedimentary and volcanic rocks in western Nevada

(King, 1878). An early study of some of these strata named a section of Miocene-Pliocene sediments "Coal Valley Formation," distinguishing them from the Truckee Formation based on the occurrence of basaltic volcanic rocks in the former (Axelrod, 1956, 1958, 1962). However, other workers argued that a distinction based on composition of intercalated volcanic rocks does not hold up in the field (Thompson and White, 1964) and concluded that the widespread rocks assigned to the Coal Valley Formation were actually deposited in a number of different basins (e.g., Bonham and Papke, 1969; Kelly, 1998). Improved age control for the Neogene sediments supports the interpretation that there were many different Neogene basins. For example, although the type section of the Coal Valley Formation and associated strata record a basin history from 13 to 9 Ma (Gilbert and Reynolds, 1973; Golia and Stewart, 1984), vertebrate remains from several other basins (e.g., Carson Valley, Nevada, and Long Valley, California) demonstrate that sedimentary rocks in these basins are late Pliocene in age (Kelly, 1998; Firby, 1979, respectively) and do not overlap in age with the type Coal Valley Formation. Our work has confirmed that the Neogene sedimentary rocks are not all correlative and—more importantly—that each basin records a different, and detailed, paleogeographic and tectonic history (e.g., Trexler et al., 2000; Muntean, 2001; Schwartz, 2001; Park, 2004; Cashman and Trexler, 2004; Mass, 2005; Mass et al., this volume; Trexler et al., this volume). (For basin locations, see Fig. 1.)

Prior to our research, sedimentary rocks of the Neogene Gardnerville Basin were mapped as "Ts", unnamed Tertiary sedimentary rocks (e.g., Noble, 1962; Moore, 1969; Kelly, 1998; Stewart, 1999), and studies were limited to quadrangle mapping (Stewart and Noble, 1979; Pease, 1980; Garside and Rigby, 1998; dePolo et al., 2000; Ramelli et al., 2003), gravity surveys (Maurer, 1984a, 1984b), and locally detailed paleontologic and paleobotanical work (Schorn, 1994; Kelly, 1994, 1997, 1998; Lindsay et al., 2002). The section was known to contain Hemphillian and Blancan vertebrates (Kelly, 1998). However, much of the information needed to decipher the tectonic evolution—duration of deposition, changes in paleogeography, fault style and faulting history, etc.—was unknown.

Structural Setting of the Neogene Gardnerville Basin

Significant dextral slip is recorded in the Walker Lane east of the Gardnerville Basin; it appears to have occurred during two different time periods. The dextral offset along northwest-striking faults in the central Walker Lane (the Walker Lake domain of Stewart, 1988) is a minimum of 48–60 km since 27 Ma; much of this offset is thought to have occurred in several episodes between 22 and 27 Ma (Hardyman and Oldow, 1991; Oldow, 1992, and references therein). In contrast, the present regime of dextral strike-slip faulting is thought to have started as recently as 10 Ma (Hardyman and Oldow, 1991) to 7 Ma (Dilles, and Gans, 1995). Farther south in the Walker Lane, the more recent transcurrent deformation is thought to have migrated west with time (e.g., Stockli et al., 2003); it is not known whether similar migration has occurred at the latitude of the Gardnerville Basin. Within the Walker Lane, the Carson domain of Stewart (1988), which borders the Walker Lake domain on the north, provides independent constraints on the timing of dextral slip. The Carson domain is characterized by east-northeast–striking sinistral faults. Paleomagnetic studies show that dextral slip in this part of the Walker Lane is accommodated by 40°–50° of clockwise vertical-axis rotation, all of which appears to have occurred since the eruption of basalts that are 9 Ma and older (Cashman and Fontaine, 2000). Smaller rotations in 4 Ma basalts from the same region indicate that about half of the rotation was accomplished before 4 Ma (Cashman and Fontaine, 2000). Notably, these timing constraints suggest that some of the dextral slip occurred prior to the opening of the Gulf of California ca. 6 Ma (Oskin and Stock, 2003) and concurrent initiation of the present tectonic regime along the western edge of the continent.

The extensional faulting history east of the Neogene Gardnerville Basin includes the dramatic, and well-dated, extension in the Wassuk and Singatse Ranges as well as more recent, less extreme, extension and tilting of the Pine Nut Mountains (see Fig. 1 for locations of these ranges). Rapid extension (>150%) occurred in the Wassuk Range between 15 and 14.4 Ma and in the Singatse Range between 15 and 12 Ma (Proffett, 1977; Dilles and Gans, 1995; Stockli et al., 2002; Surpless et al., 2002). The timing of these extension events is similar to the age of extensional faulting throughout the Basin and Range Province. The timing and amount of extension across the Pine Nut Mountains are not as well constrained, but fission-track modeling suggests fault-block tilting and footwall exhumation between ca. 10 Ma and ca. 3 Ma (Surpless et al., 2002).

The sedimentary rocks of the Neogene Gardnerville Basin depositionally overlie the western flank of the Pine Nut Mountains tilted fault block and record the tilting history of the Pine Nut block related to activity on the basin-bounding Genoa fault, of the Carson Range fault system, to the west. The Genoa fault is a normal fault that is thought to have a minor component of dextral motion (Pease, 1979; Surpless, 1999). Total vertical offset, based on gravity, is on the order of 3.2–3.7 km. Gravity modeling indicates that depth to Mesozoic bedrock 2 km east of the Genoa fault is 1.2–1.7 km, with more than 60% developed on one buried east-side-down fault ~1 km east of the Genoa fault. By including Carson Range relief (~2 km), we estimate a total of 3.2–3.7 km total vertical offset on this combined fault system. Trenching studies document two late Holocene, large-displacement events (Ramelli et al., 1999). The late Holocene slip rate based on these events is 2–3 mm/yr, which is enough to generate the structural relief along the east edge of the Carson Range in just a few million years, but several lines of evidence suggest that the longer-term slip rate may be lower (Ramelli et al., 1999).

The extensional faulting history west of the Gardnerville Basin is not as well constrained but appears to be more recent and of lower magnitude than the extension to the east. Apatite fission-track data suggest that the Carson Range block is tilted 15° west, and fault initiation occurred between 10 Ma and 3 Ma;

this requires significant offset on an east-dipping fault along the west side of the Tahoe-Truckee depression, most likely the West Tahoe fault (Surpless et al., 2002) or other faults in the Tahoe-Sierra fault system. Other workers have constrained the timing of this faulting to between 7 Ma and 2 Ma (Dalrymple, 1964), and it probably postdated andesite flows as young as 3.6 Ma (Saucedo and Wagner, 1992). The tightest age control on faulting within the Sierras is from fault-related tilting of the Neogene sedimentary section in the Boca Basin, 20 km to the north of Lake Tahoe (Fig. 1), which occurred between 2.7 and 2.6 Ma (Mass, 2005; Mass et al., this volume).

The history of deposition and syndepositional faulting preserved in the Neogene Gardnerville Basin therefore provides an opportunity to examine several questions about the timing, style, and causes of regional tectonism. The basin lies west of the Basin and Range and Walker Lane (as traditionally defined), so the age and style of faulting here can test models for regional-scale westward migration with time of: (1) extensional faulting (e.g., Surpless et al., 2002), or (2) dextral faulting (e.g., Stockli et al., 2003). The basin records motion along the eastern boundary of the Sierra Nevada microplate; it contains a significantly longer and more representative faulting history of this boundary than can be determined from trenching studies, geodetic studies, or earthquake focal mechanisms. Passage of the northward-migrating Mendocino triple junction occurred in latest Miocene or earliest Pliocene time at this latitude (Atwater and Stock, 1998), during deposition of the Sunrise Pass Formation. In addition, the sedimentary record spans the time period during which the Gulf of California started to open and plate-boundary transform motion is thought to have been transmitted to the Walker Lane by way of the Eastern California shear zone (Oskin and Stock, 2003); the Gardnerville Basin should show whether this deformation extended beyond the Walker Lane region.

DEPOSITIONAL HISTORY OF THE GARDNERVILLE BASIN

The internal stratigraphy of the Neogene Sunrise Pass Formation (Muntean, 2001) records the depositional history of the Gardnerville Basin between ca. 7 Ma and 2 Ma. The formation was deposited in a system of west-dipping half grabens, and it overlies Mesozoic crystalline rocks of the Pine Nut Mountains on the east (Fig. 2). The youngest exposed rocks are 2.0–2.3 Ma (Kelly, 1997, 2005, written commun.), near the center of the modern Carson Valley. Projecting surface exposures down to the west imply that a still-younger part of the section underlies the western half of the valley, covered by Quaternary alluvium of the Carson River. The Neogene sedimentary rocks consist of fluvial conglomerate and sandstone, and deltaic to lacustrine sandstone, siltstone, and diatomite (Fig. 3).

Age control is diverse and internally consistent throughout the area (Figs. 3 and 4). Air-fall tephras occur throughout the section; their age has been determined using tephrochronology or single-crystal $^{40}Ar/^{39}Ar$ dating. Mammal fossils also occur

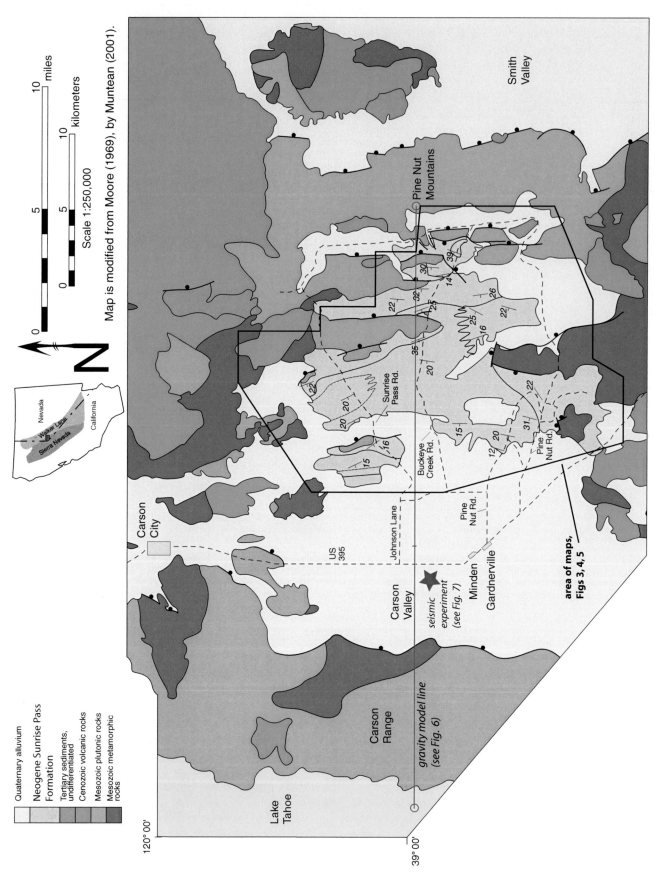

Figure 2. Geologic map of Carson Valley, Nevada, and adjacent ranges, showing the geologic setting of the Sunrise Pass Formation, in the Neogene Gardnerville Basin. The Sunrise Pass Formation depositionally overlies Mesozoic crystalline rocks of the Pine Nut Mountains on the east, and dips west. It is truncated against the Genoa fault at the west edge of the basin.

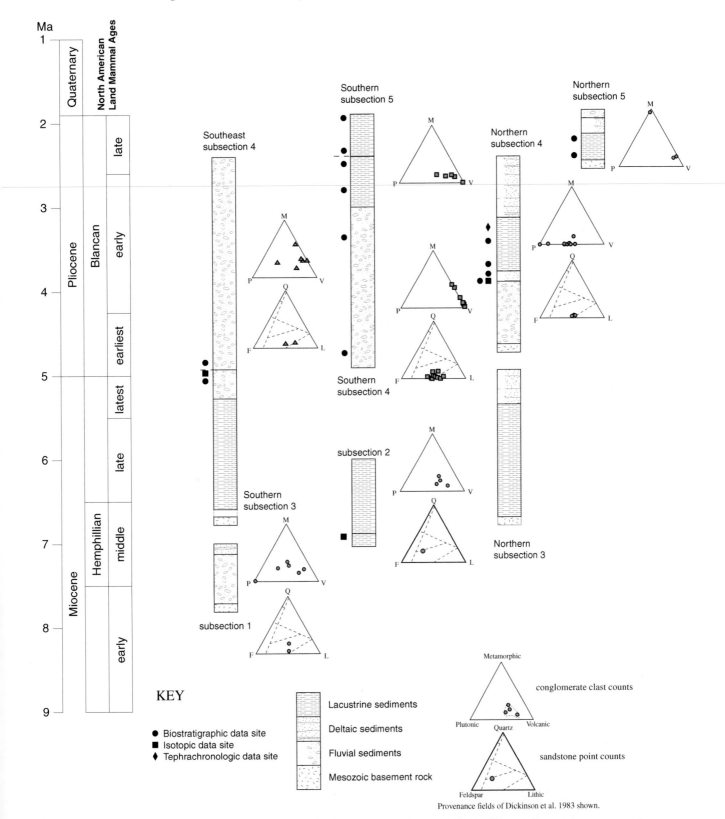

Figure 3. Composite stratigraphic sections of the Sunrise Pass Formation from Muntean (2001). The section segments are grouped as east-west transects across the basin. Since the sedimentary section dips west, older rocks crop out at the east end of each transect. The transect across the southern part of the basin is shown by the column farthest to the left, and the northernmost part of the basin is farthest to the right. These composite sections include composition data (both conglomerate clast counts and sandstone petrography), age control, and facies interpretations. Locations of subsections are shown in Figure 4.

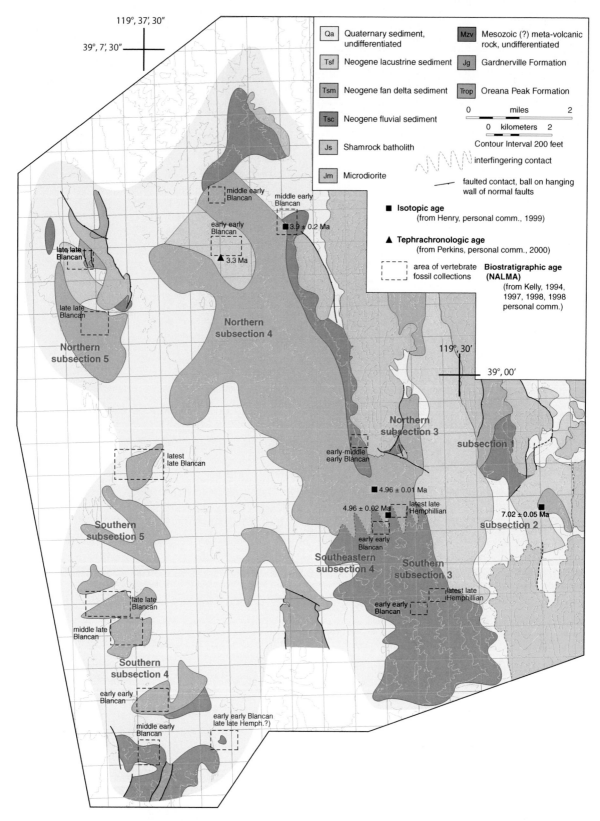

Figure 4. Geologic map of the Sunrise Pass Formation from Muntean (2001), showing facies distribution, location of samples for age control (tephrochronology, isotopic dating, and biostratigraphic dating), and locations of stratigraphic subsections used to make the composite sections in Figure 3. Boxes associated with biostratigraphic dates enclose areas in which distinctive fossil assemblages occur. NALMA—North American land mammal age.

throughout the section, and mammal biostratigraphy combined with magnetostratigraphy provide another source of age control (Kelly, 1994, 1997, 1998; Lindsay et al., 2002). It is not possible to measure a simple stratigraphic section because of intrabasin, syndepositional faulting and unconformities, but a composite section of the Sunrise Pass Formation (Fig. 3) is ~3.5 km thick (Muntean, 2001). In the following section, we briefly summarize the locations and characteristics of different facies within the Sunrise Pass Formation because these reveal important details of the Neogene tectonic evolution of the area.

Fluvial Facies

Clastic deposits interpreted to be fluvial in origin occur in all of the section segments. The conglomerates are typically clean and sand-rich. Conglomerate is interbedded with sandstone and siltstone. Sedimentary structures include tabular and trough cross-laminations, planar laminated beds, and ripple cross-laminations. Sandstone and conglomerate intervals are interpreted as fluvial channel deposits, and interbedded fine sandstone and siltstone are interpreted as overbank and floodplain sediments. Paleocurrent indicators in the fluvial deposits show flow toward the southwest, west, and northwest, i.e., drainage that is generally westward from the Pine Nut Mountains into the basin (Fig. 5), consistent with a half-graben system.

One place where coarse-grained fluvial facies rocks are prevalent is in the oldest part of the section, along the east edge of the basin (Fig. 3, subsection 4). Clast compositions here resemble subjacent crystalline rocks exposed in the foothills of the adjacent Pine Nut Mountains to the east. Sanidine grains from a reworked volcanic ash bed near the base of the section yielded a $^{40}Ar/^{39}Ar$ age of 7.02 ± 0.05 Ma (dating by Chris Henry, *reported in* Trexler et al., 2000).

Conglomerate and coarse sandstone beds also immediately overlie modern exposures of basement throughout the basin. Notably, the lowest coarse clastic sediments overlying the bedrock exposures are different ages in different parts of the basin and are progressively younger westward, as shown by biostratigraphy, tephrochronology, and isotopic dating (Figs. 3 and 4). In addition, conglomerate clast compositions vary between conglomerate localities and reflect the composition of the immediately underlying basement (Fig. 3). The basement highs are interpreted to have been long-lived topographically positive features, and to have been uplifted along active normal faults at several times during basin filling.

In addition, coarser-grained sedimentary rocks dominate throughout the Neogene section in a continuous west-trending zone through the center of the basin. The rocks in this zone are characterized by low-angle unconformities, west-trending paleocurrents, and an unusually thin (as well as coarse) sedimentary section. The central zone is unique in that it does *not* contain any basement highs exposed today. We therefore interpret these rocks to represent the trunk of an integrated drainage system that funneled sediment between the basement highs, from the topo-

graphically partitioned eastern part of the basin to the deepest part of the basin in the northwest. Age control from this higher part of the section is primarily from large mammal biostratigraphy, and it is progressively younger westward, consistent with a continuous west-dipping section (Fig. 4).

Conglomerate clast compositions change upward through the section, from more volcanic clasts low in the section to more granitic ones high in the section. This may record unroofing in the source area, waning volcanic activity, or a combination of the two.

Deltaic and Lacustrine Facies

Rocks interpreted to be deltaic are dominantly sandstone and siltstone, with local pebble conglomerate. Both laterally and vertically, they occur between the coarse-grained fluvial rocks and diatomaceous lacustrine deposits. Tabular cross-lamination and parallel lamination are common in deltaic sediments of all grain sizes. Diagnostic features include upward-coarsening intervals representing prograding delta lobes, laterally discontinuous channel deposits, and convolute lamination resulting from rapid fluid loss in water-saturated sediments. Deltaic deposits are the dominant facies in the southwestern and central parts of the basin.

Rocks interpreted to be lacustrine are dominantly diatomite and siltstone, but thin tephra and sandstone interbeds are common. Bedding in diatomite and siltstone is usually laminated or featureless. Convolute laminations occur locally, adjacent to sandstone and ash beds. Finer-grained clastic rocks and diatomite mark several long-lived depocenters within the Gardnerville Basin. Thick diatomite and siltstone deposits in two locations record lakes that were probably at least in part coeval, but may not ever have been connected.

One of the two thick lacustrine intervals occurs in the northeastern part of the present exposure of the Sunrise Pass Formation (Figs. 2 and 4), east of and possibly ponded against a granitic intrabasin structural high. The section coarsens upward, indicating progressive filling of the basin; laterally equivalent rocks are coarser to the south, indicating a southern source. This diatomite is relatively well exposed in slopes and canyons to the north of the Sunrise Pass Road. A pronounced gravity low in this area (Fig. 6) confirms the presence of diatomite where it is covered by Quaternary deposits. The silty and diatomaceous section here is ~600 m thick. Age control near the top of this section includes both a $^{40}Ar/^{39}Ar$ date of 4.96 ± 0.02 Ma on sanidine (dated by Chris Henry, *reported in* Trexler et al., 2000) and detailed biostratigraphy and magnetostratigraphy that indicate the Hemphillian-Blancan boundary in this part of the section (Kelly, 1998; Lindsay et al., 2002). Although tephras from lower in this lacustrine section did not provide plausible correlations (Perkins, 2000, written commun.), estimates based on thickness and a typical sedimentation rate suggest that lacustrine deposition was active here from ca. 7 Ma to 5 Ma.

The other thick lacustrine deposits are in the northwest part of the Carson Valley, west of most of the intrabasin highs (Figs. 3 and 4). The basin-scale paleocurrent directions and facies

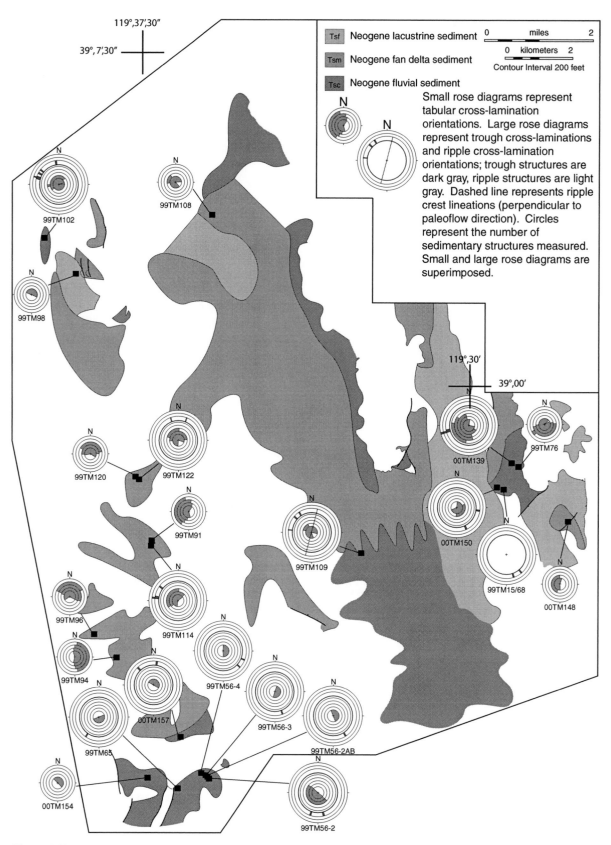

Figure 5. Simplified geologic map of the Sunrise Pass Formation from Muntean (2001), showing paleocurrent directions and locations of paleocurrent measurements. Paleocurrent data include tabular cross-laminations, trough cross-laminations, and ripple cross-laminations. See key for interpretation of rose diagrams.

181

CBA
Gravity
(mGal)

-152
-154
-155
-157
-159
-160
-162
-164
-165
-167
-168
-170
-172
-173
-177
-178
-180
-182
-183
-185
-187
-188
-190
-192
-193
-195
-197
-198
-200
-201
-203
-205
-206
-208
-210
-211
-213
-215
-216

Figure 6. Shaded complete Bouguer anomaly map of the Carson Valley and vicinity, reduced at 2.67 g/cc, with 1 mGal contours. The western half of Carson Valley corresponds with a local 25 mGal gravity low representing a 2-km-deep basin, ~8 km wide by 25 km long. National Geodetic Survey (NGS) 1999 gravity station locations are shown. Gravity coverage is not available of Lake Tahoe. See Figure 2 for location of model line.

distributions indicate that this was the main depocenter for the Neogene Gardnerville Basin. It received sediment derived from the intrabasin highs *and* sediment from farther to the east that was funneled between these highs along the central, west-trending drainage. Fossils from the lower part of this lacustrine section are middle early Blancan (Kelly, 1994, 1997), and an interbedded tephra has an age of 3.9 ± 0.2 Ma (dated by Chris Henry, *reported in* Trexler et al., 2000) (Fig. 4). Fossils in lacustrine deposits overlying a small basement high within the northwestern part of the basin are late Blancan (Kelly, 1994, 1997), indicating further uplift of basement during deposition.

Summary: Depositional Environment

The Neogene Gardnerville Basin records fluvial, deltaic, and lacustrine conditions in this area from at least 7 to 2 Ma. Along the eastern basin margin, the Neogene sedimentary rocks were deposited directly on Mesozoic plutonic and metamorphic basement; this indicates that substantial erosion occurred between the deposition of Oligocene tuffs throughout the region and the initiation of Neogene sedimentation here. The sedimentary rocks immediately overlying the basement are coarse-grained, and their compositions reflect those of the underlying basement. Drainage in the Gardnerville Basin was generally toward the west, but drainage was partitioned by syndepositional topography, with north-south sediment transport between structural highs to a central west-directed drainage system in the central part of the basin. Paleocurrent indicators are locally anomalous, showing flow away from the basement exposures, rather than directly toward the depocenters. All of these characteristics reflect the presence of crystalline basement at the ground surface (in structural highs) during deposition. The main drainage system is characterized by a relatively thin, dominantly fluvial section containing numerous unconformities. A local but long-lived depocenter persisted in the northeast part of the basin behind one of the structural highs from at least 7 Ma to 5 Ma. The main depocenter, in the northwest part of the basin, was active from at least 4 Ma and continues today to receive sediment from the Carson River.

SYNDEPOSITIONAL DEFORMATION

Fanning of dips within the Gardnerville Basin documents that faulting along the Carson Range frontal fault system to the west has been active throughout the basin history, i.e., for at least 7 m.y. In map view, bedding dips are generally steepest in the oldest rocks, i.e., along the eastern margin of the basin (Fig. 2). The same pattern can be seen in the vertical section provided by a short seismic-reflection line in the west-central part of the basin (Fig. 7). It shows three west-dipping reflections as deep as 1 km; the dip of the reflections increases with depth and age.

Crystalline rocks in structural highs within the Neogene Gardnerville Basin record syndepositional faulting. Basement-bounding faults of large displacement are typically north-striking, east-dipping normal faults. In the north part of the study area,

these faults form the east sides of a series of small, west-tilted, half-graben blocks (Fig. 2). In the southern part of the study area, faults bound two horst blocks. Displacement along both northern and southern faults decreases toward the central part of the study area. Structurally isolated subbasins between these basement highs also exhibit internal fanning of dips, recording syndepositional tilting at the scale of individual subbasins. The west-trending belt across the central part of the Gardnerville Basin has no large displacement faults or basement highs and is interpreted to be an accommodation zone between the sets of normal faults to the north and south.

The ages of the sedimentary rocks immediately overlying the basement highs document that intrabasin faulting occurred at several different times and was progressively younger westward across the basin. Isotopic dates of 4.96 Ma from sediment overlying the largest northern basement high indicate that it was exposed at the ground surface ca. 5 Ma. The adjacent basement high to the northwest was exposed ca. 4 Ma, and the farthest one to the northwest was exposed ca. 2 Ma, based on isotopic dates from the overlying sedimentary sections (Fig. 4). Age control is not as precise for the southern basement highs, but biostratigraphic dates consistently show a westward-younging pattern.

The seismic-reflection line from the western part of the basin shows abrupt changes in acoustic properties; these may record changes in depositional environment and may correspond to the tectonic uplift of the basement highs farther to the east. Each reflection appears to be the top of a sequence of layered sediments, probably lacustrine in origin. Above the reflection, the layering is discontinuous—interpreted to be coarser grained and less well bedded, so probably more fluvial—and gradually becomes more layered upward to the next reflection (Fig. 7). These vertical facies changes are interpreted as fan progradation and abandonment in response to source-area uplift or basin subsidence.

In summary, several lines of evidence point to active faulting both along the west flank and within the Neogene Gardnerville Basin during deposition. Sedimentation started by at least 7 Ma in a west-dipping half graben. The basin-bounding fault that tilts the bedding westward was active throughout the basin history, as shown by progressively steeper dips in the older sedimentary layers. In addition, there was active faulting within the basin at several times during its history. Crystalline basement rocks form several fault-bounded structural highs within the basin. These are most commonly west-tilted blocks bounded by east-dipping normal faults on the east, which are unconformably overlain by the sedimentary section on the west. The ages of the sedimentary rocks immediately overlying these basement highs are progressively younger westward. A short seismic-reflection profile in the west-central part of the basin—west of any basement highs at the surface—records abrupt changes in the character of the sedimentary rocks at three depths in the profile. These appear to represent a change from lacustrine to fluvial-deltaic deposition and may correspond to changes in internal drainage related to the uplift of basement highs within the basin to the east.

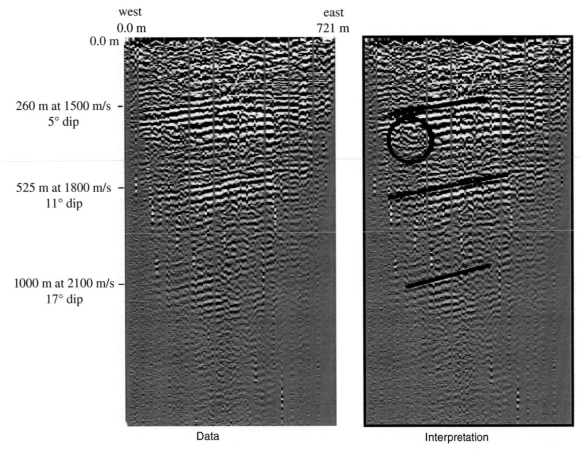

Figure 7. Reflection profile recorded in Carson Valley, interpreted at right (see Fig. 2 for location). This unmigrated stacked section is plotted at ~1:1 vertical exaggeration for the velocities found by normal-moveout (NMO) analysis. The interpretation (right) identifies three west-dipping reflections as deep as 1000 m. Their dips increase with depth and age. At the left are the depth, stacking velocity, and dip interpreted for each reflection. The circle identifies a discontinuity within continuously layered sediments, probably a fluvial structure like a channel wall, fan, or levee. See text for sedimentological interpretation. Loss of fold near the ends of this very short (720 m) profile allows interpretation only within its center.

FAULT LOCATION AND MAGNITUDE FROM GRAVITY MODELING

Geologic mapping of surface exposures does not unequivocally answer important questions about the age of initiation of the Gardnerville Basin, or age of initiation of faulting west of the Walker Lane at this latitude. The sedimentary rocks overlying basement at the eastern edge of the basin may represent progressive onlap of sediments at the basin margin, and therefore they are not necessarily the oldest rocks in the basin. Also, initiation of sedimentation is not necessarily related to initiation of faulting; rather, crosscutting relationships with sediments of known age are needed to bracket fault timing. Therefore, we did a gravity study to determine depth to basement throughout the basin. The gravity data provide new information about the location, geometry, and offset of faults, as well as about sediment thickness. The gravity modeling is constrained by rock composition information

from surface exposures and by the seismic line—an independent measure of depth to basement at one locality.

The regional complete Bouguer anomaly map (Fig. 6) and two-dimensional (2-D) model profile (Fig. 8) were developed from 869 gravity stations from the National Geodetic Survey 1999 U.S. land gravity database. These data provided useful map coverage over most of Carson Valley and east-west station spacings of 800 m. Another part of this study included collection and modeling of four new east-west gravity profiles covering ~70 km, with 219 stations at a typical station spacing of 300 m (Abbott et al., 2000).

Our new complete Bouguer anomaly map (Fig. 6) shows several features that directly correspond to the surface geology and others that contradict it. Geological and shallow seismic observations show a sediment-filled, west-dipping half graben at shallow structural levels. The gravity model is consistent with this at shallow structural levels, but the main gravity

184

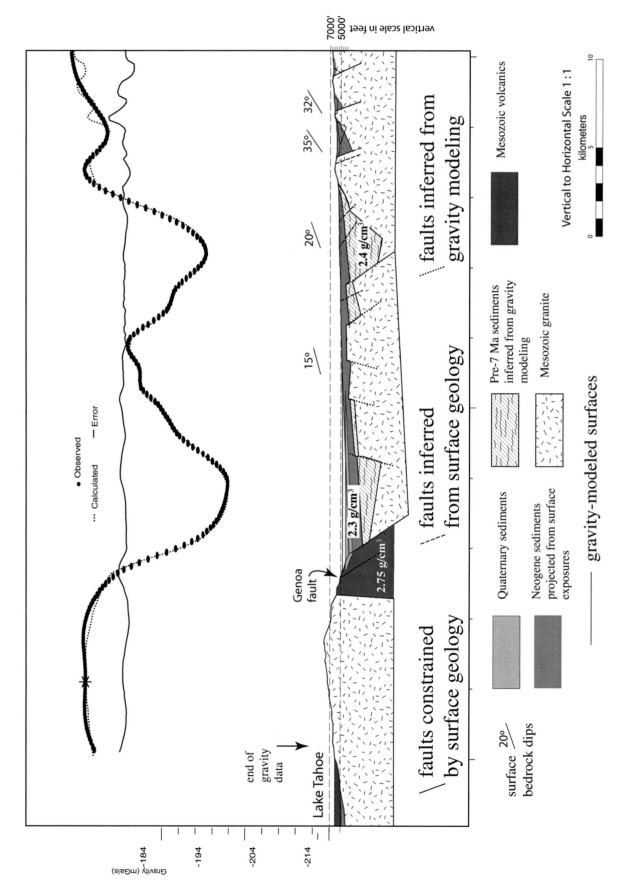

Figure 8. Gravity profile and geology model (location of profile shown on Fig. 2). The two-dimensional modeling assumed basement rocks as Mesozoic plutonic and metamorphic units averaging 2.75 g/cc, upper basin fill as Quaternary and Neogene sediments averaging 2.3 g/cc, and deeper basin fill as sediments averaging 2.4 g/cc.

expression is of a generally symmetric subbasin. This suggests a significant, early phase of basin formation that may not be related to the modern west-dipping half graben. The basement highs separate sediment-filled subbasins, as seen in the surface geology, but the gravity map identifies some subbasins as much bigger anomalies than others. The size of the gravity anomaly reflects both depth to basement and the nature of the basin fill; diatomaceous sediments exaggerate basin depth, and coarse-grained clastic sediments minimize it. Even allowing for this compositional influence, however, the Bouguer gravity map suggests faults and subbasins in addition to those recognized from the surface geology.

Integration of surface geology and gravity reveals that the geometry of the Carson Valley basin is more complex than previously recognized, and it modifies our understanding of the basin evolution based on the surface geology. Although our field mapping and seismic-reflection profile indicate that Quaternary, and probably late Neogene, basin growth has probably occurred exclusively along the *east*-dipping Sierran Range front fault system at the western edge of Carson Valley, the gravity shows an equally well-developed, but deeply buried (300–1000 m deep), *west*-dipping normal fault system centered 7 km to the east. The lack of surface morphologic features suggests this fault system is currently inactive. The fault system is expressed in the complete Bouguer anomaly gravity contour map (Fig. 6) as a 20 km zone of closely gathered north-trending contours that bisects Minden-Gardnerville. Density modeling for a gravity profile across the basin (Fig. 8) indicates this older north-south–trending normal fault system is 3 km wide and drops basement down to the west across at least two normal faults. The western of these faults adds an estimated downward displacement of 1 km, bringing the total modeled basin depth to 2 km. The density modeling also suggests a buried east-side-down normal fault east of the Genoa fault, accommodating significant additional displacement. The modeling assumed the deeper basin fill to be Cenozoic sedimentary rocks, but Cenozoic volcanic rocks (particularly Oligocene tuff or Miocene andesite) are also a possibility based on the regional geology. If volcanic rocks fill the lower part of the basin, the buried fault systems must have a still greater offset in order to compensate for the higher density of volcanic rocks than sedimentary rocks in the gravity model.

In addition, the gravity map shows an inverse relationship between offset along the Genoa fault and offset along the intrabasin normal faults to the east. The western gravity low—representing the deepest part of the basin—reaches a maximum at its midpoint, presumably also the location of maximum displacement on the Genoa fault. This occurs due west of the accommodation zone in the intrabasin fault system. The accommodation zone therefore occurs where offset on the Genoa fault is greatest. The intrabasin faults increase in offset north and south from there, accommodating more extension where the Genoa fault accommodates less (Fig. 9).

In summary, gravity studies provide evidence for a multistage faulting history for the Gardnerville Basin that started

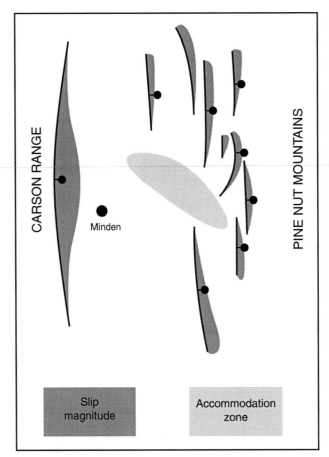

Figure 9. Schematic map of fault offset in the Gardnerville Basin, from Muntean (2001). The central accommodation zone occurs where offset on the Genoa fault is greatest. The intrabasin faults increase in offset to the north and south, accommodating more extension where the Genoa fault accommodates less.

well before 7 Ma. The older (pre–7 Ma) normal faults offset a Cenozoic sedimentary or volcanic valley fill that predates the Sunrise Pass Formation exposed at the ground surface today. The basin depth is as great as 2 km, and a previously unrecognized north-striking, west-dipping fault system with a vertical offset of at least 1 km forms the eastern boundary of this structural low. This fault system has no surface expression and appears to be no longer active. It might record middle Miocene east-west extension, as suggested by other workers in areas to the north, south, and east (e.g., Henry and Perkins, 2001; Stockli et al., 2003; Surpless et al., 2002; respectively). The gravity map also suggests a kinematic relationship between the intrabasin faults and the modern Genoa fault: displacement on the intrabasin faults increases northward and southward from a central accommodation zone, while displacement on the Genoa fault decreases away from this zone. All of these faults therefore appear to be acting together to accommodate extension across the Carson Valley.

SUMMARY AND DISCUSSION

Several new results from this study must be accounted for in any model for tectonic evolution of the northern Sierra Nevada–Basin and Range transition zone:

1. The age of the exposed Neogene sedimentary section does not overlap with type Coal Valley Formation, so a new name is required. We adopt the informal name "Sunrise Pass Formation" proposed in the detailed sedimentological and stratigraphic study by Muntean (2001), and we refer to the Neogene depositional basin as the Gardnerville Basin to avoid confusion with the modern topographic feature—the Carson Valley.

2. The Carson Range fault system along the west side of Carson Valley (including the Genoa fault) has been active since at least 7 Ma, as shown by fanning of dips in the Neogene sedimentary section. This is a longer period of activity than was projected based on trenching studies of the Genoa fault (Ramelli et al., 1999); if the known offset formed over this longer time period, the average slip rate is lower than estimated from the two-event trench record.

3. Gravity data show that a previously unrecognized, inactive, west-dipping normal fault system with a minimum of 1 km vertical offset underlies the western part of Carson Valley. It records older normal faulting than has been recognized previously.

4. Our new gravity data and modeling provide a better constraint on the total offset along the Sierran frontal fault system (Genoa fault plus buried fault to the east of it) than was previously available. We estimate a total of 3.2–3.7 km total vertical offset on this combined fault system.

5. Several intrabasin normal faults have acted in concert with the Genoa fault to accommodate extension. Significant normal offset occurred around 5 Ma along one intrabasin fault, and around 2 Ma along another one farther to the west.

6. There has been no significant *dextral* faulting within the basin throughout the ~7 m.y. history recorded in the Neogene section. Intracontinental transform motion related to the modern plate boundary and motion of the Sierra Nevada microplate do not appear to be accommodated here. In addition, there is no evidence here for westward propagation of dextral faulting.

The observed faulting history broadly supports the westward propagation of normal faulting into Sierran crystalline rocks west of the Walker Lane as suggested by Dilles and Gans (1995) and Surpless et al. (2002), but in detail, the normal faulting history is more complicated. Our estimated age for activity along the Carson Range frontal fault system here is younger than that reported in the Wassuk and Singatse Ranges to the east (e.g., Proffett, 1977; Dilles and Gans, 1995), and it is consistent with the estimate for fault-block tilting and footwall exhumation of the Pine Nut Mountains block, which underlies the sedimentary section (Surpless et al., 2002). It is older than the <5 Ma estimate for faulting in the

Tahoe Basin to the west (Surpless et al., 2002). Intrabasin normal faulting around 5 Ma and 2 Ma also stepped westward with time. However, the previously unrecognized inactive normal faults in the deeper part of the basin shown in the gravity data (Fig. 8) are of unknown age and may contradict the pattern.

In conclusion, this study documents the deformation style and history in a specific domain within the Sierra Nevada–Basin and Range transition zone—the domain west of the Walker Lane and east of the Sierra Nevada frontal fault system—and indicates the tectonic driving forces that may have influenced this intracontinental deformation. The combined depositional and deformational record in the Neogene Gardnerville basin provides a longer history than is available from geodesy or seismology and, thus, bridges an important gap between contemporary tectonics and the older geologic record. Like the Sierra Nevada to the west, this domain is underlain by crystalline rocks of the Sierra Nevada batholith. The deformation is characterized by normal faulting, which is consistent with westward propagation of Basin and Range–style extension into the Sierran crystalline rocks, possibly related to changing thermal and rheological conditions (e.g., Surpless et al., 2002, and references therein). Several intrabasin normal faults have acted in concert with the Genoa fault to accommodate extension; this deformation style could be an analog for other normal fault systems in the vicinity, e.g., the Lake Tahoe Basin, immediately west of our study area. There is no evidence in the study area for large-scale strike-slip faulting; the dextral component of relative motion of the Sierra Nevada microplate must be accommodated elsewhere. Although the timing for several deformational events in the history of the Gardnerville Basin is fairly well constrained, none of them directly corresponds to known changes in plate motions or plate-boundary configuration. The deformation appears to have continued unchanged across several tectonically significant time intervals, e.g., passage of the Mendocino triple junction and opening of the Gulf of California. This suggests it may not be valid to assume synchronous deformation over a broad area or to attribute poorly dated structures to specific tectonic events.

ACKNOWLEDGMENTS

We are grateful to many colleagues and students who have participated in various aspects of this project. Many thanks are due to Tom Kelly for mammal dates, Chris Henry for ^{40}Ar/^{39}Ar dates, Mike Perkins for tephra dates, John Louie's geophysics class (especially Jim Scott and Shane Smith) for recording the seismic data, Mike Tyler for drilling and blasting, Bently Nevada Farms for generous access to their property, and Robert Abbott and Matt Clark for collecting gravity data. All of these colleagues, as well as Larry Garside, Craig dePolo, and Jim Yount, have generously spent time with us in the field and have participated in many discussions about this project and related topics. This work was funded by National Science Foundation (NSF) grant EAR-9815122 to Cashman, Trexler, and Henry; NSF grant EAR-0001130 to Cashman, Trexler, and Louie; and EDMAP grant 99HQAG0075 to Faulds and Cashman.

REFERENCES CITED

Abbott, R.E., Cashman, P.H., Trexler, J.H., Jr., and Louie, J.N., 2000, Gravity survey constraints on the timing and style of extension of the Tertiary Gardnerville Basin, western Nevada [abs.]: Geological Society of America Abstracts with Programs, v. 32, no. 7, p. A-508.

Atwater, T.M., 1970, Implications of plate tectonics for the Cenozoic tectonic evolution of western North America: Geological Society of America Bulletin, v. 81, p. 3513–3536, doi: 10.1130/0016-7606(1970)81[3513:IOPTFT]2.0.CO;2.

Atwater, T., and Stock, J., 1998, Pacific–North American plate tectonics of the Neogene southwestern United States: An update, *in* Ernst, W.G., and Nelson, C.A., eds., Integrated Earth and Environmental Evolution of the Southwestern United States; the Clarence A. Hall, Jr. Volume: Columbia, Maryland, Bellwether Publishing, p. 393–420.

Axelrod, D.I., 1956, Mio-Pliocene floras from west-central Nevada: California University Publications in Geological Sciences, v. 33, 321 p.

Axelrod, D.I., 1958, The Pliocene Verdi flora of western Nevada: University of California Publications in Geological Sciences, v. 34, no. 2, p. 91–160.

Axelrod, D.I., 1962, A Pliocene *Sequoiadendron* forest from western Nevada: University of California Publications in Geological Sciences, v. 39, p. 195–267.

Bonham, H.F., and Papke, K.G., 1969, Geology and Mineral Deposits of Washoe and Storey Counties, Nevada: Nevada Bureau of Mines and Geology Bulletin 70, 140 p.

Cashman, P.H., and Fontaine, S.A., 2000, Strain partitioning in the northern Walker Lane, western Nevada and northeastern California: Tectonophysics, v. 326, p. 111–130, doi: 10.1016/S0040-1951(00)00149-9.

Cashman, P., and Trexler, J.H., Jr., 2004, The Neogene Verdi Basin records <2.0–2.5 Ma dextral faulting west of the Walker Lane near Reno, NV [abs.]: Geological Society of America Abstracts with Programs, v. 36, no. 4, p. 55.

Colgan, J.P., and Dumitru, T.A., 2003, Reconstructing late Miocene Basin and Range extension along the oldest part of the Yellowstone hot spot track in northwestern Nevada: Geological Society of America Abstracts with Programs, v. 35, no. 6, p. 347.

Dalrymple, G.B., 1964, Cenozoic chronology of the Sierra Nevada, California: University of California, Publications in Geological Sciences, v. 46, p. 1–41.

Deino, A.L., 1985, Stratigraphy, chemistry, K-Ar dating and paleomagnetism of the Nine Hill Tuff, California-Nevada, part I; Miocene-Oligocene tuffs of Seven Lakes Mountain, California-Nevada, part II [Ph.D. thesis]: Berkeley, University of California, 432 p.

dePolo, C.M., Ramelli, A.R., and Muntean, T., 2000, Geologic Map of the Gardnerville Quadrangle, Douglas County, Nevada: Nevada Bureau of Mines and Geology Open-File Report 00-9, scale 1:24,000.

Dickinson, W.R., Beard, S.L., Brakenridge, G.R., Erjavec, J.L., Ferguson, R.C., Inman, K.F., Knepp, R.A., Lindberg, F.A., and Ryberg, P.T., 1983, Provenance of North American Phanerozoic sandstones in relation to tectonic setting: Geological Society of America Bulletin, v. 94, p. 222–235, doi: 10.1130/0016-7606(1983)94<222:PONAPS>2.0.CO;2.

Dilles, J.H., and Gans, P.B., 1995, The chronology of Cenozoic volcanism and deformation in the Yerington area, western Basin and Range and Walker Lane: Geological Society of America Bulletin, v. 107, no. 4, p. 474–486, doi: 10.1130/0016-7606(1995)107<0474:TCOCVA>2.3.CO;2.

Dixon, T.H., Robaudo, S., Lee, J., and Reheis, M.C., 1995, Constraints on present-day Basin and Range deformation from space geodesy: Tectonics, v. 14, p. 755–772, doi: 10.1029/95TC00931.

Ducea, M.N., and Saleeby, J., 2003, Trace element enrichment signatures by slab derived carbonate fluids in the continental mantle wedge: An example from the Sierra Nevada, California: Geological Society of America Abstracts with Programs, v. 35, no. 6, p. 138.

Farmer, G.L., Glazner, A.F., and Manley, C.R., 2002, Did lithospheric delamination trigger late Cenozoic potassic volcanism in the southern Sierra Nevada, California?: Geological Society of America Bulletin, v. 114, no. 6, p. 754–768, doi: 10.1130/0016-7606(2002)114<0754:DLDTLC>2.0.CO;2.

Faulds, J.E., Henry, C.D., and Hinz, N.H., 2003, Kinematics and cumulative displacement across the northern Walker Lane: An incipient transform fault, northwest Nevada and northeast California: Geological Society of America Abstracts with Programs, v. 35, no. 6, p. 305.

Feldstein, S.N., and Lange, R.A., 1999, Pliocene potassic magmas from the Kings River region, Sierra Nevada, California: Evidence for melting of a subduction-modified mantle: Journal of Petrology, v. 40, no. 8, p. 1301–1320, doi: 10.1093/petrology/40.8.1301.

Firby, J.R., 1979, Paleogeography and biostratigraphic relationships of Late Tertiary lake beds of western Nevada, *in* Armentrout, J.M., Cole, M.R., and TerBest, H., Jr., eds., Cenozoic Paleogeography of the Western United States: Los Angeles, Pacific Section, Society of Economic Paleontologists and Mineralogists, p. 328.

Garside, L.J., and Rigby, J.G., 1998, Geologic Map of the McTarnahan Hill Quadrangle, Nevada: Nevada Bureau of Mines and Geology Open-File Map 99-5, scale 1:24,000.

Gilbert, C.M., and Reynolds, M.W., 1973, Character and chronology of basin development, western margin of the Basin and Range Province: Geological Society of America Bulletin, v. 84, p. 2489–2509, doi: 10.1130/0016-7606(1973)84<2489:CACOBD>2.0.CO;2.

Golia, R.T., and Stewart, J.H., 1984, Depositional environments and paleogeography of the Upper Miocene Wassuk Group, west-central Nevada, *in* Nilsen, T.H., ed., Fluvial Sedimentation and Related Tectonic Framework, Western North America: Sedimentary Geology, v. 38, p. 159–180.

Hardyman, R.F., and Oldow, J.S., 1991, Tertiary tectonic framework and Cenozoic history of the central Walker Lane, Nevada, *in* Raines, G.L., Lisle, R.E., Schafer, R.W., and Wilkinson, W.H., eds., Geology and Ore Deposits of the Great Basin: Reno, Nevada, Geological Society of Nevada, p. 279–301.

Henry, C.D., and Perkins, M.E., 2001, Sierra Nevada–Basin and Range transition near Reno, Nevada: Two stage development at 12 Ma and 3 Ma: Geology, v. 29, p. 719–722, doi: 10.1130/0091-7613(2001)029<0719:SNBART>2.0.CO;2.

Henry, C.D., Faulds, J.E., Garside, L.J., and Hinz, N.H., 2003, Tectonic implications of ash-flow tuffs and paleovalleys in the western US: Geological Society of America Abstracts with Programs, v. 35, no. 6, p. 138.

Hinz, N.H., 2004, Tertiary volcanic stratigraphy of the Diamond and Fort Sage Mountains, northeastern California and western Nevada—Implications for development of the northern Walker Lane [M.S. thesis]: Reno, University of Nevada, 110 p.

Hinz, N.H., Faulds, J.E., and Henry, C.D., 2003, Dextral displacement on the Honey Lake fault zone, northern Walker Lane, northeast California and westernmost Nevada: Geological Society of America Abstracts with Programs, v. 35, no. 6, p. 138.

House, M.A., Wernicke, B.P., and Farley, K.A., 1998, Dating topography of the Sierra Nevada, California, using apatite (U-Th)/He ages: Nature, v. 396, no. 5, p. 66–69, doi: 10.1038/23926.

Kelly, T.S., 1994, Two Pliocene (Blancan) vertebrate faunas from Douglas County, Nevada: PaleoBios, v. 16, no. 1, p. 1–23.

Kelly, T.S., 1997, Additional late Cenozoic (latest Hemphillian to earliest Irvingtonian) mammals from Douglas County, Nevada: PaleoBios, v. 18, no. 1, p. 1–31.

Kelly, T.S., 1998, New Miocene mammalian faunas from west central Nevada: Journal of Paleontology, v. 72, no. 1, p. 137–149.

King, C., 1878, The Comstock lode, *in* Hague, J.D., ed., Mining Industry, U.S. Geological Exploration of the 40th Parallel: Washington, D.C., U.S. Geological Survey, p. 10–91.

Lerch, D.W., McWilliams, M.O., Miller, E.L., and Colgan, J.P., 2004, Structure and magmatic evolution of the northern Blackrock Range, Nevada: Preparation for a wide-angle refraction/reflection survey [abs.]: Geological Society of America Abstracts with Programs, v. 36, no. 4, p. 37.

Lindsay, E., Mou, Y., Downs, W., Pederson, J., Kelly, T.S., Henry, C., and Trexler, J., 2002, Recognition of the Hemphillian/Blancan boundary in Nevada: Journal of Vertebrate Paleontology, v. 22, no. 2, p. 429–442, doi: 10.1671/0272-4634(2002)022[0429:ROTHBB]2.0.CO;2.

Manley, C.R., Glazner, A.F., and Farmer, G.L., 2000, Timing of volcanism in the Sierra Nevada of California: Evidence for Pliocene delamination of the batholithic root?: Geology, v. 28, no. 9, p. 811–814, doi: 10.1130/0091-7613(2000)28<811:TOVITS>2.0.CO;2.

Mass, K.B., 2005, The Neogene Boca basin, northeastern California: Stratigraphy, structure and implications for regional tectonics [M.S. thesis]: Reno, University of Nevada, 195 p.

Mass, K.B., Cashman, P.H., Trexler, J.H., Jr., Park, H., and Perkins, M.E., 2003, Deformation history in Neogene sediments of Honey Lake Basin, northern Walker Lane, Lassen County, California [abs.]: Geological Society of America Abstracts with Programs, v. 36, no. 6, p. 26.

Mass, K.B., Cashman, P.H., and Trexler, J.H., Jr., 2009, this volume, Stratigraphy and structure of the Neogene Boca Basin, northeastern California: Implications for late Cenozoic tectonic evolution of the northern Sierra Nevada, *in* Oldow, J.S., and Cashman, P.H., eds., Late Cenozoic Structure and Evolution of the Great Basin–Sierra Nevada Transition: Geological Society of America Special Paper 447, doi: 10.1130/2009.2447(09).

Maurer, D.K., 1984a, Gravity Survey and Depth to Bedrock in Carson Valley, Nevada-California: Denver, U.S. Geological Survey, 6 p.

Maurer, D.K., 1984b, Hydrogeology of Carson Valley, Nevada: Western Geological Excursions, *in* Lintz, Joseph, Jr., ed., Western geological excursions [in conjunction with 1984 annual meetings of Geological Society of America and affiliated societies]: University of Nevada, Reno, Mackay School of Mines, v. 3, p. 127–129.

Miller, E.L., Dumitru, T.A., Stockli, D.F., and Surpless, B.E., 2000, Cenozoic faulting in the northern Basin and Range Province, western U.S.: Insights from fission track studies [abs.]: Reno, Nevada, Geological Society of Nevada, Symposium 2000 Program with Abstracts, p. 63–64.

Moore, J.G., 1969, Geology and Mineral Deposits of Lyon, Douglas, and Ormsby Counties, Nevada: Reno, Nevada Bureau of Mines and Geology, 45 p.

Muntean, T.W., 2001, Evolution and stratigraphy of the Neogene Sunrise Pass Formation of the Gardnerville sedimentary basin, Douglas County, Nevada [M.S. thesis]: Reno, University of Nevada, 250 p.

Noble, D.C., 1962, Mesozoic geology of the southern Pine Nut Range, Douglas County, Nevada [Ph.D. thesis]: Stanford University, 200 p.

Oldow, J.S., 1992, Late Cenozoic displacement partitioning in the northern Great Basin, *in* Craig, S.D., ed., Walker Lane Symposium; Structure, Tectonics, and Mineralization of the Walker Lane: Reno, Nevada, Geological Society of Nevada, p. 17–52.

Oldow, J.S., Aiken, C.L.V., Hare, J.L., Ferguson, J.F., and Hardyman, R.F., 2001, Active displacement transfer and differential block motion within the central Walker Lane, western Great Basin: Geology, v. 29, no. 1, p. 19–22, doi: 10.1130/0091-7613(2001)029<0019:ADTADB>2.0.CO;2.

Oskin, M., and Stock, J., 2003, Pacific–North America plate motion and opening of the Upper Delfín Basin, northern Gulf of California, Mexico: Geological Society of America Bulletin, v. 115, no. 10, p. 1173–1190, doi: 10.1130/B25154.1.

Park, H., 2004, Honey Lake Basin: Depositional settings and tectonics [M.S. thesis]: Reno, University of Nevada, 259 p.

Park, H., Trexler, J.H., Jr., Cashman, P., and Mass, K.B., 2004, Local initiation of Walker Lane tectonism prior to 3.6 Ma recorded in Neogene sediments at Honey Lake Basin, northeastern California [abs.]: Geological Society of America Abstracts with Programs, v. 36, no. 4, p. 52.

Pease, R.C., 1979, Scarp degradation and fault history south of Carson City, Nevada [M.S. thesis]: Reno, University of Nevada, 90 p.

Pease, R.C., 1980, Geologic Map of the Genoa Quadrangle, Nevada: Nevada Bureau of Mines and Geology Map 1Cg, scale 1:24,000.

Perkins, M.E., Brown, F.H., Nash, W.P., McIntosh, W., and Williams, S.K., 1998, Sequence, age, and source of silicic fallout tuffs in middle to late Miocene basins of the northern Basin and Range Province: Geological Society of America Bulletin, v. 110, no. 3, p. 344–360, doi: 10.1130/0016-7606(1998)110<0344:SAASOS>2.3.CO;2.

Proffett, J.M., 1977, Cenozoic geology of the Yerington district, Nevada, and implications for the nature and origin of Basin and Range faulting: Geological Society of America Bulletin, v. 88, p. 247–266, doi: 10.1130/0016-7606(1977)88<247:CGOTYD>2.0.CO;2.

Ramelli, A.R., Bell, J.W., dePolo, C.M., and Yount, J.C., 1999, Large-magnitude, late Holocene earthquakes on the Genoa fault, west-central Nevada and eastern California: Bulletin of the Seismological Society of America, v. 89, no. 6, p. 1458–1472.

Ramelli, A.R., Yount, J.C., John, D.A., and Garside, L.J., 2003, Preliminary Geologic Map of the Minden Quadrangle: Nevada Bureau of Mines and Geology Open-File Report 03–13, scale 1:24,000.

Saucedo, G.J., and Wagner, D.L., 1992, Geologic Map of the Chico Quadrangle: California Division of Mines and Geology Regional Map Series 7A, scale 1:250,000.

Schorn, H.E., 1994, Floristic and vegetation changes in west-central Nevada during the later Neogene: An empirical overview [abs.]: Geological Society of America Abstracts with Programs, v. 26, p. A-521.

Schwartz, K.M., 2001, Evolution of the Middle to Late Miocene Chalk Hills Basin in the Basin and Range–Sierra Nevada transition, Virginia Range, Western Nevada [M.S. thesis]: Reno, University of Nevada, 160 p.

Stewart, J.H., 1988, Tectonics of the Walker Lane belt, western Great Basin—Mesozoic and Cenozoic deformation in a zone of shear, *in* Ernst, W.G., ed., Metamorphism and Crustal Evolution of the Western United States: Englewood Cliffs, New Jersey, Prentice Hall, p. 683–713.

Stewart, J.H., 1992, Paleogeography and tectonic setting of Miocene continental strata in the northern part of the Walker Lane Belt, *in* Craig, S.D., ed., Structure, Tectonics and Mineralization of the Walker Lane: Reno, Geological Society of Nevada, Walker Lane Symposium Proceedings, p. 53–61.

Stewart, J.H., 1999, Geologic Map of the Carson City 30 × 60 Minute Quadrangle, Nevada: Nevada Bureau of Mines and Geology Map 118, scale 1:100,000.

Stewart, J.H., and Noble, D.C., 1979, Preliminary Geologic Map of the Mount Siegel Quadrangle, Nevada-California: U.S. Geological Survey Open-File Report 79-225, 57 p.

Stockli, D.F., Surpless, B.E., and Dumitru, T.A., 2002, Thermochronological constraints on the timing and magnitude of Miocene and Pliocene extension in the central Wassuk Range, western Nevada: Tectonics, v. 21, no. 4, p. 10-1–10-19.

Stockli, D.F., Dumitru, T.A., McWilliams, M.O., and Farley, K.A., 2003, Cenozoic tectonic evolution of the White Mountains, California and Nevada: Geological Society of America Bulletin, v. 115, no. 7, p. 788–816, doi: 10.1130/0016-7606(2003)115<0788:CTEOTW>2.0.CO;2.

Surpless, B.E., 1999, Tectonic evolution of the Sierra Nevada–Basin and Range transition zone: A study of crustal evolution in extensional provinces [Ph.D. thesis]: Palo Alto, California, Stanford University, 225 p.

Surpless, B.E., Stockli, D.F., Dumitru, T.A., and Miller, E.L., 2002, Two-phase westward encroachment of Basin and Range extension into the northern Sierra Nevada: Tectonics, v. 21, no. 1, p. 2-1–2-13.

Thatcher, W., Foulger, G.R., Julian, B.R., Svarc, J., Quilty, E., and Bawden, G.W., 1999, Present day deformation across the Basin and Range Province, western United States: Science, v. 283, p. 1714–1718, doi: 10.1126/science.283.5408.1714.

Thompson, G.A., and White, D.E., 1964, Regional Geology of the Steamboat Springs Area, Washoe County, Nevada: U.S. Geological Survey Professional Paper 458-A, 52 p.

Trexler, J.H., Jr., Cashman, P.H., Henry, C.D., Muntean, T.W., Schwartz, K., TenBrink, A., Faulds, J.E., Perkins, M., and Kelly, T.S., 2000, Neogene basins in western Nevada document tectonic history of the Sierra Nevada–Basin and Range transition zone for the last 12 Ma, *in* Lageson, D.R., Peters, S.G., and Lahren, M.M., eds., Great Basin and Sierra Nevada: Boulder, Colorado, Geological Society of America Field Guide 2, p. 97–116.

Trexler, J.H., Jr., Park, H.M., Cashman, P.H., and Mass, K.B., 2009, this volume, Late Neogene basin history at Honey Lake, northeastern California: Implications for regional tectonics at 3 to 4 Ma, *in* Oldow, J.S., and Cashman, P.H., eds., Late Cenozoic Structure and Evolution of the Great Basin–Sierra Nevada Transition: Geological Society of America Special Paper 447, doi: 10.1130/2009.2447(06).

Unruh, J., Humphrey, J., and Barron, A., 2003, Transtensional model for the Sierra Nevada frontal fault system, eastern California: Geology, v. 31, p. 327–330, doi: 10.1130/0091-7613(2003)031<0327:TMFTSN>2.0.CO;2.

Wernicke, B.P., and Snow, J.N., 1998, Cenozoic tectonism in the central Basin and Range: Motion of the Sierran–Great Valley block: International Geology Review, v. 40, p. 403–410.

Whitehill, C.S., Miller, E.L., Colgan, J.P., Dumitru, T.A., Lerch, D.W., and McWilliams, M.O., 2004, Extent, style, and age of Basin and Range faulting east of Pyramid Lake: Geological Society of America Abstracts with Programs, v. 36, no. 4, p. 37.

MANUSCRIPT ACCEPTED BY THE SOCIETY 21 JULY 2008

The Geological Society of America
Special Paper 447
2009

Flattening strain during coordinated slip on a curved fault array, Rhodes Salt Marsh extensional basin, central Walker Lane, west-central Nevada

Luigi Ferranti*

Dipartimento di Scienze della Terra, Università Federico II, Napoli, 80138, Italy

John S. Oldow

Department of Geosciences, University of Texas at Dallas, Richardson, Texas 75080-3021, USA

John W. Geissman

Department of Earth and Planetary Sciences, MSC03 2040, University of New Mexico, Albuquerque, New Mexico 87131-0001, USA

Mark M. Nell

Department of Geological Sciences, University of Idaho, Moscow, Idaho 83844-3022, USA

ABSTRACT

Integrated mapping, fault-kinematic, paleomagnetic, and gravity analyses around the Rhodes Salt Marsh extensional basin, located within the east-west–trending Mina deflection of the central Walker Lane, reveal that from 8.0 to 9.0 km of late Cenozoic displacement was accommodated on a curved array of faults. The dominant slip on the faults systematically varies from left-oblique, to normal, and to right-oblique as fault strike changes from east, to north-northeast, and to north-northwest, respectively. Kinematic consistency of fault slickenline rakes, preservation of displacement budget, and paleomagnetic data from a pluton and volcanic rocks in the fault-system hanging wall indicate that the curved fault geometry is primary and not due to superposition of two fault systems nor to later vertical-axis rotation. Large-magnitude extension was localized at the apex of the curved faults and resulted in the formation of an ~3.0-km-deep prismatic basin beneath Rhodes Salt Marsh. The offset geologic structures and geophysical basin models indicate that hanging-wall displacement diverged around the curved fault array and resulted in finite flattening, with primary and secondary extensional axes oriented west-northwest and north-northeast, respectively. Fault-slip inversion yields two directions of extension consistent with the finite strain axes, and slickenlines with mutually crosscutting relations indicate formation during incremental flattening. Although broadly contemporaneous, extension parallel to the primary and secondary extension axes alternated at periods ranging from months to as much as several hundred thousand years. Large through-going structures sustained extension

*Corresponding author e-mail: lferrant@unina.it.

Ferranti, L., Oldow, J.S., Geissman, J.W., and Nell, M.M., 2009, Flattening strain during coordinated slip on a curved fault array, Rhodes Salt Marsh extensional basin, central Walker Lane, west-central Nevada, *in* Oldow, J.S., and Cashman, P.H., eds., Late Cenozoic Structure and Evolution of the Great Basin–Sierra Nevada Transition: Geological Society of America Special Paper 447, p. 189–213, doi: 10.1130/2009.2447(11). For permission to copy, contact editing@geosociety.org.

directions recorded geodetically and seismologically through multiple seismic cycles. In contrast, the alternation between primary and secondary extension directions recorded by a strainmeter suggests that, on small structures contained within fault-bounded blocks, the two extension directions alternated over time scales of as little as 2 yr.

Keywords: curved fault system, prismatic basin, finite and incremental flattening, central Walker Lane, Nevada.

INTRODUCTION

Along intracontinental fault systems of the western U.S. Cordillera, the occurrence of slip vectors with seemingly incompatible orientations on a single fault surface or subparallel faults is commonly attributed to: (1) the superposition of temporally distinct tectonic episodes with differently oriented strain fields (Angelier et al., 1985; Zoback, 1989; Bellier and Zoback, 1995); (2) the progressive rotation of crustal blocks and their bounding faults (Hudson and Geissman, 1991; Wesnousky, 2005); or (3) slip partitioning between adjacent faults with variable relative strength (Mount and Suppe, 1987; Oldow, 1992; Wesnousky and Jones, 1994). Common to these models is the a priori assumption of plane-strain conditions typical of many fault-kinematic analyses (e.g., Angelier, 1984). However, where geologic and current deformation state is demonstrably non–plane strain, as in the central Walker Lane (Oldow et al., 2001; Oldow, 2003), a more complex kinematic setting is expected both at the regional and local scale.

Constrictional strain in the Walker Lane and northern part of the Eastern California shear zone (Oldow, 2003; Murphy et al., this volume) is accommodated on a system of faults within a right-transtensional boundary zone between west-northwest extension in the Basin and Range and northwest-directed motion of the Sierra Nevada (Fig. 1). Within the boundary zone, displacement is carried on northwest- and north-northwest–striking faults of the Eastern California shear zone and transferred to the east via an east-west–striking fault array to northwest-striking transcurrent faults of the central Walker Lane. This deflection in fault strike is recorded in the pattern of seismicity (Ryall and Priestley, 1975) and is known as the Mina deflection (Fig. 1).

Several models have addressed the geometry and kinematics of the Mina deflection. Stewart (1988) proposed that east-west faults truncate northwest-striking faults and formed in discrete tectonic blocks within the Walker Lane. Oldow (1992) proposed that east-west– and northwest-striking faults are parts of a curved array that reflects a primary geometry inherited from preexisting crustal structures related to a structural stepover linking the

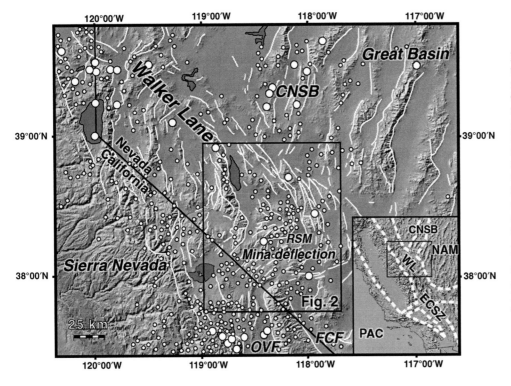

Figure 1. Digital topography of central western Great Basin, showing major Tertiary faults and seismicity. Gray filled areas are lakes. Epicenters of 1860 to 1999 earthquakes (sources: U.S. Geological Survey [USGS]–National Earthquake Information Center [NEIC]; Rogers et al., 1991) are shown by small dots (2.5 < M < 6.0) and large dots (M > 6.0). Inset shows location of the major intracontinental fault system of the western U.S. Cordillera in the context of Pacific (PAC) and North America (NAM) plate interaction. WL—Walker Lane; ECSZ—Eastern California shear zone; CNSB—central Nevada seismic belt; FCF—North Furnace Creek fault; OVF—Owens Valley fault; RSM—Rhodes Salt Marsh.

northern Eastern California shear zone and the central Walker Lane. Wesnousky (2005) viewed the geometry as the product of progressive deformation in which the east-west–striking faults form the boundaries of discrete tectonic blocks that rotated 20°–30° clockwise within a shear couple bounded by through-going northwest-striking transcurrent faults.

In this paper, we evaluate the various models for the evolution of the Mina deflection by focusing on the geometry and kinematic history of a fault system bounding a basin, the Rhodes Salt Marsh, located at the apex of east-west– and northwest-striking faults (Fig. 1). We use geologic mapping and fault-kinematic analysis to place constraints on the finite displacement budget and the slip on left-lateral, right-lateral, and dip-slip faults observed in and around the Rhodes Salt Marsh of west-central Nevada. Paleomagnetic data from volcanic and plutonic rocks provide an independent assessment of vertical-axis rotation of tectonic blocks bounded by the fault array, and gravity modeling supplies estimates of basin geometry and depth.

Our investigation documents that east-west– and northwest–striking faults are part of a kinematically coordinated system of curved faults. The slip on faults records left-oblique, dip-slip, and right-oblique displacements as strikes around the array change from east-west to north-northeast, and northwest, respectively. The primary geometry of the fault array is preserved, and it does not show appreciable rotation during progressive deformation. The fault geometry created a space problem that was accommodated by two directions of extension, which resulted in finite flattening during the evolution of a deep basin located at the apex of the curved array.

REGIONAL SETTING

During the late Cenozoic, large-magnitude right-oblique shear and extension characterized displacement within the Walker Lane in west-central Nevada (Hardyman and Oldow, 1991; Oldow, 1992), and, based on seismology and contemporary global positioning system (GPS) velocities, the structures record a history of transtensional deformation (Unruh et al., 2003; Oldow, 2003). The region is underlain by a complex system of faults that form a boundary zone between the Sierra Nevada and central Great Basin (Fig. 1) and accommodate nearly 25% of North America and Pacific plate relative motion (Argus and Gordon, 1991; Sauber et al., 1994; Dixon et al., 2000; Miller et al., 2001; Oldow et al., 2001; Oldow, 2003; Bennett et al., 2003).

Faults within the northern Eastern California shear zone, Mina deflection, and central Walker Lane are seismically active (Fig. 2) and record a pattern of displacement consistent with geologically determined offsets (Rogers et al., 1991; Oldow, 1992). Earthquakes record right-lateral and left-lateral first motions on northwest- and east-west–striking faults, respectively (Doser, 1988; Rogers et al., 1991), and correspond to the pattern of surface ruptures formed during Holocene and historical earthquakes (Gianella and Callaghan, 1934; Callaghan and Gianella, 1935; dePolo et al., 1993; Beanland and Clark, 1994; Bell et al., 1999; Caskey and Wesnousky, 1997).

In the central Walker Lane, the east-west–striking faults of the Mina deflection are part of a mid-Pliocene to Holocene structural stepover (Oldow, 1992, 2003; Oldow et al., 2008) that links structures in the Eastern California shear zone (Dokka and Travis, 1990) to faults in the central Walker Lane and Central Nevada seismic belt (Wallace, 1984; Oldow, 1992). Northwest- to north-northwest–trending faults in the Eastern California shear zone pass into a 50-km-wide belt of east-northeast– to east-west–striking faults that stretches 100 km east and merges with northwest-striking transcurrent faults of the central Walker Lane. At their transitions, northerly and easterly striking faults are linked by curved fault segments that bound deep extensional basins (Oldow, 1992; Reheis and Sawyer, 1997; Stockli et al., 2003). The basins have a prismatic geometry and gravity signatures (Saltus, 1988) that indicate substantially greater depths than elongate basins elsewhere in the central Walker Lane and within the adjacent central Great Basin.

GEOMETRY AND KINEMATICS OF THE RHODES SALT MARSH BASIN

The Rhodes Salt Marsh basin is located near the center of the Mina deflection (Fig. 1) at the junction between the east-northeast–striking faults to the southwest and northwest-striking faults of the central Walker Lane belt to the north (Fig. 3). The surface expression of the basin is a triangular depression about 12 km wide that contains the Rhodes Salt Marsh, which is localized at the lowest elevation of about 1300 m. The topographic low is ringed by highlands consisting of the Candelaria Hills to the south, the Excelsior Mountains to the northwest, and the Pilot Mountains to the east. The highlands achieve a maximum elevation of nearly 2800 m (Pilot Mountains) and are separated by 1500- and 2100-m-high topographic saddles that are spatially coincident with the southeast and southwest apexes of the triangular depression, respectively. To the north, the Rhodes Salt Marsh basin is separated from the north-northwest–trending Soda Springs Valley by a topographic sill reaching an elevation of 1400 m. Soda Springs Valley is about 5 km wide and 30 km long and separates the Pilot Mountains and Gabbs Valley Range on the east from the Garfield Hills to the west.

The highlands surrounding Rhodes Salt Marsh and Soda Springs Valley expose a Cenozoic section of Oligocene to Lower Miocene ash-flow tuff, Middle Miocene andesite, and Middle to Upper Miocene sedimentary rocks. The ash-flow tuffs are composed of several cooling units and range in thickness from 1400 to 200 m, with the greatest accumulation in the Candelaria Hills (Speed and Cogbill, 1979a), and up to 500 m along the northern flank of the Pilot Mountains (Oldow and Dockery, 1993); they are typically thin or absent in the highlands of the Excelsior Mountains, Garfield Hills, and Pilot Mountains. The tuff is locally overlain unconformably by mid-Miocene andesite and andesite lahar that range in thickness from 200 to 300 m (Speed and Cogbill, 1979a; Garside, 1979) to up to 800 m (Oldow and Dockery, 1993). The andesite unit is exposed in all of the ranges,

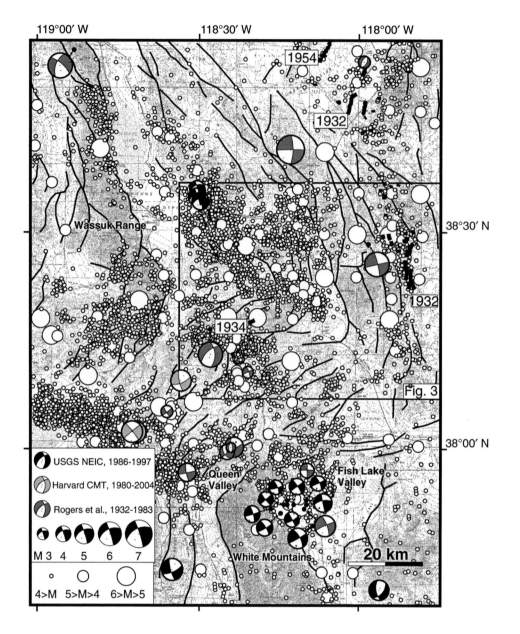

Figure 2. Seismicity map of the central Walker Lane, including major late Tertiary faults. Focal mechanisms (2.5 < M < 7.2) are after Rogers et al. (1991), U.S. Geological Survey (USGS)–National Earthquake Information Center (NEIC), and Harvard-Centroid Moment Tensor (CMT) solution. Epicenters are after dePolo and dePolo (1999). Thick discontinuous solid lines are surface breaks produced during the 1932, 1934, and 1954 earthquakes after Bell et al. (1999), Callaghan and Gianella (1935), and Caskey and Wesnousky (1997), respectively.

and it rests upon a low-relief erosional contact with underlying Cenozoic and pre-Tertiary rocks. The andesite is overlain by a poorly exposed succession of Middle to Upper Miocene sedimentary rocks that have a minimum thickness of 30 m but probably range to several hundred meters thick.

The Cenozoic rocks rest unconformably on highly deformed Paleozoic and Mesozoic rocks and postkinematic granitoid plutons. The Paleozoic rocks are volcanic and sedimentary units of the upper plate of the late Paleozoic to early Mesozoic Golconda thrust (Speed, 1977), and they form the basement to deposition of Triassic to Jurassic carbonate and clastic rocks exposed in the upper and lower plates of the late Mesozoic Luning thrust (Ferguson and Muller, 1949; Oldow, 1981). The Luning thrust is a regionally extensive structure that marks the southern mar-

gin of allochthonous rocks of the Mesozoic marine province of the northwestern Great Basin (Oldow, 1983, 1984), and it can be traced discontinuously across the Pilot Mountains, eastern Garfield Hills, across the Excelsior Mountains, and into the low hills west of the Candelaria Hills (Fig. 4).

Fault Geometry

Bedrock exposures in the bounding highlands consist of Paleozoic and Mesozoic strata, Mesozoic plutonic rocks, and Cenozoic volcanic and sedimentary units (Fig. 4) that are separated from the basin fill by a complex system of faults. The faults define a curvilinear array of structures that emanate from a single east-west fault zone in the southeast, where they bifurcate into

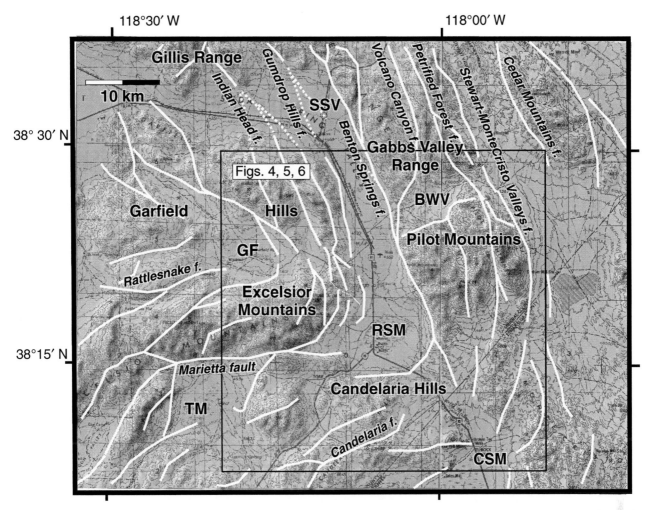

Figure 3. Fault map of the Rhodes Salt Marsh basin and environs, showing schematic traces and major fault systems, basins, and mountain ranges. BWV—Bettles Wells Valley; CSM—Columbus Salt Marsh; GF—Garfield Flat; RSM—Rhodes Salt Marsh; SSV—Soda Spring Valley; TM—Teels Marsh. f.—fault.

western and eastern fault systems forming the borders of the Rhodes Salt Marsh as they curve to the northwest and extend along the margins of Soda Springs Valley (Fig. 4). The eastern border is relatively simple; it consists of a curved fault zone bordering the northern Candelaria Hills and Pilot Mountains. In the northern Pilot Mountains and southern Gabbs Valley Range, the range-front fault bifurcates. One strand enters the mountain range and connects with northwest-striking faults along the eastern flank of the mountains (Fig. 3). The western boundary fault curves along the eastern flank of the Excelsior Mountains where it trifurcates into strands that form the range-front and two northnorthwest fault blocks within the eastern Garfield Hills (Fig. 4).

From north to south, the Indian Head, Gumdrop Hills, and Benton Springs faults project across the east-west–trending segment of Soda Springs Valley in the north and change strike from northwest to north-northwest (Fig. 3). The Benton Springs fault marks the range front of the Gabbs Valley Range; it extends south across the mouth of the east-west–trending Bettles Well Valley

and becomes the range-front fault of the Pilot Mountains. In the northern Pilot Mountains, the Volcano Canyon fault merges to the south with the Benton Springs fault and serves as a displacement relay with the northwest-striking Petrified Forest fault along the eastern part of the Gabbs Valley Range. The Gumdrop Hills fault marks the range front of the eastern Garfield Hills and extends south to the topographic sill separating Soda Springs Valley and the Rhodes Salt Marsh basin. The Indian Head fault bifurcates to the south into western and eastern strands that follow northnorthwest–trending valleys within the eastern Garfield Hills.

Along the eastern margin of Rhodes Salt Marsh basin, the north-northwest–striking Benton Springs fault (Fig. 4) progressively changes to north-south as the southeast apex of the basin is approached from the north. At the southeast corner of the basin, the Benton Springs fault bifurcates in an eastern strand, named the Jeep Trail fault, and a western strand, named the Long Canyon fault. The Long Canyon fault extends southwest and is marked by a scarp up to 1.0 m high in Quaternary basin deposits; it can be

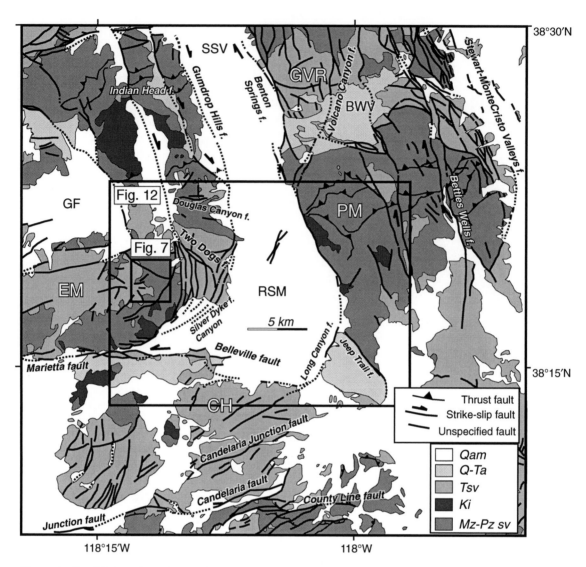

Figure 4. Simplified geological map of the Rhodes Salt Marsh basin, Excelsior Mountains, and Pilot Mountains (compiled from various sources). Rock units: Mz-Pz sv—Paleozoic to Mesozoic sedimentary and volcanic rocks of the Luning-Fencemaker thrust belt and foredeep; Ki—Cretaceous intrusive rocks; Tsv—Tertiary sedimentary and volcanic rocks; Q-Ta—Tertiary-Quaternary alluvial sediments; Qam—Quaternary alluvial and marsh sediments. Places: PM—Pilot Mountains; EM—Excelsior Mountains; GVR—Gabbs Valley Range; CH—Candelaria Hills; f.—fault. Other abbreviations are as in Figure 3.

traced southwest to the southern margin of the basin along the Candelaria Hills. The Jeep Trail fault projects to the southeast at the contact between Paleozoic strata and Cenozoic sedimentary rocks overlain by Quaternary alluvium for over 5 km, where the structure swings to the southwest for about 1 km before it is lost beneath Quaternary cover.

Faults marking the boundary between the southern margin of the Rhodes Salt Marsh basin and Candelaria Hills strike east-west and are exposed as two strands of the Belleville fault system (Fig. 4). In the east, the bounding fault is discontinuously exposed along the contact between Cenozoic volcanic rocks and Quaternary alluvium for a distance of 8 km, where the fault may enter

the Candelaria Hills along an east-west trending canyon. Farther to the west, the western strand of the Belleville fault system is located 2 km north of the mapped expression of the eastern strand of the fault system. The western part of the Belleville fault system is continuously exposed for 13 km along the contact between Cenozoic volcanic rocks and the basin fill, and it exhibits a complex pattern of anastomosing branches up to 500 m wide. The western Belleville fault system extends west and merges with the east-west–striking Marietta fault, which makes up the southern range front of the Excelsior Mountains.

The northwestern margin of Rhodes Salt Marsh basin is composed of an array of at least seven northeast-striking faults.

The more western strand, the Silver Dyke Canyon fault, has the greatest displacement and forms the eastern range-front structure for the Excelsior Mountains. Farther north, six strands of the fault array emerge from beneath Quaternary cover where they successively step bedrock down to the east toward the basin. Here, the faults change strike to north-south and are linked to the western strand of the Indian Head fault by the northwest-striking Two Dogs fault zone (Fig. 4).

Subsurface Basin Morphology

To gain insight into the subsurface geometry of the basin, 149 gravity measurements were collected in the Rhodes Salt Marsh and combined with gravity data downloaded from the Pan-American Center for Earth Sciences (PACES) gravity database for a total of nearly 1400 stations. Figure 5A depicts the distribution of stations in and around Rhodes Salt March, where several transects were measured with a nominal spacing of 300 m. All data were terrain corrected, and a complete Bouguer anomaly (CBA) was calculated with a 2.76 g/cm^3 reference density using a spreadsheet employing standard methods and constants from Holom and Oldow (2007).

A residual complete Bouguer anomaly (RCBA) was computed for the Rhodes Salt Marsh and Soda Springs Valley by subtracting minimum curvature surfaces of CBA values measured on bedrock from the regional CBA (Mickus et al., 1991). This effectively removes the regional gravity trend and isolates the gravity signal of the basin fill. The 0 mGal contour of the RCBA surface broadly follows the bedrock contact along the margins of the basin (Fig. 5B). The RCBA achieves a maximum value of –20 mGal at the center of Rhodes Salt Marsh and of –10 mGal in two subbasins beneath Soda Springs Valley. The gradient near the center of Rhodes Salt Marsh is –4 mGal/km, and the anomaly has a triangular map pattern with nearly equilateral sides that roughly parallel the bedrock boundaries of the basin. The gravity low beneath Soda Springs Valley is elongate north-south and is separated from the low beneath Rhodes Salt Marsh by an east-west–trending high (–5 mGal).

The RCBA was inverted for depth in three dimensions using the modeling program GM-SYS 3D, which uses a routine based on the Parker-Oldenburg algorithm (Oldenburg, 1974). No depth-density or independently determined depth-to-basement constraints are available for the Rhodes Salt Marsh, and we used published depth-density data from a 1.28 km well in Hot Creek Valley located 160 km to the east (Healy, 1970). The well data allowed us to calibrate a depth-density curve (Litinsky, 1989) that reflects the expected sediment density increase with depth (Maxant, 1980; Sclater and Christie, 1980; Adriansyah and McMechan, 2002; Garcia-Abdeslem, 2005; Silva et al., 2006). The surface densities for Rhodes Salt Marsh were estimated at 2.0 g/cm^3 for the saturated deposits at the surface (Coolbaugh et al., 2006), and the density increase with depth was modeled with a hyperbolic curve that achieved an upper limit of 2.67 g/cm^3. The depth inversion produced an average misfit with the observed gravity of 0.2 mGal, where the greatest basement depth was 2.9 km beneath the center of Rhodes Salt Marsh and the deepest part of Soda Springs Valley reached a depth of ~900 m.

Unlike the northerly trending basin beneath Soda Springs Valley, the Rhodes Salt Marsh basin has a prismatic geometry. Based on the gravity inversion, the basement surface has an average slope toward the basin interior of about 20° to a depth of 1.0 km with the average gradient increasing to about 40° to a depth of 2.9 km in the basin interior (Fig. 5C).

The gravity station spacing is too coarse to accurately determine the dip of the basement surface, but general patterns can be interpreted from the depth model. The shallow basement gradient along the northwestern margin is consistent with a series of down-to-the-basin normal faults mapped in bedrock exposures to the north. The origin of the basement gradient at the southeastern apex of the basin is less clear and may reflect a northwest-dipping basement ramp or a series of northeast-striking normal faults with orientations similar to the Long Canyon fault. The steep gradients in the deep center of the basin have a prismatic morphology and probably coincide with large-magnitude faults or fault zones. To the east and south, the steep gradients spatially correspond to the Benton Springs and the Belleville fault zones, respectively, and down-drop basement rocks to depths of nearly 3 km. The northeast-trending basement gradient along the northwestern margin of the deep basin interior has a vertical offset of nearly 2 km. The interpretation of this boundary as a major fault or fault zone is supported by northeast-striking fault scarps mapped along the northeastern extent of the gradient at the northern end of Rhodes Salt Marsh basin (Fig. 4).

Fault Displacement Budget

We used offset geological and geomorphologic markers across late Tertiary faults bounding the Rhodes Salt Marsh basin to estimate the finite displacement budget for the region. Specifically, we were interested in assessing how vertical and horizontal components of displacement partition as fault orientation changes around the curved array. This information provides a means of testing whether progressive displacement within the observed fault geometry can be achieved with or without deformation of the fault blocks and/or whether the fault geometry has changed during basin evolution.

The strike-slip and dip-slip components of displacement were measured along three transects across the curved fault array from the Candelaria Hills north to the Excelsior Mountains, northwesterly across Rhodes Salt Marsh from the Pilot Mountains to the Excelsior Mountains, and from the Gabbs Valley Range west across Soda Springs Valley to the eastern Garfield Hills. Several piercing points with variable accuracy were available for the slip budget estimation. For the strike-slip component of displacement, the aggregate displacement for the east-west– and north-northwest–striking faults is well constrained by offset of late Mesozoic structures of the Luning thrust belt, and no appreciable strike-slip displacement is recorded for northeast-striking

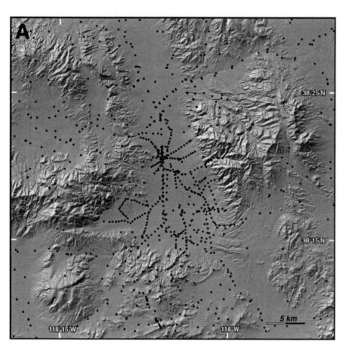

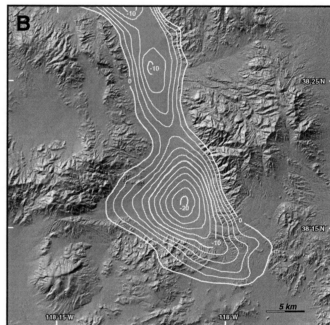

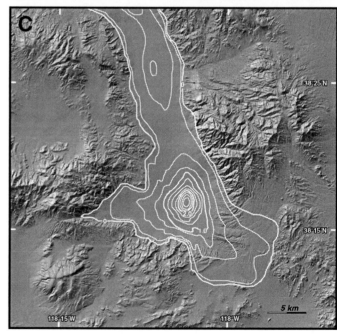

Figure 5. (A) Gravity stations acquired during this study and downloaded from the Pan American Center for Earth Sciences (PACES). (B) Residual complete Bouguer anomaly (contour interval of 2 mGal) for the
Rhodes Salt Marsh and Soda Springs Valley. (C) Depth model (in km;
250 m contour) for Rhodes Salt Marsh and Soda Springs Valley.

faults. Folds and thrust faults deformed Paleozoic through Jurassic rocks before emplacement of Cretaceous plutons (Ferguson et al., 1954; Garside, 1979; Speed and Kistler, 1980; Oldow, 1981), and they provide useful markers to assess the horizontal component of displacement on late Cenozoic faults. The vertical component of displacement for all structures is recorded by the offset of the basal unconformity beneath Tertiary andesite taken together with a depth model based on gravity data for the Rhodes Salt Marsh and Soda Springs Valley basins. The basal contact of the Tertiary andesite is mapped across the Pilot and Excelsior Mountains and into the Candelaria Hills, where, at the highest exposures, it consistently rests at altitudes of between 1900 to 2200 m. The low-relief contact is nearly planar and serves as a reasonable vertical datum for our estimates of vertical displacement on faults in and around the Rhodes Salt Marsh basin.

Displacement on Soda Springs Valley Faults

The horizontal component of displacement on north-northwest faults bounding Soda Springs Valley is based on offset of the Luning thrust fault (Ferguson et al., 1954; Oldow, 1981) and subsidiary faults in the upper plate of the fault system (Oldow and Dockery, 1993; Oldow and Speed, 1985; Oldow and Steuer, 1985). The east-west–striking Luning thrust dips steeply to moderately north and can be traced for over 15 km across the Pilot Mountains to where it is truncated by the Benton Spring fault (Fig. 6). Across Soda Spring Valley, the Luning thrust reemerges in the eastern Garfield Hills west of the eastern strand of the Indian Head fault with a northerly deflection of 7.2 km. In the northeastern Garfield Hills, offsets of thrust faults in the upper plate of the Luning thrust system indicate 1.0 km and 1.1 km of right-lateral displacement on the western and eastern strands of the Indian Head fault, respectively (Fig. 6). The aggregate right-lateral displacement across Soda Spring Valley is 8.3 km, which is consistent with estimates on the same structures measured farther north (Ekren and Byers, 1984).

The vertical component of displacement across structures bounding Soda Springs Valley basin may be determined from the offset of basal unconformity of Tertiary andesite. In the southern-central Gabbs Valley Range (Fig. 6), the basal unconformity rests at an altitude of 2200 m and steps down to ~1900 m across the Volcano Canyon fault. Across the Benton Springs fault along the range front, the base of the andesite is conservatively estimated to lie 900 m beneath the 1400 m surface elevation of Soda Springs Valley on the basis of our gravity model. The andesite reemerges along the eastern margin of the Garfield Hills at an elevation of 1650 m and again in the highlands west of the western strand of the Indian Head fault at 1900 m. The aggregate vertical displacement across the valley is 3.1 km.

Displacement on the Marietta and Belleville Faults

The horizontal component of displacement across the east-west–striking Marietta and Belleville faults is estimated by the lateral offset of the western extent of the Luning thrust, which in this location carries Permian volcanic rocks over Mesozoic clastic rocks unconformably resting on Permian strata (Garside, 1979; Oldow, 1984). In this area, the Luning thrust strikes north-south and dips moderately west and is well located in the south Excelsior Mountains north of the Marietta fault (Fig. 6). South of the Belleville fault, the Luning thrust is only approximately located in the highlands west of the Candelaria Hills where the contact between Permian volcanic and sedimentary rocks is obscured by Quaternary cover. The southern extent of the Luning thrust can be located in its intersection with the east-west Cenozoic fault to within 2.5 km, which yields an estimated left-lateral offset of 8.0 to 10.5 km.

The vertical component of displacement across the east-west late Cenozoic faults may be estimated from offset of the basal unconformity beneath Tertiary andesite exposed in the Excelsior Mountains and Candelaria Hills. In the Excelsior Mountains, the base of the andesite rests at an altitude of 2200 m. Along the southern side of Marietta fault, the base of the andesite is not exposed, but given the thickness of the unit, the basal contact is estimated at an altitude of no higher than 1500 m and may be closer to 1200 m, which provides an estimated minimum vertical displacement of between 700 to 1000 m. To the south of the Marietta and Belleville faults, the base of the andesite gains altitude and rests on Paleozoic strata at 1900 m in the central Candelaria Hills. The aggregate vertical displacement across the fault system is between 1.1 and 1.4 km.

Displacement on Rhodes Salt Marsh Basin Faults

Along a northwest-southeast transect from the southern Pilot Mountains across the Rhodes Salt Marsh basin and into the eastern Excelsior Mountains, no appreciable horizontal displacement is recorded. Rather, displacement on exposed faults is consistently down-dip. Here, as elsewhere, we used the basal unconformity beneath Tertiary andesite as a vertical datum.

The andesites rests on an erosional surface at an elevation of 2200 m in the eastern Excelsior Mountains and is successively stepped down to the east on a series of northeast-striking normal faults to below the 1400 m elevation of the surface of the basin fill at the range front. Based on our gravity model, the basin increases in depth to the east, where the basement is estimated at 2.9 km. When the thickness of the andesite is taken into account, the basal unconformity probably lies at an elevation of between 1600 to 1900 m below sea level. The aggregate vertical displacement on the western flank of the basin is between 3.8 and 4.1 km. Continuing to the east, the basement beneath the basin rises across the Benton Springs fault to exposures in the southern Pilot Mountains, where the basal unconformity below the andesite rests at an elevation of ~2000 m. Farther to the south, the same contact lies at 1800 m. The aggregate vertical offset across the eastern flank of the basin is between 3.4 and 3.9 km. The total vertical displacement across the Rhodes Salt Marsh basin ranges between 7.2 and 8.0 km.

Paleomagnetic Data, Eastern Excelsior Mountains

An important consideration in understanding the kinematic evolution of the Rhodes Salt Marsh basin and adjacent faults is

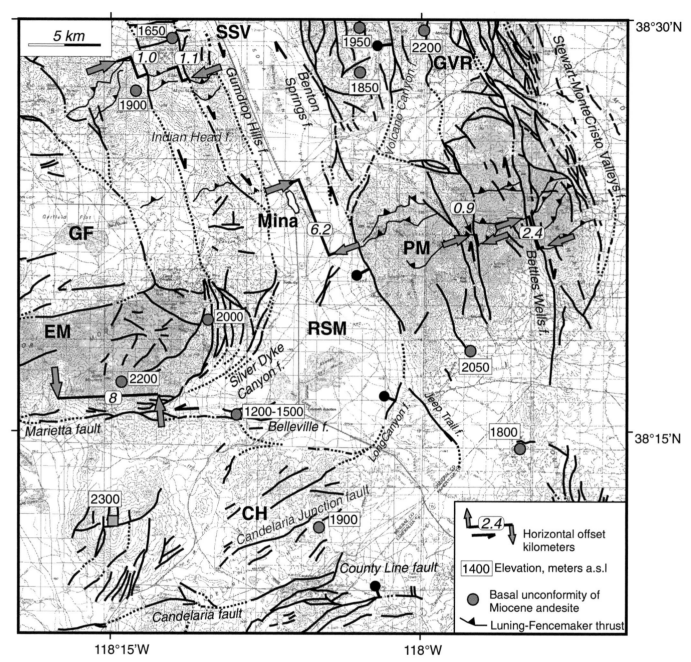

Figure 6. Horizontal and vertical geologic offset across faults in the Rhodes Salt Marsh basin area. Large arrows indicate piercing points for computing the horizontal offset. Vertical offsets are based on the present elevation of the basal unconformity of Tertiary andesites. Places are abbreviated as in Figures 3 and 4; a.s.l.—above sea level.

observed fault geometry—is there a primary configuration or has the geometry evolved during progressive deformation? Specifically, we need to explore the possibility that basin evolution is associated with large clockwise rotation of tectonic blocks and attendant reorganization of the fault geometry (Wesnousky, 2005).

To this end, we analyzed paleomagnetic data from rocks located in the southeast part of the Excelsior Mountains. Our sampling concentrated on the diorite pluton of Silver Dyke Canyon

(Garside and Silberman, 1978; Garside, 1979; Speed and Kistler, 1980), with sites distributed over as much of the exposure of the pluton as possible, typically in outcrops in well-incised drainages (Fig. 7). We also sampled host rocks to the intrusion (Permian Mina Formation and unnamed Mesozoic metasedimentary rocks) at five sites. In addition, three sites were established in Miocene rocks (a small stock that cuts the diorite and two rhyolitic ash-flow tuff deposits) (Speed and Kistler, 1980).

Sample collection and treatment were performed as follows. All samples for paleomagnetic investigations were collected as core samples drilled in the field with a portable field drill (nonmagnetic drill bits) and oriented using both magnetic and solar compasses. Specimens were cut into 2.2-cm-high cylinders. Remanence measurements were carried out on either a Schonstedt SSM-1A computer-interfaced spinner magnetometer or a 2G Enterprises Model 760R superconducting rock magnetometer. Progressive demagnetization involved either thermal (Schonstedt TSD-1) or alternating field (Schonstedt or online 2G Enterprises). Directions of magnetizations isolated in demagnetization were calculated using principal component analysis (Kirschvink, 1980), and site and grand mean directions were estimated using the method of Fisher (1953).

A total of 21 of 22 sites in the diorite of the pluton of Silver Dyke Canyon yielded interpretable demagnetization data, with sample magnetization directions exceedingly well-grouped at the site level (Table 1). Natural remanent magnetic (NRM) intensities of these rocks are typically about 0.5 A/m. The diorite is characterized by high median destructive fields (typically about 40–50 mT) and a narrow range of high laboratory unblocking temperatures below about 580 °C (Fig. 8). Demagnetization data, as well as the limited rock magnetic experiments we were able to perform (Fig. 9), suggest that fine-grained (pseudo–single domain) low-titanium magnetite is the principal remanence carrier. Limited anisotropy of magnetic susceptibility (AMS) data obtained on these rocks show no indication of a preferred fabric (Fig. 10).

The grand mean in situ pluton direction, based on 21 sites, is declination = 52.4°, inclination = 73.7°, a95 = 3.1°, and k = 109 (Fig. 11). Five sites in Jurassic and Permian metasedimentary wall rocks yielded magnetizations of a similar direction, and overall mean values of declination = 40.3°, inclination = 74.8°, a_{95} = 8.5°, and k = 82 (Fig. 11, Table 1) are thus provided by these rocks. Three sites in Miocene rocks (one small stock that crosscuts the diorite of the pluton of Silver Dyke Canyon and two overlying ash-flow tuff cooling units) yielded an in situ mean direction of declination = 16.0°, inclination = 60.4°, a95 = 4.3°, and k = 822.

These data can be used to place realistic bounds on the magnitude of post-Cretaceous deformation that has affected the southwest structural margin of Rhodes Salt Marsh basin. We assume that the magnetization characteristic of the pluton as well as its immediately adjacent wall rocks is a primary, thermoremanent magnetization acquired in the early Late Cretaceous, based on K-Ar age determinations on biotite of 95.8 ± 2.9 Ma (Garside and Silberman, 1978) and 100.0 Ma (Speed and Kistler, 1980), which provide an estimate of the minimum age of pluton emplacement. Additional geochronologic and thermochronologic investigation may refine the emplacement and cooling history of the intrusion and its immediately surrounding host rocks. The inferred age of magnetization acquisition is consistent with the exclusively normal polarity of the characteristic magnetization (acquired during the mid-Cretaceous normal polarity superchron, between about 121 and 84 Ma; Gradstein et al., 2004). The response to progressive demagnetization, as well as rock magnetic properties of the pluton (Figs. 8 and 9), is typical of intrusive igneous

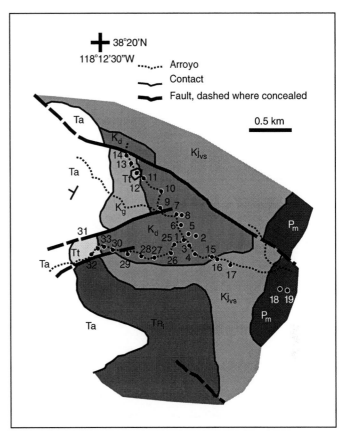

Figure 7. Simplified geologic map of a part of the eastern Excelsior Mountains (location in Fig. 4), showing locations of all sites established for paleomagnetic information. Rock units, from oldest to youngest are: P$_m$—Permian Mina Formation; TR$_l$—Triassic Luning Formation; Kj$_{vs}$—volcanic and sedimentary rocks, undifferentiated; K$_g$—granitic rocks, undivided; K$_d$—pluton of Silver Dyke Canyon; Tt—ash-flow tuffs and miscellaneous silicic intrusive rocks; Ta—predominantly andesite flows. Map is after Garside (1979).

rocks that have been demonstrated to possess geologically stable and early-acquired magnetizations. We further assume that the pluton of Silver Dyke Canyon has sufficiently averaged the geomagnetic field during the mid-Cretaceous, but we recognize that this assumption may not be fully valid.

The in situ magnetization from the diorite differs, mainly in declination, from expected mid-Cretaceous field directions for this locality derived from several estimated paleomagnetic poles for North America for this time period (Table 2). This discrepancy may reflect a combination of multiple phases of deformation since pluton emplacement. Overlying Miocene volcanic rocks exposed in the general sampling area have an approximate orientation of 225°/20°, dip to northwest (our observations and Garside, 1979). Application of a simple tilt correction about the present strike axis results in a mean paleomagnetic direction for these Miocene igneous rocks of 355°/47°. Although the data are limited in robustness, at face value, this result suggests a minimal amount of vertical-axis rotation since Miocene time, because

TABLE 1. PALEOMAGNETIC DATA FROM ROCKS OF THE EASTERN EXCELSIOR MOUNTAINS

Site	Rock type	N/N_o*	Decl.[†]	Incl.[†]	α_{95}[§]	K[##]	Comments
1	Diorite	10/10	44.3	75.7	3.3	112	
2	Diorite	10/10	37.4	64.9	3.6	99	
3	Diorite	8/8	52.1	63.0	4.6	103	
4	Diorite	6/6	61.2	68.9	6.4	80	
5	Diorite	8/8	40.5	71.4	1.4	1187	
6	Diorite	10/10	44.8	71.7	3.6	149	
7	Diorite	8/8	65.3	70.5	2.9	298	
8	Diorite	8/8	61.0	75.3	1.8	736	
9	Diorite	7/7	57.5	80.5	4.5	136	
10	Diorite	7/7	69.6	80.9	3.1	301	
11	Diorite	5/5	47.6	77.7	4.8	167	
12	Rhyolite	9/9	21.6	59.3	3.3	202	Miocene, intrusion
13	Diorite	5/5	10.5	70.4	1.9	987	
14	Diorite	5/5	40.2	69.9	2.1	891	
15	Diorite	8/8	51.7	57.4	9.8	25	
16	Metavolcanics	6/6	15.0	70.0	12.6	20	Jurassic/Cretaceous (?)
17	Metavolcanics	6/6	38.9	66.5	8.8	42	Jurassic/Cretaceous (?)
18	Metasedimentary	6/6	46.0	80.8	10.6	29	Permian Mina Fm.
19	Metasedimentary	8/8	48.6	77.9	9.4	28	Permian Mina Fm.
25	Diorite	7/7	64.2	73.9	3.6	215	
26	Diorite	7/7	69.2	77.7	10.8	24	
27	Diorite	7/7	78.1	77.7	3.5	230	
28	Diorite	6/6	58.4	77.6	3.8	233	
29a	Diorite	3/3	78.2	79.0	3.7	477	
29b	Metavolcanics	6/6	78.1	72.9	3.5	272	J/K (?) host; contact test with diorite
30	Diorite	7/7	83.3	79.3	3.1	283	
31	Rhyolite	7/7	15.0	60.7	7.0	56	Miocene, flow
32	Rhyolite	8/8	11.1	60.1	4.9	100	Miocene, flow
33	Diorite	8/8	60.1	76.9	3.9	163	

Note: Paleomagnetic data are from all rock types sampled in the eastern Excelsior Mountains. Sites are located in Figure 9.

*N/N_o—Ratio of samples used (N) to independent samples collected (N_o).

[†]Decl./Incl.—In situ declination and inclination of estimated mean magnetization direction.

[§]α_{95}—95% confidence interval about the estimated mean direction, assuming a circular distribution.

[##]K—Best estimate of Fisher (1953) precision parameter.

expected late Cenozoic field directions for North America for this area are slightly west of north and of similar inclination. The same tilt correction applied to the pre-Tertiary rocks yields mean directions of 358°/66° and 352°/64° for the mean of 21 sites in the diorite of Silver Dyke Canyon and the five sites in wall rocks to the pluton, respectively (Table 1). These directions are slightly discordant, again in declination, with most estimates of an early Late Cretaceous normal polarity reference direction (north-northwest declination and moderate positive inclination) for North America (Table 2).

Based on the available data, we infer that, in total, the pluton of Silver Dyke Canyon and its host rocks may have experienced a very modest magnitude of deformation since Cretaceous emplacement and prior to Miocene magmatism in the area. Such deformation could have involved a small (10°–15°) amount of west-side-down tilting, clockwise rotation of less than 10°, or a combination of both processes, of lesser magnitude. Notably, the good agreement between the structurally corrected data from the Miocene rocks and expected directions for this time is interpreted to indicate that little rotation of the eastern part of the Excelsior Mountains has taken place since mid-Miocene time.

Fault-Slip Analysis

The final ingredient in this integrated analysis of the kinematic history of displacement in and around the Rhodes Salt Marsh basin is supplied by a comprehensive assessment of the slip recorded by bordering faults. These include the southern part of the Benton Springs and the Jeep Trail faults to the east and southeast, the Silver Dyke Canyon fault system along the northwest, and the eastern Marietta and western Belleville faults to the southwest (Fig. 12). We supplemented the data along the southern margin of the basin, where faults have limited exposure, with an analysis of published fault-slip data from the Candelaria fault system (Speed and Cogbill, 1979b).

Fault kinematic data were collected along the main fault surface or on footwall synthetic and antithetic faults in close proximity of the range-border fault, and so they reasonably reflect the major control on basin development. Faults typically preserve lineated and decorated slickensides and the slip direction and shear sense were established following the criteria described by, among others, Gamond (1987), Hancock et al. (1987), and Petit (1987). We made a statistical analysis of the fault-slip data to determine

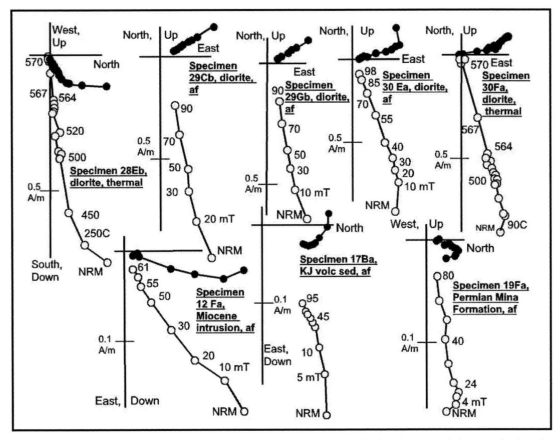

Figure 8. Examples of progressive alternating field and thermal demagnetization (orthogonal demagnetization diagrams; Zijderveld, 1967) of specimens from sites in the diorite of Silver Dyke Canyon, older Mesozoic and Permian host rocks, and a Tertiary stock. Filled (open) circles are projections on the horizontal (vertical) planes. Peak demagnetizing fields (in mT) or temperatures (°C) are given adjacent to the vertical projections. All projections are in geographic (in situ) coordinates. NRM—natural remanent magnetization.

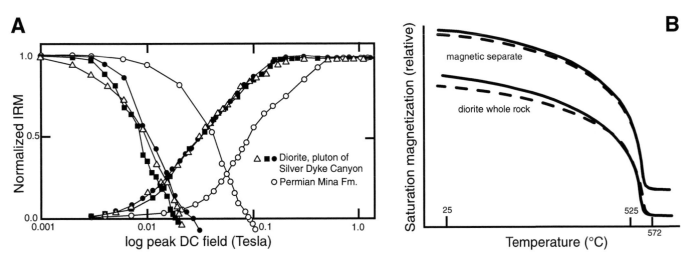

Figure 9. (A) Curves showing the acquisition of isothermal remanent magnetization (IRM) to saturation (SIRM) and backfield demagnetization of SIRM for three specimens of the diorite of Silver Dyke Canyon from three different sites and one specimen from the Permian Mina Formation. (B) Curves showing the response of saturation magnetization (Js, obtained in a 0.2 T field) to increasing (solid line) and decreasing (dashed line) temperature, which reveals the Curie temperature of the saturated or near-saturated magnetic phase. The sets of curves are for a whole rock specimen and a magnetic separate from the diorite of the Silver Dyke pluton and indicate a Curie temperature close to 560 °C.

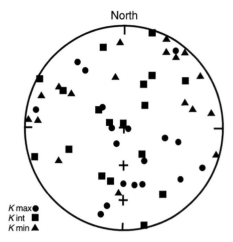

Figure 10. Anisotropy of magnetic susceptibility data (principal susceptibility directions) from 18 randomly selected samples from six sites in the diorite of Silver Dyke Canyon. All data are lower-hemisphere projections on an equal-area projection. K max, K int, and K min are the maximum, intermediate, and minimum axes, respectively, of the anisotropy of magnetic susceptibility (AMS) ellipsoid.

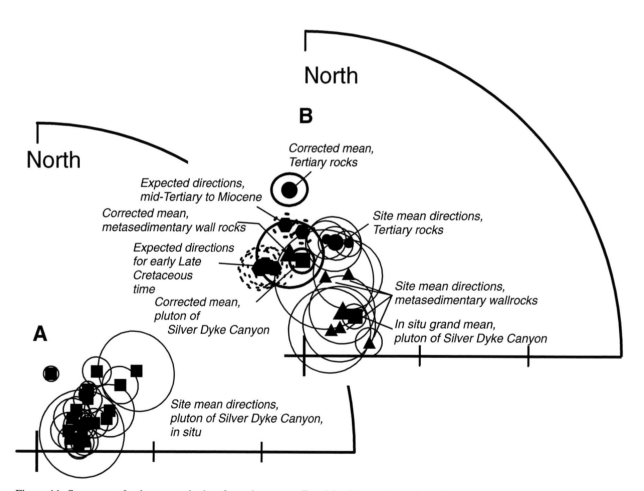

Figure 11. Summary of paleomagnetic data from the eastern Excelsior Mountains and partial equal-area projections of estimated mean directions, with all projections on the lower hemisphere. (A) In situ site mean directions for sites from the pluton of Silver Dyke Canyon. (B) Grand mean direction from the pluton of Silver Dyke Canyon (large square), compared with site mean directions from host rocks (small and large [grand mean] triangles) and site mean directions from Tertiary rocks (small and large [grand mean] circles). The three grand mean directions have been corrected for local tilt (225°/20° tilt to northwest) about present strike axis. These grand mean directions are compared with expected directions for early Late Cretaceous (Table 1) and mid-Tertiary to Miocene time (hexagons).

TABLE 2. LATE CRETACEOUS PALEOMAGNETIC POLES FOR NORTH AMERICA AND
ASSOCIATED EXPECTED DIRECTIONS

Pole latitude (°N)	Pole longitude (°W)	A95* (°)	Expected declination[†] (°)	Expected inclination[†] (°)	Reference[§]	α95* (°)
71.5	194.9	4.0	339.2	66.7	1	2.7
70.1	191.2	2.7	336.5	66.4	2	1.8
71.3	187.5	4.2	337.5	65.2	3	2.9
71.0	196.0	4.9	338.8	67.1	4	3.2
68.0	192.0	5.0	333.7	67.2	5	3.3

*95% confidence limit based on averaging specific poles.
[†]Expected declination (in degrees east of north) and inclination (in degrees positive downward) at the latitude of the eastern Excelsior Mountains (location: 37.5°N, 118°W) based on different paleomagnetic poles.
[§]References for Late Cretaceous North American poles: (1) Dickinson and Butler (1998); average of poles from Maine intrusives (McEnroe, 1996), White Mountains intrusives (Van Fossen and Kent, 1992), Monteregian Hills (Foster and Symons, 1979), Newfoundland dikes (Lapointe, 1979), Niobrara Formation (Shive and Frerichs, 1974), and Granite Mountains (Globerman and Irving, 1988). (2) Housen et al. (2003); average of 18 poles from Larochelle (1961), Shive and Frerichs (1974), Foster and Symons (1979), Lapointe (1979), Globerman and Irving (1988), Van Fossen and Kent (1992), and McEnroe (1996). (3) Van Fossen and Kent (1992); White Mountains pole based on an average of eight sites from three plutons. (4) Globerman and Irving (1988); average of four independent poles. (5) Van der Voo (1993); an average of five independent poles, assumed representative of the 67 to 97 Ma time period.

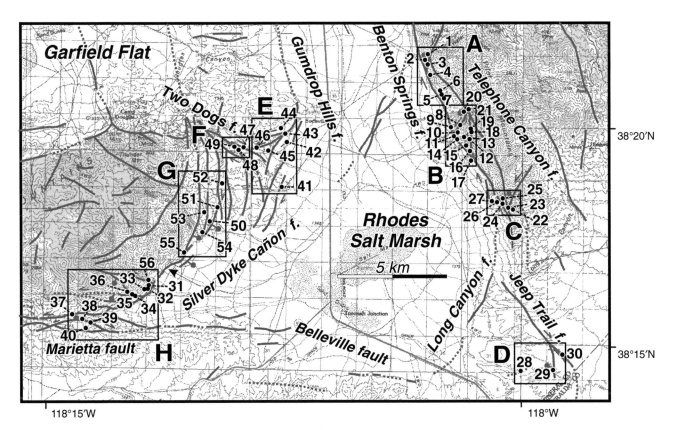

Figure 12. Fault map of the Rhodes Salt Marsh (situated in Fig. 4) showing location of slip lineation analysis sites and domains, or clusters of sites. Extension directions were computed for A to H. Domain I (where data from Speed and Cogbill [1979b] are approximately located) is shown in Figure 15.

the mean horizontal component of the extension direction for families of faults with a minimum of three orientations measured at sites located along the border of the basin. We estimated the orientation of strain axes to within five degrees, and, based on the scatter in the slip lineations together with the range of orientation of conjugate strike-slip faults, we assign a ±10° uncertainty to the determined direction of extension.

We recognized primary and secondary extensional strain axes trending west-northwest and north-northeast, respectively. The strain axes were determined from the analysis of outcrop-scale structures that make up two families of coordinated slip. Primary and secondary elongation directions were defined on the basis of the geometry of minor structures and the morphology of slickenlines and shear sense indicators. At the outcrop scale, the family of structures that accommodate slip associated with the primary extension direction consists of north-northwest and east-west right and left oblique-slip faults and north-northeast–striking dip-slip faults. These structures have a geometry akin to that of the curved fault array. Slip associated with the secondary extension direction is taken up on a system of minor structures with north-south and northeast left and right oblique transcurrent motion and west-northwest structures exhibiting dip slip. These minor structures are pervasively developed throughout the study area but are not reflected in the map-scale arrangement of faults. Although slickenlines associated with both directions of extension are found on structures of both slip systems and locally show mutually crosscutting relations, the morphology of slickenlines and shear sense indicators are preferentially developed on the slip system associated with a particular extension direction. Slickenlines and shear sense indicators associated with the primary extension direction are best developed on structures of the west-northwest slip system. Conversely, fault fabrics associated with the secondary extension direction are dominant on the north-northeast slip system.

In the following, detailed observations on fault geometry, slip lineation analysis and relations between the primary and secondary displacement fields are presented for the eastern, western, southwestern, and southern margins of the basin.

Eastern Margin

The eastern side of the Rhodes Salt Marsh is bounded by a complex fault array that changes progressively in strike from northwest on the north end to northeast on the south end (Fig. 12). The northern part of the array spans the southern termination of the Benton Springs fault system, which is characterized by a steep and nearly 500-m-high range-front scarp carved in Mesozoic rocks of the Pilot Mountains (Fig. 12). An impressive zone of shear is observed in the footwall of the Benton Springs fault, the main slip surface of which is typically concealed beneath the alluvium. A shear zone with meter-scale thickness composed of anastomosing lithons is commonly associated with secondary faults, and a 10-m-scale cataclastic breccia passing locally to a fault gouge is well developed as the border of the range is approached. In this area, minor faults associated with the Benton Springs fault are recognized in the fault footwall up to 1.0 km

east of the range front. These faults strike northwest and are connected by minor splays with variable strike.

Fault-slip analysis along the Benton Springs and subsidiary faults (domains A, B, and C, Figs. 12 and 13A) indicates a predominance of northwest-striking faults, which are generally steep (60° to nearly vertical) and, within the primary strain field, exhibit right-lateral to right-oblique slip. North-striking faults are right-oblique to normal, and east-northeast–striking faults exhibit left-oblique slip. Northeast faults have typically gentler dips (45° to 70°) and are left-oblique to normal. The horizontal component of the primary extension averaged over clusters of sites within spatially adjacent domains A, B, and C has a consistent N50°–60°W trend (Fig. 13A).

At several sites, the slip relations outlined here are locally inverted, and the northwest- to north-striking faults are normal to left-oblique. The east-northeast– and northeast-striking faults, instead, exhibit a right-oblique slip. These relations constitute the secondary displacement field for domains in this sector of the basin (Fig. 13A). The mean horizontal elongation determined for the secondary displacement field trends N35°–40°E.

Fault morphology qualitatively points to different magnitudes of aggregate slip for the two strain fields. Large tectonic grooves and Riedel-type fracture systems, 1 to 10 m scale in length, typically characterize the right-oblique slip on the northwest-striking faults and document the predominance of the primary extension. In contrast, the secondary deformation field is manifested by shorter (0.1 to 0.01 m scale) fault striations and mineral fibers.

The complicated chronological relation between the two strain fields is illustrated by slip lineation superposition observed at several sites (Fig. 13A). On an exposure of the Benton Springs fault at site 4, an early northwest right-oblique motion is overprinted by a second generation of lineations that document left-lateral slip on the same fault. The same relations are observed at site 15, where there is a more oblique trend for the second, left-slip episode, but here, a third motion event is recorded that is parallel to the first one (Fig. 13A). Similarly, north- to north-northeast–striking faults at nearby site 8, located on a minor fault in the footwall of the Benton Springs fault, show left- and right-lateral slips that are interlaced.

Toward the south, the Benton Springs fault progressively swings to a northern strike and bifurcates into the Long Canyon and Jeep Trail faults. No slip determinations are available on the Long Canyon fault, the scarp of which is carved in Quaternary alluvium and smoothly decreases in height. Few slip determinations on the Jeep Trail and ancillary faults that displace Pliocene-Quaternary sediments and basalts (domain D, Fig. 13A) are consistent with the primary and secondary displacement fields observed at sites further north.

Western Margin

The western margin of the Rhodes Salt Marsh is characterized by the complex fault array associated with the Silver Dyke Canyon fault and ancillary hanging-wall faults. Moving from

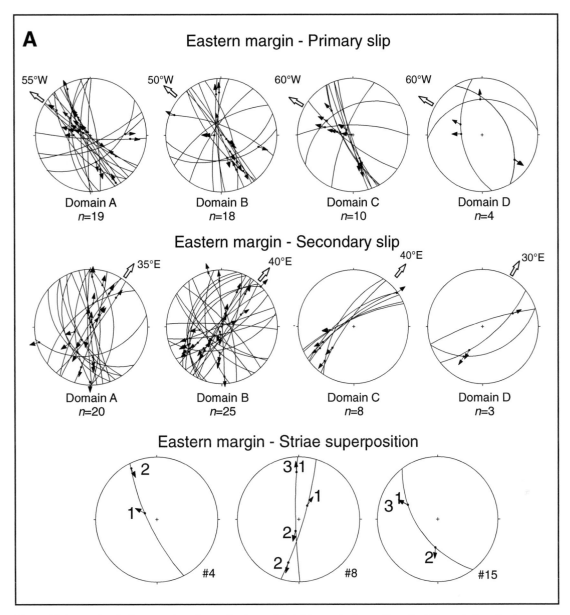

Figure 13. Lower-hemisphere, equiangular projections of fault kinematic data collected within each domain. Arrows on fault planes indicate motion of the hanging-wall block. Faults are grouped in two different projections, primary and secondary, based on slip compatibility and, as observed in some sites, striation superposition. (*Continued on following page.*)

north to south, the fault array shows a progressive curvature in strike from north to northeast. At the curvature, the range-front fault merges with the northwest-striking, ~5-km-long Two Dogs fault, which connects the range-front fault with the western strand of the Indian Head fault (Fig. 12).

Fault-slip analysis in the northern sector of the Silver Dyke Canyon fault documents right-oblique to right-lateral motion on northwest- and north-striking faults, respectively (domain E, Fig. 13B). Northeast-striking faults are left-oblique to normal. The extension direction estimated for the main strain field has a N50°W trend. To the west of the range-front fault, slip on

the northwest-striking Two Dogs fault is right-lateral and, when supplemented by determination on differently oriented subsidiary faults, yields an estimation for the primary elongation axis of N55°W (domain F, Fig. 13B). Within domains E and F, the secondary strain field is documented by right-lateral and left-oblique slip on northeast- and north-striking faults, respectively, with an estimated ~N30°E elongation axis.

To the south, the Silver Dyke Canyon fault and subsidiary hanging-wall faults strike northeast and dip more gently toward the basin. The basin-scale attitude of the fault array is reflected in the outcrop-scale fault population analysis, which documents

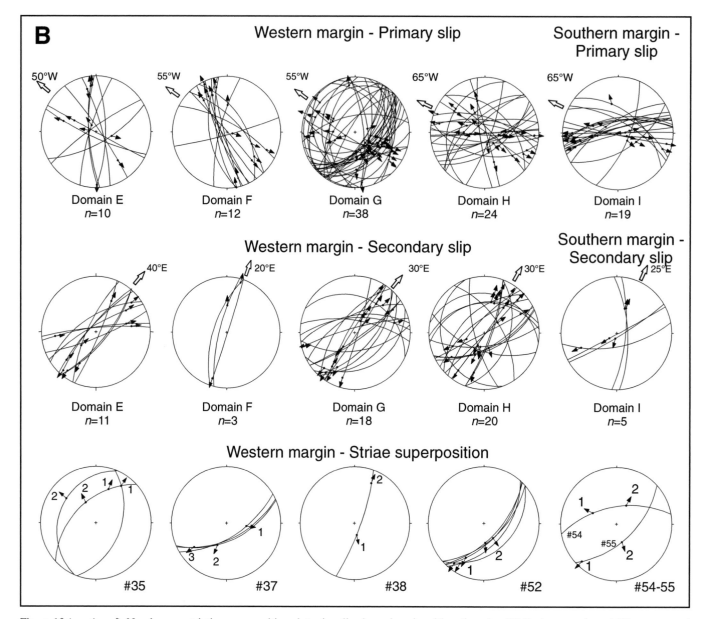

Figure 13 (*continued*). Numbers on striation superposition plots give slip chronology in arithmetic order. (A) Eastern margin and (B) western and southern margins. Domain and site locations are in Figure 12, except for domain I, which is located in Figure 15.

dips of 45°–60° for the northeast-striking faults (domain G, Fig. 13B). These faults consistently exhibit a dominant top-to-southeast normal slip, consistent with a N55°W azimuth for the horizontal component of the primary extension. Locally, the northeast-striking faults have a right-lateral slip, which, together with the normal to right-oblique slip recorded on approximately west-striking faults, document a N30°E trend for the secondary extensional axis in this sector of the basin margin.

Similar to those observed on the eastern margin of the Rhodes Salt Marsh, complex chronological relations between the main and secondary strain are recorded along the western margin on the Silver Dyke Canyon fault and ancillary faults (Fig. 13B).

At sites 52 and 55, right-lateral slip lineation on northeast-striking faults is superseded by dip-slip lineations. In contrast, at site 54, northwest slip on a east-northeast fault is followed by right-oblique displacement on the same fault surface.

Southwestern Margin

Curvilinear faults of the Silver Dyke Canyon fault array on the western side of the Rhodes Salt Marsh merge to the southwest into the west-striking Marietta fault via a complex array of interlacing secondary faults. Accordingly, fault population analysis documents a dominance of northeast- and west-striking faults (domain H, Fig. 12). As observed further north, the northeast-striking faults are normal to left-oblique. The main slip recorded on the west-striking faults is left-lateral. Additionally, right-lateral slip is observed on northwest-striking faults. Combination of the slip recorded on the three differently striking fault sets yields a N65°W azimuth for the primary extensional axis (Fig. 13B). The secondary strain field is recorded by right-lateral and right-oblique slip on northeast-striking and west-striking faults, respectively. Locally, northwest-striking faults with dip-slip lineations are observed. The secondary extension azimuth determined in this sector is N305°E, consistent with observations further north.

Development of both the main and the secondary displacements in recent times is reflected in two sets of slip lineations observed on exposures of the Marietta fault surface in Quaternary alluvium at site 37. Striation superposition criteria define early left-oblique motion followed by right-oblique slip on the east-northeast–striking fault. Whereas the fault surface has a morphology, as defined by tectonic grooves and large striations, that primarily records the first episode of left-oblique displacement and is broadly consistent with the offset recorded by geological structures (Fig. 6), the younger motion in the secondary strain field forms secondary decorations on the slickenside surface. At nearby site 35, however, the chronological relations between the two sets of lineations are reversed, with right-lateral motion followed by dip-slip on northeast-striking faults. These observations indicate that the primary and secondary deformation directions were both active in the Quaternary and alternated during this time interval.

Southern Margin

On the southern margin of the Rhodes Salt Marsh basin, Tertiary volcanic and pre-Tertiary basement rocks that underlie the Candelaria Hills are cut by the Candelaria fault system (Speed and Cogbill, 1979b). This fault array is composed of several subparallel strands (Fig. 6). Within this system, the 10-km-long Candelaria fault strikes west-southwest, dips north, and has a documented Quaternary activity (Dohrenwend, 1982; Wesnousky, 2005). Further to the south, the County Line fault overlaps with, and is locally linked with, the eastern part of the Candelaria fault (Fig. 6).

A reassessment of fault kinematic data published by Speed and Cogbill (1979b) confirms that motion on the Candelaria and County Line fault systems is essentially left-lateral. East-northeast faults form the dominant population, and minor northwest- and northeast-striking faults are also represented (domain I, Fig. 13B). Within the primary strain field, the three fault sets exhibit left-lateral, right-lateral, and normal slip, respectively. The azimuth determined for the regional elongation axis is N65°W.

In addition, a secondary N25°E extension is recorded by right-oblique and left-oblique slip on east-northeast– and north-striking structures of the County Line fault, which was attributed to gravitational sliding by Speed and Cogbill (1979b). Because northeast displacement in the area is oblique (~40°) and not orthogonal to the local geomorphic slope, we discard the interpretation by Speed and Cogbill (1979b). Rather, the northeast extension recorded on the County Line fault is in agreement with the trend of the local azimuth elsewhere along the margins of the Rhodes basin.

DISCUSSION

Origin of Basin Geometry

The Rhodes Salt Marsh basin and similar basins localized at the apex of fault curvature within the Mina deflection have an unusual geometry where bounding faults vary in strike through 120°. The geometry of the faults and basins and their relation to structures in the northern part of the Eastern California shear zone and central Walker Lane have elicited various explanations, ranging from the superposition of east-west and northwest-southeast fault systems (Stewart, 1988), to clockwise block rotation produced by a broad northwest-trending shear couple (Wesnousky, 2005), to northwesterly extension within an east-west–trending structural stepover linking northwest transcurrent faults of the Eastern California shear zone with the central Walker Lane (Oldow, 1992, 2003). The central issue addressed by these models is whether the observed fault-basin geometry is produced by the intersection of kinematically discrete fault systems or if the geometry reflects kinematically coordinated displacement on a system of curved faults. The integrated geologic and geophysical investigation of the Rhodes Salt Marsh basin and surrounding faults presented here provides the constraints needed to evaluate the merits of the alternative explanations for structures throughout the Mina deflection.

Detailed geologic mapping and kinematic analysis of the faults around the basin discount the superposed structure explanation championed by Stewart (1988) and clearly show that the structures are continuous and curved. Fault strike progressively changes around the apex of curvature, both around the northwestern and southeastern extents of the basin (Fig. 4). The transition between left-lateral and right-lateral faults around the Rhodes Salt Marsh basin is never abrupt and is manifested by a progressive change in fault strike and a concomitant reorganization of the rake of slip lineations within the displacement field (Figs. 13A and 13B). The east-west–striking faults consistently have left-lateral to left-oblique slip indicators that progressively change to a down-dip orientation as the faults pass to a northeast strike and then again progressively change slip direction to right-lateral and right-oblique as the northeast-striking structures swing to the northwest.

To resolve whether the origin of the curved fault array is a primary configuration or the product of block rotation, however, requires additional constraints. The orientations of the left-oblique and right-oblique faults are at odds with a conventional conjugate explanation (Anderson, 1951) because the primary extension direction determined from fault-slip analysis faces the acute (~60°) angle between fault arrays. This discrepancy led Wesnousky (2005) to conclude that the initial geometry of faults was formed as a conventional conjugate set of strike-slip faults with right- and left-lateral structures oriented northwest and northeast, respectively. In this scenario, the present-day orientation of the left-lateral faults was produced by a 20° to 30° clockwise rotation of tectonic blocks bound by a northwest shear

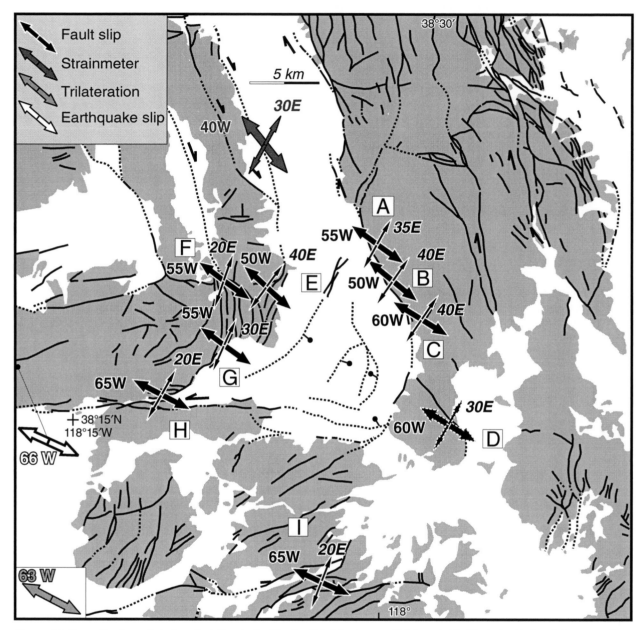

Figure 14. Synoptic view of strain orientation in the Rhodes Salt Marsh area. Thick and thin black double-headed arrows show primary and secondary extension directions, respectively, determined in composite domains from fault kinematic analysis (domains A to G from this study; domain I from Speed and Cogbill, 1979b). Heavy gray double-headed arrow shows primary and secondary extension directions (1972–1973) from a strainmeter around Mina (Ryall and Priestley, 1975). Light-gray double-headed arrow in inset is primary extension direction (1972–1979) from a trilateration network (Savage and Lisowski, 1984) for the part of the Mina deflection approximately shown in Figure 3. White double-headed arrow show tensile axis of 1934 Excelsior Mountain earthquake (Ryall and Priestley, 1975). Subsurface faults in the Rhodes basin inferred on the base of surface fault projection (Fig. 6) and Bouguer gravity anomaly model (Fig. 5C) are shown with dotted lines.

couple (Wesnousky, 2005). This rotation is a primary prediction of this model but fails on the basis of our paleomagnetic analysis of plutonic and volcanic rocks in the Excelsior Mountains. The rocks of the hanging-wall block show no significant vertical-axis rotation. Thus, if for no other reason than the process of elimination, the curved fault array is shown to be a primary configuration. Although beyond the scope of this paper, the orientations of faults bounding the Rhodes Salt Marsh basin and similar structures in the Mina deflection are thought to be controlled by long-lived crustal heterogeneity in the western Great Basin associated with continental rifting and subsequent contractional tectonics (Oldow et al., 1989; Oldow, 1992). The crustal heterogeneity localized a system of structural stepovers that evolved since the mid-Miocene (Oldow et al., 2008).

Basin Formation during Flattening

The geometry of the late Cenozoic curved fault array imparted an important boundary condition on the deformation field as displacement was transferred from left-oblique to right-oblique structures through the curved faults. The reconciliation of fault-slip data and the displacement budget for structures around the array demonstrates that displacement was accompanied by significant internal deformation of the hanging-wall blocks as the basin evolved. Specifically, divergence of the geologic displacement trajectories around the basin-bounding faults, taken together with two contemporaneous directions of extension recorded in fault-slip data (Fig. 14), testifies that progressive deformation and basin development occurred in a state of flattening strain.

We analyzed the displacement budget for the basin and curved fault array by combining the strike-slip displacement estimates and the horizontal component of vertical displacement in a two-dimensional model. The model compares the net horizontal displacement taken up across the east-west and north-northwest left and right oblique-slip faults with the horizontal component of motion accommodated across northeast dip-slip faults. We simplified the calculation by combining the distributed displacement of the five faults of the Soda Springs system as a single north-northwest–striking fault (Fig. 15) and attributed the horizontal component of vertical displacement across the Rhodes Salt Marsh basin as a single structure. The rectilinear model of fault geometry provides the basis for a simple vector summation of displacements to assess the compatibility of faults of three orientations and different slip directions. The greatest uncertainty in the analysis arises from ambiguity in the dip of faults, which substantially impacts the horizontal-vertical ratios for the dip-slip component of displacement. To bracket reasonable upper and lower bounds for fault dip, we calculated the horizontal components of displacement for structures with dips of 60°, 45°, and 30°.

For the southern margin of the curved fault array, defined by the east-west–striking Marietta and western Belleville faults, a left-lateral displacement between 8.0 and 10.5 km is established by offset of the north-south–striking segment of the Luning thrust (Fig. 15). The same structures have a vertical displacement of

between 1.1 and 1.4 km estimated by the vertical offset of the basal unconformity of Tertiary andesite. At the northern end of the curved fault array, fault slip is not taken up on a single fault system and is distributed across several north-northwest–striking faults (east and west strands of the Indian Head fault, Gumdrop Hills fault, Benton Springs fault, and Volcano Canyon fault) bordering Soda Springs Valley (Fig. 15). In this area, the strike-slip component of displacement is well constrained as 8.3 km right-lateral offset of the east-west–striking segment of the Luning thrust and subsidiary hanging-wall faults that provide redundant displacement estimates. Across the same structures, an aggregate vertical displacement of 3.1 km is recorded by offset of the basal contact of the andesite exposed in the mountains and by our depth model for Soda Springs Valley. Along a northwest-trending transect across Rhodes Salt Marsh, fault displacement is solely dip-slip and has an aggregate vertical displacement of between 7.2 and 8.0 km estimated from the offset of the basal unconformity beneath Tertiary andesite and from our depth model for the Rhodes Salt Marsh basin.

Figure 15 illustrates the range of horizontal displacements across the modeled fault system, taking into consideration uncertainties associated with geologic control and ambiguity in fault dip. The minimum displacement solution arises from 60° dips for east-west and north-northwest oblique-slip faults with displacement of about 9.0 km along an azimuth of N85°W and of about 8.4 km along an azimuth of N35°W, respectively. Displacement compatibility predicts a dip of about 45° for northeast structures bounding the Rhodes Salt Marsh basin and a displacement of about 8.0 km along an azimuth of N55°W. Divergence of the net

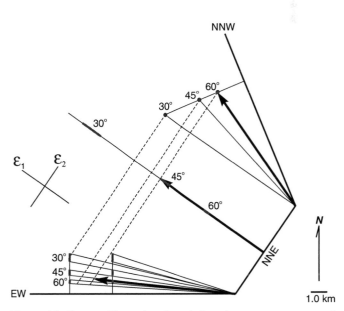

Figure 15. Cartoon illustrating the relations between geometry and horizontal displacement across the curved fault array (fault modeled as rectilinear for simplicity), and the predicted fault dips and horizontal displacement in the primary extension direction.

displacement around the curved array indicates that the geometry of the fault-basin system cannot be maintained without internal deformation of the structural blocks. Basin formation is dominated by divergent displacement and elongation toward the west-northwest but also requires a subsidiary component of orthogonal elongation toward the north-northeast. The development of two directions of elongation during basin evolution indicates finite flattening but does not discriminate between strain histories involving superposed plane strain or progressive incremental non–plane strain.

The consistency between the finite flattening recorded by the divergent displacement field and the two contemporaneous directions of extensional strain determined from fault-slip inversion indicates that basin formation occurred during a progressive history of incremental flattening and is not the product of superposed plane strain events. Throughout the region, the two directions of extensional strain are distinguished as primary and secondary on the basis of the relative prominence of fault sets and morphological features of slip lineations and shear sense indicators. The primary direction of extension has a consistent azimuth of ~N55°W, and the secondary direction is oriented ~N35°E (Fig. 14). The primary extension direction is accommodated on faults with orientations that reflect the dominant structures bounding the basin and inferred in the subsurface. In contrast, the secondary extension direction is accommodated on minor structures that do not contribute to the overall fault array and basin geometry. The contemporaneity of the two strain states is documented by mutually crosscutting relations between slip lineations associated with the two extension directions, and it is indicative of flattening deformation (Figs. 13A and 13B). The primary extensional strain axis bisects the acute (50°) angle between the diverging displacement directions measured around the curved fault array and is coincident with the direction of greatest finite elongation determined from the displacement budget. Similarly, the secondary strain axis coincides with the minor elongation axis determined from the displacement analysis.

Basin formation during progressive incremental flattening documented here still leaves unresolved the time duration between superposed displacement directions recorded in the slip data. Fault-plane solutions of moderate to large crustal earthquakes in the central Walker Lane record strain-rate axes consistent with the primary extension axis recorded in fault-slip data in the Rhodes Salt Marsh and surroundings (Figs. 2 and 14). The largest earthquakes near Rhodes Salt Marsh are the 1932 Cedar Mountain and 1934 Excelsior Mountain earthquakes, which occurred east and southwest of the Rhodes Salt Marsh basin, respectively (Fig. 2). Although the azimuth of the incremental extensional strain axes computed for these earthquakes suffers from large uncertainties (Ryall and Priestley, 1975; Doser, 1988), it nonetheless ranges between N43°–58°W and N30°–66°W for the 1932 and 1934 events, respectively. The seismologic record is consistent with the network elongation observed in geodetic trilateration (Fig. 14) south of the Rhodes Salt Marsh (Savage and Lisowski, 1984) and strain axes determined from regional GPS

geodetic velocities (Murphy et al., this volume). Although consistent with the primary direction of elongation recorded in our analysis, the contemporary strain field does not record strain-rate axes consistent with the secondary extension axis in our study.

Conceivably, this discrepancy may reflect a temporal cyclicity associated with strain accumulation and release over different length and time scales. Previous workers (Zoback et al., 1981; Zoback, 1989; Bellier and Zoback, 1995) have recognized superposed slip histories on major faults and interpreted the relations as a cyclic alternation, possibly at a time scale of few hundred thousand years, between extensional- and strike-slip–dominated deformation within the Walker Lane. Transient strain fluctuations across the Great Basin occur over durations of several years (Niemi and Wernicke, 2007), in a manner not unlike that recorded by slow earthquakes and tremors at convergent plate boundaries (Dragert et al., 2001; Miller et al., 2002; Melbourne and Webb, 2003). In the central Walker Lane, short-period cyclicity is observed in strain-axis orientation recorded by a strainmeter located in a mine 20 km north of Rhodes Salt Marsh along the western flank of Soda Springs Valley (Ryall and Priestley, 1975). Over a period of 20 mo, the strainmeter recorded a change between a primary and secondary extension direction of N40°W and N30°E, respectively (Fig. 14).

The superposition of slip lineations on structures that have accommodated two directions of extension indicates that incremental flattening in and around Rhodes Salt Marsh was accommodated by discrete episodes of displacement. The magnitude and duration of displacement episodes vary together with the different length scales of accommodating structures. Geodetic and seismological measures of the regional strain field record the primary direction of extension and reflect large-magnitude strains accommodated by a few major faults. In contrast, the strainmeter records relatively small magnitudes of extension along both primary and secondary axes and probably reflects diffuse deformation on minor structures within blocks bounded by major fault systems. Regionally extensive faults may localize strain characterized by periods of uniform displacement over time scales longer than the geodetic and seismological record over a duration of one or multiple seismic cycles. The duration of uniform displacement recorded on faults in the Mediterranean and northern Great Basin is on the order of tens of thousands of years (Marco et al., 1996; Ferranti et al., 2007; Oldow and Singleton, 2008) and, for changes in state of strain, may be on the order of a hundred thousand years (Bellier and Zoback, 1995). Small magnitudes of strain accommodated by minor structures contained within large fault-bounded blocks probably reflect the short-period alternation of strain axes seen over a period of months in the strainmeter.

CONCLUSIONS

A system of curved faults around the Rhodes Salt Marsh within the Mina deflection of west-central Nevada documents strikes varying from east, north-northeast, and north-northwest (through 120° of arc) and records left-oblique, normal, and right-oblique displacements, respectively. Detailed mapping indicates that the structures did not form at the intersection of kinematically discrete fault systems but rather are part of a kinematically coordinated system of curved faults. Paleomagnetic analysis of plutonic and volcanic rocks in the hanging wall of the fault system shows no appreciable vertical-axis rotation and indicates that the geometry of the curved array is primary and did not evolve during rotation of fault-bounded blocks within a northwest-striking shear couple. Rather, the curved faults formed in their present-day geometry as part of a regionally extensive structural stepover (Mina deflection) linking active displacement in the northern Eastern California shear zone and the central Walker Lane.

During the Pliocene to Holocene, displacement on the fault system, as determined from the offset of geologic markers and geophysical models, and large magnitude extension was localized in the formation of a 3.0-km-deep prismatic basin at the apex of the curved array beneath Rhodes Salt Marsh. Internal deformation of the hanging-wall block during divergent displacement of up to 8.0 to 9.0 km around the fault curvature resulted in finite flattening, where axes are consistent with contemporaneous incremental extensional strain axes determined from fault-slip inversion. Slip lineations on structures accommodating the primary and secondary extension directions of west-northwest and north-northeast are mutually crosscutting but clearly formed during discrete episodes of motion. Seismological and geodetic evidence suggests that large through-going structures may record alternations between primary and secondary directions of extension over durations of several seismic cycles and possibly up to several hundred thousand years. In contrast, alternation between primary and secondary extension axes recorded by a strainmeter over a period of 20 mo, suggests that small-displacement structures located within large fault blocks record a cyclicity of months to years.

ACKNOWLEDGMENTS

Ferranti acknowledges funds from the University of Naples through the Programma di Scambi Internazionali per la Mobilità di Breve Durata, for 2002 and 2003. Oldow acknowledges partial support of this research by National Science Foundation grants (EAR-0126205 and EAR-0208219). Kent Campbell is acknowledged for his participation in fault kinematic and gravity data acquisition. The careful reviews of P. Cashman and C.J. Northrup were instrumental in clarifying several aspects of the work.

REFERENCES CITED

Adriansyah, A., and McMechan, G.A., 2002, Analysis and interpretation of seismic data from thin reservoirs: Northwest Java basin, Indonesia: Geophysics, v. 67, p. 14–26, doi: 10.1190/1.1451317.

Anderson, E.M., 1951, The Dynamics of Faulting (2nd edition): Edinburgh and London, Oliver and Boyd, 206 p.

Angelier, J., 1984, Tectonic analysis of fault slip data sets: Journal of Geophysical Research, v. 89, p. 5835–5848, doi: 10.1029/JB089iB07p05835.

Angelier, J., Colletta, B., and Anderson, R.E., 1985, Neogene paleostress changes in the Basin and Range. A case study at Hoover Dam, Nevada–

Arizona: Geological Society of America Bulletin, v. 96, p. 347–361, doi: 10.1130/0016-7606(1985)96<347:NPCITB>2.0.CO;2.

Argus, D.F., and Gordon, R.G., 1991, Current Sierra Nevada–North America motion from very long baseline interferometry: Implications for the kinematics of the western United States: Geology, v. 19, p. 1085–1088, doi: 10.1130/0091-7613(1991)019<1085:CSNNAM>2.3.CO;2.

Beanland, S., and Clark, M., 1994, The Owens Valley fault zone, eastern California, and surface rupture associated with the 1872 earthquake: United States Geological Survey Bulletin, v. 1982, p. 1–29.

Bell, J.W., dePolo, C.M., Ramelli, A.R., Sarna-Wojcicki, A.M., and Meyer, C.E., 1999, Surface faulting and paleo-seismic history of the 1932 Cedar Mountain earthquake area, west-central Nevada, and implications for modern tectonics of the Walker Lane: Geological Society of America Bulletin, v. 111, p. 791–807, doi: 10.1130/0016-7606(1999)111<0791:SFAPHO>2.3.CO;2.

Bellier, O., and Zoback, M.L., 1995, Recent state of stress change in the Walker Lane zone western Basin and Range Province, United States: Tectonics, v. 14, p. 564–593, doi: 10.1029/94TC00596.

Bennett, R.A., Wernicke, B.P., Niemi, N.A., Friederich, A.M., and Davis, J.L., 2003, Contemporary strain rates in the northern Basin and Range Province from GPS data: Tectonics, v. 22, p. 3-1–3-31, doi: 10.1029/2001TC001355.

Callaghan, E., and Gianella, V., 1935, The earthquake of January 30, 1934, at Excelsior Mountains, Nevada: Bulletin of the Seismological Society of America, v. 25, p. 161–168.

Caskey, J., and Wesnousky, S.G., 1997, Static stress changes and earthquake triggering during the 1954 Fairview Peak and Dixie Valley Earthquakes: Central Nevada: Bulletin of the Seismological Society of America, v. 87, p. 521–527.

Coolbaugh, M.F., Kratt, C., Sladek, C., Zehner, R.C., and Shevenell, L., 2006, Quaternary borates as a geothermal exploration tool in the Great Basin: Geothermal Research Council Transactions, v. 30, p. 393–398.

dePolo, D.M., and dePolo, C.M., 1999, Earthquakes in Nevada 1852–1998: Nevada Bureau of Mines and Geology Map 119, scale 1:1,000,000.

dePolo, C.M., Peppin, W.A., and Johnson, P.A., 1993, Contemporary tectonics, seismicity, and potential earthquake sources in the White Mountains seismic gap, west-central Nevada and east-central California: Tectonophysics, v. 225, p. 271–299, doi: 10.1016/0040-1951(93)90302-Z.

Dickinson, W., and Butler, R., 1998, Coastal and Baja California paleomagnetism reconsidered: Geological Society of America Bulletin, v. 110, p. 1268–1280, doi: 10.1130/0016-7606(1998)110<1268:CABCPR>2.3.CO;2.

Dixon, T.H., Miller, M., Farina, F., Wang, H., and Johnson, D., 2000, Present-day motion of the Sierra Nevada block and some tectonic implications for the Basin and Range Province, North American Cordillera: Tectonics, v. 19, p. 1–24, doi: 10.1029/1998TC001088.

Dohrenwend, J.C., 1982, Surficial Geologic Map of the Walker Lake 1° by 2° Quadrangle, Nevada-California: U.S. Geological Survey Miscellaneous Field Studies Map MF-1382-C, 1 sheet, scale 1:250,000.

Dokka, R.K., and Travis, C.J., 1990, Role of the Eastern California shear zone in accommodating Pacific–North American plate motion: Geophysical Research Letters, v. 17, p. 1323–1326, doi: 10.1029/GL017i009p01323.

Doser, D.L., 1988, Source mechanisms of earthquakes in the Nevada seismic zone (1915–1943) and implications for deformation in the western Great Basin: Journal of Geophysical Research, v. 93, p. 15,001–15,015, doi: 10.1029/JB093iB12p15001.

Dragert, H., Wang, K., and James, T.S., 2001, A silent slip event on the deeper Cascadia subduction interface: Science, v. 292, p. 1525–1528, doi: 10.1126/science.1060152.

Ekren, E.B., and Byers, F.M., Jr., 1984, The Gabbs Valley Range—A well exposed segment of the Walker Lane in west-central Nevada, *in* Lintz, J., ed., Western Geologic Excursions: Boulder, Colorado, Geological Society of America Guidebook 4, p. 203–215.

Ferguson, H.G., and Muller, S.W., 1949, Structural Geology of the Hawthorne and Tonopah Quadrangles, Nevada: U.S. Geological Survey Professional Paper 216, 55 p.

Ferguson, H.G., Muller, S.W., and Cathcart, S.H., 1954, Geology of the Mina Quadrangle, Nevada: U.S. Geological Survey Quadrangle Map Series GQ 45, scale 1:125,000.

Ferranti, L., Monaco, C., Antonioli, F., Maschio, L., Kershaw, S., and Verrubbi, V., 2007, The contribution of regional uplift and coseismic slip to the vertical crustal motion in the Messina Straits, southern Italy: Evidence from raised late Holocene shorelines: Journal of Geophysical Research, v. 112, p. B06401, doi: 10.1029/2006JB004473.

Fisher, R.A., 1953, Dispersion on a sphere: Royal Society of London Proceedings, v. A217, p. 295–305.

Foster, J., and Symons, D.T.A., 1979, Defining a paleomagnetic polarity pattern in the Monteregian intrusives: Canadian Journal of Earth Sciences, v. 16, p. 1716–1725.

Gamond, J.F., 1987, Bridge structures as sense of displacement criteria in brittle fault zones: Journal of Structural Geology, v. 9, p. 609–620, doi: 10.1016/0191-8141(87)90146-5.

Garcia-Abdeslem, J., 2005, The gravitational attraction of a right rectangular prism with density varying with depth following a cubic polynomial: Geophysics, v. 70, p. J39–J42.

Garside, L.J., 1979, Geologic Map of the Camp Douglas Quadrangle, Nevada: Nevada Bureau of Mines and Geology Map 63, scale 1:24,000.

Garside, L.J., and Silberman, M.L., 1978, New K-Ar ages of volcanic and plutonic rocks from the Camp Douglas quadrangle, Mineral County, Nevada: Isochron-West, v. 22, p. 29–31.

Gianella, V.P., and Callaghan, E., 1934, The earthquake of December 20, 1932, at Cedar Mountain, Nevada, and its bearing on the genesis of Basin and Range structure: The Journal of Geology, v. 47, p. 1–22.

Globerman, B.R., and Irving, E., 1988, Mid-Cretaceous paleomagnetic reference field for North America: Restudy of 100 Ma intrusive rocks from Arkansas: Journal of Geophysical Research, v. 93, p. 11,721–11,733, doi: 10.1029/JB093iB10p11721.

Gradstein, F.M., Ogg, J.G., and Smith, A.G., eds., 2004, A Geologic Time Scale 2004: Cambridge, Cambridge University Press, 589 p.

Hancock, P.L., Al-Kahdi, A., Barka, A.A., and Bevan, T.G., 1987, Aspects of analyzing brittle structures: Annales Tectonicae, v. 1, p. 5–19.

Hardyman, R.F., and Oldow, J.S., 1991, Tertiary tectonic framework and Cenozoic history of the central Walker Lane, Nevada, *in* Raines, G.L., Lisle, R.E., Schafer, R.W., and Wilkinson, W.H., eds., Geology and Ore Deposits of the Great Basin: Reno, Nevada, Geological Society of Nevada, Symposium Proceedings, p. 279–301.

Healy, D.L., 1970, Calculated in-situ bulk density measurements from subsurface gravity observations and density logs, Nevada Test Site, and Hot Creek Valley, Nye County, Nevada: U.S. Geological Survey Professional Paper 700-B, p. 52–62.

Holom, D.I., and Oldow, J.S., 2007, Gravity reduction spreadsheet to calculate the Bouguer anomaly using standardized methods and constants: Geosphere, v. 3, p. 86–90, doi: 10.1130/GES00060.1.

Housen, B., Beck, M., Burmester, R., Fawcett, T., Petro, G., Sargent, R., Addis, K., Curtis, K., Ladd, J., Liner, N., Molitor, B., Montgomery, T., Mynatt, I., Palmer, B., Tucker, D., and White, I., 2003, Paleomagnetism of the Mount Stuart batholith revisited again: What has been learned since 1972?: American Journal of Science, v. 303, p. 263–299, doi: 10.2475/ajs.303.4.263.

Hudson, M.R., and Geissman, J.W., 1991, Paleomagnetic evidence for the age and extent of middle Tertiary counterclockwise rotation, Dixie Valley region, west-central Nevada: Journal of Geophysical Research, v. 96, p. 3979–4006, doi: 10.1029/90JB02424.

Kirschvink, J.L., 1980, The least-square line and plane and the analysis of paleomagnetic data: Geophysical Journal of the Royal Astronomical Society, v. 62, p. 699–718.

Lapointe, P.L., 1979, Paleomagnetism of the Notre Dame Bay lamprophyre dikes, Newfoundland, and the opening of the North Atlantic Ocean: Canadian Journal of Earth Sciences, v. 16, p. 1823–1831.

Larochelle, A., 1961, Application of palaeomagnetism to geological correlation: Nature, v. 192, p. 37–39, doi: 10.1038/192037a0.

Litinsky, V.A., 1989, Concept of effective density: Key to gravity determinations for sedimentary basins: Geophysics, v. 54, p. 1474–1482.

Marco, S., Stein, M., Agnon, A., and Ron, H., 1996, Long-term earthquake clustering: A 50,000-year paleoseismic record in the Dead Sea graben: Journal of Geophysical Research, v. 101, p. 6179–6191, doi: 10.1029/95JB01587.

Maxant, J., 1980, Variation of density with rock type, depth, and formation in the Western Canada basin from density logs: Geophysics, v. 45, p. 1061–1076.

McEnroe, S.A., 1996, A Barremian–Aptian (Early Cretaceous) North American paleomagnetic reference pole: Journal of Geophysical Research, v. 101, p. 15,819–15,835, doi: 10.1029/96JB00652.

Melbourne, T.I., and Webb, F.H., 2003, Slow but not quiet: Science, v. 300, p. 1886–1887, doi: 10.1126/science.1086163.

Mickus, K.L., Aiken, C.L.V., and Kennedy, W.D., 1991, Regional-residual gravity anomaly separation using the minimum curvature technique: Geophysics, v. 56, p. 279–283, doi: 10.1190/1.1443041.

Miller, M.M., Johnson, D.J., Dixon, T., and Dokka, R.K., 2001, Refined kinematics of the Eastern California shear zone from GPS observations, 1993–1998: Journal of Geophysical Research, v. 106, p. 2245–2264, doi: 10.1029/2000JB900328.

Miller, M.M., Melbourne, T., Johnson, D.J., and Sumner, W.Q., 2002, Periodic slow earthquakes from the Cascadia subduction zone: Science, v. 295, p. 2423, doi: 10.1126/science.1071193.

Mount, V.S., and Suppe, J., 1987, State of stress near the San Andreas fault: Implications for wrench tectonics: Geology, v. 15, p. 1143–1146, doi: 10.1130/0091-7613(1987)15<1143:SOSNTS>2.0.CO;2.

Murphy, J.J., Watkinson, A.J., and Oldow, J.S., 2009, this volume, Spatially partitioned transtension within the central Walker Lane, western Great Basin, USA: Application of the polar Mohr construction for finite deformation, *in* Oldow, J.S., and Cashman, P.H., eds., Late Cenozoic Structure and Evolution of the Great Basin–Sierra Nevada Transition: Geological Society of America Special Paper 447, doi: 10.1130/2009.2447(04).

Niemi, N.A., and Wernicke, B.P., 2007, Spatial and temporal distribution of strain in the Sevier Desert region from a decade of BARGEN continuous GPS observations: Eos (Transactions, American Geophysical Union), v. 88, p. T31C-0578.

Oldenburg, D.W., 1974, The inversion and interpretation of gravity anomalies: Geophysics, v. 39, p. 526–536.

Oldow, J.S., 1981, Structure and stratigraphy of the Luning allochthon and the kinematics of allochthon emplacement, Pilot Mountains, west-central Nevada: Geological Society of America Bulletin, v. 92, no. 12, p. 888–911, doi: 10.1130/0016-7606(1981)92<888:SASOTL>2.0.CO;2.

Oldow, J.S., 1983, Tectonic implications of a late Mesozoic fold and thrust belt in northwestern Nevada: Geology, v. 11, p. 542–546, doi: 10.1130/0091-7613(1983)11<542:TIOALM>2.0.CO;2.

Oldow, J.S., 1984, Evolution of a late Mesozoic fold and thrust belt, northwestern Great Basin, USA: Tectonophysics, v. 102, p. 245–274, doi: 10.1016/0040-1951(84)90016-7.

Oldow, J.S., 1992, Late Cenozoic displacement partitioning in the northwestern Great Basin, *in* Craig, S.D., ed., Structure, Tectonics, and Mineralization of the Walker Lane: Reno, Geological Society of Nevada, Walker Lane Symposium, p. 17–52.

Oldow, J.S., 2003, Active transtensional boundary zone between the western Great Basin and Sierra Nevada block, western U.S. Cordillera: Geology, v. 31, p. 1033–1036, doi: 10.1130/G19838.1.

Oldow, J.S., and Dockery, H.A., 1993, Geologic Map of the Mina Quadrangle, Mineral County, Nevada: Nevada Bureau of Mines and Geology Field Studies Map 6, scale 1:24,000.

Oldow, J.S., and Singleton, E.S., 2008, Application of Terrestrial Laser Scanning in determining the pattern of late Pleistocene and Holocene fault-displacement from the offset of pluvial lake shorelines in the Alvord extensional basin, northern Great Basin, USA: Geosphere, v. 4, no. 3, p. 536–563, doi: 10.1130/GES00101.1.

Oldow, J.S., and Speed, R.C., 1985, Preliminary Geologic Map of the Mina N.W. Quadrangle, Mineral County, Nevada: U.S. Geological Survey Miscellaneous Field Studies MF-1487, scale 1:24,000.

Oldow, J.S., and Steuer, M.R., 1985, Preliminary Geologic Map of the Mable Mountain Quadrangle, Mineral County, Nevada: U.S. Geological Survey Miscellaneous Field Studies Map MF-1486, scale 1:24,000.

Oldow, J.S., Bally, A.W., Avé Lallemant, H.G., and Leeman, W.P., 1989, Phanerozoic evolution of the North American Cordillera (United States and Canada), *in* Bally, A.W., and Palmer, A.R., eds., The Geology of North America: An Overview: Boulder, Colorado, Geological Society of America, Geology of North America, v. A, p. 139–232.

Oldow, J.S., Aiken, C.L.V., Ferguson, J.F., Hare, J.L., and Hardyman, R.F., 2001, Active displacement transfer and differential motion between tectonic blocks within the central Walker Lane, western Great Basin: Geology, v. 29, p. 19–22, doi: 10.1130/0091-7613(2001)029<0019:ADTADB>2.0.CO;2.

Oldow, J.S., Geissman, J.W., and Stockli, D.F., 2008, Evolution and strain reorganization within late Neogene structural stepovers linking the central Walker Lane and northern Eastern California shear zone, western Great Basin: International Geology Review, v. 50, p. 1–21.

Petit, J.P., 1987, Criteria for the sense of movement on fault surfaces in brittle rocks: Journal of Structural Geology, v. 9, p. 597–608, doi: 10.1016/0191-8141(87)90145-3.

Reheis, M.C., and Sawyer, T.L., 1997, Late Cenozoic history and slip rates of the Fish Lake Valley, Emigrant Peak, and Deep Springs fault zones, Nevada and

California: Geological Society of America Bulletin, v. 109, p. 280–299, doi: 10.1130/0016-7606(1997)109<0280:LCHASR>2.3.CO;2.

Rogers, A.M., Harmsen, S.C., Corbett, E.J., Priestley, K., and dePolo, D., 1991, The seismicity of Nevada and some adjacent parts of the Great Basin, *in* Slemmons, D.B., Engdahl, E.R., Zoback, M.D., and Blackwell, D.D., eds., Neotectonics of North America: Boulder, Colorado, Geological Society of America, Decade Map Volume 1, p. 153–184.

Ryall, A.S., and Priestley, K., 1975, Seismicity, secular strain, and maximum magnitude in the Excelsior Mountains area, western Nevada and eastern California: Geological Society of America Bulletin, v. 86, p. 1585–1592, doi: 10.1130/0016-7606(1975)86<1585:SSSAMM>2.0.CO;2.

Saltus, R.W., 1988, Regional, Residual and Derivative Gravity Maps of Nevada: Nevada Bureau of Mines and Geology Map 94B, scale 1:1,000,000.

Sauber, J., Thatcher, W., Solomon, S., and Lisowski, M., 1994, Geodetic slip rate for the Eastern California shear zone and recurrence time for Mojave Desert earthquakes: Nature, v. 367, p. 264–266, doi: 10.1038/367264a0.

Savage, J.C., and Lisowski, M., 1984, Deformation in the White Mountain Seismic Gap, California-Nevada 1972–1982: Journal of Geophysical Research, v. 89, p. 7671–7687, doi: 10.1029/JB089iB09p07671.

Sclater, J.G., and Christie, P.A.F., 1980, Continental stretching: An explanation of the post mid-Cretaceous subsidence of the central North Sea Basin: Journal of Geophysical Research, v. 85, p. 3711–3739, doi: 10.1029/JB085iB07p03711.

Shive, P.N., and Frerichs, W.E., 1974, Paleomagnetism of the Niobrara Formation in Wyoming, Colorado, and Kansas: Journal of Geophysical Research, v. 79, p. 3001–3009, doi: 10.1029/JB079i020p03001.

Silva, J.B.C., Costa, D.C.L., and Barbosa, B.C.V., 2006, Gravity inversion of basement relief and estimation of density contrast variation with depth: Geophysics, v. 71, p. J51–J58, doi: 10.1190/1.2236383.

Speed, R.C., 1977, Excelsior Formation, west central Nevada: Stratigraphic appraisal, new divisions, and paleogeographic interpretations, *in* Stewart, J.H., Stevens, C.H., and Fritsche, A.E., eds., Paleozoic Paleogeography of the Western United States: Pacific Section, Society Economic Paleontologists and Mineralogists, Pacific Coast Paleogeographic Symposium 1, p. 325–336.

Speed, R.C., and Cogbill, A.H., 1979a, Cenozoic volcanism of the Candelaria region, Nevada: Summary: Geological Society of America Bulletin, v. 90, Part 1, p. 143–145, doi: 10.1130/0016-7606(1979)90<143:CVOTCR>2.0.CO;2.

Speed, R.C., and Cogbill, A.H., 1979b, Candelaria and other left-oblique slip faults of the Candelaria region, Nevada: Geological Society of America Bulletin, v. 90, p. 149–163, doi: 10.1130/0016-7606(1979)90<149:CAOLSF>2.0.CO;2.

Speed, R.C., and Kistler, R.W., 1980, Cretaceous volcanism, Excelsior Mountains, Nevada: Geological Society of America Bulletin, v. 91, p. 392–398, doi: 10.1130/0016-7606(1980)91<392:CVEMN>2.0.CO;2.

Stewart, J.H., 1988, Tectonics of the Walker Lane belt, western Great Basin Mesozoic and Cenozoic deformation in a zone of shear, *in* Ernst, W.G., ed., Metamorphism and Crustal Evolution of the Western United States, Ruby Volume VII: Englewood Cliffs, New Jersey, Prentice Hall, p. 685–713.

Stockli, D.F., Dumitru, T.A., McWilliams, M.O., and Farley, K.A., 2003, Cenozoic tectonic evolution of the White Mountains, California and Nevada: Geological Society of America Bulletin, v. 115, no. 7, p. 788–816, doi: 10.1130/0016-7606(2003)115<0788:CTEOTW>2.0.CO;2.

Unruh, J., Humphrey, J., and Barron, A., 2003, Transtensional model for the Sierra Nevada frontal fault system, eastern California: Geology, v. 31, p. 327–330, doi: 10.1130/0091-7613(2003)031<0327:TMFTSN>2.0.CO;2.

Van der Voo, R., 1993, Paleomagnetism of the Atlantic, Tethys, and Iapetus Oceans: Cambridge, Cambridge University Press, 411 p.

Van Fossen, M.C., and Kent, D.V., 1992, Paleomagnetism of 122 Ma plutons in New England and the mid-Cretaceous paleomagnetic field in North America—True polar wander or large-scale differential mantle motion: Journal of Geophysical Research, v. 97, p. 19,651–19,661, doi: 10.1029/92JB01466.

Wallace, R.E., 1984, Patterns and timing of late Quaternary faulting in the Great Basin Province and relation to some regional tectonic features: Journal of Geophysical Research, v. 89, p. 5763–5769, doi: 10.1029/JB089iB07p05763.

Wesnousky, S.G., 2005, Active faulting in the Walker Lane: Tectonics, v. 24, p. TC3009, doi: 10.1029/2004TC001645.

Wesnousky, S.G., and Jones, C.H., 1994, Oblique slip, slip partitioning, spatial and temporal changes in the regional stress field, and the relative strength of active faults in the Basin and Range, western United States: Geology, v. 22, p. 1031–1034, doi: 10.1130/0091-7613(1994)022<1031:OSSPSA>2.3.CO;2.

Zijderveld, J.D.A., 1967, A.C. demagnetization of rocks: Analysis of results, *in* Collinson, D.W., Creer, K.M., and Runcorn, S.K., eds., Methods in Palaeomagnetism: Amsterdam, Elsevier, p. 254–286.

Zoback, M.L., 1989, State of stress and modern deformation of the northern Basin and Range Province: Journal of Geophysical Research, v. 94, p. 7105–7128, doi: 10.1029/JB094iB06p07105.

Zoback, M.L., Anderson, R.E., and Thompson, G.A., 1981, Cenozoic evolution of the state of stress and style of tectonism of the Basin and Range Province of the western United States, *in* Vine, F.J., and Smith, A.D., eds., Extensional Tectonics Associated with Convergent Plate Boundaries: London, Royal Society of London, p. 189–216.

MANUSCRIPT ACCEPTED BY THE SOCIETY 21 JULY 2008

The Geological Society of America
Special Paper 447
2009

Late Miocene to Pliocene vertical-axis rotation attending development of the Silver Peak–Lone Mountain displacement transfer zone, west-central Nevada

Michael S. Petronis*
Environmental Geology, Natural Resource Management Department, Ivan Hilton Science Center,
New Mexico Highlands University, Las Vegas, New Mexico 87701, USA

John W. Geissman
Department of Earth and Planetary Sciences, MSC 03 2040, University of New Mexico, Albuquerque, New Mexico 87131-0001, USA

John S. Oldow
Department of Geosciences, University of Texas at Dallas, Richardson, Texas 75080-3021, USA

William C. McIntosh
New Mexico Geochronological Research Laboratory, New Mexico Institute of Mining and Technology, Socorro,
New Mexico 87801, USA

ABSTRACT

Paleomagnetic data from three regionally extensive Oligocene ignimbrite sheets, two sequences of Miocene andesite flows, and ten sequences of Upper Miocene to Pliocene basaltic andesite flows in the Candelaria Hills and adjacent areas, west-central Nevada, provide further evidence that, since the late Miocene, and possibly between latest Miocene and earliest Pliocene time, the broad region that initially facilitated Neogene displacement transfer between the Furnace Creek and central Walker Lane fault systems experienced some 20° to 30° of clockwise vertical-axis rotation. The observed sense and magnitude of rotation are similar to those previously inferred from paleomagnetic data from different parts of the Silver Peak Range to the south. We propose that clockwise rotation within the transfer zone formed in response to horizontal components of simple and pure shear distributed between early-formed, northwest-striking right-lateral structures that initiated in mid- to late Miocene time. Notably, the spatial distribution of the early-formed transfer zone is larger and centered south of the presently active stepover, which initiated in the late Pliocene and is characterized by a transtensional deformation field and slip on east-northeast–oriented left-oblique structures that define the Mina deflection. The sense and magnitude of rotation during this phase of deformation, which we infer to be of pre–latest Pliocene age, are inconsistent with the geodetically determined regional velocity field and seismologically determined strain

†Corresponding author e-mail: mspetro@nmhu.edu.

Petronis, M.S., Geissman, J.W., Oldow, J.S., and McIntosh, W.C., 2009, Late Miocene to Pliocene vertical-axis rotation attending development of the Silver Peak–Lone Mountain displacement transfer zone, west-central Nevada, *in* Oldow, J.S., and Cashman, P.H., eds., Late Cenozoic Structure and Evolution of the Great Basin–Sierra Nevada Transition: Geological Society of America Special Paper 447, p. 215–253, doi: 10.1130/2009.2447(12). For permission to copy, contact editing@geosociety.org. ©2009 The Geological Society of America. All rights reserved.

field for this area. As a consequence, the longer-term kinematic evolution of the stepover system, and the adjoining parts of the Furnace Creek and Walker Lane fault systems, cannot be considered as a steady-state process through the Neogene.

Keywords: Walker Lane, paleomagnetism, rotation, Basin and Range, transtension.

INTRODUCTION

Modern deformation between the North American and Pacific plates is distributed across a wide zone of western North America (Fig. 1), from the San Andreas fault eastward into the western Basin and Range Province (Bennett et al., 2003; Malservisi et al., 2001; Dixon et al., 2003, 2000, 1995; Miller et al., 2001; Oldow, 2003; Oldow et al., 2001; Thatcher et al., 1999; Faulds et al., 2005). The eastern Mojave Desert and the region between the Sierra Nevada and Great Basin form a zone of distributed deformation that accommodates ~25% of the relative motion between the Pacific and North American plates (Minster and Jordan, 1987; Argus and Gordon, 1991; Bennett et al., 2003; Gan et al., 2000). Deformation to the south is localized within the Eastern California shear zone (Dokka

and Travis, 1990, 1996) and to the north in the Walker Lane belt (Locke et al., 1940; Stewart, 1980), which today is characterized by northwest-trending faults with associated earthquake focal mechanisms and Global Positioning System (GPS) velocities that are indicative of transtensional deformation (Oldow, 2003). At the latitude of the southern Sierra Nevada, north of the Garlock fault, displacement in the northern part of the Eastern California shear zone occurs in an ~250-km-wide zone of right-lateral shear that is accommodated along three major strike-slip fault zones; (1) Death Valley–Furnace Creek–Fish Lake Valley, (2) Hunter Mountain–Panamint Valley, and (3) White Mountains–Owens Valley (Fig. 1). Northward, the shear zone narrows, and east of the central Sierra Nevada, the northwest-trending strike-slip fault systems abruptly swing eastward into an array of northeast-southwest– to east-west–

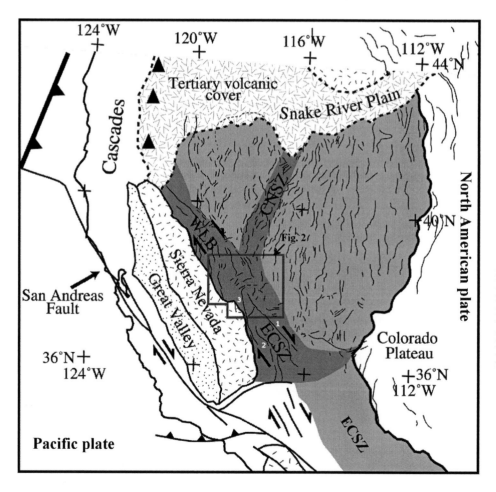

Figure 1. Simplified tectonic map of the central part of the western United States showing the major tectonic provinces, modern plate boundaries, and fault traces in the northern Basin and Range Province. Basin and Range province is in light gray; Walker Lane belt (WLB), Central Nevada seismic zone (CNSZ), and Eastern California shear zone (ECSZ) are in dark gray. Traces of Cenozoic faults in the northern Basin and Range are after Stewart (1988). Box denotes location of Figure 2. Numbers refer to major fault systems mentioned in text: (1) Death Valley–Furnace Creek–Fish Lake Valley, (2) Hunter Mountain–Panamint Valley, and (3) White Mountains–Owens Valley fault zones.

striking faults, centered on the Candelaria Hills and Excelsior Mountains, in the southern part of the central Walker Lane belt. This area is commonly referred to as the Mina deflection (Ryall and Priestly, 1975) (Fig. 2). These curvilinear faults form a "z"-shaped relay zone that currently transfers much of the slip along the Furnace Creek–Fish Lake Valley and White Mountain–Owens Valley fault zones eastward to northwest-trending transcurrent faults of the central Walker Lane and ultimately to the northern Walker Lane fault system and the Central Nevada seismic zone (Stewart, 1980; Wallace, 1984; Oldow, 1992, 2003; Oldow et al., 2001; Dickinson, 2006). Displacement transfer in this area initiated in the mid- to late Miocene, was facilitated by the development of the Silver Peak–Lone Mountain extensional complex (Oldow et al., 1994; Oldow et al., this volume), and evolved in the late Pliocene into a more focused system of discrete east-northeast–trending faults (Wesnousky, 2005). The area is the locus of moderate to large historic earthquakes (Ryall and Priestly, 1975) and late Quaternary fault scarps (Wesnousky, 2005, and references therein). Early phases of deformation related to displacement transfer, as well as more contemporary displacement on the curvilinear arrays of faults, resulted in the formation of several triangular extensional sedi-

mentary basins situated at the apices of regional curved faults (Oldow, 1992; Oldow et al., 2001; Stockli et al., 2003; Ferranti et al., this volume; Tincher and Stockli, this volume).

The way(s) in which Neogene deformation between the Sierra Nevada block and western Great Basin was accommodated and evolved with time, including the history of displacement transfer between the northern part of the Eastern California shear zone and the central Walker Lane, is a matter of considerable interest and debate (Wesnousky, 2005). To better define the nature of deformation related to the initiation phase of deformation associated with the Silver Peak–Lone Mountain extensional complex and to contemporary displacement transfer between the Furnace Creek and central Walker Lane fault systems through the Mina deflection, we used paleomagnetic analysis of Tertiary volcanic rocks to assess the degree of vertical-axis rotation accompanying regional deformation. The contemporary GPS velocity field and associated strain field recorded by earthquake focal mechanisms are not consistent with appreciable vertical-axis rotations (Oldow, 2003; Oldow et al., 2008). Prompted by previous work in the Silver Peak Range to the south, which showed that substantial vertical-axis rotation accompanied Miocene to early Pliocene deformation in the Silver Peak–Lone Mountain extensional com-

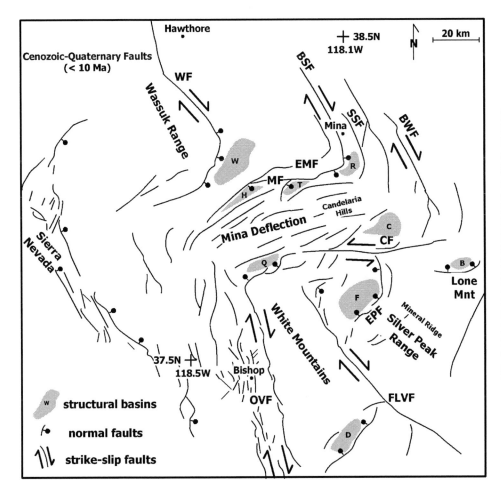

Figure 2. Simplified fault map of the Mina deflection and surrounding region. Diagram depicts late Cenozoic faults and locations of pull-apart structures in the Mina deflection area. Pull-apart structures: D—Deep Springs; F—northern Fish Lake Valley; H—Huntoon Valley; Q—Queen Valley; R—Rhodes Salt Marsh; B—southern Big Smokey Valley; T—Teels Salt Marsh; W—Whiskey Flat. C—Columbus Salt March area. Late Cenozoic faults: WF—Wassuk fault; EMF—Excelsior Mountain fault; BSF—Benton Springs fault; SSF—Soda Springs fault; BWF—Beetles Well fault; CF—Coaldale fault; EPF—Emigrant Peak fault. FLVF—Death Valley–Furnace Creek–Fish Lake Valley fault zone; OVF—White Mountains–Owens Valley fault zones. Hunter Mountain–Panamint Valley fault zone is not shown.

218

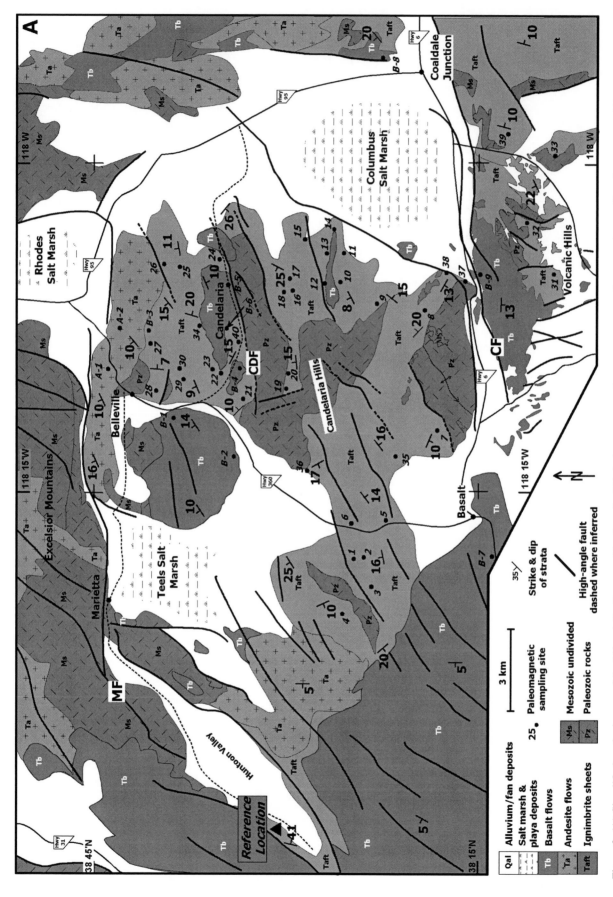

Figure 3. (A) Simplified geologic map of the Mina deflection area. Numbered solid circles are areas where one or more paleomagnetic sampling sites were established. These locations are considered "structural blocks" (see text for discussion) and correspond to localities in Table 2. Solid black triangle indicates the location of the Huntoon Valley ignimbrite section, which was used as the paleomagnetic reference location. MF—Marietta fault; CDF—Candelaria fault; CF—Coaldale fault. Dotted line with marker—state highway; solid line—dirt road; dashed line with marker—(*Continued on following page.*)

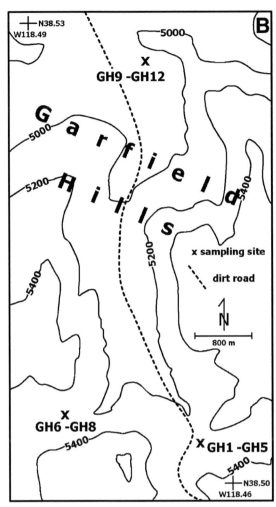

Figure 3 (*continued*). (B) Paleomagnetic sampling sites in the Garfield Hills area located 55 km north-northwest of the central Candelaria Hills.

plex (Petronis et al., 2002, 2007), we collected paleomagnetic data from broadly dispersed exposures of Oligocene- to earliest Pliocene–age volcanic rocks in the Candelaria Hills and adjoining areas to detect possible rotations and to assess their magnitude, sense, and spatial variability. Based on these new data, we infer that modest clockwise rotation affected a broad zone during the early history of displacement transfer, and we conclude that the observed rotations in this broad zone are inconsistent with the current velocity field, which developed in late Pliocene and younger time. Thus, the pattern of deformation appears to have changed since the late Miocene (Oldow et al., 2008), and Neogene deformation in this area was not a steady-state process.

TECTONIC SETTING OF THE MINA DEFLECTION

Physiographic and tectonic differences between the central Basin and Range Province and the Walker Lane belt have

long been recognized (Locke et al., 1940). The central part of the northern Basin and Range Province, with its evenly spaced north-south–trending mountain ranges, contrasts with the Walker Lane in the western part of the province (Figs. 1 and 2), which is characterized by arcuate mountain ranges and a complicated array of predominantly northwest-striking, right-lateral strike-slip and transtensional faults with lesser northeast-striking normal and left-lateral strike-slip faults (Stewart, 1980; Hardyman and Oldow, 1991; Stewart, 1991; Oldow, 1992; Dickinson, 2006; Wesnousky, 2005). Late Cenozoic faults, late Pleistocene to Holocene fault scarps, and modern seismicity indicate that the central Walker Lane belt consists of arrays of variably oriented, active faults characterized by northwest-directed right-lateral shear and dip-slip fault motion (Oldow, 1992; Wesnousky, 2005; Ferranti et al., this volume). Tertiary deformation in the Mina deflection area is hypothesized to have been strongly influenced by the pre-Tertiary geologic and structural framework of the western Basin and Range Province. The region experienced a long history of Phanerozoic continental rifting and subsequent convergent margin magmatism and tectonism (e.g., Stewart, 1972, 1991; Burchfiel and Davis, 1971; Oldow et al., 1989). In the Mina deflection, the northeast-southwest–trending structural grain appears to have been primarily controlled by pre-Cenozoic depositional and structural features that reflect the trace of the Neoproterozoic transition between attenuated continental and oceanic crust. Both Paleozoic miogeoclinal sediment distributions and structures associated with Paleozoic and Mesozoic shortening follow this Precambrian crustal boundary (Stewart, 1991; Oldow et al., 1989).

North of the Mina deflection, the northern part of the central Walker Lane is characterized by an en-echelon array of northwest-trending strike-slip faults, including the Wassuk, Soda Springs, Benton Springs, and Bettles Well fault zones (Fig. 2). To the south, these faults enter the Mina deflection, a zone of predominantly east-northeast–striking oblique-slip faults, of late Pliocene and younger age, which are characterized by extensional and left-lateral fault motion. We demonstrate here that these east-northeast–striking faults define a series of structural blocks in the Candelaria Hills and adjacent areas. The northwest-trending right-lateral strike-slip faults are not truncated by the northeast-trending faults (Stewart, 1988) but rather smoothly change orientation (Fig. 2) as they swing into the Mina deflection (Ferranti and Oldow, 2005; Ferranti et al., this volume; Tincher and Stockli, this volume). A few faults are mapped as continuous bends forming "z"-shaped releasing bends (Fig. 2) characterized by prismatic pull-apart basins that accommodate the strike-slip displacement (Oldow, 1992; Ferranti and Oldow, 2005; Ferranti et al., this volume). For example, northeast-striking faults in the Excelsior Mountains (Figs. 2 and 3) progressively change in orientation and merge with northwest-trending fault systems that bound a deep pull-apart basin in the Rhodes Salt Marsh area (Cogbill, 1979; Oldow, 1992; Ferranti and Oldow, 2005; Ferranti et al., this volume) (as well as several pull-apart basins represented by Teels Salt Marsh, Huntoon Valley, and Columbus Salt Marsh areas; Oldow, 1992). Southeast of Huntoon Valley, splays of the Excelsior fault system swing south and merge with faults that are part of the Owens Valley–White Mountain fault system (Tincher and Stockli, this volume).

The Excelsior Mountains form part of the northern Mina deflection, and they are bounded to the south by the Marietta fault and related structures (Fig. 2). The left-lateral Coaldale fault, located at the northern end of Fish Lake Valley, forms the southern boundary of the Mina deflection. This structure presently transfers displacement east from Owens Valley through Queens Valley into Big Smokey Valley (Fig. 2) and eventually northward into en-echelon north-northwest–trending structures of the central Walker Lane (Oldow, 1992; Reheis and Sawyer, 1997; Stockli, 1999; Lee et al., 2001, 2005; Schroeder et al., 2005; Stockli et al., 2003; Tincher and Stockli, this volume). In the northern Volcanic Hills, the Coaldale fault offsets an upper Pliocene basaltic andesite (3.14 ± 0.1 Ma); along strike to the west, the fault cuts basaltic andesite of earlier Pliocene age (5.28 ± 0.01 Ma, $^{40}Ar/^{39}Ar$, whole rock; and 4.7 ± 0.9 Ma, $^{40}Ar/^{39}Ar$, whole rock) (Lee et al., 2005). Apatite (U-Th)/He thermochronology data from crystalline rocks in the footwall to the Queen Valley pull-apart fault system (Tincher and Stockli, this volume) and geomorphologic and seismotectonic studies indicate that an important component of displacement along the structure occurred at ca. 3 Ma, resulting in the development of a right-lateral fault system on the west side of the White Mountains and moderate renewed uplift of the White Mountains (DePolo, 1989; Lueddecke et al., 1998; Stockli, 1999; Tincher and Stockli, this volume). The faulting at ca. 3 Ma may have been a component of the overall change in the deformational regime of the region, from an earlier one of vertical thinning and mixed horizontal components of simple and pure shear to the transtensional shear that characterizes the present-day active fault system in the Mina deflection (Oldow et al., 1994, 2001, 2008; Bennett et al., 1999; Oldow, 2003; Oldow et al., this volume). At the southwest margin of the Mina deflection, right-lateral displacement along the Fish Lake Valley fault zone appears to die out northward in a right-stepping pull-apart structure in northern Fish Lake Valley. This active pull-apart basin is structurally controlled by the northeast-trending Emigrant Peak fault (Reheis and Sawyer, 1997), which transfers slip from the Fish Lake Valley fault zone (Fig. 2) into the northern Fish Lake Valley (Oldow, 1992). The Silver Peak Range exposes part of a shallowly northwest-dipping regional detachment system (Oldow, 1992; Oldow et al., 1994), the upper plate of which consists of Mesozoic plutons and their wall rocks as well as Upper Tertiary volcanic and sedimentary rocks (Diamond and Ingersoll, 2003; Oldow et al., 2008; Oldow et al., this volume). The lower plate, exposed on Mineral Ridge, consists of metamorphosed Neoproterozoic and Paleozoic rocks, Mesozoic granitoids, and crosscutting Miocene mafic dikes (Fig. 2). Oldow et al. (1994) described northwest-directed detachment faulting and the development of the Silver Peak–Lone Mountain extensional complex as representing part of a large-magnitude regional pull-apart system active in the late Miocene between ca. 11 and 6 Ma. Displacement on the Emigrant Peak fault and

the formation of the northern Fish Lake Valley pull-apart basin postdated this earlier episode of extension (Reheis and Sawyer, 1997), and faulting migrated northward in Pliocene time, as suggested by the inception of faulting in northern Fish Lake Valley (ca. 6 Ma) and an important phase of faulting in northern Owens Valley (ca. 3 Ma) (Stockli et al., 2003).

In the northern Fish Lake Valley, Pliocene sedimentary rocks and basalt flows are deformed into northwest-trending folds, which are interpreted as the northward continuation of turtleback structures exposed in the Silver Peak Range (Oldow et al., 1994; Reheis and Sawyer, 1997). These folds are cut by east-northeast–striking high-angle faults that transfer displacement across the northern part of the Silver Peak Range to right-lateral faults of the central Walker Lane (Reheis and Sawyer, 1997). Previous paleomagnetic investigations in the Silver Peak Range (east to northeast area, Petronis et al., 2002; southwest area, Petronis et al., 2007) have shown that Miocene volcanic rocks in the upper plate of a regionally extensive, northwest-dipping detachment system (the Silver Peak–Lone Mountain extensional complex), Miocene mafic intrusions in the lower plate, and Late Cretaceous mafic dikes in the southwest Silver Peak Range have experienced modest net clockwise rotation. Major through-going east-northeast–trending faults, however, are absent in the Silver Peak Range. The observed rotations in the Silver Peak Range are thus interpreted to characterize an "early" (mid- to late Miocene to earliest Pliocene) coherent strain rotation as well as a material rotation during an initial phase of displacement transfer. The spatial distribution of the area affected by this phase of deformation is better defined with new paleomagnetic data (presented here); overall, the deformation affecting this area lacked the development of discrete structural blocks as part of the transtensional deformation that presently characterizes the Mina deflection.

GEOLOGY OF THE CANDELARIA HILLS AREA

The Candelaria Hills (Fig. 3) lie roughly at the center of the 50-km-wide zone characterized by east-northeast–striking sinistral strike-slip and oblique extensional faults in the Mina deflection (Speed and Cogbill, 1979c; Oldow, 1992). The region exposes principally mid-Tertiary volcanic rocks and minor pre-Cenozoic intrusions and Precambrian to Paleozoic metasedimentary rocks. Mid-Tertiary volcanic rocks include at least five distinct informal map units consisting of many lava flows and simple and complex pyroclastic deposits (Speed and Cogbill, 1979a). In stratigraphic order, these are "older tuff," upper tuff (Candelaria sequence), andesite flows (Gilbert Andesite), sedimentary rocks (Esmeralda Formation and volcaniclastic sedimentary rocks), and basaltic andesite flows. The available $^{40}Ar/^{39}Ar$ and K-Ar age estimates show that magmatism occurred between ca. 27.5 Ma and ca. 3 Ma (Silberman et al., 1975; Marvin and Mehnert, 1977; Robinson and Stewart, 1984; Petronis et al., 2004; Table 1). The Oligocene upper tuffs include three regionally extensive ignimbrites, as well as numerous pyroclastic deposits of more limited lateral extent and/or exposure, some of which may be the distal

TABLE 1. NEOGENE PALEOMAGNETIC POLES FOR NORTH AMERICA

Latitude (°N)	Longitude (°W)	A95	K	Age (Ma)	Expected direction Dec. (°)	Inc. (°)	Reference
84.6	164.4	3.1	107.7	10	353.2	58.3	1
83.6	163.0	3.2	84.2	15	351.9	58.3	1
81.0	156.2	4.5	68.3	20	348.6	57.5	1
82.8	165.7	5.3	57.9	25	350.9	58.6	1
81.0	211.0	13.0	6.0	24–58	353.4	63.9	2
88.3	209.0	6.3	–	17	358.8	58.7	3
85.5	108.9	4.4	–	12–20	356.0	54.3	4
81.9	143.6	4.5	17.6	20–40	350.0	55.8	5

Note: References: 1—Besse and Courtillot (2002), synthetic poles; 2—Larson and Strangeway (1966), Spanish Peak dikes; 3—Mankinen et al. (1987), Steens Mountain; 4—Calderone et al. (1990), Mojave-Sonora Desert volcanics; 5—Diehl et al. (1988), Mogollon-Datil volcanic field.

parts of voluminous ignimbrites with caldera sources 50+ km to the northeast (Speed and Cogbill, 1979a; Robinson and Stewart, 1984; C. Henry, 2006, personal commun.; Petronis and Geissman, 2008). These are all part of the Candelaria pyroclastic sequence of Speed and Cogbill (1979a). The pyroclastic rocks are overlain in angular unconformity by Miocene andesite lava and lahar flows and younger Miocene to Pliocene basaltic andesite flows erupted from numerous local vents and fissures (Ferguson et al., 1954; Silberman et al., 1975; Marvin and Mehnert, 1977; Ormerod, 1988; Lee et al., 2005).

Candelaria Pyroclastic Sequence

The Candelaria sequence consists of eleven distinct pyroclastic deposits, including three regionally extensive ignimbrites (Robinson and Stewart, 1984). The regional ignimbrites are recognized as far north as the Excelsior Mountains and Garfield Hills, as far south as the central Silver Peak Range and the west flank of the White Mountains (Krauskopf, 1971), and as far west as the southwest margin of Huntoon Valley (Robinson and Stewart, 1984; our own observations) (Figs. 2 and 3). To the east, the regional ignimbrites may extend to the edge of Big Smokey Valley. However, no conclusively identified exposures of the three regional ignimbrites are reported east of the Silver Peak Range, and to our knowledge there are no mapped occurrences or published correlations of the three regional ignimbrites east of the Candelaria Hills or in surrounding ranges to the east. The mapped exposure of the sequence is ~3200 km², and because of the vast regional distribution of the sequence, most studies have focused only on local sections in specific areas (Page, 1959; Ross, 1961; Robinson et al., 1968, 1976; Crowder et al., 1972; Robinson and Crowder, 1973; Stewart et al., 1975, 1981; Stewart, 1979, 1981a, 1981b; Garside, 1979; Speed and Cogbill, 1979a, 1979b, 1979c; Robinson and Stewart, 1984). The three regionally extensive ignimbrite sheets are readily identifiable based on their outcrop morphology and distinct mineralogy, and they are the principal Oligocene-age ignimbrites that we have studied. From oldest to youngest, the three regional ignimbrites are the Metallic City

Tuff, Belleville Tuff, and Candelaria Junction Tuff (Speed and Cogbill, 1979a; Robinson and Stewart, 1984).

New ^{40}Ar/^{39}Ar data indicate the age of emplacement of the Metallic City, Belleville, and Candelaria Junction Tuffs to be 25.64 ± 0.12 Ma, 24.09 ± 0.29 Ma, and 23.61 ± 0.13 Ma (Petronis et al., 2004), respectively, and that the numerous undifferentiated tuffs, a name we use to informally refer to the remainder of the Candelaria sequence tuffs, were emplaced between 27.54 ± 0.20 Ma and 23.16 ± 0.17 Ma[1] (see also previous published K-Ar age estimates: Gilbert et al., 1968; Robinson et al., 1968; Silberman et al., 1975; Marvin and Mehnert, 1977; Garside and Silberman, 1978). The undifferentiated tuffs exposed south of the Candelaria Hills in the Volcanic Hills and northernmost part of the Silver Peak Range (Fig. 2) consist of several simple cooling units that presently are of unknown age; however, the upper units of the sequence are likely between ca. 15 and 8 Ma in age (D. Stockli, 2005, personal commun.).

The thickest and most complete sections of the Candelaria sequence crop out in the Candelaria Hills, and individual deposits decrease in thickness, grain size, pumice and lithic content, and degree of welding away from the area (Robinson and Stewart, 1984), especially to the north, west, and southwest. Page (1959) first studied the sequence and subdivided it into 19 units, 17 of which are pyroclastic deposits. Speed and Cogbill (1979a) reinterpreted the sequence and recognized 10 major cooling units, each characterized by a distinctive ignimbrite texture and crystal mineralogy. The three regional ignimbrites (Metallic City, Belleville, and Candelaria Junction Tuffs) are voluminous and were given formal, formational status (Speed and Cogbill, 1979a). Robinson and Stewart (1984) maintained the subdivision of Speed and Cogbill (1979a) but recognized one additional unit (Tuff of Candelaria). This study follows the convention proposed by Robinson and Stewart (1984).

Candelaria Hills Andesite and Basaltic Andesite Flows

Andesite Flows

Sequences of andesite to trachyandesite lava flows, breccia lahars, flow breccia, and associated sedimentary rocks, generally less than 200 m thick, are widely distributed across the Candelaria Hills area and locally rest unconformably on the Candelaria sequence. The andesite is locally referred to as the Gilbert Andesite, and K-Ar age determinations on plagioclase, hornblende, and whole-rock separates yield age estimates between 17.4 ± 0.6 Ma and 15.7 ± 0.5 Ma (Ferguson et al., 1954; Ormerod, 1988; Hardyman and Oldow, 1991), ^{40}Ar/^{39}Ar age determinations of ca. 12 Ma, and a zircon (U-Th)/He date of 12.4 ± 1.0 Ma (Stockli et al., 2003). The andesite flows were erupted from local vents in the Monte Cristo Range, Garfield Hills, and Candelaria

[1]GSA Data Repository Item 2009073, previously unpublished ^{40}Ar/^{39}Ar age estimates from Petronis, is available at www.geosociety.org/pubs/ft2009.htm, or on request from editing@geosociety.org, Documents Secretary, GSA, P.O. Box 9140, Boulder, CO 80301-9140, USA.

Hills (Speed and Cogbill, 1979a). Lava, breccia, and lahar flows constitute the bulk of the andesite sequence. Individual flows reach a maximum thickness of 50 m, and a few can be traced for 1–2 km (Speed and Cogbill, 1979a), but most are thin (<2–3 m), lack distinctive flow markers, and cannot be correlated between discontinuous exposures. The lava flows typically contain coarse, abundant phenocrysts, and the breccia deposits contain abundant clasts of coarse porphyritic rocks. Feldspar, pyroxene, and hornblende crystals can be up to 1 cm, but typically they are less than 1 mm in length. Breccia deposits fill channels between flows and are likely debris flow in origin, and locally 1 to 3 m of andesite volcaniclastic detritus is common. The only available geochemistry on these rocks is from Ross (1961).

Basaltic Andesite Flows

Basaltic andesite flows crop out extensively throughout the central Walker Lane; thick sequences are exposed in the Candelaria Hills, Volcanic Hills, and Garfield Hills, and the Monte Cristo Range. The youngest basaltic andesite sequences in the Candelaria Hills are olivine basaltic andesites erupted from fissure-controlled vents that form small cinder cones and lava fields throughout the region. The available geochronology data indicate that the flows in the region are ca. 8.5 Ma to 3.2 Ma in age (^{40}Ar/^{39}Ar data: Johnson, 2002; Tincher and Stockli, this volume; Schroeder et al., 2005; K-Ar data: Silberman et al., 1975; Marvin and Mehnert, 1977; Ormerod, 1988; Lee et al., 2005), but, based on stratigraphic relations, some are as old as 23.5 Ma (Speed and Cogbill, 1979a; Marvin and Mehnert, 1977; Silberman et al., 1975). The Oligocene mafic lavas are dominantly olivine and hypersthene basalts, they are interbedded with the Candelaria pyroclastic sequence, and they represent some of the oldest recognized Cenozoic basaltic rocks in the western Great Basin (Silberman and McKee, 1972). The younger basaltic andesites contain sparse olivine, plagioclase, clinopyroxene, iron oxides, and fresh glass. They typically lie in angular unconformity on the Candelaria sequence and, locally, conformably on andesite flows. At a location ~0.5 km southeast of the Candelaria town site in a well-exposed outcrop at the Northern Belle open-pit mine, the youngest flows are locally draped across a paleo–fault scarp (i.e., "lava falls"), indicating that the age of last motion on this structure was before ca. 3.2 Ma.

PALEOMAGNETIC REFERENCES AND CORRELATIONS

Observed paleomagnetic declinations that deviate or are discordant from some expected value can provide rotation estimates for parts of the crust in either an absolute or a relative framework. Absolute rotation determinations require results that adequately sample the geomagnetic field over a sufficiently long time (e.g., a sequence of numerous basalt flows showing multiple polarity intervals), which are then compared with a sufficiently robust estimate of the time-averaged field based on independent paleomagnetic data (paleomagnetic poles) from the respective craton.

Relative determinations, on the other hand, require sampling a single, laterally extensive datum (e.g., Wells and Hillhouse, 1989; Byrd et al., 1994; Sussman et al., 2006) located in several different structural settings (e.g., a regionally extensive ash-flow tuff). Under ideal circumstances, where either a time-averaged geomagnetic field is well-sampled or a single datum consistently yields a very high precision result, estimates of absolute vertical-axis rotation typically have 95% confidence limits of ~4° to 10°, and relative vertical-axis rotation estimates typically have confidence limits less than 5°. The paleomagnetic data from basalt and andesite flow sequences in the Candelaria Hills area allow for an estimate of absolute rotations. We compared the group-mean directions from individual flow sequence localities (between 7 and 28 flows per sequence) with the expected reference directions for the late Cenozoic for North America. Over the past few decades, several compilations of Cenozoic paleomagnetic poles have been made to refine apparent polar wander (APW) paths for North America and other continents. The estimates of average paleomagnetic poles differ for time periods in the Cenozoic and, consequently, result in a range of estimates of vertical-axis rotation (as well as inclination shallowing or steepening). We summarize several compilations for the Oligocene and early (ca. 20 Ma) and late (ca. 10 Ma) Miocene in Table 1. If all the expected directions based on the paleomagnetic poles listed in Table 1 are averaged, they yield an expected normal polarity direction of D = 353.5° ± 3.6° and I = 58.1° ± 2.6° for the Candelaria Hills area (where D is declination and I is inclination). Our estimates of vertical-axis rotation and inclination shallowing or steepening are based on the normal and reversed polarity direction for this area, D = 353.3°, I = 58.3° and D = 173.3°, I = −58.3°, respectively (A_{95} = 3.1°, K = 107.7) using the 10 Ma synthetic pole of Besse and Courtillot (2002), which is statistically indistinguishable, at 95% confidence, from the average Miocene expected direction estimated from several compilations (Table 1).

The three regional ignimbrites were emplaced across the Candelaria Hills and adjacent areas and were later fragmented as parts of numerous internally coherent structural blocks during Tertiary extension and development of the Walker Lane. This allows for an estimate of the relative vertical-axis rotation between each block, and an evaluation of possible differential rotation across the region. Paleomagnetic investigations of ignimbrites reveal that individual cooling units typically yield an instantaneous record of the geomagnetic field (e.g., Reynolds, 1977; Geissman et al., 1982; Wells and Hillhouse, 1989; Byrd et al., 1994). Thus, the combination of results from a small number of discrete ash-flow tuffs clearly does not provide an adequate long-term average of the geomagnetic field. Consequently, our analysis of the data from these deposits involved comparing results from each of the three ignimbrites to data from an internally coherent "reference" section, where these deposits are exposed in continuous stratigraphic order, in the southern Excelsior Mountains, west of Huntoon Valley (Fig. 3A).

Each ignimbrite at the reference section has a unique characteristic remanent magnetization (ChRM) that is well defined and of low dispersion. The ChRM directional data from the reference section in the southern Excelsior Mountains are as follows: Candelaria Junction Tuff, 249.5°, −53.5°, α_{95} = 1.9° (N = 9 samples); Belleville Tuff, 044.3°, 30.5°, α_{95} = 3.6° (N = 7 samples); and Metallic City Tuff, 355.3, −32.2°, α_{95} = 2.6° (N = 15 samples). The well-defined ChRM of each ignimbrite affords the opportunity for very precise estimates of declination discrepancies between separate structural blocks; any deviation of the ChRM at individual sites compared to the reference section, when inclinations are essentially identical, is tentatively interpreted as reflecting vertical-axis rotation.

Excelsior Mountain Ignimbrite Reference Section

At this location, the average strike/dip of the ignimbrite deposits, based on the orientation of eutaxitic fabrics and contacts between units, is N14°E/41°SE. These are the westernmost exposures of the three regional ignimbrites identified, and importantly, the area is characterized by northwest-southeast–striking faults. Arguably, given the dip of the deposits, this area clearly has experienced some degree of post–late Oligocene deformation. However, given its location along the westernmost margin of the Mina deflection, we make the tentative assumption that this area experienced little, if any, vertical-axis rotation and that the current dip of the deposits is related to an early system of extensional structures formed during the inception of Basin and Range extension in this area. All paleomagnetic data from the ignimbrites, following correction for local tilt, are compared to data from this reference section.

The viability of our choice of the southwest Excelsior Mountains section as an appropriate, unrotated, reference section is based on several arguments, although we recognize that it is not an ideal reference locality. The section has been moderately tilted and it lies within a part of the western Great Basin that experienced considerable Neogene extension. The possibility certainly exists that the section has experienced some rotation. We initially chose this locality for three reasons. (1) It is a rare location where all three of the regionally extensive ignimbrites are exposed in a continuous, structurally coherent section. (2) The location of this section, on the southwest edge of Huntoon Valley, is some 35 km northwest of the termination of the Furnace Creek–Fish Lake Valley fault system and, importantly, lies west of a northwest extrapolation of this fault system. We thus assume that it lies outside of the displacement transfer zone characterized by moderate vertical-axis rotation. (3) The structural trends in the reference area are more characteristic of the broad, presumably unrotated part of the western Great Basin, which contrasts with the anomalous trends of the Mina deflection. We therefore make the assumption that there is a small likelihood of appreciable vertical-axis rotation, in this area, in response to early displacement transfer between the Furnace Creek–Fish Lake Valley and central Walker Lane fault systems. In a paleomagnetic study of the Oligocene ignimbrite section in the highly extended Yerington district located ~120 km NNW of Candelaria Hills, Geissman et

al. (1982a, 1982b) showed that, despite the fact that the Cenozoic volcanic section as well as underlying Mesozoic crystalline basement had been tilted (west-side-down) by up to 100° during Neogene extension, the volcanic sequence did not experience appreciable vertical-axis rotation. Were our proposed reference section to have experienced, for example, some appreciable clockwise (counterclockwise) rotation, then the inferred (absolute) clockwise rotations in the Candelaria Hills area would obviously be greater (smaller) than we report. We note, and this is more fully discussed later, that the average rotation estimate based on the ignimbrite data is statistically indistinguishable from the average rotation estimate determined from combining the locality mean directions for each of the basalt and andesite flow sections, which is 26.0° ± 7.8°. The similarity of independent data sets at least supports our approach of using the ignimbrite data set from Huntoon Valley as a reference section for relative, as well as absolute rotations. Because the paleomagnetic data from the ignimbrite, andesite, and basaltic andesite flows all yield similar overall rotation estimates, we hypothesize that the inferred vertical-axis rotation in the Mina deflection area affected all of these rock sequences simultaneously. The timing of this rotation is clearly no earlier than late Miocene; a key question, discussed later, is whether clockwise rotation initiated in the late Miocene and continues today, or was confined to a narrower period of time and thus no longer characterizes the deformation field.

Tilt Corrections

At most paleomagnetic localities, the orientations of the deposits, and thus stratal dip estimates, were defined principally using the orientation of eutaxitic fabrics (ignimbrites), but also, when possible, the orientations of oxidized flow tops, erosional breaks between units, and contacts with underlying or overlying sedimentary or volcaniclastic deposits (all rock types). Implicit in any correction based on eutaxitic fabrics is the assumption that pumice compaction defines a paleohorizontal datum at the time of remanence acquisition. Although this is commonly the case, significant differences between the paleohorizontal and a eutaxitic fabric can result if the deposit was emplaced onto irregular topography or confined to a high-relief channel. In addition, postemplacement flow (rheomorphism) can result in eutaxitic fabrics of variable orientation (Wolff and Wright, 1981; Tarling and Hrouda, 1993). At all locations sampled, we did not see evidence of rheomorphism, and data were collected from areas where individual ignimbrites maintained a visibly uniform thickness over as much distance as possible, to avoid potential paleo-channel margins. We assume that the compaction fabrics developed parallel to the underlying topography and that slopes were close to horizontal and thus that the current orientation of the deposits is a reflection of postemplacement local tectonic tilt. The outcrop pattern of the ignimbrites is admittedly not always ideal because of Neogene faulting and tilting of the ignimbrite sheets (Fig. 3). Although considerable effort was made to accurately define the orientation of each deposit (typically 5–15 strike and dip measurements at each locality), the maximum tilt (dip) values may have large possible errors between each deposit. The attitude of the ignimbrites at some localities was difficult to determine because exposures were discontinuous, well-exposed bases of the sequences, and contacts with underlying sedimentary-stratified rocks were uncommon, and/or compaction foliations were sometimes poorly defined. Although we attempted to minimize these potential problems, they are unavoidable, and occasional poorly defined paleohorizontal information likely contributes to the overall dispersion (specifically inclination deviation) of the paleomagnetic data from outcrops of the same ignimbrite deposit located within the same or on different structural blocks.

METHODS

Oriented samples were collected as cores using a portable gasoline-powered drill with a nonmagnetic, diamond-tipped drill bit. Each sample was individually oriented always using both solar and magnetic compasses. Each sample was cut into 2.2×2.5 cm right-circular cylinder specimens, using a diamond-tipped, nonmagnetic saw blade. Up to three specimens per sample were obtained. Remanent magnetizations were measured using a three-axis 2-G Enterprises Model 760R magnetometer with an integrated alternating field (AF) demagnetizing unit. Specimens were progressively AF demagnetized, typically in 15–20 steps, to a maximum field of 140 mT; those with high coercivity were treated with thermal demagnetization up to 610 °C. Thermal demagnetization on replicate specimens, to compare with AF behavior, was conducted with a Schonstedt TSD-1 thermal demagnetizer. Principal component analysis (PCA; Kirschvinck, 1980) was used to determine the best-fit line through selected demagnetization data points for each sample. For most samples, a single best-fit line could be fit to the data. Best-fit magnetization vectors, in most cases, involved 5–18 data points, but as few as three, and as many as 25, were used. Magnetization vectors with maximum angular deviation (MAD) values greater than 5° were not included in site mean calculations. Sample MAD values for linear data, typically, were less than 4°, but ranged from less than 1° to 7°. For less than 10% of the demagnetization results, it was necessary to anchor the magnetization vector to the origin. Individual sample directions were rejected from the site mean calculation when the angular distance between the sample direction and the estimated site mean exceeded 15°. Individual site mean directions were rejected from the group mean calculation when the angular distance between the site mean direction and the grand mean was undoubtedly an outlier; these are discussed on a case by case basis for each grand mean presented.

To test whether the site mean directions from adjacent basaltic and andesite lava flows were statistically distinct, we applied the following statistical test. For two adjacent sites, when the individual site means from each lava flow were exceptionally well defined ($\alpha_{95} \leq 5°$), and the mean of both sites fell within the 95% confidence ellipse of the adjoining site, we considered these sites to be one "cooling unit," because the two flows essentially

recorded the same paleofield direction and captured limited paleosecular variation (e.g., Calderone et al., 1990). For these sites, we combined the site means to yield a single mean field direction (Tables 2A and 2B). The complete (site level) paleomagnetic results are listed in Tables 2A and 2B.

RESULTS

Rock Magnetism

Rock magnetic analyses were conducted on the three regional ignimbrites of the Candelaria sequence. These results are discussed in greater detail in Petronis and Geissman (2008), and results are summarized in Appendix 1. Rock magnetic data indicate that the dominant magnetic mineral in the three regional ignimbrites is likely a ferrimagnetic phase, probably single-domain (SD) to pseudo–single-domain (PSD) low-titanium magnetite, although some variability was observed, and it is possible that maghemite, which may display maximum laboratory unblocking temperatures above magnetite, contributes to the remanence.

Paleomagnetism

General Demagnetization Behavior

The volcanic rocks are typified by high-quality and readily interpretable results during progressive alternating field (AF) and thermal demagnetization (Fig. 4). Duplicate specimens treated with thermal and AF demagnetization yield similar directional data (Fig. 4). In general, most samples contain a single characteristic remanent magnetization (ChRM) that is well grouped at the site level, but some samples also contain additional magnetizations, depending on rock type, which are readily removed by 20 mT or by 300 °C. We interpret these low-coercivity magnetization components as viscous overprints (VRM). After removing the VRM, the ChRM, which we interpret as the primary thermal remanent magnetization (TRM), decays along a roughly univectoral path to the origin with less than 10% of the natural remanent magnetization (NRM) intensity remaining after treatment in 140 mT fields or by 610 °C, depending on rock type (Fig. 4).

Regional Ignimbrites

Following correction for local deviation from the paleohorizontal (i.e., tilt), results from each ignimbrite have a unique ChRM that is well defined but somewhat varied in direction across the entire study area. Each deposit has a relatively unique coercivity spectra, ranging from being characterized by multidomain (MD) magnetite of intermediate coercivity to PSD/SD magnetite-dominated assemblages with moderate to high coercivity ranges. Typically, volcanic rocks, with remanence carried by populations of high coercivity magnetite/maghemite grains, have a high likelihood of yielding readily interpretable demagnetization data and very well-defined estimates of unit mean directions that are faithful records of the geomagnetic field. Those

TABLE 2A. PALEOMAGNETIC DATA FROM THE CANDELARIA REGIONALLY EXTENSIVE IGNIMBRITES

Fault/ block		Site	n\N	R	Dec. in situ (°)	Inc. (°)	α_{95}	k	VGP Lat. (°N)	VGP Long. (°W)	Strike/Dip	Dec. corrected (°)	Inc. (°)
Candelaria Junction Tuff													
1	Mean	CU53	9\9	8.98	310.3	−48.9	2.9	324.3	7.7	103.6	056, 21NW	297.0	−68.5
1	Mean	CU58	5\5	4.94	310.2	−52.3	9.7	62.9	5.2	101.8	082, 31NW	259.6	−65.9
1	Mean	CU64	6\6	5.94	327.6	−53.1	7.2	88.7	12.2	88.9	013, 22NW	0.1	−64.9
2	Ave	CU75	9\9	8.82	300.5	−12.3	7.9	43.9	19.2	126.9	020, 14NW	301.4	−26.6
2	"	CU76	9\9	8.95	291.7	−33.4	4.1	161.5	4.8	124.0	020, 14NW	292.1	−47.4
2	"	CU77	8\8	7.86	285.8	−30.7	8.0	48.5	1.7	129.1	020, 14NW	284.9	−44.7
2	Mean	CU75–77	26\26	25.16	293.3	−25.5	5.3	29.7	9.2	126.5	020, 14NW	293.9	−39.5
3	Mean	CU110	6\7	5.98	200.6	−38.0	4.0	289.1	−65.5	−170.4	068, 53SE	263.0	−56.5
4	Ave	CU37	8\8	7.97	261.1	−45.9	3.6	243.4	−23.1	134.4	349, 10SW	261.6	−55.9
4	"	CU38	8\9	7.84	264.6	−43.1	8.5	43.2	−19.2	134.6	349, 10SW	265.8	−53.0
4	Mean	Cu37–38	16\17	15.80	262.8	−44.5	4.3	75.2	−21.2	134.5	349, 10SW	263.7	−54.5
6	Mean	CH6C	8\9	7.99	291.7	−28.7	2.4	531.8	6.8	126.2	088, 26N	297.0	−57.6
6	Mean	CH10#	8\8	7.98	216.0	−47.0	3.2	302.0	−58.5	159.1	036, 16 SE	232.5	−44.7
6	Mean	CH15	9\9	8.97	285.8	−21.4	3.4	228.4	−5.3	133.3	223, 16NW	281.5	−35.4
7	Mean	MM9#	9\9	8.97	241.0	−51.8	3.0	302.8	−40.8	138.8	112, 10SW	251.1	−59.0
10	Mean	CU91	7\7	6.99	271.5	−43.7	2.7	501.2	−14.3	130.5	050, 45NW	214.5	−55.8
13	Mean	CU96	8\8	7.95	291.7	−60.7	4.7	142.0	−11.3	106.8	035, 20SE	296.4	−41.0
13	Mean	CU97	8\8	7.99	272.5	−64.0	1.8	933.0	−24.8	112.0	283, 08NE	257.7	−61.5
13	Mean	CU98	9\10	8.98	283.0	−53.6	2.4	460.4	−11.6	117.3	355, 10NE	297.8	−44.0
14	Mean	CU99	9\9	8.94	227.3	−64.7	4.5	134.1	−54.6	122.7	060, 32SE	283.8	−54.7
15	Ave	CU89	7\7	6.96	270.0	−18.4	5.1	141.0	−5.8	144.4	035, 14NW	266.2	−29.6
15	"	CU90	7\7	6.97	265.6	−41.6	4.4	185.8	−17.8	135.1	045, 12NW	255.8	−48.6
15	Mean	CU89–90	14\14	13.63	268.0	−30.0	6.8	35.4	−11.4	140.2	040, 13NW	261.6	−39.2
18	Ave	CU28	8\8	7.95	342.8	−23.5	4.6	148.3	37.0	83.0	025, 25W	355.2	−38.4
18	"	CU29	7\7	6.98	351.4	−12.9	3.9	241.5	44.5	73.9	025, 25W	359.1	−25.5
18	Mean	Cu28–29	15\15	14.83	347.0	−18.6	4.3	80.9	40.7	78.9	025, 25W	357.2	−32.4
23	Ave	BV34	Reject										
23	"	BV35	7\8	6.98	183.7	−43.7	3.2	360.8	−76.9	−133.0	085, 25S	210.3	−36.9
23	"	BV36	7\7	6.96	174.3	−32.1	4.9	151.3	−68.6	−103.0	085, 25S	194.5	−32.4
23	"	BV37	5\6	4.72	185.2	−38.8	20.8	14.5	−73.1	−135.0	085, 25S	207.6	−32.4
23	"	BV38	9\10	8.96	209.5	−30.1	3.5	190.9	−56.0	−175.9	085, 25S	220.7	−14.2
23	Mean*	BV34–38	4\4	3.91	188.4	−36.9	15.9	34.1	−71.0	−143.1	085, 25S	203.1	−60.3
24	Ave	BV66	5\6	4.99	251.5	−35.0	4.4	303.8	−26.0	147.1	045, 5SE	254.5	−32.7
24	"	BV67	5\5	4.99	252.0	−53.7	2.0	1418.0	−33.2	132.0	045, 5SE	257.6	−51.2
24	"	BV68	7\8	6.74	248.3	−53.0	12.8	23.2	−35.8	134.3	045, 5SE	254.0	−50.8
24	Mean	BV66–68	17\19	16.54	250.5	−47.9	6.1	35.1	−30.2	139.0	045, 5SE	255.2	−45.6
27	Ave	BV26	9\9	8.87	287.9	−43.2	6.7	59.9	−2.5	121.4	005, 07NW	289.7	−50.0
27	"	BV27	7\7	6.81	268.8	−58.8	10.9	31.9	−24.0	119.4	005, 07NW	267.2	−65.7
27	"	BV28	13\13	12.69	258.6	−58.8	6.8	38.2	−30.9	123.6	005, 07NW	254.4	−65.4
27	Mean	BV26–28	29\29	27.89	272.2	−54.7	5.5	25.1	−18.9	113.2	005, 07NW	271.6	−61.7
27	Mean	BV32	8\9	7.93	273.6	−60.6	5.7	94.7	−21.9	115.5	005, 07NW	273.2	−67.6
27	Mean	BV33	8\8	7.97	278.1	−39.6	3.7	220.9	−7.7	129.2	065, 20NW	254.8	−49.1
32		VH8	Reject										
32	Mean	VH9	7\9	6.70	197.7	−22.6	14.1	19.4	−59.4	−153.8	225, 33SE	214.9	−33.6
32	Mean	VH11	12\14	12.00	191.5	−54.6	2.9	231.5	−80.4	165.5	023, 36SE	237.2	−46.7
32	Mean	VH12	9\9	9.00	189.4	−54.7	1.9	749.5	−82.0	169.0	023, 36SE	236.4	−47.7
32	Ave	VH20	8\10	8.00	149.7	−59.9	4.7	141.4	−66.5	−12.0	176, 37NE	213.4	−55.6
32	"	VH21	6\9	6.00	172.3	−36.6	3.9	301.2	−71.2	−95.2	176, 37NE	197.6	−30.5
32	Mean	VH20–21	14\19	13.54	162.1	−50.5	7.6	28.5	−73.8	−47.4	176, 37NE	205.0	−45.1
34	Ave	84:a–f	6\11	5.96	224.0	−52.9	5.8	136.3	−54.4	145.6	097, 24S	259.3	−67.6
34	"	84:g–k	5\11	4.95	235.8	−64.9	8.4	84.6	−48.9	121.1	20, 15S	254.0	−54.1
34	"	BV85	10\10	9.91	240.1	−57.1	4.9	98.1	−43.5	132.8	079, 26SE	281.2	−56.3
34	Mean	BV84–85	21\21	20.72	234.1	−58.0	3.8	71.0	−48.3	133.7	079, 26SE	278.6	−59.3
34	Ave	BV86	9\9	8.96	237.3	−62.3	3.6	208.4	−47.2	125.6	015, 24SE	257.0	−43.0
34	"	BV87	9\9	8.90	241.9	−46.5	5.7	83.6	−38.0	144.0	032, 21SE	256.2	−33.7
34	Mean	BV86–87	18\18	17.69	240.1	−54.4	4.7	55.0	−42.5	136.3	032, 21SE	258.8	−41.4
34	Mean	BV89	4\7	3.92	288.7	37.9	15.1	38.8	27.2	159.2	243, 86NE	292.8	−31.8
36	Mean	CU123	9\9	8.96	270.2	−46.6	3.7	192.6	−16.6	129.1	086, 11NW	258.6	−46.3
36	Mean	CU129	10\10	9.81	273.8	−38.1	7.1	47.1	−10.1	132.4	143, 19SW	287.8	−51.0
38	Mean	CU111	8\8	7.93	220.8	−65.7	5.6	100.7	−59.0	120.1	028, 28SE	259.7	−49.6
38	Ave	CU114	6\6	5.92	197.4	−60.9	8.4	64.3	−76.1	130.1	118, 13SW	189.6	−73.5
38	"	CU115	10\10	9.99	208.4	−49.3	1.6	926.1	−65.3	161.8	118, 13SW	208.6	−62.5
38	Mean	CU114– 115	16\16	15.82	205.0	−53.7	4.1	81.2	−69.5	154.7	118, 13SW	203.5	−66.7

(continued)

TABLE 2A. PALEOMAGNETIC DATA FROM THE CANDELARIA REGIONALLY EXTENSIVE IGNIMBRITES (*continued*)

Fault/ block	Site		n/N	R	Dec. in situ (°)	Inc. (°)	α_{95}	k	VGP Lat. (°N)	Long. (°W)	Strike/Dip	Dec. corrected (°)	Inc. (°)
Candelaria Junction Tuff (*continued*)													
39	Ave	VH50	7\7	6.97	218.1	−60.1	4.3	201.5	−60.6	133.8	046, 38SE	268.6	−47.9
39	"	VH51	10\10	9.99	214.1	−60.2	1.1	1391.9	−63.7	134.1	046, 38SE	267.3	−49.7
39	Mean	VH50–51	17\17	16.96	215.7	−60.2	1.8	412.1	−62.4	134.0	046, 38SE	269.5	−47.6
39	Mean	VH58	8\9	7.97	216.2	−53.4	3.7	223.8	−60.6	148.6	059, 29S	257.8	−54.5
39	Ave	VH59	9\9	8.99	277.1	−35.7	2.2	539.1	−6.7	132.0	055, 22NW	260.4	−48.2
39	"	VH60	9\9	8.98	271.8	−27.2	2.8	345.7	−7.4	139.4	055, 22NW	259.4	−38.6
39	Mean	VH59–60	18\18	17.90	247.3	−31.5	2.7	167.3	−7.1	135.8	055, 22NW	233.1	−33.5
40	Mean	BV81	8\8	7.92	228.4	−56.0	6.0	86.9	−51.9	138.9	315, 15SW	230.8	−71.0
Belleville Tuff													
1	Mean	CU57	7\7	6.54	182.6	73.4	17.5	12.9	7.2	−119.6	136, 31NE	77.9	68.2
1	Ave	CU59	7\7	6.99	88.7	54.0	2.6	522.3	21.3	−56.1	053, 29NW	42.1	60.9
1	"	CU60	4\7	3.99	64.4	26.7	4.7	376.0	28.7	−24.0	053, 29NW	48.9	28.6
1	Mean	CU59–60	11\14	10.57	77.5	44.8	9.6	23.4	25.3	−43.1	053, 29NW	45.6	49.4
2	Mean	CU71	7\8	6.98	65.0	33.1	3.1	383.9	30.5	−28.1	007, 15W	57.8	45.3
2	Mean	CU74	9\9	8.91	97.2	53.1	5.4	92.6	15.0	−59.5	030, 17W	81.8	67.9
4	Mean	CU44	6\9	5.97	17.7	40.8	5.0	179.0	68.9	10.9	349, 10SW	9.2	45.0
8	Ave	CU103	8\8	7.98	49.4	55.4	3.0	337.6	51.0	−40.4	330, 22NE	52.7	34.6
8	"	CU104	7\7	6.98	42.9	64.6	3.9	241.7	57.7	−56.8	305, 20NE	39.8	44.7
8	Mean	CU103–104	15\15	14.90	46.8	59.7	3.2	140.6	54.2	−47.0	305, 20NE	45.1	38.7
11	Mean	CU92	7\8	6.99	347.4	60.0	2.4	649.3	79.9	171.2	060, 08NW	344.2	52.3
	Ave	CU30	7\7	6.98	56.2	27.3	3.2	360.9	35.3	−18.5	025, 25W	41.4	37.6
	"	CU31	4\5	3.87	41.1	23.6	19.8	22.5	45.3	−4.1	025, 25W	29.0	28.0
	"	CU32	10\10	9.96	57.4	35.3	3.0	252.7	37.3	−24.6	025, 25W	37.4	45.1
18	Mean	CU30–32	21\22	20.64	53.8	30.7	4.3	56.2	38.4	−18.9	025, 25W	37.0	39.6
	Ave	CU10	7\8	6.92	72.9	31.8	7.0	75.6	23.8	328.1	097, 15N	65.7	24.8
	"	CU11	6\6	5.95	61.6	36.8	6.8	98.1	34.5	−28.2	097, 15N	54.3	27.3
19	Mean	CU10–11	13\14	12.82	67.9	34.2	5.2	64.9	28.7	−30.3	097, 15N	60.5	26.0
	Ave	CU12	6\7	5.99	142.2	46.1	2.3	826.2	−15.4	276.2	060, 45N	45.8	84.4
	"	CU13	6\9	5.98	148.2	23.3	4.6	209.7	−31.5	279.0	060, 45N	145.5	68.3
	"	CU14	10\10	9.95	109.3	35.0	3.6	177.5	−2.3	304.9	060, 45N	63.6	57.6
	"	CU15	7\7	6.98	84.4	23.1	3.9	244.1	11.8	325.7	060, 45N	59.4	33.1
19	Mean*	Cu12–15	4\4	3.66	120.2	37.4	33.4	8.7	−9.8	−62.2	060, 45N	71.1	63.0
	Ave	CU16	9\9	8.98	55.4	18.7	2.3	499.2	32.9	346.7	085, 10N	52.9	13.6
	"	CU17	6\7	5.96	58.4	27.6	5.8	134.6	33.7	−20.3	"	54.3	22.8
	"	CU19	9\9	8.92	71.2	17.8	5.3	96.7	20.4	−23.8	083, 14N	67.2	14.4
	"	CU20	9\9	8.91	88.1	22.1	5.6	86.0	8.5	−36.1	083, 14N	82.3	22.6
	"	CU22	7\7	6.98	71.0	21.7	3.5	300.5	21.8	−25.5	094, 19N	65.5	13.4
	"	CU24	7\10	6.84	68.6	47.5	10.0	37.7	33.2	−40.6	094, 19N	53.9	37.1
	"	CU25	9\9	8.87	57.1	29.0	6.6	61.5	35.3	−20.2	083, 14N	51.2	22.2
20	Mean*	CU16–25	7\7	6.81	67.1	26.7	10.9	31.1	26.6	−25.6	083, 14N	61.5	21.5
20	Mean	CU27	8\9	7.98	74.0	54.4	2.9	366.6	32.1	−49.7	266, 20N	51.5	46.3
21	Mean	CU51	8\8	7.97	22.5	69.6	3.9	203.1	68.4	−79.8	In situ	22.5	69.6
23	Mean	BV48	6\7	5.90	36.0	45.4	9.4	51.5	57.9	−18.3	091, 09S	42.4	52.5
25	Mean	BV65	7\7	6.98	42.2	64.4	3.1	370.7	58.1	−56.3	208, 28SE	79.0	64.4
26	Ave	BV49'	5\8	4.91	52.2	68.8	11.8	43.4	51.7	−67.0	052, 11W	26.0	66.3
26	"	BV50'	10\10	9.96	65.1	65.9	3.1	246.3	43.0	−62.4	069, 08W	48.2	64.2
26	"	BV51'	8\8	7.74	88.9	70.7	10.8	27.3	31.0	−76.1	069, 08W	65.3	71.9
26	"	BV52'	7\9	6.98	65.2	64.4	3.4	314.9	42.4	−59.9	069, 08W	49.4	62.8
26	"	BV53'	8\8	6.80	134.3	51.0	11.3	29.5	−8.2	−80.1	298, 19E	109.9	52.4
26	Mean*	BV49'–53'	5\5	4.83	85.6	67.7	16.1	23.5	31.1	−70.5	059, 10NW	60.0	69.9
26	Ave	BV58	7\7	6.96	85.7	68.9	5.2	138.3	31.7	−72.4	326, 10NE	76.8	59.8
26	"	BV59	4\6	3.99	72.9	62.1	4.1	510.5	36.4	−58.5	326, 10NE	68.9	52.4
26	Mean	BV58–59	11\13	10.92	80.2	66.5	4.0	128.5	34.3	−71.3	326, 10NE	73.5	57.1
26	Mean	BV60	7\9	6.94	57.2	67.9	5.8	108.0	48.5	−65.1	085, 18NW	50.3	43.3
28	Ave	BV2	7\10	6.89	67.9	37.0	8.4	52.3	29.6	−32.1	In situ	67.9	37.0
28	"	BV3	8\10	7.97	87.2	9.0	3.4	262.3	5.0	−30.0	In situ	87.2	9.0
28	"	BV6	9\10	8.93	42.4	44.4	4.8	115.3	52.5	−22.0	In situ	42.4	44.4
28	Mean	BV2–3–6	24\30	22.02	67.5	31.9	9.1	11.6	28.1	−28.6	In situ	67.5	31.9
28	Ave	BV4	6\9	5.69	105.6	56.4	17.1	16.3	11.8	−66.4	In situ	105.6	56.4
28	"	BV5	7\8	6.83	89.1	43.8	10.4	34.7	16.1	−48.3	In situ	89.1	43.8
28	Mean	BV4–5	13\17	12.39	95.5	49.8	9.6	19.8	14.5	−56.0	In situ	95.5	49.8
28	*Ave*	*BV49*	*5\8*	*4.95*	*54.5*	*35.2*	*8.4*	*84.5*	*39.5*	*−22.4*	*007, 14W*	*45.9*	*44.8*
28	*"*	*BV50*	*Reject*								*007, 14W*		
28	*"*	*BV51*	*6\7*	*5.96*	*68.8*	*27.4*	*5.9*	*132.1*	*25.5*	*−26.9*	*007, 14W*	*64.1*	*39.5*

(*continued*)

TABLE 2A. PALEOMAGNETIC DATA FROM THE CANDELARIA REGIONALLY EXTENSIVE IGNIMBRITES (*continued*)

Fault/ block		Site	n/N	R	Dec. in situ (°)	Inc. (°)	α_{95}	k	VGP Lat. (°N)	Long. (°W)	Strike/Dip	Dec. corrected (°)	Inc. (°)
Belleville Tuff (continued)													
28	Mean	*BV49, 51*	11\15	10.91	61.7	31.3	7.1	108.3	23.3	−28.2	007, 14W	55.1	42.3
28	Ave	BV61	5\8	4.99	9.9	61.4	2.6	838.4	81.4	−60.8	328, 16NE	25.0	49.1
28	"	BV62	5\5	4.99	2.7	54.1	2.6	853.8	85.9	28.8	328, 16NE	16.5	43.4
28	Mean	BV61–62	10\13	9.96	6.0	57.8	3.0	256.2	85.3	−33.7	328, 16NE	20.6	46.3
28	Ave	BV91	9\9	8.95	90.3	61.8	4.1	160.2	24.8	−64.6	315, 17NE	75.3	48.2
28	"	BV92	7\7	6.97	79.1	61.5	4.0	228.5	31.9	−60.0	315, 17NE	67.9	46.5
28	"	BV93	11\11	10.91	93.3	50.4	4.4	108.1	16.2	−55.3	315, 17NE	82.0	37.8
28	"	BV94	10\10	9.89	84.6	58.1	5.4	79.9	26.4	−58.1	315, 17NE	72.8	43.9
28	Mean*	BV91–94	4\4	3.98	87.3	58.1	7.1	167.2	24.6	−59.3	315, 17NE	75.2	43.9
29	Ave	BV45	8\11	7.35	110.4	58.5	17.7	10.7	10.4	−70.6	338, 19NE	100.4	50.1
29	"	BV46	6\7	5.85	45.6	53.7	11.7	33.6	53.3	−36.2	338, 19NE	49.3	43.2
29	Mean	BV45–46	14\17	12.59	79.8	60.7	13.8	9.2	31.0	−59.1	338, 19NE	75.7	42.0
30	Ave	BV95	8\8	7.98	56.1	54.9	2.6	463.5	45.7	−42.6	338, 19NE	59.6	36.2
30	"	BV96	5\5	4.95	55.4	61.2	8.6	79.5	48.2	−52.0	338, 19NE	59.8	42.5
30	Mean	BV95–96	13\13	12.92	55.9	57.3	3.5	143.2	46.7	−45.9	338, 19NE	59.7	38.6
36	Ave	CU124	7\7	6.86	72.7	16.7	9.2	44.0	18.9	−24.3	320, 17SW	75.9	32.2
36	"	CU125	9\9	8.91	74.6	15.3	2.5	410.4	16.9	−24.9	320, 17SW	77.8	30.6
36	Mean	Cu124–125	16\16	15.84	73.8	15.9	3.8	94.4	17.8	−24.7	320, 17SW	77.0	31.3
Metallic City Tuff													
1	Ave	CU54	8\8	7.95	357.8	−17.7	4.8	134.9	42.8	64.6	246, 16NW	0.8	−32.4
1	"	CU55	5\5	4.90	26.3	−40.3	12.0	41.9	24.1	35.1	246, 16NW	39.7	−49.1
1	"	CU56	8\8	7.95	41.4	−3.3	4.5	150.1	34.9	7.9	246, 16NW	43.1	−9.8
1	Mean	CU54–56	21\21	19.11	21.4	−18.4	10.3	10.5	38.4	34.3	246,16NW	26.7	−29.1
1	Mean	CU66	7\7	6.97	352.7	−47.4	4.3	200.3	23.0	68.8	083, 25SE	352.8	−22.4
1	Ave	CU67	6\6	5.99	31.2	−64.8	2.7	609.8	0.8	40.9	215, 28NW	78.3	−54.3
1	"	CU68	Reject										
1	Mean	Cu67–68	6\6	5.99	31.2	−64.8	2.7	609.8	0.8	40.9	215, 28NW	78.3	−54.3
3	Mean	CU106	4\7	3.72	45.4	−30.9	29.5	10.7	20.6	15.1	345, 17NE	38.0	−45.2
3	Mean	CU109	Reject								350, 18NE		
5	Mean	CH4	8\8	7.95	44.1	−50.8	4.8	132.5	9.2	24.8	050, 18 SE	24.4	−45.8
5	Mean	CH4F	9\9	8.93	33.7	−52.0	5.0	109.2	12.7	33.1	160, 10SW	39.8	−43.6
7	Mean	MM1#	8\10	7.89	13.3	−18.3	7.0	62.9	41.0	44.3	112, 10SW	13.7	−8.4
9	Mean	CU100	7\9	6.32	21.8	−26.7	6.4	88.7	59.1	17.2	020, 10SE	16.8	−26.6
9	Mean	CU101	6\8	5.93	36.1	−42.0	8.2	68.1	18.9	27.2	307, 15NE	35.8	−57.0
9	Mean	CU102	6\7	5.98	68.3	−35.4	4.3	240.0	3.9	0.5	048, 36SE	39.7	−39.4
11	Mean	CU93	8\8	7.94	341.5	−17.5	5.2	114.5	39.8	86.0	060, 08NW	342.1	−25.3
13	Mean	CU94	4\8	3.90	21.7	−37.4	16.9	30.4	274.0	38.9	175, 20SW	32.4	−26.7
16	Mean	CU87	6\8	5.96	350.6	−20.1	6.2	116.8	40.6	74.1	055, 18NW	355.1	−36.1
16	Mean	CU88	6\7	5.92	47.7	−55.1	8.8	59.0	4.1	24.8	030,40SE	354.1	−47.7
18	Ave	CU33	9\9	8.96	353.2	−26.0	3.6	209.1	37.6	70.2	025, 25NW	7.3	−36.7
18	"	CU36	9\9	8.98	98.5	65.6	2.9	324.3	22.4	−72.2	025, 25NW	12.6	83.1
18	Mean	CU33–36	Reject								025,25NW		
28	Mean	BV1	Reject								007, 14W		
35	Ave	CU119	8\8	7.93	24.3	−44.6	5.4	107.5	21.8	38.4	054, 16SE	13.3	−35.3
35	"	CU120	7\7	6.95	25.6	−47.4	5.7	110.5	19.2	38.1	054, 16SE	13.3	−38.2
35	Mean	CU119–120	15\15	14.88	24.9	−45.9	3.6	112.5	20.6	38.3	054, 16SE	13.3	−36.7
35	Mean	CU122	8\8	7.76	19.3	−62.4	10.4	29.1	6.4	47.9	340, 31SW	43.1	−37.5
36	Mean	CU126	Reject										
37	Mean	CU127	Reject										
37	Ave	CU116	10\10	9.86	23.6	−35.0	6.1	64.1	28.6	36.4	171, 12NE	15.4	−40.8
37	"	CU118	8\8	7.97	23.0	−23.8	3.6	239.1	35.1	34.1	171, 12NE	17.7	−29.7
37	Mean	CU116,118	18\18	17.75	23.3	−30.0	4.3	67.2	31.6	35.4	171, 12NE	16.5	−35.8
Huntoon Valley Reference Location													
RS	Mean (TT9)	HT1	9\9	8.99	169.5	−67.9	1.9	723.2	−75.2	34.5	014, 41E	249.5	−53.5
RS	Mean (TT7)	HT3	7\9	6.98	13.8	41.6	3.6	289.3	71.5	17.8	014, 41E	44.3	30.5
RS	Mean (TT5)	HT4–5	15\17	14.93	24.4	−35.8	2.6	215.0	27.7	35.4	014, 41E	355.3	−32.2

Note: Fault block—numbers refer to locations indicated in Figures 3 and 6; RS—ignimbrite reference site located in Huntoon Valley, see text for discussion; Site—paleomagnetic sampling sites; Ave—sites that were averaged to yield a mean paleomagnetic block direction; Unit— regional ignimbrite (TT9—Candelaria Junction, TT7—Belleville, TT5—Metallic City, B—basalt, A—andesite, BA—basaltic andesite); *n/N*—ratio of samples used (*n*) to samples collected (*N*) at each site; *R*—resultant vector length, Dec/Inc in situ—in situ declination and inclination; α_{95}— 95% confidence interval about the estimated mean direction, assuming a circular distribution; *k*—best estimate of (Fisher) precision parameter; VGP Lat/Long—latitude and longitude of the virtual geomagnetic pole for the site; Strike/Dip—strike and dip of unit based on compaction fabric, flow contacts, or fiami; Dec/Inc corrected—stratigraphically corrected declination and inclination.

TABLE 2B. PALEOMAGNETIC DATA FROM ANDESITE AND BASALT FLOWS

Site	Unit	n/N	R	α_{95}	k	Dec. in situ (°)	Inc. (°)	Strike/dip (°)	Dec. corrected (°)	Inc. (°)	VGP Lat. (°N)	VGP Long. (°W)
Andesite paleomagnetic data												
A-1												
BV8	A	6\7	5.97	5.5	148.8	*36.9*	*41.5*	200, 35 W	4.3	41.9	75.5	46.0
BV9	"	7\7	6.97	4.1	220.8	*36.1*	*37.0*	"	8.0	38.3	71.9	37.1
BV10	"	7\7	6.97	4.4	190.4	*31.8*	*31.4*	"	9.4	31.8	67.4	37.8
BV11	"	4\7	3.99	3.9	565.8	*36.5*	*42.5*	"	3.0	42.3	76.0	50.5
BV12	"	7\8	6.95	5.3	131.7	*26.1*	*35.2*	"	2.3	31.5	68.7	55.8
BV13	"	7\7	6.89	8.2	55.0	*38.8*	*44.4*	"	2.5	44.8	78.0	51.0
BV14	"	7\7	6.98	3.1	388.6	*29.8*	*43.1*	"	358.0	39.1	73.8	68.5
BV15	"	7\7	6.93	6.4	88.9	*21.3*	*43.4*	"	352.3	34.9	69.8	83.4
BV16	"	7\8	6.97	4.1	214.6	*55.2*	*63.7*	"	339.9	61.7	74.1	175.1
BV17	"	6\7	5.99	3.6	342.2	*34.8*	*35.8*	"	8.1	36.7	70.9	38.1
BV29	"	8\8	7.95	4.6	143.7	*30.5*	*40.8*	046, 16 NW	18.8	34.9	64.9	15.9
BV30	"	7\8	6.96	4.9	150.3	*50.6*	*63.3*	"	20.7	60.3	73.8	314.6
BV31	"	5\5	4.93	10.5	54.0	*12.8*	*73.4*	"	346.0	61.4	78.5	178.2
Andesite volcanic neck												
BV54	A	6\7	5.91	9.0	56.1	311.8	−7.0	In situ	311.8	−7.0	29.0	120.2
BV55	"	4\7	3.99	4.1	507.8	37.3	64.5	"	37.3	64.5	61.5	303.2
BV56	"	8\8	7.96	4.3	167.0	67.7	59.1	"	67.7	59.1	38.7	307.4
A-2												
CH2A	A	8\8	7.98	3.1	323.9	98.5	58.8	047, 25 NW	*48.0*	*72.0*	53.6	284.9
CH2B*	"	3\6	3.00	5.7	464.4	138.9	43.0	"	*140.7*	*68.0*	5.7	265.4
CH2C	"	7\8	6.97	4.6	173.8	75.1	67.2	"	*12.5*	*66.6*	75.9	277.3
CH2D	"	7\7	6.96	4.8	157.7	**68.9**	**71.2**	"	*1.3*	*65.9*	80.0	246.8
CH2E	"	7\7	6.96	4.9	152.7	**60.4**	**67.7**	"	*5.7*	*61.8*	83.6	282.3
D-E Mean		*2\2*	*1.99*	*10.0*	*623.1*	*64.3*	*69.5*		*3.7*	*63.9*	*82.1*	*261.0*
CH2F	"	7\7	6.98	3.9	244.7	*81.3*	*66.1*	"	*17.6*	*68.4*	71.8	278.8
CH2G*	"	7\7	6.48	18.5	11.6	*239.5*	*75.0*	"	*288.2*	*58.4*	35.5	176.9
CH2H*	"	7\8	6.78	11.9	26.7	*321.8*	*61.3*	"	*319.1*	*36.4*	50.5	136.6
CH2I	"	6\7	5.98	4.3	249.8	**84.6**	**72.5**	"	*0.3*	*71.0*	72.8	242.4
CH2J	"	8\8	7.93	5.5	102.9	**87.9**	**75.5**	"	*350.8*	*71.7*	70.7	226.3
I-J Mean		*2\2*	*1.99*	*6.8*	*1337.3*	*86.1*	*74.0*		*355.6*	*71.4*	*71.9*	*233.9*
CH2K	"	7\7	6.98	3.5	294.6	88.8	56.9	"	*43.3*	*66.8*	57.3	297.6
CH2L	"	Conglomerate test: Positive results					~	"	~	~	~	~
CH2M	"	7\7	6.97	4.3	194.5	**83.8**	**69.5**	"	*10.7*	*69.6*	73.2	264.4
CH2N	"	8\9	7.98	2.7	432.6	**79.0**	**67.2**	"	*15.2*	*67.3*	74.0	279.3
M-N Mean		*2\2*	*1.99*	*6.3*	*1560.3*	*81.3*	*68.4*		*13.1*	*68.5*	*73.6*	*271.6*
Basalt paleomagnetic data												
B-0												
GH3	B	–	–	–	–	–	–	080, 15 N	–	–		
GH4	"	7\7	6.96	5.2	137.6	*59.1*	*50.3*	"	*44.9*	*43.2*	50.0	337.5
GH5	"	7\7	6.94	6.2	97.3	*31.9*	*44.3*	"	*24.5*	*32.5*	60.2	9.0
GH6*	"	–	–	–	–	–	–	–	–	–		
GH7	"	8\8	7.99	2.5	476.8	*244.8*	*−70.9*	"	*214.0*	*−62.9*	−62.8	122.7
GH8	"	5\6	4.99	4.4	306.9	*358.2*	*63.5*	"	*355.5*	*48.6*	80.3	85.5
GH9	"	7\7	6.93	6.5	88.3	*359.6*	*69.2*	305, 15 NE	*13.5*	*55.9*	79.1	338.3
GH10	"	7\7	6.91	7.5	65.8	*26.4*	*59.8*	"	*28.9*	*44.9*	63.0	349.4
GH11	"	7\7	6.97	4.0	225.1	**41.5**	**68.4**	"	*39.0*	*53.5*	58.5	327.5
GH12	"	7\7	6.97	4.3	201.8	**35.2**	**65.7**	"	*35.1*	*50.7*	60.6	328.4
11-12 Mean		**2\2**	**1.99**	**8.0**	**987.4**	**38.2**	**67.1**		**37.0**	**52.1**	**59.6**	**330.7**
B-1												
CH20A	B	7\7	6.94	6.1	99.1	35.4	47.3	221, 14 NW	*21.3*	*44.2*	68.3	359.9
CH20B	"	7\7	6.97	4.1	214.9	33.6	53.1	"	*16.7*	*49.2*	74.0	357.3
CH20C	"	6\7	5.92	8.4	64.7	36.1	55.0	"	*17.7*	*51.5*	74.4	349.1
CH20D	"	7\7	6.99	2.6	557.2	36.0	51.1	"	*19.9*	*47.9*	71.1	355.1
CH20E	"	5\6	5.93	8.1	69.5	32.5	55.2	"	*14.7*	*50.7*	76.2	356.7
CH20F	"	3\4	3.95	11.5	64.8	26.6	57.8	"	*8.3*	*51.9*	81.2	9.3
CH20G	"	8\8	7.97	3.4	274.7	28.8	49.8	"	*14.5*	*44.9*	73.1	11.4
CH20H	"	5\6	5.00	2.6	842.7	35.4	53.6	"	*18.1*	*49.9*	73.3	353.1
CH20I	"	5\5	4.98	5.4	203.3	43.3	52.6	"	*25.6*	*50.7*	68.0	342.0
CH20J	"	8\8	7.98	2.9	373.8	35.3	57.1	"	*15.9*	*53.1*	76.4	346.4
CH20K	"	6\6	5.99	2.6	659.8	**30.4**	**54.1**	"	*13.6*	*49.2*	76.2	3.6
CH20L	"	6\6	5.98	3.9	296.0	**34.0**	**54.3**	"	*16.5*	*50.2*	74.6	354.9
K-L Mean		**2\2**	**1.99**	**4.6**	**2935.4**	**32.2**	**54.2**		**15.0**	**49.7**	**75.4**	**359.1**
CH20M	"	5\6	4.77	19.0	17.1	34.5	56.0	"	*15.9*	*51.9*	75.9	350.6
CH20N	"	8\8	7.64	13.0	19.2	32.5	53.2	"	*15.9*	*48.9*	74.4	359.5
CH20O	"	6\7	5.98	3.8	320.7	32.8	59.3	"	*12.3*	*54.5*	79.6	345.8

(continued)

TABLE 2B. PALEOMAGNETIC DATA FROM ANDESITE AND BASALT FLOWS (*continued*)

Site	Unit	n/N	R	α_{95}	k	Dec. in situ (°)	Inc. (°)	Strike/dip (°)	Dec. corrected (°)	Inc. (°)	VGP Lat. (°N)	VGP Long. (°W)
B-2												
CH12A	B	5\6	4.88	13.5	33.0	46.3	60.1	221, 14 NW	*22.6*	*58.5*	72.4	342.2
CH12B	"	7\8	6.98	3.8	248.7	47.2	55.0	"	*27.4*	*54.1*	67.7	332.6
CH12C	"	7\7	6.99	2.4	630.9	40.3	57.2	"	*19.8*	*54.5*	73.8	337.0
CH12D	"	7\7	6.82	10.6	33.5	41.3	45.3	"	*27.5*	*43.7*	63.6	352.3
CH12E	"	7\7	6.99	2.4	628.5	40.0	51.8	"	*23.0*	*49.5*	69.5	347.4
CH12F	"	5\6	5.00	2.6	857.1	**33.0**	**50.2**	"	*17.8*	*46.4*	71.8	361.7
CH12G	"	7\7	6.98	3.5	297.5	**34.6**	**51.1**	"	*20.3*	*48.0*	70.8	354.3
F-G Mean		2\2	1.99	3.0	7140.4	**33.8**	**50.7**		*19.0*	*47.2*	**71.4**	**358.0**
B-3												
BV18	B	6\6	5.98	3.9	298.8	97.8	57.6	347, 21 E	*90.9*	*37.5*	12.1	314.6
BV19	"	5\5	5.00	1.6	2180.0	121.1	82.8	"	*88.2*	*63.4*	27.0	294.4
BV20	"	8\8	7.99	2.0	744.8	266.7	−48.8	"	*264.2*	*−28.0*	−13.6	143.5
BV21	"	7\7	6.98	3.1	372.7	**281.3**	**−52.3**	"	*274.4*	*−32.6*	−7.5	135.2
BV22	"	8\8	7.99	2.2	659.3	**281.5**	**−55.9**	"	*273.7*	*−36.2*	−9.5	133.7
21-22 Mean		2\2	1.99	3.1	1012.2	**281.4**	**−54.1**		***274.1***	***−34.4***	**−8.5**	**134.3**
BV23	"	7\7	6.94	6.2	96.4	262.6	−54.7	"	*260.9*	*−33.8*	−18.3	142.4
BV24	"	7\7	6.95	5.3	128.6	269.5	−52.8	"	*265.9*	*−32.1*	−13.8	140.4
BV25	"	7\7	6.95	5.3	129.5	263.1	−21.0	"	*262.7*	*−0.1*	−5.8	156.4
B-4 (A)												
CU2	BA	6\7	5.98	4.6	217.9	8.4	66.4	235, 20 SE	63.3	73.9	45.4	281.3
CU3	"	6\6	5.99	3.0	487.8	214.8	−64.6	"	257.9	−64.1	−33.9	117.0
CU4	"	5\6	4.97	7.1	118.0	216.0	−69.0	"	266.7	−66.5	−55.1	118.4
CU5	"	7\7	6.96	4.8	161.0	245.9	−74.1	"	289.5	−62.4	−13.7	106.4
CU6	"	5\6	4.98	5.5	195.6	201.0	−63.0	"	245.0	−67.5	−43.4	114.6
CU7	"	7\7	6.97	4.0	234.5	209.4	−67.3	"	259.5	−67.5	−34.5	111.4
CU8	"	6\7	5.96	6.2	119.8	204.2	−67.8	"	257.6	−69.4	−36.5	108.8
CU9	"	7\7	6.99	3.0	414.2	211.5	−63.7	"	254.1	−64.5	−36.6	117.6
CH14A	"	7\7	6.97	4.4	191.2	53.9	74.8	235, 20 SE	*106.1*	*65.3*	17.9	285.0
CH14B	"	6\6	5.97	5.0	181.2	61.4	77.8	"	*114.5*	*65.6*	13.8	280.9
CH14C	"	Reject						"				
CH14D	"	6\7	5.99	3.6	350.9	46.6	79.7	"	*115.7*	*69.0*	16.4	276.7
CH14E	"	7\7	6.98	3.6	289.2	357.7	64.7	"	*47.9*	*76.5*	51.7	273.1
CH14F	"	Reject, lightning						"				
CH14G	"	Reject, lightning						"				
CH14H	"	7\7	6.89	8.2	55.0	260.3	−67.8	"	*288.9*	*−54.6*	−8.5	113.3
CH14I	"	8\8	7.98	2.8	403.9	244.3	−62.8	"	*274.6*	*−54.1*	−17.4	121.1
B-5 (B)												
CH30A	B	7\7	6.98	3.3	328.8	13.9	23.4	070, 10 SE	16.9	31.6	64.3	22.1
CH30B	"	7\7	6.94	5.9	106.9	30.4	27.2	"	35.1	33.3	53.5	355.1
CH30C	"	7\7	7.00	1.5	1583.2	10.6	33.4	"	14.8	41.8	71.1	15.8
CH30D	"	6\7	5.99	2.8	574.4	16.3	33.2	"	21.0	41.0	66.9	5.1
CH30E	"	5\5	5.00	1.9	1550.6	8.6	39.5	"	13.6	48.1	75.5	6.5
CH30F	"	6\7	5.99	3.4	388.9	13.9	33.7	"	18.5	41.8	68.9	8.3
CH30G	"	5\5	4.97	6.1	157.2	20.1	37.9	"	26.2	45.2	65.2	351.6
CH30H	"	6\6	6.00	2.1	991.0	**15.1**	**38.9**	"	20.8	46.8	69.9	356.0
CH30I	"	6\7	5.98	4.2	259.7	**15.3**	**37.5**	"	20.8	45.4	69.2	358.6
H-I Mean		2\2	1.99	1.2	6616.3	**15.2**	**38.2**		**20.8**	**46.1**	**69.6**	**344.6**
CH30J	"	6\7	5.98	4.5	218.9	356.5	44.7	"	0.2	54.2	86.6	59.1
CH30K	"	Conglomerate test						"				
CH30L	"	7\7	6.99	2.4	650.8	5.5	29.4	"	8.6	38.3	71.8	35.4
CH30M	"	7\7	6.97	4.5	182.4	12.7	29.9	"	16.5	38.1	68.0	16.9
CH30N	"	6\6	5.97	5.0	183.4	18.3	33.1	"	23.2	40.7	65.2	2.1
CH30O	"	6\6	5.97	4.8	196.2	10.9	26.4	"	14.1	34.8	67.4	25.1
CH31a	"	6\7	6.89	8.2	54.8	353.3	35.3	210, 10 SE	359.9	40.8	75.2	62.2
CH31b	"	7\7	6.98	3.4	324.1	354.0	41.9	"	2.4	47.2	80.0	49.6
CH31c	"	7\7	6.98	3.5	292.7	3.9	41.2	"	12.6	44.9	74.2	15.9
CH31d	"	7\7	7.00	1.1	2835.9	8.9	35.3	"	16.1	38.3	68.4	17.5
CH31e	"	6\7	5.89	10.0	46.0	14.3	33.0	"	20.9	35.2	63.9	11.9
CH31f	"	Conglomerate test						"				
CH31g	"	7\7	6.99	2.1	826.4	9.6	45.5	"	19.9	48.1	71.2	354.6
CH31h	"	7\7	6.99	1.9	1026.8	7.8	36.9	"	15.4	40.1	69.8	16.8
CH31i	"	8\8	7.98	3.1	331.1	352.4	43.9	"	1.3	49.4	82.0	53.7
CH31j	"	7\7	7.00	1.4	1778.5	1.8	29.9	"	7.5	34.2	69.5	41.2
CH31k	"	8\8	7.99	2.3	601.0	8.2	3.4	"	9.1	7.1	54.4	46.1
CH31l	"	7\7	6.98	3.2	350.0	8.4	25.2	"	13.2	28.5	64.2	31.4
CH31m	"	Conglomerate test										

(*continued*)

TABLE 2B. PALEOMAGNETIC DATA FROM ANDESITE AND BASALT FLOWS (*continued*)

Site	Unit	n/N	R	α_{95}	k	Dec. in situ (°)	Inc. (°)	Strike/dip (°)	Dec. corrected (°)	Inc. (°)	VGP Lat. (°N)	VGP Long. (°W)
B-5 (B) (*continued*)												
CH31n	"	7\7	6.99	2.7	516.4	17.1	35.0	"	24.2	36.6	62.6	5.4
CH31o	"	7\7	6.99	2.7	492.8	4.5	35.3	"	11.6	39.1	71.1	26.7
CH31p	"	7\7	7.00	1.4	1893.3	14.2	23.1	"	18.6	25.5	60.3	23.2
B-6 (C)												
CH3A	B	7\7	6.91	7.3	68.7	*212.6*	−55.0	037, 10 SE	226.8	−54.5	−52.7	141.9
CH3B	"	7\8	7.00	1.6	1382.4	**206.5**	**−56.7**	"	221.9	−57.2	−57.3	139.2
CH3C	"	8\8	7.98	3.1	331.3	**203.7**	**−59.0**	"	220.7	−59.0	−58.6	135.9
B-C Mean		2\2	1.99	6.0	1749.2	**205.1**	**−57.9**		**221.3**	**−58.1**	**−58.0**	**137.6**
CH3D	"	5\7	4.96	7.4	107.1	*26.6*	42.3	072, 07 NW	22.7	37.1	63.8	7.1
CH3E	"	8\8	7.97	3.4	272.5	*43.8*	48.7	"	37.4	45.0	56.6	341.0
CH3F	"	5\6	5.00	1.9	1570.0	**39.4**	**37.1**	"	35.3	33.1	53.3	355.1
CH3G	"	8\8	7.99	2.0	809.6	**36.4**	**38.8**	"	32.3	34.5	56.0	357.1
F-G Mean		2\2	1.99	6.4	1547.7	**37.9**	**37.9**		**33.8**	**33.8**	**54.6**	**356.1**
CH3H	"	7\7	6.99	2.4	650.7	*27.3*	42.5	"	23.3	37.4	63.5	5.9
B-7												
CH7A	B	8\8	7.97	3.9	201.3	*195.1*	−51.6	347, 15 NE	208.5	−42.9	−62.6	171.8
CH7B	"	7\8	6.97	4.5	183.0	*179.4*	−59.1	"	200.2	−53.2	−73.1	160.1
CH7C	"	8\8	7.99	2.1	728.1	*181.6*	−53.9	"	198.7	−47.9	−72.0	176.3
CH7D	"	8\8	7.98	2.9	364.1	**187.6**	**−59.4**	"	206.7	−51.7	−67.5	158.2
CH7E	"	7\8	6.96	4.7	168.8	**189.6**	**−57.8**	"	207.3	−49.8	−66.4	161.7
D-E Mean		2\2	1.99	1.7	3602.5	**188.6**	**−58.6**		**207.0**	**−50.8**	**−67.0**	**160.5**
CH7F	"	8\8	7.98	2.8	394.9	*192.0*	−63.6	"	212.7	−54.7	−63.7	147.6
CH7G	"	9\9	8.97	3.2	255.5	*238.2*	−59.1	337, 17 SW	228.5	−75.7	−51.8	95.0
CH7H	"	4\5	3.97	9.8	88.3	*219.3*	−47.4	"	205.7	−61.5	−69.9	130.2
CH7I	"	8\8	7.96	4.1	187.7	*199.6*	−53.8	"	175.6	−62.7	−83.1	34.6
B-8												
MC16	B	7\7	6.97	4.4	188.9	33.3	44.6	045, 20 NW	16.7	37.6	67.7	17.1
MC17	"	7\8	6.99	2.3	715.8	40.2	49.8	"	19.1	44.4	69.9	3.2
MC18	"	7\8	6.95	5.3	130.8	35.0	52.0	"	13.6	44.7	73.6	13.8
MC19	"	7\7	6.98	3.8	249.7	27.9	56.5	"	5.3	46.7	79.0	36.9
MC20	"	7\7	6.97	4.5	183.9	25.7	44.5	"	10.6	35.3	69.3	32.8
MC21	"	5\6	4.99	3.6	443.0	58.3	44.3	"	38.1	45.4	56.2	340.0
MC22	"	7\7	6.98	3.2	358.9	32.2	34.1	"	20.7	27.7	60.3	18.4
MC23	"	6\7	5.98	4.2	260.6	24.4	52.9	"	5.1	42.6	75.9	42.7
MC24	"	8\8	7.97	3.8	213.8	212.8	−33.6	"	201.5	−27.4	−59.7	197.4
MC25	"	7\7	6.98	3.4	324.3	28.2	54.3	"	7.0	44.9	77.0	33.1
MC26	"	5\5	4.98	6.0	165.6	66.2	52.6	"	38.4	55.2	59.4	324.5
MC27	"	Lightning						"				
MC28	"	4\5	3.98	7.7	141.7	213.0	−15.7	"	205.4	−10.7	−50.0	200.5
B-9												
VH30	B	8\8	7.90	6.8	66.6	53.0	41.1	305, 13 SW	57.9	53.3	43.7	318.5
VH31	"	5\8	4.94	9.7	63.1	55.0	36.3	"	59.5	48.3	40.5	323.3
VH32	"	10\10	9.93	4.3	130.1	43.4	46.2	"	46.3	59.0	54.3	314.2
VH33	"	8\8	7.85	8.2	46.2	68.2	52.5	275, 13 SW	85.7	56.6	24.8	302.9
VH34	"	7\8	6.70	13.8	20.1	359.8	53.7	"	357.2	66.6	78.7	232.5
VH35	"	Lightning						"				
VH36	"	7\8	6.99	2.0	923.4	6.5	50.4	"	7.1	63.4	81.3	277.0
VH37	"	8\8	7.94	5.3	111.5	8.7	52.5	"	10.4	65.5	77.7	276.7
VH38	"	7\8	6.90	7.8	61.4	13.3	42.7	"	15.8	55.5	77.2	335.9
VH39	"	8\8	7.99	2.4	528.7	7.1	49.0	"	7.9	62.0	82.0	288.2
VH40	"	9\9	8.95	4.0	165.7	8.8	33.3	"	9.6	46.3	76.9	21.3
VH41	"	8\8	7.88	7.4	57.3	7.6	40.2	"	8.3	53.2	82.1	1.6
VH42	"	8\8	7.96	4.0	191.0	343.4	38.0	"	338.2	49.9	70.6	136.3
VH43	"	7\7	6.97	4.0	234.1	345.6	52.0	"	337.2	64.0	71.4	183.7
VH44	"	8\8	7.95	4.9	128.0	17.3	52.3	"	22.9	64.9	71.0	296.7
VH45	"	8\8	7.94	5.2	113.6	355.9	60.8	"	349.2	73.5	67.6	227.4

Note: N\N$_o$—ratio of number of sites means used in group mean to total number of site means at the section; *R*—resultant vector length, Dec/Inc—group mean declination and inclination (corrected); α_{95}—95% confidence interval of the estimated group mean direction, assuming a circular distribution; *k*—best estimate of (Fisher) precision parameter; *S*—estimated angular standard deviation of the virtual geomagnetic poles (VGPs). Asterisk—mean represents the Fisher average of four or more sites. Bold sites are adjacent flows that were combined into a single cooling unit (see text for discussion). UTM locations of all sites are given in Appendix 2 (NAD 27).

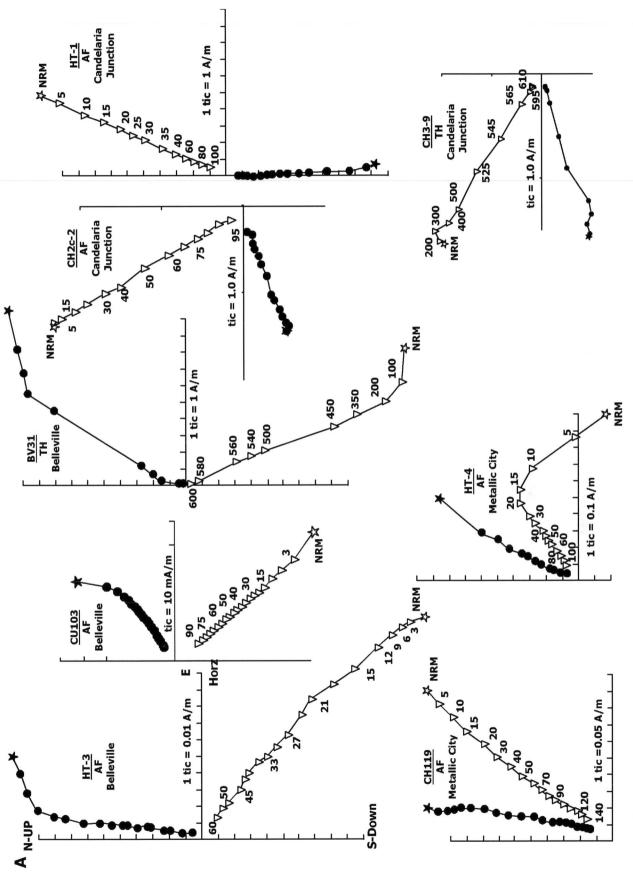

Figure 4 (*on this and following two pages*). Representative in situ modified demagnetization diagrams (Zijderveld, 1967; Roy and Park, 1974) for rocks collected in the Candelaria Hills. Solid (open) symbols represent the projection onto the horizontal (true vertical) plane. AF (alternating field) demagnetization steps are given in milliTesla and thermal demagnetization steps in degrees centigrade. Typically, AF and thermal (TH) demagnetization results from two specimens of the same sample are shown for comparison for some samples. Diagrams are designated by a site number (e.g., CU103, BV15), method of treatment (AF or TH), and rock type. Intensity (A/m) is shown along one axis for each sample, each tic equals indicated intensity. (A) ignimbrites; (B) andesite flows; (C) basalt flows. NRM—natural remanent magnetization.

232

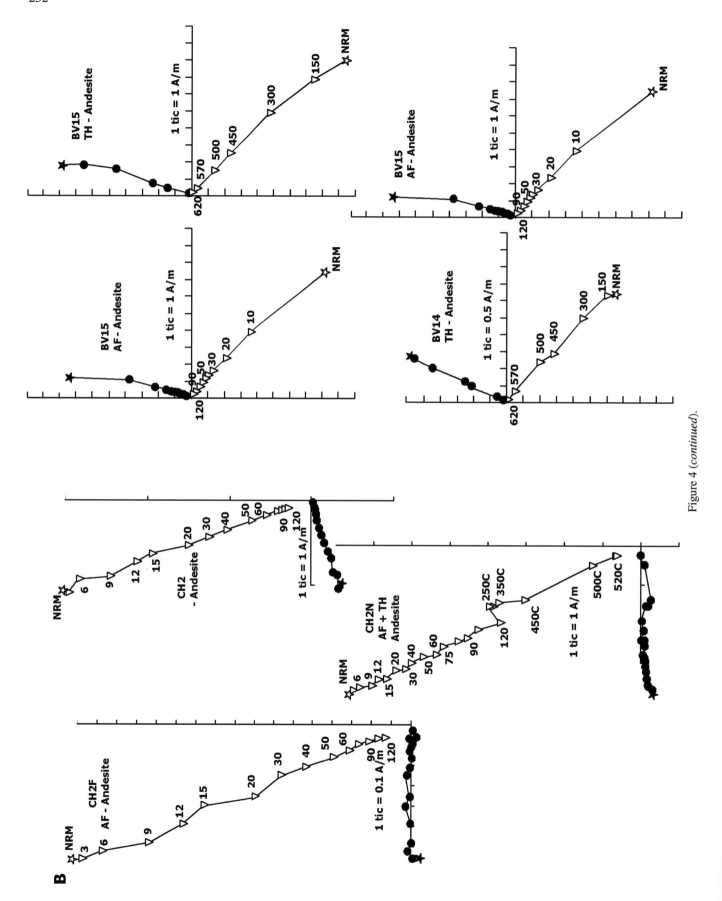

Figure 4 (*continued*).

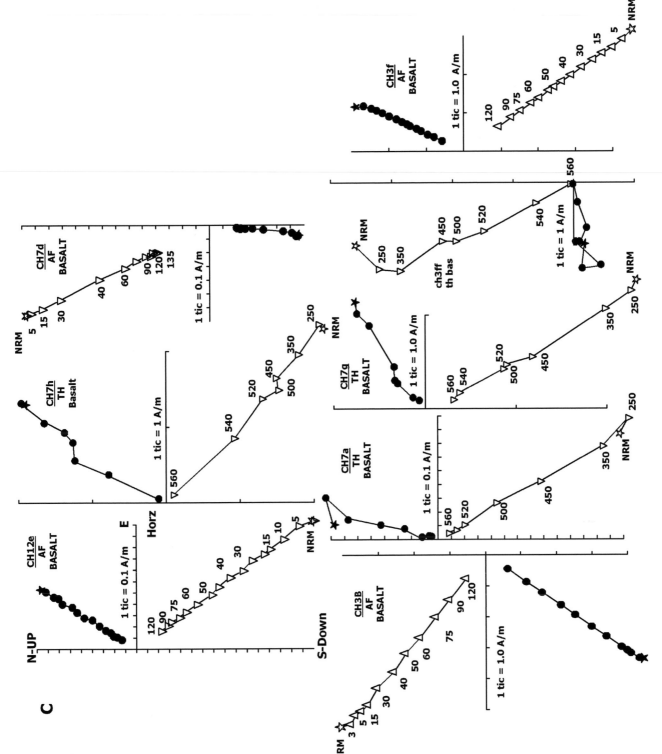

Figure 4 (*continued*).

with remanence dominated by lower coercivity, MD-dominated assemblages are often characterized by a strong viscous magnetization "overprint" and poorer overall demagnetization behavior, making accurate definition of a geomagnetic field direction less likely. The Candelaria Junction Tuff has the highest median destructive fields (MDF) values (between 40 and 50 mT), characteristic of PSD to SD grains. The Metallic City Tuff has MDF values between 10 and 15 mT, characteristic of a more MD-dominated assemblage. We note that the Metallic City Tuff typically shows a greater degree of within-site and also between-site dispersion among our collection of sites in the Candelaria Hills area, as discussed in greater detail later. The Belleville Tuff (Fig. 4A) has intermediate MDF values (~25 to ~40 mT), characteristic of mixtures of MD and PSD grains (Dunlop and Özdemir, 1997).

Paleomagnetic data were obtained from 145 sites distributed within 40 distinct structural blocks. To simplify discussion of the ignimbrite paleomagnetic data, we group sampling stations into blocks where site locations are all within ~50 m to 100 m of each other. It is rare for all three regional ignimbrites to be exposed in a single structural block, and the 40 blocks represent locations where at least one site was established in an ignimbrite. An ideal structural block is defined as a fault-bounded area with internally consistent, uniform sense of tilt that exposes coherent stratigraphic sections involving at least two tuffs. Some of the blocks are internally disrupted by small displacement faults (not shown in Fig. 3), and many blocks contain only one tuff. The area of each block varies from ~25 m^2 to 2500 m^2. For the larger blocks (>100 m^2), we sampled more than one site, typically in different zones and/or cooling units of each ignimbrite to better define the magnetization and identify subtle variations in remanence acquisition across the block. On average, sample directions from two to seven sites were averaged from the same ignimbrite on a single block. When at least four sites were available for averaging, we calculated the Fisher mean average of the site mean directions. Averaging sample data from multiple sites on a structural block from different zones and phases of the same ignimbrite provides a better estimate of the ChRM of each ignimbrite and thus the block mean direction. We note, however, that the 95% confidence limit of any block mean for a specific unit, defined by averaging samples from several sites, may artificially underestimate the true confidence limit, because the estimated α_{95} is inversely proportional to the square root of the number of specimens N, where $\alpha_{95} = 140°/(kN)^{1/2}$ (Fisher, 1953). After combining data from sites collected in the 40 structural blocks, we obtained 79 block mean paleomagnetic directions; data from the Candelaria Junction Tuff provided rotation estimates for 37 structural blocks, the Belleville Tuff provided rotation estimates for 26 structural blocks, and the Metallic City Tuff provided rotation estimates for 16 structural blocks.

Each regionally extensive tuff in the study area has a characteristic magnetization that is distinct in direction, an observation that is not inconsistent with previous paleomagnetic observations of ash-flow tuffs (e.g., Geissman et al., 1982; Wells and Hillhouse, 1989; Byrd et al., 1994; Best et al., 1995; Hudson et

al., 1994; Hudson, 1992). Therefore, the data from dispersed sites from a regionally extensive ignimbrite can be compared to estimate relative vertical-axis rotations between structural blocks. In principle, data from all three tuffs exposed in a single structural block should yield similar relative rotation estimates if rotation postdates their deposition. Paleomagnetic data from the three regionally extensive ignimbrites reveal vertical-axis rotations that vary across the study area. Typically, the data yield internally consistent results within the same structural block as well as similar magnitudes of rotation from different tuffs within the same block, although some inconsistencies are observed (Figs. 5A and 5B). The inclination of individual site mean directions from a specific tuff is similar from block to block, although some exceptions are noted (see following).

Of the 79 block mean directions determined from data from the three ignimbrites, 16 yield rotation estimates that we consider to be of relatively low confidence. Low confidence rotation estimates yield a statistically significant flattening (F) estimate that is greater than 10° of the δF error estimate (Table 3). Ideally, F values from data from the same ignimbrite should be ~0°, assuming the ChRMs at each site are appropriately referenced to the paleohorizontal and have been adequately isolated during progressive demagnetization (e.g., Wells and Hillhouse, 1989). An inappropriate structural correction is probably the most logical explanation for F values significantly greater than the δF. Other sources of error include not fully isolating the ChRM of the individual samples at a site, lightning strikes, unusual rock magnetic properties of a particular phase (Geissman et al., 1982a), or inclination flattening associated with the development of a compaction foliation (e.g., Gattacceca and Rochette, 2002; Uyeda et al., 1963). It is likely that a combination of these factors contributes to the low confidence estimates for these 16 block means (Table 3).

Ideally, statistically indistinguishable rotation estimates from several ignimbrites within the same structural block suggest that the rotation estimate for such a block is robust and well defined. Unfortunately, in this study, only four structural blocks included exposures of all three ignimbrites (blocks 1, 2, 18, and 36, see Fig. 3). With the exception of block 2, each of these localities are fragmented by small-displacement faults, and, in all four blocks, at least one of the three ignimbrites did not provide acceptable paleomagnetic results. On block 1, the three ignimbrites crop out in close proximity (Figs. 5B), but the section is not continuous. The Candelaria Junction and Belleville Tuffs are exposed in stratigraphic order, and they yield similar rotation estimates (Figs. 5B; Table 3). One site in the Metallic City Tuff yields a rotation estimate (R = +31°) that is statistically indistinguishable from those of the other two tuffs. The other site in Metallic City Tuff, located on a small fault block ~30 m from the other sites, however, yields a rotation estimate of +83° and a flattening estimate of +22°. Given the high flattening value, we attribute this unusually high rotation estimate to an erroneous structural correction based on compaction fabric orientation. For fault block 2, all three ignimbrites are in a continuous stratigraphic sequence (Figs. 5B). The data from the Metallic City Tuff are of high dispersion (α_{95} >

15° and $k < 10°$) and do not yield stable end-point behavior and were rejected. In contrast, the data from the Candelaria Junction and Belleville Tuffs are of high quality and yield similar rotation estimates (Figs. 5B; Table 3). At block 18 (Figs. 5B), the ignimbrite flows are exposed in close proximity but are not part of an intact stratigraphic section. The sites in the Metallic City Tuff did not yield acceptable results and were rejected. The Belleville Tuff is bounded by west-dipping, high-angle faults between the Candelaria Junction Tuff to the west and the Metallic City Tuff to the east. The data from the Belleville Tuff provide a questionable rotation estimate because of the high flattening value (+14°). The paleomagnetic data from the Candelaria Junction Tuff at this locality are also questionable because of the high flattening value (+21°) and an unrealistically large rotation estimate (±107°). We interpret these discrepancies in data from the Candelaria Junction and Belleville Tuffs to reflect either an error in the structural correction measurement or an unforeseen correction that was not considered, such as these sites being located on a limb of a plung-

ing fold. The rotation estimates on block 18 are of low confidence (Figs. 5B). The three regionally extensive ignimbrites crop out within block 36 (Figs. 5B); however, a continuous section of the three tuffs is not exposed. The Candelaria Junction and Belleville Tuffs crop out together and yield similar rotation estimates (Figs. 5B). The data from the Metallic City Tuff is of high dispersion ($\alpha_{95} > 15°$ and $k < 10°$) and did not yield stable end-point behavior and the site was rejected.

Several blocks preserve at least two of the tuffs in stratigraphic order (Fig. 6). For example, in block 13, the Candelaria Junction and Metallic City Tuff data yield similar-magnitude (~35° to ~40°) rotation estimates at the 95% confidence level. In block 3, the Candelaria Junction and Metallic City Tuffs are separated by ~50 m and lie on either side of an east-northeast–striking fault. Overall, both sites yield similar magnitudes of clockwise rotation (overlap at 95% confidence), yet the Metallic City result is very poorly defined ($\delta R = 33.8°$). At only two locations, opposite rotation estimates defined by different ignimbrites on the

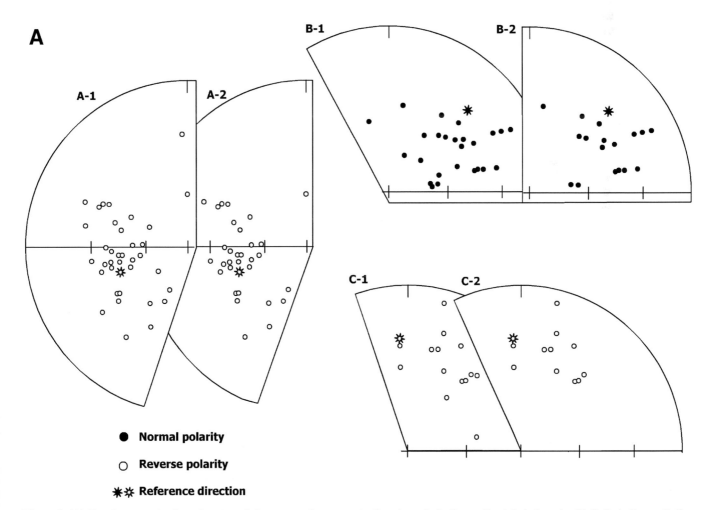

Figure 5. (A) Equal-area projection of corrected site mean paleomagnetic directions. A- indicates Candelaria Junction Tuff; B- indicates Belleville Tuff; and C- indicates Metallic City Tuff; 1 indicates all accepted paleomagnetic data; 2 indicates low confidence sites excluded (see text for discussion). (*Continued on following page.*)

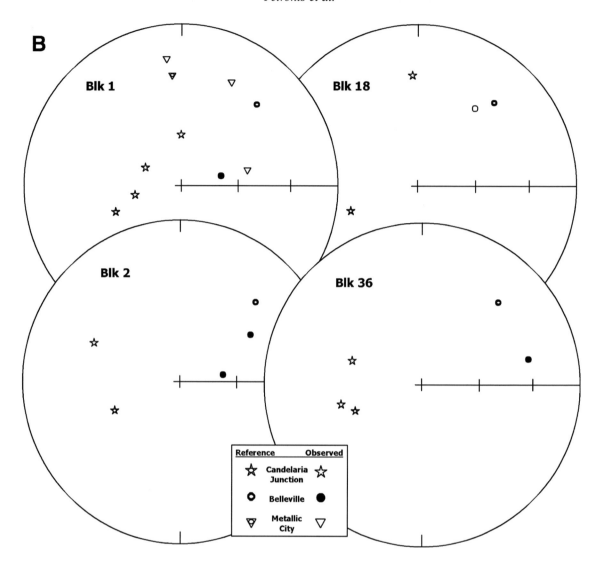

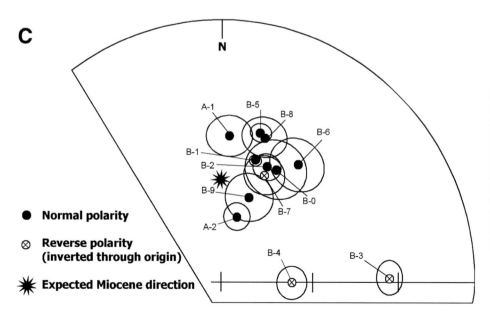

Figure 5. (B) Example of paleomagnetic data from structural blocks in which more than one tuff occurs on the same block. (C) Basalt and andesite flows from 12 sections in the Candelaria Hills. Each site number refers to abbreviations in Tables 2 and 4. Solid circle—lower-hemisphere projection; crossed circle—upper-hemisphere projection (inverted through origin); ellipse—95% confidence interval; black star—Miocene expected direction.

TABLE 3. REGIONAL EXTENSIVE IGNIMBRITES BLOCK MEAN PALEOMAGNETIC RESULTS

Fault/ block	Section	Unit	Rotation and flattening estimates			
			R	±R	F	±F
Belleville Uplift						
23	BV34-35-36-37	TT9	−46.4	10.2	−6.8	5.0
23	BV48	TT7	−1.9	12.7	−22.0	7.7
24	BV66-67-68	TT9	5.7	7.2	7.9	5.0
25*	BV60	TT7	6.0	7.1	−12.8	5.2
25	BV61-62	TT7	−23.7	5.0	−15.8	3.5
25*	BV65	TT7	34.7	7.0	−33.9	3.2
26*	BV49-53	TT7	15.7	13.2	−39.4	4.7
26*	BV58-59	TT7	29.2	7.1	−26.6	3.9
27	BV26-27-28	TT9	22.1	9.6	−8.2	4.6
27	BV32	TT9	23.7	12.0	−14.1	4.6
27	BV33	TT9	5.8	4.8	10.4	3.2
28	BV2-3-6, 4-5	TT7	31.5	8.6	−8.5	6.8
28	BV49-50-51	TT7	10.8	8.4	−11.8	6.2
28	BV91-94	TT7	30.9	4.7	−13.4	3.4
29	BV45-46	TT7	31.4	15.2	−11.5	11.2
30	BV95-96	TT7	15.4	4.9	−8.1	4.0
34	BV84-85	TT9	29.1	6.4	−5.8	3.3
34	BV86-87	TT9	9.3	5.3	12.1	4.0
34	BV89	TT9	43.3	13.7	21.7	11.6
40*	BV81	TT9	−18.7	14.7	−17.5	4.8
Candelaria Uplift						
1*	CU53	TT9	47.5	6.6	−15.0	2.5
1	CU54-56	TT5	31.4	9.7	3.1	8.8
1*	CU57	TT7	33.6	42.2	−37.7	13.8
1	CU58	TT9	10.1	18.9	−12.4	7.6
1	CU59-60	TT7	1.3	12.3	−18.9	8.0
1	CU64	TT9	110.6	13.5	−10.9	5.8
1	CU66	TT5	−2.5	4.2	9.8	5.1
1*	CU67–68	TT5	83.0	4.6	−22.1	2.7
2	CU71	TT7	13.5	4.9	−14.8	3.5
2*	CU74	TT7	37.5	12.2	−37.4	4.8
2	CU75–77	TT9	44.4	5.8	14.0	4.5
3	CU106	TT5	42.7	33.8	−13.0	22.5
3	CU109	TT5	–	–	–	–
3	CU110	TT9	13.5	6.0	−3.0	3.3
4	Cu37–38	TT9	14.2	6.2	−1.0	3.6
4	CU44	Tt7	−35.1	6.6	−14.5	4.7
5	CH4	TT5	29.1	6.0	−13.6	4.2
5	CH4F	TT5	44.5	6.0	−11.4	4.4
6	CH6C	TT9	47.5	4.1	−4.1	2.2
6	CH10#	TT9	−17.0	4.0	8.8	2.9
6*	CH15	TT9	32.0	3.7	18.1	3.2
7	MM9#	TT9	1.6	5.0	−5.5	2.6
7	MM1#	TT5	18.4	8.3	23.8	5.6
8	CU103–104	TT7	0.8	4.7	−8.2	3.9

(continued)

TABLE 3. REGIONAL EXTENSIVE IGNIMBRITES BLOCK MEAN PALEOMAGNETIC RESULTS *(continued)*

Fault/ block	Section	Unit	Rotation and flattening estimates			
			R	±R	F	±F
9	CU100	TT5	21.5	6.0	5.6	6.0
9*	CU101	TT5	40.5	12.2	−24.8	6.6
9	CU102	TT5	44.4	5.0	−7.2	4.0
10	CU91	TT9	−35.0	4.3	−2.3	2.4
11*	CU92	TT7	60.1	4.8	−21.8	3.0
11	CU93	TT5	−13.2	5.0	6.9	5.3
13	CU94	TT5	37.1	14.6	5.5	13.2
13	CU96	TT9	46.9	5.2	12.5	3.9
13	CU97	TT9	8.2	3.7	−8.0	1.8
13	CU98	TT9	48.3	3.3	9.5	2.4
14	CU99	TT9	34.3	6.5	−1.2	3.7
15	CU89–90	TT9	12.1	7.2	14.3	5.6
16	CU87	TT5	−0.2	6.5	−3.9	5.4
16	CU88	TT5	−1.2	10.6	−15.5	7.1
18*	Cu28–29	TT9	107.7	4.4	21.1	3.9
18	CU30–32	TT7	−7.3	5.7	−9.1	4.5
18	CU33–36	TT5	–	–	–	–
19	CU10–11	TT7	16.2	5.4	4.5	6.2
19*	Cu12–15	TT7	26.8	15.0	−32.5	6.8
20	CU16–25	TT7	17.2	4.3	9.0	6.4
20*	CU27	TT7	7.2	4.9	−15.8	3.4
21*	CU51	TT7	−21.8	9.7	−39.1	3.7
35	CU119–120	TT5	18.0	4.3	−4.5	3.6
35	CU122	TT5	47.8	10.6	−5.3	8.4
36	CU123	TT9	9.1	4.7	7.2	3.2
36	Cu124–125	TT7	32.7	4.7	−0.8	4.7
36	CU129	TT9	38.3	9.2	2.5	5.8
37	CU116,118	TT5	21.2	4.8	−3.6	4.1
38	CU111	TT9	10.2	7.0	3.9	4.5
38	CU114–115	TT9	−46.0	8.5	−13.2	3.4
Volcanic Hills						
32	VH8	TT9	–	–	–	–
32	VH9	TT9	−34.6	13.5	19.9	11.3
32	VH11	TT9	−12.3	3.9	6.8	2.7
32	VH12	TT9	−13.1	3.0	5.8	2.0
32	VH20–21	TT9	−44.5	8.7	8.4	6.2
39	VH50–51	TT9	20.0	2.9	5.9	1.9
39	VH56–58	TT9	8.3	5.4	−1.0	3.1
39*	VH59–60	TT9	−16.4	3.1	20.0	2.8

Note: Fault block—numbers refer to locations indicated in Figures 3 and 6; Section—paleomagnetic site locations that were averaged together to yield a fault block mean (see Table 1); Unit—regional ignimbrite (TT9—Candelaria Junction, TT7—Belleville, TT5—Metallic City); asterisk—site rejected due to a statistically significant flattening estimate. Rotation (R) and flattening (F) and associated error estimates (±R, ±F) are after Beck (1980); Demarest (1983) with respect to the Huntoon Valley ignimbrite section. See text for discussion.

238

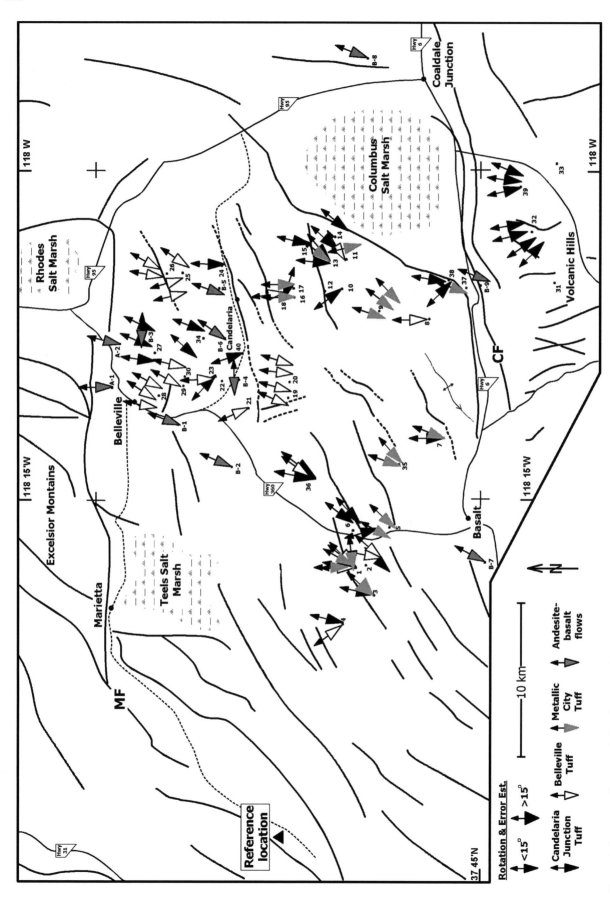

Figure 6. Summary diagram of all paleomagnetic data from the Candelaria Hills. Each arrow represents a location where several paleomagnetic sites have been averaged together to yield a fault block mean (ignimbrites) or section mean (basalts, andesites). See Tables 3 and 4 for fault block and section mean rotation estimates. Any deviation of the arrow from north indicates a vertical-axis rotation. The basalt and andesite data are referenced to the Miocene expected direction, and the ignimbrite data are given with respect to a fixed location in Huntoon Valley. See text for discussion and Table 2 for paleomagnetic data from each location. See Figure 3 for a description of symbols. Site locations are the same as in Figure 3.

same block were observed. In block 4, the Candelaria Junction and Belleville Tuffs yield data suggesting opposite senses of rotation that are statistically distinct at the 95% confidence level. In block 11, the Belleville and Metallic City Tuffs are separated by ~20 m and lie on either side of an east-northeast–striking fault. The Belleville Tuff yields data that imply clockwise rotation, and the Metallic City Tuff yields data that imply a counterclockwise rotation. The Belleville Tuff data are strongly discordant in inclination ($\delta F = 21°$), and the rotation estimate from this site is of low confidence.

Andesite Flows

Two gently dipping sections of gray to reddish gray andesite flows exposed along the east and west side of Nevada State Highway 360 in the northeast part of the Candelaria Hills (Fig. 3) yield high-quality demagnetization data (Fig. 4B). These rocks typically are of moderate to high coercivity, with MDF of 40–70 mT, and for a few high coercivity samples, thermal demagnetization up to 610 °C was required to fully isolate the ChRM, most likely at least partly carried by fine-grained maghemite (Dunlop and Özdemir, 1997). Although the two sections are likely very similar in age, they are located on different fault blocks, have different flow orientations, and therefore are discussed separately. At the eastern section (A-1, Table 4; Fig. 3), 13 sampling sites (each site in a different flow) were established, and all yield interpretable demagnetization data that are well grouped (10 sites with $\alpha_{95} < 6°$) at the site level. At the western section (A-2, Table 4; Fig. 3), we established 13 sampling sites in separate lava flows. All sites yield interpretable demagnetization data that are well

grouped (12 sites with $\alpha_{95} < 6°$) at the site level (Table 4). The paleomagnetic results from three separate adjoining lava flow sequences (sites CH2D-E, CH2I-J, and CH2M-N; Table 2B) yield site mean directions that are statistically indistinguishable. The mean of both adjacent sites falls within the 95% confidence ellipse of the adjoining site. These sites were combined into a single cooling unit. The accepted sites from each section were used to calculate group mean directions for each section after the flows were restored to the horizontal based on orientation of contacts between the individual flows. The eastern section yields a group mean of D = 3.4°, I = 43.4°, $\alpha_{95} = 6.9°$, 13/13, 13N, and the western section yields a group mean of D = 18.51°, I = 69.1°, $\alpha_{95} = 5.6°$, 7/10, 7N (Fig. 5C). The mean directions from the three sites (B, G, H; Table 2B) not included in the group mean calculation in the western section were considered outliers as the site mean was greater than 36°, 37°, and 45°, respectively, from the overall grand mean. The two group means are statistically distinct from the Miocene normal polarity expected field direction (353.2°, 58.3°; Besse and Courtillot, 2002).

Basaltic Andesite Flows

Ten sections of basaltic andesite located throughout the Candelaria Hills and in the neighboring ranges, each consisting of several lava flows having obvious cooling breaks typically indicated by oxidized flow tops, yield high-quality demagnetization data (Fig. 4C). The basaltic andesites typically have MDFs between 50 and 80 mT, and this behavior is characteristic of PSD to SD magnetite. All samples were fully demagnetized, with the ChRM isolated during thermal demagnetization between 560 °C

TABLE 4. GROUP MEAN PALEOMAGNETIC RESULTS FOR ANDESITE AND BASALT FLOWS

Location/ section	n/N	R	α_{95}	k	Dec. corrected (°)	Inc. (°)	Rotation and flattening estimates				Rev. test	S	K VGP
							R	±R	F	±F			
Basalt flows													
B-0	7\9	6.80	11.0	36.4	25.7	49.6	*32.5*	*13.7*	*–8.7*	*8.8*	i	15.6	26.8
B-1	14\14	13.97	1.9	472.0	16.6	50.0	*23.4*	*4.0*	*–8.3*	*2.6*	–	4.1	382.8
B-2	6\6	5.97	4.9	226.2	23.3	51.3	*30.1*	*6.9*	*–7.0*	*4.3*	–	4.5	321.6
B-3	5\6	4.98	5.3	260.9	267.1	–33.2	*93.9*	*5.6*	*–25.1*	*5.0*	i	4.9	271.0
B-4	14\14	13.80	5.0	69.7	270.1	–67.0	*96.9*	*5.6*	*8.7*	*4.3*	p	16.4	24.5
B-5	26\27	25.68	3.2	77.8	15.4	40.4	*22.2*	*4.5*	*–17.9*	*3.5*	–	9.2	76.9
B-6	6\6	5.89	10.10	53.9	33.1	44.6	*39.9*	*11.5*	*–13.7*	*8.1*	f	10.9	54.9
B-7	7\8	6.91	7.3	81.7	202.2	–53.8	*29.0*	*10.2*	*–4.5*	*6.0*	–	10.9	54.9
B-8	11\12	10.79	6.7	52.3	17.4	41.6	*24.2*	*7.6*	*–16.7*	*5.7*	i	11.5	49.4
B-9	13\15	12.56	8.1	29.4	16.3	60.5	*23.1*	*13.4*	*2.2*	*5.8*	–	20.6	15.5
Andesite flows													
A-1	13\13	12.68	6.9	40.1	*3.4*	43.4	*10.2*	*8.0*	*–14.9*	*5.8*		13.2	37.7
A-2	7\10	9.93	5.6	143.0	*18.5*	69.1	*25.3*	*12.9*	*10.8*	*4.7*		11.5	50.0

Note: Section locations (B-1, A-1, etc.) refer to locations indicated in Figures 3 and 6; *N*—number of accepted sites at each section, *n*—number of sites used to calculated section group mean, *R*—resultant vector length; α_{95}—95% confidence interval about the estimated mean direction, assuming a circular Fisher distribution; *k*—best estimate of (Fisher) precision parameter; Dec.—declination, Inc.—inclination. R and F are rotation and flattening with respect to the Miocene normal and reversed polarity expected direction (D = 353 [173], I = 58 [–58], $\alpha_{95} = 2.6$) based on the paleomagnetic pole of Besse and Courtillor (2002). ±R and ±F are the associated rotation and flattening errors (Beck, 1980; Demarest, 1983); Rev. test—reversal test results (McFadden and McElhinny, 1990) (i—indeterminate, f—fail, p—pass); S—estimated angular standard deviation of virtual geomagnetic pole (VGP) directions; K VGP—dispersion estimate of the virtual geomagnetic poles.

and 580 °C, implying that a cubic phase, likely low-Ti magnetite, carries the remanence. The paleomagnetic results from the 10 sections are discussed from north to south (Fig. 3; Table 4). In general, the demagnetization data from each of the sections are characterized by dominantly single-component, stable end-point response yielding well-defined magnetizations. Five sections yield single polarity magnetizations (either normal or reversed), and five sections yield dual polarity data sets. Unless otherwise noted, all paleomagnetic data from each section of basaltic andesite flows have been corrected for the dip of the flows (Fig. 5C).

The Garfield Hills basaltic andesites (B-0; Tables 2B and 4), the northernmost section sampled in this study, include 10 flows located along the north side of the Garfield Hills (Fig. 3B). Similar flows in the area have yielded K-Ar age determinations ranging from ca. 8.5 to ca. 3.2 Ma (Silberman et al., 1975; Marvin and Mehnert, 1977; Ekren et al., 1980; Ormerod, 1988). Of the 10 paleomagnetic sites, eight yield interpretable demagnetization data that are well grouped ($\alpha_{95} < 7.5°$; Tables 2B and 4) at the site level. The two excluded sites were rejected because they did not yield stable end-point behavior and had high within-site dispersion ($\alpha_{95} > 15°$, $k < 10$); these sites are not discussed further. Of the eight accepted sites, one (GH7) is of reverse polarity, and the remaining seven sites are of normal polarity. Stratigraphically, site GH7 is located near the middle of the flow sequence. The paleomagnetic results from two adjoining lava flows (sites GH11–12) yield site mean directions that are statistically indistinguishable; these were combined into a single cooling unit. After correcting for the moderate dip of the flows based on interflow contacts and inverting the reverse polarity magnetization through the origin, the seven accepted sites means provide a group mean of D = 25.7°, I = 49.6°, $\alpha_{95} = 11.0°$, 7/9, 1 reversed (R), 6 normal (N), which is statistically distinct from the Miocene normal polarity expected field direction (Fig. 5C).

We sampled two sections within the north-central part of the Candelaria Hills (B-1, B-2; Fig. 3; Tables 2B and 4). Fifteen paleomagnetic sampling sites were established at the northeast section (B-1; Table 4). All sites yield interpretable normal polarity demagnetization data that are well grouped (seven sites with $\alpha_{95} < 3°$) at the site level. The paleomagnetic results from two adjoining lava flows (sites CH20K-L) yield site mean directions that are statistically indistinguishable; these were combined into a single cooling unit. The corrected site mean data for the 14 flows provide a group mean of D = 16.6°, I = 50.0°, $\alpha_{95} = 1.9°$, 14/14, 14N (Fig. 5C), which is statistically distinct from the Miocene normal field direction. Seven paleomagnetic sample sites were established at the southeast section (B-2; Table 2B), and all sites yield interpretable normal polarity results that are well grouped (5 sites with $\alpha_{95} < 4°$) at the site level. The paleomagnetic results from two adjoining lava flows (sites CH12F–G) yield site mean directions that are statistically indistinguishable; these were combined into a single cooling unit. The corrected site mean data provide a group mean of D = 23.3°, I = 51.3°, $\alpha_{95} = 4.9°$, 6/6, 6N, which is also statistically distinct from the Miocene normal polarity field direction (Fig. 5C).

Paleomagnetic samples were collected from eight sites in a section located 3 km east of the Belleville town site (B-3, Fig. 3; Table 4). These flows are unconformably overlain by ca. 15 Ma andesite flows and are compositionally different from the other basaltic andesite flows in the area, being porphyritic with phenocrysts of pyroxene and amphibole up to 6 mm in length. All sites yield interpretable demagnetization data that are well grouped ($\alpha_{95} < 6°$) at the site level. Two sites were excluded from the group mean calculation because their mean directions were greater than 45° from the overall group mean. Of the six accepted sites, one site is normal polarity, and five sites are reversed polarity. The normal polarity site is the stratigraphically highest flow in the section. The paleomagnetic results from two adjoining lava flows (sites BV21–22) yield site mean directions that are statistically indistinguishable; these were combined into a single cooling unit. The corrected site mean data, after inverting the normal polarity magnetization through the origin, provide a group mean of D = 267.1°, I = –33.2°, $\alpha_{95} = 5.3°$, 5/7, 4R, 1N, which is statistically distinct from the Miocene reversed polarity field direction and differs in declination by ~90°. This substantial discrepancy will be discussed further later (Fig. 5C).

In the central part of the Candelaria Hills, an extensive basaltic andesite field of roughly 16 km^2 extends north for ~2 km from its southern boundary at the Candelaria fault, where several small vents and cinder cones are well exposed (Fig. 3). Sites were established along five sections. Two sections are located on the west side of the basalt field, ~2 km east of Candelaria Junction (Fig. 3), where flows rest directly on the Candelaria Junction Tuff and are petrographically distinct (i.e., altered oxide and pyroxene phenocrysts) from the other three basaltic andesite sections. We combine the two western sections into a single locality result for discussion (B-4[A]; Table 4). Seventeen sites were collected at B-4(A), and 14 sites yield interpretable demagnetization data that are well grouped (12 sites with $\alpha_{95} < 5°$) at the site level. The three sites rejected either did not yield stable end-point behavior (two sites; most likely lightning struck) or had high within-site dispersion ($\alpha_{95} > 15°$, $k < 10$) and did not yield interpretable results (one site). Of the 14 accepted sites, five are normal polarity, and nine are reverse polarity. The stratigraphic distribution of the normal and revered polarity sites indicates that this section contains at least two full reversal sequences. The corrected site mean data, after inverting the five normal polarity means, provide a group mean of D = 270.1°, I = –69.7°, $\alpha_{95} = 5.0°$, 14/14, 9R, 5N (Fig. 5C), which differs in declination from a Miocene reverse polarity reference by ~90° (see following).

Of the remaining three sections, two are located on the same structural block, ~1 km north of the Candelaria town site (Fig. 3). We combine these two sections into one composite section (B-5[B]; Fig. 3; Table 2B). For section B-5(B), all 28 sites yield interpretable exclusively normal polarity data that are well grouped (24 sites with $\alpha_{95} < 5°$) at the site level. The paleomagnetic results from two adjoining lava flows (sites CH30H–I) yield site mean directions that are statistically indistinguishable; these were combined into a single cooling unit. At three horizons in this flow sequence, we conducted modified conglomerate tests to evaluate the antiquity of the magnetization. Samples were collected (at least two samples per cobble) from deposits between flows. All three sites yield positive test results in that the magnetization directions from discrete specimens within each cobble are well-defined and similar, yet the magnetization directions are random among cobbles. The corrected site mean data provide a group mean of D = 15.1°, I = 39.2°, α_{95} = 3.8°, 26/27, 27N (Fig. 5C). The one site mean direction excluded from the group mean calculation had an inclination greater than 34° from the overall group mean.

The remaining section (B-6[C]) consists of eight flows sampled along a roughly 2 km traverse over low-relief topography 2 km west of the Candelaria town site (Fig. 3; Tables 2B and 4). The eight sites in section C yield interpretable normal and reverse polarity demagnetization data that are well grouped (seven sites with α_{95} < 5°) at the site level. Of the eight sites, five sites are normal polarity, and three sites are reverse polarity. The paleomagnetic results from adjoining lava flows (at sites CH3B–C and CH3F–G) yield site mean directions that are statistically indistinguishable; these were combined into single cooling units. The reverse polarity sites are located at the base of the flow sequence. The corrected site mean data, after inverting the reverse polarity means, provide a group mean of D = 33.1°, I = 44.6°, α_{95} = 10.1°, 6/6, 2R, 4N (Fig. 5C).

On U.S. Highway 6, ~3 km northeast of Montgomery Pass (B-7; Fig. 3; Table 4), we collected samples from a faulted sequence of basaltic andesite flows (nine flows sampled). These flows have yielded $^{40}Ar/^{39}Ar$ dates ranging from 3.14 ± 0.03 Ma to 4.08 ± 0.10 Ma (Lee et al., 2005). The nine sites all yield interpretable reverse polarity data that are well grouped (eight sites with α_{95} < 5°) at the site level. The paleomagnetic results from two adjoining lava flows (sites CH7D–E) yield site mean directions that are statistically indistinguishable; these were combined into single cooling unit. The corrected data provide a group mean, for 8/9 sites, of D = 202.2°, I = −53.8°, α_{95} = 7.3°, 7/8, 8R (Fig. 5C). The one site that was excluded from the group mean calculation has a mean direction greater than 24° from the overall group mean.

On the west flank of the Monte Cristo Range, 4.5 km northeast of Coaldale Junction, Nevada, we collected sites in a moderately west-dipping sequence of 13 basaltic andesite flows (B-8; Fig. 3; Table 2B and 4). Of the 13 sites, 12 sites yield interpretable demagnetization data that are well grouped (α_{95} < 6°) at the site level. One site appears to have been lightening struck, as it did not yield stable end-point behavior. Of the 12 accepted sites, 10 sites are normal polarity, and two sites are reversed polarity. One site was excluded from the group mean calculation because its mean is greater than 32° from the overall group mean. The corrected site mean data, after inverting the reverse polarity mean directions through the origin, provide a group mean, for 11/12 sites, of D = 17.4°, I = 41.6°, α_{95} = 6.7°, 11/12, 1R, 10N (Fig. 5C).

South of Miller Mountain, along the north slope of the Volcanic Hills, ~4 km east of Basalt, Nevada, samples were collected from a gently dipping sequence of basaltic andesite flows (B-9; Fig. 3; Table 2B and 4). Of the 16 sites, 15 sites yield interpre-

table normal polarity demagnetization data that are well grouped (13 sites with α_{95} < 8°) at the site level. One site appears to have been lightning struck, as it did not yield stable end-point behavior. The corrected site mean data provide a group mean of D = 16.3°, I = 60.5°, α_{95} = 8.1°, 13/15, 15N (Fig. 5C). The two sites excluded from the group mean calculation have mean directions that are greater than 25° from the overall group mean.

DISCUSSION

The brittle crust deforms by displacement, distortion, and rigid-body rotation. The western Great Basin, parts of which have been considerably modified by the well-recognized system of right-lateral faults defining the Walker Lane, is a logical area to expect crustal rotation to be a substantial contributor to the cumulative strain in response to different phases of Neogene deformation. Rigid-body rotations about a vertical axis are not always straightforward to identify, quantify, and explain, despite the fact that they can be a major contributor to finite strain. Paleomagnetic analysis of appropriate materials has proven to be a powerful technique for detecting and quantifying such rotations. Numerous paleomagnetic studies have documented variable degrees (or lack thereof) of vertical-axis rotation associated with strike-slip faulting and extension in the western Cordillera and some parts of the western Great Basin (e.g., Gillett and Van Alstine, 1982; Ron et al., 1984; Hudson and Geissman, 1987, 1991; Janecke et al., 1991; Hudson et al., 1994; Wawrzyniec et al., 2002; Petronis et al., 2002; Hudson, 1992; Cashman and Fontaine, 2000; Petronis et al., 2007).

The contemporary, active fault system in the Mina deflection area links northwest-striking dextral faults of the Furnace Creek–Owens Valley–Fish Lake Valley fault system to structures of similar orientation in the central Walker Lane by transferring displacement in an extensional right-step along a relatively narrow system of roughly east-west–striking left-oblique slip faults (Oldow et al., 1994, 2008; Wesnousky, 2005). The northwest-striking, right-lateral fault systems are known to have been active in the late Miocene (Niemi et al., 2001), well before initiation in the late Pliocene of the currently active east-northeast–striking fault system in the Mina deflection. This general area is well-documented by geologic, seismic, and geodetic observations to currently be in a left-transtensional state (Oldow et al., 1994, 2008; Bennett et al., 1999; Thatcher et al., 1999; Dixon et al., 2000; Oldow, 2003; Wesnousky, 2005; Oldow et al., this volume). Fundamental aspects of the mechanism of deformation in the southern part of the central Walker Lane and its evolution into the current state remain poorly understood. Our study attempts to quantify a component of deformation within the ~80–70-km-wide stepover zone between the Furnace Creek–Fish Lake Valley and the central Walker Lane fault systems prior to inception of the present-day geometry of the Mina deflection and the active transtensional system within the boundary zone between the Sierra Nevada and central Great Basin. The way in which the generally east-northeast–striking faults within the Mina deflection transfer recent to present-day slip from the Furnace Creek–Owens Valley fault system to the central Walker Lane fault system is also of considerable interest. Plausible mechanisms include (1) slip transfer dominated by left-lateral strike-slip displacement along east-northeast–oriented structures, minor normal faulting, and uniform clockwise sense, moderate-magnitude vertical-axis block rotation (Wesnousky, 2005), or (2) localized dip-slip normal faulting into extensional basins and minor strike-slip faulting (Ferranti and Oldow, 2000; Ferranti et al., this volume), or some combination of the two mechanisms. A key question regarding the long-term, Neogene evolution of this part of the western Great Basin is whether the present transtensional regime is responsible for the observed vertical-axis rotation recorded by the late Oligocene to Pliocene volcanic rocks in the Candelaria Hills and adjacent areas and whether all of the observed rotation has occurred since mid- to late Pliocene time.

The paleomagnetic data obtained in this study (Fig. 6) provide further spatial definition of what we consider to be an early zone of deformation, associated with the development of the Silver Peak–Lone Mountain stepover (Fig. 2), in the central Walker Lane. Oldow et al. (2008, p. 270) proposed that this early phase of deformation involved non–plane stain "characterized by vertical thinning and horizontal components of simple and pure shear operating along a northwest axis" and that the Miocene-Pliocene deformation differed considerably from the late Pliocene and contemporary deformation that characterizes the Mina deflection. We interpret the collective data from the ignimbrites and andesite and basalt flows (Fig. 6) to demonstrate, on average, a modest (~20° to 30°), net clockwise rotation of the sampled parts of the Candelaria Hills and nearby areas. We hypothesize that the likely timing of this rotation spanned a few million years during the late Miocene and earliest Pliocene, the early phase of deformation of the area discussed by Oldow et al. (2008), prior to the inception of the active fault system defining the Mina deflection. The paleomagnetic data are evaluated in the context of the possible identification of vertical-axis rotations and the reliability of rotation estimates, beginning with the basaltic andesite and andesite flows and followed by the results from the three regionally extensive ignimbrites.

Paleomagnetic Data

Basaltic Andesite and Andesite Flows

To assess the paleomagnetic data from the andesite and basaltic andesite flows and their viability for determining vertical-axis rotation in the Mina deflection area, two statistical tests were used to evaluate whether each group mean adequately averages paleosecular variation. Given the relatively small number of distinct flows within some of the sequences, it is unlikely that each individual section fully averages the geomagnetic field. Nonetheless, all 12 sections yield locality group mean declinations that are clockwise discordant from expected directions (Figs. 5B and 5C).

The first test compared whether the dual polarity flow mean directions from each section were drawn from populations with

means separated by 180° (McFadden and McElhinny, 1990). Five sections contained both normal and reversed polarity magnetizations. A paleomagnetic reversal test was conducted for these sections; the results are summarized in Table 4. Three sections yielded indeterminate results, one section failed the test, and one section passed at the 90% confidence level. The one section that failed and the three sections that yielded indeterminate results likely reflect the low number of reverse relative to normal polarity flow means in each of the sections (Table 4). The section that passed the reversal test likely represents a section that best averages secular variation.

Calculations of angular dispersion (Merrill and McElhinny, 1983) provide estimates of whether populations of independent directions have adequately sampled paleosecular variation. The dispersion of site virtual geomagnetic poles (VGPs)'s can be compared to latitude-dependent models to test whether a collection of VGP values sufficiently sample geomagnetic secular variation (Merrill and McElhinny, 1983; Butler, 1992). The PSV models (e.g., Merrill and McElhinny, 1983) for the average latitude of the study (38.0°N) predict a VGP angular standard deviation between 15.5° and 19.5°. If the observed angular dispersion is significantly less than that predicted by the model, then the section mean probably has not sampled an adequate time interval. If the VGP angular dispersion is greater than predicted, then an additional source of scatter is implied (e.g., inaccurate structural corrections, incorrect resolution of the ChRM). Of the 12 sections, after combining flow means from adjacent lava flows when their mean directions were found to be statistically indistinguishable (see previous; Calderone et al., 1990), three yield VGP angular standard deviation estimates that are within ±1° of the predicted PSV range (Table 4). Three sections yield low VGP dispersion values (B-1: 4.1°, B-2: 4.5°, B-3: 4.9°), and thus the inferred rotation estimates are less reliable. Yet, as argued later, the overall similarity in rotation estimates of these sections may not be a coincidence. Six sections yield VGP angular dispersion estimates that are low but within 5° of the predicted values. The individual site mean data for these sections are well defined (B-5, −7, −8, A-1, −2: $\alpha_{95} < 7.5°$; B-6: $\alpha_{95} = 10.1°$) and yield similar rotation estimates (Table 4). Four of the six sections yield exclusively normal polarity results, and the remaining two sections yield dual polarity results. For the dual polarity sections, one section fails, and the other yields an indeterminate reversal test result as discussed already. It is probable that a combination of exclusively normal polarity results in four sections and the low number of dual polarity means in the other two sections contributed to the lower than predicted dispersion estimates for these six sites. Although the dispersion associated with each of these sections is lower than predicted by PSV, we nonetheless consider the estimated group means, even though of lesser confidence, to approximate time-averaged field directions and yield reliable rotation values.

Based on the statistical tests, nine sections are considered to have adequately averaged secular variation and provide a reasonably accurate recording of the Miocene geomagnetic field and thus are expected to yield reliable absolute rotation estimates.

Three sections yield low VGP dispersion estimates and likely do not average secular variation, and their paleomagnetic data probably do not provide time averaged results, and rotation estimates are of lower confidence. Nonetheless, it seems implausibly fortuitous that these sites (B-1, B-2, B-3; Table 4) would record geomagnetic fields that are similar to the average of the other sections sampled in the area. Like Beck (1984), we propose that this correspondence suggests that the declination discrepancy from these three sections can also be explained in a manner similar to those from other sections in the area. Paleomagnetic data from the basaltic andesite and andesite flow sections yield group mean magnetizations that are, at nine localities, of north-northeast declination and moderate positive inclination (Fig. 5C). One section yields a south-southwest declination and moderate negative inclination mean, and two sections yield east declination and moderate negative inclination locality mean directions (Table 4; Fig. 5C). The paleomagnetic data from every section are statistically discordant in declination from the range of possible Miocene expected directions (Table 1). Rotation (R) and flattening (F) estimates (Beck, 1980; Demarest, 1983) for 10 of the 12 sections range from R = 10.2° to 39.9° and F = +10.8° to −25.1° (with most F values not statistically significant). For two of the sections, R is about + 90°, and the F values are +8.7° and −25.1°; these results are anomalous. The two high rotation estimates are difficult to reconcile considering that there is no field evidence for such anomalously large rotations, and ignimbrite sites in close proximity to the two sections do not yield high rotation values. Thus, it is probable that the two sections record transitional field directions or part of a short polarity event. The overall similarity of the rotation estimates from the other 10 section means (Fig. 5C), from sections distributed over an area larger than 2500 km², suggests that the mechanism controlling crustal deformation resulted in a near uniform sense and magnitude of vertical-axis rotation, affecting a large part of the Candelaria Hills and adjacent areas. If we consider the 10 results that yield modest magnitude clockwise rotation estimates, the average rotation of the basalt and andesite flows is 26.0° ± 7.8°, which is statistically indistinguishable from the average rotation results obtained from the ignimbrites (see following).

Regionally Extensive Ignimbrites

Realistic correction for postemplacement tilting at each site/ sampling locality is required to accurately interpret the paleomagnetic data from the three regionally extensive ignimbrites. In addition, any effects of magnetic fabric anisotropy on the direction of the remanence characteristic of each ignimbrite must be taken into consideration. Despite the effort to accurately define the orientation of each deposit (typically 5–15 strike and dip measurements of eutaxitic fabrics at each locality), sites with high fabric dips potentially have relatively large errors associated with the corrected data. Macdonald (1980) and Chan (1988) evaluated how apparent rotations can arise within declination data by inappropriate tilt corrections and specifically noted that the magnitude of apparent rotation may increase with the magnitude of

the tilt correction. For this study, typical dips for the ignimbrites are around 20° (±13°). A plot of inferred rotation value versus dip shows no obvious correlation (Fig. 7), suggesting that inappropriate tilt corrections have relatively little control on observed declination discrepancies. We nonetheless recognize that inaccurate definition of the paleohorizontal is a likely contributor to the inclination and declination discrepancies between or among site mean directions from the same ignimbrite deposits located on the same or different structural blocks (Fig. 7).

Magnetic anisotropy arises from alignment of the long axes of magnetic grains that define a magnetic fabric in an ignimbrite, and it can deflect the remanence from the actual geomagnetic field direction at the time of remanence acquisition (thus resulting in an anisotropy of the remanence acquisition) (Borradaile, 1988; Gattacceca and Rochette, 2002). Petronis et al. (2004) conducted artificial TRM acquisition experiments on the three regional ignimbrites. Their studies showed that the rocks were capable of acquiring a remanence essentially parallel to a weak applied field, and they also revealed a significant increase in intensity of the artificial TRM relative to the NRM. This increase in intensity is consistent with numerous previous observations that the ambient field intensity associated with unusual field directions, or non–time-averaged geomagnetic field directions, as may be recorded by ash-flow tuffs (Reynolds, 1977; Geissman et al., 1982), may be considerably less than the average ambient field during more stable normal or reverse polarity states (Tanaka et al., 1995; Prevot et al., 1985). The internally consistent remanence directions within each regional ignimbrite, despite the less than average ambient field intensity, as well as the relatively low percentages of anisotropy (Pj ranges 1.092–1.004; Jelinek, 1981) (Petronis and Geissman, 2008), argue against a significant effect of magnetic anisotropy on the remanence direction recorded in each ignimbrite.

The characteristic remanence of the three regionally extensive ignimbrites is interpreted to provide independent, essentially instantaneous Oligocene geomagnetic field records. For each of these separate ignimbrites, after structural correction, the overall dispersion of site mean directions is far greater in declination than in inclination (Figs. 5A and 6), as would be expected if a single, sheetlike deposit, with an overall uniformly directed magnetization, were fragmented and distorted by different magnitudes (e.g., Hillhouse and Wells, 1991). The paleomagnetic data from the regionally extensive ignimbrites, when compared with respective data from the reference section at the southwest end of the Excelsior Mountains, indicate typical vertical-axis rotation of structural blocks ranging between about −20° and +26°; although some exceptions exist (Figs. 5A and 6; Table 3). When compared with site mean directions from the reference section (discussed previously), 15 block means yield counterclockwise rotation estimates between −46.4° and −7.3°, 11 block means yield small rotation estimates with values less than the estimated errors (thus statistically insignificant), and 53 block means yield clockwise rotation estimates, between 110.6° and 5.8°. Four of these clockwise rotation estimates (107.7°, 110.6°, 83.0°, and

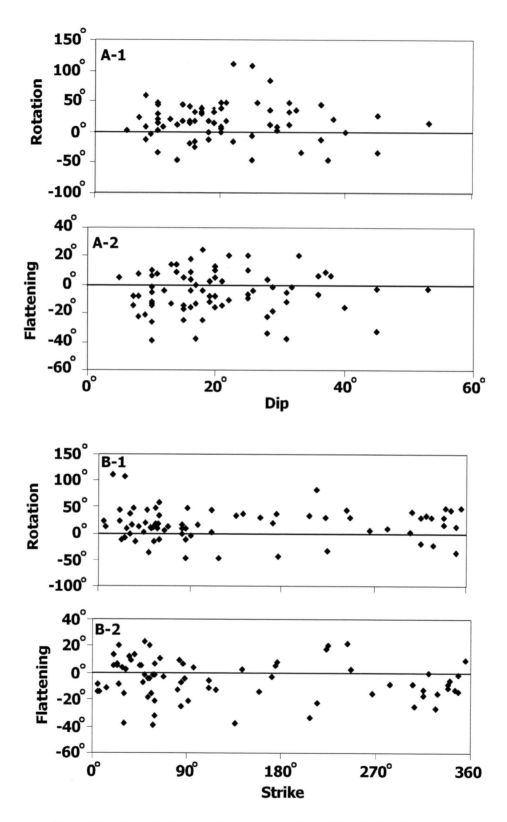

Figure 7. Rotation and flattening estimates versus strike and dip of ignimbrite sheets.

60.1°) are unrealistically high, as they would require very large strain gradients between fault blocks and no such field evidence was observed; we consider these rotation estimates anomalous and will not discuss them further.

The data from the different structural blocks show no obvious spatial pattern of rotation across the Candelaria Hills and adjacent areas (Fig. 6). The blocks yielding data suggestive of counterclockwise rotation, notably, are intermixed with other results across the study area. We suggest that, overall, the results suggesting little to modest counterclockwise rotation may reflect inaccurate tilt corrections. Alternatively, we cannot entirely discount the possibility that some blocks have experienced localized counterclockwise or very minimal rotation, and these could reflect local crustal heterogeneities and/or unusual block geometry (Lamb, 1987). If we consider only the 49 results that provide modest-magnitude clockwise rotation estimates, the average rotation of each structural block, relative to the southwest Excelsior Mountains (Huntoon Valley) reference frame, is 26.8° ± 7.8°. Whether this average rotation estimate approximates an absolute reference estimate depends on whether the southwest Excelsior Mountains section is actually an unrotated reference section.

Neogene Deformation in West-Central Nevada

Before late Pliocene development of the Mina deflection and its characteristic east-northeast left-oblique structures, active deformation east of the Furnace Creek–Fish Lake Valley fault system was concentrated in an ~85-km-wide zone along structures centered ~40 km south-southeast of the Candelaria Hills, in the Lone Mountain and Silver Peak Range area (Fig. 8A). Within the Silver Peak Range and adjacent areas to the south and north, displacement was transferred from the Furnace Creek–Fish Lake Valley and the more distributed central Walker Lane fault systems during the mid- to late Miocene, and this resulted in the development of a regionally extensive, northwest-dipping regional detachment system (Silver Peak–Lone Mountain extensional complex). The complex was ultimately abandoned, and deformation progressively stepped to the northwest (Reheis and Sawyer, 1997; Oldow et al., 1994, 2008; Petronis et al., 2002; Oldow et al., this volume). Previous paleomagnetic investigations in the Silver Peak Range (east to northeast area, Petronis et al., 2002; southwest area, Petronis et al., 2007) have shown that Miocene volcanic rocks in the upper plate and Miocene intrusions in the footwall of the northwest-dipping Silver Peak–Lone Mountain detachment system have experienced clockwise rotation of 20° to 30°. The recognition that the Candelaria Hills and adjacent areas of the central Walker Lane also experienced a similar sense and magnitude of rotation increases the known extent of the area of clockwise rotation within the central Walker Lane area. Major through-going east-northeast–trending faults (Wesnousky, 2005), which characterize the younger Mina deflection, however, are absent in the Silver Peak Range. Consequently, we argue that such structures cannot have been responsible for the observed rotations over this broad area. The data presented here support

the interpretation of Oldow et al. (2008), which states that the vertical-axis rotations in the Silver Peak Range and the Candelaria Hills and adjacent areas are the manifestation of a broad zone of "early" (late Miocene to early Pliocene) coherent strain rotation (Fig. 8B) during the initial phase of displacement transfer, which lacked the development of discrete structural blocks such as presently characterizes the Mina deflection. The regional extent of the broad zone of early coherent strain rotation, in particular to the west and south, requires further definition. We note that paleomagnetic data from the ca. 10 Ma Eureka Valley Tuff from the area north and northeast of Mono Lake (Fig. 1) have been interpreted by King et al. (2007) to indicate clockwise vertical axis rotations in this area of some 10° to 26°.

The area extending from at least the southern Silver Peak Range to possibly as far north as the Excelsior Mountains was characterized by a regional deformation field prior to the early Pliocene that was different from that of today. Incipient and distributed Walker Lane strike-slip faults to the northeast and less distributed Furnace Creek–Fish Lake Valley structures to the southwest were linked via shear couple involving pure and simple shear that affected both upper plate and lower plate rocks of the Silver Peak–Lone Mountain extensional complex (Oldow et al., 2008). The available paleomagnetic data indicate that these rocks experienced a uniform clockwise rotation of some 20° to 30°. In addition, the orientation of the principal extensional strain axis, determined from faults and structures within upper and lower plate rocks, was about N65°W, ~25° counterclockwise from the ~N40°W margins of the bounding shear couple defined by the Furnace Creek–Fish Lake Valley and central Walker Lane faults (Fig. 8B). The inferred clockwise rotation (20° to 30°) of upper and lower plate rocks is consistent with a modest combination of pure and simple shear, with cumulative right-slip displacement parallel to about N40°W of at least 25 km prior to 3 Ma. If we are correct in our inferred age of the timing, and thus duration, of clockwise rotation, we estimate integrated rotation rates that range from ~4° to 8°/m.y. This range of rotation rates is not inconsistent with previous paleomagnetic observations in actively deforming areas. For example, Sonder et al. (1994), in their study of Miocene strata in the Gale Hills area of the southern termination of the Las Vegas Valley shear zone, estimated integrated rotation rates between 1° and 12°/m.y. Rowan and Roberts (2008) inferred a rotation rate between 7° and 14°/m.y. for an early Miocene age of deformation of the Hikurangi margin of the North Island of New Zealand, and a decreased rotation rate of 3°–4°/m.y. in the Quaternary.

The localization of late Pliocene east-northeast–striking faults of the Mina deflection (Fig. 8C) beyond the northern extent of the older displacement transfer system together with deformation kinematics recorded by earthquakes and GPS velocities reflect a fundamental reorganization of the regional strain field in the Pliocene (Oldow et al., 2008). This reorganization of the strain field led to the linkage of the Owens Valley fault system and its northern termination, the Queen Valley system (Tincher and Stockli, this volume), to the central Walker Lane via the Clayton

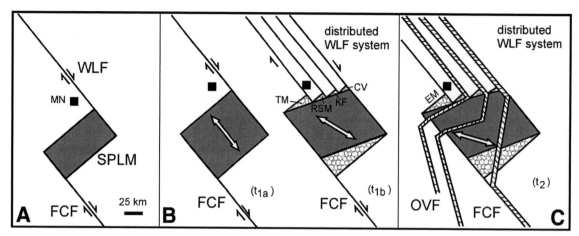

Figure 8. Schematic development of the central Walker Lane transfer system. (A) Initial and very approximate geometry of the Silver Peak–Lone Mountain (SPLM) extensional stepover system (gray rectangle) prior to significant strike-slip displacement transfer. WLF—Walker Lane fault system; FCF—Furnace Creek fault system. (B) Initiation of significant strike-slip displacement transfer (left), with a component of pure shear affecting the area as the stepover system widens. Double arrow is finite extension axis. Simultaneous simple shear deformation of the Silver Peak–Lone Mountain system (right) and the development of the distributed Walker Lane fault system and associated extensional basins (TM—Teels Marsh; RSM—Rhodes Salt Marsh; KF—Kibby Flat; CV—Cirac Valley) or areas of focused extension, in coarse detritus symbols. Double arrow is finite extension axis. Parallelogram is intended to emphasize component of homogeneous simple shear, with very approximate dimensions indicated. (C) Post ca. 3 Ma reorganization of the deformation field with localized extension on younger structures (double black lines), with slanted line interiors representing extension displacement, which form the Mina deflection. Double arrow is finite extension axis. Modified from Oldow et al. (2008). EM—Excelsior Mountains.

Valley, Coaldale, and Emigrant Peak fault systems (Fig. 2). The age of inception of the young east-northeast–trending faults at ca. 3 Ma is supported by ages of basalt flows north of the White Mountains (Dalrymple and Hirooka, 1965; Stockli et al., 2003; Lee et al., 2005), a sedimentary-basalt sequence in the Silver Peak Range (Robinson et al., 1968; Oldow et al., this volume), and the development of Queen Valley at the north end of the White Mountains (DePolo, 1989; Reheis and Dixon, 1996; Reheis and Sawyer, 1997; Stockli et al., 2003; Tincher and Stockli, this volume). In addition, in the northern Fish Lake Valley, early Pliocene sedimentary rocks and basalt flows are deformed into northwest-trending folds formed as part of the northward continuation of turtleback structures in the Silver Peak Range (Oldow et al., 1994; Reheis and Sawyer, 1997). These folds formed and are cut by east-northeast–striking high-angle faults that transfer displacement across the northern part of the Silver Peak Range to structures of the central Walker Lane (Reheis and Sawyer, 1997).

Although our estimate of the timing of the paleomagnetically determined rotations in the Candelaria Hills study area is not directly constrained, the available data are not consistent with an estimated 20°–30° of clockwise rotation taking place over the past ca. 3 Ma in association with the development of the Mina deflection. We do not preclude the possibility that localized clockwise rotation, at diminished rate, may have affected parts of the area during the past ~3 m.y., but data from the Silver Peak Range lying to the south of the Mina deflection reveal magnitudes of clockwise rotation that are similar to those from the Candelaria Hills (Petronis et al., 2002, 2007). Furthermore, data from Tertiary volcanic

rocks, a mid-Cretaceous intrusion, and host (remagnetized) Permian metasedimentary rocks in the southeast Excelsior Mountains, in the northern part of the Mina deflection, show no evidence of appreciable clockwise rotation (Ferranti et al., this volume), as has been demonstrated by data from additional Cretaceous intrusions farther to the north (Geissman et al., 1984).

CONCLUSIONS

Paleomagnetic data from regionally extensive Oligocene ignimbrites, Miocene andesitic lavas, and Miocene to earliest Pliocene basaltic andesite lavas collectively reveal that these rocks, and presumably underlying older crust, experienced a modest and statistically significant clockwise vertical-axis rotation. We interpret this rotation to have affected much of the Candelaria Hills and adjacent areas, which presently form the southern part of the Mina deflection. The inferred rotation of structural blocks is some 20° to 30° clockwise. The general agreement among results from Oligocene ash-flow tuffs (average R = 26.8° ± 7.8°), Miocene andesite flows, and Upper Miocene to Lower Pliocene basaltic andesite sequences (average R = 26.0° ± 7.8°) provides the basis supporting an early (likely late Miocene to early Pliocene) phase of coherent vertical-axis rotation during the development of a displacement transfer system. Together with additional paleomagnetic data from nearby areas, these new results support the hypothesis of Oldow et al. (2008), in which simultaneous pure and simple shear deformation of the Silver Peak–Lone Mountain extensional stepover system resulted in a zone of coherent clockwise vertical-axis rotation of rocks in upper and lower plate assemblages. This broad, stepover system linked the dextral Furnace Creek–Fish Lake Valley and the central Walker Lane fault systems. Post–early Pliocene reorganization of the deformation field involved formation of a zone of discrete, east-northeast–oriented, left-oblique structures and northwest-directed transtension. In this part of west-central Nevada, where displacement transfer is currently active, the Mina deflection has accommodated a change in regional displacement field through a combination of fragmentation of the upper crust into moderately tilted structural fault blocks, an inferred decreased rate of clockwise vertical-axis rotation of structural blocks, and localized development of deep sedimentary basins at the apex of curved fault arrays in response to northwest-directed extension. The contemporary, northwest-directed transtensional strain field (oblique divergence) in the area cannot be extrapolated back to beyond ca. 3 Ma, and thus the long-term kinematic evolution of the Mina deflection, and adjoining parts of the central Walker Lane, cannot be considered to be a steady-state process through the Neogene.

APPENDIX 1. ROCK MAGNETISM RESULTS

Methods

To characterize the magnetic mineralogy, standard rock magnetic experiments were conducted with the goal of identifying the magnetic phases carrying the remanence and the overall ability of these rocks to faithfully record an ambient field. Equipment used included a 2G Enterprise superconducting rock magnetometer, Model SSM-1A Schonstedt spinner magnetometer, Model GSD-1 Schonstedt AF demagnetizer (modified to allow tuning of input current), dTech ARM unit, and home-built static impulse magnets capable of 1–3 Tesla peak fields. The tests included in sequential order: AF demagnetization of anhysteretic remanent magnetization (ARM), DC acquisition of a saturation isothermal remanent magnetization (IRM), DC demagnetization of the saturation IRM to yield a backfield IRM (coercivity of remanence), AF demagnetization of saturation isothermal remanent magnetization (SIRM), and three component thermal demagnetization of IRM experiments. Anhysteretic remanent magnetization (ARM) was imparted in a DC field of 0.1 mT and a peak AF field of 95 mT and then AF demagnetized in ~13 steps to 95 mT. An IRM acquisition experiment involved stepwise exposure to higher fields until saturation was obtained (SIRM). Backfield SIRM demagnetization involved applying an increasing field along the –Z-axis until the +Z-remanence was reduced to zero. Finally, the samples were again saturated in a 1.33 T field and AF demagnetized in ~11 steps to 95 mT. Following the approach of Lowrie (1990), representative specimens from different parts of each of the three regionally extensive tuffs were subjected to three component IRM acquisition and progressive thermal demagnetization. One population of specimens was subjected to peak orthogonal fields of 3, 0.3, and 0.03 T; the other population was subjected to peak fields of 1.2, 0.12, and 0.02 T.

Results

Rock magnetic data indicate that the dominant magnetic mineral in the three regional ignimbrites is likely a ferrimagnetic phase, probably single-domain (SD) to pseudo–single-domain (PSD) magnetite, although some variability was observed. AF demagnetization of the NRM reveals that the three regional tuffs each have unique directions of characteristic magnetization (ChRM) as well as coercivity spectra, ranging from MD magnetite of intermediate coercivity to PSD to SD magnetite of moderate to high coercivity (Dunlop and Özdemir, 1997). Low-coercivity phases are characterized by MDF of the NRM, when it is essentially single component in nature, between 15 mT and 20 mT, and moderate to high coercivity phases are characterized by MDF of the NRM between ~25 mT and 50 mT (Fig. 4). Following the modified Lowrie-Fuller test (Johnson et al., 1975), the AF decay of normalized NRM, ARM, and SIRM was compared. The test is based on the experimental observation that normalized AF demagnetization curves of weak-field TRM (i.e., NRM) and strong-field TRM (i.e., SIRM) have different relationships for SD and MD grains of magnetite (Lowrie and Fuller, 1971; Dunlop and Özdemir, 1997). Laboratory investigations typically use weak-field ARM as a proxy for TRM. AF demagnetization of strong- and weak-field IRM showed a limited spectrum of responses by specimens of each individual ignimbrite but a wide range of response among units. For the Belleville Tuff, all curves suggest MD to PSD magnetite-dominated behavior, and for the Candelaria Junction Tuff, all curves suggest PSD to SD magnetite-dominated behavior. The Metallic City Tuff, however, yields variable demagnetization curves that suggest in some parts of the cooling unit, MD magnetite-dominated behavior of a certain grain size(s) carries the remanence, while in other parts of the cooling unit, PSD magnetite-dominated behavior is common. IRM acquisition curves show a narrow spectrum of responses within each individual ignimbrite, yet, consistent with the above observations, considerable variation among each unit. All tuff samples are likely dominated by a cubic phase (magnetite-type curves). The Metallic City and the Candelaria Junction Tuffs mostly show steep acquisition and nearly reach saturation by 0.10–0.20 T, while for the Belleville Tuff, most specimens reach complete saturation by 1.25 T, yet some did not reach a stable plateau until ~2.0 T. The acquisition curves are characteristic of MD- to PSD-magnetite–dominated behavior that is influenced by somewhat higher-coercivity SD magnetite, and, on the basis of other observations (unblocking temperature spectra during thermal demagnetization of the NRM), some maghemite, which is a characteristic restricted to the Belleville Tuff. Backfield IRM curves for the Metallic City and the Candelaria Junction Tuffs show a very narrow range of responses, with coercivity of remanence values between ~0.03 T and 0.05 T. In contrast, the Belleville Tuff shows a moderate range of responses, with values between ~0.05 T and 0.20 T. Response to thermal demagnetization of three component IRM is generally consistent within each of the three regional ash-flow tuffs, yet some variation is identified between each unit. The Candelaria Junction Tuff yields the highest IRM intensities of the three tuffs. The *Y* component has the greatest intensity and shows a uniformly distributed laboratory unblocking temperature range to ~580 °C. Behavior of Metallic City Tuff specimens is similar to those of the Candelaria Junction Tuff, although IRM intensities are slightly lower. The *Y* component is always the most intense, but several percent of the *Y* component remains unblocked above 580 °C, with complete unblocking taking place by ~625 °C. The Belleville Tuff typically shows the lowest IRM intensities and displays the most internal variability in behavior. For the most strongly magnetized specimens (CU12B, CU104C), the *Y* component dominates the other IRMs. In the more weakly magnetized specimens, the *Z* component is the most strongly magnetized IRM. These results are interpreted to indicate that a low-titanium magnetite, as well as a small fraction of maghemite, especially in the Metallic City Tuff, are the principal magnetic phases in these rocks. Some of the Belleville Tuff specimens reveal a strong contribution by a higher coercivity phase (i.e., hematite), but magnetite/maghemite are still identified in the unblocking temperature curves.

A full list of the UTM location of the paleomagnetic sampling sites can be found in Table A1.

TABLE A1. UTM LOCATION OF PALEOMAGNETIC SAMPLING SITES

Site			UTM (NAD27)		Site			UTM (NAD27)		Site			UTM (NAD27)	
			Easting	Northing				Easting	Northing				Easting	Northing
BV1	11	S	397385	4229419	BV81	11	S	403666	4226359	CU31	11	S	402092	4221041
BV2	11	S	397335	4229366	84:a-f	11	S	400662	4226577	CU32	11	S	402092	4221041
BV3	11	S	397345	4229356	84:g-k	11	S	400662	4226577	CU33	11	S	402092	4221041
BV6	11	S	397340	4229359	BV85	11	S	400662	4226577	CU34	11	S	402152	4220850
BV4	11	S	397267	4229294	BV86	11	S	400662	4226577	CU35	11	S	402180	4220814
BV5	11	S	397267	4229294	BV87	11	S	400786	4226597	CU36	11	S	402206	4220806
BV49	11	S	397266	4228840	BV89	11	S	400816	4226443	CU37	11	S	382328	4217095
BV50	11	S	397266	4228840	BV91	11	S	397032	4229867	CU38	11	S	382340	4217101
BV51	11	S	397266	4228840	BV92	11	S	397032	4229867	CU44	11	S	382495	4217392
BV26	11	S	399876	4229217	BV93	11	S	396965	4229897	CU49	11	S	382606	4217311
BV27	11	S	399876	4229217	BV94	11	S	396957	4229987	CU51	11	S	395384	4222945
BV28	11	S	399876	4229217	BV95	11	S	398524	4228025	CU53	11	S	386225	4216047
BV32	11	S	399523	4229073	BV96	11	S	398524	4228025	CU54	11	S	386184	4216112
BV33	11	S	399523	4229073						CU55	11	S	386225	4216130
BV35	11	S	398953	4225378	CU10	11	S	396814	4220215	CU56	11	S	386250	4216191
BV36	11	S	398953	4225378	CU11	11	S	396850	4220276	CU57	11	S	386261	4216226
BV37	11	S	398563	4225379	CU12	11	S	396831	4220704	CU58	11	S	385917	4216294
BV38	11	S	398563	4225379	CU13	11	S	396831	4220704	CU59	11	S	385854	4216302
BV48	11	S	397904	4225259	CU14	11	S	396891	4220719	CU60	11	S	385881	4216284
BV45	11	S	397851	4228134	CU15	11	S	396885	4220672	CU64	11	S	385540	4216205
BV46	11	S	397851	4228134	CU16	11	S	398212	4219958	CU66	11	S	385688	4215851
BV49'	11	S	405407	4228150	CU17	11	S	398195	4219970	CU67	11	S	385705	4215856
BV50'	11	S	405407	4228150	CU18	11	S	398132	4220013	CU68	11	S	385751	4215970
BV51'	11	S	405418	4228257	CU19	11	S	398105	4220035	CU70	11	S	385957	4215220
BV52'	11	S	405418	4228257	CU20	11	S	398055	4220075	CU71	11	S	385957	4215220
BV53	11	S	405418	4228257	CU21	11	S	397919	4220058	CU74	11	S	385947	4215254
BV58	11	S	405702	4228564	CU22	11	S	397898	4220110	CU75	11	S	385827	4215197
BV59	11	S	405702	4228564	CU23	11	S	398013	4220233	CU76	11	S	385827	4215197
BV60	11	S	405702	4228564	CU24	11	S	398013	4220233	CU77	11	S	385813	4215195
BV61	11	S	406011	4227408	CU25	11	S	397983	4220236	CU87	11	S	403145	4219437
BV62	11	S	406011	4227408	CU26	11	S	397558	4220049	CU88	11	S	403510	4219586
BV65	11	S	405812	4227179	CU27	11	S	397558	4220049	CU89	11	S	407414	4218525
BV66	11	S	406529	4225130	CU28	11	S	401963	4221003	CU90	11	S	407123	4218571
BV67	11	S	406529	4225130	CU29	11	S	401983	4221027	CU91	11	S	404461	4216424
BV68	11	S	406529	4225130	CU30	11	S	402061	4221047	CU92	11	S	405717	4215895
CU93	11	S	405598	4215528	VH8	11	S	406741	4203127	A-2				
CU94	11	S	403827	4217029	VH9	11	S	406777	4203124	CH2A	11	S	398812	4232837
CU96	11	S	405838	4216979	VH11	11	S	406610	4203041	CH2B	11	S	398996	4232657
CU97	11	S	405838	4216979	VH12	11	S	406972	4203217	CH2C	11	S	398613	4232604
CU98	11	S	405838	4216979	VH20	11	S	407686	4202179	CH2D	11	S	398634	4232593
CU99	11	S	407454	4215946	VH21	11	S	407797	4201973	CH2E	11	S	398634	4232593
CU100	11	S	403556	4213549	VH50	11	S	407797	4201973	CH2F	11	S	398307	4232413
CU101	11	S	404457	4213405	VH51	11	S	407797	4201973	CH2G	11	S	398511	4232546
CU102	11	S	404648	4213155	VH56	11	S	412274	4203918	CH2H	11	S	398235	4232389
CU103	11	S	403503	4210333	VH57	11	S	412274	4203918	CH2I	11	S	398136	4232382
CU104	11	S	403503	4210333	VH58	11	S	411981	4205000	CH2J	11	S	398136	4232382
CU106	11	S	383556	4214591	VH59	11	S	411981	4205000	CH2K	11	S	398253	4232393
CU109	11	S	383279	4214834	VH60	11	S	411981	4205000	CH2L	11	S	398253	4232393
CU110	11	S	383427	4214883						CH2M	11	S	398253	4232393
CU111	11	S	404587	4208064	HT1	11	S	361161	4220692	CH2N	11	S	397954	4232392
CU114	11	S	403690	4208606	HT3	11	S	360977	4220883	Basalt paleomagnetic data				
CU115	11	S	403659	4208498	HT4-5	11	S	360856	4221001	B-0				
CU116	11	S	403074	4207580	Andesite paleomagnetic data					GH3	11	S	372556	4263180
CU118	11	S	403087	4207589	A-1					GH4	11	S	372559	4263179
CU119	11	S	393164	4210263	BV8	11	S	401491	4229721	GH5	11	S	372560	4263177
CU120	11	S	393219	4210255	BV9	11	S	401491	4229721	GH6*	11	S	372561	4263175
CU122	11	S	392453	4209997	BV10	11	S	401635	4229741	GH7	11	S	372264	4263871
CU123	11	S	390007	4218233	BV11	11	S	401635	4229741	GH8	11	S	372198	4265623
CU124	11	S	390507	4218056	BV12	11	S	401874	4229711	GH9	11	S	372198	4265623
CU125	11	S	390505	4218021	BV13	11	S	401874	4229711	GH10	11	S	372198	4265623
CU126	11	S	390505	4218021	BV14	11	S	402083	4229693	GH11	11	S	372039	4265757
CU127	11	S	390505	4218021	BV15	11	S	402201	4229821	GH12	11	S	372039	4265757
CU129	11	S	390321	4217450	BV16	11	S	402201	4229821	B-1				
CH6C	11	S	387380	4214843	BV17	11	S	401635	4229741	CH20A	11	S	395195	4228275
CH10#	11	S	387482	4215584	BV29	11	S	399876	4229217	CH20B	11	S	395195	4228275
CH15	11	S	396885	4220672	BV30	11	S	399876	4229217	CH20C	11	S	398613	4232604
MM9#	11	S	394104	4209489	BV31	11	S	399876	4229217	CH20D	11	S	398613	4232604
CH4	11	S	387879	4212899	Andesite volcanic neck					CH20E	11	S	394976	4228293
CH4F	11	S	387879	4212899	BV54	11	S	405335	4228382	CH20F	11	S	394976	4228293
MM1#	11	S	393811	4209085	BV55	11	S	405335	4228382	CH20G	11	S	398307	4232413
					BV56	11	S	405335	4228382	CH20H	11	S	398511	4232546
										CH20I	11	S	398235	4232389

(continued)

TABLE A1. UTM LOCATION OF PALEOMAGNETIC SAMPLING SITES (continued)

Site			UTM (NAD27) Easting	Northing
CH20J	11	S	398136	4232382
CH20K	11	S	398136	4232382
CH20L	11	S	398136	4232382
CH20M	11	S	394674	4228268
CH20N	11	S	394674	4228268
CH20O	11	S	394674	4228268
B-2				
CH12A	11	S	394676	4224683
CH12B	11	S	394650	4224704
CH12C	11	S	394652	4224714
CH12D	11	S	394554	4224729
CH12E	11	S	394557	4224730
CH12F	11	S	394362	4224832
CH12G	11	S	394229	4224918
B-3				
BV18	11	S	400552	4228865
BV19	11	S	400552	4228865
BV20	11	S	400552	4228865
BV21	11	S	400552	4228865
BV22	11	S	400176	4228866
BV23	11	S	400176	4228866
BV24	11	S	400176	4228866
BV25	11	S	400176	4228866
B-4 (A)				
CU2	11	S	397858	4223646
CU3	11	S	397888	4223634
CU4	11	S	397864	4223578
CU5	11	S	397874	4223550
CU6	11	S	397880	4223503
CU7	11	S	397872	4223479
CU8	11	S	397885	4223461
CU9	11	S	397897	4223442
CH14A	11	S	397926	4224379
CH14B	11	S	397956	4224389
CH14C	11	S	397976	4224305
CH14D	11	S	398026	4224411
CH14E	11	S	398096	4224511
CH14F	11	S	398193	4224484
CH14G	11	S	398190	4224481
CH14H	11	S	398288	4224579
CH14I	11	S	398376	4224677

Site			UTM (NAD27) Easting	Northing
B-5 (B)				
CH30A	11	S	400959	4224244
CH30B	11	S	400995	4224269
CH30C	11	S	401034	4224330
CH30D	11	S	401053	4224330
CH30E	11	S	401148	4224343
CH30F	11	S	401153	4224377
CH30G	11	S	401153	4224377
CH30H	11	S	401193	4224417
CH30I	11	S	401242	4224466
CH30J	11	S	401293	4224530
CH30K	11	S	401314	4224549
CH30L	11	S	401314	4224549
CH30M	11	S	401366	4224555
CH30N	11	S	401442	4224602
CH30O	11	S	401469	4224609
CH31a	11	S	405039	4224702
CH31b	11	S	405024	4224729
CH31c	11	S	404950	4224723
CH31d	11	S	404920	4224733
CH31e	11	S	404829	4224736
CH31f	11	S	404829	4224736
CH31g	11	S	404739	4224795
CH31h	11	S	404739	4224795
CH31i	11	S	404715	4224800
CH31j	11	S	404675	4224813
CH31k	11	S	404668	4224811
CH31l	11	S	404620	4224803
CH31m	11	S	404585	4224827
CH31n	11	S	404585	4224827
CH31o	11	S	404567	4224843
CH31p				
B-6 (C)				
CH3A	11	S	399053	4223807
CH3B	11	S	399053	4223807
CH3C	11	S	399053	4223807
CH3D	11	S	402947	4223437
CH3E	11	S	402940	4223354
CH3F	11	S	402997	4223283
CH3G	11	S	402364	4222933
CH3H	11	S	402944	4223344

Site			UTM (NAD27) Easting	Northing
B-7				
CH7A	11	S	385999	4205759
CH7B	11	S	386052	4205847
CH7C	11	S	386100	4205968
CH7D	11	S	386227	4206035
CH7E	11	S	386536	4206128
CH7F	11	S	385940	4206262
CH7G	11	S	386120	4206293
CH7H	11	S	386751	4206315
CH7I	11	S	387099	4206330
B-8				
MC16	11	S	422309	4213153
MC17	11	S	422311	4213155
MC18	11	S	422291	4213102
MC19	11	S	422257	4213044
MC20	11	S	422383	4212984
MC21	11	S	422472	4212624
MC22	11	S	422486	4212649
MC23	11	S	422406	4212586
MC24	11	S	422388	4212460
MC25	11	S	422343	4212432
MC26	11	S	422661	4212289
MC27	11	S	422670	4212308
MC28	11	S	422671	4212309
B-9				
VH30	11	S	399394	4206241
VH31	11	S	399265	4206399
VH32	11	S	399314	4206463
VH33	11	S	399412	4206523
VH34	11	S	399722	4206535
VH35	11	S	400534	4206775
VH36	11	S	400519	4206652
VH37	11	S	400515	4206669
VH38	11	S	400539	4206637
VH39	11	S	400518	4206607
VH40	11	S	400630	4206579
VH41	11	S	400630	4206579
VH42	11	S	400810	4206490
VH43	11	S	400810	4206490
VH44	11	S	400933	4206502
VH45	11	S	401018	4206518

ACKNOWLEDGMENTS

This paper represents a major component of M.S. Petronis' Ph.D. dissertation research. This material is based upon work supported by the National Science Foundation under grant EAR-0208090 to Geissman and NSF grants EAR-0334219, EAR-0208219, and EAR-0126205 to Oldow. Additional support to Petronis was provided by Sigma XI, the Geological Society of America, University of New Mexico Research, Project, and Travel Grant, the SIPES Foundation, Monte Vista Exploration Company, and the University of New Mexico's Department of Earth and Planetary Sciences. Several individuals assisted with field sampling, including Dale Henderson, Jason Rampe, James Ashby, Melissa Pfeffer, Martine Simoes, Ian Atwell, Jamie Barnes, Leanne Scott, and Ernie Anderson. We also thank several individuals for assistance in the laboratory and numerous colleagues at the Mina Bar. Reviews by Mark Hudson and Nathan Niemi are gratefully appreciated and served to strengthen many aspects of the manuscript.

REFERENCES CITED

Argus, D.F., and Gordon, R.G., 1991, Current Sierra Nevada–North America motion from very long baseline interferometry: Implications for the kinematics of the western United States: Geology, v. 19, p. 1085–1088, doi: 10.1130/0091-7613(1991)019<1085:CSNNAM>2.3.CO;2.

Beck, M.E., 1980, Paleomagnetic record of plate-margin tectonic processes along the western edge of North America: Journal of Geophysical Research, v. 85, p. 7115–7131, doi: 10.1029/JB085iB12p07115.

Beck, M.E., Jr., 1984, Has the Washington-Oregon Coast Range moved northward?: Geology, v. 12, p. 737–740, doi: 10.1130/0091-7613(1984)12<737:HTWCRM>2.0.CO;2.

Bennett, R.A., Davis, J.L., and Wernicke, B.P., 1999, Present-day pattern of Cordilleran deformation in the western United States: Geology, v. 27, p. 371–374, doi: 10.1130/0091-7613(1999)027<0371:PDPOCD>2.3.CO;2.

Bennett, R.A., Wernicke, B.P., Niemi, N.A., Friedrich, A.M., and Davis, J.L., 2003, Contemporary strain rates in the northern Basin and Range Province from GPS data: Tectonics, v. 22, no. 2, 1008, doi: 10.1029/2001TC001355.

Besse, J., and Courtillot, V., 2002, Apparent and true polar wander and the geometry of the geomagnetic field over the last 200 Myr (v. 107, art no. 2300, 2002): Journal of Geophysical Research, v. 108, p. 2469, doi: 10.1029/2000JB000050.

Best, M.G., Christiansen, E.H., Deino, A.L., Gromme, C.S., and Tingey, D.G., 1995, Correlation and emplacement of a large, zoned, discontinuously exposed ash flow sheet: The $^{40}Ar/^{39}Ar$ chronology, paleomagnetism, and petrology of the Pahranagat Formation, Nevada: Journal of Geophysical Research, v. 100, p. 24,593–24,609, doi: 10.1029/95JB01690.

Borradaile, G.J., 1988, Magnetic susceptibility, petrofabrics and strain: Tectonophysics, v. 156, p. 1–20, doi: 10.1016/0040-1951(88)90279-X.

Burchfiel, B.C., and Davis, G.A., 1971, Clark Mountain thrust complex in the Cordillera of southern California: Geologic summary and field trip guide: University of California Riverside Museum Contribution, v. 1, p. 1–28.

Butler, R.F., 1992, Paleomagnetism: Magnetic Domains to Geologic Terranes: Oxford, England, Blackwell Scientific, 319 p.

Byrd, J.O.D., Smith, R.B., and Geissman, J.W., 1994, The Teton fault, Wyoming: Topographic signature, neotectonics, and mechanisms of deformation: Journal of Geophysical Research, v. 99, p. 20,095–20,122, doi: 10.1029/94JB00281.

Calderone, G.J., Butler, R.F., and Acton, G.D., 1990, Paleomagnetism of middle Miocene volcanic rocks in the Mojave-Sonora Desert regions of western Arizona and southeastern California: Journal of Geophysical Research, v. 95, p. 625–647, doi: 10.1029/JB095iB01p00625.

Cashman, P.H., and Fontaine, S.A., 2000, Strain partitioning in the northern Walker Lane, western Nevada and north-eastern California: Tectonophysics, v. 326, p. 111–130, doi: 10.1016/S0040-1951(00)00149-9.

Chan, L., 1988, Apparent tectonic rotations, declination anomaly equations, and declination anomaly charts: Journal of Geophysical Research, v. 93, p. 12,151–12,158, doi: 10.1029/JB093iB10p12151.

Cogbill, A.H., Jr., 1979, The relationship between seismicity and crustal structure in the western Great Basin [Ph.D. thesis]: Evanston, Illinois, Northwestern University, 294 p.

Crowder, D.F., Robinson, P.F., and Harris, D.L., 1972, Geologic Map of the Benton Quadrangle Mono County, California and Esmeralda and Mineral Counties, Nevada: U.S. Geological Survey Map GQ-1013, scale 1:62,500.

Dalrymple, G.B., and Hirooka, K., 1965, Variation of potassium, argon, and calculated age in a late Cenozoic basalt: Journal of Geophysical Research, v. 70, p. 5291, doi: 10.1029/JZ070i020p05291.

Demarest, H.H., 1983, Error analysis for the determination of tectonic rotations from paleomagnetic data: Journal of Geophysical Research, v. 88, p. 4321–4328, doi: 10.1029/JB088iB05p04321.

DePolo, C.M., 1989, Seismotectonics of the White Mountains fault system, east-central California and west-central Nevada [M.S. thesis]: Reno, University of Nevada, 354 p.

Diamond, D.S., and Ingersoll, R.V., 2003, Structural and sedimentologic evolution of a Miocene supradetachment basin, Silver Peak Range and adjacent areas, west-central Nevada, in Klemperer, S., and Ernst, W.G., eds., The George S. Thompson Volume: The Lithosphere of Western North America and Its Geophysical Characterization: Columbia, Maryland, Bellwether Publishing, Ltd., International Book Series Volume 7, p. 312–347.

Dickinson, W.R., 2006, Geotectonic evolution of the Great Basin: Geosphere, v. 2, p. 353–368, doi: 10.1130/GES00054.1.

Diehl, J.F., McClannahan, K.M., and Bornhorst, T.J., 1988, Paleomagnetic results from the Mogollon-Datil volcanic field, southwestern New Mexico, and a refined mid-Tertiary reference pole for North America: Journal of Geophysical Research, v. 93, p. 4869–4879, doi: 10.1029/JB093iB05p04869.

Dixon, T.H., Robaudo, S., Lee, J., and Reheis, M.C., 1995, Constraints on present-day Basin and Range deformation from space geodesy: Tectonics, v. 14, no. 4, p. 755–772, doi: 10.1029/95TC00931.

Dixon, T.H., Miller, M., Farina, F., Wang, H., and Johnson, D., 2000, Present-day motion of the Sierra Nevada block and some tectonic implications for the Basin and Range Province, North American Cordillera: Tectonics, v. 19, p. 1–24, doi: 10.1029/1998TC001088.

Dixon, T.H., Norabuena, E., and Hotaling, L., 2003, Paleoseismology and global positioning system: Earthquake-cycle effects and geodetic versus geologic fault slip rates in the Eastern California shear zone: Geology, v. 31, no. 1, p. 55–58, doi: 10.1130/0091-7613(2003)031<0055:PAGPSE>2.0.CO;2.

Dokka, R.K., and Travis, C.J., 1990, Late Cenozoic strike-slip faulting in the Mojave Desert, California: Tectonics, v. 9, p. 311–340, doi: 10.1029/TC009i002p00311.

Dokka, R.K., and Travis, C.J., 1996, Role of the Eastern California shear zone in accommodating Pacific–North American plate motion: Geophysical Research Letters, v. 17, p. 1323–1326, doi: 10.1029/GL017i009p01323.

Dunlop, D.J., and Özdemir, O., 1997, Rock Magnetism: Fundamentals and Frontiers: Cambridge Studies in Magnetism, Volume 3: New York, Cambridge University Press, 573 p.

Ekren, E.B., Byers, F.M., Jr., Hardyman, R.F., Marvin, R.G., and Silberman, M.L., 1980, Stratigraphy, preliminary petrology, and some structural features of Tertiary volcanic rocks in the Gabbs Valley and Gilles Ranges, Mineral County, Nevada: U.S. Geological Survey Bulletin 1464, p. 54.

Faulds, J.E., Henry, C.D., and Hinz, N.H., 2005, Kinematics of the northern Walker Lane: An incipient transform fault along the Pacific–North American plate boundary: Geology, v. 33, p. 505–508, doi: 10.1130/G21274.1.

Ferguson, H.G., Muller, S.W., and Cathcart, S.H., 1954, Mina, Nevada, Geology: U.S. Geological Survey Geological Quadrangle Map GQ-45, scale 1:62,500.

Ferranti, L., and Oldow, J.S., 2000, Displacement transfer on curved faults, central Walker Lane, western Great Basin: Geological Society of America Abstracts with Programs, v. 32, p. A104.

Ferranti, L., and Oldow, J.S., 2005, Active displacement transfer and finite flattening on curved faults, central Walker Lane, Western Great Basin, Nevada, in Lee, J., Stockli, D., Henry, C.D., and Dixon, T.H., eds., Geological Society of America Penrose Conference of Kinematics and Geodynamics of Intraplate Dextral Shear in Eastern California and Western Nevada, 21–26 April, Mammoth Lakes, California: Boulder, Colorado, Geological Society of America.

Ferranti, L., Oldow, J.S., Geissman, J.W., and Nell, M.M., 2009, this volume, Flattening strain during coordinated slip on a curved fault array, Rhodes Salt Marsh extensional basin, central Walker Lane belt, west-central Nevada, in Oldow, J.S., and Cashman, P.H., eds., Late Cenozoic Structure and Evolution of the Great Basin–Sierra Nevada Transition: Geological Society of America Special Paper 447, doi: 10.1130/2009.2447(11).

Fisher, R.A., 1953, Dispersion on a sphere: Proceedings of the Royal Society of London, v. A217, p. 295, 305.

Gan, W., Svarc, J.L., Savage, J.C., and Prescott, W.H., 2000, Strain accumulation across the Eastern California shear zone at latitude 36°30′N: Journal of Geophysical Research, v. 105, p. 16,229–16,236, doi: 10.1029/2000JB900105.

Garside, L.J., 1979, Geologic Map of the Camp Douglas Quadrangle, Nevada: Nevada Bureau of Mines and Geology Map 63, scale 1:24,000.

Garside, L.J., and Silberman, M., 1978, New K-Ar ages of volcanic and plutonic rocks from the Camp Douglas quadrangle, Mineral County, Nevada: Isochron-West, v. 22, p. 29–32.

Gattacceca, J., and Rochette, P., 2002, Pseudopaleosecular variation due to remanence anisotropy in a pyroclastic flow succession: Geophysical Research Letters, v. 29, no. 8, 1286, doi: 10.1029/2002GL014697.

Geissman, J.W., Van der Voo, R., and Howard, K.L., Jr., 1982a, A paleomagnetic study of the structural deformation in the Yerington District, Nevada: 1. Tertiary units and their tectonism: American Journal of Science, v. 282, p. 1042–1079.

Geissman, J.W., Van der Voo, R., and Howard, K.L., Jr., 1982b, A paleomagnetic study of the structural deformation in the Yerington District, Nevada: 2. Mesozoic "basement" units and their total and pre-Oligocene tectonism: American Journal of Science, v. 282, p. 1080–1109.

Geissman, J.W., Callian, J.T., Oldow, J.S., and Humphries, S.E., 1984, Paleomagnetic assessment of oroflexural deformation in west-central Nevada and significance for emplacement of allochthonous assemblages: Tectonics, v. 3, p. 179–200, doi: 10.1029/TC003i002p00179.

Gilbert, C.M., Christensen, M.N., Al-Rawi, Y., and Lajoie, K.R., 1968, Structural and volcanic history of Mono Basin, California-Nevada, in Coats, R.R., Hay, R.L., and Anderson, C.A., eds., Studies in Volcanology: A Memoir in Honor of Howel Williams: Geological Society of America Memoir 116, p. 275–329.

Gillett, S.L., and Van Alstine, D.R., 1982, Remagnetization and tectonic rotation of the upper Precambrian and lower Paleozoic strata from the Desert Range, southern Nevada: Journal of Geophysical Research, v. 87, p. 10,929–10,953, doi: 10.1029/JB087iB13p10929.

Hardyman, R.F., and Oldow, J.S., 1991, Tertiary tectonic framework and Cenozoic history of the central Walker Lane, Nevada, in Raines, G.L., Lisle, R.E., Schafer, R.W., and Wilkinson, W.H., eds., Geology and Ore Deposits of the Great Basin, Symposium Proceedings: Reno, Nevada, Geological Society of Nevada, p. 279–301.

Hillhouse, J.W., and Wells, R.E., 1991, Magnetic fabric, flow directions, and source area of the Lower Miocene Peach Springs Tuff in Arizona, California, and Nevada: Journal of Geophysical Research, v. 96, p. 12,443–12,460, doi: 10.1029/90JB02257.

Hudson, M.R., 1992, Paleomagnetic data bearing on the origin of arcuate structures in the French Peak-Massachusetts Mountain area of southern

Nevada: Geological Society of America Bulletin, v. 104, p. 581–594, doi: 10.1130/0016-7606(1992)104<0581:PDBOTO>2.3.CO;2.

Hudson, M.R., and Geissman, J.W., 1987, Paleomagnetic and structural evidence for middle Tertiary counterclockwise rotation in the Dixie Valley region, west-central Nevada: Geology, v. 15, p. 638–642, doi: 10.1130/0091-7613(1987)15<638:PASEFM>2.0.CO;2.

Hudson, M.R., and Geissman, J.W., 1991, Paleomagnetic evidence for the age and extent of middle Tertiary counterclockwise rotation, Dixie Valley region, west central Nevada: Journal of Geophysical Research, v. 96, p. 3979–4006, doi: 10.1029/90JB02424.

Hudson, M.R., Sawyer, D.A., and Warren, R.G., 1994, Paleomagnetism and rotation constraints for the middle Miocene southwestern Nevada volcanic field: Tectonics, v. 13, p. 258–277, doi: 10.1029/93TC03189.

Janecke, S.U., Geissman, J.W., and Bruhn, R.L., 1991, Localized rotation during Paleogene extension in east central Idaho: Paleomagnetic and geologic evidence: Tectonics, v. 10, p. 403–432, doi: 10.1029/90TC02465.

Jelinek, V., 1981, Characterization of the magnetic fabric of rocks: Tectonophysics, v. 79, p. T63–T70, doi: 10.1016/0040-1951(81)90110-4.

Johnson, E., 2002, Pliocene volcanism of the Excelsior Mountains, Nevada [Undergraduate honor's thesis]: Ann Arbor, Department of Geological Sciences, University of Michigan, 11 p.

Johnson, H.P., Lowrie, W., and Kent, D.V., 1975, Stability of anhysteretic remanent magnetization in fine and coarse magnetite and maghemite particles: Geophysical Journal of the Royal Astronomical Society, v. 41, p. 1–10.

King, N.M., Hillhouse, J.W., Gromme, S., Hausbeck, B.P., and Pluhar, C.J., 2007, Stratigraphy, paleomagnetism, and anisotropy of magnetic susceptibility of the Miocene Stanislaus Group, central Sierra Nevada and Sweetwater Mountains, California and Nevada: Geosphere, v. 3, p. 646–666, doi: 10.1130/GES00132.1.

Kirschvink, J.L., 1980, The least-squares line and plane and the analysis of paleomagnetic data: Geophysical Journal International, v. 62, p. 699–718, doi: 10.1111/j.1365-246X.1980.tb02601.x.

Krauskopf, K.B., 1971, Geologic Map of the Mount Barcroft Quadrangle, California-Nevada: U.S. Geological Survey Geologic Quadrangle Map GQ-0960, scale 1:62,500.

Lamb, S.H., 1987, A model for tectonic rotations about a vertical axis: Earth and Planetary Science Letters, v. 84, p. 75–86, doi: 10.1016/0012-821X(87)90178-6.

Larson, E.E., and Strangeway, D.W., 1966, Magnetic polarity and igneous polarity: Nature, v. 212, p. 756–757, doi: 10.1038/212756a0.

Lee, J., Pencer, J., and Owens, L., 2001, Holocene slip rates along the Owens Valley fault, California: Implications for the recent evolution of the Eastern California shear zone: Geology, v. 29, p. 819–822, doi: 10.1130/0091-7613(2001)029<0819:HSRATO>2.0.CO;2.

Lee, J., Stockli, D., Schroeder, J., Tincher, C., Bradley, D., and Owen, L., 2005, Fault slip transfer in the Eastern California shear zone/Walker Lane belt, with field guide, in Lee, J., Stockli, D., Henry, C.D., and Dixon, T.H., eds., Geological Society of America Penrose Conference of Kinematics and Geodynamics of Intraplate Dextral Shear in Eastern California and Western Nevada, 21–26 April, Mammoth Lakes, California: Boulder, Colorado, Geological Society of America, p. 22–46.

Locke, A., Billingsly, P.R., and Mayo, E.B., 1940, Sierra Nevada tectonic patterns: Geological Society of America Bulletin, v. 51, p. 513–540.

Lowrie, W., 1990, Identification of ferromagnetic minerals in a rock by coercivity and unblocking temperature properties: Geophysical Research Letters, v. 17, p. 159–162, doi: 10.1029/GL017i002p00159.

Lowrie, W., and Fuller, M., 1971, On the alternating field demagnetization characteristics of multidomain thermoremanent magnetization in magnetite: Journal of Geophysical Research, v. 76, p. 6339–6349, doi: 10.1029/JB076i026p06339.

Lueddecke, S.B., Pinter, N., and Gans, P., 1998, Plio-Pleistocene ash falls, sedimentation, and range front faulting along the White-Inyo Mountains front, California: The Journal of Geology, v. 106, p. 511–522.

MacDonald, W.D., 1980, Net tectonic rotation, apparent tectonic rotation, and the structural tilt correction in paleomagnetic studies: Journal of Geophysical Research, v. 85, p. 3659–3669, doi: 10.1029/JB085iB07p03659.

Malservisi, R., Furlong, K.P., and Dixon, T.H., 2001, Influence of the earthquake cycle and lithosphere rheology on the dynamics of the Eastern California shear zone: Geophysical Research Letters, v. 28, p. 2731–2734, doi: 10.1029/2001GL013311.

Mankinen, E.A., Larson, E.E., Gromme, C.S., Prevot, M., and Coe, R.S., 1987, The Steens Mountain (Oregon) geomagnetic polarity transition: 3. Its

regional significance: Journal of Geophysical Research, v. 92, p. 8057–8076, doi: 10.1029/JB092iB08p08057.

Marvin, R.F., and Mehnert, H.H., 1977, K-Ar ages of Tertiary igneous and sedimentary rocks of the Mina-Candelaria region, Mineral County, Nevada: Isochron-West, v. 18, p. 9–12.

McFadden, P.L., and McElhinny, M.W., 1990, Classification of the reversal test in palaeomagnetism: Geophysical Journal International, v. 103, p. 725, 729.

Merrill, R.T., and McElhinny, M.W., 1983, The Earth's Magnetic Field: London, Academic Press, 401 p.

Miller, M.M., Johnson, D.J., Dixon, T.H., and Dokka, R.K., 2001, Refined kinematics of the Eastern California shear zone from GPS observations, 1993–1998: Journal of Geophysical Research, v. 106, p. 2245–2263, doi: 10.1029/2000JB900328.

Minster, J.B., and Jordan, T.H., 1987, Vector constraints on western United States deformation from space geodesy, neotectonics, and plate motions: Journal of Geophysical Research-Solid Earth and Planets, v. 92, p. 4798–4804, doi: 10.1029/JB092iB06p04798.

Niemi, N., Wernicke, B., Brady, R., Saleeby, J., and Dunne, G.C., 2001, Distribution and provenance of the middle Miocene Eagle Mountain Formation, and implications for regional kinematic analysis of the Basin and Range Province: Geological Society of America Bulletin, v. 113, p. 419–442, doi: 10.1130/0016-7606(2001)113<0419:DAPOTM>2.0.CO;2.

Oldow, J.S., 1992, Late Cenozoic displacement partitioning in the northwestern Great Basin, in Craig, S.D., ed., Structure, Tectonics, and Mineralization of the Walker Lane: Reno, Nevada, Geological Society of Nevada, Walker Lane Symposium Proceedings Volume, p. 17–52.

Oldow, J.S., 2003, Active transtensional boundary zone between the western Great Basin and Sierra Nevada block, western U.S. Cordillera: Geology, v. 31, p. 1033–1036, doi: 10.1130/G19838.1.

Oldow, J.S., Bally, A.W., Avé Lallemant, H.G., and Leeman, W.P., 1989, Phanerozoic evolution of the North American Cordillera, in Bally, A.W., and Palmer, A.R., eds., The Geology of North America: An Overview: Boulder, Colorado, Geologic Society of America, Geology of North America, v. A, p. 139–232.

Oldow, J.S., Kohler, G., and Donelick, R.A., 1994, Low-angle displacement transfer system linking the Furnace Creek and Walker Lane fault zones, west-central Nevada: Geology, v. 22, p. 637–640, doi: 10.1130/0091-7613(1994)022<0637:LCETIT>2.3.CO;2.

Oldow, J.S., Aiken, C.L.V., Hare, J.L., Ferguson, J.F., and Hardyman, R.F., 2001, Active displacement transfer and differential block motion within the central Walker Lane, western Great Basin: Geology, v. 29, no. 1, p. 19–22, doi: 10.1130/0091-7613(2001)029<0019:ADTADB>2.0.CO;2.

Oldow, J.S., Geissman, J.W., and Stockli, D.F., 2008, Evolution and strain reorganization within late Neogene structural stepovers linking the central Walker Lane and northern Eastern California shear zone, western Great Basin: International Geology Review, v. 50, p. 1–21, doi: 10.2747/0020-6814.50.3.270.

Oldow, J.S., Elias, E.A., Prestia, V.I., Ferranti, L., McClelland, W.C., and McIntosh, W.C., 2009, this volume, Late Miocene to Pliocene synextensional deposition in fault-bounded basins within the upper plate of the western Silver Peak–Lone Mountain extensional complex, west-central Nevada, in Oldow, J.S., and Cashman, P.H., eds., Late Cenozoic Structure and Evolution of the Great Basin–Sierra Nevada Transition: Geological Society of America Special Paper 447, doi: 10.1130/2009.2447(14).

Ormerod, D.S., 1988, Late- to post-subduction magmatic transitions in the western Great Basin, USA [Ph.D. thesis]: Milton Keynes, UK, Open University, 313 p.

Page, B.M., 1959, Geology of the Candelaria Mining district, Mineral County, Nevada: Nevada Bureau of Mines Bulletin, v. 51, p. 513–540.

Petronis, M.S., and Geissman, J.W., 2008, Anisotropy of magnetic susceptibility data bearing on the transport direction of mid-Tertiary regional ignimbrites, Candelaria Hills area, west-central Nevada: Bulletin of Volcanology, doi: 10.1007/s00445-008-0212-3.

Petronis, M.S., Geissman, J.W., Oldow, J.S., and McIntosh, W.C., 2002, Paleomagnetic and $^{40}Ar/^{39}Ar$ geochronologic data bearing on the structural evolution of the Silver Peak extensional complex, west-central Nevada: Geological Society of America Bulletin, v. 114, no. 9, p. 1108–1130.

Petronis, M.S., Geissman, J.W., and McIntosh, W.C., 2004, Transitional field clusters from uppermost Oligocene volcanic rocks in the Central Walker Lane: Western Nevada: Physics of the Earth and Planetary Interior, v. 141, p. 207–238, doi: 10.1016/j.pepi.2003.12.004.

Petronis, M.S., Geissman, J.W., Oldow, J.S., and McIntosh, W.C., 2007, Tectonism of the southern Silver Peak Range: Paleomagnetic and geochronologic data bearing on the Neogene development of a regional extensional complex, central Walker Lane, Nevada, *in* Till, A.B., Roeske, S.M., Sample, J.C., and Foster, D.A., eds., Exhumation Associated with Continental Strike-Slip Fault Systems: Geological Society of America Special Paper 434, p. 81–106.

Prevot, M., Mankinen, E.A., Coe, R.S., and Gromme, C.S., 1985, The Steens Mountain (Oregon) geomagnetic polarity transition: 2. Field intensity variations and discussion of reversal models: Journal of Geophysical Research, v. 90, p. 10,417–10,448, doi: 10.1029/JB090iB12p10417.

Reheis, M.C., and Dixon, T.H., 1996, Kinematics of the Eastern California shear zone: Evidence for slip transfer from Owens and Saline Valley fault zones to the Fish Lake Valley fault zone: Geology, v. 24, p. 339–342, doi: 10.1130/0091-7613(1996)024<0339:KOTECS>2.3.CO;2.

Reheis, M.C., and Sawyer, T.L., 1997, Late Cenozoic history and slip rates of the Fish Lake Valley, Emigrant Peak, and Deep Springs fault zones, Nevada and California: Geological Society of America Bulletin, v. 109, p. 280–299, doi: 10.1130/0016-7606(1997)109<0280:LCHASR>2.3.CO;2.

Reynolds, R.L., 1977, Paleomagnetism of welded tuffs of the Yellowstone Group: Journal of Geophysical Research, v. 82, p. 3677–3693, doi: 10.1029/JB082i026p03677.

Robinson, P.T., and Crowder, D.F., 1973, Geologic Map of the Davis Mountain Quadrangle, Esmeralda and Mineral Counties, Nevada, and Mono County, California: U.S. Geological Survey Geologic Quadrangle Map GQ-1078, scale 1:62,500.

Robinson, P.T., and Stewart, J.H., 1984, Uppermost Oligocene and Lowermost Miocene Ash-Flow Tuffs of Western Nevada: U.S. Geological Survey Bulletin 1557, 53 p.

Robinson, P.T., McKee, E.H., and Moiola, R.J., 1968, Cenozoic volcanism and sedimentation, Silver Peak region, western Nevada and adjacent California, *in* Coats, R.R., Hay, R.L., and Anderson, C.A., eds., Studies in Volcanology: A Memoir in Honor of Howel Williams: Geological Society of America Memoir 116, p. 577–611.

Robinson, P.T., Stewart, J.H., Moiola, R.J., and Albers, J.P., 1976, Geologic Map of the Rhyolite Ridge Quadrangle, Esmeralda County, Nevada: U.S. Geological Survey Geologic Quadrangle Map GQ-1325, scale 1:24,000.

Ron, H., Freund, R., Garfunkel, Z., and Nur, A., 1984, Block rotation by strike slip faulting: Structural and paleomagnetic evidence: Journal of Geophysical Research, v. 89, p. 6256–6270, doi: 10.1029/JB089iB07p06256.

Ross, D.C., 1961, Geology and Mineral Deposits of Mineral County, Nevada: Nevada Bureau of Mines and Geology Bulletin 58, 98 p.

Rowan, C.J., and Roberts, A.P., 2008, Widespread remagnetizations and a new view of Neogene tectonic rotations within the Australia-Pacific plate boundary zone, New Zealand: Journal of Geophysical Research, v. 113, p. B03103, doi: 10.1029/2006JB004594.

Roy, J.L., and Park, J.K., 1974, The magnetization process of certain red beds: Vector analysis of chemical and thermal results: Canadian Journal of Earth Sciences, v. II, p. 437–471.

Ryall, A.S., and Priestly, K., 1975, Seismicity, secular strain, and maximum magnitude in the Excelsior Mountains area, western Nevada and eastern California: Geological Society of America Bulletin, v. 86, p. 1585–1592, doi: 10.1130/0016-7606(1975)86<1585:SSSAMM>2.0.CO;2.

Schroeder, J.M., Lee, J., Owen, L., and Finkel, R.C., 2005, Pleistocene dextral fault slip along the White Mountains fault zone, California: Implications for kinematics of fault slip in the eastern California shear zone, *in* Lee, J., Stockli, D., Henry, C.D., and Dixon, T.H., eds., Geological Society of America Penrose Conference on Kinematics and Geodynamics of Intraplate Dextral Shear in Eastern California and Western Nevada, 21–26 April, Mammoth Lakes, California: Boulder, Colorado, Geological Society of America, p. 111.

Silberman, M.L., and McKee, E.H., 1972, A summary of radiometric age determinations on Tertiary volcanic rocks from Nevada and eastern California: Part II. Western Nevada: Isochron-West, v. 4, p. 7–28.

Silberman, M.L., Bonham, H.F., Jr., Garside, L.J., and Osborne, D.H., 1975, New K-Ar ages of volcanic and plutonic rocks and ore deposits in western Nevada: Isochron-West, v. 13, p. 13–21.

Sonder, L.J., Jones, C.H., Salyards, S.S., and Murphy, K.M., 1994, Vertical axis rotations in the Las Vegas Valley shear zone, southern Nevada: Paleomagnetic constraints on kinematics and dynamics of block rotations: Tectonics, v. 13, p. 769–788, doi: 10.1029/94TC00352.

Speed, R.C., and Cogbill, A.H., 1979a, Cenozoic volcanism of the Candelaria region, Nevada: Geological Society of America Bulletin, v. 90, Part II, p. 456–493.

Speed, R.C., and Cogbill, A.H., 1979b, Deep fault trough of Oligocene age, Candelaria Hills, Nevada: Geological Society of America Bulletin, v. 90, Part I, p. 145–148, doi: 10.1130/0016-7606(1979)90<145:DFTOOA>2.0.CO;2.

Speed, R.C., and Cogbill, A.H., 1979c, Candelaria and other left-oblique slip faults of the Candelaria region, Nevada: Geological Society of America Bulletin, v. 90, Part I, p. 149–163, doi: 10.1130/0016-7606(1979)90<149:CAOLSF>2.0.CO;2.

Stewart, J.H., 1972, Initial deposits in the Cordilleran geosyncline: Evidence of a late Precambrian (850 m.y.) continental separation: Geological Society of America Bulletin, v. 83, p. 1345–1360, doi: 10.1130/0016-7606(1972)83[1345:IDITCG]2.0.CO;2.

Stewart, J.H, 1979, Geologic Map of Miller Mountain and Columbus Quadrangles, Mineral and Esmeralda Counties, Nevada: U.S. Geological Survey Open-File Report OF-79-1145, 1 sheet.

Stewart, J.H., 1980, Geology of Nevada: Reno, Nevada, Nevada Bureau of Mines and Geology, 136 p.

Stewart, J.H., 1981a, Geologic Map of the Basalt Quadrangle, Mineral County Nevada; U.S. Geological Survey Open-File Report 81-369, scale 1:24,000.

Stewart, J.H., 1981b, Geologic Map of the Jack Springs Quadrangle, Mineral County Nevada: U.S. Geological Survey Open-File Report 81-368, scale 1:24,000.

Stewart, J.H., 1988, Tectonics of the Walker Lane belt, western Great Basin: Mesozoic and Cenozoic deformation in a shear zone, *in* Ernest, W.G., ed., Metamorphism and Crustal Evolution of the Western United States: Englewood Cliffs, New Jersey, Prentice-Hall, p. 681–713.

Stewart, J.H., 1991, Latest Proterozoic and Cambrian rocks of the western United States—An overview, *in* Cooper, J.D., and Stevens, Ch., eds., Paleozoic Paleogeography of the Western United States, Volume II: Tulsa, Oklahoma, Pacific Section, Society for Sedimentary Geology (SEPM), p. 13–38.

Stewart, J.H., Robinson, P.T., Albers, J.P., and Crowder, D.F., 1975, Geologic Map of the Piper Peak Quadrangle, Nevada: U.S. Geological Survey Geologic Quadrangle Map GQ-1186, scale 1:62,500.

Stewart, J.H., Kleinhampl, F.J., Johannesen, D.C., Speed, R.C., and Dohrenwend, J.C., 1981, Geologic Map of the Huntoon Valley Quadrangle, Mineral County, Nevada, and Mono County, California: U.S. Geological Survey Open-File Report 81-274, scale 1:62,500.

Stockli, D.F., 1999, Regional timing and spatial distribution of Miocene extension in the northern Basin and Range province [Ph.D. thesis]: Palo Alto, Stanford University, 239 p.

Stockli, D.F., Dumitru, T.A., McWilliams, M.O., and Farley, K.A., 2003, Cenozoic tectonic evolution of the White Mountains, California and Nevada: Geological Society of America Bulletin, v. 115, p. 788–816, doi: 10.1130/0016-7606(2003)115<0788:CTEOTW>2.0.CO;2.

Sussman, A., Lewis, C., Soto, R., and Goteti, R., 2006, Vertical axis rotations associated with relay ramps in the Rio Grande Rift, New Mexico: Geological Society of America Abstracts with Programs, v. 38, no. 7, p. 417.

Tanaka, H., Kono, M., and Kaneko, S., 1995, Paleosecular variation of direction and intensity from two Pliocene-Pleistocene lava sections in southwestern Iceland: Journal of Geomagnetism and Geoelectricity, v. 47, no. 1, p. 89–102.

Tarling, D.H., and Hrouda, F., 1993, The Magnetic Anisotropy of Rocks: London, Chapman and Hall, 217 p.

Thatcher, W., Foulger, G.R., Julian, B.R., Svarc, J., Quility, E., and Bawden, G.W., 1999, Present-day deformation across the Basin and Range Province, western U.S.: Science, v. 283, p. 1714–1718.

Tincher, C.R., and Stockli, D.F., 2009, this volume, Cenozoic volcanism and tectonics in the Queen Valley area, Esmeralda County, western Nevada, *in* Oldow, J.S., and Cashman, P.H., eds., Late Cenozoic Structure and Evolution of the Great Basin–Sierra Nevada Transition: Geological Society of America Special Paper 447, doi: 10.1130/2009.2447(13).

Uyeda, S., Fuller, M.D., Belshe, J.C., and Girdler, R.W., 1963, Anisotropy of magnetic susceptibility of rocks and minerals: Journal of Geophysical Research, v. 68, p. 279–291, doi: 10.1029/JZ068i001p00279.

Wallace, R.E., 1984, Patterns and timing of late Quaternary faulting in the Great Basin Province and relation to some regional tectonic features: Journal of Geophysical Research, v. 89, p. 5763–5769, doi: 10.1029/JB089iB07p05763.

Wawrzyniec, T.F., Geissman, J.W., Melker, M.D., and Hubbard, M., 2002, Dextral shear along the eastern margin of the Colorado Plateau: A kinematic link between Laramide contraction and Rio Grande rifting (ca. 75–13 Ma): The Journal of Geology, v. 110, p. 305–324, doi: 10.1086/339534.

Wells, R.E., and Hillhouse, J.W., 1989, Paleomagnetism and tectonic rotation of the lower Miocene Peach Springs Tuff–Colorado Plateau, Arizona to Barstow, California: Geological Society of America Bulletin, v. 101, p. 846–863, doi: 10.1130/0016-7606(1989)101<0846:PATROT>2.3.CO;2.

Wesnousky, S.G., 2005, Active faulting in the Walker Lane: Tectonics, v. 24, TC3009, doi: 10.1029/2004TC001645.

Wolff, J.A., and Wright, J.V., 1981, Rheomorphism of welded tuffs: Journal of Volcanology and Geothermal Research, v. 10, p. 13–34, doi: 10.1016/0377-0273(81)90052-4.

Zijderveld, J.D.A., 1967, A.C. demagnetization of rocks: Analysis of results, *in* Collinson, D.W., Creer, K.M., and Runcorn, S.K., eds., Methods in Paleomagnetism: Amsterdam, Elsevier, p. 254, 286.

MANUSCRIPT ACCEPTED BY THE SOCIETY 21 JULY 2008

The Geological Society of America
Special Paper 447
2009

Cenozoic volcanism and tectonics in the Queen Valley area, Esmeralda County, western Nevada

Christopher R. Tincher
Daniel F. Stockli*
Department of Geology, University of Kansas, 1475 Jayhawk Blvd., 120 Lindley Hall, Lawrence, Kansas 66045, USA

ABSTRACT

The Queen Valley pull-apart basin is located at the northern extent of the White Mountains in western Nevada. The basin is bounded to the south by the NE-trending Queen Valley fault zone and to the north by the E-W–trending Coaldale fault zone. The curvilinear trace of the Queen Valley normal fault extends ~16 km northeast from the northern termination of the Owens Valley–White Mountain fault zone to the western Coaldale fault system. Using new (U-Th)/He and ^{40}Ar/^{39}Ar geochronology, fault kinematic data, and detailed geologic mapping (1:10,000), this study documents a three-stage late Tertiary tectonic evolution of the eastern Queen Valley area and defines the role of the Queen Valley fault system as an integral part of the right-lateral transtensional Walker Lane belt. The Queen Valley area was affected by an ignimbrite flare-up in Utah, Nevada, and California, as recorded by late Oligocene rhyolites (ca. 26 Ma). The eruption of these widespread ash flows was accompanied locally by extension, creating a series of ENE-trending half grabens. The faults are sealed by Miocene andesite (ca. 12 Ma), constraining the timing of extension to late Oligocene or early Miocene. Mid-Miocene Basin and Range extension produced E-dipping normal fault systems in the Yerington area to the north and W-dipping normal faults in the White Mountains to the south. Displacement between these fault systems with opposite polarity was accommodated by a series of right-lateral faults in the Queen Valley area. A change in extension direction from E-W extension to NW-SE during the Pliocene resulted in a transition to transcurrent and transtensional structures in the central Walker Lane belt. The beginning of transtension on the east side of the White Mountains was marked by the opening of the Fish Lake Valley pull-apart basin at ca. 6 Ma, as constrained by Upper Miocene volcanic units. Similarly, the Queen Valley pull-apart basin was a product of the reactivation of the White Mountain–Owens Valley fault zone as a right-lateral fault ca. 3 Ma, based on thermochronological data and offset Pliocene basaltic andesite (ca. 3.1 Ma) along the Queen Valley fault.

Keywords: Walker Lane, Mina deflection, transtension, volcanism, geochronology.

*Corresponding author e-mail: stockli@ku.edu.

Tincher, C.R., and Stockli, D.F., 2009, Cenozoic volcanism and tectonics in the Queen Valley area, Esmeralda County, western Nevada, *in* Oldow, J.S., and Cashman, P.H., eds., Late Cenozoic Structure and Evolution of the Great Basin–Sierra Nevada Transition: Geological Society of America Special Paper 447, p. 255–274, doi: 10.1130/2009.2447(13). For permission to copy, contact editing@geosociety.org. ©2009 The Geological Society of America. All rights reserved.

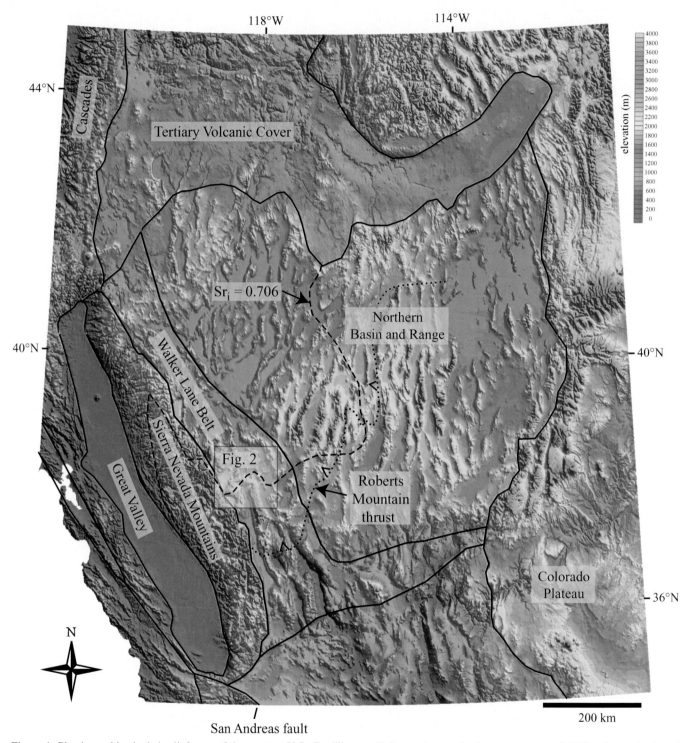

Figure 1. Physiographic shaded-relief map of the western U.S. Cordillera outlining study area in the western central Walker Lane (box) and regimes of active deformation and major pre-Tertiary structures. The Walker Lane belt represents a diffuse intracontinental transtensional belt accommodating ~15% of the modern plate boundary strain, while the San Andreas fault system marks the transform boundary between the Pacific and North American plates. Major pre-Tertiary structural belts shown are the Roberts Mountain thrust and the western margin of continental North America recorded by initial strontium isopleth, $Sr_i = 0.706$ (modified after Oldow et al., 2008).

INTRODUCTION

The interaction of the Pacific and North American plates has created a broad zone of late Cenozoic deformation that stretches from the San Andreas fault system into the western Basin and Range Province. Approximately 25% of the total displacement is accommodated through the Walker Lane belt, a 100–300-km-wide by 700-km-long zone that trends N-NW along the California-Nevada border from roughly Las Vegas, Nevada, to southern Oregon (Fig. 1). Physiographically different than the Sierra Nevada Mountains to the west and the Basin and Range Province to the east, the Walker Lane belt is characterized by N-NW–trending dextral and transtensional faults. The central portion of the Walker Lane belt, referred to as the Mina deflection, is a right-step releasing bend that links the southern and northern portions of the Walker Lane belt (Fig. 2). Previous work illustrates as much as 100 km of late Cenozoic right-lateral displacement in the northern and southern Walker Lane belt (e.g., Oldow, 1992; Stewart, 1988); however, it remains unclear how this strain is transferred through the Mina deflection.

Queen Valley, along with other fault-bound depressions, such as Rhodes Salt Marsh, Whiskey Flat, Huntoon Valley, Columbus Salt Marsh, and northern Fish Lake Valley, is a pull-apart basin characteristic of the Mina deflection (e.g., Oldow, 1992; Stewart, 1988; Stockli et al., 2003) (Fig. 2). The Queen Valley pull-apart basin is located at the northern termination of the Owens Valley–White Mountains fault zone in western Nevada. The basin is bounded to the south by the NE-trending Queen Valley fault zone and to the north by the E-W–trending Coaldale fault zone (Figs. 2 and 3). The Queen Valley fault zone marks an abrupt change from the N-S orientation of the White Mountain fault zone to NE-SW orientation as it approaches the Coaldale fault to the northeast. This study presents new detailed geological mapping, volcanic stratigraphy, geochronology, and structural data to elucidate and reconstruct the structural and magmatic evolution of the eastern portion of the Pliocene Queen Valley pull-apart structure and the role of pre-Pliocene structures on the geometry of strain transfer through the Mina deflection of the central Walker Lane.

TECTONIC SETTING OF THE CENTRAL WALKER LANE

Late Cenozoic faults, late Pleistocene to Holocene fault scarps, and modern seismicity indicate that the central Walker Lane consists of a complex array of variably oriented, active faults characterized by NW-directed right-lateral shear and dip-slip fault motion (e.g., Oldow, 1992). Tertiary deformation in the Mina deflection area is strongly influenced by the pre-Tertiary geological and structural framework of the western Basin and Range Province. The region has experienced a long-lived and complex history of Phanerozoic continental rifting and subsequent convergent margin magmatism and tectonism (e.g., Stewart, 1972; Burchfiel and Davis, 1975; Oldow et al., 1989; Stewart, 1992). In the Mina deflection, the NE-SW–trending structural

grain appears to be primarily controlled by pre-Cenozoic depositional and structural patterns that reflect the trace of the Precambrian transition between attenuated continental and oceanic crust (Fig. 1). Both Paleozoic miogeoclinal depositional facies distributions and contractional structures associated with Paleozoic and Mesozoic shortening follow this Precambrian crustal boundary (e.g., Stewart, 1992; Oldow et al., 2008).

The northern portion of the central Walker Lane belt (Fig. 2) is characterized by an en-echelon array of NW-trending strike-slip faults that, toward the south, enter the Mina deflection; this zone of predominantly NE-trending faults is characterized by extensional and left-lateral fault motion (e.g., Oldow, 1992; Stockli et al., 2003; Wesnousky, 2005; Oldow et al., 2008). The NW-trending right-lateral strike-slip faults generally do not appear to be truncated by the NE-trending faults, but, rather, they seem to smoothly change in orientation as they swing into the Mina deflection (Fig. 2). Individual faults are mapped as continuous bends as they change in orientation, forming "z"-shaped releasing bends characterized by rhomboidal pull-apart basins that accommodate the strike-slip displacement (Oldow, 1992). For example, NE-striking faults in the Excelsior Mountains progressively change in orientation and merge with the NE-trending fault systems, which bound a rhomboidal pull-apart basin in the Rhodes Salt Marsh area (e.g., Oldow, 1992; Oldow et al., 2008; Ferranti et al., this volume). To the southeast, splays of the Excelsior fault system swing south and merge with faults that are part of the distributed strain associated with the Owens Valley–White Mountain fault system. Similar smooth curvilinear fault traces can be found throughout the Mina deflection area, such as in the southern Wassuk Range, where the Wassuk fault swings southwest toward Mono Lake, structurally controlling a pull-apart structure in the Whiskey Flat area (Fig. 2).

The northern end of the White Mountains is bounded by the Queen Valley pull-apart basin, which is structurally controlled by a curvilinear NW-dipping normal fault, the Queen Valley fault (Fig. 2) (Stockli et al., 2003). The Queen Valley basin is situated at the northern termination of the Owens Valley–White Mountain right-lateral fault system and transfers slip via the Coaldale fault into the southern central Walker Lane belt (Fig. 2). Preliminary apatite (U-Th)/He thermochronological data from the northernmost White Mountains, representing the exhumed footwall block to the Queen Valley pull-apart structure, constrain the timing of inception of faulting at 3.0–2.5 Ma (Stockli et al., 2003). These data also argue for initiation of right-lateral shearing along the northern Owens Valley fault system during the Pliocene, reactivating the White Mountain normal fault as a normal-oblique right-lateral fault system (Stockli et al., 2003).

GEOLOGY OF THE QUEEN VALLEY AREA

The Queen Valley region at the northern end of the White Mountains has undergone a complex late Tertiary tectonic history characterized by the interplay between voluminous felsic to basaltic volcanism and extensional and transcurrent deformation.

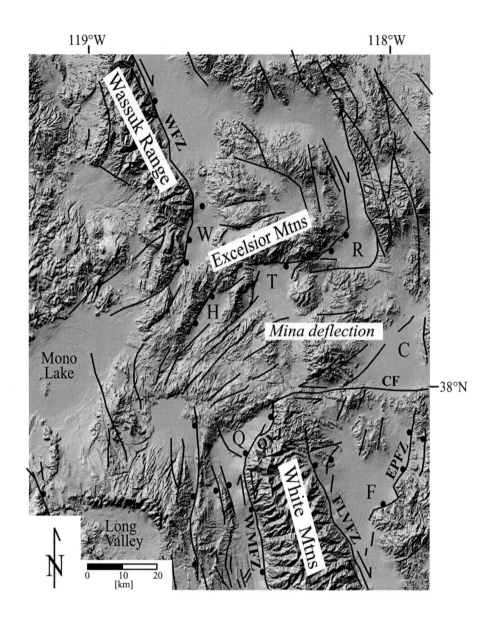

Figure 2. Physiographic map of the western Great Basin and Sierra Nevada showing major late Cenozoic faults in the greater Mina deflection region of the central Walker Lane belt. Pull-apart basins and releasing bends within the Mina deflection are: F—northern Fish Lake Valley; H—Huntoon Valley; Q—Queen Valley; R—Rhodes Salt Marsh; C—Columbus Salt Marsh; T—Teels Salt Marsh; W—Whiskey Flat area. Major faults are: WMFZ—White Mountains fault zone; FLVFZ—Fish Lake Valley fault zone; EPFZ—Emigrant Peak fault zone; QVZ—Queen Valley fault zone; CF—Coaldale fault; SSFZ—Soda Springs fault zone; and WFZ—Wassuk fault zone.

A rich record of volcanic eruptive activity temporally spans the entire Neogene structural evolution of Queen Valley, making it an ideal place to dissect the temporal, structural, and kinematic evolution of the western portion of the Mina deflection. Upper Oligocene and Lower Miocene ash-flow tuffs are found throughout the area surrounding Queen Valley (e.g., Robinson and Stewart, 1984; Petronis et al., 2004), although little is known about the local tectonic setting during eruptions of these voluminous ash-flow tuffs in the Queen Valley area. During the middle Miocene, E-W extension dominated the western Basin and Range Province. To the north of Queen Valley and the western Mina deflection, >200% extension was accommodated by E-dipping normal faults in the Wassuk Range and adjacent ranges, starting ca. 15 Ma, while the W-dipping White Mountain fault zone accommodated significant E-W extension beginning ca. 12 Ma (Proffett, 1977;

Dilles and Gans, 1995; Surpless et al., 2002; Stockli et al., 2002, 2003). Stockli et al. (2003) suggested that faults in the Queen Valley area and the western Mina deflection acted as a right-lateral accommodation zone, linking the E-dipping faults in the Wassuk Range and the W-dipping faults of the White Mountains (Fig. 2). In the following sections, this study will present new structural evidence for E-W faults acting as part of an accommodation zone during middle Miocene extension. Structural and thermochronometric studies have shown a complete tectonic reorganization and transition from extensional to NW-directed transtension by ca. 3 Ma, resulting in the formation of the Queen Valley pull-apart basin (e.g., Stockli et al., 2003; Oldow et al., 2008).

Despite this well-constrained regional tectonic framework, many outstanding questions remain regarding the detailed structural evolution, fault kinematics, timing of faulting, and

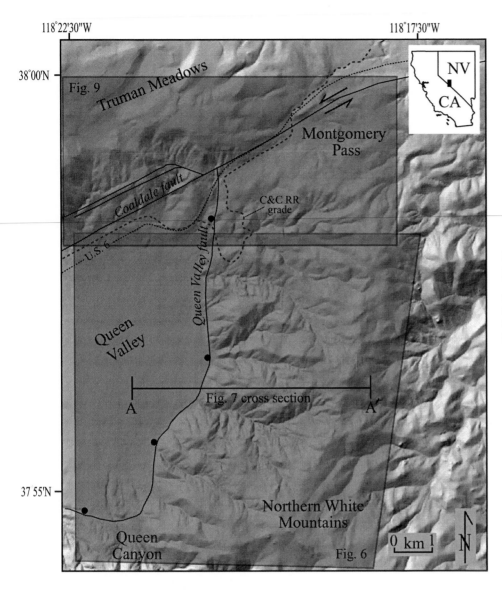

Figure 3. Shaded digital relief map of eastern Queen Valley showing the Queen Valley and Coaldale faults and locations of detailed geological maps (boxes). Map also shows U.S. Highway 6 and abandoned Carson and Colorado Railroad (C&C RR) grade as dashed lines.

interplay between volcanism and faulting in the Queen Valley area. The following sections present a detailed description of the stratigraphy and the structural evolution of the Queen Valley area aimed at gaining a better understanding of the Neogene tectonic evolution of the western Mina deflection and the transition from E-W extension to transtension in the late Miocene to Pliocene (Fig. 3).

STRATIGRAPHY OF EASTERN QUEEN VALLEY

A rich Tertiary volcanic and sedimentary succession, including Oligocene and Miocene siliceous ash-flow tuffs and associated sedimentary rocks, Miocene andesite flows, tuffs, and lahars, and Miocene and Pliocene basalts and basaltic andesite, nonconformably overlies the Paleozoic metasedimentary and

Mesozoic plutonic basement throughout the central Walker Lane belt, including the Queen Valley area (e.g., Hardyman and Oldow, 1991; Dilles and Gans, 1995; Stockli et al., 2003) (Fig. 4). The following sections describe in detail the volcanic stratigraphy of the Queen Valley area and present new $^{40}Ar/^{39}Ar$ and zircon and titanite (U-Th)/He data, providing new absolute age constraints on the volcanic history and bracketing the timing of faulting. The $^{40}Ar/^{39}Ar$ analytical data are presented in Figures 5 and 6 and Table 1, while (U-Th)/He geochronometric results are summarized in Table 2.

Pre-Tertiary Rocks of Eastern Queen Valley

The pre-Tertiary basement of Queen Valley is characterized by deformed and metamorphosed Lower Paleozoic

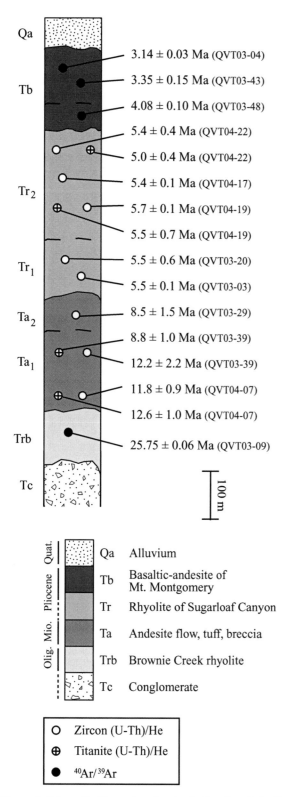

Figure 4. Schematic Cenozoic volcanic stratigraphy of the Queen Valley area constrained by new mapping and geochronological data. Dark circles designate ⁴⁰Ar/³⁹Ar ages, and open circles reflect zircon and titanite (U-Th)/He ages. Dashed lines delineate approximate boundaries of cooling units or subdivisions within the main map unit. Analytical data are shown in Table 1 (⁴⁰Ar/³⁹Ar) and Table 2 (zircon and titanite [U-Th]/He).

sedimentary rocks that are intruded by largely undated Jurassic to Cretaceous granodiorite. The Lower Paleozoic rocks consist of three marine metasedimentary units, including phyllite, marble, and slate. The oldest Paleozoic unit is a phyllite that is exposed in Queen Canyon and the area west of Montgomery Pass (Figs. 3 and 7). In Queen Canyon, the phyllite is black and gray, weathering to yellow and orange near faults, and it is generally fine-grained with large andalusite porphyroblasts restricted to contact aureoles around Mesozoic intrusive rocks. Near Montgomery Pass, the phyllites appear to be lower grade and locally transition to low-grade slates. Southwest of Montgomery Pass, along the trace of the Coaldale fault, the phyllites are overlain by fine- to medium-grained, white to gray marbles. Crowder et al. (1972) tentatively correlated the phyllites and marbles with the Cambrian Harkless and Poleta Canyon Formations, respectively.

In the lower Queen Canyon area, interbedded slates, siltstones, and marbles assigned to the Ordovician Palmetto Formation are structurally juxtaposed against (Harkless) phyllites along a low-angle thrust plane thought to be a strand of the Paleozoic Roberts Mountain thrust (Crowder et al., 1972). The same stratigraphic and structural relationships are also observed in the northwestern White Mountains (e.g., Stockli et al., 2003).

The Paleozoic metasedimentary rocks are intruded by several granodiorite, diorite, and leucogranite bodies that are part of larger intrusive suites with the northern White Mountains (Anderson, 1937; Harris, 1967; Evernden and Kistler, 1970; Crowder et al., 1972) (Fig. 7). Although mapped as a variety of Triassic and Jurassic intrusive bodies occurring in the northernmost White Mountains and eastern Queen Valley by Crowder et al. (1972), new mapping and thermal ionization mass spectrometry (TIMS) single-grain zircon U-Pb dating (Wooten and Stockli, 2004) show three major plutonic suites: (1) a voluminous medium- to coarse-grained porphyritic biotite granite with an age of ca. 162 Ma, exposed in eastern Queen Valley and the northeastern White Mountains (Pellisier Flat pluton of Crowder and Sheridan, 1972); (2) a hornblende diorite on the south side of Queen Canyon with an age ca. 160 Ma, previously mapped as Triassic (Crowder et al., 1972) and dated at ca. 157 Ma by K/Ar (Evernden and Kistler, 1970); and (3) the leucocratic Boundary Peak adamellite, with an age of ca. 88 Ma, which occupies the northwestern and northern flank of the White Mountains.

Tertiary Stratigraphy of Eastern Queen Valley

Basal Tertiary Conglomerate

The pre-Tertiary metasedimentary and intrusive rocks throughout the eastern portion of Queen Valley are unconformably overlain by a basal conglomerate (Figs. 4 and 7). The conglomerate contains (sub)rounded pebble- to cobble-size clasts of Cambrian marble and biotite granite (likely Pellisier Flat pluton) supported by a brick-red, fine-grained sandy matrix. In most localities in eastern Queen Valley, this distinct conglomerate measures <20 m in thickness and overlies a relatively planar

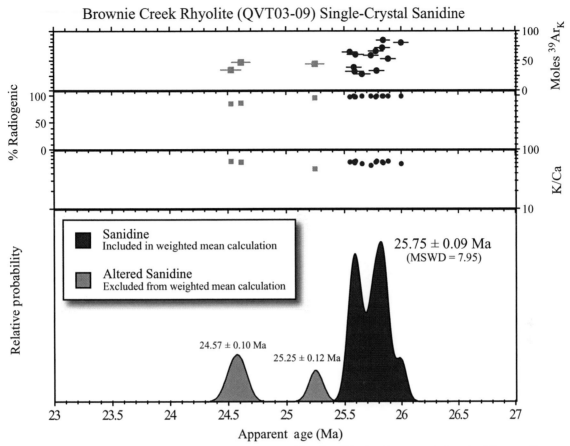

Figure 5. Diagram showing laser total-fusion–single-grain sanidine ^{40}Ar/^{39}Ar data for a sample from the Oligocene Brownie Creek rhyolite ash-flow tuff. Fifteen individual sanidine crystals from sample QVT03-09 yield a range of total fusion ages between 24.7 and 26.2 Ma. There is a correlation between low radiogenic yield and young age for some of the grains and a fairly bimodal distribution for the crystals, the radiogenic yields of which are ~100%. A weighted mean age of 25.97 ± 0.04 Ma is calculated for the oldest group of crystals (Heizler and Sanders, 2005). The data suggest a likely correction with the Metallic City tuff in the Candelaria Hills (Petronis et al., 2004). MSWD—mean square of weighted deviates.

erosional surface with very little paleorelief. It marks the base of the Tertiary volcanic section, described in detail in the following, but it is only preserved where it is directly overlain by Oligocene rhyolite, and it is commonly eroded where Miocene-Pliocene rhyolite or andesite sits directly on pre-Tertiary basement.

Oligocene Rhyolite

The oldest Cenozoic volcanic deposits consist of rhyolitic ash-flow tuffs that are up to 100 m thick and that are separated from the underlying marble conglomerate by a sharp unconformity with local erosional scours. This massive ash-flow tuff is present from Queen Canyon to near Montgomery Pass and was first mapped by Crowder et al. (1972) as the Brownie Creek rhyolite, a rhyolite unique to the Queen Valley area (Fig. 4). While previously published K-Ar age data range from 21.5 to 28.5 Ma (Gilbert et al., 1968; Robinson et al., 1968), a new ^{40}Ar/^{39}Ar laser-fusion sanidine age determination on a sample from the north side of Queen Canyon yielded an age of 25.75 ± 0.09 Ma (Table 1;

Fig. 5). Many zircon (U-Th)/He ages appear to be affected by subsequent thermal perturbations, but the oldest mode of several samples is in very good agreement with the ^{40}Ar/^{39}Ar data (Table 2). This new age constraint, combined with lithologic similarities, strongly suggests that the Brownie Creek rhyolite is correlative with the regionally extensive Metallic City Tuff (25.64 ± 0.12 Ma) of the Candelaria Hills sequence (Petronis et al., 2004). The Metallic City Tuff, along with 10 other regional extensive and voluminous ash-flow tuffs, were erupted between ca. 26 Ma and ca. 23 Ma and can be identified throughout much of the Mina deflection (Gilbert et al., 1968; Robinson et al., 1968; Silberman et al., 1975; Marvin et al., 1977; Garside and Silberman, 1978; Robinson and Stewart, 1984; Petronis et al., 2004). In the Queen Valley area, the up to 60-m-thick tuff is white to light gray in color and contains abundant phenocrysts of quartz, sanidine, and biotite. Pumice fragments are only weakly welded and make up ~15% of the flow. Many outcrops of the tuff show evidence for significant hydrothermal alteration. A sharp erosional

Tincher and Stockli

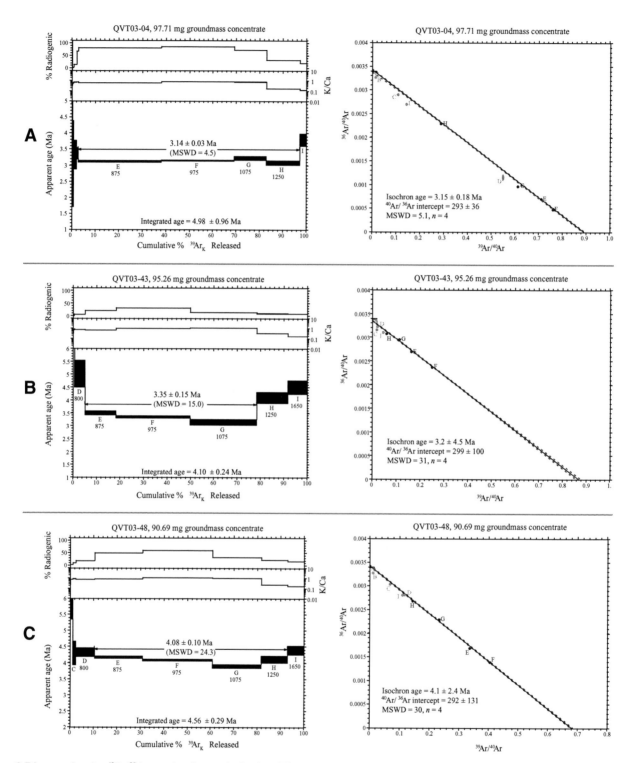

Figure 6. Diagram showing ⁴⁰Ar/³⁹Ar step-heating results for three Pliocene basaltic andesite groundmass samples from eastern Queen Valley, release spectra, and inverse isochron plots (Table 1). (A) Sample QVT03-04 exhibits a well-behaved release spectrum with a plateau age of 3.14 ± 0.03 Ma and moderately well-defined inverse isochron age of 3.15 ± 0.18 Ma. A magnetite (U-Th-[Sm])/He age of 3.1 ± 0.1 Ma (Blackburn et al., 2007) is in excellent agreement with the ⁴⁰Ar/³⁹Ar data for this sample. (B) Sample QVT03-43 exhibits a saddle-shaped age spectrum with a total gas age of 4.10 ± 0.24 Ma and a poorly constrained inverse isochron age. Most likely, the complexity of these data is related to excess argon, and thus the youngest step (step G = 3.16 ± 0.05 Ma) would represent a maximum age for eruption. (C) Sample QVT03-48 has a slightly saddle-shaped spectrum, and the youngest apparent ages are represented by intermediate heating steps. The age assigned to this sample was calculated from a pseudo–plateau segment (E to H) and is 4.10 ± 0.13 Ma, with a high mean square of weighted deviates (MSWD) value of 24.3. Similar to sample QVT03-43, the complexity of these data is related to excess argon, and thus the youngest step (step G = 3.9 ± 0.1 Ma) represents a maximum age for the eruption.

TABLE 1. ^{40}Ar/^{39}Ar GEOCHRONOLOGY DATA FROM THE QUEEN VALLEY AREA

Sample number	Unit	Type	RF/L	TFA (Ma, ±1σ)	WMPA (Ma, ±1σ)	ISO (Ma, ±1σ)	MSWD	^{40}Ar/^{36}Ar	Lat. (°N)	Long. (°W)
QTV03-04	Tb	wr	RF	4.98 ± 0.96	3.14 ± 0.03	3.15 ± 0.18	4.5	293 ± 36	37°54'20"	118°19'34"
QVT03-43	Tb	wr	RF	4.10 ± 0.24	3.35 ± 0.15	3.2 ± 4.5	15	299 ± 100	37°58'57"	118°18'56"
QVT03-48	Tb	wr	RF	4.56 ± 0.29	4.08 ± 0.10	4.1 ± 2.4	24.3	292 ± 131	37°59'05"	118°17'51"
QVT03-09	Trb	san	L	25.75 ± 0.09	–	–	8	–	37°55'07"	118°20'11"

Note: wr—basalt whole rock, san—sanidine analysis, RF—resistance furnace, L—laser analysis, TFA—total fusion age, WMPA—weighted mean plateau age, ISO—inverse isochron age, MSWD—mean square of weighted deviates. All analyses were conducted at the New Mexico Geochronological Research Laboratory (NMGRL). All analytical and methodological details can be found in NMGRL Report #NMGRL-IR-427 (Heizler and Sanders, 2005). Latitude (Lat.) and longitude (Long.) are with respect to the WGS-84 reference frame.

TABLE 2. ZIRCON AND TITANITE (U-Th)/He GEOCHRONOLOGICAL DATA, EASTERN QUEEN VALLEY, NEVADA

Sample	Age (Ma)	± (Ma)	U (ppm)	Th (ppm)	Th/U	He (ncc/mg)	mass (µg)	FT	n	Lat. (°N)	Long. (°W)
Rhyolite of Sugarloaf Canyon (Tr2)											
Z-QVT04-17	5.4	0.4	169.2	234.4	1.4	118.1	16.5	0.81	5	37°55'07"	118°20'11"
Z-QVT04-19	5.7	0.5	265.6	395.5	1.5	157.5	1.9	0.65	3	37°58'44"	118°18'10"
T-QVT04-19	5.5	0.4	3.4	24.5	7.2	7.4	8.2	1.00	4	37°58'44"	118°18'10"
Z-QVT04-22	5.4	0.4	108.3	213.7	2.0	89.6	2.5	0.68	6	37°58'24"	118°17'47"
T-QVT04-22	5.0	0.4	41.8	183.6	4.4	48.5	3.4	1.00	4	37°58'24"	118°17'47"
Rhyolite of Sugarloaf Canyon (Tr1)											
Z-QVT03-03	5.6	0.3	137.1	252.1	1.7	83.13	1.33	0.59	8	37°54'20"	118°19'29"
Z-QVT03-20	5.5	0.3	80.6	97.4	1.2	47.1	4.1	0.72	3	37°57'36"	118°19'03"
Truman Meadows Andesite (Ta2)											
Z-QVT03-29	8.3	0.6	349.7	360.2	1.0	271.34	2.38	0.64	8	37°57'07"	118°20'17"
Truman Meadows Andesite (Ta1)											
Z-QVT03-39	12.2	1.0	240.5	189.7	0.8	280.6	2.0	0.65	7	37°57'50"	118°20'09"
T-QVT03-39	8.8	0.7	85.4	463.6	5.4	226.9	2.6	1.00	3	37°57'50"	118°20'09"
Z-QVT04-07	11.8	0.9	239.9	168.7	0.7	330.4	20.0	0.82	4	37°57'07"	118°20'17"
T-QVT04-07	12.6	1.0	15.9	103.2	6.5	59.8	25.8	1.00	5	37°57'07"	118°20'17"
Brownie Creek Rhyolite (Trb)											
Z-QVT03-06	26.3	2.1	81.5	30.5	0.4	206.5	4.7	0.73	3	37°57'07"	118°20'17"
Z-QVT03-38	25.7	2.1	378.6	366.7	1.0	1092.4	8.9	0.75	2	37°57'39"	118°19'56"
Z-QVT04-14	24.8	2.0	160.9	111.5	0.7	421.3	6.5	0.74	4	37°57'39"	118°19'36"

Note: Single-grain laser zircon and titanite (U-Th)/He age determinations were carried out at the University of Kansas laboratory using laboratory procedures as described by Stockli and Farley (2004) and Biswas et al. (2007). Inclusion-free titanite and zircon grains were wrapped in acid-treated Pt foil and heated for 10 min at 1300 °C and subsequently reheated to ensure complete degassing. After laser heating, titanite was spiked using an enriched ^{235}U-^{230}Th tracer and dissolved in a HCl-HF mixture, while zircons were unwrapped from Pt foil and dissolved using HF-HNO$_3$ and HCl pressure-vessel digestion procedures. U, Th, and Sm concentrations were determined by isotope-dilution inductively coupled plasma–mass spectrometry (ICP-MS) analysis. All zircon ages were calculated using standard α-ejection corrections using morphometric analyses (Farley, 2002; Farley et al., 1996). Table shows mean ages calculated from multiple aliquot analyses (*n*). Sample location coordinates are in latitude (Lat.) and longitude (Long.) with respect to the WGS84 reference frame.

unconformity separates the Oligocene tuff from overlying middle Miocene andesite.

Miocene Andesite

The northeastern Queen Valley area is dominated by a thick succession of andesite tuffs, breccia, lahar deposits, and lavas that sit unconformably on pre-Tertiary basement (Fig. 7). Similar rocks, informally referred to as the Truman Meadows andesite, are also found between Queen Canyon and the abandoned Carson and Colorado (C&C) Railroad grade south of Montgomery Pass (Fig. 7). These andesite deposits are comparable to similar andesites from the central Walker Lane belt attributed to Miocene arc magmatism that range in age from 11 to 20 Ma (e.g., Prof-

fett, 1977; Marvin et al., 1977; Ekren et al., 1980; McKee and John, 1987; Hardyman and Oldow, 1991; Stockli et al., 2003). The Truman Meadows andesite flows are characterized by a porphyritic nature with 15%–25% plagioclase and amphibole phenocrysts set in a fine-grained, gray plagioclase matrix. Grayish-brown andesitic tuffs contain phenocrystic biotite and amphibole and are commonly interbedded with porphyritic lava flows. The entire unit, measuring up to 150 m in thickness, can be divided into two subunits: a lower unit (Ta$_1$) dominated by interlayered lavas and tuffs and an upper unit dominated by andesite breccia and lahar deposits. Most of the andesite is altered and not suitable for ^{40}Ar/^{39}Ar analysis; however, new zircon and titanite (U-Th)/He age determinations on samples from the Truman Meadows

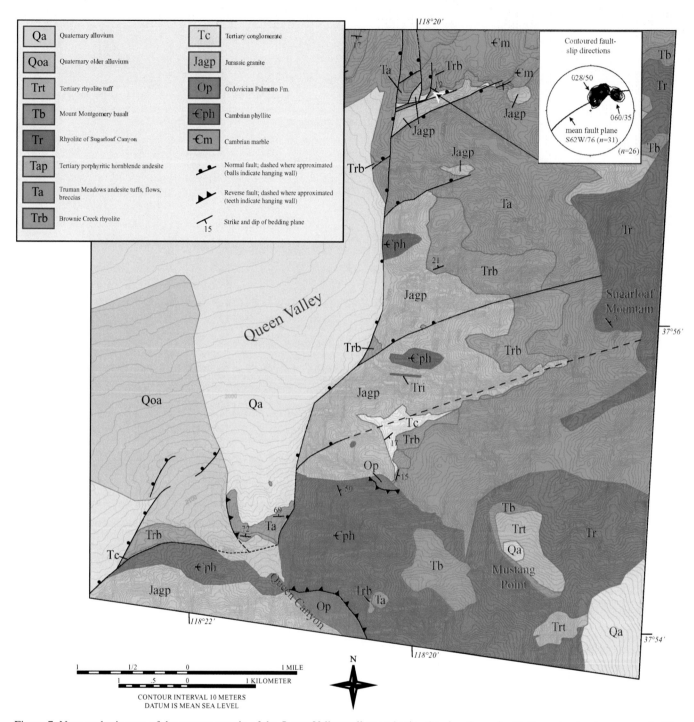

Figure 7. New geologic map of the eastern margin of the Queen Valley pull-apart basin, showing the major Tertiary structural components. The main trace of the basin-bounding Queen Valley normal fault accommodates deformation through a right step near the mouth of Queen Canyon, while several fault splays appear to transfer displacement northward into and likely across the basin onto the left-lateral Coaldale fault, as documented by Holocene fault scarps. Middle Miocene ENE-trending right-lateral normal-oblique faults with the footwall of the Pliocene Queen Valley fault (fault kinematic stereonet inset) cut the Oligocene Brownie Creek rhyolite (Trb) and middle Miocene andesite (Ta), but are sealed by late Miocene–Pliocene rhyolitic ash-flow tuffs (Tr).

andesite yielded two distinct age clusters at ca. 12 Ma (Ta_1) and ca. 8.5 Ma (Ta_2), corroborating the field-based subdivision (Table 2; Fig. 4).

Pliocene Rhyolitic Ash-Flow Tuffs

In the central and southern portions of eastern Queen Valley, the Truman Meadows andesite is overlain by massive Miocene-Pliocene rhyolites that Crowder et al. (1972) referred to as rhyolite of Sugarloaf Canyon. The rhyolite of Sugarloaf Canyon includes two distinct, mappable ash-flow tuffs (Tr_1 and Tr_2) that can be easily differentiated in terms of their welding and lithics composition.

The lower, moderately welded ash-flow tuff (Tr_1) is found exclusively at high elevations in southeastern Queen Valley near Sugarloaf Mine. This purple to light gray, intensely indurated and silicified ash-flow tuff is characterized by flow banding and, in places, abundant lithophysae (Crowder et al., 1972). Typical samples are porphyritic with ~10%–20% quartz and sanidine phenocrysts in a microcrystalline matrix. Zircon and titanite are found abundantly as accessory minerals throughout this unit.

The upper, nonwelded ash-flow tuff (Tr_2) occurs widespread and is the only Miocene-Pliocene ash-flow tuff in the central and northern portions of eastern Queen Valley, where it directly overlies the Truman Meadows andesite. It is best exposed along the old C&C railroad grade in the central portion of eastern Queen Valley and along Highway 6 east of Montgomery Pass (Fig. 3). The white to light gray, unwelded tuff contains ~15%–20% phenocrystic quartz, sanidine, biotite, and plagioclase, ~10%–15% lithic (mostly rhyolitic) fragments, and ~20%–50% pumice, ranging in size from 1 to 5 cm. Tr_1 also contains abundant obsidian (Apache tears). At an outcrop along U.S. Highway 6, the upper contact of Tr_1 displays a cooked margin caused by the overlying basaltic andesite flow.

Crowder et al. (1972) correlated these ash-flow tuffs (Tr) with ash-flow tuffs in the Silver Peak Range and northern Fish Lake Valley that have a K/Ar age of ca. 6 Ma (Robinson et al., 1968). Single-crystal laser zircon and titanite (U-Th)/He dating on five samples (33 aliquot analyses) yielded statistically indistinguishable ages of ca. 5.5 Ma (Tr_1) and ca. 5.4 Ma (Tr_2) for the two ash-flow tuffs (Table 2; Fig. 4). These ages suggest the northern Silver Peak Range as a possible source region but do not allow more precise identification of a possible origin for these eruptive products.

Pliocene Basaltic Andesite

Voluminous and widespread mafic volcanics cover much of the area east and north of Queen Valley, unconformably overlying Miocene and Pliocene andesites and rhyolites. Crowder et al. (1972) referred to these rocks as the basalt of Mount Montgomery and correlated them with basalts from the Benton Range, dated at ca. 3 Ma by K/Ar (Dalrymple and Hirooka, 1965). In the Queen Valley area, they are best exposed along Highway 6 east of Montgomery Pass, along the upper portions of the old railroad grade on both sides of the pass, and along the Coaldale

fault east of the pass. Geochemical and isotopic analyses indicate these rocks are in fact basaltic andesites with $^{87}Sr/^{86}Sr$ initial values of 0.7062–0.7067 (Winters et al., 2003; Tincher, 2005). The basaltic andesites are predominantly made up of stacked lava flows separated by highly vesicular basaltic flow-top breccias and some scoria (Winters et al., 2003). The massive, dark gray to black basaltic andesite flows contain abundant phenocrysts of clinopyroxene, olivine, and plagioclase in an aphanitic matrix composed of glass, trachytic plagioclase, and magnetite. Individual flow units are ~5–15 m thick, with a variable total unit thickness between 20 and 50 m.

New $^{40}Ar/^{39}Ar$ whole-rock age constraints on three samples from eastern Queen Valley yield $^{40}Ar/^{39}Ar$ ages of 3.14 ± 0.03 Ma (upper Queen Canyon), 3.35 ± 0.15 Ma (east Montgomery Pass), and 4.08 ± 0.10 Ma (west Montgomery Pass) (Fig. 6; Table 1). The sample from upper Queen Canyon was also dated by magnetite (U-Th-[Sm])/He dating, which gave an age of 3.3 ± 0.3 Ma (Blackburn et al., 2007). These ages corroborate the Pliocene age assigned to these rocks by Crowder et al. (1972) and are similar to basalt and basaltic andesite found throughout the western portion of the Mina deflection and the northeastern White Mountains (e.g., Stockli et al., 2003).

In the southeastern most Queen Valley area, Pliocene basaltic andesite at Mustang Point is overlain by even younger rhyolitic ash-flow tuffs (Trt on Fig. 6). Stockli et al. (2003) mapped and dated similar rhyolitic tuffs and plugs from the northeastern White Mountains with $^{40}Ar/^{39}Ar$ ages of ca. 3.0–2.8 Ma.

STRUCTURAL GEOLOGY AND FAULT KINEMATICS

The area surrounding eastern Queen Valley was mapped by the U.S. Geological Survey and published as a 15 quadrangle map by Crowder et al. (1972). Building on these efforts, we carried out detailed geological and structural mapping (1:10,000), fault kinematic analyses, and new geochronology to reconstruct the structural and kinematic evolution of the eastern Queen Valley area in the western Mina deflection. Our main structural and fault kinematic focus was the main strands of the Queen Valley fault and the Coaldale fault, in addition to subsidiary and older, truncated faults in eastern Queen Valley (Fig. 3). The Queen Valley fault is located at the northern termination of the Owens Valley–White Mountains fault zone, crossing the California-Nevada border. The smooth curvilinear trace of the Queen Valley fault marks an abrupt change from the NW-trending White Mountain fault zone to a NE-SW orientation along the northern White Mountains (Stockli et al., 2003). In eastern Queen Valley, the Queen Valley fault is characterized by prominent Holocene fault scarps and fault strands transferring displacement northward into the basin (Stockli et al., 2003; Lee et al., 2008). Along the eastern margin of Queen Valley, the basin-bounding fault changes orientation again, swinging into an ~N-S orientation as it approaches the transcurrent Coaldale fault to the west of Montgomery Pass (Fig. 7). The left-lateral Coaldale fault occupies the Montgomery Pass depression and defines the northern margin of Queen Valley and loses total fault offset westward, dissipating displacement southward into the basin (this study; Lee et al., 2008).

The following sections present the structures, fault kinematics, and timing of faulting as deduced from crosscutting relationships with Tertiary volcanic rocks for eastern Queen Valley subdivided into five geographic areas: Queen Canyon, Sugarloaf Mountain, eastern Queen Valley, northern Queen Valley, and the Mount Montgomery area (Figs. 6 and 8).

Queen Canyon Area

Along the northern margin of the White Mountains the Queen Valley fault swings from a northwesterly orientation into an ENE orientation, forming the southern margin of the Queen Valley pull-apart basin (Fig. 7). Fault-kinematic indicators show a consistent NW-directed extension direction for all segments of the fault, regardless of the absolute orientation of the fault (Stockli et al., 2003). In the Queen Canyon area, the main strand of the Queen Valley fault system continues to form the range-front fault, while several subsidiary fault strands, characterized by Holocene fault scarps, feather out into Queen Valley (Lee et al., 2008). Total displacement along the Queen Valley fault decreases from SW to NE from >1.5 km to <200 m (Stockli et al., 2003; this study), clearly illustrating that displacement is progressively transferred away from the main strand of the Queen Valley fault along these subsidiary NNE-trending faults out into and likely across the Queen Valley pull-apart basin. Farther to the east, the NE-trending main strand of the Queen Valley normal fault is linked to the more NNE-trending segment along the eastern margin of Queen Valley via an E-W–oriented right step at the mouth of Queen Canyon (Fig. 7).

In the hanging wall, steeply NE-dipping Oligocene Sugarloaf Canyon rhyolite and middle Miocene andesite are juxtaposed against Paleozoic strata along the main strand of the basin-bounding Queen Valley fault. Fault kinematic indicators preserved in the down-faulted rhyolite in the hanging wall indicate localized reverse faulting along fault planes with an orientation of ~N45°W, roughly perpendicular to the Queen Valley normal fault at this location (Fig. 7). The localized reverse faulting is easily explained either by deformation in the right-stepping geometry of the Queen Valley fault at the mouth of Queen Canyon, creating a small restraining bend, or by hanging-wall shortening attributable to a localized space problem between two normal faults with different orientations (Fig. 7).

Several NE-striking normal faults cut alluvial-fan deposits at the mouth of Queen Canyon (Crowder et al., 1972; Garwood et al., 2004; Lee et al., 2008). These faults seem to bypass the contractional right step of the Queen Valley fault, down-dropping Miocene andesite and offsetting Pleistocene alluvial-fan deposits at the mouth of Queen Canyon. Fault kinematic data from these faults in the volcanic rocks suggest a down-to-the-NW displacement, accommodating extension in the direction of ~335°, consistent with data from Stockli et al. (2003).

Sugarloaf Mountain Area

Along the eastern margin of the Queen Valley pull-apart basin, the footwall of the NNE-trending main strand of the Queen Valley fault exposes ~350 vertical meters of Jurassic intrusive rocks and an intact sequence of Oligocene Brownie Creek rhyolite, Miocene Truman Meadows andesite, and Pliocene Sugarloaf Canyon rhyolite, along with the basal conglomerate, in its exhumed footwall (Fig. 7). Reconstruction of offset Oligocene Brownie Creek rhyolite would require a minimum of ~350 m of throw across the Queen Valley fault (Figs. 7 and 8). Only sparse fault kinematic information is available from this segment of the Queen Valley fault, but it suggests predominantly normal displacement with a minor right-lateral component. At Sugarloaf Mountain, Pliocene rhyolites dip ~10°–15° toward the northeast, recording the rotational tilt that has occurred in the eastern extent of the Queen Valley fault footwall since the inception of the pull-apart basin. Near Brownie Creek, the orientation of the

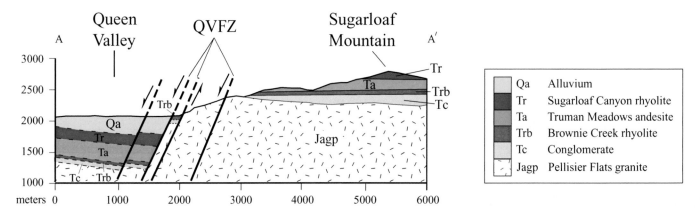

Figure 8. Geologic cross section of the southeastern margin of the Queen Valley pull-apart basin, showing offset volcanic stratigraphy and estimated Pliocene throw. Basin depth is based on gravity data (Black et al., 2005). Arrows indicate relative motion of the normal fault blocks. See Figure 2 for cross-sectional trace. QVFZ—Queen Valley fault zone.

main strand of the Queen Valley fault begins to change to a more northerly orientation.

In the footwall, we mapped several ENE-trending, top-down-to-the-NW normal faults that sequentially down-drop the Tertiary basal unconformity and repeat the Tertiary section. These faults appear to be truncated by the range-bounding Queen Valley fault. These down-to-the-NW faults offset the Mesozoic intrusive rocks and the lower volcanic sequence but are sealed by the early Pliocene Sugarloaf Canyon rhyolites (Tr_1 and Tr_2). Fault kinematic indicators preserved in the basal conglomerates and Oligocene tuffs show both normal displacement and right-lateral displacement along this fault.

Eastern Queen Valley

In eastern Queen Valley, the complexly faulted Tertiary volcanic sequence is excellently exposed along an old railroad grade (Fig. 3). Similar to the Sugarloaf Mountain area (see previous), several ENE-trending faults are truncated by the main strand of the N-S–trending, basin-bounding Queen Valley fault. However, unlike in the Sugarloaf Mountain area, here ENE-trending faults are characterized by both major down-to-the-NW and down-to-the-SE normal faults, forming graben-like structures and revealing some very interesting geological relationships (Fig. 7).

The Tertiary basalt conglomerate and the Oligocene Brownie Creek rhyolite are dissected by a multitude of ENE- to E-trending faults that tend to preserve rich fault kinematic information. While most ENE-trending faults offset both the lower and upper contact of the Oligocene rhyolite, several more E-trending faults clearly are sealed by the overlying middle Miocene andesite or die out upward with the rhyolite. These faults are characterized by largely normal displacement with mean slickenline orientations of 028°/50°. The stratigraphic relationships indicate that these faults were active prior to the deposition of middle Miocene andesite, but after eruption of the Oligocene rhyolite (<25.8 Ma).

Where observed, all ENE-trending faults offset the basal contact of the middle Miocene andesite and the andesite itself but are sealed by the early Pliocene rhyolite. Fault planes and slickenlines are well preserved along these ENE-trending faults, both in the Oligocene and Miocene volcanic rocks (Fig. 7). Fault kinematic data show an average fault-plane orientation of S62°W/76° (Fig. 7). These high-angle (~75°–80°) faults offset Oligocene Brownie Creek rhyolite and Miocene Truman Meadows andesite and yield striations (060°/35°) that indicate right-normal oblique motion. The fault kinematic information suggests that these faults are characterized by ~20% more right-lateral displacement than the older faults, which are sealed by the middle Miocene andesite. The stratigraphic relationships nicely constrain the timing of activity on these predominantly right-lateral normal oblique faults to between ca. 12 and 5.5 Ma.

The main strand of the Queen Valley fault truncates the entire faulted Tertiary volcanic section as well as the approximately ENE-trending faults. Farther north, southwest of Montgomery Pass, the exposed hanging wall of the Queen Valley fault is composed of Tertiary conglomerate overlain by Miocene andesite and Pliocene basaltic andesite (Fig. 7), exposing the same strata in the footwall, and thus limiting the apparent offset to <20 m. The Queen Valley fault continues to the north, where it is intersected or truncated by the approximately E-W–trending active left-lateral Coaldale fault (Fig. 9). The Queen Valley fault appears to have very little displacement left at the intersection with the Coaldale fault, and there is no evidence of any normal faulting associated with the Queen Valley fault on the north side of the Coaldale fault.

Northern Queen Valley (Western Coaldale Fault)

The northwestern margin of the Queen Valley depression is structurally controlled by the Coaldale fault, where it is characterized by a complex, ~500-m-wide zone of left-lateral fault strands that merge eastward into one main strand near Montgomery Pass (Fig. 9). These different fault strands offset Cambrian phyllites and marbles, as well as Ordovician slates and the overlying andesite- and basalt-dominated Tertiary cover sequence. The northernmost strand of this fault zone is well-exposed in a canyon with polished fault planes up to 5 m high, and it provides insight into the orientation of the Coaldale fault in this area (Fig. 9). Fault striations and calcite slickenfibers are well preserved with the phyllite and marbles and are found abundantly throughout a 4–5 km segment of the fault in this portion of the study area. The mean orientation of the fault plane is N57°E/89° ($n = 26$), and fault lineations (56°/28°) indicate left-lateral motion with an ~30% oblique component (Fig. 9). The normal component of slip on this left-lateral fault exposes Paleozoic units on the north side of the fault.

About 1 km southwest of Montgomery Pass, the northern strand of the fault abruptly merges with the southern strand of the Coaldale fault through a contractional right step in the fault geometry, creating a local restraining bend (Fig. 9). Cambrian phyllites and Miocene andesites are complexly folded and deformed in this zone, where small-scale fold axes plunge in the direction of N60°W. The fault strands offset both Miocene andesite and Pliocene basalt, but the stratigraphic relationships do not allow us to constrain the onset of faulting along the westernmost strands of the Coaldale fault or the timing of deformation in the restraining bend. However, fault scarps in alluvial-fan deposits along the northern margin of Queen Valley attest to the Holocene activity of the Coaldale fault (Lee et al., 2008).

Montgomery Pass Area

The NE-trending left-lateral Coaldale fault occupies the linear depression of the Montgomery Pass area, where it is characterized by a single, straight fault strand (Fig. 9). The fault cuts through Miocene andesite and Pliocene basaltic andesite and is generally poorly exposed. Farther to the east, Bradley (2005) mapped fault offsets along the central Coaldale fault and determined ~1.1 km of left-lateral offset since inception of faulting at

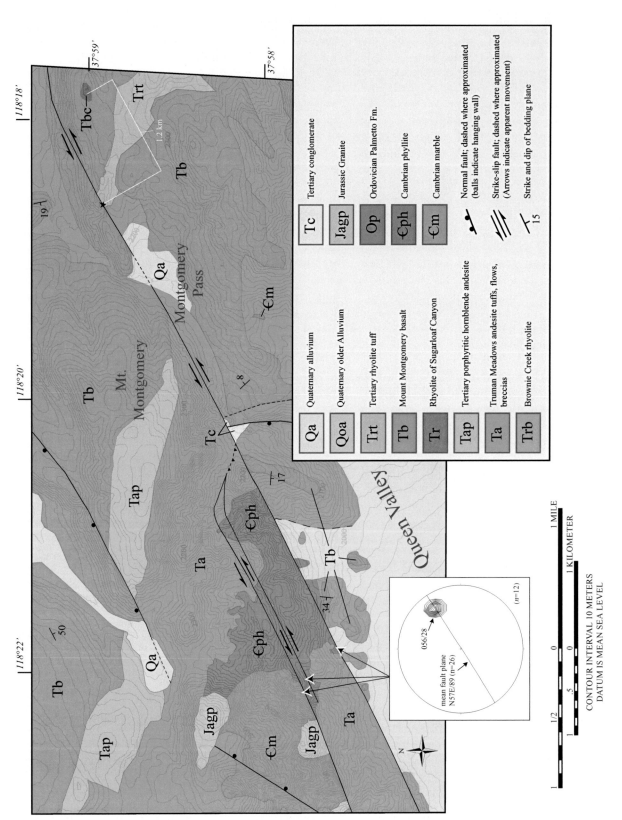

Figure 9. Geologic map of the northeastern corner of Queen Valley and the Montgomery Pass area, dominated by the western strands of the Coaldale fault and the intersection with the Queen Valley fault. While the central Coaldale fault, east of Montgomery Pass, is characterized by a single major fault strand (Bradley, 2005), it exhibits multiple NE-trending fault strands within an ~500-m-wide zone of left-lateral faults, offsetting Paleozoic metasedimentary rocks west of Montgomery Pass. Basaltic scoria is found within a small area surrounding a source vent, marked as unit Tbc. Preserved in fault gouge along the Coaldale fault, pieces of scoria suggest ~1.2 km of offset. The $^{40}Ar/^{39}Ar$ age of the basaltic andesite of Mount Montgomery is 3.14 ± 0.03 Ma, indicating an average slip rate of ~0.4 mm/yr since ca. 3 Ma.

ca. 3 Ma. In the Montgomery Pass area, the Coaldale fault cuts ca. 3 Ma basaltic andesite and appears to offset very conspicuous brick-red scoria from a local scoria vent by ~1.2 km (Fig. 9). This offset scoria represents a relatively low-confidence offset marker, but it is in remarkably good agreement with other offset constraints (Bradley, 2005; Lee et al., 2005). In order to bracket the timing of movement and constrain a minimum slip rate along the Coaldale fault, a $^{40}Ar/^{39}Ar$ age of 3.14 ± 0.03 Ma was obtained from a basaltic andesite near the scoria vent, suggesting minimum displacement rate of 0.4 mm/yr along the Coaldale fault near Montgomery Pass. However, in contrast to the western termination of the Coaldale fault, no evidence for Holocene faulting was found in Quaternary alluvium at Montgomery Pass.

CENOZOIC EVOLUTION OF THE QUEEN VALLEY AREA

Late Tertiary volcanism and faulting in the Queen Valley area constrain three main periods of tectonic activity and faulting as schematically summarized in Figure 9. These different tectonic episodes document important changes in regional kinematics and strain field through time in the western Mina deflection that could only be constrained by detailed geological and structural mapping in combination with new (U-Th)/He and $^{40}Ar/^{39}Ar$ geochronological data presented in this study. The following sections discuss the three different tectonic regimes that have controlled brittle deformation in the Queen Valley area since the Oligocene. Although evidence for structural style, fault kinematics, and timing of deformation is strongest for middle Miocene synextensional strike-slip faulting and Pliocene transtensional, pull-apart deformation, the discussion of tectonic events is chronologically organized and therefore first discusses the implications of Oligocene-Miocene faulting in the Montgomery Pass region.

Minor Oligocene-Miocene Extension

The area east of Queen Valley in the footwall of the Queen Valley fault is riddled with a series of E- to ENE-trending normal- to oblique-slip faults (Figs. 7 and 10). While most faults offset the entire Oligocene-Miocene volcanic stack, several faults only offset basal Tertiary conglomerate and Oligocene rhyolite and are sealed by the middle Miocene andesite. This represents the strongest evidence for the existence of Oligocene to early Miocene normal-oblique faulting in the area. Along the abandoned Carson and Colorado Railroad grade, normal faults with a minor right-lateral oblique component juxtapose ca. 25.8 Ma Brownie Creek ash-flow tuffs against Jurassic granites and are sealed by ca. 12 Ma Truman Meadows andesite. This observation suggests that minor normal movement along approximately ENE-trending faults occurred either synchronous with voluminous eruption of Oligocene ash-flow tuffs or that minor deformation in the right-lateral accommodation zone between the middle Miocene extensional domains in the Wassuk Range and White Mountains (Stockli et al., 2003) predates the eruptions of voluminous Miocene andesite

at ca. 12 Ma. The magnitude of faulting remains difficult to constrain given the limited exposure, but previous work by Oldow (1992) has documented the existence of Oligocene E-W–trending half grabens in the northern Walker Lane belt.

Middle Miocene Accommodation Zone

In the same area of eastern Queen Valley, a majority of the ENE-trending faults offsets both Oligocene Brownie Creek rhyolite and Miocene Truman Meadows andesite, but in turn these faults are sealed by the Miocene-Pliocene Sugarloaf Canyon rhyolite (ca. 5.5 Ma). These high-angle faults (>75°) are characterized by right-lateral displacement and a minor to moderate north-side-down normal component with ten to hundreds of meters of throw (Fig. 7). They record right-lateral strike-slip faulting at the northern end of the White Mountains between ca. 12 Ma and 5.5 Ma and appear to have been active at the same time as top-down-to-the-west normal faulting along the White Mountains fault, accommodating middle Miocene E-W extension (Stockli et al., 2003). The lack of a volcanic record between ca. 8.5 Ma and 5.5 Ma precludes a more precise determination of the cessation of right-lateral faulting in these areas. Clear structural evidence for right-lateral displacement along these faults in the middle Miocene lends strong support to the hypothesis that the Mina deflection acted as a Miocene accommodation zone (Stockli et al., 2003). The timing and kinematics are consistent with these E-W–striking faults acting as an antithetic accommodation zone in the western Mina deflection, linking middle Miocene extensional domains with opposite fault polarities in the White Mountains and the Wassuk Range region (Fig. 10). Stewart (1985) documented pre-Pliocene right-lateral displacement along E-W–trending faults in the Mina deflection that are currently characterized by left-lateral post-Miocene displacement (Stewart, 1985). Despite the lack of geological constraints, it seems possible that the ENE-trending left-lateral Coaldale fault had a history of right-lateral displacement prior to the onset of left-lateral motion after ca. 3 Ma.

Pliocene Extension and Strike-Slip Faulting

At its northern end, the White Mountains are bounded by the NW-dipping Queen Valley fault, which acts as right-stepping fault zone that transfers displacement from the Owens Valley fault system eastward via the Coaldale fault system into the southern central Walker Lane belt (Fig. 10). The geometrically right-lateral normal-oblique White Mountains fault smoothly swings around the NW corner of the White Mountain into a NE orientation, while the fault kinematics transition from right-lateral normal-oblique to normal along the Queen Valley fault (Stockli et al., 2003; Kirby et al., 2006; Lee et al., 2008). In the Queen Valley pull-apart basin, the bounding normal fault is geometrically complex with an overall en-echelon geometry or right step at the mouth of Queen Canyon, resulting in minor contraction in the hanging wall. On the eastern margin of the

A

< 25.75 Ma ~N-S extension creates normal faults that offset Oligocene rhyolitic ashflow tuffs.

B

10-15 Ma Miocene ~E-W extension requires right-lateral accommodation in the present-day Mina deflection.

C

ca. 3 Ma Pliocene reactivation of the White Mountain fault zone creates the Queen Valley pull-apart basin at the northern extent of the White Mountains.

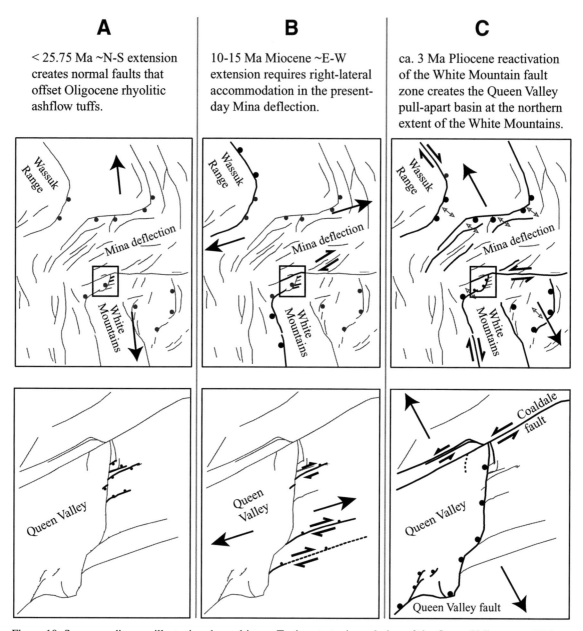

Figure 10. Summary diagram illustrating the multistage Tertiary tectonic evolution of the Queen Valley area. (A) Late Oligocene to early Miocene faulting offsets ca. 26 Ma Upper Oligocene Brownie Creek rhyolite. Previous studies suggest faulting could be related to Oligocene regional N-S extension contemporaneous with major ash-flow eruptions. (B) The Mina deflection acts as a right-lateral accommodation zone from ca. 15–10 Ma, linking normal faults to the north and south with opposite orientations. (C) Reactivation of the White Mountains–Owens Valley fault zone as a right-lateral fault ca. 3 Ma creates the Queen Valley pull-apart basin. The lower figures represent a detailed fault map of the area indicated by the box in the upper figures. The bold lines represent faults that are active during the specified time period.

basin, the Queen Valley fault swings back into a nearly N-S orientation. Along the entire trace of the fault, the total displacement along the Queen Valley fault gradually decreases from >1500 m to <100 m, where it is truncated by the Coaldale fault. We believe that the dissipation of throw is accomplished by transferring slip along splay faults out and across Queen Valley and onto the Coaldale fault (Fig. 11).

The timing of onset of faulting along the Queen Valley fault appears to postdate the entire Tertiary volcanic sequence. The fault also truncates middle Miocene ENE-trending right-lateral normal-oblique faults with no evidence for reactivation on Pliocene times. These observations are in good agreement with thermochronometric constraints, which date the onset of the Queen Valley pull-apart basin formation at ca. 3 Ma as a result of the reactivation of the White Mountains–Owens Valley fault zone as a right-lateral fault (Stockli et al., 2003).

The northern margin of Queen Valley and the linear depression across Montgomery Pass are structurally controlled by the left-lateral Coaldale fault, which transfers displacement from Queen Valley into Columbus Salt Marsh (Fig. 2). Fault kinematic data suggest almost pure left-lateral motion and a minor normal component west of Montgomery Pass. In the study area, offset Pliocene basaltic andesites suggest a post-Miocene age for the Coaldale fault with ~1.2 km of left-lateral displacement. These observations are corroborated by studies farther east along the Coaldale fault, which have determined ~1.1 km of displacement along the Coaldale fault since the onset of faulting at ca. 3 Ma (Bradley, 2005). These constraints yield a minimum time-averaged fault slip rate of ~0.4 mm/yr along the Coaldale fault.

Figure 11 displays the resolved horizontal displacement estimates of the Queen Valley and Coaldale faults since ca. 3 Ma. Stockli et al. (2003) estimated that between 1.5 and 2.5 km of net slip has occurred on the Queen Valley fault west of the study near the apex of the northwestern White Mountains. Assuming a 60° fault plane, this translates into 850–1450 m of horizontal displacement. The main strand of the Queen Valley fault near Queen Canyon accounts for less than 200 m of horizontal displacement, and this amount decreases toward the northern extent of the Queen Valley fault. This suggests that 650–1000 m of strain must be accommodated through other structures in the Queen Valley area. Lee et al. (2008) suggested that ~150 m of horizontal displacement occurred along fault splays off the Queen Valley fault as recorded by Holocene fault scarps. These intravalley faults probably feed into the western extent of the Coaldale fault, west of Montgomery Pass. The Coaldale fault, therefore, would accommodate an increasing amount of displacement toward Montgomery Pass, while the Queen Valley fault accommodates a decreasing amount of displacement (Fig. 10). Balancing resolved horizontal displacement vectors accounting for ~1.2 km of horizontal displacement along the Coaldale fault east of Montgomery Pass (Bradley, 2005; this study) requires at least 1 km of horizontal extension in Queen Valley, which is good agreement with our estimates (Fig. 11).

CONCLUSIONS

This study presents structural, fault kinematic, and geochronological evidence for three-stage tectonic evolution of the Queen Valley area since the Oligocene. More importantly, however, it elucidates the complex structural link and interplay between the extensional Queen Valley fault and the left-lateral Coaldale fault in transferring displacement from the western portion of the Eastern California shear zone (Owens Valley–White Mountain fault zone) into the regional releasing bend of the Mina deflection in the central Walker Lane belt. New structural, stratigraphic, and geochronological data are used to illustrate the volcanic and faulting history of the eastern Queen Valley area over the past ~30 m.y.

During or immediately after the eruption of voluminous Oligocene rhyolitic ash-flow tuffs, the area of eastern Queen Valley appears to have experienced minor N-S extension, based on very limited exposures. ENE-trending normal faults, with minor normal components, juxtapose basal Tertiary conglomerates and Oligocene Brownie Creek rhyolite (ca. 25.8 Ma) against the Jurassic Pellisier Flats pluton, and they are sealed by middle Miocene Truman Meadows andesite (ca. 12 Ma) (Fig. 7).

A second set of faults with an ENE orientation in eastern Queen Valley, between Montgomery Pass and Queen Canyon, is characterized by right-lateral fault kinematics and a minor northside-down normal component. These younger faults are sealed by latest Miocene to early Pliocene Sugarloaf Canyon rhyolitic ash-flow tuffs (ca. 5.5 Ma), constraining the timing of slip along these faults to between ca. 12 and 5.5 Ma. Middle Miocene E-W extension throughout the central Walker Lane belt created E- and W-dipping normal faults between 15 and 10 Ma in the Wassuk Range and White Mountain areas, respectively (e.g., Dilles and Gans, 1995; Stockli et al., 2002, 2003). During this time, the Mina deflection area acted as an accommodation zone between the two extensional domains of opposite polarity (Stockli et al., 2003) (Fig. 10).

A change in regional extension direction from approximately E-W extension to approximately NW-SE extension resulted in a transition to transcurrent and transtensional structures in the central Walker Lane. On the east side of the White Mountains, the beginning of transtension was marked by volcanism and right-lateral displacement in Fish Lake Valley at ca. 6 Ma (Reheis and Sawyer, 1997; Stockli et al., 2003). This dextral movement created the Fish Lake Valley pull-apart basin in the southern Mina deflection. Similarly, the Queen Valley pull-apart basin was a product of the reactivation of the White Mountain–Owens Valley fault zone as a right-lateral fault ca. 3 Ma (Stockli et al., 2003). Our detailed mapping documents the structural style and interaction between the extensional, pull-apart-bounding Queen Valley fault and the left-lateral Coaldale fault during the Pliocene and ongoing transtension and displacement transfer from the Eastern California shear zone into the Mina deflection of the central Walker Lane. Throw cross the Queen Valley fault decreases from SW to NW and appears to be

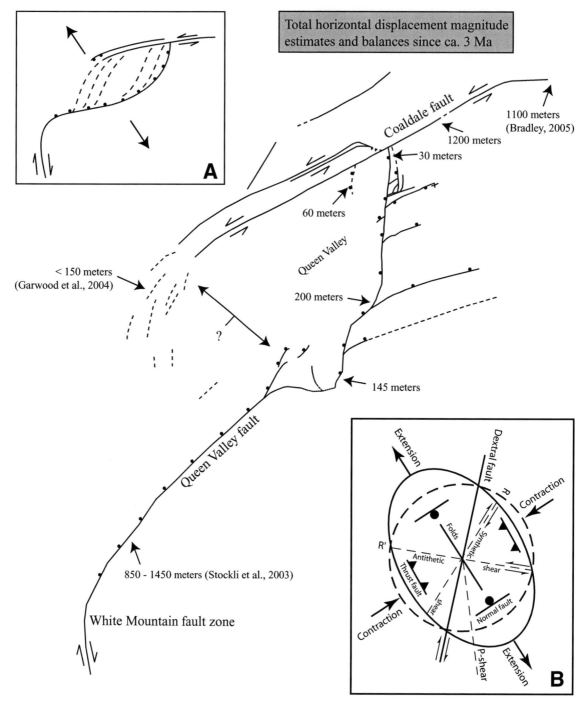

Figure 11. Schematic diagram showing the estimated resolved horizontal displacement for different portions of the Queen Valley fault and the western Coaldale fault. Stockli et al. (2003) reported an estimate of 1.5–2.5 km of net slip along the Queen Valley fault southwest of Queen Canyon, which translates into ~850–1450 m of horizontal displacement (assuming a 60° fault plane). This value agrees with horizontal displacement values of 1.1 km (this study) and 1.2 km (Bradley, 2005) reported east of Montgomery Pass along the Coaldale fault. Since the Queen Valley fault near Queen Canyon only accommodates a small percentage of this strain, a large portion must be transferred through intravalley faults, such as those reported by Garwood et al. (2004) and Lee et al. (2008). (A) A simplified model of strain accommodation through the Queen Valley area. (B) A shear diagram representing the regional extension direction of ~N35°W.

transferred into the basin by NNE-trending fault splays, relaying the displacement across the pull-apart onto the Coaldale fault. This geometry is also illustrated by the change in orientation of the main strand of the Queen Valley fault at the eastern basin margin, where the fault abruptly changes orientation from NE to nearly N-S and ultimately intersects the Coaldale fault just west of Montgomery Pass (Figs. 9 and 11).

ACKNOWLEDGMENTS

We would like to thank Kelley Wooten, Terrence Blackburn, and Stacy Rosner, who assisted with the geologic mapping and at the University of Kansas (U-Th)/He Thermochronometry Laboratory. Discussions with Jeff Lee, John Gosse, Ross Black, and David Bradley greatly contributed to this work. John Oldow, Eric Kirby, and Mike Oskin provided thoughtful reviews. Funding for this project was provided by National Science Foundation grants EAR-0125879 awarded to D. Stockli, EAR-0125782 awarded to J. Lee, and fellowships from the University of California White Mountains Research Station and University of Kansas Geology summer research to C. Tincher.

REFERENCES CITED

Anderson, G.H., 1937, Granitization, albitization, and related phenomena in the northern Inyo Range of California-Nevada: Geological Society of America Bulletin, v. 48, p. 1–74.

Biswas, S., Coutand, I., Grujic, D., Hager, C., Stockli, D.F., and Grasemann, B., 2007, Exhumation of the Shillong plateau and its influence on east Himalayan tectonics: Tectonics, v. 26, p. TC6013, doi: 10.1029/2007TC002125.

Black, R.A., Stockli, D.F., Christie, M., Desmond, J., Kueker, A., Hadley, M., Tincher, C., and Casteel, J., 2005, Gravity study of the Queen Valley pull-apart basin, western Nevada: Eos (Transactions, American Geophysical Union), v. 86, no. 52, Fall Meeting Supplement, abs. T51D-1373.

Blackburn, T.J., Stockli, D.F., and Walker, J.D., 2007, Magnetite (U-Th)/He dating and its application to the geochronology of intermediate to mafic volcanic rocks: Earth and Planetary Science Letters, v. 259, p. 360–371, doi: 10.1016/j.epsl.2007.04.044.

Bradley, D., 2005, The kinematic history of the Coaldale fault, Walker Lane belt, Nevada [M.S. thesis]: Lawrence, University of Kansas, 96 p.

Burchfiel, B.C., and Davis, G.A., 1975, Structural framework and evolution of the southern part of the Cordilleran orogen in the United States: American Journal of Science, v. 272, p. 97–118.

Crowder, D.F., and Sheridan, M.F., 1972, Geologic map of the White Mountains Peak quadrangle, Mono County, California: U.S. Geological Survey Geologic Quadrangle Map GQ-1012, scale 1:62,500.

Crowder, D.F., Robinson, P.T., and Harris, D.L., 1972, Geologic map of the Benton quadrangle, Mono County, California, and Esmeralda and Mineral Counties, Nevada: U.S. Geological Survey Geologic Quadrangle Map GQ-1012, scale 1:62,500.

Dalrymple, G.B., and Hirooka, K., 1965, Variation of potassium, argon, and calculated age in a late Cenozoic basalt: Journal of Geophysical Research, v. 70, no. 20, p. 5291–5296, doi: 10.1029/JZ070i020p05291.

Dilles, J., and Gans, P.B., 1995, The chronology of Cenozoic volcanism and deformation in the Yerington area, western Basin and Range and Walker Lane: Geological Society of America Bulletin, v. 107, p. 474–486, doi: 10.1130/0016-7606(1995)107<0474:TCOCVA>2.3.CO;2.

Ekren, E.B., Byers, F.M., Jr., Hardyman, R.F., Marvin, R.F., and Silberman, M.L., 1980, Stratigraphy, Preliminary Petrology, and Some Structural Features of Tertiary Volcanic Rocks in the Gabbs Valley and Gillis Ranges, Mineral County, Nevada: U.S. Geological Survey Bulletin 1464, 54 p.

Evernden, J.F., and Kistler, R.W., 1970, Chronology of Emplacement of Mesozoic Batholithic Complexes in California and Western Nevada: U.S. Geological Survey Professional Paper 623, 67 p.

Farley, K.A., 2002, (U-Th)/He dating: Techniques, calibrations, and applications, *in* Porcelli, D., Ballentine, C.J., and Wieler, R., eds., Noble Gases in Geochemistry and Cosmochemistry: Reviews in Mineralogy and Geochemistry, v. 47, p. 819–844.

Farley, K.A., Wolf, R.A., and Silver, L.T., 1996, The effects of long alpha-stopping distances on (U-Th)/He ages: Geochimica et Cosmochimica Acta, v. 60, no. 21, p. 4223–4229, doi: 10.1016/S0016-7037(96)00193-7.

Ferranti, L., Oldow, J.S., Geissman, J.W., and Nell, M.M., 2009, this volume, Flattening strain during coordinated slip on a curved fault array, Rhodes Salt Marsh extensional basin, central Walker Lane, west-central Nevada, *in* Oldow, J.S., and Cashman, P.H., eds., Late Cenozoic Structure and Evolution of the Great Basin–Sierra Nevada Transition: Geological Society of America Special Paper 447, doi: 10.1130/2009.2447(11).

Garside, L.J., and Silberman, M.L., 1978, New K-Ar ages of volcanic and plutonic rocks from the Camp Douglas quadrangle, Mineral County, Nevada: Isochron-West, v. 22, p. 29–32.

Garwood, J.D., Lee, J., Stockli, D.F., and Gosse, J., 2004, Fault slip rates along the Queen Valley fault system, California; slip partitioning in the eastern California shear zone/Walker Lane belt: Geological Society of America Abstracts with Programs, v. 36, no. 6, p. 16.

Gilbert, C.M., Christensen, M.N., Al-Rawi, Y., and Lajoie, K.R., 1968, Structural and volcanic history of Mono Basin, California-Nevada, *in* Coats, R.R., Hay, R.L., and Anderson, C.A., eds., Studies in Volcanology: A Memoir In Honor of Howel Williams: Geological Society of America Memoir 116, p. 275–329.

Hardyman, R.F., and Oldow, J.S., 1991, Tertiary tectonic framework and Cenozoic history of the central Walker Lane, Nevada, *in* Raines, G.L., Lisle, R.E., Schafer, R.W., and Wilkinson, W.H., eds., Geology and Ore Deposits of the Great Basin, Symposium Proceedings: Reno, Nevada, Geological Society of Nevada, p. 279–301.

Harris, D.L., 1967, Petrology of the Boundary Peak adamellite pluton, in the Benton quadrangle, Mono and Esmeralda Counties, California and Nevada [M.S. thesis]: Berkeley, University of California, 57 p.

Heizler, M., and Sanders, R., 2005, ^{40}Ar/^{39}Ar Geochronology Results for Samples from Basalt, NV: New Mexico Geochronological Research Laboratory Report NMGRL-IR-427, 26 p.

Kirby, E., Burbank, D.W., Reheis, M., and Phillips, F., 2006, Temporal variations in slip rate of the White Mountain fault zone, eastern California: Earth and Planetary Science Letters, v. 248, p. 168–185, doi: 10.1016/j.epsl.2006.05.026.

Lee, J., Stockli, D.F., Schroeder, J., Tincher, C.R., Bradley, D., and Owen, L., 2005, Fault slip transfer in the Eastern California shear zone/Walker Lane belt, *in* Geological Society of America Penrose Conference of Kinematics and Geodynamics of Intraplate Dextral Shear in Eastern California and Western Nevada, 21–26 April, Mammoth Lakes, California: Boulder, Colorado, Geological Society of America, Field Trip Guidebook, p. 1–26, doi: 10.1130/2006.FSTITE.PFG.

Lee, J., Garwood, J., Stockli, D.F., and Gosse, J., 2008, Quaternary faulting in Queen Valley, California-Nevada: Kinematics of fault-slip transfer in the Eastern California shear zone–Walker Lane belt: Geological Society of America Bulletin, doi: 10.1130/B26352.1 (in press).

Marvin, R.H., Speed, R.C., and Cogbill, A.H., 1977, K-Ar ages of Tertiary igneous and sedimentary rocks of the Mina-Candelaria region, Nevada: Isochron-West, v. 18, p. 9–12.

McKee, E.H., and John, D.A., 1987, Sample locality map and potassium-argon ages and data for Cenozoic igneous rocks in the Tonopah 1° by 2° quadrangle, central Nevada: U.S. Geological Survey Miscellaneous Field Studies Map MF-1877-1, scale 1:250,000.

Oldow, J.S., 1992, Late Cenozoic displacement partitioning in the northwestern Great Basin, *in* Craig, S.D., ed., Structure, Tectonics and Mineralization of the Walker Lane, Walker Lane Symposium Proceedings Volume: Reno, Nevada, Geological Society of Nevada, p. 17–52.

Oldow, J.S., Bally, A.W., Avé Lallemant, H.G., and Leeman, W.P., 1989, Phanerozoic evolution of the North American Cordillera (United States and Canada), *in* Bally, A.W., and Palmer, A.R., eds., The Geology of North America: An Overview: Boulder, Colorado, Geological Society of America, Geology of North America, v. A, p. 139–232.

Oldow, J.S., Geissman, J.W., and Stockli, D.F., 2008, Evolution and strain reorganization within late Neogene structural stepovers linking the central Walker Lane and northern Eastern California shear zone, western Great Basin: International Geology Review, v. 50, p. 270–290, doi: 10.2747/0020-6814.50.3.270.

Petronis, M.S., Geissman, J.W., and McIntosh, W.C., 2004, Transitional field clusters from uppermost Oligocene volcanic rocks in the central Walker Lane, western Nevada: Physics of the Earth and Planetary Interiors, v. 141, p. 207–238, doi: 10.1016/j.pepi.2003.12.004.

Proffett, J.M., Jr., 1977, Cenozoic geology of the Yerington district, Nevada, and implications for the nature and origin of Basin and Range faulting: Geological Society of America Bulletin, v. 88, p. 247–266, doi: 10.1130/0016-7606(1977)88<247:CGOTYD>2.0.CO;2.

Reheis, M.C., and Sawyer, T.L., 1997, Late Cenozoic history and slip rates of the Fish Lake Valley, Emigrant Peak, and Deep Springs fault zones, Nevada and California: Geological Society of America Bulletin, v. 109, p. 280–299, doi: 10.1130/0016-7606(1997)109<0280:LCHASR>2.3.CO;2.

Robinson, P.T., and Stewart, J.H., 1984, Uppermost Oligocene and Lowermost Miocene Ash-Flow Tuffs of Western Nevada: U.S. Geological Survey Bulletin 1557, 53 p.

Robinson, P.T., McKee, E.H., and Moiola, R.J., 1968, Cenozoic volcanism and sedimentation, Silver Peak region, western Nevada and adjacent California, *in* Coats, R.R., Hay, R.L., and Anderson, C.A., eds., Studies in Volcanology: A Memoir In Honor of Howel Williams: Geological Society of America Memoir 116, p. 577–611.

Silberman, M.L., Bonham, H.F., Jr., and Osborne, D.H., 1975, New K-Ar ages of volcanic and plutonic rocks and ore deposits in western Nevada: Isochron-West, v. 13, p. 13–21.

Stewart, J.H., 1972, Initial deposits in the Cordilleran geosyncline: Evidence of a late Precambrian continental separation: Geological Society of America Bulletin, v. 83, p. 1345–1360, doi: 10.1130/0016-7606(1972)83[1345:IDITCG]2.0.CO;2.

Stewart, J.H., 1985, East-trending dextral faults in the western Great Basin; an explanation for anomalous trends of pre-Cenozoic strata and Cenozoic faults: Tectonics, v. 4, p. 547–564, doi: 10.1029/TC004i006p00547.

Stewart, J.H., 1988, Tectonics of the Walker Lane belt, western Great Basin—Mesozoic and Cenozoic deformation in a zone of shear, *in* Ernst, W.G., ed., Metamorphism and Crustal Evolution of the Western United States, Rubey Volume VII: Englewood Cliffs, New Jersey, Prentice-Hall, p. 683–713.

Stewart, J.H., 1992, Walker Lane belt, Nevada and California; an overview, *in* Craig, S.D., ed., Structure, Tectonics and Mineralization of the Walker Lane; Walker Lane Symposium Proceedings Volume: Reno, Geological Society of Nevada, p. 1–16.

Stockli, D.F., and Farley, K.A., 2004, Empirical constraints on the titanite (U-Th)/He partial retention zone from the KTB drill hole: Chemical Geology, v. 207, p. 223–236, doi: 10.1016/j.chemgeo.2004.03.002.

Stockli, D.F., Surpless, B.E., Dumitru, T.A., and Farley, K.A., 2002, Thermochronological constraints on the timing and magnitude of Miocene and Pliocene extension in the central Wassuk Range, western Nevada: Tectonics, v. 21, 1028, doi: 10.1029/2001TC001295.

Stockli, D.F., Dumitru, T.A., McWilliams, M.O., and Farley, K.A., 2003, Cenozoic tectonic evolution of the White Mountains, California and Nevada: Geological Society of America Bulletin, v. 115, no. 7, p. 788–816, doi: 10.1130/0016-7606(2003)115<0788:CTEOTW>2.0.CO;2.

Surpless, B.E., Stockli, D.F., Dumitru, T.A., and Miller, E.L., 2002, Two phase westward encroachment of Basin and Range extension into the northern Sierra Nevada: Tectonics, v. 21, 1002, doi: 10.1029/2000TC001257.

Tincher, C., 2005, Cenozoic volcanism and tectonics in the Queen Valley area, western Nevada [M.S. thesis]: Lawrence, University of Kansas, 156 p.

Wesnousky, S.G., 2005, Active faulting in the Walker Lane: Tectonics, v. 24, p. TC3009, doi: 10.1029/2004TC001645.

Winters, N.D., Stockli, D.F., Bradley, D.B., Stockli, L.D., and Macpherson, G.L., 2003, Chemistry and geochronology of Neogene basalts along the Coaldale fault, central Walker Lane belt, Nevada: Geological Society of America Abstracts with Programs, v. 36, no. 4, p. 37.

Wooten, K., and Stockli, D.F., 2004, Mesozoic magmatism and deformation in the northern White Mountains, California and Nevada: Geological Society of America Abstracts with Programs, v. 36, no. 4, p. 90.

MANUSCRIPT ACCEPTED BY THE SOCIETY 21 JULY 2008

The Geological Society of America
Special Paper 447
2009

Late Miocene to Pliocene synextensional deposition in fault-bounded basins within the upper plate of the western Silver Peak–Lone Mountain extensional complex, west-central Nevada

John S. Oldow*
Department of Geoscience, University of Texas at Dallas, Richardson, Texas 75080-3021, USA

Elizabeth A. Elias
Department of Geological Sciences, University of Idaho, Moscow, Idaho 83844-3022, USA

Luigi Ferranti
Dipartimento di Scienze della Terra, Università di Napoli Federico II, 80138 Napoli, Italy

William C. McClelland
Department of Geosciences, University of Iowa, Iowa City, Iowa 52242, USA

William C. McIntosh
New Mexico Geochronological Research Laboratory, New Mexico Institute of Mining and Technology, Socorro, New Mexico 87801, USA

ABSTRACT

The late Miocene to Pliocene Silver Peak–Lone Mountain extensional complex in the western Great Basin is part of a structural stepover that links dextral transcurrent motion between the Furnace Creek fault system and northwest-striking transcurrent faults in the central Walker Lane. In the Silver Peak Range, the extensional complex is exposed as a west-northwest–trending turtleback structure that consists of a folded detachment fault separating a metamorphic lower-plate assemblage from unmetamorphosed upper-plate rocks. The upper plate preserves structurally attenuated lower Paleozoic carbonate and clastic rocks, upper Oligocene to lower Miocene volcanic rocks, and a synextensional mid-Miocene to Pliocene clastic and volcanic succession. The three-dimensional geometry of fault-bounded extensional basins formed during displacement on the detachment is preserved, and the synextensional units comprise five sequences separated by unconformities. The locus of deposition migrated spatially as dimensions of small basins changed through time. The entire extensional complex is deformed in two generations of late Cenozoic folds. North-northeast–trending folds formed first with axial traces oriented at a high-angle to upper-plate extension, and these are preferentially developed in proximity to thrusts and partially inverted exten-

*Corresponding author e-mail: oldow@utdallas.edu.

Oldow, J.S., Elias, E.A., Ferranti, L., McClelland, W.C., and McIntosh, W.C., 2009, Late Miocene to Pliocene synextensional deposition in fault-bounded basins within the upper plate of the western Silver Peak–Lone Mountain extensional complex, west-central Nevada, *in* Oldow, J.S., and Cashman, P.H., eds., Late Cenozoic Structure and Evolution of the Great Basin–Sierra Nevada Transition: Geological Society of America Special Paper 447, p. 275–312, doi: 10.1130/2009.2447(14). For permission to copy, contact editing@geosociety.org. ©2009 The Geological Society of America. All rights reserved.

sional faults. Younger, west-northwest–trending folds parallel the axis of the turtleback structure and involve all lithologic units, several syndepositional high-angle faults, and the basal detachment. Pliocene growth of west-northwest–trending folds marked the end of slip on the exposed parts of the basal décollement and the cessation of deposition in the fault-bounded basins. Today, upper and lower plates of the extensional complex are dissected by north- to northeast-striking normal faults that cut alluvium and cross-cut and locally reactivate earlier Cenozoic structures.

Keywords: synextensional deposition, extensional core complex, extensional stepover, central Walker Lane, extensional exhumation and deposition.

INTRODUCTION

The geologic record of near-surface deformation associated with large-magnitude extensional core complexes is rarely preserved. Typically, upper-plate assemblages are highly disarticulated, and the spatial relation between synextensional deposits and coeval structures is lost. Using remnants of sedimentary and

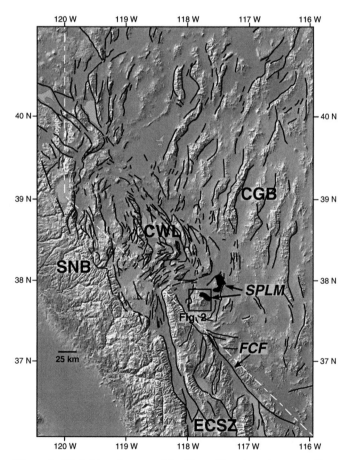

Figure 1. Digital terrain map of the Sierra Nevada and western Great Basin showing late Cenozoic faults (black). CGB—central Great Basin; CWL—central Walker Lane; ECSZ—Eastern California shear zone; SNB—Sierra Nevada block; FCF—Furnace Creek fault; SPLM—Silver Peak–Lone Mountain metamorphic core complex.

volcanic successions deposited during extension, the architecture of basins active during displacement and fragmentation of upper-plate assemblages has been pieced together (Proffett and Proffett, 1976; Gans et al., 1989; Ingersoll et al., 1996; Stewart and Diamond, 1990; Fillmore and Walker, 1996; Miller and John, 1988, 1999) to produce largely two-dimensional conceptual models of upper-plate morphology. Typically, synextensional deposition is viewed as localized in half-graben basins bounded by listric normal faults (Christie-Blick and Biddle, 1985; Gibbs, 1989, 1990). Support for these models is provided by investigations of extensional basins formed and preserved in areas of relatively modest extension (Effimoff and Pinezich, 1986; Strecker et al., 1996). Although the geometry of basins formed in low-magnitude extensional systems provides insight into the link between structures and basin growth, extrapolation to conditions found in systems of hyperextension are largely speculative. A significant disconnect remains in understanding the three-dimensional architecture and structural links of near-surface fault-controlled basins and their relation to large-displacement detachment faults related to exhumation of deep crustal rocks.

The geometry of synextensional basins is preserved in the late Miocene to Pliocene Silver Peak–Lone Mountain extensional complex (Fig. 1) of west-central Nevada. The extensional complex consists of a broad belt of metamorphic core complexes formed in a structural stepover that links right-lateral strike-slip faults of the northwest-trending Furnace Creek fault system in eastern California with faults of the central Walker Lane in Nevada (Oldow, 1992; Oldow et al., 1994, 2008). The stepover exposes an upper-plate assemblage separated from underlying metamorphic tectonites by a folded, shallowly dipping detachment fault and provides a rare view of the structural contact between lower-plate rocks and a highly extended but structurally intact extensional allochthon (Fig. 2). Within the extensional allochthon, the three-dimensional geometry of upper-plate faults and synorogenic basins is preserved.

In this paper, we present the results of several years of work in the Silver Peak Range (Prestia and Oldow, 1998, 2000; Oldow et al., 1994, 2003, 2008; Elias, 2005) and outline in detail the lithology, stratigraphy, and structure of synorogenic basin deposits exposed in the upper plate of the extensional complex. The stratigraphy of the synextensional rocks is complicated by profound spatial variation in lithology and thickness, but the exposures in the central

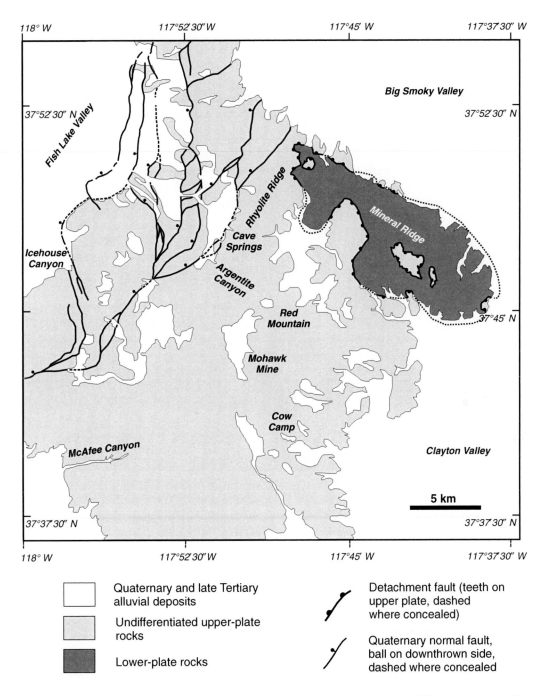

Figure 2. Generalized map of the Silver Peak Range, showing Quaternary faults, undifferentiated upper-plate and lower-plate rocks, detachment faults, and geographic locations mentioned in text.

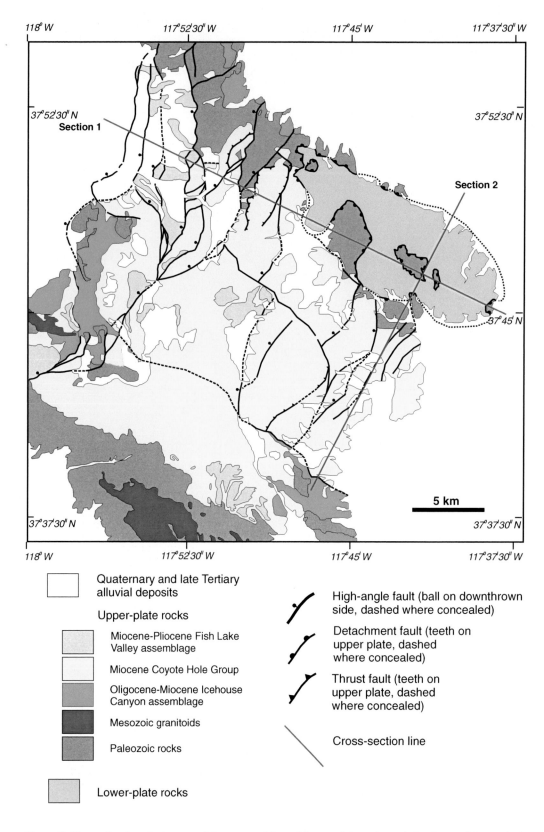

Figure 3. Generalized geologic map of the upper-plate and lower-plate rocks and structures of the central Silver Peak Range. (*Continued on following page.*)

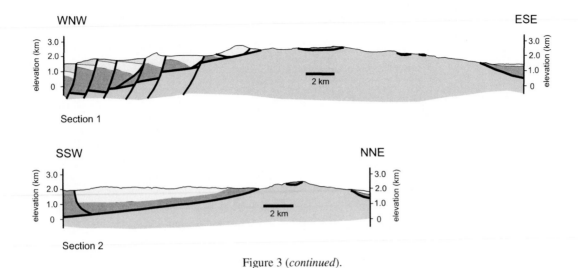

Figure 3 (*continued*).

Silver Peak Range preserve, to an unprecedented degree, the three-dimensional geometry of fault-bounded basins that formed during progressive extensional disarticulation of the upper-plate assemblage. The rocks of the upper plate are deformed both in extensional and contractional structures. We document that, over time, the locus of extension that controlled deposition migrated spatially and changed the dimensions of the upper-plate basins. Furthermore, in several instances, partial basin inversion occurred late in the depositional history, and it was accommodated by reversal of motion on pre-existing extensional faults, by the formation of new thrust faults, and by folding localized in the vicinity of the contractional faults. Deposition within the extensional basins continued through the late Miocene into the early Pliocene and ended when long-wavelength, west-northwest–trending folds that involved rocks of both upper- and lower-plate assemblages prohibited further motion on the underlying detachment.

REGIONAL GEOLOGIC FRAMEWORK

The Silver Peak–Lone Mountain extensional complex (Oldow, 1992; Oldow et al., 1994) of west-central Nevada (Fig. 1) formed in a structural stepover linking a system of Neogene faults that separate the Sierra Nevada and the central Great Basin. The array of northwest-trending faults forms part of the active structures constituting the northern part of the Eastern California shear zone (Dokka and Travis, 1990) and central Walker Lane (Locke et al., 1949), which accommodate displacement transferred inboard from the plate margin. Today, the northwest-striking transcurrent faults take up 20% to 25% of relative motion between the Pacific and North American plates (Dokka and Travis, 1990; Hardyman and Oldow, 1991; Dixon et al., 1995; Oldow, 1992; Bennett et al., 2003) in a complex zone of transtensional deformation (Unruh et al., 2003; Oldow, 2003; Dewey, 2003; Oldow et al., 2008).

At the latitude of the central Sierra Nevada, the active transtensional belt widens, and, from south to north, the northwest-

striking transcurrent faults are stepped to the east via a system of east-northeast–striking faults (Fig. 1). In the Miocene and Pliocene, the belt of deformation was more spatially restricted (Oldow, 1992), and displacement was transferred from the northern end of the Furnace Creek fault system to northwest-striking transcurrent faults of the central Walker Lane, resulting in crustal disarticulation on a shallowly north-dipping detachment fault system (Oldow et al., 1994, 2008). Extension within the stepover resulted in the exposure of metamorphic tectonites structurally overlain by a highly extended assemblage of Tertiary and pre-Tertiary volcanic, volcanogenic, and sedimentary rocks (Fig. 2).

Within the extensional complex, northwest transport of the upper-plate assemblage (Stewart and Diamond, 1990; Oldow et al., 1994, 2003; Kohler, 1994) culminated in the exhumation of lower-plate rocks exposed in doubly-plunging folds with half-wavelengths of up to 10 km. The folds constitute west-northwest–trending turtleback structures (Kirsch, 1971) that formed in response to shortening during a complex history of nonplane strain (Oldow et al., 2008) accompanied by 20° to 30° of paleomagnetically determined clockwise rotation of upper- and lower-plate rocks (Petronis et al., 2002, 2007). The growth of the turtleback structures deformed the detachment fault system separating upper and lower plates and ultimately locked the fault, ending activity within the extensional complex (Oldow et al., 1994, 2008).

Within the Silver Peak–Lone Mountain extensional complex, the structural architecture of the upper-plate assemblage is best preserved in the Silver Peak Range (Fig. 2), where Proterozoic to lower Paleozoic carbonate and clastic units, Mesozoic intrusive rocks, and an upper Tertiary volcanic and volcanogenic sedimentary succession are exposed (Fig. 3). The Proterozoic to lower Paleozoic rocks were deposited on and adjacent to the miogeosyncline of the western United States (Albers and Stewart, 1972; Stewart and Poole, 1974; Stewart et al., 1977; Oldow et al., 1989) and within an underlying late Proterozoic basin (Bally, 1989; Hoffman, 1989; Oldow et al., 1989). In west-central Nevada, the lower Paleozoic miogeosynclinal succession and underlying Proterozoic rocks have an aggregate stratigraphic thickness of more than 15 km (Albers and Stewart, 1972) and are composed of interbedded argillite, quartzite, dolomite, and limestone (Albers and Stewart, 1972). Structurally overlying the lower Paleozoic miogeosynclinal succession, there is a thrust sheet composed of Ordovician to Silurian argillite, chert, and resedimented carbonate rocks. Unlike coeval miogeosynclinal rocks, these Ordovician to Silurian rocks are of basinal facies and make up the upper plate of the Roberts Mountains thrust system, which was emplaced during mid-Paleozoic deformation (Roberts et al., 1958; Stewart, 1980; Oldow, 1984; Oldow et al., 1989; Burchfiel et al., 1992). No Mesozoic strata are preserved, but multiple generations of granitic plutons and intrusive bodies of known or inferred Jurassic to Cretaceous age are exposed in various parts of the mountains (Albers and Stewart, 1972).

The Tertiary rocks of the Silver Peak Range contain elements consistent with the regional stratigraphic pattern found in the west-central Great Basin (Stewart, 1980; Lipman, 1992). Regionally, the lower Tertiary section consists of upper Oligocene to lower Miocene siliceous to intermediate volcanic tuffs and flows with a cumulative thickness of 2–3 km (Ekren et al., 1980; Proffett and Proffett, 1976; Best, 1986; Stewart and Diamond, 1990). The tuff succession typically is overlain by middle to upper Miocene andesite flow and lahar deposits (Ferguson and Muller, 1949; Robinson et al., 1968; Ekren et al., 1980; Lipman, 1992) that exhibit substantial spatial variability in lithofacies and thicknesses, ranging from tens of meters to over 1.0 km. Following widespread andesite deposition, the region was covered by a blanket of volcanogenic rocks that are variably interbedded with intermediate to siliceous flows and tuffs. This volcanogenic-volcanic succession is characterized by substantial lateral variation in lithology and thickness and is overlain by Pliocene to Quaternary sedimentary rocks, basalt, and basaltic andesite flows (Robinson et al., 1968; Stewart et al., 1974, 1974; Reheis, 1991, 1992).

GEOLOGY OF THE CENTRAL SILVER PEAK RANGE

In the central Silver Peak Range, pre-Tertiary exposures are concentrated north and south of a region of high elevation underlain by late Cenozoic volcanic and sedimentary rocks (Fig. 3). The Tertiary rocks form a 15–20-km-wide belt along a 30 km northwest-trending axis. The southern margin of the Tertiary section consists of a thin succession of flat-lying volcanic rocks unconformably overlying deformed lower Paleozoic rocks intruded by Cretaceous and Jurassic plutons (Robinson et al., 1976; Stewart et al., 1974). Along the northern margin, Tertiary rocks unconformably overlie Paleozoic sedimentary rocks in the western part of the range, but in the east, they structurally overlie metamorphic tectonites of the Mineral Ridge turtleback structure (Kirsch, 1971). The Mineral Ridge turtleback forms a west-northwest–trending doubly-plunging antiform cored by metamorphic tectonites separated from an overlying upper-plate assemblage by a folded, shallowly to moderately dipping detachment fault. Along the eastern

flank of the mountain range, Tertiary rocks lie above a southwesterly dipping segment of the detachment. Toward the northwest, the Tertiary section rises to higher elevations and rests either on Paleozoic rocks of the upper plate or directly on metamorphic tectonites exposed beneath the underlying detachment where the trailing edge of the extensional allochthon is preserved.

The extensional allochthon exposed in the Silver Peak Range contains a Tertiary stratigraphy of several unconformably bound sequences that provide timing constraints for the evolution of structures and associated basins in the upper-plate assemblage (Stewart and Diamond, 1990; Oldow et al., 1994; Diamond and Ingersoll, 2002). Oligocene to lower Miocene volcanic rocks predate large-scale extension and, together with Paleozoic rocks, form the substratum for synextensional basins. Middle to upper Miocene clastic rocks, recognized by Stewart and Diamond (1990) as being deposited in a broad basin localized by the early stages of extension, are overlain unconformably by upper Miocene to Pliocene volcanic and sedimentary rocks. The upper Miocene to Pliocene rocks were originally interpreted as outflow from a caldera complex thought to underlie the higher elevations of the Silver Peak Range (Stewart et al., 1974; Robinson et al., 1968, 1976). As such, these rocks were viewed as an overlap succession marking the cessation of large-scale extension in this region (Stewart and Diamond, 1990; Oldow et al., 1994; Diamond and Ingersoll, 2002).

The results presented here modify earlier interpretations and document that the upper Miocene to Pliocene succession of volcanic and sedimentary rocks is not related to a postextensional caldera complex. Rather, the unconformity-bounded units of the younger Tertiary section are characterized by abrupt lateral variations in thickness associated with differential subsidence and tilt during deposition. The so-called overlap sequence does not mark the end of extensional tectonism in the region and actually was accumulated within several small fault-bounded basins that formed during the fragmentation of the upper-plate assemblage during exhumation of the Silver Peak extensional complex.

Two generations of high-angle faults and two phases of folds are recognized in the Tertiary rocks of the central Silver Peak Range (Fig. 3). First-generation faults formed a system of coeval structures oriented north-northeast and west-northwest that were active during deposition of late Cenozoic volcanic and sedimentary rocks. As will be developed later herein, timing between fault activity and deposition is documented by growth relations in the sedimentary units and by burial of the structures by younger parts of the synextensional stratigraphic sequence. Locally, upper Cenozoic synextensional rocks are deformed in north-northeast–trending folds that are spatially associated with north-northeast–striking faults. The north-northeast folds, together with the faults, are folded in broad west-northwest–trending flexures. All of these structures are crosscut by a younger system of north- to north-northeast–striking normal faults characterized by down-to-the-west displacement. The younger faults are still active and locally develop scarps in alluvium along the western range front of the Silver Peak Range and cut Quaternary deposits exposed within the western parts of the mountain range.

Quaternary Faults

Within the central and northern Silver Peak Range, a curvilinear system of north-striking high-angle faults divides the upper-plate assemblage into several structural panels. The faults form a system of splays stretching from the western range front east to the western flank of the topographic axis of the range (Fig. 2). The fault splays distribute down-to-the west displacement through a complex anastomosing pattern of faults that ramify toward the northeast. North- to northeast-striking faults are locally linked by northwest-striking fault segments, but, overall, the faults transfer displacement from Fish Lake Valley northeast across the range to southern Big Smoky Valley (Fig. 2).

The westernmost fault of the Quaternary system forms a scarp in alluvium up to ~30 m high (Fig. 3) and juxtaposes unconsolidated Quaternary deposits of the valley floor with poorly consolidated alluvial conglomerate that unconformably overlies deformed interbedded mudstone and sandstone of late Miocene to Pliocene age (Fish Lake Valley assemblage). Exposures of these Miocene-Pliocene clastic rocks and overlying Quaternary deposits form a broad belt of subdued topography that stretches along much of the western margin of the mountains. The belt is internally cut by several north-striking fault splays and is bounded on the east by a north-striking fault that exposes Paleozoic rocks and older parts of the Cenozoic section in the footwall (Fig. 3). To the south, the area of low relief is bounded by a down-to-the-west, northwest-striking fault segment that juxtaposes upper Miocene to Pliocene rocks with older Cenozoic rocks that overlie Paleozoic carbonates in the footwall. To the east at higher elevations, a wedge-shaped fault panel contains Paleozoic rocks in the north, and, toward the south, it exposes Cenozoic units. Both eastern and western faults bounding this structural panel are marked by pronounced physiographic steps associated with down-to-the-west normal displacements. The Quaternary faults merge to the south and are deflected to the southwest (Fig. 2), where they ultimately pass to the range-front fault system.

In the eastern part of the central Silver Peak Range, stretching from the range crest to Clayton and Big Smoky Valleys (Fig. 2), faults with demonstrable Quaternary displacement are absent. This region is underlain by upper-plate Paleozoic and Cenozoic rocks as well as lower-plate metamorphic tectonites exposed in Mineral Ridge (Fig. 2). The thickness of Cenozoic rocks varies substantially across the area due to activity on syndepositional faults. The syndepositional faults are preserved, in large part, because Quaternary deformation was concentrated along the western margin of the mountain range. The eastern exposures are particularly important to understanding the deformational history of the Silver Peak–Lone Mountain extensional complex because the structural contact between the lower- and upper-plate rocks is well exposed. The detachment fault that separates isolated klippen of upper-plate rocks resting on the lower plate and the highly extended but relatively coherent extensional allochthon is exposed along the northwest and southwest flanks of the Mineral Ridge turtleback (Fig. 2).

Tertiary Stratigraphy

The Tertiary stratigraphy in the Silver Peak Range and surrounding region records a complex history of volcanism and synorogenic deposition dominated by interleaved volcanic and volcanogenic rocks. Volcanic rocks have a broad range of composition, ranging from siliceous tuff, andesite flows, to basaltic lava, and sedimentary units consisting of fine- to coarse-grained siliciclastic rocks and occasional interbeds of carbonate. Although initially studied at the turn of the century (Turner, 1900) and periodically revised in subsequent years (Ferguson and Muller, 1949, Ferguson et al., 1953; Robinson et al., 1968; Albers and Stewart, 1972), details of the Tertiary stratigraphy have received only limited attention until recently (Diamond, 1990; Stewart and Diamond, 1990; Diamond and Ingersoll, 2002).

Three late Cenozoic successions (Fig. 3) are recognized in the Silver Peak Range and surrounding mountains and are named here in ascending stratigraphic order: (1) the Icehouse Canyon assemblage, (2) Coyote Hole Group, and (3) the Fish Lake Valley assemblage. The lower sequence is a lithologically heterogeneous succession of rocks, informally named Icehouse Canyon assemblage, that rest upon lower Paleozoic strata and Mesozoic plutons and that were deposited prior to large-magnitude extension. The Icehouse Canyon assemblage is unconformably overlain by rocks of several formations comprising the Coyote Hole Group. The units of the Coyote Hole Group are unconformably bounded sequences that, in ascending order, constitute the Silver Peak Formation, the Rhyolite Ridge Tuff, the Cave Springs Formation, and the Argentite Canyon Formation. The Coyote Hole Group is overlain unconformably by a succession of basalt flows and clastic rocks assigned informally to the Fish Lake Valley assemblage. The Fish Lake Valley assemblage records continued synextensional deposition up to and possibly after cessation of motion on the Silver Peak–Lone Mountain detachment system.

Icehouse Canyon Assemblage

The basal part of the Tertiary section is best exposed along the western range front near Icehouse Canyon (Figs. 2 and 3); it is composed of a lithologically heterogeneous succession of clastic and volcanic rocks assigned to the Icehouse Canyon assemblage. The assemblage is composed of a basal conglomerate containing clasts derived entirely of lower Paleozoic rock types, a middle member consisting of welded to nonwelded rhyolite tuff, and an upper unit composed of andesite lava, lahar, and debris flows (Robinson et al., 1968, 1976). The Icehouse Canyon assemblage varies in thickness laterally from a maximum of 500 m in the west-central part of the Silver Peak Range (near Icehouse Canyon) to zero ~5 km to the south and 12 km to the north. The rhyolitic rocks within the succession are dated by K-Ar as 21.5–22.8 Ma (Robinson et al., 1968) and correspond to the widespread succession of ash-flow tuffs that form the basal part of the Tertiary stratigraphy regionally.

The Icehouse Canyon assemblage contains several internal unconformities, and, locally, the rhyolitic tuff unit rests directly on Paleozoic rocks without the underlying basal conglomerate. The rhyolitic tuff is prevalent in the northern Silver Peak Range and occurs as isolated exposures along the northeastern range front near Rhyolite Ridge (Figs. 2 and 3). In the Icehouse Canyon area, the lateral thickness variations of the three units are gradual and suggest deposition in a low-relief depression. The basal conglomerate accumulated in channels overlain by a sheet of rhyolite tuff. Prior to deposition of the andesite, local erosion of the rhyolite allowed the andesitic rocks to rest directly on Paleozoic clastic and carbonate units.

Elsewhere in the central Silver Peak Range, exposures of the Icehouse Canyon assemblage are dominated by the andesite unit that, where the basal contact is exposed, rests directly on Paleozoic rocks. Small exposures of andesite are found along the northern flank of the Silver Peak Range and along the southeastern margin of Tertiary outcrops near Cow Camp (Figs. 2 and 3).

Coyote Hole Group

The Cenozoic rocks assigned here to the Coyote Hole Group (Figs. 3 and 4) were first described by Turner (1900) and named the Esmeralda Formation. Over the years, use of the name Esmeralda Formation was extended to include a succession of lithologically heterogeneous fine- to coarse-grained clastic rocks underlying much of west-central Nevada (Ferguson et al., 1953). The lack of sufficient age control and poorly understood regional stratigraphic relations amongst clastic and volcanic rocks prompted numerous revisions of the stratigraphic nomenclature, variably including and excluding interleaved volcanic rocks (Turner, 1900; Ferguson et al., 1953; Robinson et al., 1968; Albers and Stewart, 1972; Stewart et al., 1974; Robinson et al., 1976). In light of the general lack of consensus and the ambiguity surrounding usage of the Esmeralda Formation over the region, we abandon the name in our treatment of the stratigraphy of the Silver Peak Range and advocate abolishment of the term for the region.

The late Tertiary volcanogenic-volcanic succession exposed in the Silver Peak Range is a key to deciphering the progressive deformation associated with upper-plate transport and disarticulation during large-magnitude extension in the region. These units exhibit substantial differences in thickness and lithology, both along and across strike, but can be mapped throughout the western Silver Peak Range and surrounding region. Although exhibiting substantial differences in lithology, the units are genetically related and, based on well-exposed depositional and structural relations, constitute a synextensional stratigraphic succession. Rocks of the Coyote Hole Group are described in detail in later sections of this paper.

Fish Lake Valley Assemblage

The Fish Lake Valley assemblage (Fig. 3) consists of a succession of epiclastic rocks, basalt and basaltic andesite flows, and tuff. In the Silver Peak Range, the rocks exhibit significant lateral variation in grain size and composition and locally overlie the Coyote Hole Group unconformably. Along the western range front of the central Silver Peak Range, the unit is composed of

poorly lithified claystone with minor interbeds of siltstone and sandstone and local interbedded basalt flows. In these western exposures, the depositional base of the unit is buried, but in the higher elevations of the Silver Peak Range, the base is exposed. In the higher parts of the mountains, the unit reaches a thickness of 550 m and is composed of intercalated pebble to boulder conglomerate, sandstone, argillite, and basaltic lava. Conglomerate is composed predominantly of porphyritic latite and subordinate rhyolite, andesite, and basalt clasts ranging from 0.5 to 1.5 cm in diameter in a matrix of poorly sorted angular sand of quartz, biotite, and feldspar. With the exception of the basalt clasts, the clast compositions are derived from the Coyote Hole Group and Icehouse Canyon assemblage. The clastic rocks are interbedded with lava flows that occur near the top of the section. Basalt flows are dark gray, vesicular, and contain olivine phenocrysts, whereas basaltic andesite or trachyandesite is reddish brown with hornblende phenocrysts and quartz xenocrysts.

Basalt flows are dated by whole-rock K-Ar as 4.8 Ma (Robinson et al., 1968) and farther west along the northern margin of Fish Lake Valley (Fig. 2) as 3.0 Ma by (U-Th)/He thermochronology on apatite (Lee et al., 2003). Near the western flank of the central Silver Peak Range, we dated basalt flows exposed near the base of the section, where the Fish Lake Valley assemblage rests on the upper units of the Coyote Hole Group, and found whole-rock $^{40}Ar/^{39}Ar$ ages of 3.71 ± 0.01 and 3.76 ± 0.04 Ma (Appendix 1). The stratigraphic relation between the coarse- and fine-grained lithofacies assigned to the Fish Lake Valley assemblage is unclear because the units are never found in contact. Nevertheless, the lower contact of the fine-grained lithofacies is mapped as overlying rocks of the upper part of the Coyote Hole Group in the low hills west of Fish Lake Valley (Robinson et al., 1976), and, as such, they reside in the same stratigraphic position relative to the underlying rocks. It is conceivable that the fine- and coarse-grained rocks are lateral equivalents deposited in an environment of substantial topographic relief, but the alternative, that the two successions are two depositional sequences separated by an erosional unconformity, cannot be precluded.

Deposition of the Fish Lake Valley assemblage was controlled by syndepositional faults active during upper-plate extension. What remains unclear, however, is whether accumulation of the entire succession coincided with displacement and deformation of the extensional allochthon overlying the Silver Peak detachment, or if the Fish Lake Valley assemblage was deposited, at least in part, during younger extensional faulting that crosscut both upper- and lower-plate rocks.

SYNEXTENSIONAL ROCKS OF THE COYOTE HOLE GROUP

Rocks of the Coyote Hole Group rest on various units of the extensional complex exposed in the Silver Peak Range (Figs. 3 and 4). In the eastern part of the central Silver Peak Range, rocks of the Coyote Hole Group structurally overlie metamorphic tectonites of the lower-plate assemblage exposed on Mineral Ridge. Farther

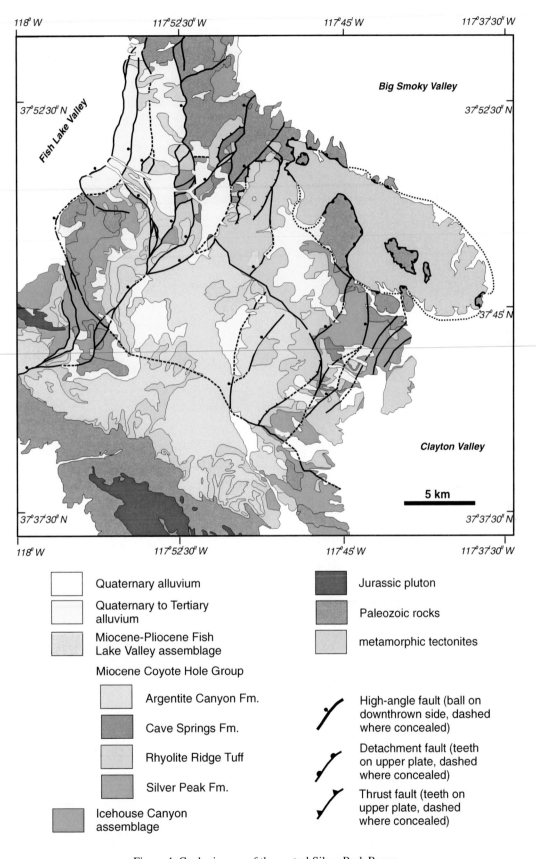

Figure 4. Geologic map of the central Silver Peak Range.

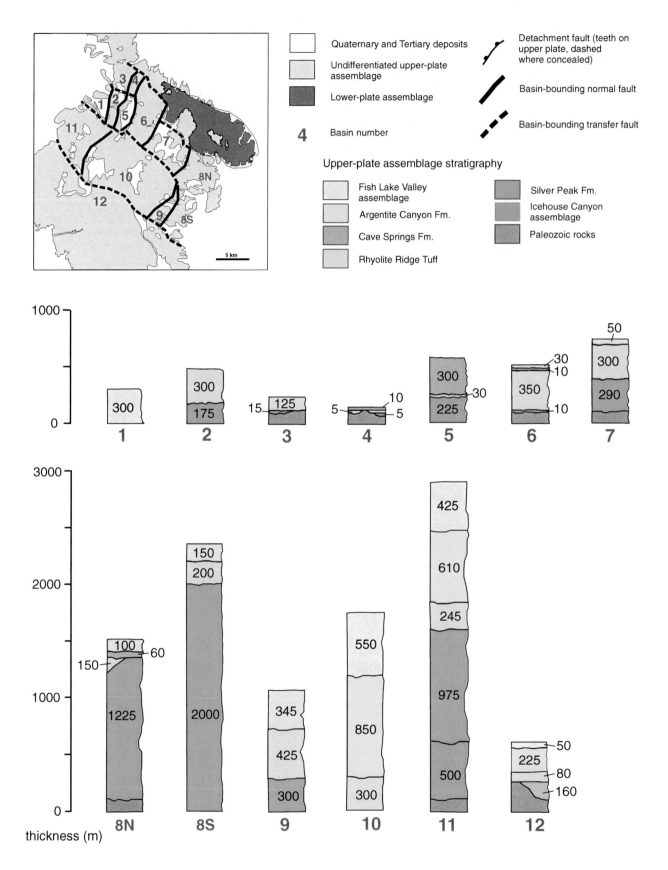

Figure 5. Stratigraphic sections of upper-plate Tertiary rocks.

west, Coyote Hole Group rocks rest with angular unconformity on different units of the upper-plate assemblage and exhibit profound variability in thickness and lateral distribution across the mountain range. Virtually all units are somewhere in depositional contact with older rocks, and local omission of units results in great differences in the stratigraphy from area to area (Fig. 5).

In most areas, a complete section of the Coyote Hole Group is not exposed, either due to local stratigraphic omission or due to disruption by faults. As described in detail in the following, many of the faults formed during deposition of the synorogenic rocks, and in several locales, the older faults were reactivated in the Quaternary. Fortunately, at least some of the formational boundaries within the group are preserved in most areas and provide the lithologic markers needed to reconstruct the stratigraphy in different parts of the central Silver Peak Range. Furthermore, lithologic differences of rocks above and below formational contacts allow assessment of the lateral heterogeneity of individual stratigraphic units in the upper-plate assemblage.

Here, we outline the spatial variability in lithology and thickness of formations comprising the Coyote Hole Group. The stratigraphy and lateral variability of each unit are discussed separately next, but they are combined in stratigraphic sections for localized basins that form the upper-plate assemblage (Fig. 5). Together with detailed geologic maps of key areas (Fig. 6), the sections provide critical stratigraphic relations used to reconstruct the geometry and history of deposition for basins formed in the extensional allochthon during tectonic transport and exhumation of the lower-plate rocks.

Silver Peak Formation

The basal unit of the Coyote Hole Group, the Silver Peak Formation, rests depositionally both on the Tertiary Icehouse Canyon assemblage and Paleozoic rocks and also is found in structural contact with the lower-plate metamorphic tectonites. Throughout the area, the Silver Peak Formation is composed of interbedded shale, sandstone, and conglomerate containing Tertiary volcanic and Paleozoic detritus. The lithology and thickness of the unit change dramatically across the central Silver Peak Range (Fig. 5), with abrupt changes occurring across high-angle faults. The Silver Peak Formation is locally omitted and, in some areas, highly dissected by Quaternary faults that obscure original stratigraphic relations. In exposures where upper and lower contacts are preserved, the unit thickness differs between 975 m and 10 m in the western Silver Peak Range and from 1225 m to 290 m in the eastern part of the mountains (Fig. 5). Although dominated by siliciclastic sedimentary rocks, two interbedded ash-flow tuffs are exposed along the range front near Icehouse Canyon (Fig. 2), and similar rocks were reported by Stewart and Diamond (1990) elsewhere in the region. A middle to late Miocene age of the Silver Peak Formation is based on dated interbedded tuffs with K-Ar ages of 13–11 Ma (recalculated by Stewart and Diamond, 1990; original dates reported by: Everden et al., 1964; Everden and James, 1964; McKee and Glock, 1984) and vertebrate and

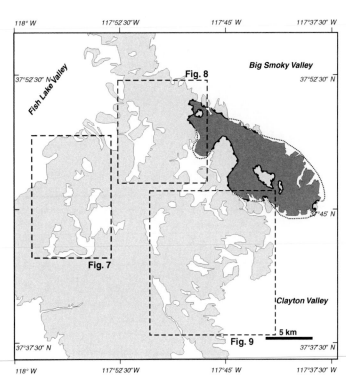

Figure 6. Location of detailed geologic maps shown in subsequent figures.

plant fossils with late to middle Miocene ages (Robinson et al., 1968; Stewart and Diamond, 1990).

Icehouse Canyon Area

A complete section of the Silver Peak Formation is exposed in the Icehouse Canyon area (Fig. 5, column 11) of the western Silver Peak Range (Figs. 6 and 7). In this area, the exposure of the Silver Peak Formation is bounded on the south by a northwest-striking fault. To the south of the fault, the Silver Peak Formation is absent, and rocks of the underlying Icehouse Canyon assemblage rest depositionally on Paleozoic strata and are overlain directly by Rhyolite Ridge Tuff (Fig. 5, column 12). North of the fault, the Silver Peak Formation reaches a thickness of ~975 m and is composed of three members that show variable thickness along strike. A lower unit of interbedded feldspathic wacke interbedded with cobble to boulder conglomerate and clasts derived from the underlying Paleozoic rocks rests depositionally on andesite of the upper part of the Icehouse Canyon assemblage. The lower part of the formation reaches thicknesses of 150–175 m and is overlain by a latite ash-flow tuff that varies in thickness from 5 to 100 m. The upper part of the Silver Peak Formation is a coarsening-upward succession of interbedded argillite, feldspathic wacke, and conglomerate that varies in thickness from 300 to 750 m.

Cave Springs Canyon Area

In the area surrounding Cave Springs Canyon (Figs. 2 and 8), the internal stratigraphy of the Silver Peak Formation is disrupted

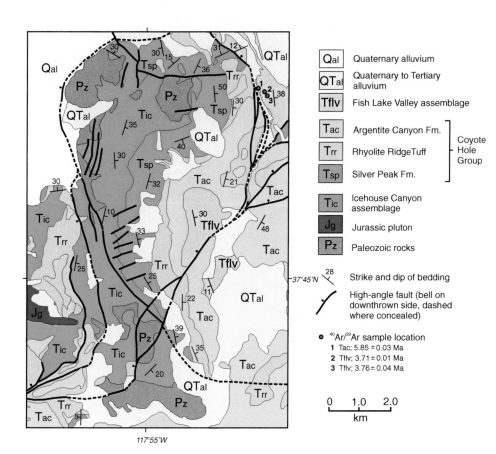

Figure 7. Geologic map of the Icehouse Canyon area, central Silver Peak Range.

by Quaternary normal faults, and the base of the formation is seldom exposed (Fig. 4). Where the base is exposed along the northern margin of Tertiary outcrops and along the western flank of Rhyolite Ridge (Fig. 2), the Silver Peak Formation rests on Paleozoic rocks with angular unconformity. The Silver Peak Formation was tilted to the southeast before deposition of the unconformably overlying Rhyolite Ridge Tuff, which is locally cut out where the Silver Peak is overlain directly by the Cave Springs Formation. Fortunately, the upper contact with younger members of the Coyote Hole Group is found in all structural blocks, allowing reconstruction of the Silver Peak stratigraphy and characterization of lateral differences in thickness and lithology, at least for the upper part of the unit (Fig. 5).

The lithology of the upper Silver Peak Formation changes from west to east across structural panels bounded by normal faults, several of which have demonstrable Quaternary displacement (Figs. 2 and 4). The western exposures preserve a >175-m-thick section (Fig. 5, column 2) composed of interbedded fine- to coarse-grained sandstone, conglomeratic sandstone, and argillite that passes upward into cobble and boulder conglomerate interbedded with feldspathic wacke and argillite as the contact with the overlying Rhyolite Ridge Tuff is approached. Farther east (Fig. 5, column 5), the Silver Peak Formation is composed of 225 m of interbedded feldspathic sandstone, argillite, and conglomerate. The lower 20 m of this section consists

of pebble to boulder conglomerate interbedded with argillite and feldspathic sandstone that passes upward into 165 m of interbedded argillite and feldspathic sandstone. The upper 35 m of the section contains interbedded pebble to cobble conglomerate and feldspathic sandstone that is unconformably overlain by the Rhyolite Ridge Tuff and locally by the Cave Springs Formation (Figs. 6 and 8).

The lower contact of the Silver Peak Formation with Paleozoic rocks is exposed in three locations (Fig. 5, columns 3, 4, and 6). Along the northern margin of the Cave Springs Canyon drainage and along the western flank of Rhyolite Ridge (Fig. 8), the Silver Peak reaches thicknesses ranging from 15 to 5 m, and it consists of thinly interbedded argillite and feldspathic wacke. In these areas, the Silver Peak Formation is overlain unconformably by Rhyolite Ridge Tuff that oversteps the underlying clastic rocks and locally rests directly on the underlying Paleozoic units.

Where the upper contact of the Silver Peak is exposed, the overlying rocks of the Coyote Hole Group dip more shallowly than the underlying clastics. The angularity of the contact varies from nearly conformable in the west, to differential tilt of the underlying rocks by 10° to 20° in eastern exposures. The across-strike heterogeneity in the lithology of the underlying Silver Peak Formation, when taken together with the dramatic differences in thickness, illustrates substantial and locally abrupt lateral changes in depositional conditions and/or preservation.

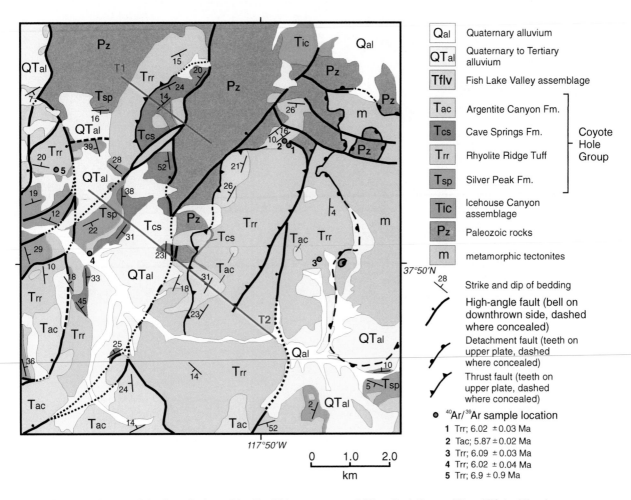

Figure 8. Geologic map of the Cave Springs–Rhyolite Ridge area, central Silver Peak Range. T1 and T2 (red lines) are transects illustrated in Figures 11 and 12, respectively.

Clayton Valley Area

Along the eastern flank of the central Silver Peak Range, the Silver Peak Formation varies substantially in thickness, exposes an internal angular unconformity, and is unconformably overlain by the Rhyolite Ridge Tuff and locally by the Cave Springs Formation (Fig. 5, columns 7, 8N, 8S, and 9). The basal contact with underlying Paleozoic rocks is locally preserved in the north, but along the southern margin of Tertiary exposures, the base of the unit is buried.

Clastic rocks exposed south of Mineral Ridge and assigned here to the Silver Peak Formation were studied in detail by Diamond (1990), where an upper and lower succession separated by an angular unconformity (Stewart and Diamond, 1990) was recognized; these have a combined thickness of 1225 m (Fig. 5, column 8N). As described by Stewart and Diamond (1990), the lower member is ~250 m thick and is composed of interbedded argillite and feldspathic sandstone with a basal conglomerate in contact with underlying Paleozoic rocks. The lower succession is variably tilted both to the east and west and cut by west-dipping

normal faults that sole into the Silver Peak detachment exposed along the northern flank of the outcrops (Fig. 9). The upper member is composed of a thick succession of conglomerate interbedded with argillite and sandstone exposed as an east-dipping homoclinal section at least 1000 m thick (Stewart and Diamond, 1990). The upper member is cut by west-facing normal faults, but the rocks locally overlie and seal similarly oriented faults found within and restricted to the lower member.

Along the western margin of Clayton Valley, the relationship between the Silver Peak Formation and younger units of the Coyote Hole Group is complicated. In the topographically high region south of Red Mountain (Figs. 2 and 9), the Silver Peak Formation reaches a thickness of at least 300 m (Fig. 5, column 9), but the base is not exposed. In this area, the unit is overlain unconformably by a thick succession of Rhyolite Ridge Tuff that passes upward into Argentite Canyon Formation.

East of the topographic front, along the southern margin of Clayton Valley (Figs. 2 and 9), the Silver Peak Formation is overlain by a thinner succession of Rhyolite Ridge Tuff in the

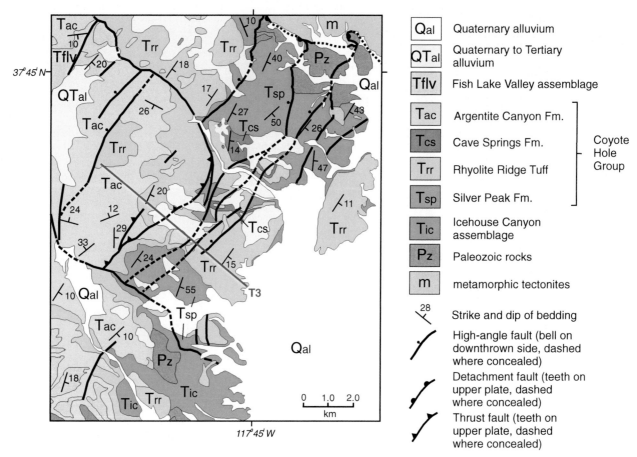

Figure 9. Geologic map of the Clayton Valley and Cow Camp area, central Silver Peak Range. T3 (red line) is transect illustrated in Figure 13.

south but by the younger sedimentary rocks of the Cave Springs Formation in northern exposures, and it is composed of a poorly exposed succession of feldspathic wacke, sandy argillite, and tuffaceous argillite. The rocks consistently dip 20° to 55° to the southeast, and the section may be over 2.0 km thick, but stratigraphic continuity is unconfirmed due to poor exposure (Fig. 5, column 8S). In the southern part of the area, the Silver Peak is unconformably overlain by Rhyolite Ridge Tuff that dips more shallowly to the southeast. In this area, the Silver Peak Formation is bound to the south by a west-northwest–striking fault that juxtaposes the tilted Silver Peak succession with nearly flat-lying andesite of the Icehouse Canyon assemblage to the south of the fault. In the areas to the south, the Silver Peak Formation is omitted, and the Icehouse Canyon assemblage is overlain directly by the Rhyolite Ridge Tuff (Fig. 5, column 12).

Rhyolite Ridge Tuff

The Rhyolite Ridge Tuff typically overlies the Silver Peak Formation, but in some areas, it rests directly on the Icehouse Canyon assemblage or Paleozoic rocks (Fig. 4). The thickness of the Rhyolite Ridge Tuff varies throughout the region, and in some

areas, the unit is absent (Fig. 5, columns 8N and 12). The unit has a maximum thickness of 425 m but typically is much thinner and exhibits lateral pinch-outs where the rocks vary from 60 m to zero over distances of less than 1 km. The unit is composed of quartz-feldspar lithic tuff containing minor biotite, phenocrystic-rich lithic tuff, and massive lithic tuff breccia. Locally, the unit is interbedded with rhyolitic lava flows, which are composed of flow-banded aphanitic rhyolite that weathers pink to gray. The flows are concentrated along two west-northwest–trending belts stretching across the center of Tertiary exposures and along their southern margin. The belt of rhyolite flows in the central part of the Tertiary outcrops extends west-northwest from Red Mountain (Fig. 2) to the western range front. The southern belt of rhyolite extends from near the southeastern margin of Tertiary exposures in the vicinity of Cow Camp northwesterly to Icehouse Canyon (Fig. 2). The rhyolite flows locally reach a maximum amalgamated thickness of 425 m, in the central part of the Silver Peak Range where they underlie Red Mountain (Figs. 2 and 9), but rapidly thin away from the west-northwest–trending outcrop belts to zero over distances of 1.5–2 km.

The age of the tuff was determined in several areas of the central Silver Peak Range (Fig. 8). Biotite separated from a sample collected at the base of the Rhyolite Ridge Tuff in the Cave Springs Canyon drainage has a $^{40}Ar/^{39}Ar$ age of 6.9 ± 0.9 Ma (P. Copeland, 1999, written commun.) and a $^{40}Ar/^{39}Ar$ age on sanidine of 6.02 ± 0.03 Ma (Appendix 1). On the eastern flank of Rhyolite Ridge, just beneath the upper contact of the tuff with the overlying Argentite Canyon Formation, a sample yielded a $^{40}Ar/^{39}Ar$ age on sanidine of 6.03 ± 0.03 Ma, and to the southeast, a sample of Rhyolite Ridge Tuff collected ~120 m below the contact with the overlying Argentite Canyon Formation provided a $^{40}Ar/^{39}Ar$ age on sanidine of 6.09 ± 0.03 Ma (Appendix 1).

Rhyolite Ridge Area

Exposures along Rhyolite Ridge (Figs. 2 and 8), the namesake of the tuff, are composed of several lithologically distinct facies that achieve a total thickness of at least 350 m (Fig. 5, column 6). The tuff is well exposed in a broad northwest-vergent anticline with a shallowly dipping eastern limb and a steeper west limb dipping up to 35° northwest. The thick accumulation of tuff is laterally continuous along the northeast-southwest–trending crest of the major fold for over 7 km. To the south, the tuff interfingers with rhyolitic lava flows that become thinner and less common to the north. To the northeast, the tuff is bounded by a west-northwest–striking, steeply dipping fault that juxtaposes the tuff succession with carbonate and siliciclastic rocks of Paleozoic age. At its western extent, the fault is overlapped depositionally by rocks of the Argentite Canyon Formation (Fig. 8). Tuff thickness varies across strike and increases from a stratigraphic pinchout in the west to 350 m in the east over a dip distance of 2 km. Although too small to be depicted on the accompanying figures, along the western flank of Rhyolite Ridge, the lower contact rests unconformably upon argillite of the Silver Peak Formation and is overlain with an angular discordance of nearly 30° by rocks of the overlying Cave Springs and Argentite Canyon Formations.

On the east flank of Rhyolite Ridge, the tuff sequence exposes two fining-upward successions of tuff breccia and matrix-rich tuff that pass into fine-grained tuffs interbedded with tuff breccia. The lower unit is between 110–130 m thick and is composed of coarse, white-weathering lithic tuff breccia that passes upward into a matrix-rich tuff. The middle unit is between 70–90 m thick, rests on an erosional surface, and consists of a medium-grained tuff breccia that passes upward into a matrix-rich tuff. The upper unit has a gradational lower contact and is composed of between 110 and 140 m of massive fine-grained pumiceous tuff and interbedded fine-grain pumiceous tuff and tuff breccia.

Cave Springs Canyon Area

Lithic tuff dominates the scattered exposures west of Rhyolite Ridge (Fig. 8) where the unit has a maximum thickness of 300 m and rests both on Silver Peak Formation and Paleozoic carbonate rocks (Fig. 5, columns 2 and 3). The lithic tuff is composed of moderately indurated nonwelded ash containing lithic clasts and subordinate phenocrysts. In the lowlands immediately west of Rhyolite Ridge, both upper and lower contacts of the unit are exposed (Fig. 5, column 5), and the tuff dips easterly at

~30° and overlies interbedded sandstone and conglomerate of the Silver Peak Formation along a sharp angular unconformity. The upper contact of the tuff is deeply eroded and incised by paleochannels filled with coarse clastic rocks and interleaved travertine localized at the base of the overlying Cave Springs Formation. From south to north beneath the Cave Springs Formation, the tuff rapidly thins and pinches out over a distance of 400 m (Fig. 8).

Clayton Valley Area

Along the western margin of Clayton Valley (Figs. 6 and 9), Rhyolite Ridge Tuff is composed of moderately indurated lithic ash-flow tuff and interleaved rhyolitic lava flows that are more prevalent along the southern margin of the Tertiary outcrop belt. The thickness of the Rhyolite Ridge Tuff varies dramatically in this area.

In the highlands south of Red Mountain, the tuff and interbedded rhyolite flows are separated from different units of the Coyote Hole Group by a east-northeast–striking fault, and they reach a minimum thickness of 425 m (Fig. 5, column 9). Lithologically, the unit is similar to exposures on Rhyolite Ridge, but it is overlain by a thick succession of the Argentite Canyon Formation.

Farther east in the lowland bordering Clayton Valley (Fig. 9), the Rhyolite Ridge Tuff rests with 30° to 40° angular discordance on the underlying Silver Peak Formation and is overlain by a thin succession of the Argentite Canyon Formation; in northern exposures, it is overlain by an intervening succession of Cave Springs Formation (Fig. 5, columns 8S and 8N). In the southern part of the Clayton Valley area, where the unit is overlain directly by the Argentite Canyon Formation, the Rhyolite Ridge Tuff is composed of nearly 50% rhyolitic flows, and it shows a gradual increase in thickness toward the east from 125 to 200 m. Toward the north from this area, the Rhyolite Ridge Tuff thins as the abundance of interleaved rhyolitic flows decreases, and the unit is overlain instead by sedimentary rocks of the Cave Spring Formation. Still farther north, the Rhyolite Ridge pinches out, and the Cave Springs Formation rests directly on the underlying Silver Peak Formation.

The southern extent of the Rhyolite Ridge Tuff coincides with the southern margin of Tertiary exposures in the central Silver Peak Range (Figs. 4 and 9). The Rhyolite Ridge Tuff is exposed both north and south of the west-northwest–striking fault that marks the southern extent of the Silver Peak Formation. North of the fault, the Rhyolite Ridge is up to 200 m thick and dips shallowly (~20°) to the east (Fig. 9). South of the fault, the Rhyolite Ridge Tuff is composed of up to 80 m of nearly flat-lying nonwelded tuff locally resting on the Icehouse Canyon assemblage, and to the northwest and southeast, on Paleozoic rocks (Fig. 5, column 12).

Cave Springs Formation

The Cave Springs Formation is composed of interbedded volcanogenic argillite, sandstone, and conglomerate and is best exposed in the west-central Silver Peak Range (Figs. 4 and 8). Fine-grained clastic rocks are composed of detritus derived largely from the underlying Rhyolite Ridge Tuff, but the coarse fraction also contains significant contributions from older Tertiary and Paleozoic units. The unit has a thickness exceeding 300 m, but it is typically thinner, ranging from 5 m to 100 m, or locally is absent. The basal contact of the formation is sharp and commonly exhibits substantial relief related to erosion and base-level change during early stages of deposition. The unit typically overlies Rhyolite Ridge Tuff, but in several locations, erosion places the Cave Springs in depositional contact with the underlying Silver Peak Formation.

Cave Springs Canyon Area

In the thickest section (Fig. 5; column 5) exposed west of Cave Springs (Fig. 2), the Cave Springs Formation reaches a thickness of at least 300 m. The bottom 30 m is lithologically heterogeneous. Possibly the most distinctive lithology, a basal ledge of travertine is up to 2 m thick. Travertine passes upward to thinly interbedded tuffaceous argillite, quartzofeldspathic sandstone, and conglomerate that are cut by paleochannels up to 20 m deep. The channels are filled with thin-bedded sandstone, argillite, and locally by claystone. The cut-and-fill deposits are overlain by thin-bedded argillite, sandstone, and conglomerate exposed as laterally continuous beds that constitute the upper 270 m of the Cave Spring Formation. The dominant lithology of the upper part of the section is thinly interbedded tuffaceous argillite and immature quartzofeldspathic wacke, which contains rare rock fragments and biotite grains. Thin bedded quartzofeldspathic arenite and sandy pebble conglomerate are subordinate constituents of the Cave Springs rocks, but they are found throughout the section. The lower 45 m of the upper section is distinguished by thin beds of matrix-supported pebble to boulder conglomerate derived from Paleozoic and older Tertiary volcanic sources. The upper parts of the section occasionally contain layers of thin-bedded tuffaceous argillite reaching several meters thick.

Along the western margin of Rhyolite Ridge (Fig. 2), and east of a north-northeast–striking normal fault buried beneath the Argentite Canyon Formation (Fig. 8), the Cave Springs Formation is only 10 m thick (Fig. 5, section 6). Here, the unit consists of light-gray, fine-grained quartzofeldspathic wacke and tuffaceous argillite. The argillite is very fine-grained, weathers white, and constitutes ~30% of the lithology. The formation rests upon the Rhyolite Ridge Tuff with an angular discordance of ~30° and is laterally continuous over distances of nearly 3.5 km.

Clayton Valley Area

In the low hills west of Clayton Valley (Fig. 9), the Cave Springs Formation rests unconformably on the Rhyolite Ridge Tuff and, where the tuff is stratigraphically omitted, directly on the Silver Peak Formation (Fig. 5, column 8N). The unit reaches a maximum thickness of 60 m and locally is overlain by volcanic rocks of the Argentite Canyon Formation. Overall, the clastic rocks form a fining-upward succession ranging from

matrix-supported boulder conglomerate at the base to laminated claystone at the top. The basal conglomerate is not laterally continuous, and in some locations, the basal section is composed of coarse- to medium-grained quartzofeldspathic wacke.

In the easternmost exposures, the Cave Springs Formation is underlain by the Rhyolite Ridge Tuff and rests on a basal conglomerate that is 10 m thick (Fig. 9). The matrix-supported conglomerate contains subangular to rounded pebbles and boulders derived from the underlying Rhyolite Ridge Tuff and Silver Peak Formation. The basal conglomerate fines upward into a section 20 m thick composed of quartzofeldspathic wacke and cobble to pebble conglomerate that grades upward into thinly bedded coarse- to medium-grained quartzofeldspathic wacke. The upper 30 m of the section is composed of thin-bedded wacke interleaved with laminated sandy argillite layers.

Where the underlying Rhyolite Ridge Tuff is thinner, or where the underlying Rhyolite Ridge is eroded and the Cave Springs Formation rests on the Silver Peak Formation, the Cave Springs unit is composed of interbedded wacke and claystone. The basal section is composed of ~20 m of thin-bedded (1–5 cm) quartzofeldspathic wacke that passes upward into ~30 m of laminated claystone and interbedded fine-grained wacke.

Argentite Canyon Formation

The Argentite Canyon Formation forms the upper unit of the Coyote Hole Group (Fig. 4), and it consists of latite tuff and lava with a total unit thickness ranging from 30 m to in excess of 850 m. At lower elevations along the northern margin of Tertiary outcrops in the central Silver Peak Range (Fig. 4), the Argentite Canyon Formation is exposed as thin (10–30 m) erosional remnants composed dominantly of welded, porphyritic ash-flow tuff that in some areas is underlain by nonwelded tuff up to 10 m thick. In the lowlands west of Clayton Valley, the Argentite Canyon Formation is composed of tuff reaching a thickness of 150 m. The Argentite Canyon Formation rests depositionally on the Cave Springs Formation, the Rhyolite Ridge Tuff, and locally on Paleozoic rocks (Fig. 4), and, in several locales, it depositionally overlies high-angle faults that controlled the depositional pattern of older units of the Coyote Hole Group. At higher elevations in the southwestern part of the range, the Argentite Canyon Formation is composed of a thick succession of feldspar porphyry flows and ash-flow tuff with subordinate interleaved tuff breccias. Here, the unit overlies the Rhyolite Ridge Tuff and locally rests upon Paleozoic rocks and reaches an exposed thickness of 850 m but may exceed 1000 m in some areas.

Robinson et al. (1976) reported K-Ar ages of 5.9 and 6.1 Ma on two of the flows exposed along the southern margin of the Cenozoic outcrops where the unit is overlain by rocks of the Fish Lake Valley assemblage (Fig. 4). We dated the Argentite Canyon Formation in two locations. On the northern flank of Rhyolite Ridge, a sample from the base of the unit, where it rests unconformably on the Rhyolite Ridge Tuff, provided a $^{40}Ar/^{39}Ar$ age on sanidine of 5.87 ± 0.02 Ma (Appendix 1). Along the western range, a sample from the top of the Argentite Canyon Formation, just below the contact with the overlying Fish Lake Valley assemblage, yielded a $^{40}Ar/^{39}Ar$ age on sanidine of 5.85 ± 0.03 Ma (Appendix 1).

The thick accumulations of Argentite Canyon Formation in the southern highlands are composed of tuff and lava that are divided into two units by an unconformity. The thick succession of tuff and lava in the southern highlands dips to the east by up to 40° and is separated from the thin accumulations of tuff farther north by a west-northwest–striking fault that bisects Cenozoic deposits in the central Silver Peak Range (Fig. 4). In the highlands south of the central fault, the lower member of the Argentite Canyon Formation is composed of latite tuff from 345 m up to 370 m thick, and it rests on rocks of the Rhyolite Ridge Tuff (Fig. 5, columns 9 and 10). The lower member contains sporadic interbeds of pumiceous tuff breccia reaching thicknesses of 30 m that can be mapped along strike for up to 2 km. Unconformably overlying the lower member, there is a feldspar porphyry latite of the upper member that reaches a thickness of up to 480 m, for a total unit thickness of 850 m (Fig. 5, column 10). South of the west-northwest–striking fault traced from Cow Camp to Icehouse Canyon (Figs. 2 and 4), the Argentite Canyon Formation reaches a thickness of 225 m (Fig. 5, column 12) and is composed of thinned upper and lower members. The lower member is composed of nearly flat-lying latite tuff, which reaches a thickness of ~100 m and overlies thin deposits of the Rhyolite Ridge Tuff. The upper member has a thickness of ~125 m and is composed of a vesicular andesite that passes upward into a lithic-rich latite ash-flow tuff (Robinson et al., 1976).

PRE-QUATERNARY STRUCTURES

Tertiary rocks of the central Silver Peak Range experienced a complex history of late Cenozoic deformation prior to disruption by Quaternary normal faults. During deposition of the Coyote Hole Group, north-northeast– and west-northwest–striking, high-angle faults were active and formed a nearly orthogonal system of basin-margin structures. In many areas, high-angle faults juxtapose different lithologic units or different parts of the same unit, and in several areas are themselves depositionally overlain by younger members of the volcano-sedimentary sequence. Preserved depositional contacts that locally seal the older faults provide clear evidence that fault displacement and deposition within the basin were coeval. The synextensional rocks are deformed in two generations of folds and are cut by north-northeast–striking faults with reverse motion.

Folds

At least four generations of folds and associated penetrative fabrics are recorded in the pre-Tertiary rocks of the Silver Peak Range (Oldow, 1984; Oldow et al., 2003). Of these structures, two generations are observed in the Tertiary rocks of the Icehouse Canyon assemblage, the Coyote Hole Group, and Fish Lake Valley assemblage. Here, we do not discuss pre-Tertiary structures and focus only on the Tertiary folds, designated D_1 and D_2 in order of decreasing age, and their relation to the structural evolution of the upper plate of the Silver Peak extensional complex.

First-Generation Folds

First-generation folds (D_1) have upright, northwest-vergent axial surfaces and shallowly plunging, north-northeast–trending fold axes. In most areas, folds are gentle to open and have half-wavelengths of 100 m to 5 km and amplitudes ranging from a few meters to 400 m. Locally, the limb appression of first-generation folds becomes tight and half-wavelengths decrease to 1.0–2.5 m. The north-northeast–trending folds are prevalent in a belt stretching from Rhyolite Ridge and the flanking lowlands to the west toward the southeast into the topographic highlands of the central Silver Peak Range (Figs. 2 and 10A).

Rhyolite Ridge has a prominent topographic expression that reflects a northeast-trending, asymmetric, west-facing anticline (Fig. 8). This northwest-vergent fold has a half-wavelength of ~1.5 km and an amplitude of nearly 400 m. The western limb dips up to 35° northwest and exposes a thick section of Rhyolite Ridge Tuff overlain by the Cave Springs and Argentite Canyon Formations. In the eastern limb, rocks of the Rhyolite Ridge Tuff dip shallowly to the east (10° or less) and are truncated by a west-dipping high-angle fault that most recently experienced ~100 m of reverse motion (Fig. 8). Along the steeper northwest limb, the fold locally is cut by a southeast-dipping thrust fault that exhibits a few tens of meters of displacement. The offset along this thrust fault decreases along strike to the northeast and southwest over several hundred meters to where the anticline passes continuously to the west into a low-amplitude syncline.

In the lowlands west of Rhyolite Ridge (Figs. 8 and 10), a broad north-northeast–trending anticline-syncline pair with a wavelength of ~1 km is exposed and preserves rocks of the Cave Springs Formation in the core of the syncline. The eastern limb of the syncline dips from 20° to 60° toward the west and is truncated along its strike length by a Quaternary normal fault. The western limb of the syncline dips easterly between 20° to 40° and passes to the west into an anticline exposing the underlying Rhyolite Ridge Tuff and Silver Peak Formation. The Rhyolite Ridge Tuff rests unconformably on the Silver Peak Formation, which had an easterly dip of 10° to 15° prior to tuff deposition and subsequent folding. To the northeast in a deep canyon, a north-northeast–trending fold train is localized in the footwall of a thrust fault with a steep westerly dip (Fig. 8). In the footwall, minor fold limb appression increases to tight as the fault is approached, and the structures have half-wavelengths of 2–7 m and amplitudes of 3–5 m.

The belt of north-northeast–trending folds broadens to the southeast (Fig. 10), and in the topographic highlands of the central Silver Peak Range, rocks of the upper Coyote Hole Group are deformed in a series of long-wavelength, north-northeast–trending folds. The folds have half-wavelengths of 2–5 km and amplitudes up to 400 m. The folds are truncated by a west-northwest–striking fault exposed near the southeastern margin of Tertiary

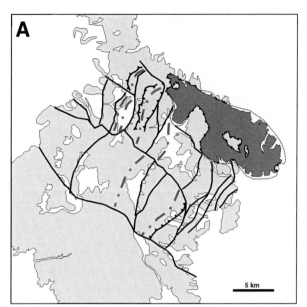
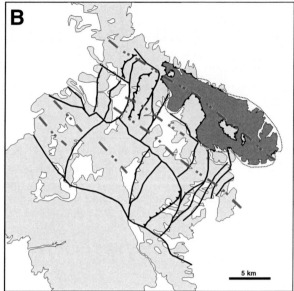

Figure 10. Reconstructed pre-Quaternary faults of the upper-plate assemblage and major D_1 (A) and D_2 (B) fold traces (in red). See Figure 4 for fault symbols.

outcrops. South of the fault, the Tertiary section is nearly flat-lying and rests on a low-relief contact with Paleozoic rocks.

Second-Generation Folds

Second-generation folds (D_2) are open, northwest-trending structures with shallowly plunging axes and upright axial planes. Major D_2 folds have half-wavelengths of 0.5–1.5 km and amplitudes of 0.5–1.0 km, and the map-scale anticline and syncline pairs control the distribution of Tertiary rocks in the west-central Silver Peak Range (Fig. 10B). The turtleback structure underlying Mineral Ridge is a doubly-plunging D_2 antiform. Smaller second-generation folds, with half-wavelengths of 70–100 m and amplitudes of 30–40 m, are found throughout the area (Fig. 10B). The second-phase folds mildly deform earlier northeast-trending folds and several extensional and contractional faults.

Faults

Although disrupted by Quaternary displacement on new and reactivated faults in the western part of the Silver Peak Range, the structural architecture of the extensional allochthon is preserved in several areas. In these areas, two sets of faults, striking north-northeast and west-northwest, are intimately associated with the geometry and distribution of synextensional stratigraphic units. In the following sections, observations from selected areas in the Silver Peak Range are outlined in some detail and provide constraints on the timing and displacements of the pre-Quaternary faults.

Icehouse Canyon Area

In the Icehouse Canyon area (Figs. 2 and 7), faults with Quaternary displacement cut across deformed Tertiary rocks and their basement. The western range-front fault is curved and enters the mountain range at the mouth of Icehouse Canyon, and to the east, a few kilometers inboard of the range front, a major northeasterly striking system of west-side-down normal faults merges to the south and cuts across older structures. In this part of the western Silver Peak Range, the southern extent of Coyote Hole Group rocks is marked by a west-northwest–striking fault with pre-Quaternary displacement. The fault is well located in the west where Paleozoic and Tertiary rocks are juxtaposed, but farther east, where rocks of the Icehouse Canyon assemblage are exposed on both sides of the structure, the fault is located with more difficulty. Nevertheless, the fault juxtaposes a thin Tertiary section resting on Paleozoic strata and Jurassic plutonic rocks above a subhorizontal contact to the south with a thick accumulation of Tertiary rocks and their dipping Paleozoic basement to the north (Fig. 7). South of the fault, the rocks of the Icehouse Canyon assemblage have laterally variable thicknesses, reaching a maximum of ~200 m, and they are overlain by an incomplete section of nearly flat-lying Coyote Hole Group rocks that does not exceed a thickness of 200 m. North of the fault, the Tertiary section is 2755 m thick, dips 30° to 50° to the east, and is broadly folded in northwest-trending map-scale folds.

In the Tertiary rocks north of the southern boundary structure, a series of relatively small-displacement syndepositional faults controlled the thickness of the lower part of the Silver Peak Formation (Fig. 7). The faults were folded together with the Tertiary rocks, and they reside along the southern limb of a northwest-trending, easterly-plunging D_2 syncline. The faults strike northeasterly, and differential down-to-the-east displacement controlled the thickness of the sedimentary rocks and latite tuff composing the lower, middle, and parts of the upper member of the Silver Peak Formation in this area. The displacement on individual faults varies from a few to as much as 200 m and has provided accommodation space for the deposition of the lower parts of the Silver Peak Formation. The faults are traced upsection into clastic rocks of the upper part of the Silver Peak Formation, where their displacement is lost. The faults do not involve the basal contact of the overlying Rhyolite Ridge Tuff.

Cave Springs Canyon Area

The exposures in the Cave Springs Canyon area west of Rhyolite Ridge (Figs. 2 and 8) are highly dissected by Quaternary faults, but segments of older faults are preserved and locally buried by different units of the Coyote Hole Group. Locally, the older faults are broadly folded in west-northwest–trending D_2 folds.

North of Cave Springs Canyon, a west-northwest–striking fault marks the boundary between a 175-m-thick section of Silver Peak Formation on the south and Paleozoic rocks overlain by 15 m of Silver Peak clastics to the north (Fig. 8). Although largely concealed by Quaternary deposits, the high-angle fault truncates the along-strike continuity of the thick section of Silver Peak rocks to the north. The fault is overlapped by 125 m of Rhyolite Ridge Tuff, which, to the north, rests on Paleozoic rocks and the thin (15 m) Silver Peak section and, to the south, overlies the thick sequence of Silver Peak rocks (Fig. 8). Along strike to the east, across a system of northeast-striking faults with Quaternary displacement, the easterly continuation of the west-northwest fault juxtaposes a thick succession (300 m) of Cave Springs Formation on the south with a thin succession of Rhyolite Ridge Tuff (5 m) overlain by 10 m of Cave Springs Formation to the north. The west-northwest structure is lost father east, where it is truncated by a northeast-striking Quaternary fault that exposes Paleozoic rocks in the footwall (Fig. 8).

Where overlapped by the Rhyolite Ridge Tuff, the west-northwest–striking fault is crosscut by a north-northeast fault with reverse displacement (Fig. 8). The thrust fault is deformed by a broad D_2 fold and is truncated along strike to the north and south by Quaternary normal faults. The thrust fault carries 125 m of Rhyolite Ridge Tuff in the hanging wall easterly over a thin succession of Coyote Hole Group rocks. In the footwall, the Tertiary section consists of Cave Springs Formation (10 m) that depositionally overlaps Paleozoic rocks and locally rests upon thin erosional remnants of the Silver Peak Formation (~5 m) and Rhyolite Ridge Tuff (~5 m).

Rhyolite Ridge Area

Along the western flank of Rhyolite Ridge (Fig. 2), a north-northeast–striking normal fault with pre-Quaternary displacement is exposed to the east and within the footwall of a westerly dipping Quaternary fault (Fig. 8). The older fault juxtaposes Cave Springs Formation rocks exposed in the western hanging wall with Paleozoic rocks in the footwall and is depositionally overlapped by rocks of the Argentite Canyon Formation. Several hundred meters to the east of the old normal fault, the western

flank of Rhyolite Ridge is underlain by a west-directed thrust fault. The thrust fault carries a thick section of Rhyolite Ridge Tuff (350 m) overlain by thin Cave Springs Formation (10 m) and Argentite Canyon Formation in the hanging wall westerly over a footwall containing tuff of the Argentite Canyon Formation that rests unconformably on Paleozoic rocks and a thin sliver of east-dipping Silver Peak clastic rocks. The thrust has a few tens of meters of westerly displacement and cuts the west-dipping limb of a map-scale D_1 fold. Fault displacement diminishes along strike, and the fault disappears in the axis of a syncline mapped along the base of Rhyolite Ridge (Fig. 8).

One of the most dramatic exposures of a pre-Quaternary fault system occurs in high-relief outcrops along the northeastern flank of Rhyolite Ridge (Figs. 2 and 8). In this area, Paleozoic and Tertiary rocks of the upper-plate assemblage are in structural contact with the underlying metamorphic tectonites of the lower plate. The detachment fault between the upper and lower plates is broadly folded about a west-northwest axis and dips shallowly (10° or less) to the northwest. In the upper plate, 350 m of Rhyolite Ridge Tuff is juxtaposed with Paleozoic carbonate and clastic rocks by a west-northwest–striking fault that dips steeply to the south. The fault is well exposed along strike for 2 km to the northwest, where it is buried beneath a depositional contact with the Argentite Canyon Formation. To the southeast, the west-northwest–striking fault does not intersect the basal detachment but swings to the south-southwest, parallel to and along the eastern flank of Rhyolite Ridge. Here, the fault merges with a north-northeast–striking thrust fault that uplifts internal units of the Rhyolite Ridge Tuff in the western hanging wall ~150 m over coeval units to the southeast.

Red Mountain Area

A system of north-northeast–striking faults with clear evidence of displacement during deposition of the Coyote Hole Group is preserved near the eastern range front of the Silver Peak Range, east of Red Mountain (Figs. 2 and 9). Normal faults dip steeply to the west, cut the Silver Peak Formation, and merge with the underlying detachment fault exposed to the north (Stewart and Diamond, 1990). Several fault splays record displacements of hundreds of meters during deposition of the Silver Peak Formation (Stewart and Diamond, 1990) but are overlapped locally by younger stratigraphic units of the Coyote Hole Group. In one location, a north-northeast–striking splay offsets the lower part of the Rhyolite Ridge Tuff but is buried beneath upper units of the same tuff. Elsewhere, fault splays were active during Rhyolite Ridge Tuff deposition but are overlapped by the Cave Springs and Argentite Canyon Formations.

Northwest of Red Mountain (Figs. 2 and 9), the tuffs of the Argentite Canyon and Rhyolite Ridge are separated by a steeply dipping, west-northwest–striking fault. The fault is mapped for over 12 km along strike to the west, where it is truncated by a northeast-striking Quaternary fault (Fig. 4). The west-northwest–striking fault marks a major change in the thickness and lithology of the Argentite Formation, and, more significantly, several north-northeast–striking faults mapped in areas to the north and south are truncated by the structure.

In the vicinity of Red Mountain at the east end of the west-northwest–striking fault (Fig. 2), the structure swings southwest and merges with a thrust fault that forms the eastern margin of a topographic highland east of the Mohawk Mine (Fig. 2). The north-northeast–striking thrust has top-to-the-southeast displacement and juxtaposes coeval units of the upper Coyote Hole Group that have substantial differences in thickness. The hanging-wall assemblage is composed of at least 425 m of Rhyolite Ridge Tuff overlain by at least 345 m of Argentite Canyon Formation. In contrast, the thickness of the topographically lower Rhyolite Ridge Tuff in the footwall does not exceed 200 m, and the Argentite Canyon Tuff is only 150 m thick.

Cow Camp Area

In the Cow Camp area, near the southeastern margin of Tertiary exposures (Figs. 2 and 9), a steeply north-dipping, west-northwest–striking fault separates a thick succession of southeasterly tilted rocks of the Coyote Hole Group on the north from a thin, nearly flat-lying succession of Tertiary volcanic rocks resting on Paleozoic substratum to the south. The fault is traced with confidence from the eastern range front to the northwest for over 10 km, but in the eastern part of the topographic highlands south of the Mohawk Mine (Fig. 2), the location of the structure is less certain due to poor exposure.

Near Cow Camp (Fig. 2), the fault juxtaposes a thick, easterly dipping succession of Coyote Hole Group rocks on the north with a thin, nearly flat-lying succession of partially coeval units to the south. North of the west-northwest–striking fault, rocks of the Silver Peak Formation are cut by a series of east-northeast–striking normal faults, and the strata dip 20° to 55° to the southeast. In this area, the thick (300 m to 2.0 km) Silver Peak Formation is unconformably overlain by the Rhyolite Ridge Tuff. The Rhyolite Ridge Tuff is locally involved in the east-northeast–striking faults, but in several areas, it depositionally overlies faults mapped over several kilometers in the Silver Peak Formation rocks. In this area, the Rhyolite Ridge Tuff reaches a maximum thickness of 200 m and is overlain by 150 m of Argentite Canyon Formation. South of the fault, the prominent east-northeast–striking normal faults are absent, and the Silver Peak Formation is missing. The nearly flat-lying Tertiary section is less than 200 m thick. Andesite of the Icehouse Canyon assemblage is locally absent but reaches thicknesses of up to 160 m where it rests unconformably on a shallowly dipping, low-relief Paleozoic substratum. The Icehouse Canyon andesite is unconformably overlain by 80 m of Rhyolite Ridge Tuff and the Argentite Canyon Formation, the dips of which do not exceed 10°.

Northwest from Cow Camp and south of the Mohawk Mine (Fig. 2), the west-northwest–striking fault separates the thin Tertiary section on the south from a thick accumulation of Rhyolite Ridge (425 m) overlain by Argentite Canyon (850 m). Unlike the relatively flat-lying Tertiary rocks south of the fault, the Neogene rocks to the north are folded and involved in a southeast-directed

thrust that borders the highlands south of Red Mountain (Figs. 2 and 9). The folds and thrust fault do not cross the west-north-west–striking fault.

In the highlands southwest of the Mohawk Mine (Fig. 2), the fault trace is located with more difficulty because much of the area is covered by forest. Nevertheless, the fault can be traced west-northwest across the southern part of Tertiary exposures to the headwaters of Icehouse Canyon (Fig. 2) and is located by the juxtaposition of a thick accumulation (850 m) of deformed Argentite Canyon rocks on the north from a thin (225 m) succession of nearly flat-lying Argentite Canon to the south. Folds (D_1) and east-northeast–striking normal faults do not cross the west-northwest–striking structure, which serves as a major structural boundary in the central Silver Peak Range.

BASIN EVOLUTION

As discussed later, the internal stratigraphy and geometry of Coyote Hole Group rocks preserve the history of basin development during disarticulation of the extensional allochthon exposed in the Silver Peak Range. In this part of the upper-plate assemblage, preserved basin margins were localized by coeval displacement on north-northeast– and west-northwest–striking faults. Half grabens formed along west-facing normal faults created multiple basins with north-northeast long axes that were offset and linked by west-northwest–striking faults.

The coeval development and kinematic coordination of the north-northeast– and west-northwest–striking faults is clearly established by the geometry of synextensional basin deposits in several areas. The rate of accommodation space made during fault displacement was not constant, and fault activity waxed and waned as displacement was transferred from one fault system to another. In some instances, displacement on faults ceased, and structures were buried beneath units of the Coyote Hole Group. Where not disrupted by younger Quaternary faults in the eastern central Silver Peak Range, east-northeast faults are truncated along strike (Fig. 4) by west-northwest–striking faults. In two locations, west-northwest–striking faults curve and merge with east-northeast structures (Figs. 8 and 9).

There are only a few instances where the pre-Quaternary faults preserve kinematic indicators, and their history of displacement is inferred from the dip of upper-plate rocks and the geometry of synextensional deposits. The northwest extension direction within the Silver Peak–Lone Mountain extensional complex was initially inferred from the systematic southeasterly tilt of upper-plate units found as klippen resting on metamorphic tectonites (Stewart and Diamond, 1990; Oldow et al., 1994; Kohler, 1994). The direction of extension is supported by the geometry and depositional growth relations observed in clastic rocks of the Silver Peak Formation (Stewart and Diamond, 1990; Diamond and Ingersoll, 2002) and younger rocks of the Coyote Hole Group.

During upper-plate breakup, volcanic and sedimentary depocenters migrated both in space and time. The migration of active extension did not track a simple advance toward the direction of upper-plate motion as is commonly portrayed in two-dimensional models (Proffett, 1977; Wernicke, 1981; Stewart and Diamond, 1990). Rather, as the three-dimensional basin geometry changed through time, the depocenter axes moved forward and backward with respect to northwesterly displacement of the extensional allochthon. Faults that controlled along-strike segmentation of the basins were not active throughout synorogenic deposition, and basin geometries changed along strike as well in the dip direction. Furthermore, basin evolution did not maintain a simple history of extension, and, in several instances, bounding extensional faults reversed their motions during late-stage basin inversion and, locally, thrust faults were formed.

Basin Geometry

Reconstruction of basin geometry is possible because of the degree of exposure and preservation found in synorogenic deposits in the central Silver Peak Range. Here, we discuss specific instances that illustrate the relation between sedimentation and faults oriented normal (north-northeast) and parallel (west-northwest) to upper-plate extension.

Normal Faults

The internal stratigraphy of the Coyote Hole Group documents the development of northeast-trending basins that experienced progressive easterly tilts during their evolution. The synextensional basins in the Silver Peak Range have prismatic geometries that thicken eastward toward the location of known or inferred basin-bounding normal faults. The basin fill contains angular unconformities that separate depositional sequences with progressively steeper easterly dips downsection. Taken together, the basin stratigraphy and tilt indicate progressive syndepositional displacement on basin-bounding faults with a listric geometry (Bally and Oldow, 1983; Bally et al., 1985; Gibbs, 1987, 1989, 1990; McClay and Ellis, 1987).

Differential thicknesses of coeval units in the Coyote Hole Group in different parts of the extensional allochthon show that growth-fault displacements were not constant in space or time. In several areas, sufficient stratigraphic range is preserved in adjacent basins to document the spatial and temporal pattern of displacement on basin-bounding faults. In this discussion, we focus on three transects across a northerly trending belt of prominent outcrops that mark the crest in the central part of the range (Fig. 11).

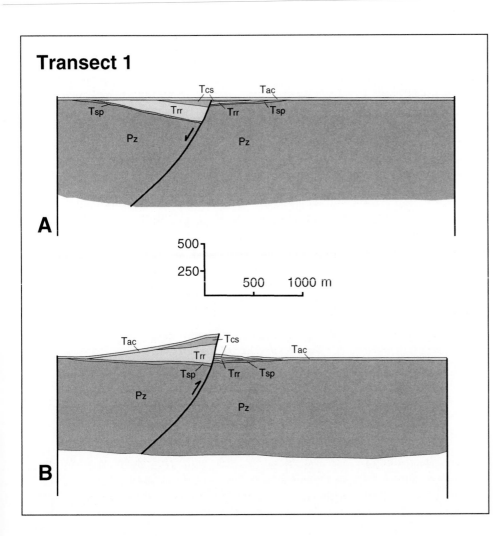

Figure 11. Schematic model for extension (A) and partial basin inversion (B) for transect 1. See Figure 8 for line of section.

In these exposures, the differential displacement on growth faults and internal geometry of extensional basins are particularly well preserved because Tertiary rocks are uplifted by long-wavelength D_1 folds and faults with reverse motion (discussed next).

Transect 1. West of Rhyolite Ridge, a north-northeast–striking fault separates two successions of Rhyolite Ridge Tuff of vastly different thickness (Figs. 2 and 8). Today, the fault has west-side-up reverse displacement and is broadly folded in a northwest-trending D_2 fold. The stratigraphic relations preserved in the footwall and hanging-wall assemblages, however, indicate that the fault originally experienced normal displacement during deposition of the lower parts of the Coyote Hole Group (Fig. 11). In the fault hanging wall to the west, the Rhyolite Ridge Tuff is 125 m thick and rests unconformably on a thin (~15 m) layer of Silver Peak Formation clastic rocks, which locally are cut out where the tuff lies in direct depositional contact with Paleozoic carbonate strata. In the footwall block, the Rhyolite Ridge is up to ~5 m thick and rests either on a laterally discontinuous layer of Silver Peak clastics that reach a maximum thickness of ~5 m or upon Paleozoic carbonate rocks. In the footwall, the Rhyolite Ridge is unconformably overlain by ~10 m of Cave Springs clastic rocks, which are not preserved in the hanging-wall assemblage.

Transect 2. In the Cave Springs area (Figs. 2 and 8), along Rhyolite Ridge and the lowlands immediately to the west, the internal geometry of two basins separated by a normal fault that was capped by Argentite Canyon Formation tuff is well exposed (Fig. 12). Today, the basins are disrupted by late-stage contractional structures associated with basin inversion and are separated by a normal fault with Quaternary motion (Fig. 8). Nevertheless, rocks along this transect preserve a history of differential subsidence and tilt.

From west to east, rocks of the Coyote Hole Group are involved in an anticline-syncline pair truncated on the east by a Quaternary fault (Figs. 8 and 12). West of the fault, the north-northeast–trending anticline exposes an easterly dipping succession of Silver Peak Formation of more than 225 m thick that is overlain by a thin layer of Rhyolite Ridge Tuff. The Rhyolite Ridge Tuff reaches a maximum thickness of 30 m and is overlain by Cave Springs clastic rocks that reach a thickness of at least 300 m in the core of the syncline. The basal contact of the Rhyolite Ridge Tuff is an angular unconformity where the underlying Silver Peak clastic rocks dip more easterly by 10°. East of the Quaternary fault marking the western flank of Rhyolite Ridge (Fig. 8), a syndepositional normal fault juxtaposes Cave Springs clastic rocks in the hanging wall with Paleozoic carbonate rocks in the footwall. The fault marks the eastern margin of the western basin and is unconformably overlain by Argentite Canyon Formation tuff.

The Paleozoic rocks in the footwall of the syndepositional normal fault and overlying units of the Coyote Hole Group are deformed in a major north-northeast–trending anticline that is underlain locally by a small-displacement, west-directed thrust fault (Fig. 8). The small displacement of the thrust fault allows reconstruction of the internal stratigraphy of the rocks now exposed in the core of the anticline. In the western part of the exposure, Argentite Canyon Tuff overlies the Paleozoic rocks. To the east, a thin interval of Cave Springs clastic rocks that reaches a maximum thickness of 10 m underlies Argentite Canyon Tuff, which from west to east progressively rests on Paleozoic rocks and easterly thickening wedges of Silver Peak Formation and Rhyolite Ridge Tuff (Fig. 12). The underlying Silver Peak clastic rocks dip easterly 35° and reach a maximum thickness of 10 m to the east where they intersect the ground surface. The base of the Rhyolite Ridge Tuff dips east 30°, and the unit thickens to the southeast over a horizontal distance of 2 km from zero to over 350 m.

The thick accumulation of coarse clastic rocks of the Silver Peak Formation exposed in the western basin is separated from deposition in the eastern basin by the up-thrown block of the intervening syndepositional fault. In the eastern basin, only a thin succession of fine-grained Silver Peak Formation rocks is found resting on the Paleozoic substratum, recording either limited deposition or differential preservation. No exposures of the Silver Peak Formation are observed along the eastern flank of Rhyolite Ridge.

The accumulation of the Rhyolite Ridge Tuff in the two basins was also controlled by the intervening fault. In the lowlands west of Rhyolite Ridge, the easterly dipping Rhyolite Ridge Tuff thins toward the west and locally is missing in the stratigraphic section where Cave Springs Formation rocks rest unconformably on Silver Peak clastic rocks (Figs. 8 and 12). The unit thickens to the east beneath exposures of the Cave Springs Formation and is lost in the subsurface. In the eastern basin now exposed as the western flank of the north-northeast–trending anticline, the Rhyolite Ridge Tuff rests on more steeply dipping rocks of the Silver Peak Formation and forms a prismatic geometry that thickens to the east to over 350 m. Development of accommodation space within the western and eastern basins continued during deposition of the Rhyolite Ridge Tuff, but the two basins were separated by the structural high formed by the tilted footwall block of the intervening normal fault.

During deposition of the overlying Cave Springs Formation, the western and eastern basins experienced different subsidence histories. In the western basin, at least 300 m of Cave Springs clastic rocks were deposited, whereas in the eastern basin, accommodation space was not developed, and the unit reached a maximum thickness of 10 m. West-side-down motion continued on the fault bounding the western basin during Cave Springs deposition, but the fault inferred along the eastern flank of the east basin apparently was inactive (Fig. 12).

Transect 3. South of Red Mountain (Figs. 2 and 9), a north-northeast–striking thrust fault juxtaposes a thick succession of Rhyolite Ridge Tuff and Argentite Canyon Formation on the western up-thrown block with a stratigraphically thinned succession of coeval rocks to the east. The differential thicknesses of the Rhyolite Ridge Tuff and Argentite Canyon assemblages across the fault indicate that substantially greater accommodation space was available to the west, in the present-day up-thrown block. The implication is that reverse displacement was superposed on an earlier west-side-down normal fault (Fig. 13). No base is observed for the Rhyolite Ridge Tuff exposed on the up-thrown

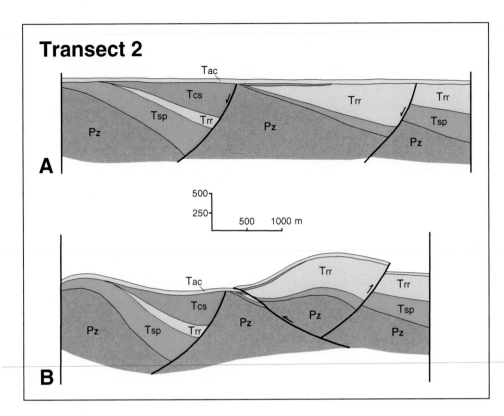

Figure 12. Schematic model for extension (A) and partial basin inversion (B) for transect 2. See Figure 8 for line of section.

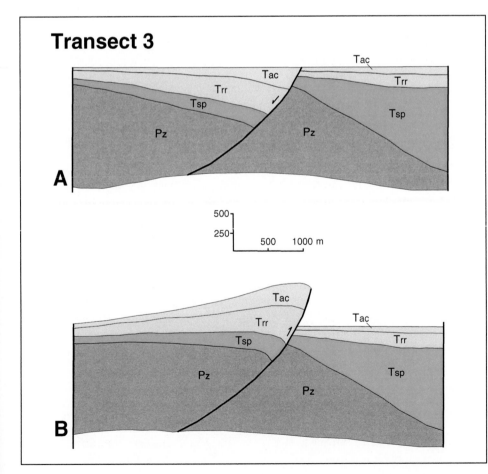

Figure 13. Schematic model for extension (A) and partial basin inversion (B) for transect 3. See Figure 9 for line of section.

block, and the unit exceeds 425 m in thickness, in contrast to a total thickness of coeval rocks to the east, which themselves thicken from 125 m to 200 m over a distance of 5 km eastward. Similarly, the Argentite Canyon Formation to the west has no depositional top, but it reaches a thickness of 345 m as opposed to a thickness of 150 m for coeval rocks to the east.

Longitudinal Faults

The lithology and thickness of the Coyote Hole Group not only vary across strike but also along the axes of the north-northeast–trending basins. Along basin axes, transitions in thickness and lithology are, in many areas, abrupt and occur across steeply dipping west-northwest–trending faults (Fig. 4). The pattern of basin deeps is strongly three dimensional, and basin morphology is controlled along strike by differential tilt across longitudinal faults oriented nearly orthogonal to the north-northeast–striking normal faults. The longitudinal faults have acted as transfer zones (Faulds and Varga, 1998), accommodating differential extension in basins linked by displacement transfer structures in a manner similar to that described for the North Sea (Gibbs, 1987, 1989, 1990).

The west-northwest–striking longitudinal faults are locally well exposed along the northern and southern margins of Tertiary exposures and along the central axis of Tertiary rocks, which divides the synextensional basins into northern and southern belts. The faults are well documented over distances of up to 12 km, but the lateral continuity of the structures across the western part of the mountain range is disrupted by younger, northerly striking Quaternary faults. The central and southern longitudinal faults are the locus of rhyolitic intrusions and rhyolitic flows interbedded within the Rhyolite Ridge Tuff. These intrusive rocks and flows apparently used the structures as conduits.

The northern margin of Rhyolite Ridge (Fig. 8) exposes a spectacular example of a longitudinal fault striking west-north-west and dipping steeply to the south. The fault lies in the trailing edge of the extensional allochthon, just above the westernmost exposure of the underlying detachment fault. The transfer fault juxtaposes 350 m of east-dipping Rhyolite Ridge Tuff on the south against Paleozoic carbonate and clastic rocks on the north. The fault zone is up to 15 m wide and contains an anastomosing system of splays locally intruded by basalt, possibly related to flows in the Fish Lake Valley assemblage. The fault does not intersect the underlying detachment at the surface and toward the east swings to the south and merges with a west-dipping, north-northeast–striking basin-bounding fault. Along strike to the west-northwest, the fault is sealed depositionally by tuff of the Argentite Canyon Formation. The fault accommodated differential extension and facilitated accumulation of the thick succession of Rhyolite Ridge Tuff during displacement on the north-northeast–striking basin-bounding fault. Extensional activity on the structures clearly was older than the onset of Argentite Canyon Formation deposition, dated at this location as 5.87 ± 0.02 Ma.

In the Cave Springs Canyon area (Fig. 8), the thickness of the Silver Peak Formation is controlled by a west-northwest

transfer fault that preserves a history of motion earlier than that of the structures exposed on Rhyolite Ridge. From south to north, Silver Peak clastic rocks abruptly thin from more than 175 m to 15 m. At the western reaches of the fault, the structure is sealed by the basal contact of the Rhyolite Ridge Tuff, giving the a minimum age of activity on the structure at this location. The eastern extension of this fault, however, continued activity during deposition of the Cave Springs Formation and, as pointed out in the discussion of transect 2, shows how the locus of displacement was accommodated by segments of west-northwest–striking faults as younger north-northeast–striking faults crosscut segments of the earlier longitudinal structures during progressive deformation.

A long-lived longitudinal fault is well exposed in the Cow Camp area (Fig. 9) along the southeastern margin of Tertiary exposures. The fault strikes west-northwest and juxtaposes east-tilted rocks of the Silver Peak Formation and overlying Rhyolite Ridge Tuff against nearly flat-lying andesite of the Icehouse Canyon assemblage. South of the fault, the Silver Peak Formation is absent, and a relatively thin sequence of Rhyolite Ridge Tuff rests directly on the Icehouse Canyon assemblage or Paleozoic rocks. This structure marks the southern boundary of a basin that accommodated deposition of a thick sequence of Silver Peak clastic rocks. North of the fault, the overlying Rhyolite Ridge Tuff is tilted 15° to the east, less than the 20° to 55° tilts of the underlying Silver Peak Formation, and it reaches a maximum thickness of 200 m. South of the fault, the Rhyolite Ridge Tuff is flat-lying and achieves a maximum thickness of 80 m. Activity on the fault continued after Silver Peak accumulation and controlled the deposition of the Rhyolite Ridge Tuff. The fault served as a conduit for rhyolitic magma, as expressed by the abundance of flows interbedded with the tuff to the north and the existence of rhyolitic intrusives along the structure.

Along strike to the northwest, the fault is traced across the central Silver Peak Range to Icehouse Canyon on the western range front (Fig. 4). South of the fault trace, units of the Coyote Hole Group are thin, nearly flat-lying, and rest variably on Paleozoic and Icehouse Canyon assemblage rocks. To the north, rocks of the Coyote Hole Group show substantial differences in thickness and tilt across north-northeast–trending faults that do not cross the transfer fault. Similarly, folds in Coyote Hole Group rocks are found only to the north of the transfer fault. The folds involve rocks of the Coyote Hole Group and the Fish Lake Valley assemblage, indicating continued activity into the early Pliocene.

The trace of the southern longitudinal fault is not straight. Locally, the fault is deflected by large-displacement, north-northeast–striking normal faults, as at the basin-bounding fault exposed south of Mohawk Mine (Figs. 2 and 4). Here, the longitudinal fault steps to the north before continuing west-northwest toward Icehouse Canyon, and there is no southern continuation of the north-northeast–striking normal fault to the southwest. In the highlands southwest of Red Mountain near the Mohawk Mine (Fig. 2), the longitudinal fault forms the boundary between different members of the Argentite Canyon Formation.

To the north, the Argentite Canyon Formation is at least 850 m thick and deformed in north-northeast–trending folds. South of the fault, the Argentite Canyon Formation reaches a maximum thickness of 225 m and is nearly flat-lying. From the highlands, the fault enters the headwaters of Icehouse Canyon and swings northwest (Fig. 7). Here, the fault is located by the juxtaposition of a thick sequence of Silver Peak Formation to the north and a thin sheet of Icehouse Canyon assemblage overlying Paleozoic rocks to the south. The fault trace continues northwest through Icehouse Canyon and merges with younger faults at the western extent of the range.

Another long trace-length longitudinal fault system follows the west-northwest–trending central axis of the belt of Tertiary rocks in the central Silver Peak Range (Fig. 4). The fault is well preserved along the eastern trace, but to the west, the fault is disrupted by younger Quaternary faults that cross it at a high angle. The stratigraphy of the Coyote Hole Group differs sharply north and south of a west-northwest–trending belt of rhyolitic intrusives and flows of the Rhyolite Ridge Tuff that vented along the fault system. Discontinuous, west-northwest–trending fault segments juxtapose thick accumulations of Rhyolite Ridge Tuff and Argentite Canyon Formation across steeply dipping contacts characterized by zones of hydrothermally altered breccia. To the south, Argentite Canyon Formation rocks consist of interbedded porphyritic flows, tuffs, and tuff breccias that reach a thickness in excess of 850 m. To the north, the same unit is composed of welded and nonwelded tuff, with few interbedded flows, and it reaches a maximum thickness of 30 m. In the south, over 550 m of Fish Lake Valley assemblage rests on the Argentite Canyon Formation, and related rocks are found to the north only along the western range front. The fault trace coincides with the locus of along-strike changes in lithology and thickness of the Rhyolite Ridge Tuff and Silver Peak Formation. Near the western range front, a young west-northwest–trending fault juxtaposes a thick accumulation of Coyote Hole Group rocks resting on Icehouse Canyon assemblage and Paleozoic rocks against Fish Lake Valley assemblage to the north. The fault lies along the western projection, and probably is a reactivated a segment of, the central longitudinal fault system.

Basin Inversion

Although the locus of deposition within the basins was controlled by extensional faults, the final distribution of synorogenic units was not controlled solely by extensional deformation. In at least three instances, basins were partially inverted by thrust faults that either reactivated the earlier basin-bounding extensional structures or broke through the synextensional succession near basin-bounding faults. The contractional faults were accompanied by the formation of northeast-trending folds, both in the footwall and hanging-wall blocks of the contractional faults (Fig. 10). The inverted basins within the central Silver Peak Range lie within a spatially restricted belt trending north-south across the mountain range.

Cave Springs Canyon and Rhyolite Ridge Areas

Along the northern margin of the Cave Springs Canyon area (Fig. 8), an east-directed thrust is exposed, and it has a 3.5-km-long curved trace that is truncated on both ends by younger normal faults. The fault trace curvature is not a primary feature and was produced by a superposed northwest-trending fold. In the hanging-wall block, the west-dipping thrust exposes 125 m of Rhyolite Ridge Tuff that rests depositionally on Paleozoic rocks and clastic rocks of the Silver Peak Formation. In the adjacent footwall, the Rhyolite Ridge Tuff is only ~5 m thick and is overlain by ~10 m of clastic rocks of the Cave Spring Formation. The footwall assemblage is exposed in the bottom of a northerly trending canyon and resides topographically below hanging-wall rocks exposed to the west. The stratigraphic and structural relations between footwall and hanging-wall assemblages indicate slip reversal on a basin-bounding normal fault (Fig. 11). Close to tight D_1 folds are localized in the footwall and are truncated by the overlying thrust fault but folded together with the fault in a broad northwest-trending D_2 fold. The folds and reverse motion of the fault postdate deposition of the Cave Springs Formation and formed relatively late in the history of Coyote Hole Group deposition. Earlier motion on the fault was normal, down-to-the west, which was necessary to accommodate the differential thickness of the Rhyolite Ridge Tuff between footwall and hanging-wall blocks.

In the Rhyolite Ridge area (Fig. 8), two faults with reverse motion are recognized and have opposing directions of upper-plate motion. The easternmost fault records easterly directed hanging-wall displacement of ~150 m (needed to accommodate observed juxtaposition of stratigraphic units). The east-dipping reserve fault carries folded synorogenic rocks westward over a shallowly dipping footwall (Fig. 12). The structure of Rhyolite Ridge is dominated by a northeast-trending anticline (D_1) underlain by a west-directed thrust fault (Fig. 8) that is well exposed on the walls of a steep canyon crossing the western margin of Rhyolite Ridge. The fault has a shallow easterly dip, strikes north-northeast, and has a surface expression traceable for several hundred meters. The fault has several tens of meters of throw and cuts all formations of the Coyote Hole Group, indicating development late in the deformational history. In the hanging-wall block, the Rhyolite Ridge Tuff thickens rapidly to the east but overlying rocks of the Cave Springs and Argentite Canyon Formations are relatively thin and do not change thickness. All of the hanging-wall rocks are folded in the Rhyolite Ridge anticline, the western limb of which dips 35° to the west (Fig. 12). In the footwall, a thin wedge of Silver Peak Formation rests on Paleozoic carbonate rocks and, together with Rhyolite Ridge Tuff of the footwall, is unconformably overlain by flat-lying ignimbrite of the Argentite Canyon Formation. The thrust fault dips east 30° and essentially follows the easterly dipping contact between the Silver Peak Formation and Rhyolite Ridge Tuff in the footwall. The fault offsets the subhorizontal basal contact of the overlying Argentite Canyon Formation and projects to the surface along the boundary between the west-dipping steep limb of the hanging-wall anticline and the nearly flat-lying rocks of the footwall. In contrast to the inferred inversion of an earlier basin-bounding normal fault outlined previously for the northern Cave Springs Canyon area, the fold-fault system along the western margin of Rhyolite Ridge did not reactivate an earlier normal fault. Rather, the fold and fault formed along the western margin of an easterly thickening basin filled with rocks of the Rhyolite Ridge Tuff. Most of the shortening was accommodated by the anticline that apparently formed late as a fault-propagation fold (Suppe and Medwedeff, 1990; Erslev, 1991; Allmendinger, 1998).

Red Mountain Area

Along the topographic escarpment south of Red Mountain (Fig. 9), an inferred basin-bounding normal fault was reactivated with reverse motion following deposition of the Argentite Canyon Formation. The fault separates 125–200 m of Rhyolite Ridge Tuff in the footwall block to the east from at least 425 m of Rhyolite Ridge Tuff to the west in the hanging-wall block (Fig. 13). Both successions of Rhyolite Ridge are overlain depositionally by rocks of the Argentite Canyon Formation. In the western block, the Argentite Canyon Formation has an exposed thickness of 345 m, but farther west, where overlain by the Fish Lake Valley assemblage, it is 850 m thick. To the east of the fault, the Argentite Canyon Formation has no observed top but does not exceed 150 m in thickness. Based on the thicknesses of coeval rock units, the initial motion on the fault was down-to-the west, allowing at least 225 m more Rhyolite Ridge Tuff to be accumulated in the western highland. West-side-down displacement on the fault may have continued during deposition of the Argentite Canyon Formation, which may have been thicker by as much as 700 m.

Today the thicker western exposure of Rhyolite Ridge Tuff and Argentite Canyon Formation sits 350 m topographically above the thinner exposures to the east (Fig. 13). This discrepancy is explained by reversal of motion on the basin-bounding normal fault separating the two successions. Such reserve slip was accompanied by the formation of a long-wavelength, north-northeast–trending fold in the hanging-wall block (Fig. 10). The fold involves rocks of the Fish Lake Valley assemblage, indicating late-stage formation, and does not extend south of the west-northwest–striking longitudinal fault.

Basin Inversion Mechanism

Basin inversion formed late in the development of the synextensional basins and involved rocks of the Argentite Canyon Formation throughout the study area and the Fish Lake Valley assemblage in the southern highlands. The contractional structures (D_1 folds and thrust faults) are spatially associated with preexisting east-northeast–striking normal faults that controlled basin deposition and are arrayed as a northerly trending belt across the central Silver Peak Range (Fig. 10). The structures are bounded to the north and south by west-northwest–striking longitudinal faults, and the central longitudinal fault truncates and offsets the contractional structures.

The spatially restricted distribution of contractional structures in the upper-plate assemblage in the Silver Peak Range is not easily reconciled with contraction developed during an episode of regional shortening. Rather, these observations suggest that contractional structures formed in the upper plate during progressive extension and tectonic transport above a shallowly dipping detachment with a ramp-flat geometry. Hanging-wall folding above a footwall ramp is well known in thrust systems where the change in dip of the underlying fault induces buckling (Suppe, 1983). Similarly, changes in the underlying fault geometry of an extensional detachment are also capable of producing local shortening during extensional transport (Buchanan and McClay, 1991; McClay and Scott, 1991; Mancktelow and Pavlis, 1994). A local flat in the underlying detachment will provoke shortening as the upper-plate passes from a steeper-dipping segment of the basal fault. Local shortening is accommodated by reversal of slip on existing normal faults during basin inversion and may be accompanied by folding of upper-plate rocks and the formation of new thrust structures.

Three-Dimensional Depocenter Migration

Changes in the distribution of late Tertiary units in the central Silver Peak area provide insight into the temporal and spatial variability of depocenters in this part of the extensional complex. Unique models for the basins are difficult to achieve in many locations because structural and depositional relations are complicated by disruption by younger Cenozoic faults. The younger faults truncate earlier structures and, in several areas, reactivate older basin-bounding faults. Nevertheless, the spatial control is adequate to document the gross aspects of the migration of the three-dimensional geometry through time. The amount of synorogenic accumulation and the degree to which the units are tilted help to constrain the depth of the basin. Although specifics regarding the geometry of the underlying detachment are beyond our reach, basin depths and deformation within synorogenic units give clues to the shape and position of the underlying structure.

Basin development within the upper plate of the Silver Peak extensional complex formed in response to half-graben subsidence along northwest-facing normal faults that, based on differential tilts within the synextensional stratigraphy, had listric geometries and merged with the detachment at depth. The northeast-trending long axes of the extensional basins were offset along strike or terminated by longitudinal faults that strike subparallel to the overall west-northwest extension direction. In general, the half-graben geometry of syndepositional faults conforms to accepted two-dimensional models of extensional basin formation (Stewart and Diamond, 1990; Ingersoll et al., 1996; Diamond and Ingersoll, 2002), but the timing relations and three-dimensional geometry preserved in the Silver Peak Range document a more complex spatial-temporal pattern of deposition than is generally thought to occur within extensional terranes.

We reconstructed the history of basins in the central Silver Peak Range using lateral variations in formational thickness and lithology. The reconstruction is illustrated in a forward model that is depicted in Figure 14 as a series of time slices defined

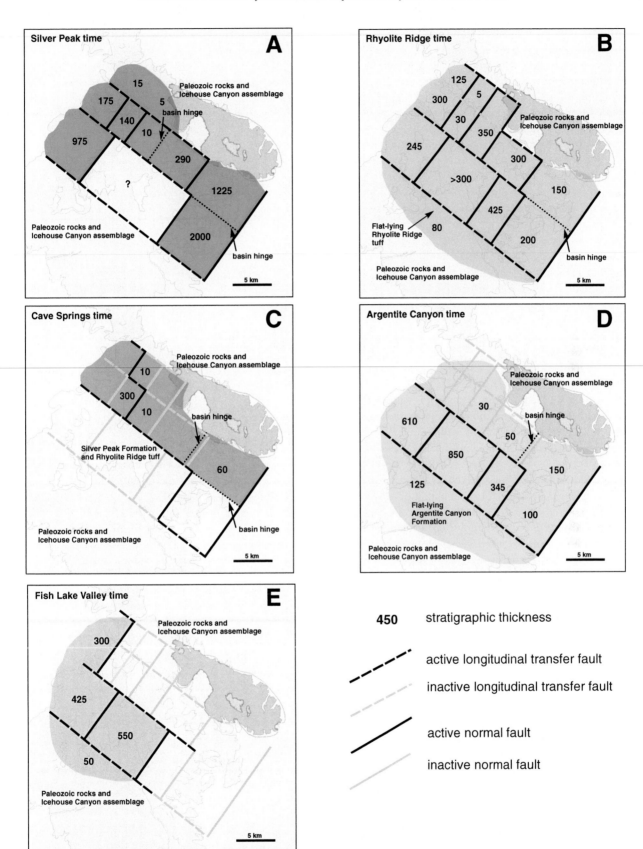

Figure 14. Reconstruction of basin activity showing spatial and temporal migration of depocenters during extension of the upper-plate assemblage.

by the stratigraphic units composing the Coyote Hole Group. The model illustrates progressive changes in basin geometry and the role of basin-bounding faults in the distribution of the synextensional units.

Silver Peak Time

Upper-plate basin development began during deposition of the Silver Peak Formation (Fig. 14A). The distribution of the lower member of the Coyote Hole Group is exposed locally throughout the region, but in the Silver Peak Range, the variations in unit thickness and lithology were clearly controlled by faults active during deposition. Longitudinal faults striking parallel to the northwest extension direction form the northern and southern boundaries of the Silver Peak deposits within the mountain range and, together with the central longitudinal fault system, subdivide the basin into northern and southern sectors. The longitudinal faults juxtapose spatially restricted basins that contain different facies and thicknesses of the Silver Peak Formation. Stratigraphic accommodation space varied from basin to basin and ranged from a few hundred meters to over a kilometer and was provided by west-dipping, north-northeast–striking faults that did not extend north or south beyond longitudinal faults.

The longitudinal faults show a complex pattern of displacement and interaction with north-northeast–striking normal faults. Along strike, the northern longitudinal fault system marks an abrupt change in unit thickness. The eastern segment of the fault systems is poorly located, but it juxtaposed a relatively thick succession (290 m) of Silver Peak Formation rocks with Paleozoic rocks now residing as klippen on lower-plate metamorphic rocks exposed to the north. To the west, the thickness of the Silver Peak Formation diminishes to 10 m, probably across a basin hinge line localized in the footwall of a north-northeast normal fault farther to the west (Fig. 14A). To the west, the northern boundary fault zone juxtaposed a thick Silver Peak clastic accumulation (175 m) with thin (15 m) Silver Peak Formation rocks resting on Paleozoic strata. Similarly, the southern longitudinal fault zone clearly controlled the distribution of the Silver Peak Formation exposed in the Cow Camp and Icehouse Canyon areas at the southeastern and northwestern ends of the structure, respectively. The continuation of the southern longitudinal fault across the southern margin of the central highland cannot be confirmed by direct observation for Silver Peak time but is inferred from the lack of Silver Peak rocks to the south of the through-going structure. Depocenters controlled by north-northeast–striking faults are offset across the western part of the central longitudinal fault (Fig. 14A), where a thick succession of Silver Peak Formation rocks south of the fault in Icehouse Canyon area (975 m) is separated from a thinner accumulation of the same unit farther to the north in Cave Springs area. Activity on the central part of the fault is inferred for much of Silver Peak time, and farther to the east, the structure swings south where it merges with a north-northeast–striking basin-bounding fault.

Rhyolite Ridge Time

All basins were active during deposition of the Rhyolite Ridge Tuff. During deposition of the Rhyolite Ridge Tuff, basin geometries changed from those experienced during Silver Peak time, and accumulation thicknesses varied spatially from 30 m to 425 m (Fig. 14B). The pattern of subsidence for individual basins reversed between the deposition of the Silver Peak Formation and the Rhyolite Ridge Tuff. Basins that exhibited minor thicknesses of Silver Peak Formation were the sites of substantial Rhyolite Ridge Tuff accumulation, and, conversely, sites of thick Silver Peak Formation deposition were overlain by thin accumulations of Rhyolite Ridge Tuff. For example, in the Icehouse Canyon area, 975 m of Silver Peak Formation is overlain by only 245 m of Rhyolite Ridge Tuff. In the easternmost exposures of Tertiary rocks near Clayton Valley, the Rhyolite Ridge Tuff achieves a maximum thickness of 200 m, but the underlying Silver Peak Formation is over a kilometer thick. In the Rhyolite Ridge area, the Silver Peak Formation is only 5–10 m thick, whereas the overlying Rhyolite Ridge Tuff reaches a thickness of 350 m. With the advent of Rhyolite Ridge deposition, the location of the northern longitudinal fault stepped north from the position of the structure during Silver Peak deposition. The western trace of the older northern longitudinal fault was truncated by a north-northeast–striking normal fault that provided accommodation space for northern exposures of Rhyolite Ridge Tuff, but the eastern segment of the northern longitudinal fault remained active (Fig. 14B). The central and southern longitudinal faults were active during Rhyolite Ridge deposition, as evidenced by the truncation of depocenters controlled by north-northeast–striking normal faults.

In the easternmost basin along the western margin of Clayton Valley, the thickness of the Rhyolite Ridge Tuff also varies along the north-northeast axis of the basin. In this area, the thickness of the tuff is greatest (200 m) in proximity to the southern longitudinal fault but thins to zero near the southern flank of Mineral Ridge (Fig. 2), where the underlying detachment fault and lower-plate rocks are exposed. The transition in Rhyolite Ridge Tuff thickness coincides with the eastern projection of the central longitudinal fault. There is no evidence that the central longitudinal fault extended beyond the mapped eastern limit near Red Mountain (Fig. 2), but we cannot conclusively preclude the possibility that the structure continued into and segmented the eastern basin before deposition of younger units of the Rhyolite Ridge Tuff. Alternatively, the easterly projection of the central longitudinal fault may have acted as a basin hinge line during Silver Peak time and certainly controlled the thickness of the Rhyolite Ridge Tuff. The tuff thins to the north and is absent near the underlying detachment and lower-plate rocks now exposed on Mineral Ridge. The uplift along the northern flank of the basin may mark the onset of differential uplift associated with the initiation of turtleback structures (Figs. 2 and 9).

Cave Springs Time

The Cave Springs Formation is found in only three basins: along the western margin of Clayton Valley, in the Rhyolite Ridge area, and in the Cave Springs area (Fig. 14C). The thickness and distribution of Cave Springs clastics highlight the spatial variability of subsidence and the role of longitudinal faults in controlling deposition. In the three basins, Cave Springs thicknesses range from >300 m to less than 10 m, and the greatest thickness differences occur across a single north-northeast–trending normal fault and part of the northern longitudinal fault (Fig. 14C). Accumulations of Cave Springs Formation did not extend south across the central longitudinal fault, indicating that the structure was the southern boundary for basin subsidence to the north. Along the western margin of Clayton Valley, Cave Springs deposition was localized in the northern part of the basin and did not extend south beyond the eastern projection of the central longitudinal fault. This boundary was active during deposition of the underlying Rhyolite Ridge Tuff. The accumulation of Cave Springs clastic rocks in this area records a reversal in depositional pattern within the basin, with an area of uplift becoming the site of subsequent deposition.

Argentite Canyon Time

Flows and tuffs of the Argentite Canyon Formation achieve a maximum thickness of 850 m in the southern highland, south of the central longitudinal fault. In the basins to the north of the fault, the Argentite Canyon Formation consists of tuffs that do not exceed 30 m in thickness. Farther to the east along the flanks of Clayton Valley, the Argentite Canyon Tuff reaches a maximum thickness 150 m (Fig. 14D). Major thickness differences were controlled by the central longitudinal fault and the north-northeast–striking fault along the western margin of the eastern basin. Within the southern basins, north-northeast–striking faults were active during Argentite Canyon time and controlled deposition. These faults separate deposits of the Argentite Canyon Formation with thicknesses that vary by as much as 500 m. In the Icehouse Canyon area, the Argentite Canyon Formation is preserved as a prism that thickens eastward to a maximum of 610 m. In Argentite Canyon area (Fig. 2), thick accumulations of the Argentite Canyon Formation to the south of the central longitudinal fault are juxtaposed with tuff of the Rhyolite Ridge to the north (Figs. 4 and 8). North of the fault, only thin deposits of the Argentite Canyon Formation are found, and they locally overlap both the northern longitudinal fault and north-northeast–striking normal faults. This pattern of deposition suggests that extension and subsidence were greater in the southern part of the allochthon than in the north during this time interval.

Fish Lake Valley Time

The upper contact of the Fish Lake Valley assemblage is not preserved in the study region, so synextensional stratigraphic relations for the entire succession are difficult to assess. Nevertheless, continued activity on basin-bounding faults is apparent during Fish Lake Valley time (Fig. 14E). The Fish Lake Valley assemblage is preserved in four different locations within the Silver Peak Range. The southernmost deposits lie south of the southern longitudinal fault, and in this locale, the Fish Lake Val-

ley assemblage is composed of thin basalt flows devoid of interbedded sedimentary rocks. North of the southern longitudinal fault, the Fish Lake Valley assemblage is exposed in two basins (Fig. 14E), and it consists of boulder conglomerate and sandstone with interbedded basalt flows ranging in thickness from a few tens of meters to over 550 m. The northernmost deposits outcrop along the western range front of the Silver Peak Range, have no exposed base, and are made up of at least 300 m of poorly lithified claystone with minor interbeds of sandstone and siltstone. The lithologic heterogeneity of the Fish Lake Valley assemblage from basin to basin suggests that depocenters were isolated by active basin-bounding faults.

Summary

The overall pattern of deposition within the extensional allochthon indicates that activity on basin-bounding faults controlled deposition, the locus of which migrated spatially through time. The thickness variability of all units of the Coyote Hole Group clearly points to a complex pattern of differential extension and subsidence on north-northeast–trending normal faults separated by west-northwest–striking faults that served as longitudinal transfer structures.

The north-northeast–striking normal faults that provided significant accommodation space during Silver Peak time were less active during Rhyolite Ridge Tuff time, and, conversely, the faults that exhibited minor displacement during Silver Peak deposition provided deep depocenters during accumulation of Rhyolite Ridge deposits. The relative paucity of thick Cave Springs clastic rocks suggests that accumulation of this unit occurred over a short time interval or that extensional was slow and little basin subsidence occurred during this time. During deposition of the Argentite Canyon Formation, only basins south of the central transcurrent fault experienced significant extension, and subsidence to the north was greatly reduced or inactive. The lack of accommodation space developed in the northern basins may indicate uplift of the region, possibly related to turtleback structure development. The latest episode of extension was concentrated in the western and southern parts of the central Silver Peak Range, in basins that accumulated rocks of the Fish Lake Valley assemblage. Activity on syndepositional faults after Cave Springs time was concentrated in the southern and western parts of the Tertiary exposures.

Formation of Turtleback Structures

Synextensional deposition and displacement of the upper-plate assemblage ceased with the growth and emergence of lower-plate structural culminations that core the turtleback structures. Although it is conceivable that displacement on the detachment continued during initial development of the turtleback structure, continued northwesterly displacement of the extensional allochthon was not possible after fold growth reached some critical state.

In the Silver Peak Range, the Mineral Ridge turtleback structure has a half-wavelength of ~10 km and an amplitude of ~1 km. The doubly-plunging fold has a geometry consistent with west-northwest–trending D$_2$ folds, which are not restricted to upper-plate rocks and are pervasively developed in the lower-plate tectonites as well (Oldow et al., 2003; Kolher, 1994). The folded detachment separating upper- and lower-plate assemblages was deflected from an initial, shallow, northwest-dipping geometry during progressive deformation that followed cooling of the lower plate through apatite fission-track closure at ca. 6 Ma (Oldow et al., 1994). Growth of the turtleback structures continued after deposition of the Coyote Hole Group and Fish Lake Valley assemblage, but it may have been initiated during deposition of the Rhyolite Ridge Tuff.

The late-stage, west-northwest–trending folds reflect shortening localized within the Silver Peak–Lone Mountain extensional stepover and were accompanied by paleomagnetically determined clockwise rotation of rocks in the upper- and lower-plate assemblages by 20° to 30° (Petronis et al., 2002, 2007). Displacement transfer through the stepover was not accommodated solely by slip on the basal detachment and extension within the upper-plate assemblage. Rather, a substantial component of distributed nonplane strain is envisioned as having contributed both to folding and rotation of upper- and lower-plate rocks (Oldow et al., 1994, 2008; Petronis et al., 2002, 2007).

CONCLUSIONS

The Silver Peak extensional complex preserves relationships between deposition and deformation during extensional transport and fragmentation of the upper plate during movement on an underlying, shallowly dipping detachment. Northnortheast– and west-northwest–striking syndepositional faults soled into the underlying detachment and accommodated extension within the upper plate. These syndepositional faults were oriented parallel and perpendicular to the inferred upper-plate extension direction and acted synkinematically to create isolated half-graben basins that provided accommodation space for synextensional deposition of sedimentary and volcanic rocks. The three-dimensional geometry of the basins evolved as the rate of displacement on bounding faults changed through time and as some faults became inactive and were buried beneath synextensional deposits. The locus of deposition migrated spatially and through time and records a complex pattern related to tectonic thinning of the trailing edge of the extensional allochthon. Synextensional stratigraphic units were deformed by late-stage folds, and contractional displacement of several basin-bounding faults probably formed in response to extensional displacement across a ramp-flat geometry of the underlying detachment. The upper-plate structures of the extensional allochthon were overprinted by west-northwest–trending folds that involved both upper- and lower-plate rocks. The west-northwest–trending folds produced the characteristic turtleback structures of the Silver Peak–Lone Mountain extensional complex and marked the cessation of displacement on the regional detachment. In the late Pliocene and/or Quaternary, younger normal faults related to active displacement transfer from the Furnace Creek fault system and the central Walker Lane cut across the western part of the Silver Peak Range and cut or locally reactivated older structures formed during earlier extension.

APPENDIX 1. ^{40}Ar/^{39}Ar ANALYTICAL METHODS AND RESULTS

Rocks being prepared for groundmass concentrates were crushed, sieved, and cleaned with hydrochloric acid. Those being prepared as sanidine separates were crushed, sieved, and treated with dilute hydrofluoric acid. The groundmass concentrate and sanidine separates were loaded into aluminum discs and irradiated for 7 h at the Nuclear Science Center in College Station, Texas.

The groundmass concentrates were analyzed with the furnace incremental heating age spectrum method, while the sanidine crystals were fused with a 50 W CO$_2$ laser. The age data for the sanidine analyses are displayed on a probability distribution diagram (cf. Deino and Potts, 1992). Abbreviated analytical methods for the dated samples are given in Table A1.

102803-1A

Weighted mean age = 3.71 ± 0.01 Ma, n/n_{total} = 5/9, mean square of weighted deviations (MSWD) = 1.27.

Groundmass concentrate from 102803-1a yielded a well-behaved age spectrum (Fig. A1). The initial 8.7% of the ^{39}Ar released was slightly disturbed, while the following 89.3% of the ^{39}Ar released was nearly concordant, and a weighted mean age of 3.71 ± 0.01 Ma was calculated from this portion of the age spectrum. The final heating step contained 2% of the ^{39}Ar released and yielded old apparent ages. Inverse isochron analysis of steps A–H yielded an isochron age of 3.70 ± 0.02 Ma with a ^{40}Ar/^{36}Ar intercept of 296.9 ± 4.1, within error of atmosphere (Fig. A1B). The radiogenic yields increased from 1.3% to 89.4% over the initial 58% of the ^{39}Ar released and then decreased to 26.8% over the remainder of the age spectrum. The K/Ca values increased from 0.18 to 1.1 over the initial 58% of the age spectrum and then decreased to 0.048.

102803-1B

Weighted mean age = 3.76 ± 0.04 Ma, n/n_{total} = 9/9, MSWD = 1.22.

Groundmass concentrate from 102803-b yielded a well-behaved age spectrum, with 100% of the ^{39}Ar released yielding a nearly concordant population (Fig. A2A). A weighted mean age of 3.76 ± 0.04 Ma was calculated from all the heating steps. Inverse isochron analysis of steps A–I yielded an isochron age of 3.73 ± 0.05 Ma with a ^{40}Ar/^{36}Ar age of 296.5 ± 0.8 Ma (Fig. A2B). The radiogenic yields increased from 0.5% to 47.3% over the initial 75.1% of the ^{39}Ar released and then decreased to 21.8% radiogenic over the remainder of the age spectrum. The K/Ca values increased from 0.05 to 1.2 over the initial 35.6% of the ^{39}Ar released and then decreased to 0.34 over the remainder of the age spectrum.

102803-2

Weighted mean age = 5.85 ± 0.03 Ma, n/n_{total} = 15/15, MSWD = 5.88.

Fifteen sanidine crystals were used to calculate a weighted mean age of 5.85 ± 0.03 Ma (Fig. A3A). It is noted that this population has a MSWD of 5.88, larger than the other sanidine samples. The radiogenic yields were similar to the other samples, ranging from 76.0% to 97.8%

TABLE A1. SUMMARY OF $^{40}Ar/^{39}Ar$ RESULTS AND ANALYTICAL METHODS

Sample	Location	Unit	Map reference	Mineral	Age analysis	Steps	Age	±2σ	MSWD
102801-1a	37°47.91′N, 117°53.55′W	Tflv	Fig. 7, site 2	Groundmass concentrate	Furnace step-heat	5	3.71	0.01	1.27
102803-1b	37°47.91′N, 117°53.55′W	Tflv	Fig. 7, site 3	Groundmass concentrate	Furnace step-heat	9	3.76	0.04	1.22
102803-2	37°48.30′N, 117°53.83′W	Tac	Fig. 7, site 1	Sanidine	Laser total fusion	15	5.85	0.03	5.88
102703-5	37°50.46′N, 117°40.01′W	Tac	Fig. 8, site 2	Sanidine	Laser total fusion	13	5.87	0.02	1.29
102703-4	37°50.46′N, 117°40.01′W	Trr	Fig. 8, site 1	Sanidine	Laser total fusion	8	6.03	0.03	1.48
102403-3	37°49.17′N, 117°47.75′W	Trr	Fig. 8, site 3	Sanidine	Laser total fusion	14	6.09	0.03	3.43
102903-1	37°48.95′N, 117°51.62′W	Trr	Fig. 8, site 4	Sanidine	Laser total fusion	13	6.02	0.03	1.99

Note: Sample preparation and irradiation: Minerals were separated with standard heavy liquid, Franz Magnetic, and handpicking techniques. Sanidine and groundmass concentrates were loaded into a machined Al disc and irradiated for 7 h in D-3 position, Nuclear Science Center, College Station, Texas. Neutron flux monitor Fish Canyon Tuff sanidine (FC-2) was used, and the assigned age was 28.02 Ma (Renne et al., 1998). The muscovite was loaded into a machined Al disc and irradiated for 24.83 h at the McMaster Nuclear Reactor in Hamilton, Ontario. Neutron flux monitor Fish Canyon Tuff sanidine (FC-2) was used, and the assigned age was 28.02 Ma (Renne et al., 1998). Instrumentation: Mass Analyzer Products 215-50 mass spectrometer on-line with automated all-metal extraction system. Groundmass concentrate and muscovite were step-heated for 8 and 10 min, respectively, using a Mo double-vacuum resistance furnace. Reactive gases were removed during furnace analysis by reaction with three SAES GP-50 getters, two of which were operated at ~450 °C and one at 20 °C. Gas also exposed to a W filament operated at ~2000 °C. Sanidine was fused by a 50 W Synrad CO_2 laser. Reactive gases were removed during a 2 min reaction with two SAES GP-50 getters, one of which was operated at ~450 °C and one at 20 °C. Gas also was exposed to a W filament operated at ~2000 °C and a cold finger operated at −140 °C. Analytical parameters: Electron multiplier sensitivity averaged 4.19×10^{-16} moles/pA for furnace analyses. Electron multiplier sensitivity averaged 1.02×10^{-16} moles/pA for laser analyses. Total system blank and background averaged 1260, 5.3, 0.72, 5.1, and 2.3×10^{-18} moles at masses 40, 39, 38, 37, and 36, respectively, for the furnace analyses. Total system blank and background averaged 233, 6.1, 0.93, 0.93, and 1.38×10^{-18} moles at masses 40, 39, 38, 37, and 36, respectively, for the laser analyses. J-factors were determined by CO_2 laser-fusion of six or 12 single crystals from each of three or six radial positions around the irradiation tray. Correction factors for interfering nuclear reactions were determined using K-glass and CaF_2 and are as follows: $(^{40}Ar/^{39}Ar)_K$ = 0.0000 ± 0.0004; $(^{36}Ar/^{37}Ar)_{Ca}$ = 0.00028 ± 0.00001; and $(^{39}Ar/^{37}Ar)_{Ca}$ = 0.0007 ± 0.00005. MSWD—mean square of weighted deviations.

radiogenic. The K/Ca values are very uniform, with values ranging from 12.1 to 17.4.

102703-5

Weighted mean age = 5.87 ± 0.02 Ma, n/n_{total} = 13/15, MSWD = 1.29.

Thirteen of the analyzed sanidine crystals were used to calculate a weighted mean age of 5.87 ± 0.02 Ma (Fig. A3B). These thirteen crystals yielded a near Gaussian population. An anomalously young crystal with a very low radiogenic yield and an older possible xenocrystic crystal were eliminated. The radiogenic yields of the included crystals range from 77.1% to 96.6%. The K/Ca values of the sanidine range from 9.2 to 16.2.

102703-4

Weighted mean age = 6.03 ± 0.03 Ma, n/n_{total} = 8/8, MSWD = 1.48.

The mineral separate from sample 102703-4 contained a mix of sanidine and plagioclase as identified by the K/Ca values (Fig. A4A). K/Ca values for sanidine are typically over 10. We eliminated seven plagioclase analyses (K/Ca values of 1.4 and lower) from the data set before calculating a weighted mean age of 6.03 ± 0.03 Ma. The K/Ca values of the sanidine range from 13.5 to 112.3. The radiogenic yields of the sanidine grains are between 72.1% and 91.2%.

102403-3

Weighted mean age = 6.09 ± 0.03 Ma, n/n_{total} = 14/14, MSWD = 3.43.

Fourteen sanidine crystals from sample 102403-3 were used to calculate a weighted mean age of 6.09 ± 0.03 Ma, with a nearly Gaussian distribution (Fig. A4B). The somewhat elevated MSWD values (3.43) indicated that scatter in ages was greater than expected from analytical error. The radiogenic yields range from 95.2% to 100.3%. The K/Ca values range from 8.8 to 153.2.

102903-1

Weighted mean age = 6.02 ± 0.03 Ma, n/n_{total} = 13/13, MSWD = 1.99.

A single crystal sanidine from 102903-1 yielded a well-behaved nearly Gaussian population, with all 13 of the analyzed crystals yielding a weighted mean age of 6.02 ± 0.03 Ma (Fig. A5). The radiogenic yields range from 75.5% and 97.7%, while the K/Ca values range from 19.1 to 43.1.

ACKNOWLEDGMENTS

This research was supported by grants to Oldow from the American Chemical Society Petroleum Research Fund (PRF 28854-AC8) and the National Science Foundation (EAR-0126205) and to Oldow and McClelland in a grant from the National Science Foundation (EAR-0208219). We thank James Trexler and Scott Minor for the careful and insightful reviews that contributed greatly to improving the quality of the manuscript.

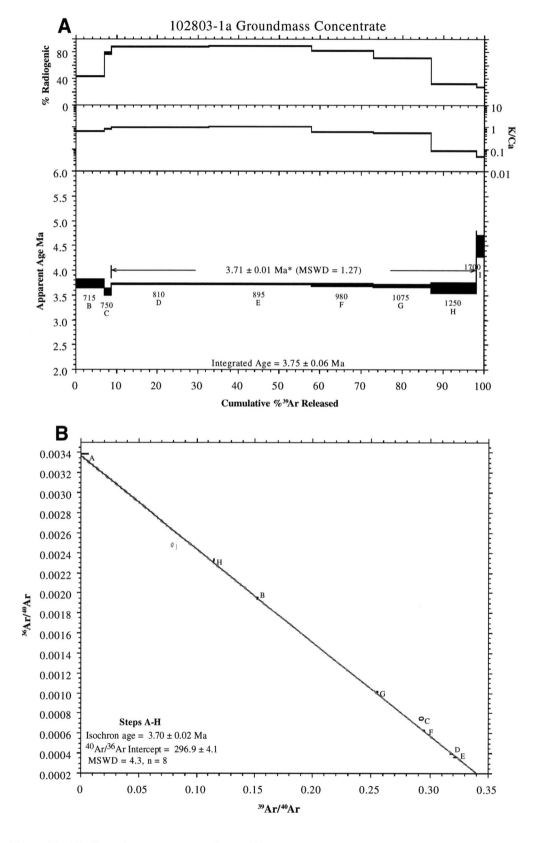

Figure A1. (A) Groundmass concentrate from 102803-1a yielded a well-behaved age spectrum, with a weighted mean age of 3.71 ± 0.1 Ma. (B) Inverse isochron analysis of steps A–H yielded an isochron age of 3.70 ± 0.02 Ma with a $^{40}Ar/^{39}Ar$ age of 296.9 ± 4.1 Ma. MSWD—mean square of weighted deviations.

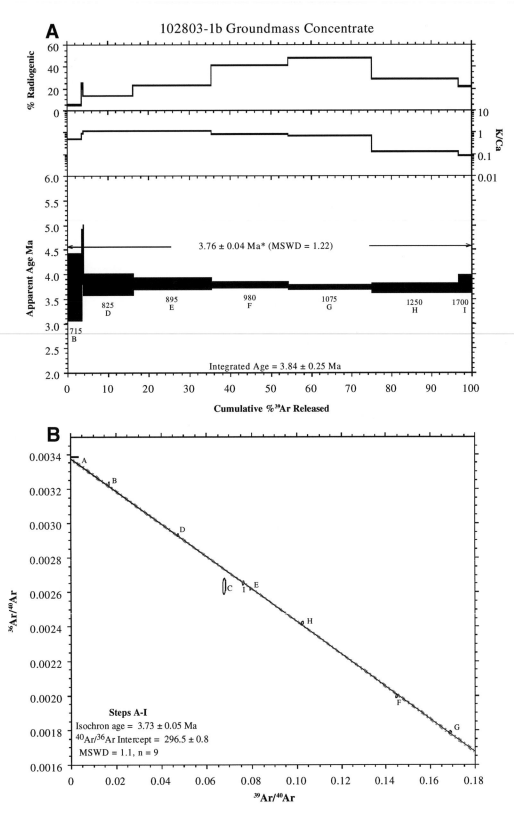

Figure A2. (A) Groundmass concentrate from 102803-b yielded a well-behaved age spectrum, with 100% of the ^{39}Ar released yielding a nearly concordant population. (B) Inverse isochron analysis of steps A–I yielded an isochron age of 3.73 ± 0.05 Ma with a ^{40}Ar/^{36}Ar age of 296.5 ± 0.8 Ma. MSWD—mean square of weighted deviations.

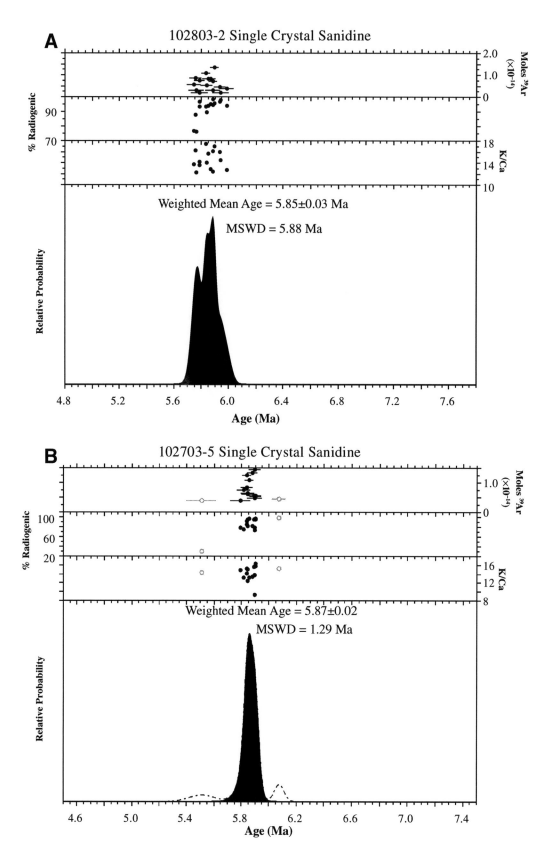

Figure A3. (A) Fifteen sanidine crystals were used to calculate a weighted mean age of 5.85 ± 0.03 Ma. (B) Thirteen of the analyzed sanidine crystals were used to calculate a weighted mean age of 5.87 ± 0.02 Ma. MSWD—mean square of weighted deviations.

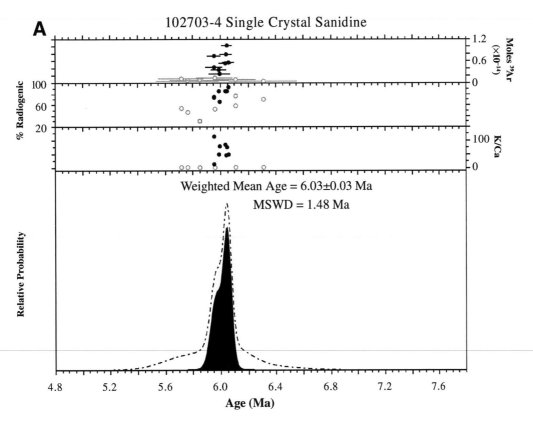

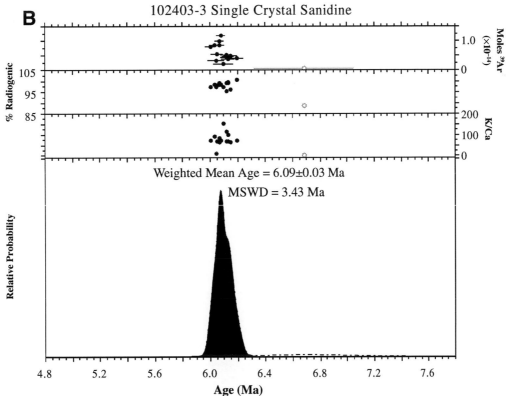

Figure A4. (A) The mineral separate from sample 102703-4 contained a mix of sanidine and plagioclase as identified by the K/Ca values. (B) Fourteen sanidine crystals from sample 102403-3 were used to calculate a weighted mean age of 6.09 ± 0.03 Ma, with a nearly Gaussian distribution. MSWD—mean square of weighted deviations.

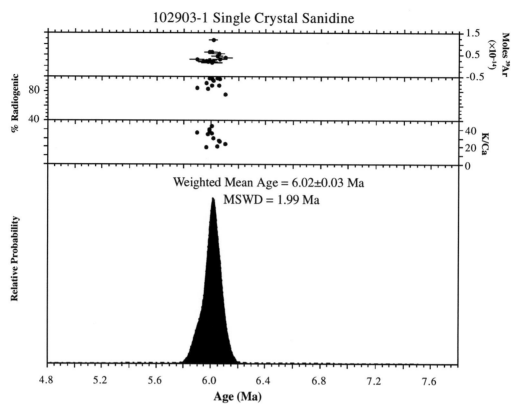

Figure A5. A single crystal sanidine from 102903-1 yielded a well-behaved nearly Gaussian population, with all 13 of the analyzed crystals yielding a weighted mean age of 6.02 ± 0.03 Ma. MSWD—mean square of weighted deviations.

REFERENCES CITED

Albers, J.P., and Stewart, J.H., 1972, Geology and Mineral Deposits of Esmeralda County, Nevada: Nevada Bureau of Mines and Geology Bulletin 78, 88 p.

Allmendinger, R.W., 1998, Inverse and forward numerical modeling of tri-shear fault-propagation folds: Tectonics, v. 17, p. 640–656, doi: 10.1029/98TC01907.

Bally, A.W., 1989, Phanerozoic basins of North America, *in* Bally, A.W., and Palmer, A.R., eds., The Geology of North America—An Overview: Boulder, Colorado, Geological Society of America, Geology of North America, v. A, p. 397–446.

Bally, A.W., and Oldow, J.S., 1983, Plate tectonics, structural styles, and the evolution of sedimentary basins: Houston, Houston Geological Society, 238 p. Reprinted in 1985 as Cordilleran Section, Geological Association of Canada Short Course No. 7, 238 p.

Bally, A.W., Catalano, R., and Oldow, J.S., 1985, Elementi di Tettonicia Regionale: Bologna, Pitagora Editrice, 276 p.

Bennett, R.A., Wernicke, B.P., Niemi, N.A., Friedrich, A.M., and Davis, J.L., 2003, BARGEN continuous GPS data: Tectonics, v. 22, no. 2, 1008, doi: 10.1029/2001TC001355.

Best, M.G., 1986, Some observations on late Oligocene–early Miocene volcanism in Nevada and Utah: Geological Society of America Abstracts with Programs, v. 18, p. 86.

Buchanan, P.G., and McClay, K.R., 1991, Sandbox experiments of inverted listric and planar fault systems: Tectonophysics, v. 188, p. 97–115, doi: 10.1016/0040-1951(91)90317-L.

Burchfiel, B.C., Coward, D.S., and Davis, G.A., 1992, Tectonic overview of the Cordilleran orogen in the western United States, *in* Burchfiel, B.C., Lipman, P.W., and Zoback, M.L., eds., The Cordilleran Orogen; Conterminous U.S.: Boulder, Colorado, Geological Society of America, Geology of North America, v. G-3, p. 407–480.

Christie-Blick, N., and Biddle, K.T., 1985, Deformation and basin formation along strike-slip faults, *in* Biddle, K.T., and Christie-Blick, N., eds., Strike-slip Deformation, Basin Formation, and Sedimentation: Society of Economic Paleontologists and Mineralogists Special Publication 37, p. 1–34.

Deino, A., and Potts, R., 1992, Age probability spectra from examination of single-crystal ^{40}Ar/^{39}Ar dating results: Examples from Olorgesailie, southern Kenya: Quaternary International, v. 13, p. 47–53.

Dewey, J.F., 2003, Transtension in arcs and orogens, *in* Klemperer, S., and Ernst, W.G., eds., The George Thompson Volume: The Lithosphere of Western North America and Its Geophysical Characterization: International Book Series Volume 7: Columbia, Maryland, Bellwether Publishing Limited, p. 105–142.

Diamond, D.S., 1990, Structural and sedimentologic evolution of an extensional orogen, Silver Peak Range and adjacent areas, West-Central Nevada [Ph.D. thesis]: Los Angeles, University of California–Los Angeles, 360 p.

Diamond, D.S., and Ingersoll, R.A., 2002, Structural and sedimentologic evolution of a Miocene supradetachment basin, Silver Peak Range and adjacent areas, west-central Nevada: International Geology Review, v. 44, p. 588–623, doi: 10.2747/0020-6814.44.7.588.

Dixon, T.H., Lee, J., and Reheis, M.C., 1995, Constraints on present-day Basin and Range deformation from space geodesy: Tectonics, v. 14, p. 755–772, doi: 10.1029/95TC00931.

Dokka, R.K., and Travis, C.J., 1990, Role of the Eastern California shear zone in accommodating Pacific–North American plate motion: Geophysical Research Letters, v. 17, p. 1323–1326, doi: 10.1029/GL017i009p01323.

Effimoff, I., and Pinezich, A.R., 1986, Tertiary structural development of selected basins: Basin and Range Province, northeastern Nevada, *in* Mayer, L., ed., Extensional Tectonics of the Southwestern United States: A Perspective on Processes and Kinematics: Boulder, Colorado, Geological Society of America Special Paper 208, p. 31–42.

Ekren, E.B., Byers, F.M., Jr., Hardyman, R.F., Marvin, R.F., and Silberman, M.L., 1980, Stratigraphy, Preliminary Petrology, and Some Structural

Features of Tertiary Volcanic Rocks in the Gabbs Valley and Gillis Ranges, Mineral County, Nevada: U.S. Geological Survey Bulletin B1464, 54 p.

Elias, E.A., 2005, Structural control of Late Miocene to Pliocene volcanic and volcaniclastic deposition during upper-plate fragmentation in the Silver Peak extensional complex, west-central Great Basin [M.S. thesis]: Moscow, Idaho, University of Idaho, 62 p.

Erslev, E.A., 1991, Trishear fault-propagation folding: Geology, v. 19, p. 617–620, doi: 10.1130/0091-7613(1991)019<0617:TFPF>2.3.CO;2.

Everden, J.F., and James, G.T., 1964, Potassium-argon dates and the Tertiary floras of North America: American Journal of Science, v. 262, p. 945–974.

Everden, J.F., Savage, D.E., Curtis, G.H., and James, G.T., 1964, Potassium-argon dates and the Cenozoic mammalian chronology of North America: American Journal of Science, v. 262, p. 145–198.

Faulds, J.E., and Varga, R.J., 1998, The role of accommodation zones and transfer zones in the regional segmentation of extended terranes, in Faulds, J.E., and Stewart, J.H., eds., Accommodation Zones and Transfer Zones: The Regional Segmentation of the Basin and Range Province: Geological Society of America Special Paper 323, p. 1–45.

Ferguson, H.G., and Muller, S.W., 1949, Structural Geology of the Hawthorne and Tonopah Quadrangles, Nevada: U.S. Geological Survey Professional Paper 216, 53 p.

Ferguson, H.G., Muller, S.W., and Cathcart, S.H., 1953, Geology of the Coaldale Quadrangle, Nevada: U.S. Geological Survey Geologic Quadrangle Map GQ-23, scale 1:100,000.

Fillmore, R.P., and Walker, J.D., 1996, Evolution of a supradetachment extensional basin: The lower Miocene Pickhandle Basin, central Mojave Desert, California, in Beratan, K.K., ed., Reconstructing the History of Basin and Range Extension Using Sedimentology and Stratigraphy: Geological Society of America Special Paper 303, p. 107–126.

Gans, P.B., Mahood, G., and Schermer, E., 1989, Synextensional Magmatism in the Basin and Range Province—A Case Study from the Eastern Great Basin: Geological Society of America Special Paper 233, 53 p.

Gibbs, A.D., 1987, Deep seismic profiles in the northern North Sea, in Brooks, J., and Glennie, K.W., eds., Petroleum Geology of North West Europe: 3rd Conference on Petroleum Geology of North West Europe, London, United Kingdom: London, Graham and Trotman, p. 1025–1028.

Gibbs, A.D., 1989, A model for linked basin development, around the British Isles, in Tankard, A.J., and Balkwill, H.R., eds., Extensional Tectonics and Stratigraphy of the North Atlantic Margins: American Association of Petroleum Geologists Memoir 46, p. 501–509.

Gibbs, A.D., 1990, Linked fault families in basin formation: Journal of Structural Geology, v. 12, p. 795–803, doi: 10.1016/0191-8141(90)90090-I.

Hardyman, R.F., and Oldow, J.S., 1991, Tertiary tectonic framework and Cenozoic history of the central Walker Lane, Nevada, in Raines, G.L., Lisle, R.E., Schafer, R.W., and Wilkinson, W.H., eds., Geology and Ore Deposits of the Great Basin: Reno, Nevada, Geological Society of Nevada, Symposium Proceedings, p. 279–301.

Hoffman, P.F., 1989, Precambrian geology and tectonic history of North America, in Bally, A.W., and Palmer, A.R., eds., The Geology of North America—An Overview: Boulder, Colorado, Geological Society of America, Geology of North America, v. A, p. 447–512.

Ingersoll, R.V., Devaney, K.A., Geslin, J.K., Cavazza, W., Diamond, D.S., Heins, W.A., Jagiello, K.J., Marsaglia, K.M., Paylor, E.D., II, and Short, P.F., 1996, The Mud Hills, Mojave Desert, California: Structure, stratigraphy, and sedimentology of a rapidly extended terrane, in Beratan, K.K., ed., Reconstructing the History of Basin and Range Extension Using Sedimentology and Stratigraphy: Geological Society of America Special Paper 303, p. 61–84.

Kirsch, S.A., 1971, Chaos structure and turtleback dome, Mineral Ridge, Esmeralda County, Nevada: Geological Society of America Bulletin, v. 82, p. 3169–3176, doi: 10.1130/0016-7606(1971)82[3169:CSATDM]2.0.CO;2.

Kohler, G., 1994, Structural evolution of metamorphic tectonites beneath the Silver Peak–Lone Mountain detachment fault, west-central Nevada [M.S. thesis]: Houston, Texas, Rice University, 126 p.

Lee, J., Stockli, D., Schroeder, J., Tincher, C., Bradley, D., and Owen, L., 2003, Fault slip transfer in the Eastern California shear zone and Walker Lane, in Lee, J., Stockli, D., Henry, C., and Dixon, T., conveners, Penrose Conference: Kinematics and Geodynamics of Intraplate Dextral Shear in Eastern California and Western Nevada: Boulder, Colorado, Geological Society of America, p. 22–44.

Lipman, P.W., 1992, Magmatism in the Cordilleran United States; Progress and problems, in Burchfiel, B.C., Lipman, P.W., and Zoback, M.L., eds., The Cordilleran Orogen: Conterminous U.S.: Boulder, Colorado, Geological Society of America, Geology of North America, v. G3, p. 481–514.

Locke, A., Billingsly, P.R., and Mayo, E.B., 1949, Sierra Nevada tectonic patterns: Geological Society of America Bulletin, v. 51, p. 513–540.

Mancktelow, N.S., and Pavlis, T.L., 1994, Fold-fault relationships in low-angle detachment systems: Tectonics, v. 13, p. 668–685, doi: 10.1029/93TC03489.

McClay, K.R., and Ellis, P.G., 1987, Geometries of extensional fault systems developed in model experiments: Geology, v. 15, p. 341–344, doi: 10.1130/0091-7613(1987)15<341:GOEFSD>2.0.CO;2.

McClay, K.R., and Scott, A.D., 1991, Experimental models of hanging wall deformation in ramp-flat listric extensional fault systems: Tectonophysics, v. 188, p. 85–96, doi: 10.1016/0040-1951(91)90316-K.

McKee, E.H., and Glock, P.R., 1984, K-Ar ages of Cenozoic volcanic rocks; Walker Lane 1° by 2° quadrangle, eastern California and western Nevada: Isochron-West, v. 40, p. 9–11.

Miller, J.M.G., and John, B.E., 1988, Detached strata in a Tertiary low-angle normal fault terrane, southeastern California: A sedimentary record of unroofing, breaching, and continued slip: Geology, v. 16, p. 645–648, doi: 10.1130/0091-7613(1988)016<0645:DSIATL>2.3.CO;2.

Miller, J.M.G., and John, B.E., 1999, Sedimentation patterns support seismogenic low-angle normal faulting, southeastern California and western Arizona: Geological Society of America Bulletin, v. 111, p. 1350–1370, doi: 10.1130/0016-7606(1999)111<1350:SPSSLA>2.3.CO;2.

Oldow, J.S., 1984, Spatial variability in the structure of the Roberts Mountains allochthon, western Nevada: Geological Society of America Bulletin, v. 95, p. 174–185, doi: 10.1130/0016-7606(1984)95<174:SVITSO>2.0.CO;2.

Oldow, J.S., 1992, Late Cenozoic displacement partitioning in the northwestern Great Basin, in Craig, S.D., ed., Structure, Tectonics, and Mineralization of the Walker Lane: Reno, Geological Society of Nevada, p. 17–52.

Oldow, J.S., 2003, Active transtensional boundary zone between the western Great Basin and Sierra Nevada block, western U.S. Cordillera: Geology, v. 31, p. 1033–1036, doi: 10.1130/G19838.1.

Oldow, J.S., Bally, A.W., Avé Lallemant, H.G., and Leeman, W.P., 1989, Phanerozoic evolution of the North American Cordillera (United States and Canada), in Bally, A.W., and Palmer, A.R., eds., The Geology of North America—An Overview: Boulder, Colorado, Geological Society of America, Geology of North America, v. A, p. 139–232.

Oldow, J.S., Kohler, G., and Donelick, R., 1994, Late Cenozoic extensional transfer in the Walker Lane strike-slip belt, Nevada: Geology, v. 22, p. 637–640, doi: 10.1130/0091-7613(1994)022<0637:LCETIT>2.3.CO;2.

Oldow, J.S., Payne, J.D., Prestia, V.I., McClelland, W.C., and Ferranti, L., 2003, Stratigraphic and structural evolution of the Silver Peak extensional complex, western Great Basin, USA, in Regional Geology and Gold Deposits in the Silver Peak Area, Mineralization Hosted by Metamorphic Core Complexes: Geological Society of Nevada Special Publication 38, p. 51–78.

Oldow, J.S., Geissman, J.W., and Stockli, D.F., 2008, Evolution and strain reorganization within late Neogene structural stepovers linking the central Walker Lane and northern Eastern California shear zone, western Great Basin: International Geology Review, v. 50, p. 1–21, doi: 10.2747/0020-6814.50.3.

Petronis, M.S., Geissman, J.W., Oldow, J.S., and McIntosh, W.C., 2002, Paleomagnetic and 40Ar/39Ar geochronologic data bearing on the structural evolution of the Silver Peak Range, west-central Nevada: Geological Society of America Bulletin, v. 114, p. 1108–1130.

Petronis, M.S., Geissman, J.W., Oldow, J.S., and McIntosh, W.C., 2007, Tectonism of the southern Silver Peak Range: Paleomagnetic and geochronologic data bearing on the Neogene development of a regional extensional complex, central Walker Lane, Nevada: Geological Society of America Special Paper 434, p. 81–106, doi: 10.1130/2007.2434(05).

Prestia, V.I., and Oldow, J.S., 1998, Synorogenic deposition of volcanic and volcaniclastic rocks during upper-plate disarticulation of the Silver Peak–Lone Mountain extensional complex, west-central Nevada: Geological Society of America Abstracts with Programs, v. 30, p. 60.

Prestia, V.I., and Oldow, J.S., 2000, Upper-plate fragmentation in the Silver Peak extensional complex during displacement transfer from the southern to central Walker Lane, western Great Basin: Geological Society of America Abstracts with Programs, v. 32, p. A166.

Proffett, J.M., Jr., 1977, Cenozoic geology of the Yerington district, Nevada, and implications for the nature and origin of Basin and Range faulting: Geological Society of America Bulletin, v. 88, p. 247–266, doi: 10.1130/0016-7606(1977)88<247:CGOTYD>2.0.CO;2.

Proffett, J.M., Jr., and Proffett, B.H., 1976, Stratigraphy of the Tertiary Ash-Flow Tuffs in the Yerington District, Nevada: Nevada Bureau of Mines Report 27, 28 p.

Reheis, M.C., 1991, Geologic Map of Late Cenozoic Deposits and Faults in the Western Part of the Rhyolite Ridge 15 Minute Quadrangle, Esmeralda County, Nevada: U.S. Geological Survey Miscellaneous Investigations Series Map I-2183, scale 1:24,000.

Reheis, M.C., 1992, Geologic Map of Late Cenozoic Deposits and Faults in Parts of the Soldier Pass and Magruder Mountain 15 Minute Quadrangles, Inyo and Mono Counties, California, and Esmeralda County, Nevada: U.S. Geological Survey Miscellaneous Investigations Series Map I-2268, scale 1:24,000.

Renne, P.R., Swisher, C.C., Deino, A.L., Karner, D.B., Owns, T.L., and DePaolo, D.J., 1998, Intercalibration of standards, absolute ages and uncertainties in $^{40}Ar/^{39}Ar$ dating: Chemical Geology, v. 145, p. 117–152, doi: 10.1016/S0009-2541(97)00159-9.

Roberts, R.J., Hotz, P.E., Gilluly, J., and Ferguson, H.G., 1958, Paleozoic rocks of north-central Nevada: American Association of Petroleum Geologists (AAPG) Bulletin, v. 42, p. 2813–2857.

Robinson, P.T., McKee, E.H., and Moiola, R.J., 1968, Cenozoic volcanism and sedimentation, Silver Peak region, western Nevada and adjacent California, in Coats, R.R., Hay, R.L., and Anderson, C.A., eds., Studies in Volcanology; A Memoir in Honor of Howel Williams: Geological Society of America Memoir 116, p. 577–611.

Robinson, P.T., Stewart, J.H., Moiola, R.J., and Albers, J.P., 1976, Geologic Map of the Rhyolite Ridge Quadrangle, Esmeralda County, Nevada: U.S. Geological Survey Map GQ-1325, scale 1:62,500.

Stewart, J.H., 1980, Geology of Nevada: Nevada Bureau of Mines and Geology Special Publication 4, 136 p.

Stewart, J.H., and Diamond, D.S., 1990, Changing patterns of extensional tectonics; overprinting of the basin of the middle and upper Miocene Esmeralda Formation in western Nevada by younger structural basins, in Wernicke, B.P., ed., Basin and Range Extensional Tectonics Near the Latitude of Las Vegas, Nevada: Geological Society of America Memoir 176, p. 447–475.

Stewart, J.H., and Poole, F.G., 1974, Lower Paleozoic and uppermost Precambrian Cordilleran miogeocline, Great Basin, western United States, in Dickinson, W.R., ed., Tectonics and Sedimentation: Society of Economic Paleontologists and Mineralogists Special Publication 22, p. 28–57.

Stewart, J.H., Robinson, P.T., Albers, J.P., and Crowder, D.F., 1974, Geologic Map of the Piper Peak Quadrangle, Nevada-California: U.S. Geological Survey Map GQ-1186, scale 1:62,500.

Stewart, J.H., Stevens, C.H., and Fritsche, A.E., eds., 1977, Paleozoic Paleogeography of the Western United States: Pacific Section, Society of Economic Paleontologists and Mineralogists, Pacific Coast Paleography Symposium 1, 502 p.

Strecker, U., Smithson, S.B., and Steidtmann, J.R., 1996, Cenozoic basin extension beneath Goshute Valley, Nevada, in Beratan, K.K., ed., Reconstructing the History of Basin and Range Extension Using Sedimentology and Stratigraphy: Geological Society of America Special Paper 303, p. 15–26.

Suppe, J., 1983, Geometry and kinematics of fault-bend folding: American Journal of Science, v. 283, p. 684–721.

Suppe, J., and Medwedeff, D., 1990, Geometry and kinematics of fault-propagation folding: Eclogae Geologicae Helvetiae, v. 83, p. 409–454.

Turner, H.W., 1900, The Esmeralda Formation, a Fresh-Water Lake Deposit in Nevada: U.S. Geological Survey 21st Annual Report, Part 2, p. 191–226.

Unruh, J., Humphrey, J., and Barron, A., 2003, Transtensional model for the Sierra Nevada frontal fault system, eastern California: Geology, v. 31, p. 327–330, doi: 10.1130/0091-7613(2003)031<0327:TMFTSN>2.0.CO;2.

Wernicke, B.P., 1981, Low-angle normal faulting in the Basin and Range Province: Nappe tectonics in an extending orogen: Nature, v. 291, p. 645–648, doi: 10.1038/291645a0.

MANUSCRIPT ACCEPTED BY THE SOCIETY 21 JULY 2008

The Geological Society of America
Special Paper 447
2009

Deformation of the late Miocene to Pliocene Inyo Surface, eastern Sierra region, California

A.S. Jayko*

U.S. Geological Survey, White Mountain Research Station, Bishop, California 93514, USA

ABSTRACT

A middle and late Miocene erosion surface, the Inyo Surface, underlies late Miocene mafic flows in the White Mountains and late Miocene and (or) early Pliocene flows elsewhere in the eastern Sierra region. The Inyo Surface is correlated with an erosion surface that underlies late Miocene mafic flows in the central and northern Sierra Nevada. The mafic flows had outpourings similar to flood basalts, although of smaller volume, providing paleohorizontal and paleolowland indicators. The flows filled and locally topped the existing landscape forming broad plateau-like flats. Topographic relief in the region was characterized by weathered and rounded slopes prior to late Miocene mafic magmatism. Relicts of the older landscape lie adjacent to late Miocene and early Pliocene basalt-covered lowlands that now occur within the crests of ranges that have 2500–3000 m relief and dramatically steep escarpments. Late Miocene mafic flows that lie on the crest of the Sierra Nevada adjacent to the White Mountains predate significant activity on the Sierra Nevada frontal fault zone. These deposits and accompanying erosion surfaces provide excellent strain markers for reconstructing part of the Walker Lane north of the Garlock fault and west of the Amargosa drainage, here referred to as the eastern Sierra region.

The Inyo Surface is a compound erosional surface that records at least four major erosion events during the Cenozoic. These four surfaces were first recognized on the Kern Plateau and named from oldest to youngest, the Summit Upland, the Subsummit Plateau, the Chagoopa Plateau, and the Canyon. The three older surfaces have also been subsequently modified by Pleistocene glaciation. The compound erosion surface, which is locally overlain by late Miocene mafic flows in the northern and central Sierra Nevada, is here referred to as the Lindgren Surface. Correlatives in the eastern Sierra region are found in the White Mountains, Inyo Mountains, Darwin Plateau, Coso Range, and nearby ranges.

Keywords: late Miocene erosion surface, Inyo Surface, Sierra Nevada frontal fault, eastern Sierra region, Lindgren Surface, Sierra Nevada block, Inyo Range.

*Corresponding author e-mail: ajayko@usgs.gov.

Jayko, A.S., 2009, Deformation of the late Miocene to Pliocene Inyo Surface, eastern Sierra region, California, *in* Oldow, J.S., and Cashman, P.H., eds., Late Cenozoic Structure and Evolution of the Great Basin–Sierra Nevada Transition: Geological Society of America Special Paper 447, p. 313–350, doi: 10.1130/2009.2447(15).

INTRODUCTION

The eastern Sierra region, which lies east of the Sierra Nevada along the southwestern edge of the Great Basin, is structurally and topographically distinctive, characterized by extremely high-relief and narrow elongate basins (Fig. 1). It contrasts in both morphology and structural character with the adjacent bounding regions of the Mojave and Amargosa Deserts to the south and east. The uplift and consequent Miocene-Pliocene record of the eastern Sierra region are strongly linked to the late Miocene to Pleistocene tectonics of the Sierra Nevada block (Fig. 1; Wright, 1976), which lies to the west. The major faults that bound the region discussed here are the Sierra Nevada frontal fault on the west, the Death Valley fault zone on the east, and the Garlock fault zone on the south (Fig. 1).

The elevation of the eastern Sierra region prior to onset of the late Miocene transtensive deformation has been a question of debate for decades. Although the vertical slip rate and thus the growth rate of relief have been constrained for many of the range fronts of the area (cf. compilation in Reheis and Sawyer, 1997; Table 1), the uplift history of the ranges is of current intrigue. There have been lingering questions about the elevation of the region and extent of a high plateau prior to the advent of Basin and Range extension and displacement of the Sierra Nevada block.

Subhorizontal to very gently inclined remnants of an irregular, late Miocene erosion surface are part of a relict landscape here collectively referred to as the Inyo Surface. The Inyo Surface, along with the overlying mafic volcanics, provides a datum that constrains the uplift, tilting, and structural geometry of mainly Pliocene-Pleistocene strain in the eastern Sierra region. Steeply inclined remnants of the Inyo Surface are step-faulted, tilted, or warped along range front structures including relay ramps and rollovers (Hopper, 1939, 1947; Dalrymple, 1963, 1964; Larsen, 1979; Stinson, 1977a, 1977b; Schweig, 1989; Sternlof, 1988; Coleman et al., 1987; Duffield and Bacon, 1981; Bacon et al., 1982; Bateman, 1965; Krauskopf, 1971). The erosion surface is irregular and characterized by broad flats with low relief and monadnocks that occur as relicts within mountain highlands. It is locally deeply dissected by late Pliocene and Quaternary drainages. The limited occurrence of a paleosol and sparse clastic deposition along segments of the disconformity, although not its entirety, suggest a broad, low-relief upland that was largely denuded of Tertiary rock prior to eruption of basalts across the region (Schweig, 1989; Jayko, 2008).

Much of the evolution of the eastern Sierra region is linked to displacement of the Sierra Nevada block (Fig. 1). The Sierra Nevada, which forms the east side of the Sierra Nevada block, has been recognized as gently west-tilted since geologists first arrived in California (LeConte, 1880, 1886, 1889; Lawson, 1904; Lindgren, 1911). Subsequent work has focused on the uplift and rotation history of the block (cf. Christensen, 1966; Huber, 1981, 1990; Unruh, 1991; Wakabayashi and Sawyer, 2001). Considerable recent attention has also been given to constraining the topographic history using a variety of isotopic and cosmogenic

dating methods (cf. House et al., 1998, 2001; Chamberlain and Poage, 2000; Stock et al., 2004, 2005; Clark et al., 2005; Cecil et al., 2006; Mulch et al., 2006). Far less attention has been given to the adjacent eastern Sierra region, which locally reaches similar elevation, has a similar upland or "mountain-valley" geomorphic history, and similar topographic relief along the range fronts.

Lindgren Surface, Sierra Nevada Block

Two lines of evidence, stratigraphic and geomorphic, indicate that the modern topographic edifice of the Sierra Nevada in excess of ~1500 m is relatively young, primarily late Miocene to Quaternary in age, and formed, at least in part, as a result of a mainly rigid rotation of a large crustal block, albeit with some local faulting (Lindgren, 1911; Christensen, 1966; Huber, 1981; Fig. 1). These observations have been supported by much subsequent work. The most important observations were detailed in Lindgren's (1911) report on Tertiary gravels of the Sierra Nevada; thus, the late Miocene erosion surface and overlying deposits in the Sierra Nevada are herein referred to as the Lindgren Surface.

The results of Lindgren's (1911) work showed that late Miocene strata, including interbedded volcanic flows, extended uninterrupted from the Sierra Nevada crest to the Great Valley. Mafic flows of the Mehrten Formation and trachyandesite of Kennedy Table, which were subsequently dated to ca. 9.5–10.2 Ma, provided constraints on the initiation and amount of westward tilt (Christensen, 1966; Dalrymple, 1964; Huber, 1981, 1990; Table 1). These units and the underlying erosion surface extend across the ~100 km width of the exposed block and have rotated ~1.4° westward since late Miocene time (Christensen, 1966; Huber, 1981; Unruh, 1991; Table 1). The flows and underlying erosion surface delineate an accordant ridge envelope that characterizes the upper surface of the Sierra Nevada block (Christensen, 1966; Wakabayashi and Sawyer, 2001; Fig. 1). Wahrhaftig (1965) described contrasting morphology in the southern Sierra where "stepped" topography was present rather than the fairly simple, gently west-dipping surface found in the northern Sierra Nevada. Christensen (1966), Matthes (1930), and Huber (1981) also noted this.

Lindgren (1911) and Christensen (1966) recognized that the axis of Sierra Nevada block rotation was about subhorizontal and trended ~330° to 340° near the western foothills of the Sierra Nevada (Fig. 1). Stratigraphic and seismic-reflection results show late Neogene syntectonic rotation and sedimentation in the Great Valley (Marchand and Allwardt, 1981; Unruh, 1991), which was previously recognized in the geomorphology (cf. Lindgren, 1911; Matthes, 1930, 1937; Huber, 1981). In addition, Lindgren (1911), Matthes (1930), and Huber (1981) described paleodrainages near the crest and westward that formerly had carried sediment from the east, and that had been offset by the Sierra Nevada frontal fault zone. These paleodrainages locally contain volcanic cobbles of Miocene age (ca. 11 Ma) from volcanic centers to the east, indicating truncation from former headwaters in late Miocene or younger time (Huber, 1981).

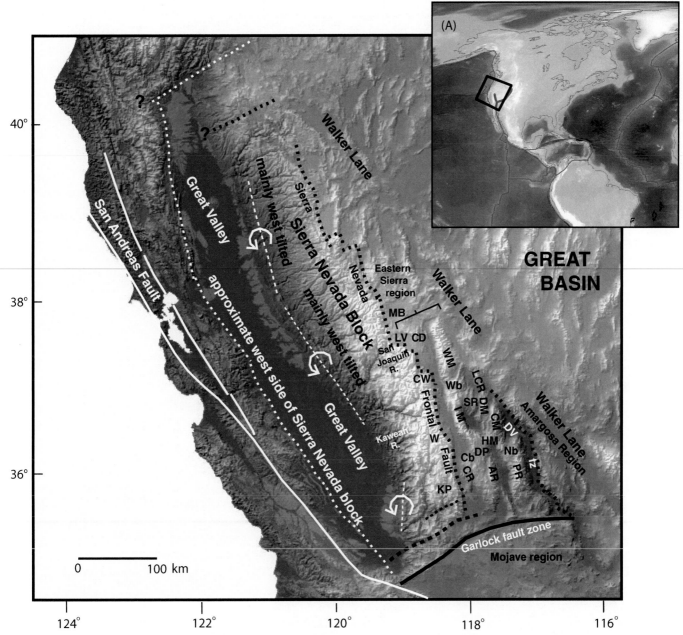

Figure 1. Map showing location of Sierra Nevada block, eastern Sierra region, and localities including: AR—Argus Range, CD—Casa Diablo, CM—Cottonwood Mountains, CW—Coyote Warp, DM—Dry Mountain, DP—Darwin Plateau, DVfz—Death Valley fault zone, HM—Hunter Mountain, KP—Kern Plateau, IM—Inyo Mountains, LCR—Last Chance Range, LV—Long Valley, MB—Mono Basin, SR—Saline Range. Paleobasins: Cb—Coso Basin; CR—Coso Range, Nb—Nova Basin, PR—Panamint Range, W—Mount Whitney, Wb—Waucobi Basin, WM—White Mountains. Inset map shows (A) western Pacific plate boundary and initiation of Pacific plate capture of a large crustal fragment of western North American plate associated with plate-boundary reorganization and cessation of Gorda plate subduction.

TABLE 1. CHARACTERISTICS OF THE LINDGREN SURFACE, SOUTHERN SIERRA NEVADA

Location	Unit	Age (Ma)	Reference	Elevation (m)	Morphology
Lindgren Surface, Kern Plateau*					
Summit-Upland		Pre–late Miocene relief (Eocene)	Dalrymple (1963)	3505	Minimum late Miocene paleorelief
Subsummit Plateau, Kern Plateau		Oligocene–late Miocene relief (?)	Matthes (1937); Dalrymple (1963)	2865	Late Miocene baseline
Ramshaw Meadow, Chagoopa S.		Hempillian-Clarendonian? 4.75–11.0	Axelrod and Ting (1961)	2435	Eroded during late Miocene
Bakeoven Meadow, Chagoopa S.		Hempillian	Axelrod and Ting (1961)	2440	Eroded during late Miocene
Kennedy Meadow, Chagoopa S.		Hempillian, 4.75 to 9.0	Axelrod and Ting (1961)	1890	Eroded during late Miocene
Chagoopa equivalent (?)		≥4.72	Stock et al. (2005)	1860	Late Miocene–early Pliocene baseline
Chagoopa surface		Late Miocene–late early Pliocene (?)	Matthes (1937); Dalrymple (1963)		Late Miocene–early Pliocene baseline
Chagoopa surface, confluence of Kern and Little Kern Rivers	Tb	≥3.8	Dalrymple (1963)		Late Miocene–early Pliocene baseline
Chagoopa surface, confluence of Kern and Little Kern Rivers	Tv	>3.5	Bacon and Duffield (1981)		Late Miocene–early Pliocene baseline
Alabama Hills		Hempillian, 4.75 to 9.0	Axelrod and Ting (1961)		Late Miocene baseline
Depocenter, Owens Valley			Pakiser (1964)		Late Miocene baseline (?)
Lindgren Surface, Coyote Warp, correlative surfaces and deposits[†]					
Cloudripper, The Hunchback		Summit-Upland	cf. Dalrymple (1963)	4120	Summit-Upland
Table Mountain		Subsummit Plateau, Kern Plateau	Figure 8B	3540	Subsummit Plateau
Table Mountain	Ts	Subsummit Plateau, Kern Plateau	Bateman (1965)	3550	Clastic deposit on Subsummit Plateau
Round Mountain	Tb	Subsummit Plateau, Kern Plateau	Bateman (1965)	3200	Subsummit Plateau
Near Little Egypt Prospect	Tb	9.6 ± .02	Dalrymple (1963)	2990	Subsummit Plateau tilted north, northeast
Coyote Flat	Tb	9.6 ± .02	Dalrymple (1963)	2070	Subsummit Plateau tilted north, northeast
Tungsten Hills	Tb	9.6 ± .02	Figures 8C and 8D	1840	Erosional relict, Subsummit Plateau
Red Hill	Tb	9.6 ± .02	Figures 8C and 8D	1400	Erosional relict, mainly buried by Qa
Unnamed, near Bishop Creek	Tb	9.6 ± .02	Figures 8C and 8D	1450	Erosional relict, mainly buried by Qa
Near Mill Creek pond	Tb	9.6 ± .02	Figures 8C and 8D	1350	Erosional relict, mainly buried by Qa
Casa Diablo			Figure 3A		Summit-Upland

Note: Tb—Tertiary basalt; Ts—Tertiary sedimentary rock; Tv—Tertiary volcanic rock; Qa—Quaternary alluvium.

*Minimum displacement across Sierra Nevada frontal fault, Lone Pine section from offset Chagoopa surface overlying the Alabama Hills to the depocenter ~2.2 km. Minimum uplift rate with respect to late Miocene: 0.22 mm/yr, and with respect to late Pliocene: 0.55 mm/yr. Minimum paleorelief in late Miocene: 640 m (1900 ft).

[†]Minimum displacement across Sierra Nevada frontal fault, Wheeler Crest section, from Coyote Warp basalts: ~2.2 km. Minimum uplift rate with respect to late Miocene: 0.22 mm/yr, and with respect to late Pliocene: 0.55 mm/yr. Minimum paleorelief in late Miocene: 680 m (1900 ft).

Uplift of the Sierra Nevada crest relative to middle Miocene and early Pliocene elevations is primarily a consequence of block rotation (Lindgren, 1911; Christensen, 1966; Huber, 1981, 1990; Unruh, 1991). This contrasts with a model of gravitational collapse from a much higher Andean-type altiplano of thickened continental crust (Wernicke et al., 1996; House, et al., 1998, 2001; Dilek and Moores, 1999). Gravitational collapse may represent part of a viable working model for the late Mesozoic and early Cenozoic, but it does not agree very well with the late Neogene geologic constraints. The information summarized in this report concerning the extent and elevation of an erosion surface east of the Sierra Nevada, provides additional scaling constraints for the extent of the surface that preceded development of modern relief along the Sierra Nevada frontal fault zone, and for the extent of the region affected by the mainly middle and late Pliocene lithospheric processes such as mantle delamination (cf. Ducea and Saleeby, 1996).

Paleoelevation of the Sierra Nevada

The first estimate for the minimum elevation of the Sierra Nevada before late Miocene uplift was ~762 m (2500 ft) based on the fossil flora (Axelrod, 1957, 1962). Huber (1981) estimated the minimum elevation in the late Miocene as ~900 m by subtracting the estimated Miocene and Pliocene uplift from the present range elevation. Recent (U/Th)/He isotopic studies in the southern and central Sierra Nevada indicate the elevation was around 850–2150 m (~2600 to 6600 ft) in Late Cretaceous time (Clark et al., 2005; House et al., 2001). Hydrogen isotopes from Eocene deposits in the northern Sierra Nevada indicate higher early Cenozoic elevations of ~2200 m (Mulch et al., 2006), somewhat higher than the (U/Th)/He results.

Pollen and other paleofloral remains indicate the elevation of the Sierra Nevada was subalpine by early–late Pliocene time (ca. 3.5 Ma), a range of elevations that is in part climate dependent (Axelrod and Ting, 1960, 1961; Christensen, 1966). Several different lines of evidence indicate that erosion and incision rates have been quite rapid during the Pliocene and Pleistocene relative to earlier times, which is considered indicative of a recent tectonic uplift signal (Lawson, 1904; Dalrymple, 1963; Stock et al., 2004; Clark et al., 2005; Ross, 1986). Early workers also noted the presence of significant late Miocene paleorelief, as much as 365–610 m (1200–2000 ft), where paleocanyons were inset on an erosion surfaces near the Sierran crest. This suggests nonstatic tectonic conditions prior to onset of late Miocene uplift (Bateman and Wahrhaftig, 1966; Christensen, 1966; Huber, 1981, 1990). Dalrymple (1963) noted at least ~1220 m (4000 ft), nearly double the relief in the Kern Canyon of the southern Sierra Nevada, by Pliocene time, ca. 3.5 Ma.

Timing of Sierran Uplift

The uplift that formed the modern topography is fairly well constrained, albeit occasionally attracting controversy over

the decades (Lindgren, 1911; Matthes, 1937; Axelrod, 1957; Dalrymple, 1963, 1964; Christensen, 1966; Gilbert and Reynolds, 1973; Huber, 1981; Rood et al., 2005). The late Miocene Table Mountain Latite, trachyandesite of Kennedy Table, and Mehrten Formations (Lindgren, 1911; Piper et al., 1939; Matthes, 1937, 1960; Bateman et al., 1971; Huber, 1981), deposited around 9.3–10.2 Ma (Dalrymple, 1963, 1964, corrected ages; Huber, 1981), constrain the initiation and amount of Sierra Nevada block tilting (Table 2). These units extend from the Sierra Nevada crest to the Great Valley. Stratigraphic studies from basins adjacent to the southeastern corner of the Sierra Nevada block show arrival of Sierran detritus ca. 8 Ma and extensional faulting around 9 Ma, which is interpreted as the timing of Sierra Nevada uplift or emergence (Loomis and Burbank, 1988). Likewise, stratigraphic studies from the Great Valley (Table 2) indicate that the Sierra Nevada block was relatively stable and not noticeably tilting between Oligocene and late Miocene time, a duration of ~25 m.y. (Lindgren, 1911; Marchand and Allwardt, 1981; Unruh, 1991; Huber, 1981, 1990). About 1.4° tilting of the Sierra Nevada block followed deposition of the Mehrten Formation (Dalrymple, 1963; Marchand and Allwardt, 1981; Huber, 1981), resulting in ~2100–2400 m of uplift near the Sierran crest (Christensen, 1966; Huber, 1981; Fig. 2; Table 2).

Recent (U/Th)/He isotopic results on apatite suggest that the modern Sierra Nevada topography is Late Cretaceous and early Eocene in age (House et al., 1998, 2001), based on assumptions about erosion rate and source of He in the rocks. However, sedimentologic studies of Jurassic to Cretaceous deposits in the Great Valley have shown that Mesozoic plutons of the Sierra Nevada were exhumed and being eroded by Late Cretaceous time (Ingersoll, 1982; Linn et al., 1992). Late Cretaceous exhumation of the batholith is also supported by fission-track cooling ages that indicate rapid cooling between 65 and 75 Ma (Dumitru, 1990). Also, (U/Th)/He data on zircon from the northern Sierra Nevada show rapid cooling between 90 and 66 Ma, consistent with Cretaceous exhumation rates of 0.2–0.8 mm/yr followed by a much slower rate during the Cenozoic (Cecil et al., 2006). Therefore, a model that assumes a long-term, steady-state erosion rate of ~0.1 mm/yr since the Late Cretaceous does not fit with the late Miocene to Pleistocene topographic history of the Sierra Nevada (House et al., 1998, 2001), as discrete tectonic uplift events are indicated by the geologic and geomorphic record (cf. Lawson, 1904; Christensen, 1966; Marchand and Allwardt, 1981; Huber, 1981, 1990).

Uplift Rates, Sierra Nevada Block

The Sierra Nevada block extends from the Sierra Nevada frontal fault in the east to the Coast Range fault in the west (cf. Wright, 1976; Wakabayashi and Sawyer, 2001; Fig. 1). Stratigraphic studies from the west side of the Sierra Nevada indicate the Sierra Nevada block tilted ~0.3° to 0.6° around an axis trending ~310° to 325° between the middle Eocene and Oligocene (Marchand and Allwardt, 1981; Table 2; Fig. 2). This early Cenozoic tilting occurred within an ~25 m.y. span and resulted in

TABLE 2. STRATIGRAPHIC UNITS AND GRADIENTS THAT INDICATE CENOZOIC ROTATION OF THE SIERRA NEVADA BLOCK

Unit	Stratigraphic age	Unit characteristics and (or) depositional environment*	Reference for age control*	Approx. age picks (Ma)*	Dip (°) converted from published values of gradient (m/km)						Strike of beds (°)	Estimated average age (Ma)	Average dip (°)	Uplift (m)
					Oroville†	American River§	San Joaquin§	Merced River#	San Joaquin River#	Kennedy Table**				
Alluvium	Holocene	Alluvial fan; initial dip (?)	Unruh (1991)	0.01–0				0.05	0.045	0.057		0.01	0.004	0
Modesto Formation	Late Pleistocene, last major aggradation event, therefore last glacial?	Alluvial fan (dip may be primary slope?)	Marchand and Allwardt (1981); Burow et al. (2004)	0.05–0.01		0.05–0.12	0.08	0.10–0.07	0.07–0.05		330	0.025	0.05	
Riverbank Formation	Middle to late Pleistocene	Fluvial	Marchand and Allwardt (1981); Burow et al. (2004)	0.781–0.126?	0.09–1.5 (local deform.?)			0.11–0.16	0.09–0.11		330	0.45	0.1	
Turlock Lake Formation	Early Pleistocene	Lower part—Corcoran Clay (proglacial lake)—is conformably overlain by "Friant pumice member," which is distal Bishop Ash ca. 0.760 Ma	Sarna-Wojcicki et al. (2000); Marchand and Allwardt (1981); Burow et al. (2004)	Upper part ca. 0.76; lower part ca. 1 Ma or older			0.2	0.18	0.16		330	0.76	0.17	290
Arroyo Seco Gravel	Early Pleistocene	Glacial outwash (~Sherwin till age inferred, this study)	Cherven (1984)	ca. 1(?) from interpolation and correlation with glacial event		0.22						1	0.22	
North Merced Gravel	Early Pleistocene or Pliocene	Locally overlies Laguna Formation and locally erosional pediment surface, the North Merced pediment (interglacial age inferred here)	Marchand and Allwardt (1981); Burow et al. (2004)	ca. 1.1 to 1.8(?) and from interpolation and correlation with interglacial event			0.17–0.27	0.2–0.46			330–335	1.5	0.33	
China Hat Gravel; upper member of Laguna Formation (Marchand and Allwardt, 1981)	Pliocene	Glacial outwash (?); may be equivalent to parts of Laguna Formations (?); (~McGee till age inferred, this study)	Marchand and Allwardt (1981); Cherven (1984)	ca. 2.5 (?) from interpolation and correlation with glacial event	0.6		0.6	0.6			320–325	2.5	0.6	1100
Laguna Formation (northern San Joaquin Valley)	Late Pliocene	Unconformably overlies the Mehten Formation; correlated with deposits containing the 3.4 Ma Nomlaki Tuff (Busacca et al., 1982) near the Yuba River; correlated with gravels containing 3.76 Ma basalt clasts north of Kings River.	Busacca et al. (1982); Marchand and Allwardt (1981); Page and Balding (1973); Burow et al. (2004)	ca. 5.0–3.0(?)		0.98	0.9					3.6	0.94	1750

(continued)

TABLE 2. STRATIGRAPHIC UNITS AND GRADIENTS THAT INDICATE CENOZOIC ROTATION OF THE SIERRA NEVADA BLOCK (continued)

Unit	Stratigraphic age	Unit characteristics and (or) depositional environment*	Reference for age control*	Approx. age picks (Ma)*	Dip (°) converted from published values of gradient (m/km)						Strike of beds (°)	Estimated average age (Ma)	Average dip (°)	Uplift (m)
					Oroville†	American River§	San Joaquin§	Merced River§	San Joaquin River§	Kennedy Table**				
Laguna Formation (continued)		Lower unit is magnetically reversed; upper unit is normal, suggesting the boundary includes the Gilbert-Gauss reversal at ca. 3.4 Ma; also has been correlated with the Turlock Fm. (Cherven, 1984). The strike and dip suggest the Pliocene correlation is more probable. Either Pliocene age or, if Pleistocene, then locally deformed.												
Mehrten Formation (northern San Joaquin Valley)	Late Miocene	Fluvial; dominantly andesitic volcanic detritus (ca. 9.8 Ma)	Dalrymple (1964); Burow et al. (2004)	9.3–9.8 (recalculated)		1.0–2.0	1.3–1.4	1.1	1.26–1.54		325	9.5	1.4	2600
Trachyandesite site of Kennedy Table	Late Miocene	A single trachyandesite flow (10.2 Ma)	Huber (1981)	10.2						1.28–1.55 (1.22)	320	10.2	1.4	2600
Lovejoy Basalt (northern Sacramento Valley)	Miocene	Basalt	Coe et al. (2005)	14–17		1.4						16	1.4	2600
Valley Springs Formation (northern San Joaquin Valley)	Oligocene and (or) early Miocene	Mainly fluvial and wetlands deposits consisting of rhyolitic ash and nonmarine clastic sediments	Bartow (1994)	36–19		1.5–2	1.2–1.35	1.0–2.0			325–310	23	1.5	2600
Ione Formation (northern San Joaquin Valley)	Middle Eocene	Fluvial, lagoonal, marine; (ca. 50 Ma) Ulatisian stage	Marchand and Allwardt (1981)	52–40, most likely (50) Ulatisian stage		5 (not included in average)		1.72	1.74	2	310	50	2	3500

Notes: Early Pleistocene: 1.806 to 0.781 Ma; Middle Pleistocene: 0.781 to 0.126 Ma; Late Pleistocene: 0.126 to 0.011 Ma. Dip values are from Unruh (1991). Only units with published stratigraphic age control and/or dated tuffs are included here. Strike values are from Marchand and Allwardt (1981).
*Compiled from this study.
†Unruh (1991).
§Sources cited in Unruh (1999).
#Marchand and Allwardt (1981).
**Huber (1981).

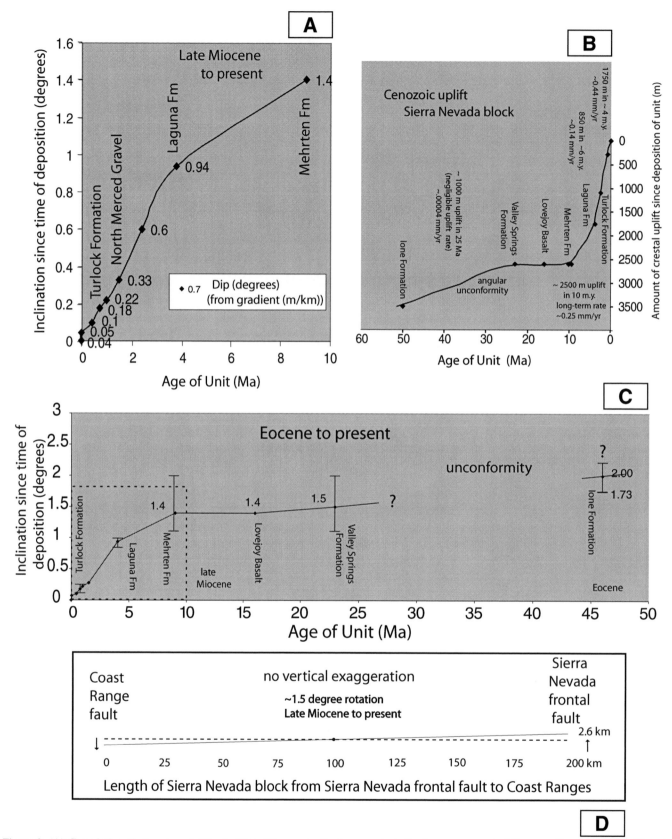

Figure 2. (A) Cumulative dip (westward tilting) of late Miocene to Quaternary strata (in degrees, converted from gradients [m/km]); (B) estimated cumulative uplift of the Sierra Nevada from block rotation during the Cenozoic; (C) cumulative dip of Eocene to Quarternary strata (in degrees, converted from gradients [m/km]); and (D) scaled schematic drawing showing the approximate length of Sierra Nevada block, tilting, and estimated uplift; no vertical exaggeration.

~1000–1300 m of uplift (Christensen, 1966; Huber, 1981; Table 2; Fig. 2), with a fairly negligible uplift rate of ~0.0004 mm/yr. The uplift and uplift rates (Table 3; Fig. 2) presented here were calculated assuming an ~200-km-wide rigid block with a centrally located axis of rotation (Christensen, 1966; Huber 1981; Fig. 2). The axis of rotation lies near the interface between the Great Valley and Sierran foothills (Christensen, 1966). It presently trends ~330°, about normal to the gradient (km/m) of the late Cenozoic strata exposed in the western Sierran foothills (cf. Marchand and Allwardt, 1981). The rotation axis shifted clockwise to its more northerly 330° position from ~320° to 325° during the late Miocene and Pliocene uplift event (Marchand and Allwardt, 1981; Table 2). About 6° ± 8° of clockwise rotation corroborates paleomagnetic results from Cretaceous plutons in the south-central part of the Sierra Nevada, which indicate a shift of the block after ca. 90 Ma (Frei, 1986).

About 1.4° of late Miocene (ca. 10 Ma) and younger tilting followed deposition of extensive mafic flows that are widely distributed on the perched erosion surfaces of the Sierra Nevada block (Christensen, 1966; Huber, 1981; Marchand and Allwardt, 1981). This tilting resulted in ~2100–2600 m of uplift, which gives an average uplift rate of ~0.25 mm/yr (Huber, 1981). However, the stratigraphic record in the Great Valley (Marchand and Allwardt, 1981; Huber, 1981) and the timing of volcanic activity in the Sierra Nevada indicate that the tilt rate, and so also the uplift rate, increased significantly at ca. 4.0–3.5 Ma. The uplift estimated from rigid block rotation is ~850 m during the late Miocene and early Pliocene (ca. 10–4 Ma), and ~1750 m during the late Pliocene and Pleistocene (ca. 4 Ma to present), suggesting an initial uplift rate of ~0.14 mm/yr that increased to ~0.44 mm/yr in the late Pliocene (Table 2; Fig. 2). The units that provide important constraints for the differential tilting rates during the last 10 m.y. are from: (1) the Friant pumice member of the Turlock Lake Formation, which contains the Bishop Tuff (ca. 760 ka) (Sarna-Wojcicki et al., 2000; Marchand and Allwardt, 1981); (2) part of the Laguna Formation, which is correlated with gravels near the Yuba River containing the Nomlaki Tuff (ca. 3.4 Ma) (Busacca et al., 1982, 1989), (3) gravels near the King River containing a ca. 3.76 Ma basalt clast, and Laguna Formation deposits in Merced County that contain a magnetic reversal interpreted as the Gilbert-Gauss reversal at ca. 3.4 Ma (Marchand and Allwardt, 1981); (4) mafic flows and clastic rocks of the Mehrten Formation at ca. 9.8 Ma (Dalrymple, 1963); and (5) trachyandesites of Kennedy Table at ca. 10.2 Ma (Huber, 1981) (Table 2; Fig. 2). The early Tertiary stratigraphic and geomorphic record indicates that major river systems occupied many of the areas that are now in the crestal highlands, as well as the entire west side of the block (Matthes, 1930; Huber, 1981).

Sierra Nevada Frontal Fault Zone

The east side of the Sierra Nevada was recognized as a fault-controlled escarpment bounding the west-tilted "Sierran" block in the late 1800s and early 1900s by Fairbanks (1898), Reid (1906), and Lindgren (1911). Lindgren (1911) concluded that uplift and

incision of the erosion surface preceded initiation of the Sierra Nevada frontal fault based on offset drainages and canyons with Miocene deposits. Regional mapping in the Owens Valley area by Knopf (1918) included a geomorphic description of the faceted spurs along the Sierran Range front west of Owens Valley, along with a detailed description of numerous fault scarps on Quaternary deposits over the length of the valley. The range front fault, which also represents the western boundary of the Great Basin, has subsequently been referred to as the Sierra Nevada frontal fault (cf. Evernden, 1980; Greensfelder et al., 1980; Unruh, 1991; Wakabayashi and Sawyer, 2001; Slemmons et al., 2008) and as the Sierra Nevada fault zone (Jennings, 1994). The Sierra Nevada frontal fault is part of a zone of active seismicity referred to as the Eastern California Seismic Zone, which includes the northern part of the Walker Lane and the region that lies between Death and Owens Valleys (cf. Wallace, 1984).

Several studies along the central part of Sierra Nevada frontal fault, including Huber (1981) and Gilbert and Reynolds (1973), have indicated that significant vertical displacement initiated around 8–10 Ma. Voluminous basaltic outpouring around 8–10 Ma is also interpreted as an indication of the break-up of the late Miocene landscape (Dalrymple, 1963, 1964; Larsen, 1979; Huber, 1981; Sternlof, 1988; Schweig, 1989). Although the Sierra Nevada frontal fault appears to have developed significant activity around late Miocene time, it is also recognized as having greater Pliocene-Pleistocene displacement by Matthes (1930, 1960) and Blackwelder (1931) and many subsequent workers. These observations are supported by recent erosion and incision rate studies of the Sierra block that are constrained by cosmogenic (U/Th)/He data (Stock et al., 2004, 2005; Clark et al., 2005). These isotopic results indicate high Pliocene rates relative to Miocene and earlier time. A late Pliocene lithospheric decoupling event was deduced as the likely driving mechanism from studies of xenoliths in late Miocene and Pliocene basalts (cf. Ducea and Saleeby, 1996).

The largest displacement across the Sierra Nevada frontal fault occurs near the highest elevations of the Sierran crest near Mount Whitney (3230 m; Fig. 3). In southern Owens Valley, gravity studies indicate that ~2440 +610 m of basin fill lie below the playa floor (Pakiser et al., 1964), suggesting that displacement across the frontal fault is locally ~5.7 ± 0.6 km. Approximately 2–3 km of displacement can be accommodated by rigid block rotation (Huber, 1981; Unruh, 1991; Table 2; Fig. 2), suggesting the additional ~3 km of has been accommodated by other displacement mechanisms. Late Quaternary vertical slip rates of ~0.4–0.5 mm/yr from late Pleistocene alluvial-fan and lacustrine deposits in southern Owens Valley, and Pleistocene cinder cones of the Big Pine area in Owens Valley requires ~8–10 m.y. of fault-slip activity along the Sierra Nevada frontal fault to develop the present relief (Table 3).

Apparent fault slip and uplift rates range from ~0.3 to 1.0 mm/yr along different parts of the Sierra Nevada frontal fault (see references cited in Table 3). The highest displacement rate of ~1.0 mm/yr from both offset Pliocene deposits and from late

TABLE 3. REPRESENTATIVE UPLIFT AND VERTICAL SLIP RATES, SIERRA NEVADA FRONTAL FAULT AND ADJACENT EASTERN SIERRA REGION

Uplift rates, eastern Sierra	Reference	Average uplift/dip-slip rate (mm/yr)	Duration (m.y.)	Constraining deposit
Eastern Sierra				
Owens Gorge	Curry (1971)	0.3	3.2	Owens Gorge basalt
San Joaquin crest	Curry (1971)	0.25	2.7	Two Teats quartz latite
San Joaquin crest	Curry (1971)	0.34	0.76	Bishop Tuff
San Joaquin crest	Curry (1971)	0.37	3.1	Andesite of Deadman Pass
Mono Basin	Gilbert et al. (1968)	0.45	3	Mono Lake Rocks
Sierra escarpment, Mono quadrangle	Kistler (1966)	0.41	0.94	Sherwin till?
Bloody Canyon, Mono Basin	Phillips et al. (1990)	0.77	0.2	Mono Basin moraines
Long Valley	Clark and Gillespie (1981)	1.1	0.2	Hilton Creek fault
White Mtns.	DePolo (1989)	0.9		Marble Creek
East of Big Pine	Bachman (1978)	0.5	2.3	Waucobi Lake Beds
Sierra Nevada frontal fault zone				
Owens Valley fault zone	Beanland and Clark (1994)	0.35	300	Lake deposits
Owens Valley fault zone	Bacon et al. (2005)	0.12	9	Lake and delta deposits
Fish Springs fault	Martel et al. (1987)	0.2	314	Quaternary cinder cone
Fish Springs fault	Zehfuss et al. (2001)	0.24	300	Alluvial fan
Lone Pine	Lubetkin and Clark (1988)	1.0	10–21	Alluvial fan
Lone Pine area	Martel et al. (1987)	0.4	~0.4	Cinder cone
Sierra Nevada frontal fault, Independence	Gillespie (1982)	~0.25	0.46	Quaternary basalt
Sierra Nevada frontal fault, Independence	Le et al. (2007)	0.5	0.025	Alluvial fan
Sierra Nevada frontal fault, Hartley Springs	Bursik and Sieh (1989)	0.9	0.2	Tahoe moraine
Sierra Nevada frontal fault, Reversed Peak	Bursik and Sieh (1989)	0.5	0.2	Tahoe moraine
Sierra escarpment, Mono quadrangle	Kistler (1966)	0.41	0.84	Sherwin till?
Sierra Nevada frontal fault, Mono Basin	Huber (1981)	0.25	10	Late Miocene volcanic rock
Adjacent faults, eastern Sierra region				
Fish Lake Valley fault zone	Reheis and Sawyer (1997)	0.2–0.5	10	
Deep Springs fault	Reheis and Sawyer (1997)	0.35		Alluvial fan
Emigrant Peak fault zone	Reheis and Sawyer (1997)	0.35		Alluvial fan
Dyer Section	Reheis and Sawyer (1997)	0.2		Alluvial fan
Chiatovich Creek	Reheis and Sawyer (1997)	0.3		Alluvial fan
Death Valley fault zone, Willow Creek	Klinger (1999)	0.5	Holocene	Alluvial fan
Dry Mountain fault	This study	0.4	4	Pliocene basalt
Ash Hill fault, Panamint Valley	Densmore and Anderson (1997)	0.1	~0.2	Shoreline
Panamint Valley fault	Vogel et al. (2002)	0.35	0.9	Alluvial fan
Darwin Plateau	Gillespie et al. (2002)	0.5	6	Miocene basalt

Note: Rates for the Eastern Sierra represent uplift. Rates for the Sierra Nevada frontal fault zone and adjacent faults represent dip.

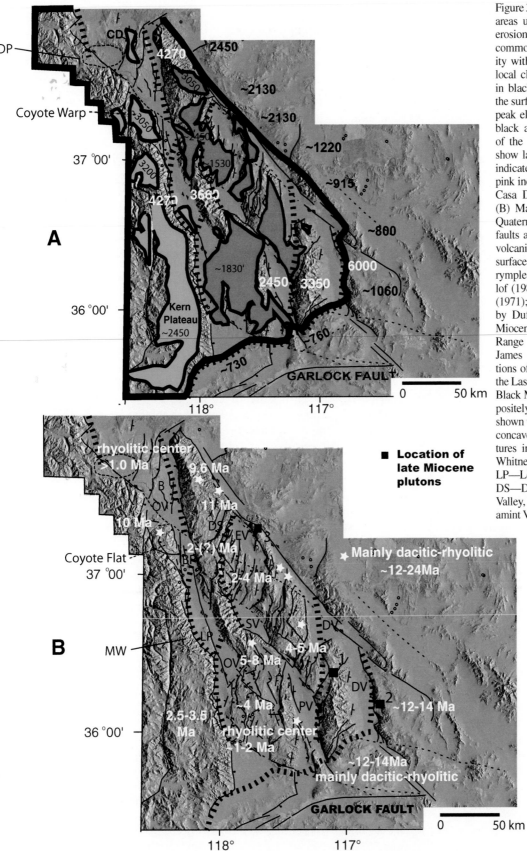

Figure 3. (A) Map showing generalized areas underlain by the late Miocene erosion surface. The erosion surface is commonly overlain by, or in continuity with, extensive basaltic flows and local clastic basin deposits. Numbers in black are generalized elevation of the surface, numbers in white are local peak elevations, and numbers in bold black are baseline elevations outside of the high-relief area. Dashed lines show large breakaway zones. Orange indicates the Lindgren Surface, and pink indicates the Inyo Surface. CD—Casa Diablo, DP—Devils Post Pile. (B) Map showing location of major Quaternary and Pliocene-Quaternary faults and crustal blocks. Star—dated volcanic rocks overlying the erosion surface; basalts were dated by Dalrymple (1963), Larsen (1979), Sternlof (1988); Schweig (1989), and Hall (1971); Coso volcanic field was dated by Duffield and Bacon (1981); and Miocene volcanics in southern Argus Range were dated by Evernden and James (1964). Filled squares—locations of three late Miocene plutons in the Last Chance, Panamint Range, and Black Mountains. Very large-scale oppositely verging breakaway zones are shown with hachure; vergence is in the concave direction with rollover structures in the footwalls. MW—Mount Whitney, B—Bishop, BP—Big Pine, LP—Lone Pine, OV—Owens Valley, DS—Deep Spring Valley, SV—Saline Valley, EV—Eureka Valley, PV—Panamint Valley, DV—Death Valley.

Quaternary fault scarps was determined along the scarp in the Mono Basin and Long Valley area. The lower, more characteristic rates of ~0.2–0.5 mm/yr are capable of developing the escarpment in the 9–10 m.y. duration suggested by the geologic indicators. It is impressive that the measured fault-slip and uplift rates determined from offset Quaternary to Pliocene horizons are in good agreement with the uplift rates estimated from rigid block rotation based on the dips of Cenozoic units along the west edge of the exposed block.

Tectonic Setting of the Sierra Nevada Block and Eastern Sierra Region

Late Miocene and Pliocene rotation and uplift of the Sierran crest has been concurrent with northwestward displacement of the Sierra Nevada block resulting from Pacific plate motion (cf. Hay, 1976; Frei, 1986; Unruh et al., 2003; Jones et al., 2004). The transtensive tectonics of the eastern Sierra region are a consequence of Pacific–North American plate motion east of the San Andreas system (cf. Hay, 1976; Wallace, 1984; Frei, 1986). This transtensive tectonic regime controls the structural character of the region and has been noted since the 1960s and 1970s (Shaw, 1965; Burchfiel and Stewart, 1966; Hay, 1976; Slemmons et al., 1979; Stewart, 1985, 1992). More recently, crustal displacements have been determined by global positioning system (GPS) and paleoseismic studies (Savage and Lisowski, 1995; Bennett et al., 1999; Dixon et al., 2000; Miller et al., 2001; Reheis and Sawyer, 1997; Unruh et al., 2003).

Early workers recognized that westward tilting of the Sierran block was controlled by deep crustal or lithospheric processes due both to the dimensions of the block (LeConte, 1886; Christensen, 1966) and to consideration of the rheological strength of the crust (Christensen, 1966). Subsequent work has verified this concept and resulted in the currently held model developed from petrologic observations (Ducea and Saleeby, 1996) and geophysical studies (Jones et al., 2004) that that uplift is in part driven by Pliocene lithospheric delamination.

Objective

The objective of this paper is to provide documentation for the extent and deformation of the late Miocene erosion surface in the eastern Sierra region, herein named the Inyo Surface. The Inyo Surface was mapped at a scale of 1:100,000 in the southern part of the eastern Sierra region during an investigation of the Darwin Hills 30′ × 60′ quadrangle (Jayko, 2008). It has also been observed in reconnaissance and compiled from previous mapping elsewhere in the area (Fig. 3). Cross sections and topographic profiles were used to delineate the structural character and to estimate vertical separation rates. The Lindgren Surface in the Sierra Nevada and the Inyo Surface in the eastern Sierra region are mapped where unusually flat or gently dipping surfaces and arrays of accordant ridges lie adjacent to late Miocene deposits perched in mountain upland areas and along range fronts. These

surfaces generally form broad, low-relief areas within the mountain highlands (Figs. 3 and 4). In some places, it can be shown that the Lindgren Surface of the Sierra Nevada was formerly continuous with part of the Inyo Surface west of the Sierran crest, as described next.

THE INYO SURFACE

Four prominent late Miocene to Pliocene erosion surfaces that include a broad mountain uplands, the Lindgren Surface, have been described in the southern and central Sierra Nevada (Lawson, 1904; Matthes, 1930). The mountain uplands is overlain by ca. 10 Ma mafic flows in the central and northern Sierra Nevada (Lindgren, 1911; Christensen, 1966; Huber, 1981). The best chronologic link between the late Neogene erosion surfaces recognized by Lawson (1904) and mountain uplands overlain by late Miocene basalt flows can be established between the Kern Plateau and Coyote Flat areas in the south and southeastern Sierra Nevada (Dalrymple, 1963). Therefore, the description of the Inyo Surface in the eastern Sierra region will be prefaced by an overview of fluvial strath or erosion surfaces overlain by late Miocene and Pliocene mafic lava flows in the southeastern Sierra Nevada.

A relict of the late Miocene paleolandscape is best preserved on the Kern Plateau, southeastern corner of the Sierra Nevada block, but the erosion surface is largely stripped of Cenozoic cover rock that would provide chronostratigraphic control. It is, however, an excellent place to observe the landscape relicts with vestiges of the late Miocene morphology and older uplands. This morphology and its relative position in the modern landscape can be traced north to the Coyote Flat area, where late Miocene mafic flows constrain the relative age of these mountain uplands. Same-age mafic flows and relict landscape can be found in the White Mountains east of the Sierra Nevada. From the White Mountains, the Inyo Surface can be traced more or less continuously to the Darwin Plateau and northern Panamint Range, as described in the following sections.

The Inyo Surface commonly occurs adjacent to, or in close proximity to late Miocene and (or) early Pliocene basin sediments, pediment surfaces, and volcanic flow deposits, most commonly basalts and basaltic andesites (see following discussion). In addition, at many localities, rare resistant lag deposits that commonly consist of quartzite and basaltic cobbles, or other resistant clast types, overlie the relict erosion surface. These erosion surfaces are generally irregular, tilted, or slightly undulatory and cap accordant ridges and flats. Areas of paleohighlands, with generally gently sloping and rounded monadnocks indicating paleorelief up to 700 m, are described in the following sections. There are only a few isolated occurrences where early and middle Cenozoic deposits are preserved in the ranges west of the Death Valley fault zone (Fig. 5). Deposits of mainly middle Miocene and locally of late Oligocene age, which represent relicts of the earlier subdued pre–Sierra Nevada landscape, occur as small pockets only in the southern and eastern part of the eastern Sierra region (Fig. 5).

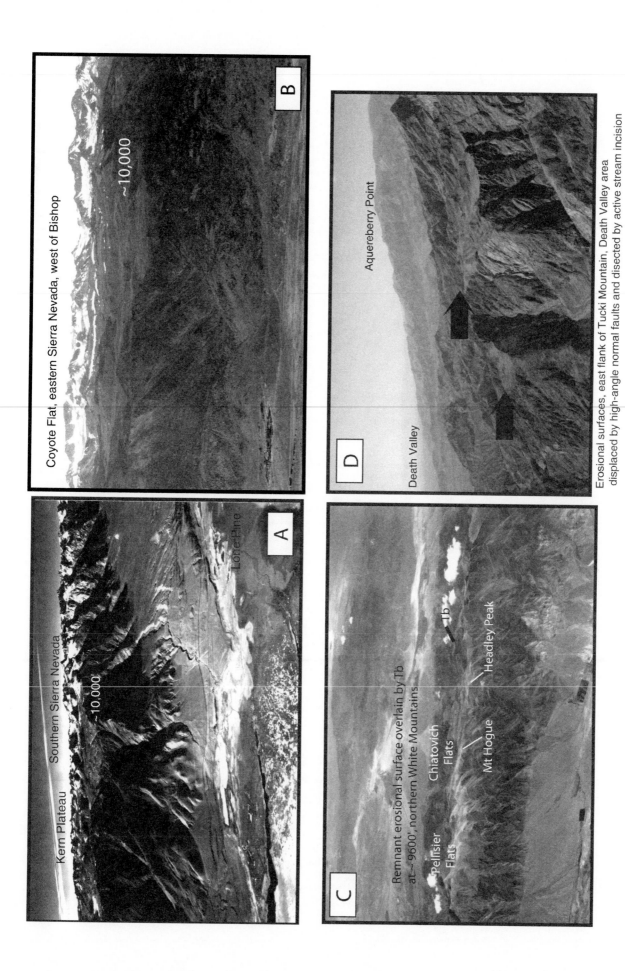

Figure 4. Oblique aerial photographs showing relicts of the Lindgren and Inyo Surfaces: (A) the Kern Plateau south of Mount Whitney in the Sierra Nevada; (B) Coyote Flat in the Sierra Nevada east of Bishop; (C) Chiatovich Flats in the northern White Mountains; and (D) the east side of Tucki Mountain in the northern Panamint Range, west of Death Valley. Tb—Tertiary basalt flows.

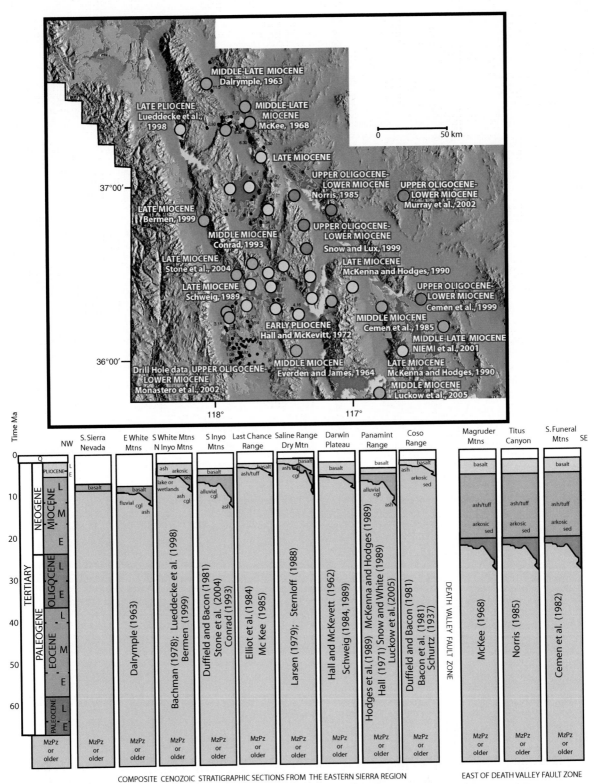

Figure 5. Map showing the location of late Miocene deposits shaded with respect to the oldest deposit at the base of the section that overlies Mesozoic or older basement rock. Selected references are given for description of sections at locations indicated. Schematic stratigraphic sections show age range of Tertiary deposits overlying the Inyo Surface. Time is on the left. Sections are meant to highlight age of the oldest Tertiary shown shaded according to time. Sections are tapered to schematically show the relative abundance of Tertiary-age rocks. The sections are drawn to the age of youngest strata in the area. Sections are generally late Miocene and (or) Pliocene west and northwest of Death Valley and middle Miocene or older east of Death Valley. MzPz—Mesozoic-Paleozoic.

The Kern Plateau

In the southern Sierra Nevada, an area commonly referred to as the Kern Plateau (Figs. 1, 4A, and 6) is eroded and largely denuded of Cenozoic rocks except for sparse early late Pliocene volcanic rocks that are ca. 3.5–2.4 Ma in age (Dalrymple, 1963; Bacon and Duffield, 1981). The broad expanse of fairly flat-lying, eroded terrain is recognized as an erosion surface that preserves a paleolandscape formed during three notable Cenozoic events of different uplift magnitude, originally described by Lawson (1904) and subsequently by Knopf (1918) and Dalrymple (1963).

The oldest relict landscape, called the Summit Upland by Lawson (1904), extends from the crest at ~4400 m to ~3500 m and defines the minimum relief and incision of mainly Paleogene relict landscape (Figs. 7 and 8A). The Subsummit Plateau, which generally lies at about ~3550 m elevation, was first described by Lawson (1904) and also was well documented by Knopf (1918). It forms a broad erosion plain, mainly a broad fluvially eroded strath within this mountainous landscape. There was considerable paleorelief from an older phase of uplift that gradually diminished during the middle Cenozoic. The relief above the Subsummit Plateau, effectively above a fluvially eroded slope break, to the Summit Upland is often also referred to as the "Eocene" surface (cf. Matthes, 1930, 1960; Dalrymple, 1963). The Chagoopa Plateau (Figs. 7 and 8A) is inset into the Subsummit Plateau, and both are highly incised by deep canyons, resulting in a compound paleoerosion surface (Lawson, 1904).

Discontinuous relicts of the Summit Uplands, Subsummit Plateau, and Chagoopa Plateau erosion surfaces are present north of the Kern Plateau along the Sierra Nevada crest. These relicts can be tracked discontinuously to the Coyote Flat area, near Bishop (Figs. 4C, 8B, and 8C) where the Subsummit Plateau surface at ~3550 m is overlain by basalt that erupted ca. 9.8 Ma (Dalrymple, 1963). These classic erosion surfaces of Knopf (1918) and Lawson (1904) can be interpreted with respect to the available radiometric control and episodes of tilting (Tables 1 and 4), giving first-order approximations for ages of the relict geomorphic surfaces of the Kern Plateau (Figs. 7 and 8; Table 5), which are here also extended to the eastern Sierra region (Fig. 3; Tables 1 and 4) using age-correlative basalts in the White Mountains.

Dalrymple concluded that the broad Chagoopa surface, Chagoopa Plateau of Lawson (1904), Axelrod and Ting (1961), and Dalrymple (1963), which underlies late Pliocene (3.5 Ma and younger) volcanic rocks, formed during the late Miocene ca. 10–15 Ma. However, recent work indicates it may be more narrowly constrained to ca. 5–10 Ma from the time that tilting of Mehrten Formation and trachyandesite of Kennedy Table occurred (Table 1; Fig. 2). Cosmogenic exposure ages suggest the erosion surface at ~1900 m (6200 ft), the Chagoopa equivalent(?) in the South Fork Kaweah is older than 4.7 Ma (Stock et al., 2005), therefore predating the late Pliocene ca. 4.0–3.5 Ma volcanic event in the Sierra Nevada and eastern Sierra region (Dalrymple, 1963, 1964; Hall, 1971; Sternlof, 1988; Larsen, 1979; Duffield and Bacon, 1981).

In addition, the deep canyons of both the Kern and upper San Joaquin drainages had incised significantly, to almost half their present depth by ca. 3.5 Ma, when basalts partially filled the gorges (Dalrymple, 1963; Huber, 1981). The Pine Flat basalt at Devils Postpile in the Upper San Joaquin River (ca. 3.5 Ma) filled the upper 305 m, and the river subsequently eroded an addition 400 m (Huber, 1981). A similar observation was made by Dalrymple (1963) of ca. 3.5 Ma basalt flows in the Kern River that filled deep paleochannels, indicating that the "Canyon stage" was well established prior to 3.5 Ma (Fig. 7).

Axelrod's reports in the 1960s concerning the late Neogene flora of the Sierra Nevada are of value if his interpreted ages are reconsidered in light of new constraints from a variety of sources since 1960, including significant changes in designation of stage boundaries based on new radiometric dating. Axelrod and Ting (1961) had interpreted the age of his cold/wet flora on the Chagoopa surface as equivalent to the first recorded glaciation (McGee-Glacier Point tills), which would therefore be Kansan, i.e., Pleistocene. However, the oldest dated glaciation is now recognized at ca. 3.2–2.8 Ma, or late Pliocene (cf. Curry, 1971; Huber, 1981, corrected ages; Phillips et al., 2000). Therefore, the conclusions about the timing and magnitude of differential uplift based on floral changes inferred in Axelrod and Ting (1961) report become more consistent with modern results from isotopic exposure dating when the ages are reevaluated.

Also, at the time his reports were published, the Hempillian and Clarendonian both were considered to be Pliocene ages (cf. Dalrymple, 1963). Hempillian is now dated as 4.75–9.0 Ma, mainly early Pliocene and late Miocene, and the Clarendonian is 9.0–11.8 Ma, i.e., mainly late Miocene. Axelrod and Ting (1961) stated in a footnote (p. 140) "It should be noted that if we follow the age-curve analysis proposed by Barghoorn (1951) and revised by Wolfe and Barghoorn (1960) then the Ramshaw Meadows flora would be ~Early Pliocene (thus ~Clarendonian)....; the Bakeoven Meadows would be Middle Pliocene (Hempillian)..; and the Alabama Hills Middle Pliocene (Hempillian)." Therefore, the deposits on the Kern Plateau and the correlative offset surface at the top of the Alabama Hills (Figs. 3B and 6) are most likely ca. 5–11 Ma in age.

Coyote Flat and the Coyote Warp

The most important chronostratigraphic and geomorphic link between the late Miocene Lindgren Surface of the Sierra Nevada block and the equivalent Inyo Surface of the eastern Sierra region is located at Coyote Flat, south of Long Valley and west of Bishop. Basaltic rocks ca. 10.0 ± 0.2 Ma (whole-rock K/Ar, corrected age) overlie a late Miocene erosion surface in both the eastern Sierra Nevada at Coyote Flat and in the White Mountains (uncorrected ages of 9.6 ± 0.2 Ma and 10.8 ± 1.0 Ma, respectively; Dalrymple, 1963; Figs. 3 and 8D). These mafic rocks on Coyote Flat and the adjacent "Coyote Warp" (Von Huene et al., 1963; Pakiser et al., 1964), here informally referred to as the Coyote Warp basalts, are equivalent in

328

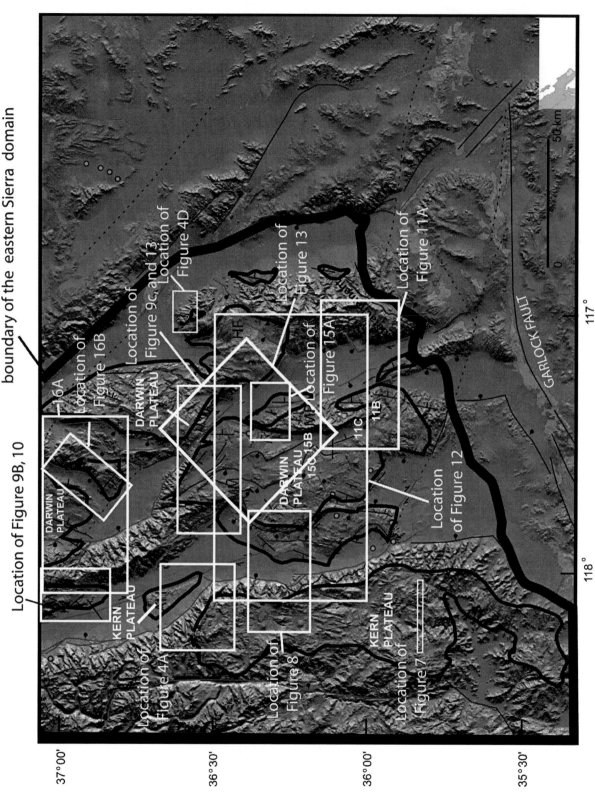

Figure 6. Shaded relief map showing the location of the Kern and Darwin Plateaus in the southeastern Sierra region. AH—Alabama Hills, which was formerly continuous with the Kern surface and is now separated by the Sierra Nevada frontal fault, HF—Harrisburg Flat, MC—Mazourka Canyon, MM—Malpais Mesa, MP—Mazourka Peak. Major Quaternary faults are in gray lines with ball on the down-thrown block. Light gray indicates the location of early Pliocene and (or) late Miocene flood-like basalt flows. Darker gray indicates location of Pliocene bimodal volcanic rocks. Dark black line outlines the extensive plateau areas that include the surfaces described by Lawson (1904) and mapped in the Darwin Plateau area by Jayko (2008). Other figure locations are also indicated.

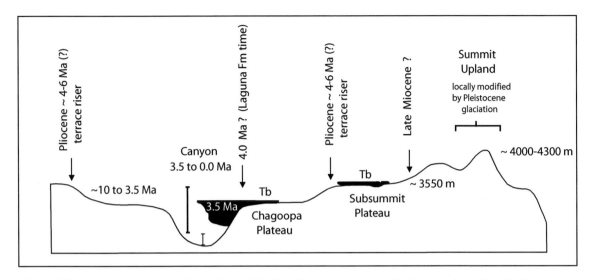

Figure 7. Drawing showing the relative position of the Kern Plateau surfaces originally mapped and described by Lawson (1904) and subsequently dated by Dalrymple (1963). The erosion surfaces include: the Summit Uplands, the Subsummit Plateau, the Chagoopa Plateau, and the Canyon. Correlative surfaces or close approximations are also widely distributed in the White Mountain, Inyo Range, and adjacent areas. See Figure 6 for approximate location of schematic profile. Pliocene volcanic rocks are present at this location; however, late Miocene volcanic rocks are found at Coyote Warp to the north. Tb—Tertiary basalt flows.

TABLE 4. CORRELATION OF STRATIGRAPHIC UNITS IN THE GREAT VALLEY WITH EROSION SURFACES ON THE KERN PLATEAU AND SAN JOAQUIN DRAINAGE

Great Valley units*	Kern Plateau[†]	San Joaquin drainage[§]	White Mountains	Inyo Mountains area	Age
Lone Formation	Summit Uplands				Eocene
Valley Springs Formation	Subsummit Plateau	Boreal, broad valley			Oligocene
Mehrten Formation	Subsummit Plateau	Boreal, broad valley	White Mountain basalt	Bonham Gravels	Late Miocene
Trachyandesite of Kennedy Table	Subsummit Plateau	Boreal, broad valley	White Mountain basalt	Darwin Plateau basalt	Late Miocene
	Chagoopa Plateau	Mountain-valley		Coso Formation	Late Miocene to early Pliocene
				Mazourka Canyon beds	Late Miocene to early Pliocene
Laguna Formation	Chagoopa Plateau	Mountain-valley		Waucobi Lake beds, Saline Range basalt	Early to late Pliocene
China Hat Gravel	Canyon	Canyon		Waucobi Lake beds and upper Coso Fm	Late Pliocene
North Merced Gravel	Canyon	Canyon		Glacial moraines and alluvial fan	Late Pliocene
Tulare	Canyon	Canyon		Glacial moraines and lake deposits	Pleistocene

*From Marchand and Allwardt (1981).
[†]From Lawson (1904).
[§]From Matthes (1930, 1937).

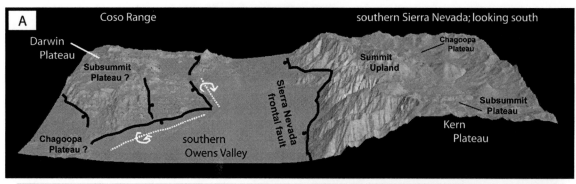

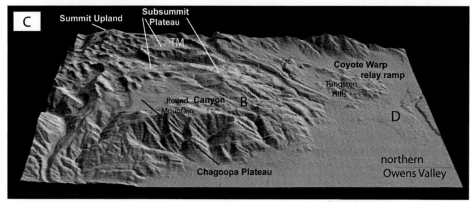

Figure 8. (A) Oblique shaded-relief view of Kern Plateau in the southern Sierra Nevada and lower-elevation but equivalent-age Darwin Plateau in the Coso Range area. View is looking from north to south. Lawson's (1904) erosion surfaces are indicated. Sierra Nevada frontal fault is in black. Rollover hinges on edge of Coso block are shown schematically. (B) Photograph showing Summit Upland, Subsummit Plateau, Chagoopa Plateau, and Canyon surface near the north end of Coyote Flat. View is looking to the west toward the Sierra Nevada crest from ~3000 m elevation. The relict late Miocene topography is referred to as the Lindgren Surface in the Sierra Nevada. The Subsummit Plateau lies at ~3300 m elevation. (C) Oblique shaded-relief view of the Coyote Warp relay ramp at the north end of Owens Valley. Image shows the four surfaces in B, the north and west dipping ramps, and locations where photos in B and D were taken. The Coyote Warp rolls over toward the east above a listric, west-dipping normal fault. To the north, the Lindgren Surface is partly step-faulted and tilted northward toward the Owens Valley floor. TM—Table Mountain. (D) Photo of the Coyote Warp looking from north to south (location labeled D in part C). Basalt (Tb) was dated at 9.6 Ma (Dalrymple, 1963) on the Subsummit Plateau near the ridge crest. The same basalt unit is faulted lower on the range front and is also exposed on the valley floor.

TABLE 5. CHARACTERISTICS OF THE INYO SURFACE, WHITE-INYO MOUNTAINS

Location	Unit	Age (Ma)	Reference for dated unit	Elevation (m)	Morphologic unit
Inyo Surface, White Mountain block*					
White Mountain		Late Miocene (?)		4342	Summit Upland
Mount Dubois		Late Miocene (?)		4145	Summit Upland
Pellisier Flats		Late Miocene (?)	Figure 4C	3870	Subsummit Plateau
Mt. Hogue		Late Miocene (?)	Figure 4C	3808	Subsummit Plateau
Chiatovich Flats		Late Miocene (?)	Figure 4C	3808	Subsummit Plateau
Mt. Barcroft	Tb	Late Miocene (?)	Figure 9A	3600	Subsummit Plateau
Tres Plumas Flat	Tb	Late Miocene (?)	Krauskopf (1971)	3350–2800	Subsummit Plateau
Bucks Peak	Tb	Late Miocene (?)	Nelson (1966)	3310	Subsummit Plateau
Red Peak	Tb	Late Miocene (?)	Nelson (1966)	3075	Subsummit Plateau
Perry Aikee Flat	Tb	9.8 ± 0.3	Krauskopf (1971); Reheis et al. (1995)	2926	Subsummit Plateau
Sage Hen Flat	Tb	Late Miocene (?)	Nelson (1966)	2883	Subsummit Plateau
Black birch Canyon	Tb	Late Miocene (?)	Nelson (1966)	2540	Subsummit Plateau
Chocolate Mountain	Tb	11.3 ± 1.0*	Dalrymple (1963); 10.8 + 1.0**	2200	Subsummit Plateau
Lower Furnace Creek	Tb	11.5 ± 0.4	Krauskopf (1971); Reheis et al. (1995)	1700	Subsummit Plateau
Lower Cottonwood Creek	Tb	Late Miocene (?)	Nelson (1966)	1700	Subsummit Plateau
Northern Deep Springs Valley	Tb	Late Miocene (?)	McKee and Nelson (1966)	1700	Subsummit Plateau
Inyo Surface, northern and northwestern Inyo Range block†					
Chocolate Mountain	Tb	11.3 ± 1.0*	Dalrymple (1963); 10.8 + 1.0**	2200	Subsummit Plateau
Harkless Flat		Late Miocene (?)		2250	Subsummit Plateau
Crooked Road Flat		Late Miocene (?)		2440	Subsummit Plateau
Nohorn Ridge flat		Late Miocene (?)		2645	Subsummit Plateau
Juniper Flat		Late Miocene (?)		2732	Minimum late Miocene paleorelief
Papoose Flat		Late Miocene (?)	Figure 9B	2620	Subsummit Plateau
Squaw Flat		Late Miocene–Pliocene (?)	Figure 9B	2500	Chagoopa surface (?)
Waucobi Mountain		Late Miocene (?)	Figure 9B	3377	Summit Upland
Inyo crest, blue		Late Miocene (?)	Figure 9B	3270	Summit Upland
Inyo crest, south gradient		Late Miocene (?)	Figure 9B	2900	Subsummit Plateau
Bager Flat		*Late Miocene (?)*	Figure 9B	2600	Subsummit Plateau
Mazourka Peak		*Late Miocene (?)*	Figure 10	2870	Summit Upland
Santa Rita Flat	Ts	*Late Miocene (?)*	Ross (1965)	2100	Subsummit Plateau
Mazourka Canyon	Ts	*>5.3*	Berman (1999)	1400	Subsummit Plateau
Inyo Surface, southern and southwestern Inyo Range and Darwin Plateau§					
Inyo crest, Keynot Peak		Late Miocene (?)	Figure 16B	3352	Summit Upland
New York Butte		Late Miocene (?)	Figure 16B	3260	Summit Upland
Burgess Mine Flat		Late Miocene (?)	Figures 9C and 16B	2835	Subsummit Plateau
Cerro Gordo Peak		Late Miocene (?)	Figure 9C	2800	Summit Upland
Santa Rosa Flat	Ts	<13 ± 0.5	Conrad (1993)	1825	Bonham gravels
Conglomerate Mesa		Late Miocene (?)	Conrad (1993)	2285	Subsummit Plateau
Cerro Gordo fans	Ts	6.0–6.7	Stone et al. (2004)	1280	Subsummit Plateau or Chagoopa surface
Malpais Mesa	Tb	Ca. 6.0	Bacon et al. (1982)	2350	Subsummit Plateau (?)
Southeastern Lee Flat	Tb	Ca. 7.7	Schweig (1989)	1770	Subsummit Plateau
Inyo Surface, greater Darwin Plateau, Coso to Argus Ranges					
Matarango Peak	Tgor	Late Miocene (?)	Figure 12	2695	Summit Upland
Coso Peak	Tb	Late Miocene–Pliocene	Duffield and Bacon (1981)	2890	Subsummit Plateau
Upper Cactus Flat		Late Miocene–Pliocene	Figure 12	1525	Subsummit Plateau
Cole Flat	Tb	Late Miocene (?)	Figure 12	1860	Subsummit Plateau
Etcheron Flat	Tb	Late Miocene (?)	Figure 12	1950	Subsummit Plateau
Darwin Wash		Late Miocene (?)	Hall (1971)	1465	Chagoopa surface
Ash Hill	Tb	Early Pliocene	Hall (1971)	670	Chagoopa surface
Inyo Surface, Panamint Range					
Telescope Peak		Late Miocene (?)		3366	Summit Upland (?)
Panamint Butte	Tb	Late Miocene (?)	Hall (1971)	2040	Subsummit Plateau
Harrisburg Flats		Late Miocene (?)	Figure 12	1525	Subsummit Plateau (?)
Flats, east of Tucki Mtn.		Early Pliocene	Figure 4D	1220	Subsummit Plateau (?)
Hunter Mountain	Tb	Early Pliocene	Hall (1971)	2255	Subsummit Plateau
Manly Peak		Late Miocene (?)	Figure 3B	2195	Summit Upland (?)
Butte Valley		Early Pliocene	Figure 3B	1220	Chagoopa or canyon (?)

Note: Units as in Figure 9.
*Displacement across White Mountain frontal fault, from offset basalts and gravity data ~4.2 km (Pakiser et al., 1964; Bateman, 1965). Minimum uplift rate with respect to late Miocene: 0.4 mm/yr, and with respect to late Pliocene: 1.0 mm/yr. Minimum paleorelief in late Miocene: 680 m (1900 ft). ~4.6° dip based on gradient in Mt. Blanco quadrangle.
†Italics indicate flats of the Mazourka Canyon relay ramp. Minimum paleorelief in late Miocene from Inyo Crest surface: 477 m (1580 ft) Waucoba Mountain to Inyo Crest, south.
§Minimum paleorelief in late Miocene from Inyo Crest surface: 548 m (1800 ft) Keynot Mountain to flat below Burgess Mine.
*Corrected K/Ar age.
**Uncorrected K/Ar age.

age to the Mehrten Formation (Dalrymple, 1963; Huber, 1981) and trachyandesite of Kennedy Table (Huber, 1981), described previously (Table 1).

Basaltic flows that overlie Coyote Flat in the Sierra Nevada erupted prior to significant activity on the Sierra Nevada frontal fault and development of the graben block that now underlies northern Owens Valley. The flows are tilted and step-faulted from near the range crest at ~3200 m elevation on Coyote Flat to the valley north of Bishop at ~1350 m elevation by an active relay ramp, the Coyote Warp. Coyote Flat is step-faulted, down to the west from Table Mountain, a flat at ~3540 m, ~5–6 km to the west (Figs. 4C, 8C, and 8D). The northern extent of the Summit Upland surface that is faulted into northern Owens Valley by the Coyote Warp is buried by the Bishop Tuff and exposed west of Long Valley at Casa Diablo (Fig. 3). The Coyote Warp is one of the most important sites in the eastern Sierra area south of the Mono Basin for constraining the maximum age of Owens Valley and offset of the Sierra Nevada frontal fault zone (Fig. 1; Table 3).

The White Mountain Block

The late Miocene paleotopography of the White Mountains lies due east of its late Miocene corollary at Coyote Flat, Table Mountain, and the adjacent Coyote Warp in the Sierra Nevada (Figs. 3, 4C, and 4D). The White Mountains are also overlain by ca. 9.8 Ma and older basalt that is tilted east and northeast (Fig. 9A). The local Summit Uplands, Subsummit Plateau, and Chagoopa Plateau surfaces east of Owens Valley are collectively referred to as the Inyo Surface. Localities where equivalents of the Kern morphology, in particular, the Subsummit Plateau, can be seen include Pellisier, Chiatovich, and Sage Hen Flats (Figs. 3 and 4C), which represent the late Miocene and older landscape relicts near the crest and east side of the range. Basement rocks near the range crest at ~3550 m elevation east of Mount Barcroft are locally overlain by late Miocene basalt (Krauskopf, 1971; Reheis et al., 1995; Table 5; Fig. 9A).

Mount Barcroft is here considered correlative with the Summit Upland surface, and the erosion surface underlying the basalt is correlated with the Subsummit Plateau. At lower elevation, the basalts locally overlie a thin veneer of rhyolitic tuff, suggesting the land surface had a gentle eastward-facing gradient at the time of eruption (Nelson, 1966). K/Ar whole-rock ages of ca. 9.8–11.4 Ma on basalts have been determined at three sites on the east side of the White Mountains (Dalrymple, 1963, 1964; Reheis et al., 1995; Table 5). The youngest age, radiometrically dated ca. 9.8 + 0.3 Ma (Reheis et al., 1995), is from the northernmost and highest elevation site at ~2926 m (Table 5). The central and southern White Mountains provide the strongest evidence for former continuity and parallel topographic evolution of the Coyote Warp on the Sierra Nevada block and the eastern Sierra region. This location also provides the best chronologic and geomorphic correlation between the classic tilted late Miocene surfaces of the central and western Sierra Nevada studied by Lindgren (1911) and Matthes (1937, 1960).

The Inyo Surface is tilted south, southeastward, and displaced downward in the extensional relay that separates the White and Inyo crustal blocks in the vicinity of Westgard Pass (Fig. 9A). Between the White Mountains and Inyo Range, flat-topped ridges that represent relicts of the Subsummit Plateau and Chagoopa surfaces, and possibly the Summit Upland, are preserved south of the White Mountains at Chocolate Mountain on the southeast corner of Deep Springs Valley, the ridges north and south the Westgard Pass area, the intervening flats including Cedar Flat at Westgard Pass, and mountain uplands to the southeast (Table 5; Fig. 9A).

These paleosurfaces and their basaltic cover rocks constrain the late Miocene and early Pliocene deformation of the White Mountain block (Dalrymple, 1963; Bateman, 1965). The White Mountains are part of a block that has been gently east-tilted ~4.6° since the late Miocene, concurrent with displacement along the White Mountain fault zone (dip gradient determined on basalt flows mapped by Nelson, 1966). Relicts of this paleolandscape now extend between 3870 m in the White Mountains to 1350 m in Owens Valley, indicating at least ~2520 m of vertical separation by both faulting and warping during the last 10 m.y. This gives a minimum ~0.25 mm/yr rate for the formation of the relief. Bateman (1965) estimated depth to basement at the depocenter of the northern part of the Owens Valley between Bishop and Big Pine from gravity data. The gravity data suggest that a buried normal fault with ~1800 m of displacement lies along the east side of the southern White Mountain block. Thus, displacement along the White Mountain fault zone may be as great as ~4300 m, nearly double the offset along the Wheeler Crest section of the Sierra Nevada frontal fault zone.

The White Mountain fault may be significantly older than the Sierra Nevada frontal fault at this latitude. Cooling ages (Stockli et al., 2003) indicate uplift of the White Mountains began around late middle Miocene and or early late Miocene, whereas the Wheeler Crest section of the Sierra Nevada frontal fault may be mainly Pliocene and Pleistocene in age between the east side of the Coyote Warp and Mono Basin (Blackwelder, 1931; Curry, 1971; Phillips et al., 2000; Huber, 1981, 1990). Significant early work has noted the remarkable apparent youthfulness of the Sierra Nevada frontal fault zone along this section (cf. Axelrod and Ting, 1961; Curry, 1971; Huber, 1981, 1990), but has not emphasized the relative difference in age of the White Mountains fault zone, which also bounds Owens Valley. The Owens Valley apparently developed in a broad axial graben-type structural setting. East-tilted basalt flows presently lie adjacent to flat-topped ridges underlain by steeply dipping basement rocks. The accordant, flat-topped ridges are part of a broad, eroded, formerly continuous, low-gradient, probably arched, late Miocene surface that has been dissected throughout Pliocene-Pleistocene time.

The Former White-Inyo Mountains Block

Across large extents of the White Mountains, Inyo Mountain, Saline Range, Darwin Plateau, Coso Range, and locally in the

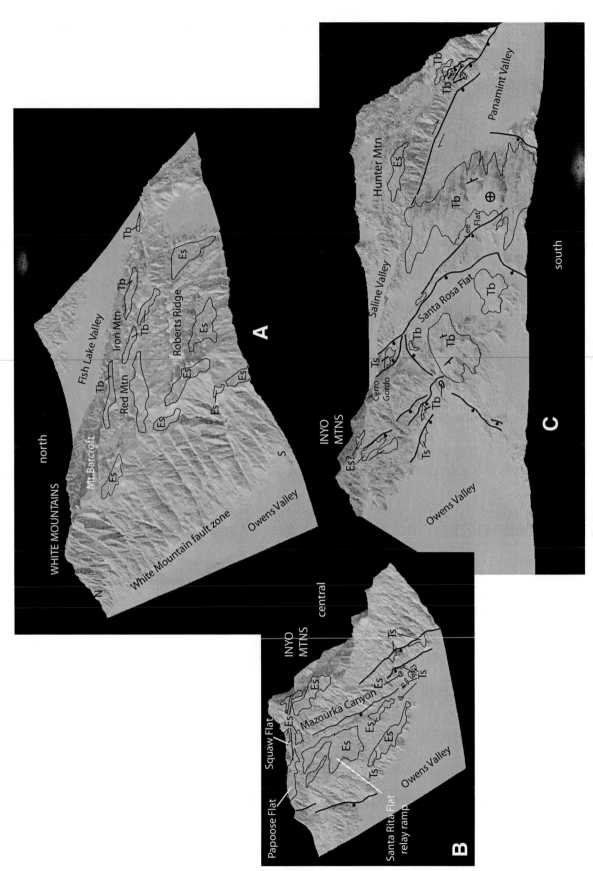

Figure 9. Oblique shaded-relief views looking from south-southwest to north-northeast. (A) Southern White Mountains, Es—Inyo Surface (erosion surface). In this area, it is mainly equivalent to the Subsummit Plateau and is locally overlain by flood-like basalt flows (Tb) with ages of ca. 9.6–10.6 Ma. Mount Barcroft is topographically higher than the highest elevation flat at ~3500 m, which is overlain by Tb near the base of Mount Barcroft, so it is considered Summit Upland. The Subsummit Plateau surface is inset into the Summit Uplands in the areas shown as Es. (B) Santa Rita Flat and Mazourka Canyon area on the west side of the central Inyo Range. The Inyo Surface (Es) and locally over-lying late Miocene deposits (Ts) are scissor-faulted down-to-the-west along the range front by a relay ramp, the Mazourka Canyon fault. Photos of the Santa Rita fault surface are shown in Figure 10. (C) Southern Inyo Mountains, northern Darwin Plateau, and Hunter Mountain area showing the location of the Inyo Surface (Es), late Miocene clastic deposits (Ts) with interbedded ca. 4–6 Ma tuff (Stone et al., 2004), and ca. 6 Ma flood-like basalt flows (Tb) that drape relicts of the late Miocene topography.

Last Chance Range, late Miocene and (or) early Pliocene basaltic flows filled formerly low-relief topography that now occupies mountain summits or the flanks of large, tilted mountain blocks. These flows were commonly extruded directly onto basement or onto thin pediment veneers ranging from 1 to 15 m in thickness. Stratigraphic sections consisting of basin fill deposits, 20–350 m or more thick, are also present, but these are much more spatially restricted. The mafic flood-like basaltic flows range in age from ca. 9.6 to 11 Ma in the north (Dalrymple, 1963; Reheis et al., 1995) and ca. 4 to 8 Ma in the south (Hall, 1971; Larsen, 1979; Bacon et al., 1982; Sternlof, 1988; Schweig, 1989). This record of basaltic volcanic activity is coeval with the initiation of fault activity that formed the basins.

The paleohorizontal and paleolowland indicators provided by the formerly laterally extensive basalt flows and the relics of late Miocene landscape, the Inyo Surface, suggest that the White and Inyo Mountains were formerly a large contiguous crustal block during early late Miocene time at the time the Owens Valley began forming. The White Mountain fault initiated around 10–11 Ma (Dalrymple, 1963; Stockli et al., 2003). The White Mountain and southern Sierra Nevada frontal fault were oppositely facing during initiation of the Owens Valley graben (Jayko, 2008). The White-Inyo block was coherent until middle to late Pliocene time, as described later. Summit Upland and Subsummit Plateau surfaces are preserved in both ranges. The White-Inyo crustal block floundered into several smaller fault blocks and associated half grabens during middle Pliocene time, giving rise to the Waucobi Embayment (Walcott, 1897; Bachman, 1978; Lueddecke et al., 1998), Deep Springs Valley, Eureka Valley (Reheis et al., 1995), Saline Valley (Sternlof, 1988; Larsen, 1979; Zellmer, 1980), and much of Panamint Valley (Hall, 1971; Burchfiel et al., 1987). These topographic basins appear to be younger than ca. 4.0 Ma and to have formed subsequent to initiation of Death Valley and Owens Valley.

Pliocene inversion of late Miocene to early Pliocene lowland and basin topography is also well documented in several mountain ranges and range divides where alluvial, fluvial, and lacustrine deposits of the Coso (Bacon et al., 1982), Nova (Harding, 1988; Snow and White, 1990), and Waucobi (Bachman, 1978) Formations are perched 914–1829 m (3000–6000 ft) above the present valley floors (Figs. 3, 4, and 5).

Inyo Mountain Block

The Inyo Mountain block is not significantly tilted and is flanked on both sides by oblique normal faults (Stone et al., 1989, 2004; Bacon et al., 2005; Oswald and Wesnousky, 2002). Relict upland surfaces include Harkless Flat, Papoose Flat, Squaw Flat, and adjacent rolling uplands at the north end of the block (Table 5; Fig. 9B). The Summit Upland and relics of the Subsummit Plateau are also present on the crest of the Inyo block. The crest is locally ~5 km wide and has very gently sloping, low-gradient, well-preserved "mountain valley" topography or "flats" that extend continuously for at least 30 km. This mountain-valley

relict is flanked by extremely steep, fault-controlled, east and west slopes.

The Summit Upland and Subsummit Plateau surfaces, which occur along the crest of the Inyo Mountains, are step-faulted to Badger Flat (Subsummit Plateau) and Mazourka Peak (Summit Upland), and step-faulted down to Santa Rita Flat (Subsummit Plateau) (Table 5; Figs. 9B and 10). Santa Rita Flat, at ~2100 m elevation, is scissor-faulted and southwest-tilted to ~1400 m elevation along a relay ramp on the western range front in the northern and southern part of the range. Late Miocene to early Pliocene volcanic ash, fluvial, and lacustrine deposits (5.3 Ma or older) overlie the flat and also are step-faulted along the range front north and south of Mazourka Canyon (Ross, 1965; Berman, 1999). The base of the fluvial section dips 20° to 25° westward. These deposits are in structural and topographic continuity with the Santa Rita Flat surface north of Mazourka Canyon, which rises along a gentle gradient to the elevation of the Subsummit Plateau on the Inyo Range crest (Figs. 6 and 11A). Emergence of a "bouldery" landscape from under a late Miocene paleosol and overlying deposits occurs on the lower part of Santa Rita Flat (Figs. 9B and 10). This landscape morphology is also found in the Alabama Hills. Adjacent surfaces are tilted westward into Owens Valley along a footwall tear fault or relay ramp that extends from the crest of the range to the Bee Flat area south of Mazourka Canyon (Figs. 9B and 10). The north end of Santa Rita Flat is locally overlain by basalt (Ross, 1965).

Pliocene-Pleistocene erosion and tectonic denudation (Figs. 9C and 10) largely modify the southern Inyo Range. However, relics of the late Miocene Inyo erosion surface are present north and south of New York Butte at ~3070–2880 m elevation, continuing along a southward-trending gradient to lower elevations, and under the late early Miocene basalts (ca. 6–7 Ma; Bacon et al., 1982; Schweig, 1989; Conrad, 1993) and associated clastic deposits at Malpais Flat in the southernmost Inyo Range (Figs. 9C and 12). The elevation difference between the high peaks at ~3353 m and nearby relict flats indicates a minimum late Miocene paleorelief of ~550 m.

A late Miocene alluvial-fan deposit with about ca. 6.0–6.7 Ma tephra (Stone et al., 2004; Fig. 9C) overlies an erosion surface that is also overlain by ca. 6 Ma basalt further south along the Inyo Mountains range front. The erosion surface in this area is probably correlative with the Subsummit and (or) Chagoopa Plateau. These late Miocene deposits on the west side of the southern Inyo Range indicate the minimum age for the southern Owens Valley and Inyo Mountain fault zone (Bacon et al., 2005).

Darwin Plateau

The Darwin Plateau lies east of the Sierra Nevada frontal fault, adjacent, but downthrown relative to the Kern Plateau (Figs. 6 and 12). The greater Darwin Plateau area discussed here is a large, 35–40-km-wide highland cut by several widely spaced faults forming gently tilted blocks and shallow half grabens south

Figure 10. Photos of Santa Rita Flat in the west-central Inyo Range. (A) Photo taken from the Inyo Surface, which here is most likely correlative with the Subsummit Plateau on Santa Rita Flat, looking up the south-tilted surface toward the Inyo Range crest at ~3500 m elevation. (B) View from about the same area looking down the surface, showing contrast in morphology between denuded granitic bedrock and the surface with preserved paleosol and late Miocene (?) deposits. (C) View from lower west edge of Santa Rita Flat looking across the Owens Valley toward the Sierra Nevada Range. The Alabama Hills in the foreground are overlain by either the Chagoopa Plateau and (or) Subsummit Plateau surfaces (Axelrod and Ting, 1961).

336

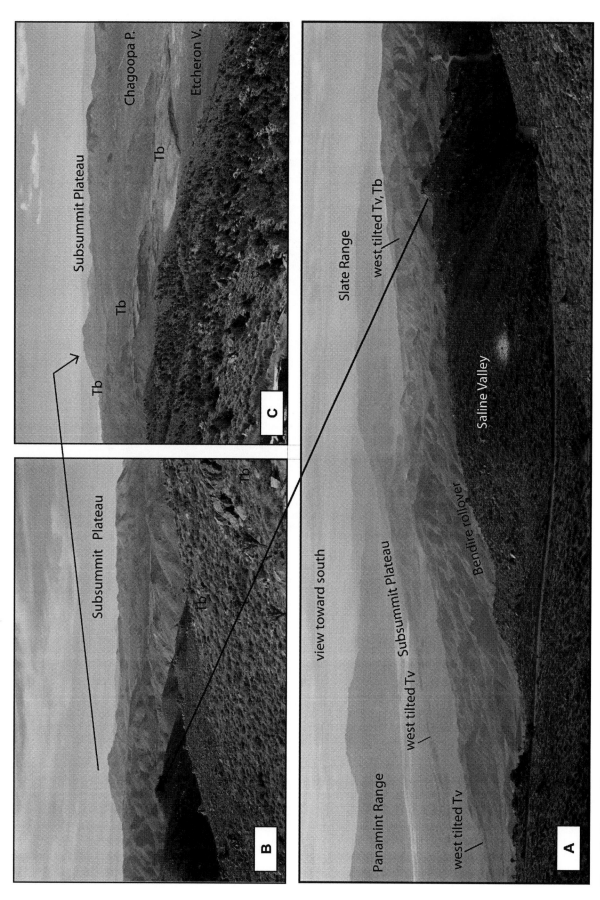

Figure 11. Photos of the greater Darwin Plateau. (A) Note east-tilted surface on the range front. View from the Argus Range south of Maturango Peak looking toward Panamint Valley and the Slate Range in the southeast east. Dip-slopes of east-tilted Tertiary are noted. (B) View from the Argus Range south of Maturango Peak looking south, showing accordant ridges, erosional flats, and remnants of Tb (basalt flows) on the Subsummit Plateau, nearly horizontal in this area. (C) View from the Argus Range south of Maturango Peak looking southwest toward Etcheron Valley. Tb is step-faulted into Etcheron Valley. Chagoopa surface is inset into Subsummit Plateau. Tv—Tertiary volcanic rock.

of the Inyo Mountains, including the mountain upland flats of the Argus and Coso Ranges. The Inyo Surface occurs as irregular, low-gradient, step-faulted, and gently tilted flats between 1525 and 1830 m elevation (Jayko, 2004a, 2004b, 2008) on the plateau. Geomorphic units correlative with the Kern Plateau surfaces of Lawson (1904) are present.

A clastic deposit (Ts) with a ca. 13 Ma detrital biotite age interpreted as reworked tephra (Conrad, 1993) lies on the east side of the southern Inyo Mountains, at the northern end of the Darwin Plateau at Santa Rosa Flat (Figs. 9C). This surface is contiguous, albeit step-faulted (Jayko, 2008), with the surface that underlies ca. 6–8 Ma basalts (Larsen, 1979; Bacon et al., 1982; Schweig, 1989) on the northern Darwin Plateau, so is most likely correlative with the Subsummit Plateau. The Summit Upland and Subsummit Plateau surfaces can be traced continuously from the Inyo Range crest to the northern Darwin Plateau, and then discontinuously to the southern edge of the plateau. Therefore, the tilted and faulted erosion surface that underlies late Miocene and (or) early Pliocene basalt in the northern plateau is tentatively correlated with the Subsummit rather than the Chagoopa Plateau.

The margins of the greater Darwin Plateau are bounded on the west, northwest, and east by rollover structures that overlie listric normal or normal-oblique slip faults emanating from the Sierra Nevada on the west and Panamint Range on the east (Jones et al., 2004; Jayko, 2008; Figs. 12–14). Late Miocene to early Pliocene deposits and relicts of the Inyo Surface overlie the rollover structures. Circa 4–6 Ma strata of the Coso Formation (Fig. 12) (Duffield and Bacon, 1981) are tilted ~14° to 20° west-northwestward (Stinson, 1977a; Duffield and Bacon, 1981) on a west-facing rollover hinge, the Haiwee rollover (Jayko, 2008), west of Cactus Flat (Fig. 12), indicating significant late Pliocene and Pleistocene displacement on a listric Sierra Nevada frontal fault. The northwest side of the Darwin Plateau (Figs. 8 and 12) is overlain by ca. 6 Ma and younger fluvial-lacustrine and coeval, largely bimodal volcanic deposits (Duffield and Bacon, 1981; Bacon et al., 1982) tilted northwest on the Red Hill rollover. A thin veneer of late Miocene(?) tuff and fine-grained paludal or lacustrine sediment tentatively correlated with the Coso Formation also overlies the Inyo Surface in the central part of the plateau (St. Amand and Roquemore, 1979; Jayko, 2008). The highest peak, Coso Mountain, in the central uplands of the greater Darwin Plateau overlies the top of a horst block that reaches ~2430 m elevation. It is capped by early Pliocene basalt flows (Bacon et al., 1982; Duffield and Bacon, 1981) that overlie the Subsummit or Chagoopa Plateaus. The thin-bedded, flat-lying basalt flows are step-faulted to adjacent lower-lying flats (Fig. 12).

A thick accumulation of widespread 5–6 Ma flood-like basalt flows overlies Paleozoic and Mesozoic basement rocks on the southernmost part of the Inyo Range, and these flows are west-tilted and step-faulted into Owens Valley west of Malpais Mesa (Bacon et al., 1982; Figs. 9C and 12). Basalt flows, ca. 5–8 Ma (Schweig, 1989), which overlie Paleozoic basement rocks, are east-tilted on a rollover, the Rainbow Canyon rollover,

into Panamint Valley on the east side of the northern Darwin Plateau (Figs. 12 and 13). The basement rocks at the north end of the Argus Range and lower Darwin Wash are overlain by ca. 4 Ma basalt flows (Hall, 1971) and locally preserve underlying pockets of late Miocene and (or) early Pliocene rhyolitic ash, arkosic clastic rocks, and ash-flow tuff (Fig. 15). Nearly flat-lying, late Miocene to early Pliocene basalt also caps the crest of the Argus Range. Early Pliocene basalt overlies late Miocene (?) poorly welded rhyolitic ash and basement rocks on the Ash Hill horst (Hall, 1971; Figs. 5 and 15) and is east tilted above the Revenue rollover into Panamint Valley (Jayko, 2008; Figs. 12 and 15A).

Very small isolated pockets of late middle Miocene clastic rocks and intermediate and siliceous volcanic rocks from the Miocene arc (cf. Rood et al., 2005) are found east and southeast of Ash Hill. Late Miocene, ca. 12–13 Ma arc-related rhyolitic and intermediate volcanic rocks (Evernden and James, 1964) are present on the southeast flank of the Argus Range, north of Water Canyon on the east side of the Darwin Plateau, where they are tilted east and overlie basement rock (Figs. 6 and 11). Detrital biotite from reworked tephra (?) of similar age is interbedded with conglomerate at the north end of the Darwin Plateau (Conrad, 1993) and faulted against the southeastern corner of the Inyo Range. These are the oldest basin-filling volcanic rocks and associated sediments of the greater Darwin Plateau (Figs. 5 and 12). These older Miocene deposits are here inferred to have been deposited in the arc or back-arc landscape that preceded development of the present, north-northwest–trending Basin and Range topography associated with evolution of the Sierra Nevada frontal fault. Their sparse preservation under late Miocene and early Pliocene flows on the east side of the Darwin Plateau and absence on the west side support the inference that the late Miocene landscape had an east-southeastward–sloping gradient in the late Miocene.

The Darwin Plateau in Panamint and Cottonwood Mountain Areas

The northeastern continuation of the late Miocene Darwin Plateau block can be traced where basalt-covered flats and adjoining erosion surfaces are present within the northern Panamint Range, Panamint Butte, Cottonwood Mountains, Dry Mountain, and parts of the Saline Range east of the Hunter Mountain–Panamint Valley faults (Hall, 1971; Figs. 8 and 14). The elevation and morphologic characteristics of the Inyo Surface, and the dominantly ca. 4 Ma mafic volcanic rock cover, are correlative with those of the greater Darwin Plateau area sensu strictu, except to the east, where the erosion surface is less continuous and has been more extensively faulted, tilted, and incised during middle Pliocene to Pleistocene extension.

The crustal block, or the northeastern continuation of the late Miocene Darwin Plateau that occurred between the Death Valley and Panamint–Hunter Mountain–Saline Valley fault zones, largely underwent extension during the last 2–4 m.y., based on K/Ar dates on basalt flows (Larsen, 1979; Sternlof, 1988) that

338

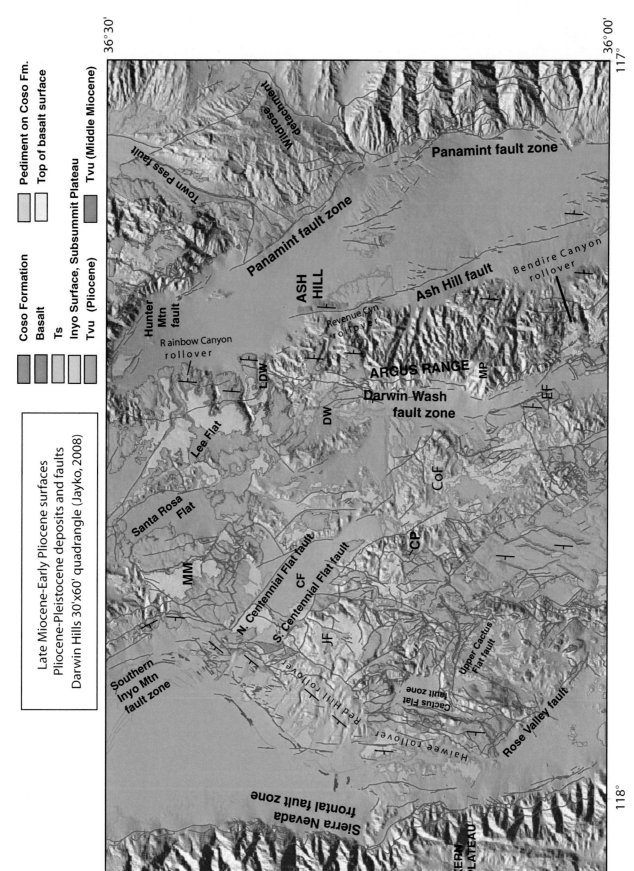

Figure 12. Map showing mapped areas of the Inyo Surface and late Miocene and (or) early Pliocene basalt on the Darwin Plateau and adjacent areas from Jayko (2008). This erosion surface and overlying basalts once formed a continuous lowland landscape that is inferred to have been continuous with the Kern Plateau to the west. Pliocene to Quaternary faults are in red. MM—Malpais Mesa, CF—Centennial Flat, CoF—Cole Flat, CP—Coso Peak, DW—Darwin Wash, EF—Etcheron Flat, JF—Joshua Flat, LDW—Lower Darwin Wash, MP—Maturango Peak. Tvu (Miocene) occurs along the southwest side of the Argus Range.

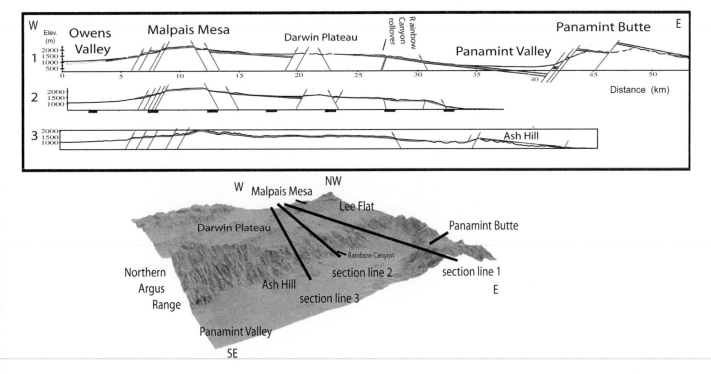

Figure 13. Structure profiles drawn across the Darwin Plateau showing dip and offset of early Pliocene basalts used to constrain vertical separation and slip rates.

are offset across northeast-trending normal faults linked to northwest-trending transfer faults (Figs. 1, 3, and 16). Likewise, high extensional activity after 3.6–3.8 Ma in the Harrisburg Flats–Wildrose Canyon area of the northern Panamint Range tectonically stranded basalt that overlies basement rocks (Hodges et al., 1989a, 1989b; Fig. 14). A long-term, ~0.4 mm/yr vertical separation rate across the Dry Mountain fault is constrained by offset basalts (Figs. 4A, 16A, and 16B).

Pliocene Breakup of the Darwin-Panamint Crustal Block

The southwestern corner of the crustal block that underlies the Darwin Plateau and Coso Range is overlain by an extensive section of ca. 3–4 Ma basalt flows that range in thickness from one to several meters (Duffield and Bacon, 1981; Bacon et al., 1982). The northern end of the block is overlain by ca. 5–7 Ma basalt flows (Schweig, 1985, 1986, 1989), and one paleocanyon-filling flow is as old as ca. 7.7 Ma on the east side of the Darwin Plateau (Schweig, 1989). Basin deposits along the northern end of the Argus Range have been correlated with the Nova Formation of the northern Panamint Range and inferred to be ca. 5–6 Ma in age (Hall, 1971; Larsen, 1979; Schweig, 1989). Basalt overlies a very thin section of fluvial clastic sediments and late Miocene rhyolitic tuffs adjacent to the northern end of the Argus Range and overlies conglomerate and alluvial gravels with late Miocene tuffs at the southern end of the Inyo Range. Geochemically correlative basalts that have ages of ca. 4–5 Ma are found capping

adjacent ridges, including Hunter Mountain and Panamint Butte, and are tilted westward into Death Valley north of Townes Pass (Figs. 4D and 14) (Hall, 1971; Coleman and Walker, 1990). The late Miocene and early Pliocene basalts, which erupted prior to formation of the present topography, indicate that the Panamint Basin largely formed during and subsequent to early Pliocene time, and mainly after ca. 4 Ma (Hall, 1971; Burchfiel et al., 1987; Schweig, 1985, 1989; Coleman and Walker, 1990).

Late Miocene volcanic rocks (Larsen, 1979; Hodges et al., 1989a, 1989b) flanking the northern and central parts of the Panamint block indicate that the block is east-tilted and detached from the Amargosa region along the Death Valley fault zone (Hodges et al., 1989a, 1989b). Tertiary volcanic rocks on the east flank of the Panamint Mountains that face into Death Valley have K/Ar ages of ca. 9.0–10.7 Ma (Hodges et al., 1989a, 1989b). These strata correlate with a section in the Black Mountains west of southern Death Valley (i.e., Sheephead Andesite and Rhodes Tuff—10.4 and 10.8 Ma; Shoshone volcanics—8.2–8.7 Ma; McKenna and Hodges, 1990; Wright et al., 1984). The Panamint block is structurally denuded in the north (Wernicke et al., 1988a, 1988b; Labotka and Albee, 1990) and flanked by 30°S-dipping Tertiary rocks (Johnson, 1957) that are underlain by a detachment fault at the south end. The Inyo Surface has largely been tectonically denuded in the central part of the Panamint Range (cf. Fig. 4D). Most of the deformation in this region is younger than ca. 8 Ma (McKenna and Hodges, 1990). However, displacement stepped from central Death Valley to the northwest ca. 3.0–4.0 Ma when

340 *Jayko*

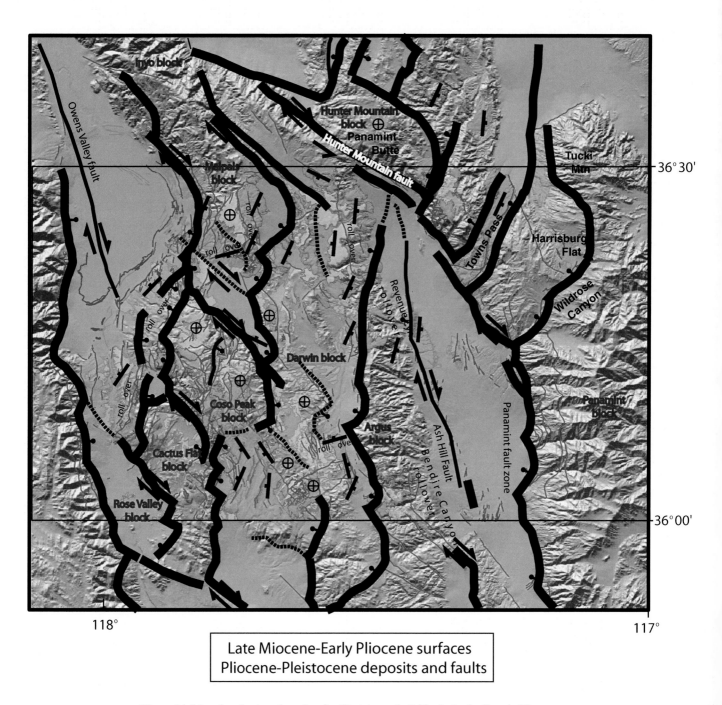

Late Miocene-Early Pliocene surfaces
Pliocene-Pleistocene deposits and faults

Figure 14. Map showing location of major Quaternary fault blocks in the Darwin Plateau area.

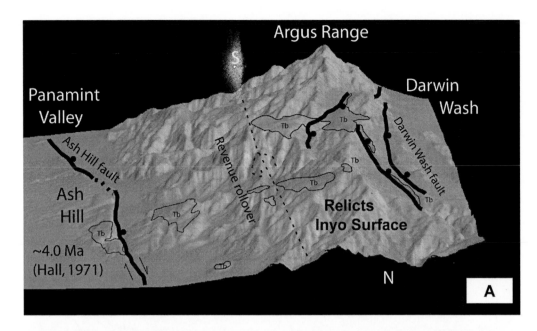

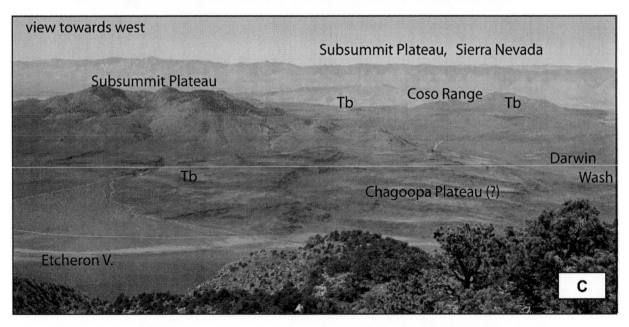

Figure 15. (A) Oblique view of the northern Argus Range and Ash Hill horst block; Tertiary basalt is labeled Tb. The Argus Range is locally capped by flat-lying erosion surfaces with Tb; east of the range crest is a rollover hinge, the Ash Hill rollover. Tb is gently east-tilted downslope from the hinge, and the ramp is faulted by the younger Ash Hill fault. (B) Gently south-tilted basalt at the north end of Darwin Wash and southern Etcheron Valley. (C) View looking west from the Argus Range across the greater Darwin Plateau toward the Sierra Nevada, showing nearly flat-lying and gently tilted Tb and Chagoopa (?) surface in lowland area, accordant hilltops of the Subsummit Plateau surface in the Coso Range, and accordant crest of the Sierra Nevada in the vicinity of the Kern Plateau.

Pliocene flood basalts offset by normal faults
and folded on rollover hinges, Saline Range

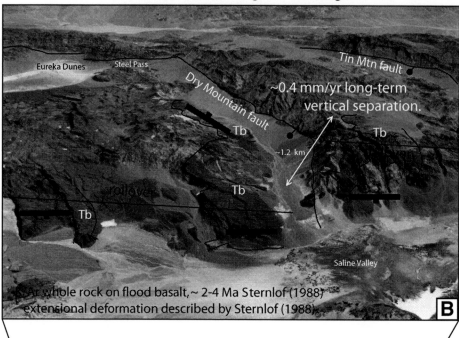

Figure 16. (A) Oblique aerial photograph showing the southern Inyo Range near Keynot Peak and New York Butte relics of the Summit Upland modified by Pleistocene weathering. Burgess Mine Flat is a relict of the Subsummit Plateau surface. (B) Oblique aerial photograph showing faulted rollover hinges on the hanging-wall blocks that deform early Pliocene basalt flows, which overlie Chagoopa (?) surface on the east side of Saline Valley. Basaltic flows provide excellent strain markers for offset and tilted blocks. Tb—late Miocene and (or) early Pliocene basalt, Es—Inyo Surface (erosion surface).

the Panamint and Hunter Mountain fault zones became active (Walker et al., 1987) and the former block that now underlies the White–Inyo–Darwin Plateau–Panamint Range floundered and extended.

Vertical displacement rates were estimated from three section lines, ~35–60 km long, that were constructed between Owens Valley and the northern Panamint Range. The section lines were made where radiometrically dated flood basalts locally cap ridges and are tilted into the adjacent basins (Fig. 13). Vertical separation of Darwin Plateau basalts, ~2.5 km on the west side of the plateau and ~2.7 km on the east side of the plateau, has occurred since late Miocene–early Pliocene time. About 3.7 km of dip-slip separation has occurred across the northern Panamint–Hunter Mountain fault zone where the range front faults dip ~45–55° west. A dip-slip rate of 0.93 mm/yr and vertical separation rates of 0.65 mm/yr result, if most of the displacement postdates the youngest flood basalts ca. 4 Ma (Hall, 1971; Schweig, 1989). If displacement began synchronously with the first eruption of basalt ca. 8 Ma, the dip-slip rate is ~0.48 mm/yr, and the vertical separation rate is ~0.35 mm/yr (Gillespie et al., 2002).

DISCUSSION

The average elevation of the present topography east of the southern Sierra Nevada in the greater Darwin Plateau at ~1525–1830 m may approximate the minimum baseline elevation of the late Miocene paleosurface prior to tilting of the Sierra Nevada block. This elevation for the late Miocene Lindgren and Inyo Surfaces is equivalent to the baseline in the Mono Basin area (Huber, 1981) and to recent paleoelevation results using helium isotopes (Clark et al., 2005). If this approximate paleosurface elevation is valid, then ~1220–1830 m of uplift has occurred along the frontal fault of the Sierra Nevada. Uplift of the erosion surface likely accompanied the late Miocene and (or) early Pliocene regional arching followed by development of the range front fault systems that bound the large crustal blocks of Owens Valley and the rest of the eastern Sierra region. The Miocene plateau surface was either simultaneously or subsequently down-faulted by ~305–1525 m to its current elevation at ~300–1200 m in valley floors (Fig. 17). Thus, the time-averaged vertical separation rate is a minimum, and the higher, Pliocene-Pleistocene displacement rates determined along the Sierran escarpment (Table 3) are more reasonable for the long-term slip rate. If the interpretation that the Kern and Darwin Plateaus are both relics of a more continuous surface is correct, then there has been ~915 m of separation in the last ~8 m.y. This gives a long-term Sierran uplift rate at this latitude of ~0.08–0.17 mm/yr, which is lower than the 0.25–1.0 mm/yr constrained from Holocene and late Pleistocene slip rates to the north, where the range is on average higher in elevation (Lubetkin and Clark, 1988; Beanland and Clark, 1994). Pliocene volcanic activity ca. 3–5 Ma was coincident with uplift of the southern part of the Sierra Nevada, higher uplift rates compared to the late Miocene, and the breakup of the White-Inyo and Darwin Plateau crustal blocks in the eastern Sierra region (Larsen,

1979; Hall, 1971; Sternlof, 1988; Burchfiel et al., 1987; Ducea and Saleeby, 1996; Jones et al., 2004).

The late Miocene landscape of the eastern Sierra region broke into several fault-bounded blocks coeval with evolution of the Sierra Nevada frontal fault zone, and uplift and westward tilting of the Sierra Nevada block (Figs. 14 and 18). Initiation of the Sierra Nevada frontal fault by at least 10.5–9.6 Ma is suggested by the age of the oldest basalt flow on Coyote Warp in the Sierra Nevada and the youngest east-tilted basalt in the White Mountains. This timing is also recognized to the north in the Sonora Pass area (Rood et al., 2005). Initiation of the southern part of the Sierra Nevada frontal fault by at least ca. 7.8 Ma is also suggested by the age of the oldest basalt flow on the Darwin Plateau in Rainbow Canyon (Schweig, 1989).

Mafic volcanic activity ca. 9.5–10.5 Ma constrains initiation of the Sierra Nevada frontal fault and early west tilting of the Sierra Nevada block (Dalrymple, 1963; Bateman, 1965; Christensen, 1966; Bateman et al., 1971; Huber, 1981, 1990). Development of the frontal fault was accompanied by widespread normal or oblique-normal faulting and block rotation in the eastern Sierra region (Figs. 3B and 14). The Last Chance Range (Fig. 1), east of the Inyo Mountains and west of the Death Valley fault zone, is also locally overlain by ca. 10 Ma basalt (Reheis, 1992; Reheis et al., 1995). Basalt extrusion in the Last Chance Range is effectively concurrent with 10.8 Ma basalt found in Deep Springs Valley and 9.6 Ma basalt in the White Mountains (Dalrymple, 1963) and Coyote Warp in the Sierra Nevada. Late Miocene basaltic activity was also concurrent with shallow siliceous plutonic activity of the same age in the eastern Sierra region straddling Death Valley.

Three small plutons now exposed in ranges within highly extended parts of the eastern Sierra region give ca. 7.5–11 Ma ages, supporting the conclusion that the transtensive (or extensional) deformation began around the time that mafic or basaltic extrusion was occurring at the surface along the Death Valley and Panamint Valley fault zones. These three plutonic localities (Fig. 3B) are: (1) The Little Chief stock in the northern Panamint Range (McDowell, 1974, 1978), which gives Ar/Ar biotite ages of 10.8 and 10.4 Ma (McKenna and Hodges, 1990) and a Rb/Sr age of 10.6 (McKenna, 1986) (the hanging-wall volcanic units including the Trail Canyon Volcanic sequence are 10.3 and 10.0 Ma; McKenna and Snee, 1990); (2) Willow Spring diorite, western Black Mountains, with ages of 10.3, 9.9, 9.2, and 8.4 Ma (McKenna et al., 1989); and (3) intrusive dioritic and granitic rocks of the Last Chance Range, which give ages of 6.3, 7.0, 7.3, 7.5, and 7.9 Ma (McKee, 1985; Miller and Wrucke, 1995).

The late Miocene activity along the Sierra Nevada frontal fault zone and Owens Valley area is not consist with a model that proposes simple east to west younging of deformation between Death Valley and the Sierra Nevada block, as suggested by Jones et al. (2004). Both Death Valley and Owens Valley experienced significant late Miocene and early Pliocene deformation, although the intervening region now underlain by Panamint, Saline, Eureka, and Deep Springs Valleys did not begin extending significantly until early Pliocene time. Late Miocene alluvial and

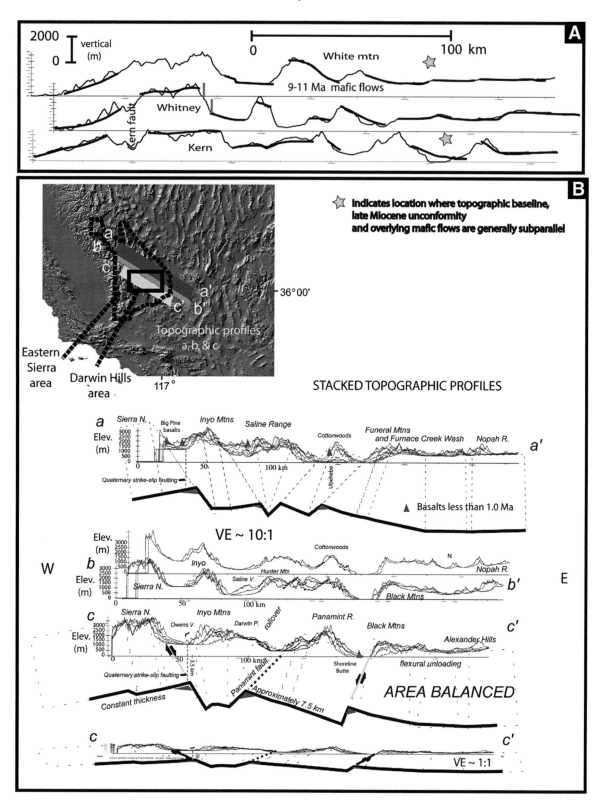

Figure 17. (A) Inset shaded-relief maps show stacked topographic profiles from areas a–a′, b–b′, and c–c′ on inset map constructed approximately perpendicular to the Quaternary and Pliocene normal and (or) normal-oblique slip faults. The stacked topographic profiles are subparallel with ~25–20 km spacing. Sections a–a′ and c–c′ are area balanced with respect to a constructed ~7.5 km depth line. (B) The three section lines perpendicular to the Sierra Nevada frontal fault constructed across the southern White Mountains, the Sierra Nevada at the latitude of Mount Whitney, and the Kern Plateau at the latitude of Maturango Peak (MP) in the Argus Range (see Fig. 6 for location of MP). The lines show a broad faulted arch and baseline elevations to the east. Heavy black lines characterize the offset Lindgren and Inyo Surfaces.

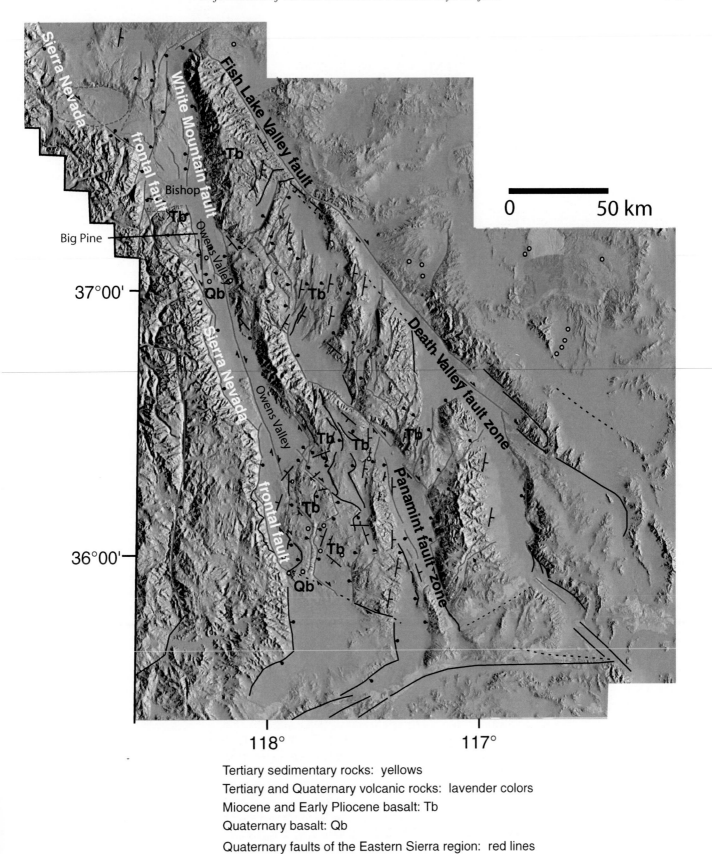

Tertiary sedimentary rocks: yellows
Tertiary and Quaternary volcanic rocks: lavender colors
Miocene and Early Pliocene basalt: Tb
Quaternary basalt: Qb
Quaternary faults of the Eastern Sierra region: red lines

Figure 18. Map of eastern Sierra region showing prominent Pliocene-Pleistocene faults and highly extended areas.

fluvial-lacustrine deposits with late Miocene ca. 6 Ma interbedded tephra overlie basement on step-faulted blocks along the Inyo Range front, indicating the presence of a confining basin on a graben block (see topical sections).

The Lindgren and Inyo Surfaces

The Inyo Surface in the eastern Sierra region, like its correlative in the Sierra Nevada, was relatively stable during the Oligocene and Miocene relative to late Miocene and Pliocene time. Deformation rates increased considerably by late Miocene time, and extensive uplift and canyon incision occurred during the Pliocene and Pleistocene. The evolution of relief in the eastern Sierra region resulted from similar processes driving uplift of the Sierra Nevada Range. The distribution and thickness of sparse late Miocene deposits, which are deposited directly on Mesozoic basement, suggest that the eastern Sierra region was broadly arched prior to initiation of the Sierra Nevada and White Mountain fault zones. The symmetry of tilt and distribution of late Miocene lowland deposits suggest that Owens Valley developed within the axial graben of a broad arch concurrent with regional transtensive deformation.

The suggestion by Lindgren (1911), Matthes (1937), Axelrod (1957), Christensen (1966) and Huber (1981) that flexure or warping may have preceded development of the Sierra Nevada frontal fault is supported by: (1) the development of an upland erosional surface in the high mountains east of the Sierran crest; (2) the absence of much of an early and middle Tertiary record, which, however, is preserved in the Amargosa and Mojave regions to the east and south, respectively; (3) the absence or only thin veneer of late Neogene clastic deposits underlying extensive areas of mafic, generally basaltic flows that define paleohorizontal and fill topographic lows; and (4) the gradual eastward thickening of late Miocene deposits, for example, east of the White Mountains (Crowder and Sheridan, 1972).

Long-wavelength, regional warping or arching may have accompanied or preceded initiation of the Sierra Nevada frontal fault zone. Regional topographic profiles (Fig. 17) show the general character an arch, which is evident in both the topography and deformed late Miocene mafic flows. This arch apparently formed concurrent with the transtensive structural regime that characterizes the Walker Lane and Eastern California shear zone (cf. Wallace, 1984; Unruh et al., 2003; Fig. 18).

CONCLUSIONS

(1) The late Miocene landscape that was truncated by the Sierra Nevada frontal fault is widely preserved in the eastern Sierra region. The late Miocene landscape was part of an erosional baseline that was relatively stable for 10–20 m.y. before it was truncated. It provides an important strain marker for relative vertical displacements in the eastern Sierra part of the Walker Lane belt. It is far more extensive than late Miocene basin deposits, which are only locally preserved.

(2) The late Miocene erosion surface is a mappable geomorphic unit that helps to define important structural features along fault-bounded blocks that would not otherwise be well constrained, including relay ramps and rollovers on the hanging wall of listric faults (Fig. 12). The late Miocene erosion surface that underlies the ca. 9–10 Ma mafic flows in the Sierra Nevada is herein referred to as the Lindgren Surface; its equivalent in the eastern Sierra region is referred to as the Inyo Surface. The erosion surface is compound and represents a relict mountainous landscape that was surrounded by vast rivers that flowed from the eastern Sierra area during the Miocene. The rivers became inset, forming a broad terrace during the late Miocene, which was succeeded by canyon cutting in the Pliocene (Lawson, 1904; Dalrymple, 1963; Matthes, 1930).

(3) Relicts of the erosion surfaces originally described by Lawson (1904), including the Summit Upland, Subsummit Plateau, and Chagoopa Plateau, are widely preserved throughout the eastern Sierra region and are here referred to as the Inyo Surface. Large areas of the compound erosion surface remain in high mountain crestal areas or are gently tilted on rotated crustal blocks. The Inyo Surface provides a strain marker that helps delineate relay ramps and rollover structures along the edges of fault-bounded blocks.

(4) Much of the erosion surface that forms the Subsummit Plateau surface was eroded during Oligocene and Miocene time (Dalrymple, 1963). Uplift of the Subsummit Plateau near the Sierran crest was probably under way at the time of basalt eruption, based on the absence of Tertiary deposits under the basalt flow near the crest. The Subsummit Plateau surface was inset by the Chagoopa Terrace during the late Miocene, when the Sierra Nevada frontal fault became active as the Sierra Nevada block became kinematically detached from the continent east of the Walker Lane area. The large-scale "terrace riser" that separates the Subsummit and Chagoopa Plateaus probably formed ca. 9–10 Ma. Initiation of canyon cutting probably occurred during the early Pliocene, when uplift rates appear to have increased (Table 2; Fig. 2).

(5) The initiation of the grabens that now form Death Valley and Owens Valley began in late Miocene time. The Sierra Nevada frontal fault was active by ca. 9.5–10.5 Ma. A relatively coherent structural block that extended between what is now occupied by the White–Inyo–Cottonwood–Panamint Ranges began floundering in mainly late early Pliocene time.

(6) The relative uplift that characterizes the region (Fig. 18) was concurrent with transtensive deformation. The Inyo Surface and its thin Miocene-Pliocene cover sequence provide an excellent strain marker for reconstructing vertical displacements and structural characteristics of Pliocene-Pleistocene transtensive deformation in the Walker Lane east of the Sierra Nevada. This surface provides a regional datum that helps to constrain long-term, i.e., 4–6 m.y., vertical separation rates that range on the order of 0.2–0.4 mm/yr (Table 2). Long-term vertical separation rates determined from ca. 4–9.6 Ma basaltic rocks also yield displacement rates that range from ~0.2 to 0.4 mm/yr. The late

Pleistocene to Holocene vertical separation rates range from ~0.2 to 0.9 mm/yr but are dominantly around 0.2–0.5 mm/yr, consistent with the higher tilting rates of the Sierra Nevada block in the Pliocene (Table 2).

ACKNOWLEDGMENTS

I thank John Oldow for reviewing the revised version of this manuscript and making significant suggestions that very greatly changed and improved an early draft and resulted in this paper. Constructive and insightful reviews by Jack Stewart and Dave Harwood, U.S. Geological Survey, Menlo Park, are also very much appreciated. I thank A.J. Gillespie, U.S. Geological Survey intern in 2001, for help with constructing the structure profiles across the Darwin Plateau. Discussions with Fred Phillips, New Mexico Tech., Steve Bacon, Desert Research Institute, and Collin Amos, University of California–Santa Barbara, have added breadth to discussions in this paper. All errors in content are my own oversights. This paper is dedicated to the memory of N. King Huber, U.S. Geological Survey (1926–2007).

REFERENCES CITED

Axelrod, D.I., 1957, Late Tertiary floras and the Sierra Nevada uplift, California-Nevada: Geological Society of America Bulletin, v. 68, p. 19–45, doi: 10.1130/0016-7606(1957)68[19:LTFATS]2.0.CO;2.

Axelrod, D.I., 1962, Post-Pliocene uplift of the Sierra Nevada, California: Geological Society of America Bulletin, v. 73, p. 183–198, doi: 10.1130/0016-7606(1962)73[183:PUOTSN]2.0.CO;2.

Axelrod, D.I., and Ting, W.S., 1960, Late Pliocene floras east of the Sierra Nevada, California-Nevada: University of California Publications in Geological Sciences, v. 39, 117 p.

Axelrod, D.I., and Ting, W.S., 1961, Early Pleistocene floras from the Chagoopa Surface, southern Sierra Nevada: University of California Publications in Geological Sciences, v. 39, p. 119–194.

Bachman, S.B., 1978, Pliocene-Pleistocene break-up of the Sierra Nevada–White–Inyo Mountains block and formation of Owens Valley: Geology, v. 6, p. 461–463, doi: 10.1130/0091-7613(1978)6<461:PBOTSN>2.0.CO;2.

Bacon, C.R., and Duffield, W.A., 1981, Late Cenozoic rhyolites from the Kern Plateau, southern Sierra Nevada, California: American Journal of Science, v. 281, p. 1–34.

Bacon, C.R., Giovannetti, D.M., Duffield, W.A., Dalrymple, G.B., and Drake, R.E., 1982, Age of the Coso Formation, Inyo County, California: U.S. Geological Survey Bulletin 1527, 18 p.

Bacon, S.N., Jayko, A.S., and McGeehin, J., 2005, Holocene and latest Pleistocene oblique dextral faulting on the Southern Inyo Mountains fault, Owens Lake Basin, California: Bulletin of the Seismological Society of America, v. 95, p. 2472–2486, doi: 10.1785/0120040228.

Barghoorn, E., 1951, Age and environment: A survey of North American Tertiary floras in relation to paleoecology: Journal of Paleontology, v. 25, p. 736–744.

Bartow, J.A., 1994, Tuffaceous ephemeral lake deposits on an alluvial plain, middle Tertiary of central California: Sedimentology, v. 41, p. 215–232, doi: 10.1111/j.1365-3091.1994.tb01402.x.

Bateman, P.C., 1965, Geology and Tungsten Mineralization of the Bishop District, California: U.S. Geological Survey Professional Paper 470, 208 p.

Bateman, P.C., and Wahrhaftig, C., 1966, Geology of the Sierra Nevada, in Bailey, E.G., ed., Geology of Northern California: California Division of Mines and Geology, v. 190, p. 107–172.

Bateman, P.C., Lockwood, J.P., and Lydon, P.A., 1971, Geologic Map of the Kaiser Peak Quadrangle, Central Sierra Nevada, California: U.S. Geological Survey Quadrangle Map GQ-894, scale 1:62,500.

Beanland, S., and Clark, M.M., 1994, Owens Valley Fault: U.S. Geological Survey Bulletin 1982, 29 p.

Bennett, R.A., Davis, J.L., and Wernicke, B.P., 1999, Present-day pattern of Cordilleran deformation in the western United States: Geology, v. 31, p. 327–330.

Berman, F.A., 1999, The paleoclimate and tectonic significance of the Bee Springs alluvial fan sequence, west-central Inyo Mountains, east-central California [M.S. thesis]: Arcata, California, Humboldt State University, 77 p.

Blackwelder, E., 1931, Pleistocene glaciation in the Sierra Nevada and Basin ranges: Geological Society of America Bulletin, v. 42, p. 865–922.

Burchfiel, B.C., and Stewart, J.H., 1966, 'Pull-apart' origin of the central segment of Death Valley, California: Geological Society of America Bulletin, v. 77, p. 439–441.

Burchfiel, B.C., Hodges, K.V., and Royden, L.H., 1987, Geology of Panamint Valley–Saline Valley pull-apart system, California; palinspastic evidence for low-angle geometry of a Neogene range-bounding fault: Journal of Geophysical Research, v. 92, p. 10,422–10,426, doi: 10.1029/JB092iB10p10422.

Burow, K.R., Shelton, J.L., Hevesi, J.A., and Weissmann, G.S., 2004, Hydrogeologic Characterization of the Modesto Area, San Joaquin Valley, California: U.S. Geological Survey Scientific Investigations Report 2004-5232, 54 p.

Bursik, M., and Sieh, K.E., 1989, Range front faulting and volcanism in the Mono Basin, eastern California: Journal of Geophysical Research, v. 94, p. 15,587–15,609, doi: 10.1029/JB094iB11p15587.

Busacca, A.J., Verosub, K.L., and Singer, M.J., 1982, Late Cenozoic geologic and soil geomorphic history of the Feather and Yuba River areas, Sacramento Valley, California: Geological Society of America Abstracts with Programs, v. 14, p. 153.

Busacca, A.J., Singer, M.J., and Verosub, K.L., 1989, Late Cenozoic stratigraphy of the Feather and Yuba Rivers area, California, with a section on soil development in mixed alluvium at Honcut Creek: U.S. Geological Survey Bulletin B 1590-G, p. G1–G132.

Cecil, M.R., Ducea, M.N., Reiners, P.W., and Chase, C.G., 2006, Cenozoic exhumation of the northern Sierra Nevada, California, from (U-Th)/He thermochronology: Geological Society of America Bulletin, v. 118, p. 1481–1488, doi: 10.1130/B25876.1.

Chamberlain, C.P., and Poage, M.A., 2000, Reconstructing the paleotopography of mountain belts from the isotopic composition of authigenic minerals: Geology, v. 28, p. 115–118, doi: 10.1130/0091-7613(2000)28<115:RTPOMB>2.0.CO;2.

Cherven, V.B., 1984, Early Pleistocene glacial outwash deposits in the eastern San Joaquin Valley, California; a model for humid-region alluvial fans: Sedimentology, v. 31, p. 823–836, doi: 10.1111/j.1365-3091.1984.tb00889.x.

Christensen, M.N., 1966, Late Cenozoic crustal movements in the Sierra Nevada of California: Geological Society of America Bulletin, v. 77, p. 163–182, doi: 10.1130/0016-7606(1966)77[163:LCCMIT]2.0.CO;2.

Clark, M.K., Maheo, G., Saleeby, J., and Farley, K.A., 2005, The non-equilibrium landscape of the southern Sierra Nevada, California: GSA Today, v. 15, p. 4–9, doi: 10.1130/1052-5173(2005)015[4:TNLOTS]2.0.CO;2.

Clark, M.M., and Gillespie, A.R., 1981, Record of late Quaternary faulting along the Hilton Creek Fault in the Sierra Nevada, California: Earthquake Notes, v. 52, p. 46.

Coe, R.S., Stock, G.M., Lyons, J.J., Beitler, B., and Bowen, G.J., 2005, Yellowstone hotspot volcanism in California? A paleomagnetic test of the Lovejoy flood basalt hypothesis: Geology, v. 33, p. 697–700, doi: 10.1130/G21733.1.

Coleman, D.S., and Walker, J.D., 1990, Geochemistry of Miocene-Pliocene volcanic rocks from around Panamint Valley, Death Valley area, California, in Wernicke, B.P., ed., Tertiary Extensional Tectonics Near the Latitude of Las Vegas: Geological Society of America Memoir 176, p. 391–411.

Coleman, D.S., Walker, J.D., Bickford, M.E., and Dickinson, W.R., 1987, Geochemistry of Mio-Pliocene volcanic rocks from northern Panamint Mountains and Darwin Plateau, Basin and Range Province; implications for extension and regional geology: Geological Society of America Abstracts with Programs, v. 19, p. 623.

Conrad, J.E., 1993, Late Cenozoic tectonics of the southern Inyo Mountains, eastern California [M.Sc. thesis]: San Jose, San Jose State University, 84 p.

Crowder, D.F., and Sheridan, M.F., 1972, Geologic Map of the White Mountain Peak Quadrangle, Mono County, California: U.S. Geological Survey Quadrangle Map GQ-1012, scale 1:62,500.

Curry, R.R., 1971, Glacial and Pleistocene History of the Mammoth Lakes Sierra Nevada; a Geologic Guidebook: Missoula, Department of Geology, University of Montana, Geological Series, v. 2, 47 p.

Dalrymple, G.B., 1963, Potassium-argon dates of some Cenozoic volcanic rocks of the Sierra Nevada, California: Geological Society of America

Bulletin, v. 74, p. 379–390, doi: 10.1130/0016-7606(1963)74[379:PDOSCV] 2.0.CO;2.

Dalrymple, G.B., 1964, Cenozoic Chronology of the Sierra Nevada, California: University of California Publications in Geological Sciences, v. 47, 41 p.

Densmore, A.L., and Anderson, R.S., 1997, Tectonic geomorphology of the Ash Hill fault, Panamint Valley, California: Basin Research, v. 9, no. 1, p. 53–63, doi: 10.1046/j.1365-2117.1997.00028.x.

dePolo, C.M., 1989, Seismotectonics of the White Mountains fault system, east-central California and west-central Nevada [M.S. thesis]: Reno, University of Nevada, Reno, 354 p.

Dilek, Y., and Moores, E.M., 1999, A Tibetan model for the early Tertiary western United States: Journal of the Geological Society of London, v. 156, p. 929–941, doi: 10.1144/gsjgs.156.5.0929.

Dixon, T.H., Miller, M., Farina, F., Wang, H., and Johnson, D., 2000, Present-day motion of the Sierra Nevada block and some tectonic implications for the Basin and Range Province, North American Cordillera: Tectonics, v. 19, p. 1–24, doi: 10.1029/1998TC001088.

Ducea, M.N., and Saleeby, J.B., 1996, Buoyancy sources for a large unrooted mountain range, the Sierra Nevada, California: Evidence from xenolith thermobarometry: Journal of Geophysical Research, v. 101, p. 8229–8241, doi: 10.1029/95JB03452.

Duffield, W.A., and Bacon, C.R., 1981, Geologic Map of the Coso Volcanic Field and Adjacent Areas, Inyo County, California: U.S. Geological Survey Miscellaneous Investigations I-1200, scale 1:50,000.

Dumitru, T.A., 1990, Subnormal Cenozoic geothermal gradients in the extinct Sierra Nevada magmatic arc; consequences of Laramide and post-Laramide shallow-angle subduction: Journal of Geophysical Research, v. 95, p. 4925–4941, doi: 10.1029/JB095iB04p04925.

Evernden, J.F., 1980, Earthquake Hazards along the Wasatch and Sierra-Nevada Frontal Fault Zones: U.S. Geological Survey Open-File Report OF 80-801, 688 p.

Evernden, J.F., and James, G.T., 1964, Potassium-argon dates and the Tertiary floras of North America: American Journal of Science, v. 262, p. 945–974.

Fairbanks, H.W., 1898, The Great Sierra Nevada Fault Scarp: Popular Science, v. 52, p. 609–621.

Frei, L.S., 1986, Additional paleomagnetic results from the Sierra Nevada: Further constraints on Basin and Range extension and northward displacement in the western United States: Geological Society of America Bulletin, v. 97, p. 840–849, doi: 10.1130/0016-7606(1986)97<840:APRFTS>2.0.CO;2.

Gilbert, C.M., and Reynolds, M.W., 1973, Character and chronology of basin development, western margin of the Basin and Range Province: Geological Society of America Bulletin, v. 84, p. 2489–2509, doi: 10.1130/0016-7606(1973)84<2489:CACOBD>2.0.CO;2.

Gilbert, C.M., Christensen, M.N., Al-Rawi, Y., and Lajoie, K.R., 1968, Structural and volcanic history of Mono Basin, California-Nevada, *in* Coats, R.R., Hay, R.L., and Anderson, C.A., eds., Studies in Volcanology: A Memoir in Honor of Howel Williams: Boulder, Colorado, Geological Society of America Memoir 116, p. 275–329.

Gillespie, A.R., 1982, Quaternary glaciation and tectonism in the southeastern Sierra Nevada, Inyo County, California [Ph.D. thesis]: San Marino, California Institute of Technology, 695 p.

Gillespie, A.J., Jayko, A.S., and Thompson, R.A., 2002, Estimate of Plio-Pleistocene dip-slip rates using offset flood basalts in the northern Panamint Range and Darwin Plateau, California: Eos (Transactions, American Geophysical Union), v. 83, p. 1313–1314.

Greensfelder, R.W., Kintzer, F.C., and Somerville, M.R., 1980, Seismotectonic regionalization of the Great Basin, and comparison of moment rates computed from Holocene strain and historic seismicity: Geological Society of America Bulletin, v. 91, p. 518–523, doi: 10.1130/0016-7606(1980)91<518:SROTGB>2.0.CO;2.

Hall, W.E., 1971, Geology of the Panamint Butte Quadrangle, Inyo County, California: U.S. Geological Survey Bulletin B-1299, 67 p.

Harding, M.B., 1988, Geology of the Wildrose Peak area, Panamint Mountains, SE CA: Geological Society of America Abstracts with Programs, v. 20, p. 166.

Hay, E.A., 1976, Cenozoic uplifting of the Sierra Nevada in isostatic response to North American and Pacific plate interactions: Geology, v. 4, p. 763–766, doi: 10.1130/0091-7613(1976)4<763:CUOTSN>2.0.CO;2.

Hodges, K.V., McKenna, L.W., Stock, J., Knapp, J., Page, L., and Sternlof, K., 1989a, Evolution of extensional basins, and Basin and Range topography west of Death Valley, California: Tectonics, v. 8, p. 453–467, doi: 10.1029/TC008i003p00453.

Hodges, K.V., McKenna, L.W., Stock, J., Knapp, J., Page, L., Sternlof, K., Wuest, G., and Walker, J.D., 1989b, Evolution of extensional basins and Basin and Range topography west of Death Valley, California: Tectonics, v. 8, p. 453–467, doi: 10.1029/TC008i003p00453.

Hopper, R.H., 1939, Paleozoic section in the Argus and Panamint Ranges, Inyo County, California: Geological Society of America Bulletin, v. 50, p. 1952.

Hopper, R.H., 1947, Geologic section from the Sierra Nevada to Death Valley, California: Geological Society of America Bulletin, v. 58, p. 393–432, doi: 10.1130/0016-7606(1947)58[393:GSFTSN]2.0.CO;2.

House, M.A., Wernicke, B.P., and Farley, K.A., 1998, Dating topography of the Sierra Nevada, California, using apatite (U-Th)/He: Nature, v. 396, p. 66–69, doi: 10.1038/23926.

House, M.A., Wernicke, B.P., and Farley, K.A., 2001, Paleo-geomorphology of the Sierra Nevada, California, from (U-Th)/He ages in apatite: American Journal of Science, v. 301, p. 77–102, doi: 10.2475/ajs.301.2.77.

Huber, N.K., 1981, Amount and Timing of Late Cenozoic Uplift and Tilt of the Central Sierra Nevada, California—Evidence from the Upper San Joaquin River: U.S. Geological Survey Professional Paper 1197, 28 p.

Huber, N.K., 1990, The late Cenozoic evolution of the Tuolumne River, central Sierra Nevada, California: Geological Society of America Bulletin, v. 102, p. 102–115, doi: 10.1130/0016-7606(1990)102<0102:TLCEOT>2.3.CO;2.

Ingersoll, R.V., 1982, Initiation and evolution of the Great Valley forearc basin of northern and central California, U.S.A., *in* Leggett, J.K., ed., Trench-forearc geology; sedimentation and tectonics on modern and ancient active plate margins, conference: Geological Society of London Special Publication 10, p. 459–467, doi: 10.1144/GSL.SP.1982.010.01.31.

Jayko, A.S., 2004a, Plio-Pleistocene extension of a late Miocene–early Pliocene erosion surface in the southern Inyo, Argus and Coso Ranges, Eastern California shear zone: Geological Society of America Abstracts with Programs, v. 36, no. 4, p. 17.

Jayko, A.S., 2004b, Comparison of the Plio-Pleistocene and Quaternary transtensive deformation structures along the Pacific–North American plate boundary, eastern Sierra region, southwestern Inyo County, California: Eos (Transactions, American Geophysical Union), v. 85, no. 47, fall meeting supplement, p. F567, abstract G11A-0782.

Jayko, A.S., 2008, Surficial Geology of the Darwin Hills 30′ × 60′ Quadrangle, Inyo County, California: U.S. Geological Survey Scientific Investigations Map 3040, scale 1:100,000, color map, 38 p.

Jennings, C.W., 1994, Fault activity map of California and adjacent areas with locations and ages of recent volcanic eruptions: California Division of Mines and Geology Data Map Series 6, 92 p., 2 plates, scale 1:750,000.

Johnson, B.K., 1957, Geology of part of the Manly Peak Quadrangle, southern Panamint Range: University of California Publications in Geological Sciences, v. 30, p. 353–424.

Jones, C.H., Farmer, G.L., and Unruh, J., 2004, Tectonics of Pliocene removal of lithosphere of the Sierra Nevada: Geological Society of America Bulletin, v. 116, p. 1408–1422, doi: 10.1130/B25397.1.

Kistler, R.W., 1966, Structure and metamorphism in the Mono Craters Quadrangle, Sierra Nevada, California: U.S. Geological Survey Bulletin B 1221-E, p. E1–E53.

Klinger, R.E., 1999, Tectonic geomorphology along the Death Valley fault system; evidence for recurrent late Quaternary activity in Death Valley National Park, *in* Slate, J.L., ed., Proceedings of Conference on Status of Geologic Research and Mapping, Death Valley National Park: U.S. Geological Survey Open-File Report 99-153, p. 132–140.

Knopf, A., 1918, A Geologic Reconnaissance of the Inyo Range and the Eastern Slope of the Southern Sierra Nevada, California: U.S. Geological Survey Professional Paper 110, 130 p.

Krauskopf, K.B., 1971, Geologic Map of the Barcroft Quadrangle, California-Nevada: U.S. Geological Survey Quadrangle Map GQ-960, scale 1:62,500.

Labotka, T.C., and Albee, A.L., 1990, Uplift and exposure of the Panamint metamorphic complex, California, *in* Wernicke, B., ed., Basin and Range Extensional Tectonics near the Latitude of Las Vegas, Nevada: Geological Society of America Memoir 176, p. 345–362.

Larsen, N.W., 1979, Chronology of Late Cenozoic basaltic volcanism: The tectonic implications along a segment of the Sierra Nevada and Basin and Range province boundary [Ph.D. thesis]: Provo, Utah, Brigham Young University, 95 p.

Lawson, A.C., 1904, The geomorphology of the upper Kern Basin, California: Geological Society of America Bulletin, v. 3, p. 291–376.

Le, K., Lee, J., Owen, L.A., and Finkel, R., 2007, Late Quaternary slip rates along the Sierra Nevada frontal fault zone, California: Evidence for slip partitioning across the western margin of the Eastern California shear zone/Basin and Range province: Geological Society of America Bulletin, v. 119, p. 240–256, doi: 10.1130/B25960.1.

LeConte, J., 1880, The old river beds of California: American Journal of Science, v. 38, p. 176–190.

LeConte, J., 1886, A post-Tertiary elevation of the Sierra Nevada shown by the river beds: American Journal of Science, v. 38, p. 167–181.

LeConte, J., 1889, On the origin of normal faults and the structure of the basins region: American Journal of Science, v. 38, p. 257–263.

Lindgren, W., 1911, The Tertiary Gravels of the Sierra Nevada of California: U.S. Geological Survey Professional Paper 73, 225 p.

Linn, A.M., DePaolo, D.J., and Ingersoll, R., 1992, Nd-Sr isotopic, geochemical, and petrographic stratigraphy and paleotectonic analysis; Mesozoic Great Valley forearc sedimentary rocks of California: Geological Society of America Bulletin, v. 104, p. 1264–1279, doi: 10.1130/0016-7606(1992) 104<1264:NSIGAP>2.3.CO;2.

Loomis, D.P., and Burbank, D.W., 1988, The stratigraphic evolution of the El Paso Basin, Southern California; implications for the Miocene development of the Garlock fault and uplift of the Sierra Nevada: Geological Society of America Bulletin, v. 100, p. 12–28, doi: 10.1130/0016-7606 (1988)100<0012:TSEOTE>2.3.CO;2.

Lubetkin, L.K., and Clark, M.M., 1988, Late Quaternary activity along the Lone Pine fault, eastern California: Geological Society of America Bulletin, v. 100, p. 755–766, doi: 10.1130/0016-7606(1988)100<0755: LQAATL>2.3.CO;2.

Luckow, H.G., Pavlis, T.L., Serpa, L.F., Guest, B., Wagner, D.L., Snee, L.W., Hensley, T.M., and Korjenkov, A., 2005, Late Cenozoic sedimentation and volcanism during transtensional deformation in Wingate Wash and the Owlshead Mountains, Death Valley: Earth-Science Reviews, v. 73, p. 177–219.

Lueddecke, S.B., Pinter, N., and Gans, P., 1998, Plio-Pleistocene ash falls, sedimentation and range front faulting along the White-Inyo Mountains front, California: The Journal of Geology, v. 106, p. 511–522.

Marchand, D.E., and Allwardt, A., 1981, Late Cenozoic Stratigraphic Units, Northeastern San Joaquin Valley, California: U.S. Geological Survey Bulletin B-1470, 70 p.

Martel, S.J., Harrison, T.M., and Gillespie, A.R., 1987, Late Quaternary vertical displacement rate across the Fish Springs fault, Owens Valley fault zone, California: Quaternary Research, v. 27, p. 113–129, doi: 10.1016/0033-5894(87)90071-8.

Matthes, F.E., 1930, Geologic History of the Yosemite Valley: U.S. Geological Survey Professional Paper 160, 137 p.

Matthes, F.E., 1937, The geologic history of Mount Whitney: Sierra Club Bulletin, v. 22, p. 1–18.

Matthes, F.E., 1960, Reconnaissance of the Geomorphology and Glacial Geology of the San Joaquin Basin, Sierra Nevada, California: U.S. Geological Survey Professional Paper 329, 62 p.

McDowell, S.D., 1974, Emplacement of the Little Chief Stock, Panamint Range, California: Geological Society of America Bulletin, v. 85, p. 1535–1546, doi: 10.1130/0016-7606(1974)85<1535:EOTLCS>2.0.CO;2.

McDowell, S.D., 1978, Little Chief Granite porphyry; feldspar crystallization history: Geological Society of America Bulletin, v. 89, p. 33–49, doi: 10. 1130/0016-7606(1978)89<33:LCGPFC>2.0.CO;2.

McKee, E.H., 1985, Geologic Map of the Magruder Mountain Quadrangle, Esmeralda County, Nevada and Inyo County, California: U.S. Geological Survey Quadrangle Map GQ-1587, scale 1:62,500.

McKenna, L.W., III, 1986, New Rb-Sr constraints on the age of detachment faulting in the Panamint Range, Death Valley, CA: Geological Society of America Abstracts with Programs, v. 18, no. 2, p. 156.

McKenna, L.W., and Hodges, K.V., 1990, Constraints on the kinematics and timing of late Miocene–Recent extension between the Panamint and Black Mountains, southeastern California, *in* Wernicke, B., ed., Basin and Range Extensional Tectonics near the Latitude of Las Vegas, Nevada: Geological Society of America Memoir 176, p. 363–376.

McKenna, L.W., and Snee, L.W., 1990, Timing of movement and strain rate of Cenozoic extensional faulting in the Death Valley area, California: Geological Society of America Abstracts with Programs, v. 22, p. 227.

McKenna, L.W., Hodges, K.V., and Deaps, K.V., 1989, Constraints on middle Miocene–Recent extension, Death Valley region, Basin and Range Province: Geological Society of America Abstracts with Programs, v. 21, p. 353.

Miller, M.M., Johnson, D.J., Dixon, T.H., and Dokka, R.K., 2001, Refined kinematics of the Eastern California shear zone from GPS observations, 1993–1998: Journal of Geophysical Research, v. 106, p. 2245–2263, doi: 10.1029/2000JB900328.

Miller, R.J., and Wrucke, C.T., 1995, Age, chemistry, and geologic implications of Tertiary volcanic rocks in the Last Chance Range and part of the Saline Range, northern Death Valley region, California: Isochron-West, v. 62, p. 30–36.

Monastero, F.C., Walker, D.J., Katzenstein, A.M., and Sabin, A.E., 2002, Neogene evolution of the Indian Wells Valley, east-central California, *in* Glazner, A.F., Walker, J.D., and Bartley, J.M., eds., Geological Evolution of the Mojave Desert and Southwestern Basin and Range: Geological Society of America Memoir 195, p. 199–228.

Mulch, A., Graham, S.A., and Chamberlain, C.P., 2006, Hydrogen isotopes in Eocene river gravels and paleoelevation of the Sierra Nevada: Science, v. 313, p. 87–89, doi: 10.1126/science.1125986.

Nelson, C.A., 1966, Geologic Map of the Blanco Mountain Quadrangle, Inyo and Mono Counties, California: U.S. Geological Survey Geological Quadrangle GQ-529, scale 1:62,500.

Oswald, J.A., and Wesnousky, S.G., 2002, Neotectonics and Quaternary geology of the Hunter Mountain fault zone and Saline Valley region, southeastern California: Geomorphology, v. 42, p. 255–278, doi: 10.1016/S0169-555X(01)00089-7.

Page, R.W., and Balding, G.O., 1973, Geology and Quality of Water in the Modesto-Merced Area, San Joaquin Valley, California, with a Brief Section on Hydrology: U.S. Geological Survey Water-Resources Investigations 6-73, 85 p.

Pakiser, L.C., Kane, M.F., and Jackson, W.W., 1964, Structural Geology and Volcanism of Owens Valley Region, California: A Geophysical Study: U.S. Geological Survey Professional Paper 438, 64 p.

Phillips, F.M., Zreda, M.G., Smith, S.S., Elmore, D., Kubik, P.W., and Sharma, P., 1990, Cosmogenic chlorine-36 chronology for glacial deposits at Bloody Canyon, eastern Sierra Nevada: Science, v. 248, p. 1529–1532, doi: 10.1126/science.248.4962.1529.

Phillips, F.M., McIntosh, W.C., and Dunbar, N.W., 2000, Extensional collapse of the Owens Valley during the late Pliocene: Geological Society of America Abstracts with Programs, v. 32, p. 507.

Piper, A.M., Gale, H.S., Thomas, H.E., and Robinson, T.W., 1939, Geology and Ground Water Hydrology of the Mokelumne Area, California: U.S. Geological Survey Water-Supply Paper 780, 230 p.

Reheis, M.C., 1992, Geologic Map of Late Cenozoic Deposits and Faults in Parts of the Soldier Pass and Magruder Mountain 15′ Quadrangles, Inyo and Mono Counties, California and Esmeralda County, Nevada: U.S. Geological Survey Miscellaneous Investigations I-2268, scale 1:24,000.

Reheis, M.C., and Sawyer, T.L., 1997, Late Cenozoic history and slip rates of the Fish Lake Valley, Emigrant Peak, and Deep Springs fault zones, Nevada and California: Geological Society of America Bulletin, v. 109, p. 280–299, doi: 10.1130/0016-7606(1997)109<0280:LCHASR>2.3.CO;2.

Reheis, M.C., Slate, J.L., and Sawyer, T.L., 1995, Geologic Map of the Late Cenozoic Deposits and Faults in Parts of the Mt. Barcroft, Piper Peak and Soldier Pass 15′ Quadrangles, Esmeralda County, Nevada, and Mono County, California: U.S. Geological Survey Miscellaneous Investigations I-2464, scale 1:24,000.

Reid, J.A., 1906, A detail of the Great Fault Zone of the Sierra Nevada: Geological Society of America Bulletin, v. 16, p. 593.

Rood, D.H., Busby, C.J., Jayko, A.S., and Luyendyk, B.P., 2005, Neogene to Quaternary kinematics of the central Sierran frontal fault system in the Sonora Pass region; preliminary structural, paleomagnetic, and neotectonic results: Geological Society of America Abstracts with Programs, v. 37, no. 4, p. 65.

Ross, D.C., 1965, Geology of the Independence Quadrangle, Inyo County, California: U.S. Geological Survey Bulletin 1181-O, scale 1:64,500, 64 p.

Ross, D.C., 1986, Basement-Rock Correlations across the White Wolf–Breckenridge–Southern Kern Canyon Fault Zone, Southern Sierra Nevada, California: U.S. Geological Survey Bulletin 1651, 25 p.

Sarna-Wojcicki, A.M., Pringle, M.S., and Wijbrans, J., 2000, New Ar^{40}/Ar^{39} age of the Bishop Tuff from multiple sites and sediment rate calibration for the Matuyama-Brunhes boundary: Journal of Geophysical Research, v. 105, p. 21,431–21,443.

Savage, J.C., and Lisowski, M., 1995, Strain accumulation in Owens Valley, California, 1974 to 1988: Bulletin of the Seismological Society of America, v. 85, p. 151–158.

350

Jayko

Schweig, E.S., III, 1985, Neogene tectonics and paleogeography of the southwestern Great Basin, California [Ph.D. thesis]: Palo Alto, California, Stanford University, 200 p.

Schweig, E.S., III, 1986, The inception of Basin and Range tectonics in the region between Death Valley and the Sierra Nevada, California: Geological Society of America Abstracts with Programs, v. 18, p. 181–182.

Schweig, E.S., III, 1989, Basin-Range tectonics in the Darwin Plateau, southwestern Great Basin, California: Geological Society of America Bulletin, v. 101, p. 652–662, doi: 10.1130/0016-7606(1989)101<0652:BRTITD>2.3.CO;2.

Shaw, D.R., 1965, Strike-slip control of Basin-Range structure indicated by historical faults in western Nevada: Geological Society of America Bulletin, v. 76, p. 1362–1378.

Slemmons, D.B., Van Wormer, D., Bell, E.J., and Silberman, M.L., 1979, Recent crustal movements in the Sierra Nevada–Walker Lane region of California-Nevada: Part 1. Rate and style of deformation: Tectonophysics, v. 52, p. 561–570.

Slemmons, D.B., Vittori, E., Jayko, A.S., Carver, G.A., and Bacon, S.N., 2008, Quaternary fault map of Owens Valley, Inyo County, Eastern California: Geological Society of America Map and Chart Series 96, 25 p., 2 sheets, scale 1:100,000, doi: 10.1130/2008.MCH096.

Snow, J.K., and White, C., 1990, Listric normal faulting and synorogenic sedimentation, northern Cottonwood Mountains, Death Valley region, California, in Wernicke, B.P., ed., Tertiary Extensional Tectonics Near the Latitude of Las Vegas: Geological Society of America Memoir 176, p. 413–445.

St. Amand, P., and Roquemore, G.R., 1979, Tertiary and Holocene development of the southern Sierra Nevada and Coso Range, California: Tectonophysics, v. 52, p. 409–410, doi: 10.1016/0040-1951(79)90255-5.

Sternlof, K.R., 1988, Structural style and kinematic history of the active Panamint-Saline extensional system, Inyo County, California [M.Sc. thesis]: Cambridge, Massachusetts Institute of Technology, 30 p.

Stewart, J.H., 1985, East-trending dextral faults in the western Great Basin; an explanation for anomalous trends of pre-Cenozoic strata and Cenozoic faults: Tectonics, v. 4, p. 547–564, doi: 10.1029/TC004i006p00547.

Stewart, J.H., 1992, Walker Lane belt, Nevada and California—An overview, in Craig, S.D., ed., Structure, Tectonics and Mineralization of the Walker Lane: Walker Lane Symposium Proceedings Volume: Reno, Geological Society of Nevada, p. 1–17.

Stinson, M.C., 1977a, Geology of the Haiwee Reservoir 15′ Quadrangle, Inyo County, California: California Division of Mines and Geology Map Sheet 37, scale 1:62,500.

Stinson, M.C., 1977b, Geologic Map and Sections of the Keeler 15′ Quadrangle, Inyo County, California: California Division of Mines and Geology Map Sheet 38, scale 1:62,500.

Stock, G.M., Anderson, R.S., and Finkel, R.C., 2004, Pace of landscape evolution in the Sierra Nevada, California, revealed from cosmogenic dating of cave sediments: Geology, v. 32, p. 193–196, doi: 10.1130/G20197.1.

Stock, G.M., Anderson, R.S., and Finkel, R.C., 2005, Rates of erosion and topographic evolution of the Sierra Nevada, California, inferred from cosmogenic ^{26}Al and ^{10}Be concentrations: Earth Surface Processes and Landforms, v. 30, p. 985–1006, doi: 10.1002/esp.1258.

Stockli, D.F., Dumitru, T.A., McWilliams, M.O., and Farley, K.A., 2003, Cenozoic tectonic evolution of the White Mountains, California and Nevada: Geological Society of America Bulletin, v. 115, p. 788–816, doi: 10.1130/0016-7606(2003)115<0788:CTEOTW>2.0.CO;2.

Stone, P., Dunne, G.C., Stevens, C.H., and Gulliver, R.M., 1989, Geological Map of Paleozoic and Mesozoic Rocks in Parts of the Darwin Plateau and Adjacent Areas, Inyo County, California: U.S. Geological Survey Miscellaneous Field Investigations Map I-1932, scale 1:31,250.

Stone, P., Dunne, G.C., Conrad, J.E., Swanson, B.J., Stevens, C.H., and Valin, V.C., 2004, Geologic Map of the Cerro Gordo Peak 7.5′ Quadrangle, Inyo County, California: U.S. Geological Survey Geological Science Investigation Map I-2851, scale 1:24,000.

Unruh, J.R., 1991, The uplift of the Sierra Nevada and implications for late Cenozoic epeirogeny in the western Cordillera: Geological Society of America, v. 103, p. 1395–1404, doi: 10.1130/0016-7606(1991)103<1395:TUOTSN>2.3.CO;2.

Unruh, J., Humphrey, J., and Barron, A., 2003, Transtensional model for the Sierra Nevada frontal fault system, eastern California: Geology, v. 31, p. 327–330, doi: 10.1130/0091-7613(2003)031<0327:TMFTSN>2.0.CO;2.

Vogel, M., Jayko, A.S., Wooden, J., and Smith, R.S.U., 2002, Quaternary exhumation rate central Panamint Range, California from U-Pb zircon ages: Geological Society of America Abstracts with Programs, v. 34, p. 249.

Von Huene, R., Bateman, P.C., and Ross, D.C., 1963, Guidebook for Seismological Tour: Part 4. Indian Wells, Owens, Long Valley and Mono Basin: Karlsruhe, Germany, International Union of Geodesy and Geophysics, 109 p.

Wahrhaftig, C., 1965, Stepped topography of the southern Sierra Nevada, California: Geological Society of America Bulletin, v. 76, p. 1165–1190, doi: 10.1130/0016-7606(1965)76[1165:STOTSS]2.0.CO;2.

Wakabayashi, J., and Sawyer, T.L., 2001, Steam incision, tectonics, uplift and evolution of the topography of the Sierra Nevada, California: The Journal of Geology, v. 109, p. 539–562, doi: 10.1086/321962.

Walcott, C.D., 1897, The post-Pleistocene elevation of the Inyo Range, and the lake beds of Waucobi Embayment, Inyo County, California: The Journal of Geology, v. 5, p. 340–348.

Walker, J.D., Coleman, D.S., and Dickinson, W.R., 1987, Correlation of Mio-Pliocene rocks of the northern Panamint Mountains and Darwin Plateau; implications for normal-fault development and the opening of Panamint Valley: Geological Society of America Abstracts with Programs, v. 19, p. 878.

Wallace, R.E., 1984, Patterns and timing of late Quaternary faulting in the Great Basin Province and relation to some regional tectonic features: Journal of Geophysical Research, v. 89, p. 5763–5769, doi: 10.1029/JB089iB07p05763.

Wernicke, B.P., Axen, G.J., and Snow, J.K., 1988a, Basin and Range extension at the latitude of Las Vegas, Nevada: Geological Society of America Bulletin, v. 100, p. 1738–1757, doi: 10.1130/0016-7606(1988)100<1738:BARETA>2.3.CO;2.

Wernicke, B.P., Snow, J.K., and Walker, D., 1988b, Correlation of early Mesozoic thrusts in the southern Great Basin and their possible indication of 250–300 km of Neogene crustal extension: Boulder, Colorado, Geological Society of America, Field Trip Guide Book, p. 255–268.

Wernicke, B.P., Clayton, R.W., Ducea, M.N., Jones, C.H., Park, S.K., Ruppert, S.D., Saleeby, J.B., Snow, J.K., Squires, L.J., Fliedner, M.M., Jiracek, G.R., Keller, G.R., Klemperer, S.L., Luetgert, J.H., Malin, P.E., Miller, K.C., Mooney, W.D., Oliver, H.W., and Phinney, R.A., 1996, Origin of high mountains in the continents; the southern Sierra Nevada: Science, v. 271, p. 190–193, doi: 10.1126/science.271.5246.190.

Wolfe, J.A., and Barghoorn, E.S., 1960, Generic changes in Tertiary floras in relation to age: American Journal of Science, v. 258A, p. 388–399.

Wright, L., 1976, Late Cenozoic fault patterns and stress fields in the Great Basin and westward displacement of the Sierra Nevada block: Geology, v. 4, p. 489–494, doi: 10.1130/0091-7613(1976)4<489:LCFPAS>2.0.CO;2.

Wright, L.A., Kramer, H., Thornton, C.P., and Troxel, B.W., 1984, Type sections of two newly named volcanic units of the central Death Valley volcanic field, eastern California: California Division of Mines and Geology Map Sheet 34, p. 21–24.

Zehfuss, P.H., Bierman, P.R., Gillespie, A.R., Burke, R.M., and Caffe, M.W., 2001, Slip rates on the Fish Springs fault, Owens Valley, California, deduced from cosmogenic ^{10}Be and ^{26}Al and soil development on fans: Geological Society of America Bulletin, v. 113, p. 241–255, doi: 10.1130/0016-7606(2001)113<0241:SROTFS>2.0.CO;2.

Zellmer, J.T., 1980, Recent deformation in the Saline Valley region, Inyo County, California [Ph.D. thesis]: Reno, University of Nevada, 224 p.

MANUSCRIPT ACCEPTED BY THE SOCIETY 21 JULY 2008

The Geological Society of America
Special Paper 447
2009

Seismotectonics of an evolving intracontinental plate boundary, southeastern California

Jeffrey Unruh*

William Lettis & Associates, Inc., Walnut Creek, California 94596, USA

Egill Hauksson

Seismology Laboratory, California Institute of Technology, Pasadena, California 91125, USA

ABSTRACT

An analysis of seismicity in the northern Mojave block, southern Walker Lane belt, and southern Sierra Nevada provides a detailed snapshot of the kinematics of active deformation within a young and probably evolving intracontinental plate boundary in eastern California. Earthquakes in this region were relocated using joint hypocentral inversion, double-difference, and waveform cross-correlation techniques. Groups of focal mechanisms were inverted for the components of a reduced deformation rate tensor, and the inversion results were synthesized in maps of the regional seismogenic deformation field. In general, seismogenic deformation in the Eastern California shear zone and Walker Lane belt is characterized by distributed simple shear and primarily accommodates northwest translation of the Sierra Nevada–Central Valley (Sierran) microplate relative to stable North America. Crustal thinning in the southern Walker Lane belt is subordinate to northwest-directed dextral shear and typically is associated with releasing geometries among dextral faults. There is no significant variation in strain geometry in the transition from the Eastern California shear zone to the Walker Lane belt across the Garlock fault; we interpret this to indicate that this structure is being sheared and rotated clockwise by distributed northwest dextral shear. In contrast, seismogenic deformation in the southeastern Sierra Nevada near latitude 36°N is characterized by horizontal elongation and locally by flattening, i.e., extension in two perpendicular horizontal directions. The transition from wrench tectonics in the southern Walker Lane to flattening in the High Sierra occurs across a 25- to 35-km-wide transtensional zone west of the Sierran range front. The elongation and flattening occur directly east of the "Isabella anomaly," a high-velocity anomaly in the upper mantle interpreted to be lower Sierran lithosphere that is foundering or convectively descending into the asthenosphere. The seismogenic thinning of the southern Sierra may be related to thickening above the descending lithosphere to the west and/or may be driven by tensile buoyancy forces. The seismotectonics of this region may provide insights into the process by which the Walker Lane belt has progressively expanded westward into the Sierra microplate during late Cenozoic time.

Keywords: neotectonics, seismicity, seismotectonics, lithospheric dynamics.

---

*Corresponding author e-mail: unruh@lettis.com.

Unruh, J., and Hauksson, E., 2009, Seismotectonics of an evolving intracontinental plate boundary, southeastern California, *in* Oldow, J.S., and Cashman, P.H., eds., Late Cenozoic Structure and Evolution of the Great Basin–Sierra Nevada Transition: Geological Society of America Special Paper 447, p. 351–372, doi: 10.1130/2009.2447(16). For permission to copy, contact editing@geosociety.org. ©2009 The Geological Society of America. All rights reserved.

INTRODUCTION

Geodetic measurements of the secular velocity field in southern California demonstrate that northwest-directed dextral shear along the eastern edge of the Pacific plate splits into two branches north of the Salton trough (Fig. 1). The main branch continues to the northwest as the San Andreas fault system, and ancillary structures in western California, and accommodates ~75% of the total motion between the Pacific plate and stable North America (Argus and Gordon, 1991, 2001; Dixon et al., 2000). An eastern branch of dextral shearing, called the Eastern California shear zone by Dokka and Travis (1990), trends north-northwest from the Salton trough and crosses the central and eastern Mojave block. Geodetic data indicate that ~25% of Pacific–North American plate motion is accommodated within the Eastern California shear zone (Savage et al., 1990; Sauber et al., 1994; Miller et al., 2001). Distributed dextral shear in the Eastern California shear zone continues northward in the Walker Lane belt, a 100-km-wide zone of active seismicity and late Cenozoic faulting east of the Sierra Nevada mountain range (McClusky et al., 2001; Bennett et al., 2003; Monastero et al., 2005). Active strike-slip and normal faults within the Eastern California shear zone and Walker Lane belt accommodate northwest translation of the western Mojave block and the Sierra Nevada–Central Valley block as discrete microplates relative to stable North America (Argus and Gordon, 1991, 2001; Dixon et al., 2000; Fig. 1).

Although geodetic data show that the Sierra Nevada–Central Valley block (hereafter called the Sierran microplate) effectively moves as a rigid block (Dixon et al., 2000), relatively high levels of background seismicity in the southeastern Sierra east of longitude 118.5°W indicate that some deformation is occurring within the interior of the microplate (Fig. 2). This part of the Sierra Nevada produced moderate to large earthquakes in 1868 and 1946 (Jones and Dollar, 1986; Bawden et al., 1999), and a swarm of small earthquakes in the early 1980s characterized by normal faulting (Fig. 3). The region of elevated seismicity in the southeastern Sierra is characterized by thinner crust (~35 km) and higher heat flow (32–57 mW/m²) than the Sierran foothills to the west, where crustal thickness is ~41 km and heat flow values range from 18 to 21 mW/m² (Ruppert et al., 1998; Saltus and Lachenbruch, 1991; Fig. 1).

The study area outlined in Figure 1 also is east of the "Isabella anomaly," a narrow, vertically elongated zone of high Pn velocities in the upper mantle beneath the western Sierran foothills (Benz and Zandt, 1993). The Isabella anomaly is interpreted to be lower lithosphere that has detached from the base of the crust and is descending into the asthenosphere (Saleeby et al., 2003; Zandt et al., 2004; Jones et al., 2004; Boyd et al., 2004). The region of elevated seismicity in the southern Sierra occurs directly east of the Isabella anomaly (Fig. 1). Zandt et al. (2004) interpreted a local depression in the Moho above the Isabella anomaly (documented by Ruppert et al., 1998; Fliedner et al., 2000) to be crustal thickening above the descending lithosphere. A late Cenozoic basin in the eastern Central Valley above the

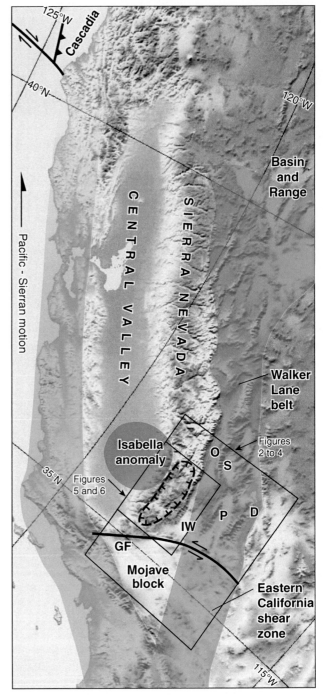

Figure 1. Oblique Mercator map of California using the Pacific–Sierra Nevada Euler pole (Argus and Gordon, 2001) as a basis for projection. The blue band illustrates distributed strike-slip deformation in western California that accommodates ~75% of the motion between the Pacific plate and stable North America. The green band encompasses the Eastern California shear zone and Walker Lane belt, which in southern California accommodate the remaining 25% of Pacific–North American motion. The detailed study area discussed in this paper is outlined by the black box. The elliptical area in the southern Sierra Nevada defined by black barbs adjacent to the "Isabella anomaly" is characterized by thin crust and high heat flow relative to the Sierran foothills to the west. O—Owens Valley; S—Saline Valley; P—Panamint Valley; IW—Indian Wells Valley; D—Death Valley; GF—Garlock fault.

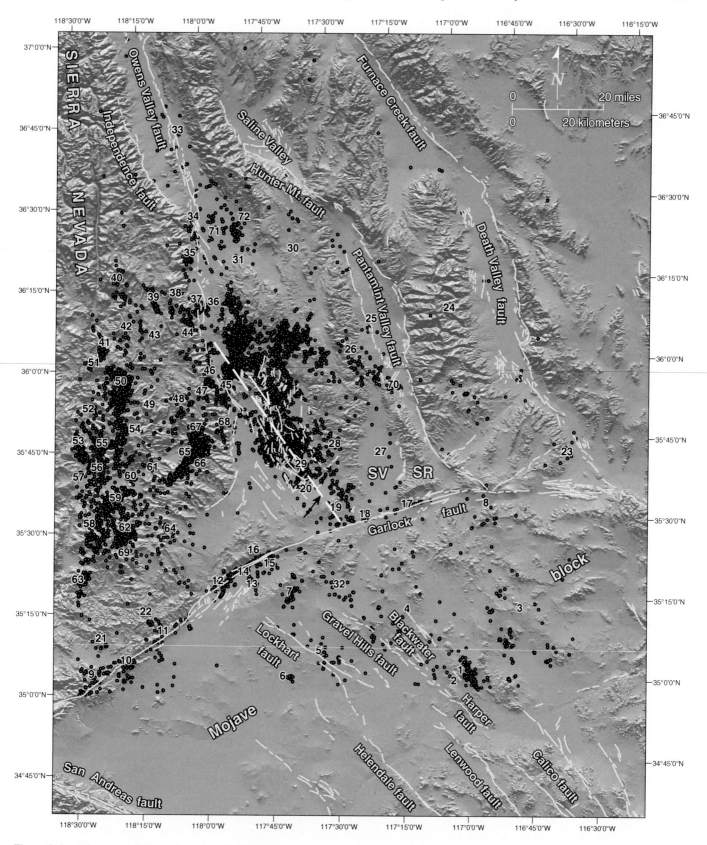

Figure 2. Location map of the study region with Quaternary faults (in yellow) from Jennings (1994) and epicenters of earthquakes located via joint hypocentral inversion. Numbers indicate individual groups of earthquakes selected for kinematic analysis. Dashed white line shows interpreted western margin of the Walker Lane belt. SV—Searles Valley; SR—Slate Range.

354

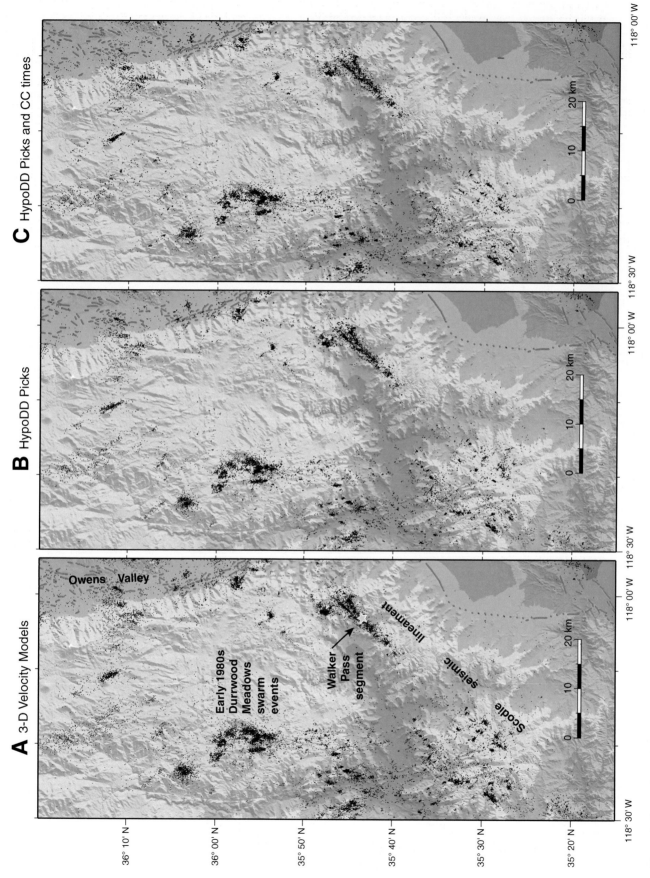

Figure 3. Comparison of epicentral locations in the southern Sierra Nevada derived from: (A) Three-dimensional (3-D) velocity models (yellow star—epicenter of 1946 Walker Pass earthquake); (B) double-difference (DD) analyses; and (C) waveform cross-correlation (CC) techniques.

Isabella anomaly may represent surface subsidence driven by the negative buoyancy of the descending lithosphere (Saleeby and Foster, 2004). The seismogenic normal faulting in the southern Sierra adjacent to thickened crust above downwelling lithosphere suggests a causative relationship between these processes.

This paper presents an analysis of seismicity in the region that encompasses the tectonic boundaries among the western Mojave block, Sierran microplate, Eastern California shear zone, and Walker Lane belt (Fig. 1). We assume that distributed seismicity recorded in this region during the last several decades provides a reasonably accurate snapshot of late Quaternary time-averaged deformation, and we use earthquake focal mechanisms to evaluate and map the regional seismogenic deformation field. We assess variations in deformational style across the boundary between the Walker Lane belt and Sierran microplate and adjacent to the Isabella anomaly. In particular, we focus on the kinematics and dynamics of active extension in the southern Sierra as the expression of deformation associated with foundering lithosphere adjacent to a young and probably evolving intracontinental plate boundary.

EARTHQUAKE DATA

The Southern California Seismic Network (SCSN), jointly operated by Caltech and the U.S. Geological Survey, recorded the waveforms of more than 100,000 earthquakes in the study region from 1981 to 2003, and it provided the P and S arrival time data used in our analysis.

We used the method of Thurber (1993) and detailed approach described in Hauksson (2000) to invert for the regional velocity structure. Using the 15 km grid mode from Hauksson (2000), we interpolated for a 10 km horizontal grid and determined three-dimensional (3-D) Vp and Vp/Vs models for the southern Sierra, Walker Lane, and northern Mojave region. The Vp and Vp/Vs models include arrival times from more than 11,500 earthquakes, and thus there are ample data to constrain the models. The final Vp model was used to relocate the seismicity in the area from 1981 through December 2002 (Fig. 2).

To further refine locations of hypocenters in the southern Sierra Nevada, we applied waveform cross-correlation to the seismograms for over 20,000 events located in that region between 1981 and 2002 (Hauksson and Shearer, 2005). Waveforms recorded by the SCSN were first extracted from the Southern California Earthquake Data Center (SCEDC) data center in 50 s windows that include both P and S waves. The traces then were resampled to a uniform 100 Hz sample rate and band-pass filtered to between 1 and 10 Hz. Next, we applied time-domain waveform cross-correlation for P and S waves between each event and 100 neighboring events that were identified from the catalog based on the 3-D velocity model of Hauksson (2000). We determined differential times from the peaks in the cross-correlation functions and used a spline interpolation method to achieve a nominal timing precision of 0.001 s. These differential times, together with existing P and S phase picks, were input to the double-difference

program of Waldhauser and Ellsworth (2000) to determine the hypocenters. The results of the combined cross-correlation and double-difference (CC/DD) methods are compared in Figure 3 to locations obtained from the 3-D model. In general, the CC/DD epicenters are more tightly clustered and linear alignments have sharper definition than those produced by the 3-D model (Fig. 3).

We used the method of Hardebeck and Shearer (2002) to determine 12,484 first-motion focal mechanisms for events with 12 or more reported first motions. For earthquakes located in the southern Sierra Nevada, we used the hypocenters determined with the double-difference method (Fig. 3) as starting locations. The Hardebeck and Shearer (2002) method takes into account possible errors in the earthquake location, velocity model, and polarity observations. The algorithm performs a grid search to find a set of all mechanisms for each event that are acceptable, using an expected polarity error rate and the range of allowed hypocenters and velocity models. Outliers are removed, and the remaining solutions are averaged to find the best focal mechanism.

KINEMATIC ANALYSIS OF SEISMOGENIC DEFORMATION

We inverted focal mechanisms from groups of earthquakes in the Sierra Nevada, southern Walker Lane belt, and northern Mojave block (Fig. 2) to evaluate the regional seismogenic deformation field. Our analytical approach is similar to that of previous studies (e.g., Unruh et al., 1996, 1997, 2002, 2003). Using focal mechanisms from earthquakes located with the 3-D velocity model (Fig. 3A), we first identified individual groups of earthquakes based on spatial clustering, discrete alignments of epicenters and association with active faults or other tectonic structures. We derived seismic P and T axes from the focal mechanisms of events in each group and inverted these data for the components of a reduced deformation rate tensor using a micropolar continuum model (Twiss et al., 1993; Twiss and Unruh, 1998). Inversions were performed using an automated grid search algorithm called FLTSLP_2K6 (see Appendix D in Guenther [2004] for user's manual). The best-fit micropolar model is parameterized by the orientations of the principal strain rates ($d_1 > d_2 > d_3$; lengthening positive); a scalar parameter (D), which is formed by a ratio of the differences in the principal strain rates, and which characterizes the shape of the strain rate ellipsoid; a scalar parameter (W), which characterizes the relative vorticity of rigid, fault-bounded blocks about an axis parallel to d_2; and a measure of the misfit between the best-fit model and the data (Twiss et al., 1993; Unruh et al., 1996). Because the period of time over which the earthquakes occurred is very small and essentially instantaneous relative to geologic time, during which significant deformation accumulates, the strain rates that form the basis of micropolar theory can be thought of as incremental strains for the purpose of interpretation.

To characterize the net vertical deformation accommodated by the earthquakes, we calculated the ratio V of the vertical deviatoric deformation rate to the maximum deformation rate (Unruh et al., 2002). Positive values of V indicate net crustal thickening, negative values indicate net crustal thinning, and a value of zero indicates horizontal plane strain. The range of values from $-1 \leq V \leq 1$ characterizes gradations from extensional, transtensional, plane strain, transpressional, to shortening deformations (Lewis et al., 2003). Values of $-1 > V > -2$ indicate a flattening deformation, i.e., vertical shortening accompanied by elongation in two perpendicular horizontal directions.

In analyzing the focal mechanisms, we used two different approaches to evaluate the quality of the best-fit model and homogeneity of the data. For the regional data set in Figure 2, which included over 70 individual seismicity domains, we performed multiple grid search inversions of each data set. Each grid search was begun from a randomly selected set of initial parameters, and this process was repeated 25 times to generate a total of 26 best-fit models per data set. Results of the random restart process for individual domains fell into one of three general categories:

(1) The majority of the best-fit models for a given domain are very similar, both in terms of the model parameters and misfit values—We interpret these results to indicate that the best-fit solution occupies a well-defined minimum in the solution space, and that there are no significant false minima. We infer that the given data set is relatively homogeneous.

(2) The best-fit solutions fall into two or more groups, each of which is characterized by a distinct, limited range of misfit values—We interpret these data sets to be heterogeneous, i.e., the focal mechanisms represent more than one strain geometry within the sampled crustal volume. In previous studies, we have found that strain heterogeneity may be expressed by multiple maxima of seismic P and T axes when plotted on lower-hemisphere Kamb contour plots (e.g., Figs. 8 and 9 in Unruh et al., 2002), and we used these criteria to assess heterogeneity for the present work.

(3) The best-fit solutions from the random restart analysis do not fall into one or more well-defined groups—In these cases, the solutions obtained from the random restart analysis do not define a clear minimum (or minima) in solution space. Most of these solutions come from data sets with small numbers of earthquakes that are not associated with a discrete fault and that are not organized in a discrete cluster or linear alignment.

In general, we accepted best-fit models in the first category as reasonable solutions. Data sets with multiple minima (the second category) were reexamined and subdivided if there were enough focal mechanisms to produce two or more data sets with a minimum of ~25–30 focal mechanisms each, which we have found from past experience to be the minimum data necessary to obtain a well-constrained solution. We evaluated the homogeneity of data from the subdivided domains through visual inspection of Kamb plots of seismic P and T axes and varied the horizontal and/or vertical boundaries of the domains as necessary to obtain data sets characterized by single, well-defined maxima of P and T axes. The best-fit models and their misfits obtained in this manner are provided in Table 1.

The third category is represented almost exclusively by data sets with fewer than 25 focal mechanisms, and/or domains with relatively few events distributed across large areas. We used

standard bootstrap methods to evaluate these data sets and find the best-fit model parameters and their 95% confidence intervals (Guenther, 2004). The results of the bootstrap analyses for selected domains are listed in Table 2. For comparison, the orientations of d_1 and d_3 from the best-fit bootstrap models are presented along with the mean d_1 and d_3 orientations for all models within the 95% confidence intervals. Similarly, the best-fit values of D and W are reported along with the mean values of D and W within their respective 95% confidence intervals. Inspection of Table 2 shows that the best-fit bootstrap model parameters generally are equivalent or very close to the mean values within the 95% confidence intervals.

Because the bootstrap inversions are time intensive, we limited their application to domains with small and possibly heterogeneous data sets when analyzing seismicity throughout the entire study region shown in Figures 1 and 2. As discussed later, this analysis reveals that seismogenic deformation in the southern Sierra Nevada is very different in character from that in the Walker Lane belt (Fig. 4). To characterize the seismotectonics of the southern Sierra in greater detail, we analyzed a refined subset of Sierran earthquakes located via double-difference and waveform cross-correlation techniques (Fig. 3C), and for which focal mechanisms could be determined using the Hardebeck and Shearer (2002) method. The earthquakes were sorted into spatially distinct groups (Fig. 5), and focal mechanisms for all groups were inverted using bootstrap methods to obtain the mean best-fit values for the model parameters and respective 95% confidence intervals. The southern Sierra inversion results are reported in Table 3 and are discussed in detail in a following section.

The inversion results for the regional data set (Fig. 2) and the refined Sierran data set (Fig. 4) are synthesized in maps of the seismogenic deformation field (Figs. 4, 5, and 6). The horizontal components of the deformation are depicted by trajectories drawn parallel to the trends of axes d_1 (maximum extension) and d_3 (maximum shortening). If one or both of d_1 and d_3 are moderately plunging, then they may not appear orthogonal in plan view. In cases where d_1 or d_3 is steeply plunging to subvertical, their respective trajectories are not plotted on the map. If one or both of the principal strain rate axes is plunging, then the deformation likely has a net vertical component. For example, if only d_1 axes are shown, then the deformation is characterized by horizontal extension parallel to the d_1 trajectories, and d_3 is steeply plunging to subvertical, indicating net crustal thinning. Similarly, if only d_3 axes are shown, then d_1 is steeply plunging to subvertical and the deformation locally is characterized by net vertical crustal thickening. For the regional overview in Figure 4, individual domains where the absolute value of V is greater than 0.5 (Tables 1 and 2) are delineated with blue and orange colors for net vertical thickening and thinning, respectively.

RESULTS

At a regional scale, the inversion results reveal a transition from dextral wrench faulting in the Walker Lane belt to

TABLE 1. MICROPOLAR INVERSION RESULTS, REGIONAL SEISMOTECTONICS

Domain number	Domain tag	d_1 (trend, plunge) (°)	d_3 (trend, plunge) (°)	D	W	V	Misfit	n	Domain description
1	NEMH	111, 0	201, 7	0.5	0.3	−0.1	7.2821	213	NW-trending alignment of epicenters in the Mud Hills
2	CMHC	125, 34	221, 8	0.6	−0.4	0.4	5.9065	33	Cluster of events in central Mud Hills
3	NEMO	300, −7	211, 8	0.6	0.1	0.1	9.3928	84	Broadly distributed seismicity in the NE Mojave block near Ft. Irwin
4	REVNCMO	274, 0	185, −1	0.5	0	−0.1	7.5706	57	Scattered events SW of Blackwater fault
5	LOCK	111, −7	202, −9	0.3	0.1	−0.2	6.3041	51	Distributed seismicity in the vicinity of the Lockhart fault
7	SERM	102, 1	194, 59	0.6	−0.1	−0.8	7.9627	126	Seismicity southeast of the Rand Mts.
9	GFET	106, −12	197, −3	0.5	0	0.1	7.5078	42	Garlock fault in the eastern Tehachapi Mts.
10	GFTP	75, 8	164, −13	0.5	−0.1	−0.1	8.9966	78	Garlock fault at Tehachapi Pass
11	GWFV	265, −8	172, −12	0.5	−0.1	−0.1	5.9134	42	Garlock fault north of western Fremont Valley
12	GEFV	286, 0	196, −2	0.6	0.4	0.1	8.1872	98	Garlock fault north of eastern Fremont Valley
13	NRMP	272, 22	247, −66	0.7	−0.3	−0.9	3.2446	22	Cluster of events north of Rand Mts. piedmont
14	KLCV	323, −12	328, 78	0.7	−0.4	−1.3	4.1989	36	Seismicity near Koehn Lake, in vicinity of Cantil Valley fault
16	GFSE	108, −4	18, 5	0.6	0.5	0.1	8.3295	82	Garlock fault south of the El Paso Mts.
18	SWSV	264, 15	219, −69	0.4	0.2	−0.7	3.8087	23	NNW-trending alignment of events north of Garlock fault, SW of Searles Valley
19	SPHL	95, 5	186, 9	0.6	0.4	0.1	8.329	74	Seismicity in the Spangler Hills
20	SECL	95, −16	6, 2	0.6	0.3	0.2	4.5741	41	Seismicity near Lone Butte, southeast of China Lake
22	SNSP	67, 8	158, 8	0.5	0.5	0	5.2863	69	Cluster of events near springs at southern end of Sierran block
23	SDVC	267, 26	18, 36	0.4	−0.3	−0.1	2.96	13	Scattered events south of the Confidence Hills and eastern Owlshead Mts.
24	REVPANA	282, −6	195, 27	0.5	0.3	−0.2	7.8432	34	Southern Panamint Range
25	REVCPAN	288, −8	193, −30	0.1	−0.5	−0.4	5.7832	17	Central Panamint Valley
26	REVSRTC	274, 4	11, 64	0.5	0.1	−0.8	4.8449	40	Eastern Slate Range
28	SARG	103, 1	193, −1	0.5	−0.2	0.1	8.168	172	Seismicity in the southern Argus Range
29	SRID	99, −13	8, −1	0.5	0.3	0.1	5.266	190	NE-trending alignment of epicenters south of Ridgecrest
32	ERED	105, −7	17, 10	0.5	−0.2	0	6.5592	24	Area E-SE of Red Mountain
35	SEAC	144, 13	232, −9	0.4	−0.2	−0.1	6.9741	72	E-W alignment of epicenters along Sierran escarpment near Ash Creek
37	SEDM	156, 14	3, 75	0.5	−0.3	−0.9	7.207	41	Seismicity along the Sierran escarpment near Deer Mountain
39	WMON	264, −12	311, 72	0.7	−0.4	−1.1	9.9961	106	NW alignment of epicenters at Monache Meadows
42	SBRC	301, −5	91, −84	0.6	−0.3	−1.1	4.175	26	Seismicity near Rattlesnake Creek, southern Sierra
43	SBSM	112, 0	24, 83	0.7	−0.4	−1.2	6.6877	29	Seismicity near Smith Mt., southern Sierra
45	SELL	110, 7	199, −6	0.4	0	−0.1	6.5016	65	Sierran escarpment west of Little Lake
46	SLLC	110, −2	14, −71	0.5	0.2	−0.9	8.3482	141	Sierran escarpment at Little Lake
47	SBBP	316, 20	13, −57	0.5	0.6	−0.6	3.6821	22	Seismicity near Big Pine Meadows, southern Sierra
48	SBLV	273, 31	252, −57	0.7	−0.3	−0.6	6.9308	75	Seismicity near Long Valley, southern Sierra
50	REVSBDW	296, 4	246, −83	0.7	−0.4	−1.3	5.7648	114	Durrwood Meadows area, locus of swarms in 1980s
51	SWDW	110, −19	91, 70	0.4	0.5	−0.7	2.064	11	Distributed seismicity west of Durrwood Meadows
52	SBBR	282, −4	94, −86	0.4	−0.2	−0.9	3.78	43	Seismicity near Baker Ridge, southern Sierra
53	SBBM	108, −4	180, 77	0.6	−0.6	−1	6.1311	188	Seismicity in the vicinity of Black Mt., southern Sierra
54	SBCM	87, 13	52, −74	0.5	0.1	−0.9	8.3302	309	Seismicity near Cannell Meadow, southern Sierra
55	REVSBCC	265, −14	260, 76	0.4	−0.1	−0.8	5.3223	67	Seismicity near Red Mt., southern Sierra
56	ISAB	71, −4	217, −86	0.5	0	−1	6.1323	100	Lake Isabella region
57	KNVL	261, −13	317, 68	0.5	−0.1	−0.8	5.0716	27	Seismicity near Kernville, southern Sierra
58	REVSBRM	66, 22	123, −53	0.5	0.2	−0.5	7.2528	97	Seismicity near Red Mt., southern Sierra
59	REVSBPM	109, 20	131, −69	0.5	0.2	−0.7	5.8878	64	Seismicity near Piute Mt., southern Sierra
60	KELS	95, 23	179, −13	0.6	0.1	0.2	7.4087	57	Seismicity in the vicinity of Kelso Creek, southern Sierra
61	SBOP	84, −4	88, 86	0.4	0	−0.9	4.3881	22	Seismicity near Onyx Peak, southern Sierra
62	REVSBWM	270, −20	169, −25	0.4	−0.6	−0.1	6.5268	64	Seismicity near Waterhole Mine, southern Sierra
63	SHAR	284, 9	13, −3	0.4	−0.1	−0.1	7.969	56	Seismicity near Harper Peak, southern Sierra
64	SBKV	108, 0	198, 0	0.5	0.4	0	5.0389	53	Seismicity near Kelso Valley, southern Sierra
65	REVNEKM	293, 2	87, 87	0.6	−0.2	−1.1	6.0522	59	NE-trending alignment of events in Kiavah Mountains, southern Sierra
66	SEMP	116, −10	210, −22	0.5	−0.1	−0.1	7.9367	76	Sierran escarpment near Morris Peak
67	REVSBOW	98, −16	112, 73	0.6	−0.3	−1	7.2523	128	Seismicity near Owens Peak, southern Sierra
68	SEIW	113, 13	207, 17	0.5	−0.6	0	4.7656	34	Sierran escarpment bordering Indian Wells Valley
69	SBGC	105, 9	195, −2	0.4	0	−0.1	4.5067	38	Seismicity near Gallup Camp, southern Sierra
70	MPFS	287, −5	198, 5	0.6	0.3	0.2	5.5286	43	Manly Pass fault system
71	WOLK	280, −10	187, −14	0.7	0	0.2	2.9473	19	Western Owens Lake basin

Note: See text for explanation of *D, W,* and *V.*

Unruh and Hauksson

TABLE 2. RESULTS OF BOOTSTRAP INVERSIONS FOR SELECTED DOMAINS

Domain number	Domain tag	Best-fit d_1 (trend, plunge) (°)	Mean d_1 (trend, plunge) (°)	Std. dev. (°)	Best-fit d_3 (trend, plunge) (°)	Mean d_3 (trend, plunge) (°)	Std. dev. (°)	Best-fit D	Mean D	95% C.L.	Best-fit W	Mean W	95% C.L.	V (best-fit model)	Misfit (best-fit model)	n	Domain description
6	BORO	143, 19	143, 19	15	279, 65	279, 64	14	0.4	0.4	0.3/−0.2	0.1	0.2	0.6/−0.5	−0.6	3.486	16	Distributed seismicity near Boron
8	GFNG	109, 15	108, 15	9	208, 28	208, 28	8	0.8	0.8	0.2/−0.4	−0.1	−0.1	0.6/−0.6	0.1	6.106	15	Garlock fault north of the Granite Mts.
15	GFEF	62, −2	64, −3	14	147, 65	148, 63	25	0.6	0.6	0.2/−0.3	0.3	0.4	0.6/−0.4	−0.9	7.768	32	Garlock fault northeast of Fremont Valley
17	GFSS	90, −25	96, −30	7	209, −34	202, −38	8	0.2	0.3	0.3/−0.2	0.2	0.4	0.5/−0.4	−0.2	4.445	18	Garlock fault south of the Slate Range
21	NETM	104, 6	105, 6	24	9, 39	10, 40	29	0.6	0.6	0.3/−0.4	0.4	0.4	0.6/−1.4	−0.4	12.183	27	Seismicity in the northeastern Tehachapi Mts.
27	SEAR	91, −7	89, −6	12	181, −1	179, −2	19	0.5	0.5	0.3/−0.4	−0.2	−0.2	0.6/−0.5	0	4.212	14	Distributed seismicity in Searles Valley
30	DARW	123, 67	124, 68	13	205, −3	202, −5	18	0.2	0.3	0.6/−0.2	0.1	0.2	0.4/−0.5	0.8	5.946	18	Seismicity in the Darwin Plateau area
31	NCRP	83, −65	89, −62	14	158, 7	178, 1	26	0.4	0.4	0.6/−0.4	0.4	0.4	0.5/−0.7	0.8	7.599	13	Sparse seismicity along the northern Coso Range piedmont
33	SOWN	93, −51	93, −56	17	99, 39	97, 34	17	0.5	0.5	0.3/−0.4	0.7	0.7	0.3/−0.7	0.2	11.366	23	Distributed seismicity in southern Owens Valley
34	SEBT	90, −10	95, −10	22	108, 79	118, 79	13	0.6	0.6	0.4/−0.5	0.1	0.2	0.7/−1.1	−0.9	8.772	19	Seismicity along the Sierran piedmont near Bartlett
36	NWHR	167, 7	165, 8	14	3, 83	355, 82	19	0.8	0.7	0.2/−0.5	−0.1	−0.1	0.6/−0.5	−1.4	8.918	30	Distributed seismicity northwest of Haiwee Reservoir
38	SBDM	217, −18	218, −19	10	283, 51	283, 51	7	0.5	0.5	0.1/−0.2	0.1	0.1	0.5/−0.4	−0.5	4.446	22	Seismicity near Deer Mt., southern Sierra
40	SBHP	93, −15	99, −13	15	45, 69	53, 71	7	0.6	0.6	0.3/−0.2	−0.2	−0.3	0.3/−0.4	−0.9	4.571	46	Seismicity near Hockett Peak, southern Sierra
41	SBLM	166, 10	163, 11	6	297, 75	299, 75	4	0.6	0.6	0.1/−0.1	0	0	0.4/−0.4	−1	6.248	132	Seismicity near Lookout Mt., southern Sierra
44	SERV	98, −16	99, −16	10	39, 60	40, 62	15	0.6	0.6	0.2/−0.2	−0.2	−0.2	0.6/−0.6	−0.8	9.0248	63	Sierran escarpment west of Rose Valley
49	SBSK	101, 3	104, −1	22	243, 86	148, 89	17	0.6	0.6	0.3/−0.4	0.3	0.3	0.5/−0.8	−1.1	7.493	24	Seismicity in the vicinity of the south fork of the Kern River
72	EOLK	40, −28	41, −27	6	337, 41	337, 41	4	0.9	0.9	0.1/−0.1	0.8	0.8	0.1/−0.2	−0.2	5.8296	18	Distributed seismicity, eastern Owens Lake basin

Note: C.L.—confidence limit. See text for explanation of *D*, *W*, and *V*.

Figure 4. Regional seismogenic deformation field in the northern Mojave block, southern Sierra Nevada, and southern Walker Lane belt derived from kinematic analyses of earthquakes (data in Tables 1 and 2). Inversion results from a previous seismotectonic analysis of the Indian Wells Valley and Coso Range (Unruh et al., 2002) are included in the map. Solid purple lines are trajectories drawn parallel to the trends of the maximum extensional strain rates (d_1), and dashed purple lines are drawn parallel to the trends of the maximum shortening strain rates (d_3). Colors indicate areas where the deformation includes significant vertical thinning (orange) or thickening (blue). Geographic features mentioned in text (north to south): OVF—Owens Valley fault; SV—Saline Valley; HMF—Hunter Mountain fault; DP—Darwin Plateau; CR—Coso Range; DV—Death Valley; PR—Panamint Range; PV—Panamint Valley; SR—Slate Range; DM—Durrwood Meadows; ALF—Airport Lake fault; AR—Argus Range; SV—Searles Valley; CH—Confidence Hills; IWV—Indian Wells Valley; RV—Rose Valley; RM—Rand Mountains.

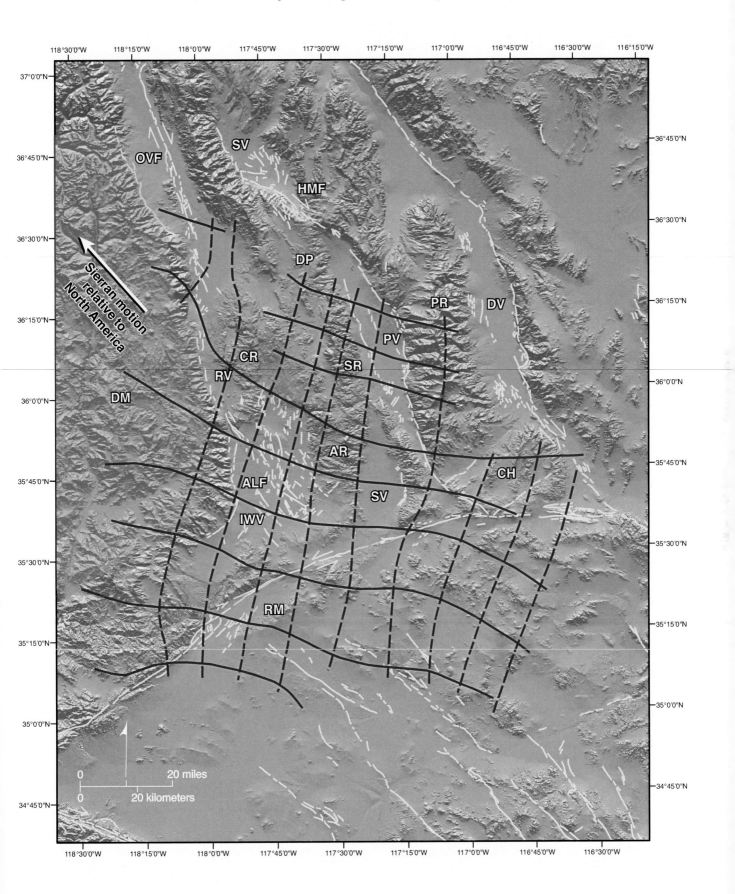

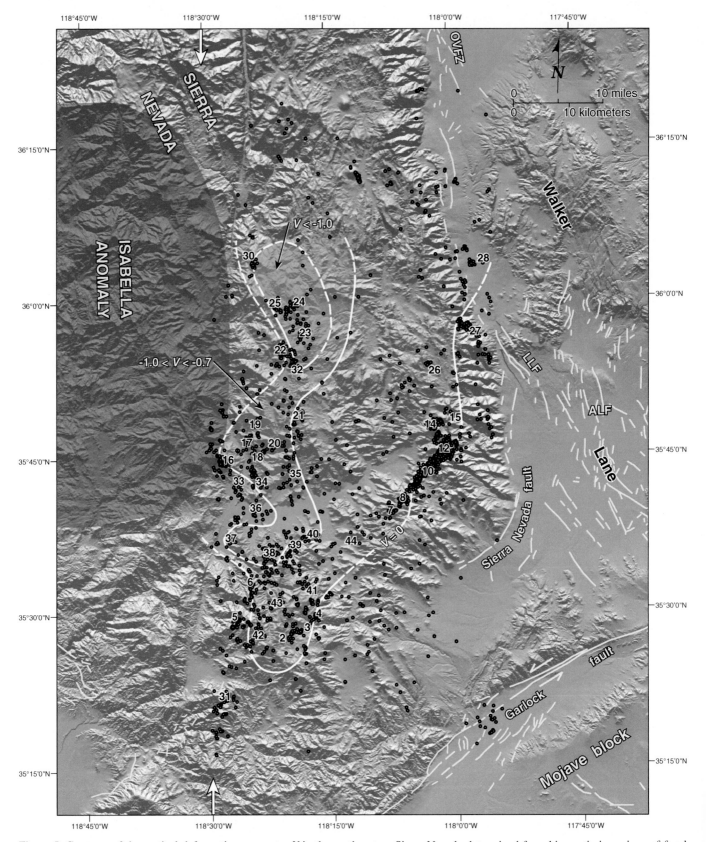

Figure 5. Contours of the vertical deformation parameter *V* in the southeastern Sierra Nevada determined from kinematic inversions of focal mechanisms (see Table 3 for tabulated inversion results). Numbers refer to individual domains selected for analysis. Blue-colored region represents the location (in plan) of the "Isabella anomaly." ALF—Airport Lake fault; LLF—Little Lake fault; OVFZ—Owens Valley fault zone.

TABLE 3. BOOTSTRAP INVERSIONS FOR DOMAINS IN THE SOUTHERN SIERRA NEVADA

Domain tag	Mean d_1 (trend, plunge) (°)	Std. dev. (°)	Mean d_3 (trend, plunge) (°)	Std. dev. (°)	Mean D	95% C.L.	Mean W	95% C.L.	V	95% C.L.	n
SS01	73, −4	12	162, 16	13	0.5	0.4/−0.2	0.5	0.3/−0.4	−0.1	0.7/−0.7	22
SS02	77, −17	4	55, 72	7	0.3	0.2/−0.1	0.4	0.3/−0.2	−0.6	0.3/−0.3	32
SS03	73, 1	6	161, −63	13	0.3	0.1/−0.1	0.5	0.3/−0.3	−0.6	0.4/−0.2	34
SS04	106, −12	6	197, −7	5	0.5	0.2/−0.2	−0.1	0.6/−0.5	0.1	0.3/−0.2	34
SSO5	62, −11	8	150, 11	16	0.5	0.2/−0.2	−0.1	0.5/−0.4	0	0.4/−0.9	63
SSO6	77, −14	9	134, 65	31	0.5	0.2/−0.2	0.7	0.3/−1.0	−0.6	0.9/−0.6	49
SSO7	99, −9	5	192, −15	9	0.4	0.3/−0.3	−0.1	0.3/−0.3	−0.1	0.4/−0.3	28
SS08	85, −8	4	47, 79	5	0.3	0.1/−0.1	0.7	0.2/−0.3	−0.8	0.2/−0.2	33
SS09	96, −6	4	158, 78	8	0.4	0.1/−0.1	0.1	0.3/−0.3	−0.8	0.2/−0.2	130
SS10	87, 3	7	179, 23	13	0.3	0.3/−0.2	−0.4	0.3/−0.4	−0.2	0.4/−0.4	70
SS11a	99, −1	8	189, −3	9	0.5	0.2/−0.2	−0.4	0.3/−0.4	0	0.2/−0.3	56
SS11b	80, −4	8	160, 66	13	0.4	0.2/−0.2	0.2	0.5/−0.5	−0.7	0.5/−0.3	66
SS12	84, −2	6	13, 84	20	0.4	0.1/−0.2	0.1	0.3/−0.6	−0.8	0.7/−0.2	193
SS13	98, 2	11	238, 88	9	0.5	0.2/−0.4	0.3	0.4/−0.7	−1	0.5/−0.3	47
SS14	88, 1	7	351, 81	26	0.4	0.2/−0.1	0.2	0.6/−0.5	−0.7	0.8/−0.3	242
SS15	86, −1	5	6, 82	10	0.4	0.1/−0.2	0.6	0.3/−0.6	−0.8	0.4/−0.2	28
SS16	82, −4	12	163, 69	13	0.4	0.2/−0.2	0.4	0.3/−0.4	−0.7	0.5/−0.4	73
SS17	82, −7	6	148, 73	46	0.4	0.2/−0.1	−0.2	0.7/−0.5	−0.6	0.7/−0.6	37
SS18	86, −9	4	65, 81	3	0.5	0.1/−0.2	0.8	0.1/−0.2	−0.9	0.3/−0.2	90
SS19	266, 17	6	133, 66	6	0.4	0.1/−0.1	−0.8	0.3/−0.2	−0.7	0.4/−02	27
SS20	73, −2	7	161, 38	6	0.4	0.1/−0.1	0.3	0.6/−0.6	−0.4	0.2/−0.2	67
SS21	101, 1	3	191, 14	4	0.7	0.2/−0.3	−0.6	0.3/−0.3	0.2	0.3/−0.7	61
SS22	276, 30	4	110, 60	3	0.4	0.2/−0.2	−0.3	0.3/−0.3	−0.4	0.2/−0.3	48
SS23	222, −23	4	353, −57	2	1	0/−0	0.7	0.1/−0.1	−1.1	0.3/−0.3	46
SS24	60, 1	5	346, −85	2	0.7	0.1/−0.1	0	0.2/−0.3	−1.4	0.2/−0.2	67
SS25	111, −9	9	64, 78	5	0.6	0.2/−0.1	−0.6	0.3/−0.2	−1	0.1/−0.2	34
SS26	103, 37	3	18, −6	3	0.6	0.1/−0.1	0.9	0.1/−0.1	−0.5	0.1/−0.2	22
SS27	113, 2	5	203, −4	18	0.6	0.1/−0.2	−0.6	0.4/−0.3	0.2	0.3/−0.9	106
SS28	119, −2	6	208, 43	19	0.2	0.2/−0.2	0.1	0.4/−0.6	−0.5	0.3/−0.2	14
SS29	72, −3	6	161, 35	32	0.4	0.4/−0.3	0	0.6/−0.4	−0.4	0.8/−0.8	46
SS30	314, −1	5	173, −88	2	0.7	0.2/−0.1	0.7	0.2/−0.3	−1.3	0.2/−0.3	21
SS31	88, −16	14	173, 19	19	0.5	0.2/−0.2	0.1	0.6/−0.8	−0.2	0.8/−0.6	27
SS32	26, 15	12	37, −75	6	0.7	0.2/−0.2	−0.5	0.3/−0.2	−1.1	0.4/−0.3	30
SS33	81, −21	4	113, 66	5	0.4	0.1/−0.1	0.5	0.2/−0.2	−0.6	0.5/−0.3	37
SS34	86, −15	5	91, 75	5	0.4	0.2/−0.1	0.6	0.3/−0.5	−0.7	0.3/−0.3	34
SS35	80, 76	5	230, 83	12	0.3	0.2/−0.2	−0.4	0.4/−0.2	−0.7	0.2/−0.2	35
SS36	74, −15	8	152, 38	16	0.3	0.3/−0.1	0.2	0.5/−0.6	−0.3	0.6/−0.5	45
SS37	62, −8	12	119, 75	19	0.5	0.2/−0.2	0.3	0.6/−0.8	−0.9	0.9/−0.4	39
SS38	80, −13	4	72, 77	7	0.4	0.1/−0.1	0.2	0.3/−0.3	−0.8	0.3/−0.2	101
SS39	90, −1	5	181, −37	29	0.3	0.2/−0.2	0	0.4/−0.6	−0.4	0.4/−0.5	73
SS40	83, −10	5	144, 70	7	0.4	0.1/−0.1	0.3	0.4/−0.4	−0.7	0.3/−0.2	55
SS41	96, −10	6	151, 73	16	0.4	0.2/−0.2	0.2	0.4/−0.6	−0.8	0.7/−0.3	122
SS42	79, −18	8	197, −56	11	0.3	0.2/−0.1	−0.3	0.7/−0.5	−0.5	0.5/−0.4	62
SS43	76, −21	4	27, 60	13	0.3	0.1/−0.1	0.5	0.2/−0.3	−0.5	0.3/−0.3	37
SS44	92, −5	6	180, 21	13	0.4	0.2/−0.1	0.1	0.6/−0.6	−0.2	0.2/−0.7	37
SS45	124, −10	14	119, 80	5	0.8	0.1/−0.2	0.4	0.3/−0.5	−1.3	0.4/−0.2	20

Note: C.L.—confidence limit. See text for explanation of D, W, and V.

transtension and local crustal flattening in the southern Sierra Nevada (Fig. 4; Table 1). At the latitude of Indian Wells Valley, the transition occurs west of the Sierran range front. At the latitude of the Coso Range north of Indian Wells Valley, negative "transtensional" values of the vertical deformation parameter V extend east of the Sierran range-front fault into Rose Valley. In general, there is a gradient rather than a sharp boundary between dextral wrenching in the southern Walker Lane belt and extensional deformation in the interior of the Sierra Nevada, characterized by increasingly negative values of V from east to west across a distance of several tens of kilometers (Fig. 5). In contrast to the east-to-west gradation from simple shear to transtension to elongation expressed in the seismogenic deformation, geodetic data published by Monastero et al. (2005) document a well-defined western limit of active northwest dextral shear in Indian Wells Valley and Rose Valley (dashed white line, Fig. 2). Global Positioning System (GPS) observation stations west of this boundary do not move significantly with respect to the interior of the southern Sierra, and thus we interpret it to be the effective eastern margin of the Sierran microplate at this latitude. The seismogenic deformation in the southern Sierra is occurring at a much lower rate than shearing in the Walker Lane belt. In the following sections, we describe the seismotectonics of the Walker Lane and southern Sierra in detail.

Seismogenic Deformation East of the Southern Sierra Nevada

Seismogenic deformation in the Eastern California shear zone and southern Walker Lane belt primarily is characterized by dextral wrench tectonics (i.e., d_1 and d_3 subhorizontal, V close to or equal to zero; Fig. 4). This includes areas with fault-bounded extensional basins like Panamint Valley (domain 25) and Searles

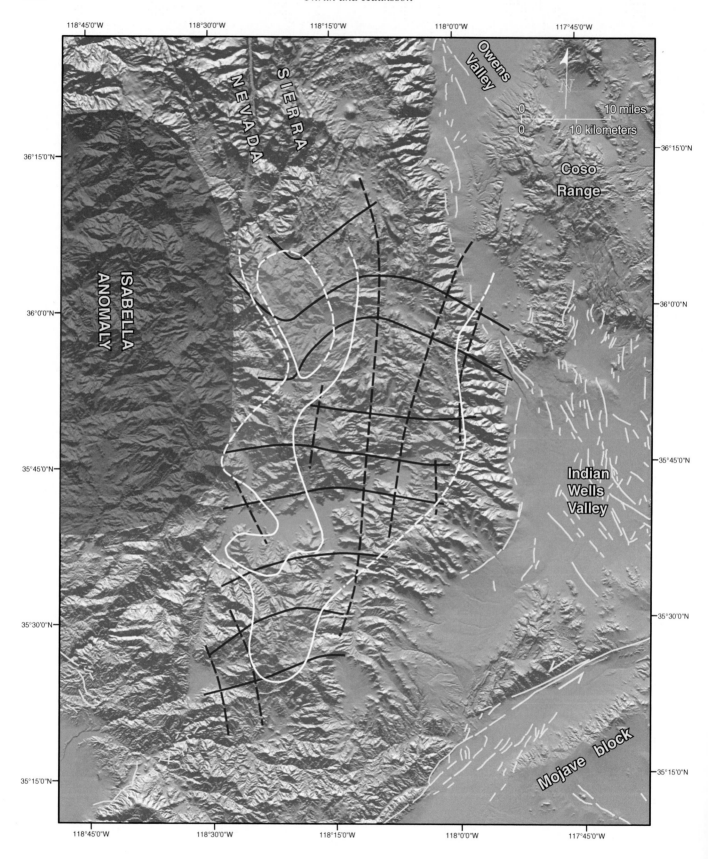

Figure 6. Trajectories of maximum extensional strain rate (d_1) and maximum shortening strain rate (d_3) in the southeastern Sierra Nevada derived from inversion of focal mechanisms (data in Table 3).

Valley (domain 27). Horizontal elongation with significant vertical thinning ($V \leq -0.5$) occurs only locally east of the Sierra Nevada, e.g., domains along the Garlock fault (domains 14, 15, 18), in the Rand Mountains (domain 13), the Slate Range (domain 26), and in the Argus Range (domain 7) (Tables 1 and 2). These local areas of extension are isolated within much larger areas of wrench deformation, and we attribute them to second-order kinematic complexities in the regional distribution of northwest-directed dextral shear.

For example, extension in the Coso Range is driven by a releasing right step between the dextral Airport Lake and Owens Valley faults (Unruh et al., 2002; Monastero et al., 2005; Hauksson and Unruh, 2007), which are the major structures associated with the geodetic boundary between the Sierran microplate and Walker Lane belt at this latitude. Quaternary normal faults in the Coso Range generally strike normal to local d_1 trajectories, indicating that they are optimally oriented to accommodate elongation in the regional dextral shear regime. Values of the parameter V are moderately negative (i.e., transtensional) within the releasing stepover area (Hauksson and Unruh, 2007). Similarly, normal faults that bound eastern Searles Valley and east-central Panamint Valley are at high angles to the local d_1 trajectories (Fig. 4) and are optimally oriented to accommodate normal slip in the regional plane strain regime. Although seismicity data are too sparse to evaluate the deformation geometry east of the Panamint Range with confidence, the normal fault bounding east-central Death Valley appears to be at a high angle to the d_1 trajectories in the Panamint Mountains to the west, suggesting that active extension here also is a component of regional northwest dextral shear in the Walker Lane belt.

In addition to localized extension, several small domains in the northern part of the study region are characterized by transpression and net crustal thickening. The northern Coso Range piedmont (domain 31) and Darwin Plateau (domain 30) are transpressional regions ($V > 0$) surrounded by transtension and dextral shearing (Fig. 2; Table 1). Some of the net shortening in the Darwin Plateau area may be related to transfer of dextral slip from Panamint Valley to Saline Valley along the dextral Hunter Mountain fault, which strikes more toward the west than other dextral faults in the southern Walker Lane belt and thus has a restraining geometry relative to the direction of regional macroscopic northwest dextral shear (Fig. 4).

Strain geometries along the Garlock fault are consistent with observed left-lateral slip on this structure. The orientations of the principal strains systematically rotate counterclockwise from east to west along strike so that the fault is everywhere subparallel to the direction of maximum resolved sinistral shear (Fig. 4). Although the Garlock fault is the principal tectonic boundary between the Eastern California shear zone and Walker Lane belt, there is no obvious or abrupt change in strain geometry from southeast to northwest across the structure. The orientations of the principal strains near the Blackwater fault in the northern Eastern California shear zone are similar those in Indian Wells Valley, southern Walker Lane belt. These relations indicate that the Garlock fault is not a discontinuous strain boundary and that dextral shear is approximately uniform across the Eastern California shear zone–Walker Lane belt transition. This in turn implies that the Garlock fault is being sheared and rotated clockwise by distributed northwest dextral shear in eastern California.

Previous workers have interpreted the strike of the Garlock fault east of the Sierra Nevada to have rotated clockwise in late Cenozoic time based on palinspastic reconstruction of the Eastern California shear zone (Dokka and Travis, 1990) and models for progressive westward expansion of the Walker Lane belt (Gan et al., 2003). Although a detailed kinematic analysis of the Garlock fault is beyond the scope of this paper, the inversion results prompt the following observations:

(1) The Garlock fault exhibits a uniform N54°E strike along the southern margin of the Sierra Nevada. The strike of the fault abruptly rotates clockwise by ~4° northwest of the dextral Lockhart fault, which is the westernmost Quaternary active fault of the Eastern California shear zone at this latitude (Fig. 2). The change in strike is marked by a complex pattern of splays, en echelon steps, and a local left stepover (domains 12–16; Fig. 2).

(2) South of Indian Wells Valley, Searles Valley, and the Slate Range (between domains 15 and 17 on Fig. 2), the strike of the Garlock fault is a relatively uniform N72°E. The strike of the Garlock fault abruptly rotates ~18° clockwise to an east-west orientation near its intersection with the southern end of the Panamint Valley fault zone, and it is uniform along the southern margin of the Panamint Range between the Panamint Valley and Death Valley fault zones.

Following Gan et al. (2003), we suggest that the discrete changes in strike represent division of the Garlock fault into a series of structural segments to accommodate localization of regional simple shear on the major strike-slip faults in the Walker Lane belt to the north. The Garlock fault is the southern structural boundary of the blocks bounded by the faults (i.e., the Panamint Range and Slate Range). Differential northwest motion of the blocks has progressively deformed and rotated the trace of the fault. The individual structural segments that comprise the Garlock fault may behave as independent rupture segments during earthquakes. Compatibility problems associated with abrupt changes in strike may be resolved by distributed deformation and aftershock activity at the segment boundaries.

Seismotectonics of the Southern Sierra Nevada

In contrast to the Eastern California shear zone and Walker Lane belt, seismogenic deformation in the interior of the southern Sierra Nevada regionally includes components of horizontal extension and vertical crustal thinning (Figs. 4, 5, and 6; Tables 1, 2, and 3). In some areas, vertical thinning is the dominant style of deformation. For example, the value of V in the Durrwood Meadows region (domain 50, Fig. 4; Table 1) is -1.3, which indicates that d_1 and d_2 both are subhorizontal and positive; thus, they accommodate extension in two orthogonal horizontal directions (i.e., "flattening"). For a constant-volume deformation, the

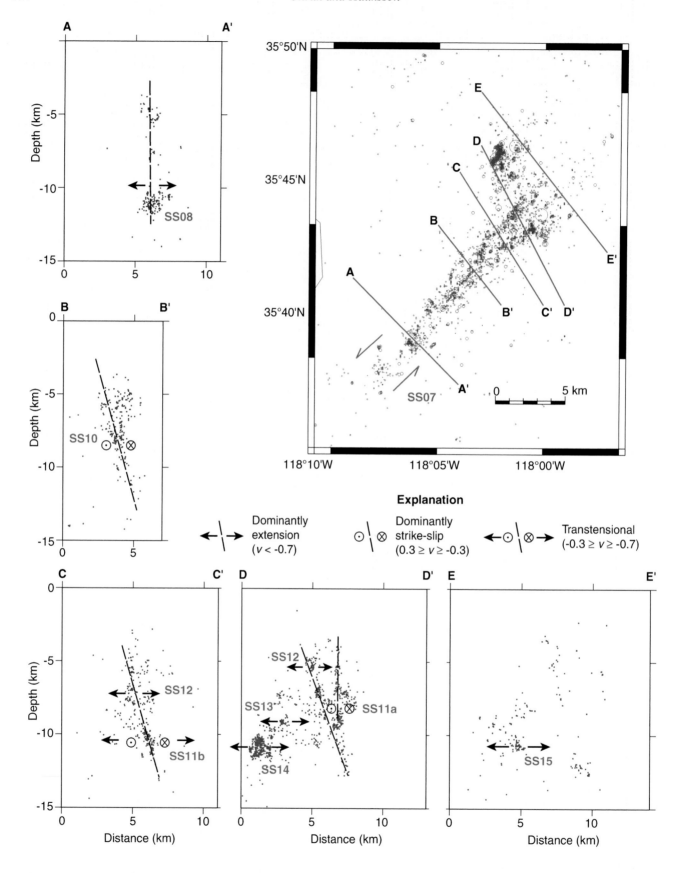

Figure 7. Kinematics of seismogenic deformation in the Walker Pass segment of the Scodie seismic lineament. See text for discussion.

horizontal flattening must be balanced by vertical crustal thinning (i.e., d_3 vertical). This deformation is essentially an asymmetric "pancaking" of the brittle crust in the horizontal plane.

Transtension in the Southeastern Sierran Microplate

The transition from northwest dextral shear in the Walker Lane belt to crustal thinning in the interior of the southern High Sierra occurs across a 25- to 35-km-wide zone that borders the eastern Sierran escarpment, and that also encompasses the southern Sierra north of the Garlock fault (Fig. 5). Values of the parameter V in this region range from zero to –0.7, indicating that the seismogenic deformation accommodates components of both horizontal shearing and vertical thinning.

A seismogenic feature in the transtensional zone is the Scodie seismic lineament, a northeast-trending band of epicenters in the southern Sierra that extends for ~60 km from the eastern end of the White Wolf fault to the Scodie Mountains area of the southeastern Sierra Nevada (Bawden et al., 1999) (Fig. 3). The most seismically active part of the Scodie lineament in the present study area is the 20-km-long, northeast-trending alignment of earthquakes in the epicentral region of the 1946 Walker Pass earthquake. For convenience, we refer to this as the Walker Pass segment of the Scodie lineament (Fig. 3A). Bawden et al. (1999) studied the Walker Pass segment in detail and concluded that the seismicity was associated with development of a young, anastomosing left-lateral strike-slip fault that lacks both surface expression and integrated structure.

Partitioning of normal, dextral, and oblique slip in the Walker Pass segment is illustrated in Figure 7 by a map and series of cross sections. Deformation accommodated by an isolated group of earthquakes at the southwest end of the Walker Pass segment (domain SS07; Fig. 7) is characterized by plane strain, consistent with sinistral shear on a northeast-striking high-angle fault. Moving northeast along the segment, cross-section A–A′ shows that the next discrete group of epicenters (SS08) accommodates a dominantly extensional deformation ($V = –0.8$; Table 3). Section B–B′, near the center of the Walker Pass segment, shows a steeply southeast-dipping zone of hypocenters (SS10). Inversions of focal mechanisms from these events suggest that this part of the segment primarily accommodates strike-slip faulting with minor extension ($V = –0.2$; Table 3).

As previously recognized by Bawden et al. (1999), the northeast part of the Walker Pass segment is characterized by multiple shallow branches that splay upward from a single fault at depth (sections C–C′ and D–D′, Fig. 7). Section C–C′ is near the branch point, and it shows two vertically segregated clusters of earthquakes. Based on the kinematic inversions, the upper cluster (SS12) primarily accommodates vertical thinning ($V = –0.8$), and the lower cluster (SS11b) has a slightly more transtensional solution ($V = –0.7$; Table 3). Section D–D′ is near the north end of the Walker Pass segment, and it shows at least one and possibly two shallow antithetic faults branching upward from a steeply southeast-dipping fault. The upper part of the main southeast-dipping fault between 3 and 6 km is the northern part of domain SS12, which accommodates vertical

thinning ($V = –0.8$; Table 3). The southwest-dipping antithetic splay (SS11a) accommodates sinistral shearing ($V = 0$; Table 3). Section D–D′ also reveals two spatially segregated clusters of earthquakes in the 7–11 km depth range, northwest of the main Walker Pass segment fault. The shallower cluster (SS13) accommodates extension ($V = –1.0$), and the deeper cluster (SS14) includes a component of shearing ($V = –0.7$). The small cluster adjacent to cross section E–E′ at the northernmost end of the Walker Pass segment (SS15) primarily accommodates horizontal elongation ($V = –0.8$; see Fig. 7; Table 3).

Our kinematic interpretation of the Walker Pass segment is illustrated in a simplified plan view diagram that shows the local orientation of d_1 for individual domains (Fig. 8). The main seismogenic fault at the southwest end of the segment is a vertical structure in the 3–13 km depth range. We interpret a small releasing left step in this fault northeast of domain SS07 to account for the negative value of V in domain SS08. At about domain SS10, the vertical fault merges with or rotates into a fault that dips steeply southeast, and this fault accommodates sinistral-normal

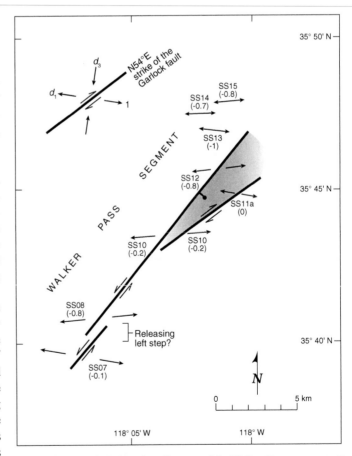

Figure 8. Simplified plan-view diagram of the Walker Pass segment of the Scodie seismic lineament with kinematic interpretation (see also Figs. 3 and 7). Outward-pointing arrows labeled with domain numbers indicate local d_1 trends. Numbers in parentheses are values of the vertical deformation parameter V (see Table 3 for full inversion results).

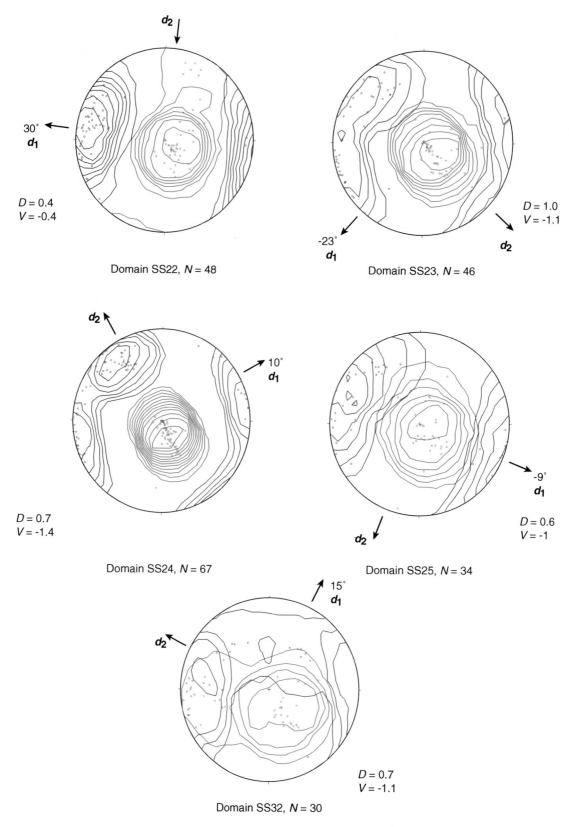

Figure 9. Kamb contour plots of seismic P and T axes from focal mechanisms in domains in the southern Sierra that accommodate flattening of the crust (see Fig. 5 for locations). Orientation of the maximum extensional principal strain rate (d_1) from kinematic inversions is plotted for each domain. The approximate orthogonal orientation of the intermediate principal strain rate (d_2) also is shown.

oblique shear. The southeast-dipping fault has a more northerly strike than the vertical fault, which has a releasing geometry relative to the N54°E strike of the Garlock fault, which we assume is parallel to the direction of macroscopic sinistral shear. A vertical sinistral fault in the upper 9–10 km of the hanging wall of the southeast-dipping fault (domain SS11a) creates a local strike-slip duplex. Inversions of focal mechanisms from isolated clusters of seismicity north of the northeast end of the Walker Pass segment (domains SS13, 14, and 15) indicate horizontal east-west elongation, but the fault geometry is not apparent from the hypocenters. Values of V decrease northward toward the interior of the Sierran block (Fig. 8), consistent with regional trends in Figure 5.

Crustal Thinning in the Southern High Sierra

Two areas dominated by vertical thinning (here defined as $V < -0.7$) west and north of the transtensional transition zone lie along a roughly north-trending belt directly east of 118.5°W longitude (Fig. 5). The best-fit micropolar models for several domains in the Durrwood Meadows area at the northern end of the extensional belt are characterized by flattening in the horizontal plane (i.e., $V < -1.0$). Bootstrap statistics indicate that the flattening deformation in two of the domains (SS24 and SS30; Fig. 5) can be distinguished from a biaxial strain accommodating horizontal elongation and vertical thinning (i.e., $V = -1.0$) at the 95% confidence interval (Table 3). The flattening deformation is associated with a pronounced rotation of d_1 to the northeast-southwest, from its regional northwest-southeast orientation in the transtensional domain and Walker Lane belt to the east (Fig. 6).

In their analysis of the 1983–1984 Durrwood Meadows swarm, Jones and Dollar (1986) reported that seismic T axes fall into two distinct groups: (1) one with T axes oriented WSW-ENE, consistent with normal slip on NNW-striking faults; and (2) one with T axes oriented NNW-SSE, consistent with normal slip on northeast-southwest–striking faults. These clustering trends are well expressed in Kamb plots of T axes from domains SS23 and SS24 (Fig. 9). The northwest-southeast clusters of T axes in these two domains probably account for the northwest-southeast orientation of d_1 in the inversion solutions. Extension in the northwest-southeast direction, expressed by concentrations of T axes in the northwest-southeast quadrants, is reflected in the inversion solutions by values of $D > 0.5$ (Table 3), which imply that the intermediate principal strain d_2 is positive and extensional. Earthquakes from adjacent domains SS25 and SS32 do not exhibit well-defined multiple clusters of T axes along these trends but rather an azimuthal variation from WSW to northwest (Fig. 9). This dispersion of T axes is evidence that there are components of horizontal elongation resolved in two orthogonal directions.

Domain SS22 exhibits a relatively well-defined transtensional solution ($V = -0.4$), with d_1 parallel to regional d_1 trends in the Walker Lane belt to the east. The intermediate principal strain rate (d_2) in domain SS22 is a horizontal shortening parallel to the regional trend of d_3 (maximum shortening) in the Walker Lane belt. This strain geometry suggests that northwest-directed simple shear extends into the region of flattening, and that shearing

is interacting with an independent tectonic process in the southern High Sierra to produce the observed deformation. Given the pronounced local rotation of d_1 to a northeast-southwest orientation, the flattening could be a combination of northwest-directed dextral shear (d_1 horizontal and oriented northwest-southeast; d_3 horizontal and oriented northeast-southwest), and an independent northeast-southwest–directed elongation (d_1 horizontal and oriented northeast-southwest; d_3 vertical), where the absolute value of d_1 associated with the local northeast-southwest elongation is greater than the value of northeast-southwest–oriented d_3 associated with shearing in the Walker Lane belt (Fig. 10).

KINEMATICS AND DYNAMIC PROCESSES

Seismogenic deformation in the southern Walker Lane belt is dominated by horizontal plane strain and northwest-directed dextral shear, consistent with the presence of large strike-slip structures in this region like the Airport Lake, Owens Valley, Panamint, and Death Valley–Furnace Creek fault zones (Fig. 2). These structures primarily accommodate northwest translation of

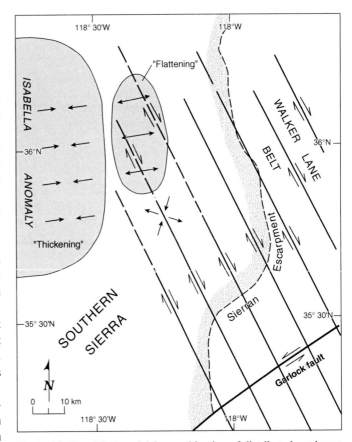

Figure 10. Simplified model for combination of distributed northwest dextral shear in the southern Sierra and horizontal northeast-southwest stretching adjacent to the Isabella anomaly to produce the unusual flattening deformation observed in the Durrwood Meadows area (i.e., $V < -1.0$; Fig. 5).

the Sierran microplate relative to stable North America (Unruh et al., 2003) and comprise an intracontinental transform boundary. Magnetotelluric soundings reveal that major strike-slip faults in the southern Walker Lane belt are associated with steeply dipping conductive features that extend to the Moho (Park and Wernicke, 2003). From these data, we interpret major strike-slip faults in the Walker Lane belt as shear zones that penetrate the full thickness of the crust (Fig. 11).

Based on the orientations and relative magnitudes of the principal strain rates in the southeastern Sierra Nevada, we interpret that northwest-directed dextral shear extends a minimum of ~30 km into the eastern Sierran microplate. The orientations of d_1 and d_3 west of the Sierran escarpment are similar to those in the Walker Lane belt, indicating that the direction of macroscopic dextral shear is uniform across the Sierran–Walker Lane tectonic boundary. We interpret that the shearing component of deformation in the transtensional domain is driven by shear stresses associated with strike-slip deformation along the eastern margin of

the Sierran microplate. Small earthquakes in the Sierra Nevada transtensional domain may be releasing some of the regional elastic strain accumulating on major strike-slip faults bordering the eastern Sierran microplate, such as the Airport Lake and Owens Valley faults.

The transition from wrench-dominated deformation to transtension occurs along or west of the geodetically defined eastern boundary of the Sierran microplate at this latitude. North of the present study area, Oldow (2003) observed a similar east-to-west kinematic transition within the Walker Lane belt, well to the east of the Sierran microplate. It is possible that the distribution of wrenching and transtension adjacent to the Sierran microplate is more heterogeneous than previously observed at regional scales. Alternatively, the transition from wrenching to transtension documented herein may be directly related to the tectonics of the southern Sierra, and thus may be unique to the present study area.

Horizontal flattening in the southern High Sierra is distinct kinematically from wrench faulting in the Walker Lane belt

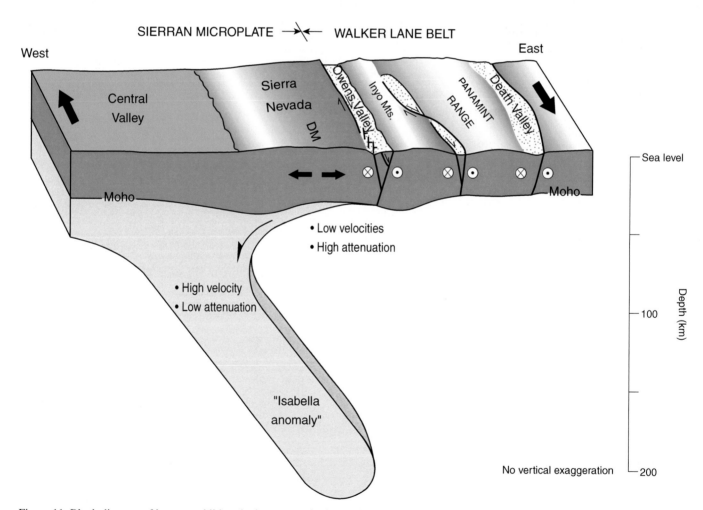

Figure 11. Block diagram of interpreted lithospheric structure in the southeastern California study area. Relief and structure on the base of the crust are from Fliedner et al. (2002). Depiction of the Isabella anomaly is modified from Boyd et al. (2004). Location and geometry of strike-slip shear zones are from data and interpretations in Park and Wernicke (2003). DM—Durrwood Meadows.

and from transtension in the transitional zone. We suggest two hypotheses to account for the flattening deformation:

Hypothesis 1: Flow Associated with Down-Welling Lithosphere

Seismogenic extension in the southern High Sierra may represent thinning of the upper crust above lithosphere that is flowing toward the Isabella anomaly. Fleitout and Froixdevaux (1982) discussed hypothetical patterns of lithospheric flow in the presence of buoyancy forces as a function of lateral variations in viscosity (Fig. 12A). Buoyancy forces, which are proportional to the vertical moment of density in the lithosphere, could be very large and negative in the southwestern Sierran foothills, given the depth to which the Isabella anomaly has been imaged (~200 km; Benz and Zandt, 1993; Boyd et al., 2004). A negative buoyancy force ($-\gamma$; Fig. 12) would produce horizontal compressive stresses in the lithosphere that act to drive vertical thickening, consistent with observations of a thickened crustal welt and depressed Moho above the Isabella anomaly (Fliedner et al., 2000; Zandt et al., 2004). Fleitout and Froixdevaux (1982) envisioned a simple model (Fig. 12A) where a lithospheric "drip" is surrounded by an annulus of lower viscosity, resulting in radial stretching around the thickening drip. In the case of the Isabella anomaly, however, the thinner lithosphere (Fliedner et al., 2000) and higher heat flow (Saltus and Lachenbruch, 1991) in the southeastern Sierra relative to the Sierran foothills and Central Valley to the west may concentrate deformation on the eastern margin of the Isabella anomaly (Fig. 12B), producing asymmetric flow and localized stretching on the eastern margin of the drip rather than the radial deformation envisioned by Fleitout and Froixdevaux (1982). Numerical experiments by Houseman et al. (2000) and Le Pourheit et al. (2006) suggest that deformation adjacent to downwelling lower lithosphere may be much more complex than the simple models in Figure 12.

Hypothesis 2: Tensile Buoyancy Forces

Alternatively, tensile buoyancy forces associated with the topographically high and thin Sierran lithosphere east of the Isabella anomaly may drive transtension and crustal thinning within the southern Sierran microplate. Buoyant uplift of the southern Sierra Nevada lithosphere following removal of its ecologitic root in the late Neogene predictably would increase the total gravitational potential energy, which in turn would be balanced by an increase in horizontal tensile buoyancy forces. Jones et al. (2004) estimated that the ~1 km of late Cenozoic surface uplift would increase the average potential energy of the Sierran lithosphere by ~1.2×10^{12} N/m. This is equivalent to an increase in tensile stress of ~340 bars averaged over the 35-km-thick southeastern Sierran lithosphere, which would act primarily to thin the lithosphere. In this model, the flattening deformation is attributed to spreading of the Sierran lithosphere under its own weight. The belt of transtensional deformation between the Walker Lane belt and High Sierra may result from the combination of shear tractions from the Walker Lane belt and tensile buoyancy forces (Fig. 13).

Preferred Hypothesis

Of these two hypotheses, we favor the first model to account for the flattening on the eastern margin of the Isabella anomaly, primarily because it can be explained by a combination of northwest dextral shear, which clearly extends several tens of kilometers westward into the southern Sierra block from the Walker

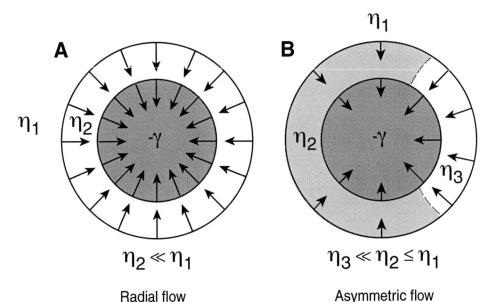

Figure 12. (A) Model for buoyancy-driven deformation associated with downwelling lithosphere (after Fleitout and Froixdevaux, 1982). Central region is characterized by thick, downwelling lithosphere subject to compressive buoyancy forces ($-\gamma$). Adjacent region (η_2) is characterized by much lower effective viscosity than surrounding area (η_1). Lithospheric thickening above downwelling lithosphere is accommodated by horizontal stretching in adjacent region. Radial patterns of flow result from the symmetry of the model. (B) Model for localized high rate of stretching along the margin of a downwelling lithospheric drip produced by asymmetric variations in effective viscosity (η_3).

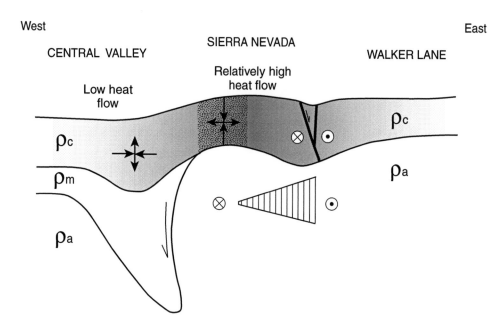

Figure 13. Cartoon of east-west variations in thickness of the crust (c), lithospheric mantle (m), and asthenosphere (a) (relative densities: $\rho_m > \rho_a > \rho_c$), and inferred buoyancy forces (arrows) across the Walker Lane belt–southeastern Sierra Nevada boundary. Westward narrowing pattern of vertical lines beneath the eastern Sierra represents encroachment of dextral shearing from the Walker Lane belt.

Lane belt, and localized northeast-southwest stretching along the northeastern margin of the Isabella anomaly (Figs. 10 and 12B). Although tensile buoyancy forces predictably would increase with elevation, which could account for the observed decrease in *V* west of the Walker Lane belt, it is not obvious why these forces would have a strong preferred northeast-southwest orientation sufficient to cause the observed rotation of the d_1 strain rate trajectories (nearly 90°) in the vicinity of Durrwood Meadows from their regional orientations (Fig. 6). It is also important to note that buoyancy forces may be modified by and compete in importance with flexural stresses at horizontal scales of several tens of kilometers (i.e., the dimensions of the seismically active areas in the southern Sierra), and thus the net effect of tensile buoyancy forces on the observed deformation cannot be confidently assessed without a more complex analysis.

DISCUSSION

Jones and Dollar (1986) and Saltus and Lachenbruch (1991) interpreted the Durrwood Meadows earthquake swarm to represent westward encroachment of Basin-and-Range–style normal faulting and extension into the southern Sierra Nevada. We favor the alternative hypothesis that strike-slip deformation, rather than extension, is propagating into the southeastern Sierran microplate from the dextral Walker Lane belt. Workers have proposed that the eastern margin of the relatively undeformed Sierran microplate at ca. 5 Ma was located tens of kilometers east of the present eastern front (see summary in Jones et al., 2004), implying that mixed strike-slip and normal faulting of the Walker Lane belt has propagated westward during late Cenozoic time.

At the latitude of our study area, the western limit of dextral shearing associated with the Walker Lane belt stepped abruptly

west from the Panamint Valley area to the Indian Wells Valley area ca. 3.4 Ma or later (Gan et al., 2003). In the period between 3.4 and 5 Ma, prior to this westward expansion of the Walker Lane belt, Indian Wells Valley was as an active extensional basin in the hanging wall of a north-south–striking, east-dipping normal fault system (Monastero et al., 2002). The ancestral Indian Wells Valley graben became inactive after ca. 3.5 Ma (possibly as recently as 2 Ma, per Monastero et al., 2002), and the basin was subsequently deformed by the Airport Lake fault, Little Lake fault, and other active dextral faults of the western Walker Lane belt at this latitude. The westward shift of northwest dextral shear from Panamint Valley to Indian Wells Valley created a new structural segment of the Garlock fault south of Searles Valley (Gan et al., 2003). Distributed dextral shear in Indian Wells Valley and Searles Valley since ca. 2–3.5 Ma has rotated this new segment clockwise from the Garlock fault's regional strike of N54°E south of the intact Sierran block.

We suggest that active seismogenic extension in the southern High Sierra Nevada may be analogous to the late Neogene tectonics of Indian Wells Valley prior to or during westward expansion of the Walker Lane belt into this region around 2.0–3.5 Ma. Although there is presently no large extensional basin in the southern Sierra Nevada comparable to Indian Wells Valley, workers have reported evidence of Quaternary surface rupture along normal faults in the region (Nadin and Saleeby, 2001), and the seismogenic extension may represent an early stage in the process of basin formation. If the Walker Lane belt widens through westward encroachment into the Sierran microplate, then the geology of the southern Sierra several million years hence predictably would show evidence of a period of extensional deformation (i.e., the current seismotectonic regime) overprinted by strike-slip faulting, similar to the late Cenozoic history of Indian Wells Valley (Monastero et al., 2002). The modern east-to-west variation in strain geometry across the Sierra Nevada–Walker Lane belt boundary thus could be a snapshot of the process whereby the Sierran block has been progressively narrowed by westward expansion of the Walker Lane belt, which occurred in Indian Wells Valley between 2.0 and 3.5 Ma (Monastero et al., 2002).

CONCLUSIONS

Kinematic analysis of seismogenic deformation in southeastern California reveals a well-defined, east-to-west transition from distributed northwest dextral shear in the Walker Lane belt, to transtensional shearing within the eastern Sierran microplate, to flattening in the southern High Sierra. We interpret this variation in deformation style to reflect the influence of different classes of tectonic forces: shearing in the Walker Lane is driven primarily by far-field forces that are responsible for large-scale motions of the plates, and deformation in the southern Sierra Nevada primarily is driven by local forces associated with late Cenozoic foundering of lower lithosphere represented by the Isabella anomaly. The observed transition from extension to strike-slip faulting in Indian Wells Valley after ca. 2.0–3.5 Ma effectively shifted the eastern edge of the Sierran microplate westward to its present location. A similar transition from extension to strike-slip faulting may be presently occurring in the southeastern Sierra Nevada, thus providing insights into the process whereby the Sierran microplate has been narrowed through westward encroachment of the Walker Lane belt.

ACKNOWLEDGMENTS

We performed this study under contract N68936-02-C-0207 from the U.S. Navy. We thank Frank Monastero, U.S. Navy Geothermal Program Office, for supporting this work and engaging us in many stimulating discussions about the seismotectonics of the study region. Hauksson also was supported by NEHRP/U.S. Geological Survey grants 04HQGR0052 and 05HQGR0040. Steve Thompson provided constructive criticism of an early draft of this paper. Comments by volume editor John Oldow and formal reviews by Basil Tikoff and an anonymous reviewer significantly improved the final version. This is contribution number 10,001 of the Division of Geological and Planetary Sciences, California Institute of Technology, Pasadena.

REFERENCES CITED

Argus, D.F., and Gordon, R.G., 1991, Current Sierra Nevada–North America motion from very long baseline interferometry: Implications for the kinematics of the western United States: Geology, v. 19, p. 1085–1088, doi: 10.1130/0091-7613(1991)019<1085:CSNNAM>2.3.CO;2.

Argus, D.F., and Gordon, R.G., 2001, Present tectonic motion across the Coast Ranges and San Andreas fault system in central California: Geological Society of America Bulletin, v. 113, p. 1580–1592, doi: 10.1130/0016-7606(2001)113<1580:PTMATC>2.0.CO;2.

Bawden, G.W., Michael, A.J., and Kellogg, L.H., 1999, Birth of a fault: Connecting the Kern County and Walker Pass, California, earthquakes: Geology, v. 27, p. 601–604, doi: 10.1130/0091-7613(1999)027<0601:BOAFCT>2.3.CO;2.

Bennett, R.A., Wernicke, B.P., Niemi, N.A., Friedrich, A.M., and Davis, J.L., 2003, Contemporary strain rates in the northern Basin and Range Province from GPS data: Tectonics, v. 22, no. 2, 31 p.

Benz, H.M., and Zandt, G., 1993, Teleseismic tomography: Lithospheric structure of the San Andreas fault system in northern and central California, *in* Iyer, H.M., and Hirahara, K., eds., Seismic Tomography; Theory and Practice: London, UK, Chapman and Hall, p. 440–465.

Boyd, O.S., Jones, C.H., and Sheehan, A.F., 2004, Foundering lithosphere imaged beneath southern Sierra Nevada, California, USA: Science, v. 305, p. 660–662, doi: 10.1126/science.1099181.

Dixon, T.H., Miller, M., Farina, F., Wang, H., and Johnson, D., 2000, Present-day motion of the Sierra Nevada block and some tectonic implications for the Basin and Range Province, North American Cordillera: Tectonics, v. 19, p. 1–24, doi: 10.1029/1998TC001088.

Dokka, R.K., and Travis, C.J., 1990, Late Cenozoic strike-slip faulting in the Mojave Desert: Tectonics, v. 9, p. 311–340, doi: 10.1029/TC009i002p00311.

Fleitout, L., and Froixdevaux, C., 1982, Tectonics and topography for a lithosphere containing density anomalies: Tectonics, v. 1, p. 21–56, doi: 10.1029/TC001i001p00021.

Fliedner, M., Klemperer, S.L., and Christensen, N.I., 2000, Three-dimensional seismic model of the Sierra Nevada arc, California, and its implications for crustal and upper mantle composition: Journal of Geophysical Research, v. 105, p. 10,899–10,921.

Gan, W., Zhang, P., Shen, Z.-K., Prescott, W.H., and Svarc, J.L., 2003, Initiation of deformation of the Eastern California shear zone: Constraints from Garlock fault geometry and GPS observations: Geophysical Research Letters, v. 30, no. 10, p. 1496, doi: 10.1029/2003GL017090.

Guenther, L.D., 2004, Tectonics of the northern California coast ranges, past and present, as inferred from analysis of kinematic data [M.S. thesis]: Davis, University of California, 312 p.

Hardebeck, J.L., and Shearer, P.M., 2002, A new method for determining first-motion focal mechanisms: Bulletin of the Seismological Society of America, v. 92, no. 6, p. 2264–2276, doi: 10.1785/0120010200.

Hauksson, E., 2000, Crustal structure and seismicity distribution adjacent to the Pacific and North American plate boundary: Journal of Geophysical Research, v. 105, no. B6, p. 13,875–13,903, doi: 10.1029/2000JB900016.

Hauksson, E., and Shearer, P., 2005, Southern California hypocenter relocation with waveform cross-correlation: Part 1. Results using the double-difference method: Bulletin of the Seismological Society of America, v. 95, p. 896–903, doi: 10.1785/0120040167.

Hauksson, E., and Unruh, J.R., 2007, Regional tectonics of the Coso geothermal area along the intracontinental plate boundary in central eastern California: Three-dimensional Vp and Vp/Vs models, spatial-temporal seismicity patterns, and seismogenic deformation: Journal of Geophysical Research, v. 112, B06309, doi: 10.1029/2006JB004721.

Houseman, G.A., Neil, E.A., and Kohler, M.D., 2000, Lithospheric instability beneath the Transverse Ranges of California: Journal of Geophysical Research, v. 105, p. 16,237–16,250, doi: 10.1029/2000JB900118.

Jennings, C.W., 1994, Fault activity map of California and adjacent areas: California Department of Conservation, Division of Mines and Geology, Geologic Data Map No. 6, scale 1:750,000.

Jones, C.H., Farmer, G.L., and Unruh, J.R., 2004, Tectonics of Pliocene removal of lithosphere of the Sierra Nevada, California: Geological Society of America Bulletin, v. 116, p. 1408–1422, doi: 10.1130/B25397.1.

Jones, L.M., and Dollar, R.S., 1986, Evidence of Basin and Range extensional tectonics in the Sierra Nevada: The Durrwood Meadows swarm, Tulare County, California (1983–1984): Bulletin of the Seismological Society of America, v. 76, p. 439–461.

Le Pourheit, L., Gurnis, M., and Saleeby, J., 2006, Mantle instability beneath the Sierra Nevada mountains in California and Death Valley extension: Earth and Planetary Science Letters, v. 251, no. 1–2, p. 104–119, doi: 10.1016/j.epsl.2006.08.028.

Lewis, J.C., Unruh, J.R., and Twiss, R.J., 2003, Seismogenic strain at the Cascadia convergent margin: Geology, v. 31, p. 183–186, doi: 10.1130/0091-7613(2003)031<0183:SSAMOT>2.0.CO;2.

McClusky, S.C., Bjornstad, S.C., Hager, B.H., King, R.W., Meade, B.J., Miller, M.M., Monastero, F.C., and Souter, B.J., 2001, Present day kinematics of the eastern California shear zone from a geodetically constrained block model: Geophysical Research Letters, v. 28, no. 17, p. 3369–3372, doi: 10.1029/2001GL013091.

Miller, M.M., Johnson, D.J., Dixon, T.H., and Dokka, R.K., 2001, Refined kinematics from the eastern California shear zone from GPS observations, 1993–1998: Journal of Geophysical Research, v. 106, no. B2, p. 2245–2263, doi: 10.1029/2000JB900328.

Monastero, F.C., Walker, J.D., Katzenstein, A.M., and Sabin, A.E., 2002, Neogene evolution of the Indian Wells Valley, east-central California, *in* Glazner, A.F., Walker J.D., and Bartley, J.M., eds., Geologic Evolution of the Mojave Desert and Southwestern Basin and Range: Geological Society of America Memoir 195, p. 199–228.

Monastero, F.C., Katzenstein, A.M., Miller, J.S., Unruh, J.R., Adams, M.C., and Richards-Dinger, K., 2005, The Coso geothermal field: A nascent metamorphic core complex: Geological Society of America Bulletin, v. 117, p. 1534–1553, doi: 10.1130/B25600.1.

Nadin, E.S., and Saleeby, J.B., 2001, Relationship between the Kern Canyon fault (KCF) and the proto–Kern Canyon fault (PKCF), southern Sierra Nevada, CA: Eos (Transactions, American Geophysical Union), v. 82, no. 47, p. F1183.

Oldow, J.S., 2003, Active transtensional boundary zone between the western Great Basin and Sierra Nevada block, western U.S. Cordillera: Geology, v. 31, no. 12, p. 1033–1036, doi: 10.1130/G19838.1.

Park, S.K., and Wernicke, B., 2003, Electrical conductivity images of Quaternary faults and Tertiary detachments in the California Basin and Range: Tectonics, v. 22, no. 4, p. 1030, doi: 10.1029/2001TC001324.

Ruppert, S., Fliedner, M.M., and Zandt, G., 1998, Thin crust and active upper mantle beneath the southern Sierra Nevada in the western United States: Tectonophysics, v. 286, p. 237–252, doi: 10.1016/S0040-1951(97)00268-0.

Saleeby, J., and Foster, Z., 2004, Topographic response to mantle lithosphere removal in the southern Sierra Nevada region, California: Geology, v. 32, no. 3, p. 245–248, doi: 10.1130/G19958.1.

Saleeby, J.B., Ducea, M.N., and Clements-Knott, D., 2003, Production and loss of high-density batholithic root, southern Sierra Nevada, California: Tectonics, v. 22, no. 6, 24 p., doi: 10.1029/2002TC001374.

Saltus, R.W., and Lachenbruch, A.H., 1991, Thermal evolution of the Sierra Nevada: Tectonic implications of new heat flow data: Tectonics, v. 10, p. 325–344, doi: 10.1029/90TC02681.

Sauber, J., Thatcher, W., Solomon, S.C., and Lisowski, M., 1994, Geodetic slip rate for the Eastern California shear zone and recurrence time of Mojave Desert earthquakes: Nature, v. 367, p. 264–266, doi: 10.1038/367264a0.

Savage, J.C., Lisowski, M., and Prescott, W.H., 1990, An apparent shear zone trending north-northwest across the Mojave Desert into Owens Valley, eastern California: Geophysical Research Letters, v. 17, p. 2113–2116, doi: 10.1029/GL017i012p02113.

Thurber, C.H., 1993, Local earthquake tomography: Velocities and V_p/V_s theory, *in* Iyer, H.M., and Hirahara, K., eds., Seismic Tomography: Theory and Practice: London, Chapman and Hall, p. 563–583.

Twiss, R.J., and Unruh, J.R., 1998, Analysis of fault-slip inversions: Do they constrain stress or strain rate?: Journal of Geophysical Research, v. 103, p. 12,205–12,222, doi: 10.1029/98JB00612.

Twiss, R.J., Souter, B.J., and Unruh, J.R., 1993, The effect of block rotations on the global seismic moment tensor and patterns of seismic P and T axes: Journal of Geophysical Research, v. 98, p. 645–674, doi: 10.1029/92JB01678.

Unruh, J.R., Twiss, R.J., and Hauksson, E., 1996, Seismogenic deformation field in the Mojave block and implications for the tectonics of the Eastern California shear zone: Journal of Geophysical Research, v. 101, no. B4, p. 8335–8362, doi: 10.1029/95JB03040.

Unruh, J.R., Twiss, R.J., and Hauksson, E., 1997, Kinematics of post-seismic relaxation from aftershocks of the 1994 Northridge, California, earthquake: Journal of Geophysical Research, v. 102, p. 24,589–24,603, doi: 10.1029/97JB02157.

Unruh, J.R., Hauksson, E., Monastero, F.C., Twiss, R.J., and Lewis, J.C., 2002, Seismotectonics of the Coso Range–Indian Wells Valley region, California: Transtensional deformation along the southeastern margin of the Sierran microplate, *in* Glazner, A.F., Walker, J.D., and Bartley, J.M., eds., Geologic Evolution of the Mojave Desert and Southwestern Basin and Range: Geological Society of America Memoir 195, p. 277–294.

Unruh, J., Humphrey, J., and Barron, A., 2003, Transtensional model for the Sierra Nevada frontal fault system, eastern California: Geology, v. 31, p. 327–330, doi: 10.1130/0091-7613(2003)031<0327:TMFTSN>2.0.CO;2.

Waldhauser, F., and Ellsworth, W.L., 2000, A double-difference earthquake relocation algorithm: Method and application to the northern Hayward fault: Bulletin of the Seismological Society of America, v. 90, no. 6, p. 1353–1368, doi: 10.1785/0120000006.

Zandt, G., Gilbert, H., Owens, T.J., Ducea, M., Saleeby, J., and Jones, C.H., 2004, Active foundering of a continental arc root beneath the southern Sierra Nevada, California: Nature, v. 431, p. 41–46, doi: 10.1038/nature02847.

MANUSCRIPT ACCEPTED BY THE SOCIETY 21 JULY 2008